STUDIES IN BAYESIAN
ECONOMETRICS AND STATISTICS

CONTRIBUTIONS TO ECONOMIC ANALYSIS

86

NORTH-HOLLAND PUBLISHING COMPANY – AMSTERDAM • OXFORD
AMERICAN ELSEVIER PUBLISHING COMPANY, INC. – NEW YORK

STUDIES IN BAYESIAN ECONOMETRICS AND STATISTICS

IN HONOR OF LEONARD J. SAVAGE

Edited by

STEPHEN E. FIENBERG
University of Minnesota

and

ARNOLD ZELLNER
University of Chicago

1975

NORTH-HOLLAND PUBLISHING COMPANY – AMSTERDAM • OXFORD
AMERICAN ELSEVIER PUBLISHING COMPANY, INC. – NEW YORK

Library of Congress Catalog Card Number: 73-86697

ISBN North-Holland for this series 0 7204 3100 x
ISBN North-Holland for this volume 0 7204 3188 3
ISBN American Elsevier for this volume 0 444 10579 4

Publishers:

NORTH-HOLLAND PUBLISHING COMPANY—AMSTERDAM
NORTH-HOLLAND PUBLISHING COMPANY, LTD.—OXFORD

Sole distributors for the U.S.A. and Canada:

AMERICAN ELSEVIER PUBLISHING COMPANY, INC.
52 VANDERBILT AVENUE, NEW YORK, N.Y. 10017

PRINTED IN THE NETHERLANDS

INTRODUCTION TO THE SERIES

This series consists of a number of hitherto unpublished studies, which are introduced by the editors in the belief that they represent fresh contributions to economic science.

The term *economic analysis* as used in the title of the series has been adopted because it covers both the activities of the theoretical economist and the research worker.

Although the analytical methods used by the various contributors are not the same, they are nevertheless conditioned by the common origin of their studies, namely theoretical problems encountered in practical research. Since for this reason, business cycle research and national accounting, research work on behalf of economic policy, and problems of planning are the main sources of the subjects dealt with, they necessarily determine the manner of approach adopted by the authors. Their methods tend to be "practical" in the sense of not being too far remote from application to actual economic conditions. In addition they are quantitative rather than qualitative.

It is the hope of the editors that the publication of these studies will help to stimulate the exchange of scientific information and to reinforce international cooperation in the field of economics.

THE EDITORS

CONTENTS

PART 1

INTRODUCTION

S. E. Fienberg and A. Zellner (eds.)
Studies in Bayesian Econometrics and Statistics.

1.1 DEDICATION TO LEONARD JIMMIE SAVAGE

On October 14 and 15, 1971, Leonard Jimmie Savage attended the Third Meeting of the National Bureau of Economic Research–National Science Foundation Seminar on Bayesian Inference in Econometrics in Cambridge, Massachusetts. Just over two weeks later, on November 1, he died suddenly at his home in Connecticut.

To those of us who were present at the Cambridge meeting, Savage's passing came as a special shock. He participated in the discussions with such energy and verve that it was hard for us to accept the fact that we would never again be privileged to experience the full force of his towering intellect. In particular, we recall Savage's discussion of a paper presented by Jacques Drèze regarding the problem of identifiability, and some comments he made to James Dickey and one of us regarding Kolmogorov's definition of conditional probability and the apparent paradoxes resulting from its use. On the second day of the meeting, Savage presented a talk entitled "Elicitation of personal probabilities and expectations", based in part on his paper which we have included in this volume. His talk, a product of over ten years' work, was both illuminating and provocative, and its published version, reproduced in the present volume, serves as a fitting testimonial to Savage's contributions to the theory of personal probabilities and Bayesian inference.

It was at the Cambridge meeting that we had discussions regarding the present volume. Jimmie Savage strongly supported the preparation of a collection of papers based on presentations at meetings of the Seminar, and was the first to say he would be happy to have one of his papers included in the volume.

At the time of his death, Savage was the Eugene Higgins Professor of Statistics and Chairman of the Department of Statistics at Yale University. Earlier in his life he was on the faculty of the University of Michigan and the University of Chicago (where he also served as Chairman of the Department of Statistics). His book, "The foundations of statistics" (originally published in 1954 by John Wiley and Sons and recently reissued in a revised paperback edition by Dover Publications), was a pathbreaking effort and the first modern exposition of the rationale behind subjective probability and its use in statistical inference. Without a doubt, he was the moving force and leading exponent behind the resurgence of the theory of personal probability. His collaboration with Bruno de Finetti, and his efforts to promote de Finetti's work among American statisticians, have culminated in the publication of de Finetti's book "Probability, induction and statistics" (Wiley, 1972), which appeared shortly after Savage's death. Savage continued to lecture on the use of subjective probabilities throughout his career, and some of these lectures appeared in "The foundations of statistical inference" (Wiley, 1962). In December of 1970, Savage delivered the R. A. Fisher Memorial Lecture, entitled "On re-reading Fisher", at the annual meeting of the American Statistical Association. He held an overflow audience spellbound for close to two hours, giving his personal reflections on Fisher's contributions to statistics. Those of us who attended this lecture could not help emerging with the belief that Fisher was in fact a Bayesian, though he never admitted to it. Savage's writing will undoubtedly serve to inspire future generations of statisticians.

Jimmie Savage was a teacher of several contributors to this volume, a colleague of others, and an inspiration to all. In the words of a former student of Savage: "To his colleagues and students each moment with him was cherished—around him the atmosphere tingled; the tingles emanating not only from his sparkling intelligence and wit, but also from his empathy and warm-heartedness. He was ever willing to discuss a subject at any level, and ever capable of leading one's fallacious argument gracefully and elegantly over a nearby logical cliff. I cannot imagine the existence of another who combines raw intelligence and teaching ability as effectively as did Jimmie Savage."[1] We will miss Jimmie Savage, and we hope that this collection of papers will serve in a small way as a testimonial to the enormous contributions of his lifetime's work.

[1] An excerpt from a letter to the editor of the American Statistician written by Donald A. Berry, Volume 26, Number 1, 1972.

S. E. Fienberg and A. Zellner (eds.)
Studies in Bayesian Econometrics and Statistics.

1.2 OVERVIEW OF VOLUME

The papers in this volume reflect recent and on-going research on Bayesian methods and applications of them. Many of the papers were presented to meetings of the National Bureau of Economic Research–National Science Foundation Seminar on Bayesian Inference in Econometrics.[1] Comments of Seminar participants were often extremely helpful in clarifying issues and providing technical suggestions. Indeed, some of Savage's remarks had a profound impact on several research projects that were reported at the Cambridge meeting of the Seminar in October, 1971.

Since interaction between theory and applications is thought to be extremely important, both theoretical and applied papers have been included in the present volume. While different models, problems and applications are treated, the unity of the Bayesian approach, emphasized by many, is apparent in all the papers. This is not to say that all Bayesians are in agreement about all aspects of Bayesian analysis. In fact, there are differences, some of them nontrivial, among Bayesians;[2] however, these differences are more akin to variations on a common theme rather than disagreements about the common theme or about the superiority of the Bayesian approach vis-à-vis non-Bayesian approaches. Hopefully, the present volume will serve to introduce a broad audience to a range of Bayesian analyses that will enhance general understanding of the approach and be useful in generating further development and applications of it.

The first set of contributions relates econometrics and Bayesian analysis. While some early quantitative economists, particularly Cournot and Edgeworth, contributed significantly to the development of Bayesian analysis, in more recent times most econometric work has involved use

[1] See Appendix for a listing of the programs of the first five meetings of the Seminar. Reports of the Seminar meetings can be obtained on request from Mrs. Anna Tremblay, National Bureau of Economic Research, 575 Technology Square, Cambridge, Massachusetts, 02139, USA.

[2] In this connection see the delightful piece by I. J. Good, 46656 varieties of Bayesians, American Statistician 25 (1971), 62–63.

of non-Bayesian techniques. In the last two decades, there has been a resurgence of interest and work on Bayesian statistical methods and their application in econometrics. To give readers perspectives regarding these developments, the first section of the volume presents contributions that discuss various aspects of the Bayesian approach.

Drèze's contribution, a published version of his presidential address to the Econometric Society, is a lucid and penetrating discussion of the role of Bayesian analysis in "the formal analysis of decision problems in economics, resting on a specification of ends and means firmly rooted in economic theory, incorporating a probabilistic treatment of econometric information, and making use of the possibilities offered by mathematical programming techniques to compute optimal policies". His detailed consideration of potentially fruitful problem areas and the attention that he devotes to analysis of the simultaneous equation model permit readers to obtain a valuable orientation with respect to a wide range of problems in econometrics to which Bayesian analysis can contribute useful and meaningful solutions.

Rothenberg and Zellner discuss comparative features of Bayesian and non-Bayesian approaches to inference and decision problems in econometrics. Zellner considers Bayesian and non-Bayesian approaches to a broad range of basic inference and decision problems and discusses their relative performance in specific terms. He concludes that the learning and decision techniques of Bayesian analysis have much to commend them for use in econometrics. Rothenberg presents a thoughtful discussion of some of Sir Ronald A. Fisher's reservations about the Bayesian approach, a consideration of Bayesian decision and canonical theorists' possible reactions to Fisher's points, and finally his views on how science progresses.

In comments on Rothenberg's and Zellner's papers, Kaufman and Pratt find arguments for the Bayesian approach compelling. However, they point to several important problems requiring additional work in order to make the Bayesian approach applicable to a broader range of problems and more fruitful in econometric and statistical work. Zellner in a rejoinder makes some remarks about prior distributions and points out that there is nothing in the Bayesian approach that logically implies an incremental theory of scientific progress that is charming and peaceful. Changes in beliefs produced by new information can and have been dramatic, disturbing and even revolutionary. He urges that Jeffreys' historical account of many major developments in science be considered before readers reach a final decision on how science "really" progresses.

The next section, Bayes' theorem and economic theory, is included to illustrate some uses of Bayesian analysis in economic theory. In this regard the reader should note that Bayes' theorem can be viewed as a formal learning model that represents how new information is processed to reach a more informed position, that is to reduce uncertainty. The information considered may relate to parameter values, future values of random variables, alternative models, etc. In the Cyert–DeGroot paper, an analysis of a duopoly problem is presented in which two firms learn about each other's decision-making behavior in a variety of ways. By using the formal Bayesian learning model to represent this process, Cyert and DeGroot have introduced a basic innovation in duopoly theory.[3] Similarly, Turnovsky's paper, illustrating that a Bayesian learning model can rationalize the ad hoc assumption of adaptive expectations, is an innovation that will undoubtedly be followed by many more contributions linking the theory of expectations and Bayesian learning processes.[4]

The problem of assessing prior distributions is a central one in Bayesian analysis. Also, since many non-Bayesians employ a good deal of prior information in their work, progress on understanding the nature of prior information and on methods for representing it will be of value not only to Bayesians but also to many non-Bayesians. Savage's paper, "Elicitation of personal probabilities and expectations", appearing in the section on Assessing Prior Distributions, represents an extremely thoughtful and detailed exposition of the theory of subjective prior information and a penetrating discussion of a broad range of unresolved problems. That the presentation uses many concepts and terms from the economic theory of choice under uncertainty and examples from economics makes it particularly relevant for econometricians as well as statisticians. The extensive bibliography should be of great value to readers who wish to pursue the many issues raised by Savage.

The next section, Identification and Bayesian Analysis, deals with the

[3] Additional work on the duopoly problem, utilizing Bayesian analysis, appears in the following papers by Cyert and DeGroot: Interfirm learning and the kinked demand curve, Journal of Economic Theory 3 (1971), 272–287; An analysis of cooperation and learning in a duopoly context, paper presented to the Second NBER–NSF Seminar on Bayesian Inference in Econometrics, May 1971; and Rational expectations and Bayesian analysis, manuscript, Carnegie–Mellon Univ., 1972.

[4] See, e.g., Sanford Grossman, A Bayesian approach to static, stochastic equilibria, manuscript, Univ. of Chicago, presented to the Fifth NBER–NSF Seminar on Bayesian Inference in Econometrics, 1972.

concept of identification in Bayesian analysis. Drèze considers the Bayesian theory of identification in the context of simultaneous equation models. He emphasizes that very often identifying information is uncertain and can be represented by a prior distribution. With such a representation, he presents a theorem indicating the implications of underidentification for posterior distributions. He goes on to extend the classical conditions for identification to necessary conditions for the existence of a proper posterior distribution for parameters of a model. Kadane continues the analysis of the concept of identification in Bayesian theory in a paper that stresses the view that "identification is properly a part of the study of the design of experiments", a remark that he attributes to Leonard J. Savage, and that "identification is a property of the likelihood function, and is the same whether considered classically or from the Bayesian approach". He goes on to discuss real-valued identified functions and characterization of identified functions in the contexts of modification of opinion and of making decisions. Finally, he presents a general result on Bayesian analysis of a possibly unidentified model. There can be no question but that these two papers constitute a fundamental contribution to the further understanding and clarification of the concept of identification.

The next section includes several applications of Bayesian approaches to regression analysis. In the first paper, Varian analyzes a real estate assessment problem within the context of Bayesian decision theory. He shows how prior information and a flexible class of loss functions can be employed in conjunction with a regression model for sales prices to obtain estimates of the market prices for properties that can serve as a basis for assessed valuations. Then his approach is applied in analyses of actual data to illustrate a potentially important advance in the methodology of real estate assessment. In the second paper, Hogarth provides a useful review of some previous work that attempts to separate the effects of genetic and environmental influences on intelligence as measured by IQ scores. Then he goes on to analyze data relating to monozygotic twins reared apart. In these analyses, a bivariate "seemingly unrelated" regression model is employed to describe twins' IQ's and measures of socio-economic status relating to the homes in which they were reared. The results will be of interest to a broad spectrum of social scientists. In the last paper in this section, Geisel analyzes the problem of comparing alternative models. After a valuable general discussion of a Bayesian procedure for comparing parametric stochastic models, he proceeds to apply the procedure to the comparison of simple

'Quantity Theory and Keynesian macroeconomic regression models. These models were considered earlier by Friedman and Meiselman in analyses that were widely discussed in the economic literature. Geisel's analyses of these models and elaborations of them from the Bayesian point of view illustrate well the fruitfulness and flexibility of Bayesian techniques for comparing models.

In the next section, several missing observation problems are analyzed. Since missing observation problems are frequently encountered in applied work, it is fortunate that they can be dealt with in an effective and operational manner by use of Bayesian methods. In the context of normal regression models, Press, Scott and Dagenais present solutions to missing observation problems under various assumptions. The Press–Scott paper includes interesting comparisons of Bayesian and non-Bayesian solutions, explicit Bayesian estimators for missing observations, and techniques for joint estimation of regression parameters and missing observations on dependent and independent variables. In Dagenais' paper, sample information on regression variables and on variables related to independent variables is employed to solve the problem of missing observations. Posterior distributions for parameters are derived and used in a numerical example to illustrate the importance of taking proper account of missing observations in analyses. In the Mehta–Swamy paper, Bayesian methods for making inferences about means of a bivariate normal distribution when some observations are missing are developed. This work involves a generalization of the Behrens–Fisher problem to the case of correlated observations. In a computed example, Mehta and Swamy provide an interesting comparison of the exact posterior distribution of the difference of the means with conditional and approximate posterior distributions.

That usual Bayesian methods can be utilized in the analysis of time series problems is indeed fortunate since such problems are of great importance in econometrics and statistics. In the first paper in the section on Time Series Problems, Chow takes up the problem of obtaining multi-period predictions from stochastic difference equations. Using Bayesian methods, he provides an analytical solution to the problem for a univariate first-order autoregressive process and then generalizes the results to apply to systems of stochastic difference equations. In particular, Chow emphasizes that usual non-Bayesian "estimation methods may be inadequate if the resulting estimates are to be used to make ex ante predictions for more than one period ahead, and if the accuracy of the predictions is measured, as it usually is, by the mean square errors". The following papers by Cleveland and Shiller analyze distributed lag models that have been widely used in econometrics. In Cleveland's paper, methods of parameter estimation using a class of prior distributions that includes members placing large and small prior weights on smooth values of the parameters are developed. There is a close connection between the methods for choosing priors advocated by Cleveland and the use of data-based priors as discussed by Sutherland, Holland, and Fienberg in a paper in the last section of this volume. Similarly, Shiller derives Bayesian estimators for distributed lag parameters utilizing prior distributions incorporating information regarding the "smoothness" of the lag curve. Both Cleveland and Shiller present interesting numerical examples to illustrate applications of their methods and provide comparisons of their methods with widely-used non-Bayesian techniques. The flexible manner in which prior information is introduced in Cleveland's and Shiller's work is noteworthy.

The next three papers deal with the analysis of simultaneous equation models from the Bayesian point of view. Chetty shows how Bayesian techniques can be employed to analyze prediction and robustness of assumptions problems in the context of two simple models put forward by Haavelmo. For the second model, he analyzes data to show how sensitive inferences about the marginal propensity to consume are to assumptions about a serial correlation parameter and to an assumption that a variable is exogenous. In both cases it is found that inferences are sensitive to incorrect assumptions, a finding that underlines the importance of appraising specifying assumptions carefully. In the next paper, Kaufman discusses features of a Bayesian analysis of a simultaneous equation system when both time series and cross-section data are available. The importance of combining data from various sources in order to improve the precision of inferences has been recognized in the literature. Building on previous work of Drèze, Kaufman's analysis provides Bayesian techniques for accomplishing this objective. In an empirical application, he computes posterior distributions for parameters of Tintner's model of the U.S. meat market. In the last paper of this section, Rothenberg presents a comprehensive survey of work on the simultaneous equation model from the Bayesian point of view. While much has been accomplished to date, there are some problems requiring additional research. By summarizing what has been done, Rothenberg's paper represents an excellent point of departure for those interested in pursuing further work on simultaneous equation models.

The next two papers involve Bayesian analyses of control problems that arise in using econometric models for policy purposes. Prescott's paper with its simple illustrative examples compares the performance of linear decision rules that neglect uncertainty about parameter values with that of Bayesian adaptive and optimal decision rules that take account of such uncertainty and experimentation. His analyses reveal a marked superiority of the Bayesian adaptive decision rules, particularly when there is a fair amount of uncertainty regarding parameters' values. In the paper by Bowman and Laporte a one-period control problem is analyzed in relation to a fully recursive econometric model and a quadratic criterion function. An application of the techniques using a version of the St. Louis econometric model of the U.S. economy is provided. As in Prescott's study, it is found that for a broad range of conditions the Bayesian solution that takes account of uncertainty about parameter values provides considerably lower expected loss than a certainty equivalence solution that neglects such uncertainty. These findings are encouraging with respect to possibilities of improving policy-making by use of Bayesian techniques. Further, it is fortunate that much additional research on Bayesian control theory and applications of it is in progress.[5]

Many critics of the Bayesian approach to statistical inference claim that it is not within the scope of the scientific tradition to report the results of a statistical analysis based on a personal prior assessment. In the section on Reporting Results, Dickey and Roberts give attention to the important yet often neglected problem of reporting the procedures and results of statistical investigations. Dickey discusses how investigators can use graphical reports of inference by probabilities to enhance the quality of reporting. He illustrates his procedures in analyses of several central statistical problems. Roberts' paper is a thoughtful and extensive survey of the problem of reporting results. Much statistical reporting could be improved considerably if investigators followed the suggestions put forward by Dickey and Roberts. Improved reporting will generally provide readers with more accurate and meaningful accounts of the ways in which researchers' initial views have been modified by sample information and better understanding of how initial views and sample information have been used in reaching decisions.

The last section, New Developments in Estimation, includes several

[5] See, e.g., Special issue on control theory, Annals of Economic and Social Measurement 1 (1972).

papers that deal with recent innovations in estimation theory. Many of these papers have their origins in work done, in a classical statistical decision theory framework, by Stein on the inadmissibility of the multivariate normal mean. A recent paper by Lindley and Smith[6] gives a multi-stage Bayesian analysis of the general multivariate normal mean problem, and complements the present contributions.

In Dickey's paper, Bayesian analogues are derived for the Stein-type and Hoerl–Kennard-ridge-regression-type improvements on the traditional least-squares estimates. In addition, three practical decision-theoretic alternative tests are provided for the traditional *F*-test. Several examples are analyzed to illustrate the procedures. The analyses clearly illustrate how prior information that is often used informally in applied work can be represented formally by prior distributions and used to produce useful estimates and tests.

In the second paper, Hill is concerned with the general problem of making inferences about parameters, perhaps many in number. After a cogent discussion of the concepts of coherence and admissibility, he goes on to emphasize that it is extremely important for analysts "to study carefully the behavior of the inferential procedure for various data, before actually taking the data, until one is confident that the particular model he has chosen or is tempted to choose, yields to within an adequate approximation the kind of behavior he regards as wise". As an example of these considerations, he presents an analysis of a random means model. His analysis shows clearly how prior information about the means can be employed to obtain improved estimators in situations in which means are thought to be related a priori. However, when the means are considered independent a priori, Hill shows that the usual estimator is admissible relative to the usual quadratic loss function and a certain class of estimators. These results, along with others included in the paper, should help to enhance our understanding of "improved" estimators that have appeared in the literature in the past few years.

In the next paper in this section, Sutherland, Holland and Fienberg take up the problem of estimating the vector of cell probabilities for a multinomial random variable. Using a combination of Bayesian models and frequency calculations (i.e., risk functions), they develop a class of estimators that resemble the James–Stein estimator for the multivariate

[6] D. Lindley and A. F. M. Smith, Bayes estimates for the linear model (with discussion), Journal of the Royal Statistical Society (B) 34 (1972), 1–41.

normal mean vector, and can be viewed as approximations to Bayesian point estimators generated using two-stage prior distributions. Introducing a novel asymptotic theory where the number of observations per parameter is moderate but the number of parameters is large, they approximate the risk functions of various estimators, and prove a "first-order" optimality result for one of the estimators in their class. An application of the results to the analysis of a set of occupational mobility data is provided. The paper represents valuable progress in the analysis of a multi-parameter problem, where it is extremely difficult to completely specify prior distributions for the parameters involved.

The paper by Tiao and Box raises questions about the usefulness of point estimators in relation to inference problems. They emphasize that "inferences about a parameter are provided not by the sampling properties of some arbitrarily chosen characteristic of the posterior distribution, but by that distribution itself". Very fundamentally, they raise the issue of the relevance of sampling properties, including admissibility, for the task of making inferences from *given* data. Finally, Tiao and Box present some interesting examples to illustrate their point of view. The reader should note that the paper by Hill in this section, using the subjectivist notion of coherency, does show the relevance of admissibility from the Bayesian point of view.

In the paper by Zellner and Vandaele, a brief review of Stein and Stein-like estimators for the normal-mean and regression models is presented. Various Bayesian and non-Bayesian interpretations are provided for these estimators. It is pointed out that the normal-mean problem, as well as others, has three components, namely a model for the observations, a prior distribution and a loss function. In the fixed parameter case with a uniform prior and usual quadratic loss function, the posterior mean, the usual estimate, minimizes posterior expected loss and is optimal in this respect. However, since average risk is not finite, there is no assurance that the posterior mean is in general admissible. In this circumstance, it is shown how an improved point estimator can be obtained without changing prior assumptions, the model or the loss function. Other procedures for obtaining improved point estimators that involve changes in the components of the problem are described. Among these are analyses that utilize standard informative prior distributions for the parameters and prior distributions that lump some probability on specific parameter values. In connection with the latter class of problems, Zellner and Vandaele comment that "in this approach the complete posterior pdf for the parameter vector is available and thus there is no need to rely on only a point estimate in making inferences about parameters' probable values".

In concluding this introductory section, it is the editors' hope that the present volume will be useful not only to Bayesian researchers but also to others interested in learning about Bayesian inference and decision procedures. Success in this regard will be a fitting testimonial to Savage who was ever-ready to work hard to enhance students' and colleagues' understanding of the bases of statistical inference and decision.

PART 2

ECONOMETRICS AND BAYESIAN INFERENCE

S. E. Fienberg and A. Zellner (eds.)
Studies in Bayesian Econometrics and Statistics.

2.1 ECONOMETRICS AND DECISION THEORY*

Jacques H. DRÈZE

University of Louvain

This is a written version of the Presidential Address delivered at the Second World Congress of the Econometric Society, Cambridge, September 10, 1970.

The oral presentation was introduced by the following remark: There is a substantial difference between the science of writing a cookbook and the art of cooking a good dinner. A similar difference exists between the task of writing a presidential address and that of presenting it at a Congress. I did consult many econometric cookbooks over the summer in preparing this address, but today I will present it in the spirit in which a dinner is served. So I begin by announcing the

MENU

The address was dedicated to William Vickrey, who introduced me to econometrics. The footnotes were not included in the oral presentation.

* Reprinted from Econometrica 40 (1972), 1–17, with permission of the Econometric Society.

[1] Served under limited information.

1.

This address will review a development which lies in the future as much as in the past, and still calls for a tone of sober and subdued confidence. It is a development which I foresaw as my friends and colleagues at Carnegie Tech were "Planning production, inventories and work force" [25], while French engineers were putting "Marginal cost pricing in practice" [36]; the picture became sharper after I became acquainted with Savage's view of "The foundations of statistics" [43], and then discovered how Arrow could deduce from it "The role of securities in the optimal allocation of risk bearing" [1] and how Schlaifer could extend it to "Probability and statistics for business decisions" [44].

My theme will be that we should now regard as a realistic challenge the formal analysis of decision problems in economics, resting on a specification of ends and means firmly rooted in economic theory, incorporating a probabilistic treatment of econometric information, and making use of the possibilities offered by mathematical programming techniques to compute optimal policies.

2.

"Knowledge is useful if it helps to make the best decisions", noted Marschak in the opening sentence of "Studies in econometric method" [35, p. 1]. The decisions of concern to us are the economic decisions of households, firms, and public authorities. These decisions are typically made under uncertainty. Formal analysis of decisions under uncertainty rests on logical foundations that were clearly set forth in Savage's book [43] in terms of three basic concepts: the states, which describe the environment; the consequences, which describe what happens to the decision maker; and the acts, which are functions assigning a consequence to each state (or to sets of states called events). A decision problem calls for choosing a most preferred element from a given set of acts.

Consistency of preferences is usually identified with the existence of a preordering, among the objects of choice, which is simple—that is, complete and transitive. A natural extension of this requirement to the realm of uncertainty leads Savage to postulate: (i) a simple preordering among acts, conditionally on any non-null event; (ii) a simple preordering among consequences; and (iii) a simple preordering among events.

Under appropriate technical precautions, these postulates imply the powerful "moral expectation theorem": the preordering among consequences is representable by a real-valued function, defined up to a linear transformation—call it "utility" or "loss"; the preordering among events is uniquely representable by a probability measure; and the preordering among acts is then representable by their expected utility, so that an optimal act maximizes expected utility over the feasible set.[2]

In the light of this theorem, a decision problem under uncertainty can be logically decomposed into the following four steps: (i) define the set of acts; (ii) define a utility on the consequences; (iii) define a probability measure on the events; and (iv) find an act which is maximal with respect to expected utility.

This decomposition, permitting each of the four steps to be attacked independently, is useful for applications and research strategy alike. At the cost of oversimplification, let me suggest that positive economics should assist in defining the set of acts (that is, the range of alternatives and their consequences conditionally upon the states); that normative economics should assist in defining the utility; that statistics and econo-

[2] This theory is usually presented with reference to "games against nature" as opposed to "games of strategy" or "games of strength and skill". One reason for this distinction is the following: (i) the events must be defined with interpersonal objectivity, if the theory is to remain operational; (ii) the choice of strategy by a player, in a game of strategy or in a game of strength and skill, is not always observable (especially when mixed strategies are used); (iii) consequently, the probability measure associated with the events used in describing these games may depend upon the strategy chosen by the decision maker. I personally hold the view that the theory outlined in the text can be generalized in a natural way so as to cover the three (collectively exhaustive) types of games. The generalization consists in (i) providing an operational test whereby one can identify sets of acts having at least one *common* optimal strategy; (ii) restricting the postulates that imply a qualitative probability to such sets of acts. One then defines probability measures on the events *conditionally* upon the choice of a strategy by the decision maker, and one verifies that the moral expectation principle governs the choices among strategies as well as the choices among acts. The weaker postulates underlying this reformulation also allow the ordering among consequences to become conditional upon the state that obtains, a useful generalization in itself. This viewpoint is developed in my unpublished Ph.D. dissertation [8]; a summary presentation of the generalized theory is given in [10]; a nontechnical presentation is given in [9]. An application of this approach to choices affecting the decision maker's probability of survival is given in [11].

metrics should bring empirical observations to bear upon the assessment of probabilities; and that techniques of mathematical programming should assist in finding an optimal act.

I may now restate my theme more specifically. Recent advances in statistical and econometric methodology enable us to derive from empirical observations a probability measure on the relevant events for many decision problems; this new possibility challenges us to state our decision problems formally; the directions in which normative economics should be extended in order to meet that challenge and the specific mathematical programming problems with which we are confronted can be sharply defined. This is a limited but well-defined area of research that deserves to be systematically explored.

By way of illustration, let me rise to my natural level of incompetence, and invade that most serious-minded of our substantive fields, agricultural economics.

In response to short-term random fluctuations in yields and long-term trends in productivity, and of course under due political pressure, many countries have engaged in programs of price support for basic agricultural commodities, accompanied by government-held buffer stocks. Typical policies consist in choosing a level for the price paid to farmers and a level for the wholesale price (or for the buffer inventory). The consequences of policy choices include the direct costs of the support and buffer program, its incidence on the welfare of consumers and agricultural producers, and all kinds of indirect effects, through the prices and outputs of substitutes, migration between farms and cities, and so on. The inventory levels resulting from one year's decisions are part of the initial conditions affecting the following year's decisions. Imperfect predictability characterizes not only agricultural yields but also the responses of farmers and consumers to price changes, the future states of the economy, and so on.

Such programs may be continued, or redesigned, or discontinued upon disposal of the existing inventories; in any case, rational decisions would call for an analysis in four steps: (i) specification of the range of feasible policies and of a model linking the policy instruments to their consequence; (ii) definition of a criterion function (utility) encompassing the various dimensions of the consequences; (iii) specification of the joint probability density function of the unknown parameters in the model; and (iv) sequential stochastic maximization of the expected value of the criterion function over the set of feasible policies.

Of course, this is not the way in which we proceed nowadays. In some cases, there will be no formal analysis whatever; in those cases where formal analysis is attempted, it will typically consist in computer simulation of some simple policy, using a random generator of annual yields reflecting historical experience, but proceeding conditionally as regards other parameters (such as trends, demand elasticities, and the like). The criterion function will typically be some discounted net sum of consumers surplus plus producers surplus minus direct costs of the program. Such simulations are clearly valuable (see, e.g., [41]), and they are a significant step in the right direction. But we should prepare ourselves to do better: first, by relating more tightly the specification of the criterion function to general economic theory (recognizing, for instance, the role of private speculators, the efficiency aspects of the program, and so on); second, by treating in probabilistic terms all the unknown parameters of the model, instead of proceeding conditionally on point estimates for some of these; and third, by searching for optimal decision rules by mathematical programming techniques.

My remarks will successively point to the directions for research in normative economics connected with the first task; to the new possibilities offered by Bayesian statistics as regards the second task; and to some of the specific problems in stochastic sequential optimization associated with the third task.

3.

By way of introduction to my first topic, I will recall some relevant propositions from welfare economics under certainty. We start from consumers' sovereignty, and take consumers' tastes as given. In the absence of externalities and of consumer saturation, a competitive equilibrium is a Pareto-optimum; under additional convexity assumptions on preferences and production possibilities, one can associate with every Pareto-optimum a price system and a distribution of initial resources for which the allocation is also competitive. Accordingly, profit maximization at given prices by firms (producers) is compatible with productive and allocative efficiency; equilibrium prices reflect the marginal values of consumers and can, for instance, guide evaluations of costs and benefits, at least locally. These two propositions hold in any institutional framework. But when there are externalities, or when the prevailing allocation is not a competitive equilibrium, they lose their generality. In some cases, one may resort to the methodology of second-best, or constrained Pareto-optimality, to define efficiency conditions. This has been done successfully for a few specific situations, in which it has typically been found that the possibility of carrying out lump-sum transfers among consumers would preserve the separation of the efficiency problems from the distributive problems. In other cases, it seems more meaningful to approach group decisions by means of solution concepts that are not reducible to decentralization through prices, but have distinct claims to realism. An example of such a concept is the core, which has been studied extensively; for some specific models, the analysis has interesting implications for allocative efficiency.

This is still a narrow basis from which to proceed towards policy decisions in industrial states and developing countries alike. But narrow as it may be, that basis has the merit to exist; it is a rigorous and well-articulated theoretical model. The underlying assumptions are clearly defined; as Koopmans remarked [31, p. 28], "their importance is emphasized by the strength of the conclusions that can be derived from them". The frontiers of our knowledge are sharply defined, inviting "the construction of economic knowledge through a sequence of models".

The possibility of extending these results to a world of uncertainty was first suggested in Arrow's pioneering paper "The role of securities in the optimal allocation of risk bearing" [1]. Arrow's results for an exchange economy were extended by Debreu [6] to a private ownership economy with production, and then by Radner [38] to the case in which different agents have different information. Letting again the uncertainties be described by a set of states, a consumption plan for a consumer now specifies the quantity of each commodity to be consumed under each state of the world: it is a vector in an enlarged commodity space, obtained as the cartesian product of the traditional commodity space and the state space. A production plan for a producer is similarly defined. Taking again as given the preferences of consumers among consumption plans, the propositions of welfare economics under certainty carry over directly to this model. In particular, the correspondence between Pareto-optima and competitive equilibria is preserved, with price systems now defined as vectors in the enlarged commodity space. The interpretation of such price systems is straightforward: the price of commodity c under state s, in terms of today's numeraire, is the price to be paid today for one unit of commodity c, to be claimed if and only if state s obtains; it is the price of a claim to one unit of commodity c contingent on the occurrence of state s, and may be interpreted as an insurance premium. Given such a price system, that is, given a full set of insurance markets for all commodities and all states, profit maximization at given prices by firms is again efficient, and prices provide appropriate marginal valuations of costs and benefits.[3]

Of course, in an economy extending over time and with continuous random variables entering the definition of the states, a full system of prices for contingent claims can only be an idealization that ignores the costs of computations, communication, and market organization. In real economies, the possibilities of exchanging contingent claims are limited, and the corresponding prices are neither unique nor directly observable: they exist only in the latent form of marginal rates of substitution for consumers or of transformation for producers. The welfare economics of uncertainty are accordingly the welfare economics of second-best, or constrained Pareto-optimality, under restricted exchange opportunities. The time has come when this type of normative economics could, and should, be developed. Some beginnings have already been made, as in some work of Diamond [7], Foley [20], Hahn [22], and Radner [39].

In the specific context of uncertainty, recent work on consumption decisions [16, 23] and portfolio selection [34] provides some of the required building blocks.[4] But the analysis of production decisions raises

[3] Thus, in the Arrow–Debreu model, the information provided by the price system reduces the decision problems of firms and public authorities to mere optimization under certainty. There is, however, an alternative formulation of this conclusion, which runs as follows. It is readily verified that, under appropriate normalization, the prices for contingent claims to any given commodity have all the formal properties of a probability measure on the events (including the rules of conditional probability). Let us for simplicity abstract from time considerations. Choose a numeraire, normalize the prices for contingent claims to the numeraire so that they add up to unity, and call these prices "market probabilities". The price of a claim to commodity c contingent on state s may then be viewed as the "market probability" of state s times the price at which commodity c would exchange against the numeraire, should state s obtain; call this last price the "utility of commodity c under state s". The value of a plan, at the market prices for contingent claims, has become the "expected utility" of that plan, in terms of the "market probabilities" and "utilities" just defined. Under this alternative formulation, the information provided by the price system amounts to the definition of a "utility function" for firms and public authorities. The utility function so defined is the same for all firms; it is linear in profit for any given state, but with a different proportionality constant for each state. For further details, see [12].

[4] Of particular interest to me, in this connection, has been the search for conditions under which the decisions about consumption can be separated from the portfolio choices, with current consumption chosen first, on the basis of a "certainty-equivalent" rate of return, and the portfolio choices coming thereafter. See [15, 16].

altogether new problems. By contrast to the certainty theory, efficiency of production decisions under uncertainty with limited insurance opportunities seems hard to analyze without reference to the prevailing institutional framework. For instance, in private ownership economies with a stock market, many risks accepted by firms are ultimately borne by stockholders, and one would like to relate explicitly the decisions of the firms to the preferences of stockholders, a feature that is specific to private ownership economies, and may well call for appropriate criteria of group decision within the firms.

These remarks may seem quite remote from my illustrative decision problem about price support and buffer stocks. But in fact they are not remote at all. The only reason that such programs are considered, and are not trivial to manage efficiently, is the absence of markets where agricultural output could be traded "contingently" on yields and other determinants of supply or demand (notice that futures markets do *not* fulfill the same functions). A rigorous normative analysis of the problem would require an assessment of the available opportunities for allocating risk bearing, of the additional opportunities offered by the program, and of the efficiency conditions for a constrained Pareto-optimum under the limited exchange opportunities. Pending such an analysis, any discussion of the merits of the program remains somewhat vague, and only sub-optimal policies can be defined.

Before leaving this topic, I would like to draw attention to the dual nature of the tasks just assigned to normative analysis: the inductive aspect, concerned with discovering the functioning and shortcomings of existing institutions, is just as important as the deductive aspect, concerned with discovering the corresponding second-best efficiency conditions. Theoretical analysis is needed to clarify the testable implications of restricted exchange opportunities, but empirical analysis is indispensable to bring the theory to bear upon realistic decision problems.

4.

In the formal analysis of realistic decision problems, we will typically work with a model specifying the relationship among decision variables, exogenous variables, and dependent variables; typically again, these relationships will not be exact, but will contain random terms. We may be uncertain about the structure of the model, the numerical values of its parameters, and the future values of exogenous or random variables.

In terms of the moral expectation theorem, all these sources of uncertainty combine to define the states. We would like to summarize our imperfect information in a probability measure on the states. We would also like to bring empirical observations to bear upon the numerical assessment of that probability measure. The tool which has been developed over the past fifteen years to accomplish precisely this purpose goes under the name of Bayesian statistics. Its logical basis is very simple.

Consider a set of states, a partition of that set into events B_i, and a probability measure which assigns to B_i the probability $P(B_i)$. Consider next an observation x, and a statistical model specifying, for all x and B_i, the conditional probability $P(x|B_i)$. It then follows from Bayes' theorem that:

$$P(B_i|x) = \frac{P(B_i)\cdot P(x|B_i)}{\Sigma_j P(B_j)\cdot P(x|B_j)}\,.$$

Bayes' theorem is the formula through which an empirical observation x is brought to bear upon the assessment of a probability measure on the states. There are two inputs to this formula. The first input is $P(B_i)$, the probability measure defined on the states *prior* to the observation; if $P(B_i)$ is the same for all i, then $P(B_i|x)$ is proportional to $P(x|B_i)$, and we say that our prior measure is noninformative. The second input is $P(x|B_i)$, the likelihood of the observation; if $P(x|B_i)$ is the same for all i, then $P(B_i|x)$ is equal to $P(B_i)$, and the observation is noninformative; when this property holds for all x, the B_i's are called "observationally equivalent." These are the two inputs. The output of the formula is a posterior probability measure which combines our prior information with that supplied by the observation; it is the probability measure that will be used for decision purposes.

This procedure is very general; it enables us to combine various types of information, like cross-section and time-series observations or judgmental probabilities and sample evidence; and it enables us to define probability measures on the states on the basis of observations alone, starting from prior ignorance.

It is of course a long distance from these simple principles to operational statistical procedures, but substantial contributions have already been made towards bridging the gap. The work of Schlaifer, Raiffa, Pratt and several of their associates has been particularly significant in this respect [44, 40, 37]. For a number of standard data generating processes (like binomial, Poisson, and normal sampling), they have defined families of prior densities on the process parameters having the useful property that posterior densities belong to the same families and are characterized by parameters that can be readily computed as functions of the corresponding parameters of the prior densities on the one hand and of sufficient statistics summarizing the sample information on the other hand. With these families of "natural conjugate densities", "the kernel of the prior density combines with the sample kernel in exactly the same way that two sample kernels combine" [40, p. 49]. Typically, a limiting member of the natural conjugate family may be regarded as noninformative about the process parameters. When one wishes to incorporate in the analysis prior information which is not equivalent to what might have been learned from a hypothetical previous sample generated by the same process, it is sometimes possible to extend the natural conjugate family by introducing additional parameters. Otherwise, one must resort to specific algebraic or numerical analysis; in such cases, one loses the reproductive properties which make natural conjugate analysis so attractive; and one faces the practical limitations of numerical analysis, especially when many variables are involved.

A further step consists of preposterior analysis, which is concerned with the design of experiments. By studying the attributes of the posterior density prior to the observation, and relating these attributes to optimal decisions, it is sometimes possible to compute the expected value of sample information and thereby to encompass data collection in the formal analysis.

5.

I would like to review some features of this approach in the specific econometric context of simultaneous equations models. I will do this mostly on the basis of my own work [13, 14, 17], but I wish to acknowledge the decisive stimulation provided by others who have worked in the same area, in particular V. K. Chetty [5], R. Harkema [24], G. Kaufman [29], T. Kloek [30], J. Kmenta [52], T. J. Rothenberg [42], and A. Zellner [51]; and I wish to thank my closer associates, Juan-Antonio Morales, Michel Mouchart, and Jean-François Richard, who frequently provided specific assistance as well as intellectual encouragement.

My interest in Bayesian analysis of simultaneous equations arose naturally in 1961, as I taught econometrics to a group of students trained in Bayesian statistics. My starting point consisted of the paper by Walter Fisher on "Estimation in the linear decision model" [19] and the paper by Theil and Goldberger on "Pure and mixed statistical estimation in economics" [47]. In addition, a paper by Frank Fisher, "On the cost of approximate specification in simultaneous equation estimation" [18], pointed to a natural application of Bayesian methods: in case of doubt about the prior specification, why not introduce stochastic constraints on the structural parameters, instead of deterministic constraints?

Consider the structural system $By+\Gamma z = u$, with reduced form $y = \Pi z+v$, and let Σ and Ω denote the parameters of the structural and reduced form disturbances respectively. The structural parameters are thus (B, Γ, Σ) and the reduced form parameters (Π, Ω). The identification problem arises because the likelihood can be written as a function of reduced form parameters only, whereas the correspondence between structural and reduced form parameters is many-to-one. That problem is traditionally solved by restricting a priori the range of the parameters in such a way that a one-to-one correspondence emerges. More generally, one can define a prior density function $f(B, \Gamma, \Sigma)$ on the space of structural parameters, and combine it with the sample likelihood by means of Bayes' theorem to obtain the posterior density $f(B, \Gamma, \Sigma | Y, Z) \propto f(B, \Gamma, \Sigma)L(Y, Z | B, \Gamma, \Sigma)$. *This operation is well defined, whether the model is identified or not. But conditionally on a set of values for the reduced form parameters, the prior and posterior densities of B are identical:*

$$f(B|\Pi, \Omega, Y, Z) \underset{Y,Z}{\equiv} f(B|\Pi, \Omega).$$

This identity reveals that all nonsingular matrices B are observationally equivalent: the observations are informative about the reduced form parameters alone. The identity provides a simple and general statement of the identification problem. Indeed, exact identifying restrictions correspond to the special case where $f(B|\Pi, \Omega)$ has all its mass concentrated at a single point.

From the Bayesian viewpoint, classical identification theory is really concerned with local uniqueness of posterior modes. For a system of m equations identified by means of exclusion restrictions, a necessary condition for identification of a given equation is that at least $m-1$ variables be excluded from it, and a sufficient condition is that a certain submatrix from Π should have full rank. The Bayesian counterpart of these conditions states that the coefficients of a given equation will have locally unique posterior modes if and only if the prior density has locally unique modes for at least $m-1$ of these coefficients and the *very same* rank condition is fulfilled. Thus, a prior density with finitely many modes carries essentially the same identification power as it would with all its mass concentrated at a single point.[5]

This remark should bring comfort to those who regard underidentification as the natural state of an econometric model: indeed, stochastic prior information, even if imprecise, should enable them to proceed with structural estimation. Turn now to overidentification. In principle, deterministic restrictions are maintained, even if they conflict with the observations: the prior information dominates the sample information. In contrast, stochastic prior information gets combined with sample information through Bayes' theorem; neither type of information dominates the other, but both count in proportion to their respective precisions.

As a final remark about identification, let me note that locally unique posterior modes need not be relevant for decision purposes. Consider a decision problem in which the criterion function involves the square of a structural coefficient; examples are easy to find. We are then interested in the second moments of the posterior density. But we know, for instance, that even moments of two-stage least-squares estimators exist only up to an order not exceeding the degree of overidentification of the equation being estimated; and even less can be claimed about limited information maximum likelihood estimators [33]. We shall keep this problem in mind as we review some more specific results about Bayesian inference.

Space limitations impose selectivity, so I shall only discuss a limited information approach. When we study a single equation from a system under limited information, we ignore all the a priori information about the remaining equations, except for the classification of variables into endogenous ones and exogenous ones. In a Bayesian approach, this means using a noninformative prior density about the parameters of these remaining equations. Noninformative prior densities on variances and covariances are a persistent headache, the classical work of Jeffreys [28] notwithstanding. Fortunately, the nature of our problem suggests

[5] Unlike the other statements in the text, this result is not quoted from my papers [13, 14, and 17]. I have only recently become aware of this close analogy between the classical conditions for identification through exclusion restrictions and their Bayesian counterpart. This analogy was clarified through discussion with Messrs. Morales, Mouchart, and Richard. More complete results on this point will be presented in a forthcoming paper.

unambiguously how we should proceed. Indeed, let there be m endogenous and n exogenous variables, and let the structural disturbances be jointly normally distributed with zero expectation and covariance matrix Σ, identically and independently for all observations. If we want our prior density to be constant-valued over the space of structural regression coefficients (B, Γ), and to remain constant-valued if we perform the nonsingular integrand transformation from Γ to Π, then the joint prior density $f(B, \Gamma, \Sigma)$ is *uniquely* defined by its kernel

$$|\Sigma|^{-(m+n+1)/2}$$

and

$$f(B, \Gamma, \Sigma^{-1}) \propto |\Sigma|^{(m+1-n)/2}.$$

Combining this prior density with the likelihood leads to a posterior density which can be factored into: a normal density for Γ, conditionally on B and Σ^{-1}; a Wishart density for Σ^{-1}, conditionally on B; and a constant-valued marginal density for B (as was to be expected in the absence of identifying prior information). This factorization permits analytical integration with respect to all the parameters except those of a single equation. In order to identify that equation, we may combine at will deterministic and stochastic types of prior information. In particular, the extended natural conjugate prior family is conditionally normal for the regression coefficients given the residual variance, and marginally gamma for that variance; this is also the form in which information would arise from, say, a regression analysis of cross-section data immune from simultaneity bias.

For illustrative purposes, I shall consider the case of identification through exclusion restrictions; the more general case of a natural conjugate prior is both logically and computationally analogous. And I shall discuss only the marginal posterior density for the regression coefficients of the endogenous variables appearing in one equation, say β_Δ. That marginal density, $f(\beta_\Delta | Y, Z)$, is obtained upon further analytical integration, first with respect to the nonzero regression coefficients of exogenous variables—let there be $n - n_*$ of these—and next with respect to the residual variance; it is defined by

$$f(\beta_\Delta | Y, Z) \propto \frac{(\beta_\Delta W_{\Delta\Delta} \beta'_\Delta)^{(T-m+1)/2}}{(\beta_\Delta W^*_{\Delta\Delta} \beta'_\Delta)^{(T-m+1+n-n_*)/2}} \, d\beta_\Delta,$$

where $W_{\Delta\Delta}$ and $W^*_{\Delta\Delta}$ are the familiar matrices of cross-products of least-squares residuals appearing in the variance ratio from which limited information maximum likelihood estimators are computed. Actually, our posterior kernel differs from the variance ratio to be maximized for LIML estimation in one respect only, namely the different exponents in the numerator and denominator.

The difference in exponents results from the fact that our density is a marginal one: had we instead looked for the conditional density of β_Δ, given that the other parameters in our equation take their modal or mean values, then the two exponents would have been equal, as in the LIML case. The difference in exponents has an important implication: the posterior moments of β_Δ exist up to the degree of overidentification, whereas the small sample moments of LIML estimators of β_Δ do not exist. When the likelihood function has a complicated algebraic expression and involves many nuisance parameters, or when there are strong interdependences among parameters, marginal densities may be quite different from conditional densities; the marginal densities are usually the relevant ones, and will be more reliable—but they are sometimes difficult to obtain analytically. The posterior mode of β_Δ can be computed in almost the same way as LIML estimators, but so far moments can only be evaluated by numerical integration techniques. The numerical integration involves quadratic forms which are easy to evaluate. In general, a posterior density based directly upon the likelihood function is much more tractable than the sampling distribution of a modal value of that function, which involves the maximization operator.

These points are vividly illustrated by the overidentified version of the demand equation in Tintner's two-equation model of the U.S. meat market [48]. After normalization, β_Δ consists of a single element, β_{12}, the price coefficient in the demand function. The model uses three exogenous variables, so that $m+1-n = 0$, and our prior density embodies no other information than the exclusion restrictions. The sample consists of twenty-three observations. Fig. 1 depicts, as a function of β_{12}, the variance ratio in a form to be maximized for limited information maximum likelihood estimation; notice first that the integral of this function does not converge: *conditionally* on the modal (or mean) values of Γ and Σ^{-1}, β_Δ does not have a proper density; notice also that the function is constant, up to computer accuracy, over the whole range from 4 to 7. (Tintner has computed a LIML estimator of 5.8, Goldberger has recomputed that estimator as 4.8, with an asymptotic standard error of 5.7; the two-stage least-squares estimator is 1.6, with an asymptotic standard error of 0.6; the ordinary least-squares estimator is 1.0.) Fig. 2 gives the posterior marginal density of β_{12}, obtained under the *very same* prior information. It has a mode of 2.46, a mean of 2.96, and a standard deviation of 2.9 (over the range of integration). The contrast between the two figures is the contrast between a conditional measure and a marginal density; algebraically, the whole difference has consisted in raising the exponent in the denominator from 11 to 12. . . .

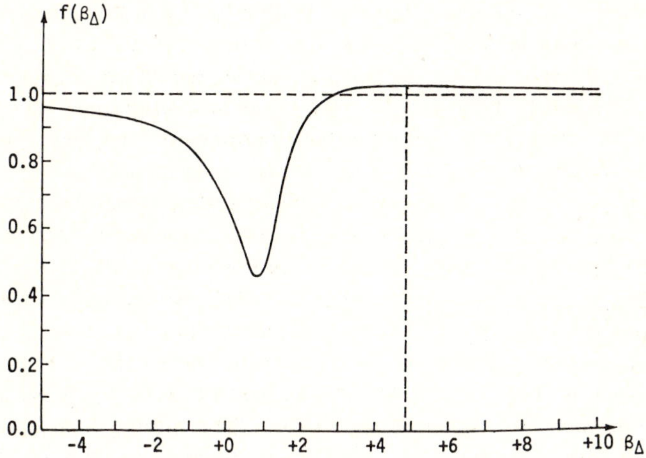

Fig. 1

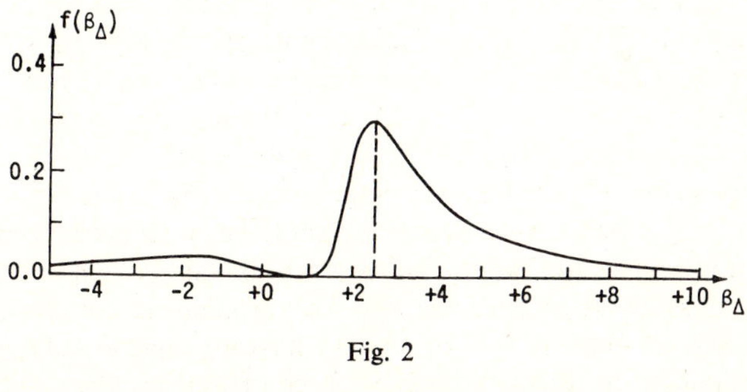

Fig. 2

I have also used some results of Wold's analysis of Swedish time-series and cross-section data [50] to define a prior density on the coefficients of Tintner's demand function. My prior is much less concentrated than suggested by Wold's standard errors; it relaxes somewhat Tintner's exclusion restrictions by making them stochastic; and it is noninformative about the residual variance. The resulting posterior density of β_{12}, shown in fig. 3, has a mode of 1.7, a mean of 1.6, and a standard deviation of 1.1; the gain in precision is, of course, due to the use of prior information.

In summary, I would say that limited information analysis can now be carried out under a Bayesian approach with no more computational effort than under a maximum likelihood approach; but the Bayesian approach is more flexible, leads to sharper finite sample results, and is better suited for decision purposes.

The relevance of Bayesian econometric methods for such decision problems as my illustrative example is obvious enough. The price

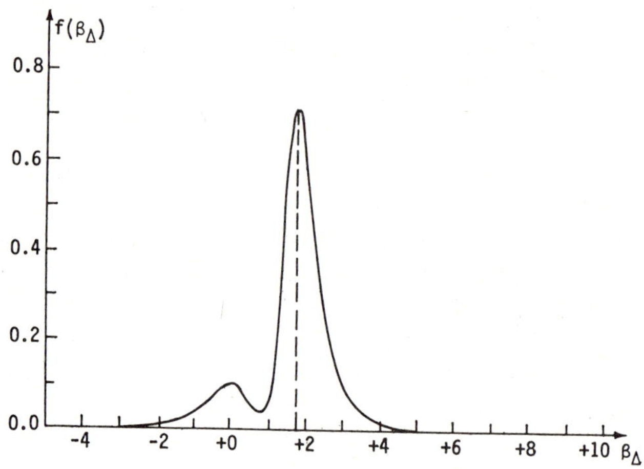

Fig. 3

elasticity of demand for some agricultural product comes in, both to evaluate consumers' surplus and to forecast inventory levels associated with specific price policies; information about a price elasticity will typically come from time-series data, and a simultaneous equations model will be required to identify the demand function; additional information about other parameters of that function, like the income elasticity, will typically be available from other sources, such as cross-section data, and more efficient use will be made of the time-series data if that additional information is taken into account. Bayesian methods permit summarizing our information in a posterior marginal density on the price elasticity; that density may be used for calculating expected values associated with specific policies; this procedure is much more satisfactory than the alternative of proceeding conditionally on some point estimate of the price elasticity of demand.

6.

There remains now to comment briefly about the prospects for computing optimal policies. My example may serve as a starting point. To be specific, let us assume that demand functions and storage costs are linear, let us take agricultural inputs as given, and let us further ignore indirect effects. In the suboptimization problem outlined above, the criterion function combining consumers' and producers' surpluses with direct costs of the program will be a quadratic function of the policy variables, with stochastic coefficients. The range of feasible policies will be defined by inequality constraints, like the nonnegativity and storage constraints on inventories.

The resulting optimization problem may seem relatively simple, but real difficulties are associated with its sequential nature. Indeed, assume for a moment that we were only concerned with a one-period problem, some value being attached to terminal inventory; the total value of end-of-period inventory could for instance be given as a quadratic function of inventory level, with stochastic coefficients. We would then need only to solve a quadratic programming problem, a rather simple one at that. But how could the value of terminal inventory be given to us, when terminal inventory for the first period is simply the starting inventory for the second period? It is only by solving the second period's problem that we could specify the terminal value for the first period's problem; and in order to solve the second period's problem, we need a value for the second period's terminal inventory, which can only be obtained by solving the third period's problem, and so on. This is the horizon problem. We start with an indefinite future; in looking for an optimal solution, we must either solve the problem over the infinite future, or we must show that if the problem is truncated in a certain way at a certain point, and an optimal solution to the finite problem is found, then there will always exist a continuation to the finite solution which leads to a global optimum.

Solving the problem over the infinite future would be easy in the absence of constraints on the policy variables: the Simon–Theil certainty equivalence theorem for unconstrained quadratic optimization [45, 46] could be invoked and would justify replacing all stochastic parameters by their expected values; stationarity assumptions might then permit computing optimal decision rules. But there are constraints: buffer stocks cannot become negative, and one certainly hopes that they will vanish at some point!

Truncating the problem would be easy if the price elasticity of demand were zero and holding costs were linear: the problem would then reduce to a stochastic warehousing problem, for which horizon rules and forward-working algorithms have been defined.[6] But certainly the price elasticity is different from zero.

To the best of my knowledge, standard algorithms for sequential stochastic problems exist neither in the quadratic case with constraints nor in the linear case with more than two periods; the model of two-period stochastic linear programming, or linear programming with recourse [49], is an interesting one, of genuine mathematical interest, and some of the results obtained for that problem may eventually help in attacking N-stage problems [27]; but as of now, N-stage algorithms are not available.

Unless we had the rare good fortune to discover that our specific problem is discussed in the inventory literature [21] there would seem to be few avenues worth exploring. The main ones are Markov programming [3] and stochastic control theory [32] (sometimes also referred to under other names). The first avenue clearly holds some promise; of course, it would require a discrete formulation of our problem; but the approximations involved, if any, might be tolerable. Of greater concern would be the realism of the stationarity assumptions on which that theory rests. Stochastic control theory seems less promising in terms of operational algorithms, but it might help with analytical characterizations of optimal policies.

Before generating too much enthusiasm about these possibilities, however, we should realize fully the implications of imperfect knowledge

[6] Thanks to its simplicity, the warehousing model provides a neat illustration of sequential stochastic problems. Rules for optimal truncation (horizon rules) can be defined in terms of the densities of the unknown parameters (prices and costs, in this case). The rules consist in placing bounds on the expected value of terminal inventory for the relevant period. Bounds can also be placed on the expected value of the first period's terminal inventory by considering only the first n periods, $n \geq 2$. As n becomes larger, the bounds are brought towards each other in a monotonic way. This is the basis for defining forward-working algorithms. For further details, see [4].

about the model parameters in sequential stochastic problems. The density function of such parameters will frequently be the posterior marginal density of some statistical or econometric investigation. At each stage, additional observations will be available to revise these densities, and the revisions will seldom be given by simple formulae. The marginal densities, as of the first period, of parameters for more distant periods can only be specified through preposterior analysis. More importantly, the conditional densities of the stochastic coefficients pertaining to each period will depend upon all the intervening observations, even if the random terms themselves have a Markovian property. To top it all, these intervening observations will themselves depend upon the decisions made in earlier pericds. Consequently, full optimization would call for taking into consideration the expected value of the information generated by the decisions, in addition to the expected value of the direct consequences of the decisions. To be concrete, a monopolist may wish to depart from the price which maximizes expected profit, simply to learn more about his demand function. Our experience with the mathematical programming problems associated with such situations is too meager to permit further considerations now, but it is obvious that the difficulties will be great and the challenge lasting. In the meantime, we will live with suboptimization and continue to solve proximate problems, following the suggestion already voiced by Arrow in his presidential address fourteen years ago [2].

7.

To have identified certain problems is only a first step towards solving them, and I realize that my remarks have covered more problems than results. But my main purpose was to review the broad picture as I see it tcday. To repeat once more, we are still a long way from operational routines for a formal analysis of economic decision problems under uncertainty, but the goal is in sight, and the task is challenging. The challenges come on several fronts: normative economics, statistical methods of econometrics, and mathematical programming. I have no doubt that these challenges will be answered; this is just a matter of time. When problems of practical relevance and scientific importance are clearly defined, they attract talent. The problems of decision making under uncertainty have all these attributes.

I thus feel tempted to conclude this address with a note of optimism.

But optimism about an exciting subject is too easily wishful. So, to those who will extend the work in this area, I had better refrain from promising success. I will simply offer the modest advice that they are sure to learn a great deal from looking at real-life problems; and I will make the modest promise that it will be a lot of fun.

References

[1] K. Arrow, Le rôle des valeurs boursières pour la répartition la meilleure des risques. In: Econométrie. CNRS, Paris, 1953, 41–48; translated as: The role of securities in the optimal allocation of risk-bearing. Review of Economic Studies 31 (1964), 91–96.

[2] K. Arrow, Statistics and economic policy. Econometrica 25 (1957), 523–531.

[3] D. Blackwell, Discrete dynamic programming. The Annals of Mathematical Statistics 33 (1962), 719–726.

[4] A. Charnes, J. H. Drèze and M. H. Miller, Decision and horizon rules for stochastic planning problems: a linear example. Econometrica 34 (1966), 307–330.

[5] V. K. Chetty, Bayesian analysis of Haavelmo's models, Econometrica 36 (1968), 582–602. (Reprinted in the present volume.)

[6] G. Debreu, Theory of value. New York, Wiley, 1959.

[7] P. Diamond, The role of a stock market in a general equilibrium model with technological uncertainty. American Economic Review 57 (1967), 759–776.

[8] J. H. Drèze, Individual decision making under partially controllable uncertainty. Unpublished Ph.D. Dissertation, Columbia University, 1958.

[9] J. H. Drèze, Les probabilités subjectives ont-elles une signification objective?. Economie Appliquée 13 (1960), 55–70.

[10] J. H. Drèze, Les fondements logiques de l'utilité cardinale et de la probabilité subjective. In: La Décision. CNRS, Paris, 1961, 73–87.

[11] J. H. Drèze, L'utilité sociale d'une vie humaine. Revue Française de Recherche Opérationnelle 23 (1963), 93–118.

[12] J. H. Drèze, Market allocation under uncertainty. European Economic Review 2 (1971), 133–165.

[13] J. H. Drèze, The Bayesian approach to simultaneous equations estimation. ONR Research Memorandum 67, The Technological Institute, Northwestern University, 1962.

[14] J. H. Drèze, Limited information estimation from a Bayesian viewpoint. CORE Discussion Paper 6816.

[15] J. H. Drèze and F. Modigliani, Epargne et consommation en avenir aléatoire. Cahiers du Séminaire d'Econométrie 9 (1966), 7–33.

[16] J. H. Drèze and F. Modigliani, Consumption decisions under uncertainty. Journal of Economic Theory 5 (1972), 308–335.

[17] J. H. Drèze and J. A. Morales, Bayesian full information analysis of the simultaneous equations model. CORE Discussion Paper 7031.

[18] F. Fisher, On the cost of approximate specification in simultaneous equation estimation. Econometrica 31 (1961), 139–170.
[19] W. Fisher, Estimation in the linear decision model. International Economic Review 3 (1962), 1–29.
[20] D. K. Foley, Economic equilibrium with costly marketing. Working Paper 52, Department of Economics, Massachusetts Institute of Technology, 1970.
[21] G. Hadley, and T. H. Whitin, Analysis of inventory systems. Englewood Cliffs, Prentice-Hall, 1963.
[22] F. Hahn, Equilibrium with transaction costs. Econometrica 39 (1971), 417–439.
[23] N. Hakansson, Optimal investment and consumption strategies under risk for a class of utility functions. Econometrica (forthcoming).
[24] R. Harkema, A class of tractable prior distributions on structural parameters of simultaneous equation systems. Report 6919, Econometric Institute, Rotterdam.
[25] C. C. Holt, F. Modigliani, J. F. Muth and H. A. Simon, Planning production, inventories and workforce. Englewood Cliffs, Prentice-Hall, 1960.
[26] W. C. Hood and T. C. Koopmans, Studies in econometric method. New York, Wiley, 1953.
[27] R. Jagannathan, A solution procedure for a class of multi-stage linear programming problems under uncertainty. Management Science Research Report 146, Carnegie-Mellon University, 1968.
[28] H. Jeffreys, Theory of probability, 3rd ed. Oxford, Clarendon Press, 1961.
[29] G. Kaufman, Posterior inference for structural parameters using cross-section and time-series data. Paper presented at the Second World Congress of the Econometric Society, Cambridge, September, 1970. (A revised version of this paper appears in the present volume.)
[30] T. Kloek and R. Harkema, A limiting Bayesian approach to simultaneous equation systems. Report 7004, Econometric Institute, Rotterdam.
[31] T. C. Koopmans, Three essays on the state of economic science. New York, McGraw-Hill, 1957.
[32] H. J. Kushner, Stochastic stability and control. New York, Academic Press, 1967.
[33] R. S. Mariano, On distributions and moments of single-equation estimators in a set of simultaneous linear stochastic equations. Econometric Series Report 2, Institute for Mathematical Studies in the Social Sciences, Stanford University, 1969.
[34] H. M. Markowitz, Portfolio selection. New York, Wiley, 1959.
[35] J. Marschak, Economic measurements for policy and prediction, 1–26 in [26].
[36] J. R. Nelson (ed.), Marginal cost pricing in practice. Englewood Cliffs, Prentice-Hall, 1964.
[37] J. Pratt, H. Raiffa and R. Schlaifer, Introduction to statistical decision theory. New York, McGraw-Hill, 1965.
[38] R. Radner, Competitive equilibrium under uncertainty, Econometrica 36 (1968), 31–58.
[39] R. Radner, Problems in the theory of markets under uncertainty. American Economic Review 60 (1970), 454–460.
[40] H. Raiffa and R. Schlaifer, Applied statistical decision theory. Cambridge, M.I.T. Press, 1961.

[41] S. Reutlinger, A simulation model for evaluating national buffer stock programs. Paper presented at the Second World Congress of the Econometric Society, Cambridge, September, 1970.

[42] T. J. Rothenberg, A Bayesian analysis of simultaneous equation systems. Report 6315, Econometric Institute, Rotterdam. (A revised version of this paper appears in the present volume.)

[43] L. J. Savage, The foundations of statistics. New York, Wiley, 1954.

[44] R. Schlaifer, Probability and statistics for business decisions; an introduction to managerial economics under uncertainty. New York, McGraw-Hill, 1959.

[45] H. A. Simon, Dynamic programming under uncertainty with a quadratic criterion function. Econometrica 24 (1956), 74–81.

[46] H. Theil, Optimal decision rules for government and industry. Amsterdam, North-Holland, 1964.

[47] H. Theil and A. Goldberger, Pure and mixed statistical estimation in economics. International Economic Review 2 (1961), 65–78.

[48] G. Tintner, Econometrics. New York, Wiley, 1952.

[49] R. J-B. Wets, Stochastic programs with recourse: A survey. Mathematical Note 614, Boeing Scientific Research Laboratories, 1969.

[50] H. Wold and L. Jureen, Demand analysis. New York, Wiley, 1953.

[51] A. Zellner, Bayesian and non-Bayesian analysis of simultaneous equation models. Paper presented at the Second World Congress of the Econometric Society, Cambridge, September, 1970.

[52] A. Zellner, J. Kmenta and J. H. Drèze, Specification and estimation of Cobb–Douglas production function models. Econometrica 34 (1966), 784–795.

S. E. Fienberg and A. Zellner (eds.)
Studies in Bayesian Econometrics and Statistics.

2.2 THE BAYESIAN APPROACH AND ALTERNATIVES IN ECONOMETRICS—I*

Arnold ZELLNER

University of Chicago

Introduction

The Bayesian approach and alternative approaches to the problems of inference and decision have been discussed at length in books and journals for quite a long time.[1] The subjects of these discussions are of fundamental importance for science in general since they involve consideration of conceptual frameworks and methods for analyzing data and making decisions. If, as Karl Pearson suggests, the unity of science is in the main unity in the methods employed in analyzing and learning from data, then the aforementioned discussions may have important implications for work in econometrics.[2]

Ideally, it would be desirable to have a unified set of principles for making inferences and decisions which can be readily applied in a broad

* Reprinted from M. D. Intriligator (ed.), Frontiers of quantitative economics. Amsterdam, North-Holland Publishing Company, 1971, with permission of the North-Holland Publishing Company. This work was financed in part by the National Science Foundation, Grant GS-2347, and by income from the H. G. B. Alexander Endowment Fund, Graduate School of Business, Univ. of Chicago.

[1] See, e.g., H. Jeffreys, Theory of probability. Oxford, Clarendon Press, 1939, 1948, 1961 and 1966, especially Chapters I, VII and VIII.—L. J. Savage et al., The foundations of statistical inference, London, Methuen, 1962.—D. V. Lindley, The use of prior probability distributions in statistical inferences and decisions. In: Proceedings of the fourth Berkeley symposium on mathematical statistics and probability, edited by J. Neyman, Vol. 1 (1961), 453–468.—J. W. Pratt, Bayesian interpretation of standard inference statements, J. Royal Stat. Soc. B 27 (1965), 169–203 (with discussion).—I. J. Good, A subjective evaluation of Bode's law and an "objective" test for approximate numerical rationality. J. Am. Stat. Assoc. 64 (1969), 23–49 (with discussion).

[2] K. Pearson, The grammar of science. Everyman edition, 1938, 16.

range of circumstances and fields to yield good results. One of the main points of this paper, and hardly a novel one, is that the Bayesian approach approximates this ideal much more closely than do non-Bayesian approaches currently in use in econometrics.[3] In fact, many non-Bayesian approaches in econometrics and elsewhere are rather good examples of what has been called a collection of "ad hockeries".[4] Some examples illustrating this point will be provided below.

The plan of the paper is as follows: In section 1 several key features of the Bayesian approach will be considered with main emphasis on illustrating its unity, generality and usefulness. Section 2 provides comparisons with other approaches utilized in econometrics. Last, section 3 contains some concluding remarks.

1. Aspects of the Bayesian approach[5]

The following are some key points which are relevant for comparing Bayesian and non-Bayesian approaches in econometrics.

(1) As Jeffreys and others have emphasized, the Bayesian approach to inference complements very nicely the activities of researchers.[6] A researcher is often concerned with the problem of how information in data modifies his beliefs about hypotheses and parameter values. In the Bayesian approach to inference, an investigator has formal and operational techniques for determining how sample information modifies his beliefs. That is, initial beliefs, represented by prior probabilities, are combined by means of Bayes' Theorem with information in data, incorporated in the likelihood function, to yield posterior probabilities relating to parameters or hypotheses. In a fundamental sense, this Bayesian procedure for changing initial beliefs is a learning model of great value in accomplishing what Jeffreys and others consider to be a major objective of science, namely learning from experience.

[3] This point has been emphasized by Jeffreys, de Finetti, and others in connection with the Bayesian approach in general vis-à-vis other approaches.

[4] This phrase, used by B. de Finetti to describe non-Bayesian approaches at a meeting on Statistical Methods of Econometrics, sponsored by the Institute of Mathematics, Univ. of Rome, Frascati, June 1968, is probably due to I. J. Good.

[5] A number of these points were put forward earlier in A. Zellner, Bayesian inference and simultaneous equation econometric models. Paper presented to the First World Congress of the Econometric Society, Rome, September 1965.

[6] Cf. H. Jeffreys, Theory of probability, cit. supra.

(2) As regards statistical estimation, Bayes' Theorem (sometimes also called Bayes' Rule or the Principle of Inverse Probability) can and has been applied in analyses of all kinds of statistical models including regression models, time series models, Markov transition probability models, simultaneous equation models, errors in the variables models, etc. *In every instance*, the posterior probability density function (pdf) for parameters is proportional to the prior pdf times the likelihood function with the factor of proportionality being a normalizing factor. That one simple principle has such wide applicability is indeed appealing. Further, it should be noted that by appropriate choice of prior pdf an investigator can introduce as much or as little prior information in his analysis as he chooses. The likelihood function in Bayes' Theorem is known to incorporate all the sample information. Thus the posterior pdf for parameters of a model incorporates all the available information, prior and sample. In addition, the posterior pdf so obtained is an exact finite sample pdf which can be used to make exact finite sample posterior probability statements about parameters of a model. There is in general no need to rely on large sample approximations.[7]

Since the entire posterior pdf for a model's parameters, which incorporates both prior and sample information, is generally available, posterior beliefs about parameters are fully represented. If for some reason an investigator wishes to obtain a point estimate, the Bayesian prescription, choose the point estimate which minimizes expected loss, is a general, operational principle in accord with the expected utility hypothesis. For example, if the loss function is quadratic, the mean of a posterior pdf is optimal in general. Given an absolute error loss function, the optimal estimate is the median of a posterior pdf. The Bayesian point estimate is thus tailored to the particular loss function which is deemed appropriate and is an exact finite sample solution to the problem of point estimation. That Bayesian estimators are known to be admissible and consistent, and to minimize average risk are additional features which some point to in commending their use in practice.[8] Finally, in large samples under rather general conditions, posterior pdf's assume a normal form with mean equal to the maximum likelihood (ML) estimate. This dovetailing of Bayesian and ML results in large samples

[7] Of course, if the sample size is large, large sample approximations can be useful and are often convenient.

[8] For consideration of these properties, see, e.g., T. S. Ferguson, Mathematical statistics: a decision theoretic approach. New York, Academic Press, 1967. Note that these results are based on the assumed use of a *proper* prior pdf for parameters.

is indeed noteworthy. Jeffreys interprets this fact as providing a justification for the ML approach in large samples.[9]

(3) Bayesian methods for analyzing prediction problems are simple, operational, and generally applicable. Whatever the model, the predictive pdf for future observations is obtained.[10] This predictive pdf serves as a basis for making probability statements about future observations. Also, for a given loss function involving prediction errors, it is generally possible to obtain a point prediction which minimizes expected loss. As with point estimation, the solution to the point prediction problem is in accord with the expected utility hypothesis and is tailored to be appropriate for the loss function which is employed.

(4) In the area of control theory, Bayesian methods are particularly valuable in that their application yields a combined solution to the control and estimation problems.[11] For single period control problems, a Bayesian solution incorporates both prior and sample information and involves due allowance for uncertainty about parameter values. For

[9] See H. Jeffreys, Theory of probability, cit. supra, 193–194. Further work on the asymptotic properties of Bayesian procedures appears in D. V. Lindley, The use of prior probability distributions in statistical inferences and decisions, cit. supra, and L. Le Cam, On some asymptotic properties of maximum likelihood estimates and related Bayes estimates. Univ. of California Publications in Statistics 1 (1953), 277–330.

[10] If $\tilde{\boldsymbol{y}}$ denotes a vector of future observations generated by a model with parameter vector $\boldsymbol{\theta}$, and the joint pdf for $\tilde{\boldsymbol{y}}$ and $\boldsymbol{\theta}$, given the sample information, $\boldsymbol{y}$, and prior information, I_0, is denoted $p(\tilde{\boldsymbol{y}}, \boldsymbol{\theta}|\boldsymbol{y}, I_0)$, the predictive pdf is obtained by integrating this joint pdf with respect to the elements of $\boldsymbol{\theta}$. See, e.g., H. Jeffreys, Theory of probability, cit. supra.—H. Raiffa and R. Schlaifer, Applied statistical decision theory, Cambridge, Mass., Harvard Univ. Press, 1961.—D. V. Lindley, Introduction to probability theory from a Bayesian viewpoint, Part Two: Inference. Cambridge, 1965, and A. Zellner, Bayesian inference in econometrics. New York, Wiley, 1971, for examples of predictive pdf's for a number of models.

[11] See M. Aoki, Optimization of stochastic systems. New York, Academic Press, 1967, for a valuable introduction to topics in the control area and an extensive list of references to earlier work. Some works by econometricians include W. D. Fisher, Estimation in the linear decision model, Int. Econ. Review 3 (1962), 1–29.—A. Zellner and V. K. Chetty, Prediction and decision problems in regression models from the Bayesian point of view. J. Am. Stat. Assoc. 60 (1965), 608–616.—A. Zellner and M. S. Geisel, Sensitivity of control to uncertainty and form of the criterion function. In: The future of statistics, edited by D. G. Watts, New York, Academic Press, 1968, 269–289.—E. C. Prescott, Adaptive decision rules for government and industry. Doctoral dissertation, Grad. Sch. of Ind. Administration, Carnegie-Mellon Univ., 1967.

multi-period control problems, Bayesian principles can and have been employed to obtain optimal, computable solutions which not only take account of prior and past sample information and uncertainty about parameter values but also take account of new information as it becomes available and how settings of control variables affect the precision of information about parameter values. Thus solutions to adaptive control problems employing Bayesian methods are solutions to the joint problems of control, estimation and design of experiments.

(5) In the Bayesian approach, prior information about parameters or models can be flexibly and formally incorporated in analyses of estimation, prediction, control, and hypothesis testing problems. This flexibility of the Bayesian approach with respect to the incorporation of prior information in analyses contrasts markedly with currently available non-Bayesian techniques for using prior information in econometric analyses. (See the next section for examples.) Of course non-Bayesians use prior information extensively in their work. For example, exact prior restrictions on parameters are introduced to identify parameters of simultaneous equation models. In the errors in the variables model, prior information about parameter values, say the ratio of error terms' variances, is required to identify parameters. Often in analyses of Cobb–Douglas production function models, the prior assumption of constant returns to scale is introduced in an effort to deal with multicollinearity. What should be appreciated in these and other examples is that Bayesians can introduce such required information in a flexible manner which can more accurately reflect the prior information that we may have about parameter values.[12]

(6) The problem of nuisance parameters is solved quite straightforwardly and neatly in the Bayesian approach. Parameters not of interest to an investigator, that is nuisance parameters, can be integrated out of a posterior pdf to obtain the marginal pdf for the parameters of interest. This marginal posterior pdf can then be employed to make inferences about the parameters of interest.[13]

(7) The Bayesian approach is convenient for the analysis of effects of

[12] This point has been made by J. Drèze on many occasions. See, e.g., The Bayesian approach to simultaneous equations estimation. ONR Research Memo No. 67, The Technological Institute, Northwestern Univ., 1962.

[13] For an example, see A. Zellner and G. C. Tiao, Bayesian analysis of the regression model with autocorrelated errors. J. Am. Stat. Assoc. 59 (1964), 763–768.

departures from specifying assumptions. That is, use of conditional posterior pdf's enables an investigator to determine how sensitive his inferences about a particular subset of parameters are to what is assumed about other parameters. Such an approach has been used on a variety of problems.[14]

(8) In the Bayesian approach, inferences about parameters, etc., can be made on the basis of the prior and sample information which we have. There is no need to justify inference procedures in terms of their behavior in repeated, as yet unobserved, samples. This is not to say that properties of procedures in repeated samples are not of interest and, in fact, e.g., Bayesian estimators have several good sampling properties in that they are admissible and constructed so as to minimize average risk. However, in an inference situation, what is most relevant is the information at hand in the sample and in our prior beliefs.

(9) In the area of comparing and testing hypotheses and models, the Bayesian approach is distinguished from non-Bayesian approaches in that it associates probabilities with hypotheses and provides formal, operational techniques for modifying such probabilities in the light of new information. These posterior probabilities associated with hypotheses and models incorporate prior and sample information and are viewed by many Bayesians as representing degrees of belief. Then too, if one has explicit losses associated with actions, such as accepting or rejecting a particular hypothesis, and posterior probabilities associated with possible states of the world, he can act so as to minimize expected loss in testing hypotheses.[15]

[14] See, e.g., G. E. P. Box and G. C. Tiao, A Bayesian approach to the importance of assumptions applied to the comparison of variances. Biometrika 51 (1964), 153–167.—A. Zellner and G. C. Tiao, Bayesian analysis of the regression model with autocorrelated errors, cit. supra.—V. K. Chetty, Bayesian analysis of Haavelmo's models. Econometrica 36 (1968), 582–602 (reprinted in the present volume).

[15] Some works dealing with comparing and testing hypotheses from the Bayesian point of view include H. Jeffreys, Theory of probability, cit. supra.—D. V. Lindley, Introduction to probability theory from a Bayesian viewpoint. Part Two: Inference, cit. supra.—and A statistical paradox. Biometrika 44 (1957), 187–192 (see also M. S. Bartlett's comment on this paper in the same journal and volume, p. 533).—G. E. P. Box and W. J. Hill, Discrimination among mechanistic models. Technometrics 9 (1967), 57–91.—H. Thornber, Applications of decision theory to econometrics. Doctoral dissertation, Univ. of Chicago, 1966.—A. Zellner, Bayesian inference in econometrics, cited above.—M. S. Geisel, Comparing and choosing among parametric statistical models: A Bayesian analysis with macroeconomic applications. Doctoral dissertation, Univ. of Chicago, 1970.

Having reviewed several important aspects of the Bayesian approach, a comparative review of the characteristics of some non-Bayesian approaches in econometrics will now be presented.

2. *Non-Bayesian approaches in econometrics*

Comparison of the Bayesian (B) approach with non-Bayesian (NB) approaches in econometrics is difficult since there are many formal and informal NB approaches utilized in econometrics. Some of these NB approaches are designed to handle specific types of problems and do not represent a unified set of principles which can be applied to a broad range of problems. Also, it is recognized that good researchers can sometimes obtain reasonable results even if they are not operating with an explicitly formulated set of principles. While this latter fact is recognized, what will emerge from the considerations presented below *is that there is no alternative set of principles currently being employed in econometrics which is as unified and as generally applicable as those embedded in the B approach.* To provide some structure to the discussion, a review of properties of NB approaches in econometrics to the issues and problems listed in section 1 will be presented.

(1) Above, it was pointed out that the Bayesian approach incorporates an explicit learning model which appears to complement the activities of researchers rather nicely. As of the present writing, it is a fact that NB approaches in econometrics do not incorporate learning models as fundamentally and as explicitly as is done in the B approach. For example, in NB approaches it is usually not considered meaningful to associate probabilities with hypotheses or models. Thus in NB approaches, it is impossible to quantify a statement of the type, "The permanent income hypothesis is probably true." Further, they can not formally allow for the modification of such probabilities in the light of new sample information. If the revision of such probabilities is regarded as a representation of the learning process in research, then, of course, NB approaches do not formally provide what is needed for this process. However, this is not to say that non-Bayesians do not engage in considerations concerning whether a particular hypothesis is probably true. They do so, but only in an informal and subjective manner.

That the B approach incorporates a formal and explicit learning model is not necessarily an advantage if, e.g., by doing so the B approach could not be applied fruitfully in practice. That this latter circumstance is not the case is easily established by noting that a wide range of problems has been analyzed from the B point of view without undue difficulty and, in fact, in many instances with less difficulty, both practically and theoretically, than is the case with NB approaches.

(2) In the area of econometric estimation, it was stated above that Bayes' Theorem is a simple and generally applicable principle which can and has been used to analyze estimation problems for a very wide range of models. In NB approaches in econometrics, the situation is quite different. There is a plethora of estimation principles, some more general than others. The following are a few principles which have been put forward: maximum likelihood, least squares, best linear unbiased estimation, generalized least squares, minimum absolute deviations, minimum chi-square, indirect least squares, instrumental variable methods, generalized classical linear estimation, two- and three-stage least squares, simultaneous equation least-squares, *k*-class estimation, double *k*-class estimation, minimal mean square error estimation, minimum variance unbiased estimation, best quadratic unbiased estimation, etc. While some of these methods produce satisfactory results in particular circumstances and for particular models, it is clearly the case that many are not generally applicable and are not related in any obvious way to a simple set of unified and generally applicable principles. This is one reason why many Bayesians regard current practice in econometrics and elsewhere to be rather ad hoc.

One of the more general principles listed above is the maximum likelihood (ML) method of estimation. It can and has been employed to produce estimates of parameters in many different kinds of econometric models. Usually in econometrics, estimators generated by ML are justified in terms of their sampling properties in large samples, e.g., consistency and large sample efficiency. As pointed out above, a ML estimate will be approximately equal to the mean of the Bayesian posterior pdf in large samples under general conditions. Thus, for many models, there is a compatibility between large sample Bayesian and ML results, although there is an important difference in interpretation of them. However, in "small" samples the situation is fundamentally different. ML estimators do not necessarily have good small sample sampling properties. For example, in the standard linear normal regression model, the ML estimator for the disturbance variance is usually discarded in favor of one with "better" properties. Recent work of Stein indicates that if there are three or more regression coefficients to be estimated, the ML estimator for the regression coefficients is inadmissible relative to a quadratic loss function.[16] Further, with regard to the log-normal distribution, it appears that there are estimators with better finite sample properties than possessed by the ML estimator. Last, in connection with parameters of simultaneous equation models, it is a fact that many econometricians use other than ML estimators for a variety of reasons.

Thus while the ML method of estimation is a general one and useful in quite a few circumstances, it is a fact that for quite a few important problems in econometrics it produces unsatisfactory small sample results according to sampling theory criteria and is often replaced by alternatives. This stands in contrast to Bayesian estimators which can be generated by simple principles and which have good sampling properties in that they are known to be admissible and minimize average risk (when it exists).

While not emphasized in much of the econometric literature, the likelihood approach of Barnard and others deserves some comment. In this approach great emphasis is placed on characterizing the shape and general features of the likelihood function without bringing in sampling properties of procedures.[17] That is, the likelihood function is considered in connection with data on hand and not with respect to other possible samples which could have arisen. The location of the mode of the likelihood function is just one feature, and perhaps not the most important feature, to be studied according to proponents of this approach. In this respect, most Bayesians would agree that a thorough study of the likelihood function is desirable. However, without Bayes' Theorem, it appears necessary to entertain not only the concept of probability but also the concept of likelihood. By use of Bayes' Theorem the analysis can go forward in terms of the concept of probability alone which appears to be advantageous and which does not rule out the desirable practice of studying the form of the likelihood function.

While some Bayesians do not emphasize the sampling properties of B estimators, it should be recognized that B estimators have rather good

[16] C. Stein, Multiple regression. In: Contributions to probability and statistics: essays in honor of Harold Hotelling, edited by I. Olkin. Stanford, Calif., Stanford Univ. Press, 1960, 424–443. See also S. L. Sclove, Improved Estimators for Coefficients in Linear Regression. J. Am. Stat. Assoc. 63 (1968), 596–606.

[17] See, e.g., G. A. Barnard, G. M. Jenkins and C. B. Winsten, Likelihood inference and time series. J. Roy. Statistical Society A 125 (1962), 321–372 (with discussion).

sampling properties both in small and large samples. As pointed out above, the B estimator is the estimator which minimizes average risk. This is a well-known exact finite sample property of B estimators which should be of great interest to NB's. Several Monte Carlo experiments have been performed to compare properties of B and NB estimators.[18] In all experiments to date, B estimators have performed as well or better than leading NB estimators. In one particularly outstanding study of an autoregressive model, risk functions and average risk associated with alternative B and NB estimators were estimated.[19] It was established that the properties of estimators for this "simple" model are quite different in small samples and that the B estimator's average risk is quite a bit smaller than of the ML estimator and several other popular NB estimators. Of course point estimation is just a part of the problem of inference. That the Bayesian approach provides the complete, finite sample posterior pdf for models' parameters and can be employed to make other kinds of inferences should not be overlooked.

(3) As pointed out above, the B approach provides a unified and operational approach to the problem of prediction. In NB approaches various principles are employed to generate "optimal" predictors which are not generally applicable. For example, in some situations NB econometricians use minimum variance linear unbiased predictors. It is well known that this principle can not be applied in the case of many important models used in econometrics. Further, the restrictions that a predictor be *linear* and *unbiased* can result in sub-optimal predictors in a number of cases.[20] Bayesians do not place such restrictions on their predictors and as a result can usually obtain point predictions which minimize expected loss. In addition, since the complete predictive pdf is in general available in the B approach, various probability statements

[18] See, e.g., A. Zellner, Bayesian inference and simultaneous equation econometric models, cit. supra.—V. K. Chetty, Bayesian analysis of some simultaneous equation models and specification errors. Doctoral dissertation, Univ. of Wisconsin, Madison, 1966.—H. Thornber, Finite sample Monte Carlo studies: an autoregressive illustration. J. Am. Stat. Assoc. 62 (1967), 801–818.—J. B. Copas, Monte Carlo results for estimation in a stable Markov time series. J. Roy. Stat. Soc. A (1966), 110–116.—T. C. Lee, G. G. Judge and A. Zellner, Maximum likelihood and Bayesian estimation of transition probabilities. J. Am. Stat. Assoc. 63 (1968), 1162–1179.

[19] H. Thornber, op. cit.

[20] See, e.g., A. Zellner, Decision rules for economic forecasting. Econometrica 31 (1963), 111–130.

about future observations can be made for a variety of models. These probability statements, made conditional upon the given sample and prior information, appear to be the kinds of statements which are of great value to economic forecasters. Needless to say, forecasters use a good deal of outside or prior information in their work. That the B approach provides a means of formally incorporating such information in forecasts and in probability statements regarding future outcomes is indeed fortunate and contrasts with what is currently available in NB approaches for accomplishing this objective.

(4) With respect to control problems in econometrics, it is probably accurate to state that approaches currently in use do not allow for uncertainty about values of parameters appearing in the equations of econometric models. This is the case in applications of "certainty equivalence" and "linear decision rule" principles which have appeared in the literature. Several studies have appeared which indicate that Bayesian solutions to control problems, which take account of uncertainty about parameter values, are different from certainty-equivalence solutions and provide lower expected loss particularly when sample information is not very extensive or precise.[21] In addition, B solutions to control problems reflect both sample and prior information (as much or as little of the latter as is judged reasonable to use). As regards multiperiod control problems with unknown model parameters which must be estimated, it appears that there is no alternative to the B adaptive control solution in the econometric literature.

(5) On the question of introducing prior information in analyses, often this is done in an informal manner in NB approaches. For example, in regression problems some use the following procedure: If an independent variable's coefficient estimate turns out to have the "wrong" algebraic sign, the variable is dropped from the regression. This ad hoc procedure can yield good results in terms of, for example, a mean square error criterion. However, it is apparent that the resulting estimator is no longer a minimum variance linear unbiased estimator. Rather it is a biased, nonlinear estimator having a mixed distribution that is part discrete (some probability piled up at zero) and part in a

[21] In particular see the work of Aoki, Fisher, Geisel, Prescott and Zellner cited in footnote 11 above.

truncated normal form.[22] Thus, obviously, there is no secure basis for the usual t tests and for constructing usual confidence intervals. This is then a good example of an ad hoc procedure which departs from what some regard as a fairly general principle (minimum variance linear unbiased estimation).

Another special method for incorporating subjective prior information in NB analyses which has appeared in econometrics texts and the literature is the "mixed linear estimation" procedure for the linear regression model,[23] $\tilde{\boldsymbol{y}} = X\boldsymbol{\beta}+\tilde{\boldsymbol{u}}$. The subjective prior information about the nonstochastic, true parameter vector $\boldsymbol{\beta}$ is represented by $\tilde{r} = R\boldsymbol{\beta}+\tilde{v}$ where R is a matrix to be assigned by an investigator and a tilde denotes random variables. It is further assumed that $E\tilde{v} = 0$ and $E\tilde{v}\tilde{v}' = \Omega$, a matrix whose elements are assigned by an investigator. Then generalized least squares (GLS) is employed to combine the sample and prior information with the common variance of the elements of $\tilde{\boldsymbol{u}}$ replaced by a sample estimate thereby getting an approximation to the GLS estimator. What appears not to have been recognized in this approach is that a rather stringent condition has been placed on the prior information. That is $E\tilde{r} = R\boldsymbol{\beta}$, or if $R = I$, $E\tilde{r} = \boldsymbol{\beta}$, the *true parameter* vector. That prior subjective information be unbiased is a severe restriction on the nature of such information, a restriction that is not imposed in the B approach. Further in analyzing the large sample properties of the linear mixed estimator, it has been customary to assume that $R'\Omega^{-1}R$ is of order T and thus that the prior subjective information grows with the sample size, T, and is not dominated by the sample information as the sample size grows large. That the precision of the prior subjective information depends on the sample size and that the sample information does not dominate in large samples are features of this approach which appear unsatisfactory from the B point of view.

Prior information in the form of exact constraints and/or inequality constraints can of course be introduced in B and NB approaches. In

[22] If $\hat{\beta}$ is the unrestricted least squares estimator, then a "sign-test" estimator, denoted $\hat{\beta}^*$, is given by

$$\hat{\beta}^* = \begin{cases} \hat{\beta} \text{ for } \hat{\beta} > 0 \\ 0 \text{ for } \hat{\beta} \leq 0. \end{cases}$$

[23] See, e.g., A. S. Goldberger, Econometric theory. New York, Wiley, 1964, 261, and the references cited there.

connection with inequality constraints, the criterion function (likelihood function, sum of squared errors, etc.) can be maximized (or minimized) subject to inequality constraints.[24] The resulting point estimators often have extremely complicated, unknown distributions. Testing and interval estimation procedures for the inequality constraint case remain to be worked out. Reliance is often placed on large sample approximations. However, if the sample is truly a large one, the prior information incorporated in the inequality constraints will often be unimportant. In the B approach inequality constraints can and have been introduced via an appropriate choice of prior pdf.[25] Posterior pdf's have been obtained in the usual manner and, as always, summarize the complete information (sample and prior) about parameters. If a point estimate is desired, it can be computed and has a finite sample justification, as explained above.

NB analyses of random parameter models superficially appear to resemble B analyses of fixed parameter models in that a pdf for parameters appears in a NB analysis. However, this pdf is not given the interpretation of representing subjective prior information and, further, B analyses of random parameter models involve placing a prior pdf on the parameters of the pdf for the random parameters.[26]

(6) The problem of nuisance parameters is a particularly thorny one in NB approaches in econometrics and elsewhere. For example, an optimal estimator, say a generalized least squares estimator, may depend on a disturbance autocorrelation parameter which is not of special interest to an investigator. Often a sample estimate of the nuisance parameter (or parameters) is inserted in the expression for an optimal estimator. This procedure produces an approximation to the optimal

[24] Examples of such an approach appear in G. G. Judge and T. Takayama, Inequality restrictions in regression analysis. J. Am. Stat. Assoc. 61 (1966), 166–181. —T. C. Lee, G. G. Judge and A. Zellner, Maximum likelihood and Bayesian estimation of transition probabilities, cit. supra.—A. Zellner and M. S. Geisel, Analysis of distributed lag models with applications to consumption function estimation. Invited paper presented to the Econometric Society, Amsterdam, Sept. 1968, and published in Econometrica 38 (1970), 865–888.

[25] See, e.g., the last two references in footnote 24.

[26] These issues have been discussed by Lindley in connection with the paper D. V. Lindley and G. M. El-Sayyad, The Bayesian estimation of a linear functional relationship. J. Roy. Stat. Soc. B 20 (1968), 190–202. See also P. A. V. B. Swamy, Statistical inference in random coefficient models. Doctoral dissertation, Univ. of Wisconsin, Madison, 1968.

estimator which is usually justified in terms of large sample theory. If the sample size is not large, there is a question about how good the approximation is. As mentioned above, Bayesians handle the nuisance parameter problem quite simply: such parameters are integrated out of a posterior pdf to yield the marginal posterior pdf for the parameter (or parameters) of interest.

(7) The effects of departures from specifying assumptions have been analyzed to some extent using B and NB approaches. If, e.g., disturbance terms in a regression model are thought to be autocorrelated, the model can be broadened to include a stochastic process, say a first-order autoregressive process, for the disturbance terms. Then under alternative assumptions about the value of the autoregressive parameter, both B and NB approaches can be utilized to assess how sensitive inferences about regression coefficients are to what is assumed about the autoregressive parameter. Similar calculations, involving introduction of a new parameter or two, can be performed in connection with possible departures from normality, linearity and homoscedasticity in regression models although in these cases it is difficult to obtain the exact finite sample distributions of NB unconditional estimators.[27] Since Bayesians and some likelihood advocates take the sample as given (not random), the relevance of these sampling distributions for the problems of inference and of investigating effects of departures in a particular analysis is unclear. Bayesians, in any event, can compute conditional posterior pdf's for parameters of interest given values of a parameter (or parameters) introduced to allow for a departure from specifying assumptions. This sort of analysis can generally be carried through without a need for approximations.

(8) The relevance for inference from given data of the criterion of performance in repeated samples, which is featured in many NB approaches in econometrics and elsewhere, has not received much attention in the econometric literature. Bayesians and some likelihood advocates emphasize that inferences from given data should be based on the information that is in the given data and, for Bayesians, on given prior information. The relevance of other possible samples for the problem of analyzing a given set of data is not clear. However, as

[27] See, e.g., G. E. P. Box and D. R. Cox, An analysis of transformations. J. Roy. Stat. Soc. B 26 (1964), 211–243.

pointed out above, Bayesian estimation procedures do have good properties in repeated samples. It appears that the question of how procedures perform in connection with random data samples is an important issue before the data are drawn, for example in designing a survey. Once the data have been obtained, the problem of inference appears to be that of bringing the information in the given data to bear on beliefs about parameter values, etc.

(9) In the area of analyzing alternative hypotheses and models, it was pointed out above that Bayesians can compute probabilities associated with alternative hypotheses or models which reflect both prior and sample information and which can be regarded as measures of belief in alternative hypotheses or models. Posterior odds relating to two mutually exclusive hypotheses or models are given *in general* by the product of prior odds times a likelihood ratio factor. Further, if a decision has to be made, namely accept or reject, given a loss structure specifying the consequences of these acts, it is possible to act so as to minimize expected loss. NB approaches for analyzing alternative hypotheses and models do not involve introduction of probabilities associated with hypotheses or models. Rather what is involved in testing is the computation of a test statistic. If the test statistic assumes an "unusual" value under the null hypothesis, an investigator's view of the null hypothesis is somehow affected. The jump in logic from an unusual event under the null hypothesis to the conclusion that the null hypothesis is suspect is an issue that concerns many Bayesians. This is not to say that NB test procedures always or usually yield meaningless results. In fact under vague prior information, Lindley has provided a B rationale for usual tests (likelihood ratio, t tests, F tests, etc.).[28] On the other hand, as emphasized by Jeffreys and others, in testing situations our prior information is often not vague and it is important to take account of this fact in order to get meaningful results. Most, if not all, NB approaches to testing in econometrics do not appear at present to have the capability of formally introducing prior subjective information in testing hypotheses. Rather, such information tends to be used informally in viewing final results and, perhaps, in the choice of a significance level.

[28] D. V. Lindley, Introduction to probability theory from a Bayesian viewpoint. Part Two: Inference, cit. supra.

3. Conclusions

In a short paper, it is difficult to do justice to the delicate and deep issues involved in a comparison of alternative approaches in econometrics. What has been presented is an overview of some distinctive characteristics of the B approach and a comparison of NB approaches in relation to these characteristics of the B approach. Further, it must be recognized that the B approach is in a stage of rapid development with work going ahead on many new problems and applications. While this is recognized, it does not seem overly risky to conclude that the B approach, which already has had some impact on econometric work, will have a much more powerful influence in the next few years. The most important consideration underlying this prediction is the fact that the B approach rests on a unified and relatively simple set of principles which are broadly applicable and produce good results. In addition, the considerations in sections 1 and 2 of this paper point up specific features of the B approach which commend it for use in econometrics and other areas of science.

S. E. Fienberg and A. Zellner (eds.)
Studies in Bayesian Econometrics and Statistics.

2.3 THE BAYESIAN APPROACH AND ALTERNATIVES IN ECONOMETRICS—II*

Thomas J. ROTHENBERG

University of California at Berkeley

Should economists in the course of their empirical research use Bayesian methods of inference? This explicitly normative question is at the core of much current discussion in econometric methodology. Indeed, the question of the role of Bayesian techniques has been raised in nearly every application of statistics and has caused an enormous amount of controversy in each of them. I do not intend in this paper to give a complete survey of this controversy over Bayesian methods—such a survey would take many pages and in any case would not be very fruitful; nor will I try to present a convincing defense of my own answer to the question. For, in fact, I have none—neither a convincing defense, nor even an answer at all. It seems, however, that much of the controversy over Bayesian methods has been directed at false issues and has been based on misconceptions as to where the real difficulties lie. I would like, therefore, to spend a few moments indicating what are for me the crucial aspects of the debate between Bayesians and non-Bayesians.[1]

Before beginning, it is only fair that I state my current feelings about the role of Bayesian techniques in econometrics. Frankly, they are not altogether free from contradiction. I think that there is a set of important problems where Bayesian techniques are clearly superior to any traditional alternatives. Some examples will be given later on in the paper.

* Reprinted with minor changes from M. D. Intriligator (ed.), Frontiers of quantitative economics. Amsterdam, North-Holland Publishing Company, 1971, with permission of the North-Holland Publishing Company.

[1] My thinking on these questions has been greatly influenced by the following: L. J. Savage, The foundations of statistics reconsidered. In: Studies in subjective probability, edited by H. Kyburg and H. Smokler. New York, Wiley, 1964.—D. V. Lindley, Statistical inference, Journal of the Royal Statistical Society B 15 (1953).—R. A. Fisher, Statistical methods and scientific inference. Hafner, 1956.—Karl Popper, Conjectures and refutations. Basic Books, 1962.

I would suggest, however, that for many (perhaps most) statistical problems which arise in practice the difference between Bayesian methods and traditional methods is too small to worry about, and that when the two methods differ it is usually a result of making strongly different assumptions about the problem that is being answered. Yet, I would also suggest that for certain key problems involving major changes in thinking, the difference between the Bayesian and traditional approaches is enormous and here the Bayesian approach is very unsatisfactory. These views will, I hope, become somewhat more clear as I proceed.

As a starting point in the discussion, I would like to quote some passages from a famous critic of an earlier version of Bayesian statistics (which then went under the name of "inverse probability"). In attacking the Bayesian approach in his 1935 book, "Design of experiments", R. A. Fisher gave three basic arguments:

> (1) ... advocates of inverse probability seem forced to regard mathematical probability, not as an objective quantity measured by observable frequencies, but as measuring merely psychological tendencies, theorems respecting which are useless for scientific purposes.
>
> (2) ... it is the nature of an axiom that its truth should be apparent to any rational mind which fully apprehends its meaning. The axiom of Bayes has certainly been fully apprehended by a good many rational minds, including that of its author, without carrying this conviction of necessary truth.
>
> (3) ... inverse probability has been only very rarely used in the justification of conclusions from experimental facts, although the theory has been widely taught, and is widespread in the literature of probability. Whatever the reasons are which give experimenters confidence that they can draw valid conclusions from their results, they seem to act just as powerfully whether the experimenter has heard of the theory of inverse probability or not.[2]

Fisher, of course, was talking about a very primitive form of Bayesian analysis. It is of some interest to see how the modern mid-twentieth-century version of Bayesianism stands up to his attacks. A first problem is that it is not altogether clear what one means by the modern Bayesian approach. The term Bayesian is used in a number of different ways by different people. All of these uses, however, have some common elements: Hypotheses about the state of nature are assigned probabilities. Sample evidence is viewed as a random variable whose probability distribution is conditional on the state of nature. By Bayes' formula one calculates a new probability distribution for the hypotheses conditional on the sample data. This process of going from a prior (before the data)

[2] Pages 6, 7 in the seventh edition, printed in 1960 by Oliver and Boyd.

probability distribution to a posterior (after the data) probability distribution is the basis of inductive inference for a Bayesian.

This simple description leaves out two crucial steps in the inference procedure. First, how does one choose the prior distribution and, second, what does one do with the posterior distribution once one has it? Here is where there are differences of opinion among the Bayesians. One may distinguish two major schools of thought. The *decision theorists* (like Raiffa and Savage) argue that any specific inference problem involves a decision maker choosing among alternative actions where the degree of preference for the various possible actions can be expressed by a utility function which depends on the unknown state of nature. Given the posterior distribution the best act (and hence implicitly the best inference) is the one which maximizes expected utility. As for the prior distribution, the decision theorists argue as follows: people who make consistent choices in uncertain situations always behave as though they had a subjective probability distribution. Since decision makers do make decisions even in the absence of data, some a priori information is already available before experiments are made and this can be expressed as a prior probability distribution over the possible states of nature.

In contrast to the decision theorists, however, there are other Bayesians who do not assume a decision problem at all. These Bayesians argue that in those cases where there is little previous evidence available one should use as a prior distribution a uniform distribution over alternative hypotheses. A somewhat more sophisticated version of this rule is provided by Jeffreys[3] who suggests an invariance procedure for obtaining prior distributions. In either case, there is a canonical prior distribution determined by the likelihood function alone and not by any subjective beliefs. Since no decision problem is posed, the posterior distribution on the alternative hypotheses is the final product of this type of inference procedure.

Although there are great similarities between these two versions of Bayesianism (and it is possible to be a decision theorist for some problems and a canonical theorist for others), I shall treat these variants as quite separate. For they differ in one very important respect. Two decision theorists, looking at the same data and agreeing on the stochastic process which produced the data, could come to very different conclusions if they started with different prior beliefs or utility functions.

[3] Harold Jeffreys, Theory of probability, 3rd edition, Oxford, 1961, especially chapter 1.

Two followers of Jeffreys' invariance rule of induction, however, would always come to the same inference in this circumstance. Furthermore a Jeffreys Bayesian can argue (along with non-Bayesians like Fisher) that inference and knowledge are quite different from behavior and decision, whereas a decision theorist treats the former as special cases of the latter.

The conclusion of this brief aside is that there are two sorts of Bayesians: (1) decision theorists who argue that inference should be thought of as action and that any prior probability which reflects one's beliefs can be used in deriving a posterior distribution, and (2) canonical theorists who argue that inference is meaningful even without action and that any specific problem gives rise to a specific set of prior probabilities to be used in Bayes' formula.

How, then, would these two types of Bayesians respond to Fisher's attacks? In regard to the first point—that Bayesians must give up a frequency interpretation of probability—both schools would agree. Indeed, both Savage and Jeffreys would argue that there are fundamental weaknesses in the frequency interpretation and that it has to be dropped in any case. Fisher's proposition that non-frequency interpretations are useless for scientific purposes is of course denied by Bayesians. Both decision theorists and canonical theorists would argue that statements like "it will probably rain tomorrow" are used constantly in scientific discourse and that such statements follow a precise logical structure. In some sense the Bayesians are clearly right: it *is* possible to develop a system of logic obeying the axioms of probability which is not based on relative frequencies. Both Savage and Jeffreys produce such systems. Whether one should *use* such a system in scientific procedures, however, cannot be answered on purely logical grounds. After all, both Euclidean and non-Euclidean geometries are consistent logical systems. The usefulness of such a system must be judged at a different level.

Fisher's second argument—that the Bayesian axiom on the existence of a prior probability distribution over states of nature is not readily apparent—is also answerable, at least by some Bayesians. The decision theorist can point to the fact that this axiom can be derived from a set of more basic postulates on the consistency of individual choices under uncertainty.[4] These latter postulates—although not accepted by every-

[4] See, for example, J. Pratt, H. Raiffa, and R. Schlaifer, The foundations of decision under uncertainty: An elementary exposition, J.A.S.A. (1964), or L. J. Savage, The foundations of statistics. New York, Wiley, 1954.

one—are quite convincing to many rational minds (including this one). The canonical Bayesian theorists, however, have no clear answer to Fisher's objection. They simply argue that the axiom is indeed self-evident; that there is a valid primitive idea expressing the degree of confidence that we may reasonably have in a proposition and further that there is a simple mathematical form for expressing the notion of complete ignorance. Here, unfortunately, I share Fisher's skepticism. Jeffreys' axioms do not possess the sure ring of truth, even though they may sometimes give rise to sensible statistical procedures. It is for this reason I believe that few people have been able to accept the Jeffreys analysis.

Finally, Fisher's third point—that in practice researchers do not defend their conclusions by Bayesian arguments—is in my opinion the crucial one. It is an argument that avoids tautologies and endless epistemological debate. It is an argument consistent with a dictum of modern methodological research: to gain an understanding of the philosophy of science, one should study the history of science. Thus I would like to spend the rest of this paper exploring the question: is there compelling evidence in the history of science—in economics and elsewhere—that tends to support or contradict the Bayesian approach to learning from experience?

In this context it is possible to consider the original question in three parts. First, can most problems in scientific research be usefully classified as statistical decision problems? Second, if they can be so classified, is the Bayesian solution to the decision problem an accepted one? Third, if they cannot be so classified, do scientists use nondecision theoretic Bayesian arguments to justify their conclusions?

I would now like to take up these three questions, starting with the second one because it is in many ways the easiest. Suppose we argue that a given estimation problem does fall into the structure of statistical decision theory. That is, we have to guess the value of some parameter θ and that the cost of being wrong can be described by a loss function (say a quadratic function of the estimation error). We observe a sample which has a distribution completely known to us except for the parameter θ. Any estimator (i.e., any function of the sample values) has associated with it a mean squared error which depends on θ. This expected loss, expressed as a function of θ, is called the risk function for the estimator. In general, the risk functions for various estimators will cross; that is, no risk function is, for all θ, uniformly below all the others. Thus we have a problem of choosing among the alternative risk functions.

The traditional solution is to restrict ourselves to unbiased estimators and hope that among estimators in the restricted class there is one which has uniformly smallest risk. Luckily, for many problems such a minimum variance unbiased estimate exists. But would anyone in a real decision context choose such a solution? It is hard to see why anyone should. Does a plant manager when making ordering decisions care that in some sense he will be right "on the average"? Won't he naturally use his rough information about the world, combine it with current data and essentially act as a Bayesian? Surely people involved in real-world decision-making do possess enormous amounts of prior information. Furthermore, they very likely behave—or at least would like to behave—in a manner consistent with the "coherence" axioms of Savage. Thus I find very convincing the arguments of Savage, Raiffa, and others that Bayesian analysis is consistent with—indeed implied by—sensible decision making in practice. I will not try to defend this position here since it is persuasively done in the published literature. I shall simply state my view that, to the extent statistical problems truly are applied decision problems, I find the Bayesian approach more attractive than the traditional ones.

Up until just a few years ago the decision aspect of economic research had been largely ignored. In an unpublished paper written in 1963 I argued that a Bayesian approach to econometrics was needed so that this decision aspect could be incorporated into practical statistical procedures.[5] Since that time enormous progress has been made in developing the tools for solving econometric decision problems. The excellent work of Drèze, Kaufman, Zellner and many others has brought us to the point where practical application of Bayesian methods in economics is now possible. Although many technical problems remain to be solved—convenient prior distributions when there are nuisance parameters have still to be developed, for example—the basic framework for Bayesian econometrics now exists. Textbooks in elementary economic statistics now treat Bayesian methods routinely. The view that Bayesian decision theory is useful in economic research is widely accepted and no longer open to serious controversy.[6]

[5] T. J. Rothenberg, A Bayesian analysis of simultaneous equation systems. Report 6315, Econometric Institute, Netherlands School of Economics, 1963. A revised version of this paper appears as chapter 6 of T. Rothenberg, Efficient estimation with a priori information. Yale, 1973.

[6] This is perhaps an exaggeration. See for example, J. Wolfowitz, Bayesian inference and axioms of consistent decision. Econometrica 30 (1962).

Why then is there controversy over Bayesian methods at all? Clearly because there is serious disagreement among statisticians over the usefulness of considering *all* inference problems as decision problems. Here we must be rather careful in defining our terms. Surely most textbook examples of point estimation and hypothesis testing problems do fit easily into the framework of decision theory. The modern textbook writers like Mood and Graybill and Lindgren specifically present these classical problems as special cases of the general decision problem. But does scientific research really proceed in the fashion described by these decision models? Is there not something rather artificial in this textbook treatment? I would suggest that recent work in the history and philosophy of science casts considerable doubt on the usefulness of the simple decision-theoretic model of scientific inference.

Let me suggest two different views on how scientists work.[7] Needless to say, these are highly stylized. One view, which I will call the incremental theory, sees intellectual development as a smooth piecemeal process by which facts and theories are added to the ever-growing stockpile that constitutes scientific techniques and knowledge. The scientist learns from experience; and a natural way to describe the combining of past knowledge with current sample information is by Bayes' theorem. Thus each economist makes his small, but useful contribution by modifying his subjective probability distribution over the possible hypotheses concerning Nature. With more and more data posterior distributions become more and more concentrated and uniformity of opinion among economists is approached.

This charming and peaceful picture of progress through hard work and nice priors is of course a caricature. Yet it seems to fit the story told by some Bayesians. For example, Savage writes, "Inference means for us the change in opinion induced by evidence on the application of Bayes' theorem".[8] And de Finetti says, "Everything that does not reduce to a simple statement, to an isolated historical truth, . . . constitutes a judgement of probability which is based, perhaps unconsciously and indistinctly, on the principles of the calculus of probability.

[7] The following arguments are based on ones presented by Thomas Kuhn in The structure of scientific revolution. University of Chicago Press, 1962, and by Paul Feyerabend, in Problems of empiricism, included in Beyond the edge of certainty, edited by R. Colodny. Prentice-Hall, 1965. These authors do not draw the same conclusions from their arguments that I do; they are not, of course, responsible for what I say.

[8] The foundations of statistics reconsidered, op. cit., 178.

This calculus thus constitutes the foundation of the greatest part of our thought, and we can well repeat with Poincare, 'Without it, Science would be impossible'."[9]

If scientific progress is made by adapting prior beliefs as a result of new experimental evidence, one would expect development to occur smoothly. Intellectual life would be very peaceful; over time divergent opinions come closer together as scholars observe common new evidence. This view of scientific development seems to have some validity if one looks crudely over the general trend of history. Observational evidence has modified old beliefs to form new ones. Conflicting theories are merged into consistent ones as a result of cumulative evidence. Yet, as a description of the day-to-day work of scientists this incremental theory seems very misleading. It does not capture the arguments which researchers actually use in their discussions with colleagues.

A very different view on how science works is given by Karl Popper.[10] It is not continuity and accumulation which is emphasized, but quite the opposite. His dictum is: *We learn from our mistakes.* Progress is made by attack, by criticism, by attempted refutation of proposals. New truths are not produced by combining old ones with data. New truths are unjustified guesses, conjectures, which happen to survive severely critical tests. This point of view is well expressed by the quote:

> ... whenever we try to propose a solution to a problem, we ought to try as hard as we can to overthrow our solution, rather than defend it. Few of us, unfortunately, practice this precept; but other people, fortunately, will supply the criticism for us if we fail to supply it ourselves.[11]

Anyone who has had a lunchtime economics discussion with our good colleagues at the University of Chicago will see the truth—if not the fortune—of this last sentence!

Stronger versions of the Popper view are given by Kuhn and Feyerabend. Kuhn makes the (now famous) distinction between *normal science*—"a strenuous and devoted attempt to force nature into the conceptual boxes supplied by professional education"—and *scientific revolutions*—"the extraordinary investigations that lead the profession at last to a new set of commitments, a new basis for the

[9] Bruno de Finetti, Foresight: Its logical laws, its subjective sources. In: Studies in subjective probability, op. cit., 156.

[10] The classic source is The logic of scientific discovery. Harper Torchbooks, 1965 (a translation of the 1934 German edition). A clearer presentation can be found in Conjectures and refutations, cited earlier.

[11] K. Popper, The logic of scientific discovery, op. cit., 7.

practice of science".[12] Normal science is a slow, incremental process of examining data and fitting them into the traditional paradigm. But scientific progress is not simply the cumulative result of normal scientific activity. The scientific revolutions, the great crises that occur when one paradigm collapses and another one takes its place, clearly do not fit into the incrementalist view of the world. But discontinuities in scientific progress are not restricted to revolutions. As Feyerabend puts it:

> Now it is commonly assumed ... that crises are, or at least should be, *transitory stages* in the history of thought, that they are periods of disorder and embarrassment which are void of knowledge and provide no suitable basis for methodological discussions ... The realization that alternatives precipitate progress suggests an evaluation of the relative merits of normal science and periods of crises that differs radically from [the above view]. Normal science, extended over a considerable time, now assumes the character of stagnation, of a lack of new ideas; it seems to become a starting point for dogmatism and metaphysics. Crises, on the other hand, are not accidental disturbances of a desirable peace; they are periods where science is at its best, exhibiting as they do the method of progressing through the consideration of alternatives.[13]

These scientific crises do not occur peacefully. The new paradigms are not related to the old ones by a generalized Bayes formula. As Kuhn quotes Max Planck, "a new scientific truth does not triumph by convincing its opponents and making them see the light, but rather because its opponents eventually die, and a new generation grows up that is familiar with it".[14] Although this is surely an exaggeration it does emphasize the crucial role played by criticism and intellectual antagonism in science. To summarize this crises theory in brief: Science at its best consists of two activities—imaginative, indefensible conjectures and ruthless attempts at refutation.

What does all this have to do with Bayesian econometrics? I think it has the following implications: If we accept the view that scientific progress results from criticism and attack, that scientific method consists of conjecture and refutation, then the naive view of a smooth transition via induction from prior probability to posterior probability must be rejected as historically invalid. Furthermore, the simple-minded decision models described in the textbooks must be rejected as a

[12] T. Kuhn, op. cit., 5–6.

[13] P. Feyerabend, Problems of empiricism, op. cit., 167, 172.

[14] T. Kuhn, op. cit., 150. The citation is in: Max Planck, Scientific autobiography and other papers. New York, 1949, 33–34.

general framework for scientific work. Although Bayesian methods are clearly useful for actual decision problems, they are not necessarily useful for fundamental economic research. Here Fisher's third argument has considerable force. It is, I think, factually correct to say that economists involved in basic empirical research do not justify their arguments on Bayesian grounds. The above arguments suggest there is no compelling reason why they should.

This position, however, does not constitute a general argument against the Bayesian approach. Even if science does consist of refutation rather than induction, rules for refutation must be established. Even if science is viewed as warfare among competing theories, some method for choosing among them is needed. It is simple enough to say that a theory should be rejected if it is inconsistent with the facts. But, as Kuhn emphasizes, all historically significant theories have agreed with the facts—more or less. The Bayesian framework does give us a basis for deciding which of two theories fits the facts best. Thus one may argue that Bayesian methods can play an important role even if one accepts Popper's view of science. This interpretation, however, has two weaknesses. First, it is very hard to place this role of Bayesian analysis into a decision context. I would have no idea how to specify a loss function or a prior probability distribution. One can resort, of course, to the Jeffreys' type of analysis—but that has all the weaknesses discussed earlier. Second, most disagreements about fundamental ideas involve noncommensurabilities. "Neither side will grant all the nonempirical assumptions that the other needs in order to make its case", argues Kuhn.[15] Rarely can we specify completely and impartially the essential ideas underlying fundamental disagreements. Simple likelihood functions and smooth prior densities cannot capture the deep and often subtle differences of economic opinion.

Although in principle Bayesian arguments might be useful in choosing among alternative theories, in practice they seem to play no important role. I would suggest the following analogy. In the theory of perfect competition under constant returns to scale, equilibrium is characterized by zero profits although profit seeking is the dynamic mechanism which produces this result. That is, the motivation of the individual participants is not reflected in the historical observations of equilibria. In a similar fashion, we may view the motivations of individual research workers. Each scientist behaves in the manner described by Popper,

[15] Ibid., p. 147.

dogmatically asserting his positions and attacking the positions of others. The end result is a merging of views, a development which may, when viewed from afar, be described as learning from experience. Bayes' theorem may apply in the aggregate even though there are no Bayesian researchers![16]

The above arguments are relevant for basic disagreements among economists. In practice, however, many econometric problems involve inference about parameters in well-defined models where the basic theory of the phenomena and the process generating the data are not open to serious debate. Data analysis is performed to gain a more precise measure of a parameter, not to resolve some fundamental disagreement about the way the economy works. In such cases, the problem of inference is straightforward. The classical statistician will calculate some unbiased estimator and note its standard error. If the standard error is small and the estimate "plausible", he will accept (tentatively) the new estimate as an improvement over past ones obtained with larger standard errors. If the standard error is large or the estimate is "implausible", he will search for new methods of learning about the parameter. The Bayesian statistician will behave in a slightly different way. He is likely to have a prior density which puts roughly uniform weight over the range of parameter values indicated by past experiments and puts low weight on values far away from past results. The posterior distribution will be concentrated at the unbiased estimate as long as the experimental precision is high and the estimate is in the range of past estimates. Otherwise the posterior will be rather dispersed and the statistician will be led to reconsider the whole problem. Thus, in practice, the classical and the Bayesian statisticians will come to essentially the same conclusions based on the data. Whether it is more convenient to report the results in terms of a classical confidence region or in terms of how a class of priors gets transformed by the data is an interesting open question. I doubt that it is of great econometric importance, however, since such cases of normal science with well-defined models rarely involve serious problems of communication.

The really important economics problems, I fear, are badly specified and very difficult. They involve serious disagreements about the world and are only partially solvable by the data at hand. The proper inference method for these problems is not at all obvious. It would seem,

[16] The original 1969 version of the paper ended at this point. The material which follows was added for the present volume.

however, that any theoretical basis for econometrics must face up to the following facts:

(1) Econometrics is to a large extent a group or social activity. Data are collected and used in conversation and argumentation by people with diverse opinions. Rather than characterizing statistics as a method of "learning from experience", it would be more natural to characterize it as a method for resolving (or maybe even instigating) controversies. It may be usefully seen as a decision game, but it is often a game against other people, not a game against nature.

(2) There must be an important sociological and psychological component in any model of inference. What types of arguments are convincing? What interplay among economists is most conducive to progress? Econometrics cannot be separated from the day-to-day activities of economists.

(3) Statistics is indeed the language of economic science; but it is an extraordinary complex one. How in fact does this language function in practice? Subjective probabilities may be a natural part of this language, but the simple one-person learning model seems inadequate. Statements about probabilities may represent degrees of belief (just as statements about values represent preferences), but scientific discourse (like ethical discourse) requires a common grammar for integrating subjective vocabularies.

(4) There is no simple, unified model which seems to capture all of the diverse activities that we tend to label as econometrics. Attempts to force research into a narrow mould is likely to be counterproductive. Much actual economic research is exploratory, tentative, casual—more like storytelling than formal analysis. The simple proposition that observations that are implausible under an hypothesis tend to discredit the hypothesis seems to be at the heart of good storytelling. It would be a pity to have to give it up.

Econometrics at the moment does seem to consist of a large number of ad hoc procedures. It would surely be desirable to have a general theory of inference which would serve as an organizing principle. This theory, it seems to me, will have to be based on two fundamental concepts: individual beliefs (represented perhaps by subjective probabilities) and interpersonal discourse. Current Bayesian models emphasize the former concept; the traditional models can be interpreted in terms of the latter. Professor Savage's "The foundations of statistics" (especially chapters 8 and 10) represents a pathbreaking, but unsuccessful attempt to develop a unified theory incorporating both ideas. Until a successful theory is developed, econometrics will have to make do with its ad hoc, but often surprisingly useful, methods.

S. E. Fienberg and A. Zellner (eds.)
Studies in Bayesian Econometrics and Statistics.

2.4 COMMENTS ON PAPERS 2.2 and 2.3*

Gordon M. KAUFMAN

Massachusetts Institute of Technology

John W. PRATT

Harvard University

and

Arnold ZELLNER

University of Chicago

Comments, by Gordon M. Kaufman

Zellner and Rothenberg have presented two thoughtful, provocative papers. Rothenberg has done an admirable job of putting into philosophical perspective a number of fundamental dilemmas that cannot be ignored—even by those of us who are firmly convinced that Bayesian methods have an important role to play in the future development of econometrics. Zellner has effectively summarized a battery of pro-Bayesian arguments that I find, in the main, compelling.

As Rothenberg points out, the Bayesian model of inference will not do as a vehicle for capturing the act of creativity underlying what he terms "scientific revolution": conceptualization of a testable model or hypothesis. But neither do any of the alternate methodologies currently available. Until an implementable global model of inference that jives with all of the rigorous precepts governing a well constructed scientific philosophy is created—an unlikely event—we are forced to be pragmatic in the methodological tools we use.

None of this implies that a (non-Jeffreys) Bayesian approach is necessarily unattractive in implementing choice among alternative

* Reprinted with minor changes from M. D. Intriligator (ed.), Frontiers of quantitative economics. Amsterdam, North-Holland Publishing Company, 1971, with permission of the North-Holland Publishing Company.

theories that are representable as *parametric* models. The posterior odds that Model A is true as against Model B has meaning and there is nothing incommensurable about such an expression of *relative* degree of belief after the objective evidence is in. Rather than reach a conclusion now about the viability of Bayesian methods in this context, I personally choose to defer judgment at this time and to wait and see how the Bayesian approach functions in some particular cases where such a choice is to be made. The work done by Geisel and Zellner on comparing and choosing among parametric statistical models is a promising start.

There are, however, two closely related *methodological* questions addressed by neither Zellner nor Rothenberg that loom large given the present state of development of the Bayesian approach. Before posing them let us assume that we are in the realm of *statistical decision*, so that we can all agree on the usefulness (in principle) of analysis à la Savage et al. The first is:

> What methods should one use to elicit prior judgements about parameters of an econometric model?

This question has a normative ring about it. In order to make it a valid one, then, we should prescribe the goals of such methods. A reasonable one suggested by Howard Raiffa is *empirical validity*. To see what it means consider a weather forecaster who, each day, states a probability of the event "snow on the morrow".

Suppose the forecaster restricts himself to asserting that the probability of snow is 0, 0.1, 0.2, . . ., 1.0; i.e. to whole tenths and records for *each* of these eleven probabilities (a) the number of days that he asserted that probability and (b) the number of days among all such days that it *actually* snowed. When the total number of days he asserted that this probability is 0.3, say, becomes very large, we would hope that the proportion of these days that it *actually snowed* is very close to 0.3. If this is so for each of the eleven probabilities, then we assert that this forecaster *validates empirically*. This is very desirable.

In an econometric setting empirical validity has strong appeal. It can be interpreted as a measure of the degree to which an assessor is able to say what he really feels—provided that he has no clear motivation to state assessments different from those he believes. Experiments done by Raiffa at Harvard and replicated by others of us show that it is very difficult in general for an assessor to achieve empirical validity. His experiments show that most of us—even when of truly honest intent—

are *strikingly* not empirically valid. A practitioner of Bayesian econometrics ought to take heed!

The second question is also about assessment of priors. It has to do with the tyranny of large numbers: to motivate it suppose we are interested in doing a Bayesian analysis of a simultaneous equation system with ten equations—not an unduly large number. If one takes into account the normalization rule, but imposes no exact a priori constraints the number of functionally independent parameters in the 10×10 nonsingular matrix premultiplying the exogenous variables is 90. One also has to worry about the remainder of the structure. I do not think anyone here seriously believes he can responsibly assess a joint density over 90 random variables. An appeal to natural conjugate analysis or its extensions helps somewhat, but not much. The question is then:

> How can a Bayesian responsibly assess a prior for parameters of a large-scale econometric model?

One might argue that we are usually seriously interested in only a small subset of all such parameters. But this begs the question. Parenthetically, Jacques Drèze's work on a Bayesian analogue of limited information maximum likelihood estimation is indirectly addressed to this question in one special case.

I am not aware of any clear-cut general answer to the latter question.

These observations about practical problems that arise in assessing priors mirror, I believe, what may be a truly substantial barrier to widespread application of Bayesian methods in econometrics, except in the simplest of cases.

Comments, by John W. Pratt

Ordinarily when I am invited to participate in a session of this kind I assume that I am supposed to represent the far left—or is it the far right—Bayesian view. When I was preparing my remarks, however, with only Zellner's paper in hand, I thought the other speakers would be all quite pro-Bayesian, so I planned to try to restore balance to the session by indicating some difficulties with Bayesian methods in practice. I was certainly wrong about the other speakers but I will make the remarks I had planned to make, anyway. Specifically, I will mention some problems which are important, but which are hard or impossible to handle by the Bayesian approach at present and whose future handling may involve us Bayesians in "ad hockeries" of our own.

(1) *Problems related to flat (diffuse) prior distributions.* Prior distributions are flexible, as Zellner said, but it is an inflexible requirement to have one. There is a great tendency to use "improper", flat prior distributions—even Zellner has this tendency. But the nice admissibility features Zellner mentions in several places hold in general only for proper prior distributions. In particular, Stein's inadmissibility results unfortunately apply to the usual (flat prior) Bayesian estimates. In fact, the usual Bayesian estimates coincide with classical estimates typically, and hence are neither better nor worse.

Choosing between models requires proper prior distributions, and the posterior probabilities of the models may be sensitive to their choice. What is crucial is something like the ratio of the heights of the prior densities of the parameters in the relevant regions under the two models. Here uncomfortable assessment problems cannot be avoided.

In this connection, in relation to Rothenberg's paper, it seems only fair to point out that the corresponding classical procedures, hypothesis tests and *P*-values, are not important in revolutionary science either. Regarding other difficulties with testing, I have already had my say to the Econometric Society in the previous paper mentioned by Zellner. That was two years ago, and I was not saying anything new even then.

One problem of choice of models receiving much attention nowadays is the problem of choice of subset regression. The advantage of parsimony, in terms of preferring a simpler explanation if it is almost as good, can be captured in a Bayesian analysis in various ways. But another aspect of this problem, expressed in classical terms, is that, when there are r parameters and n observations, the average over the n points of the variance of the usual (classical or Bayesian) estimate of the regression is $(r/n)\sigma^2$ and the corresponding average forecast variance is $(1+r/n)\sigma^2$. Adding more parameters to a model can therefore hurt. It cannot hurt a Bayesian in theory, if he assesses his "true" prior distributions. It can, however, if he uses flat prior distributions. Here I think we might look toward some kind of compromise, using prior distributions which are largely but not wholly conventional. This will inevitably be rather ad hoc.

(2) *Robustness problems.* It is generally agreed that randomization is nice if you can do it. In fact, the inability to randomize makes econometricians worry about their models much more than experimenters

who randomize. (Notice, incidentally, that it is correspondingly hard to find a physically realizable frequency interpretation of classical statistical inference statements in typical practical econometric studies.) Classical methods appear to be able to make strong use of randomization if it has occurred, and generally, to permit inference statements which hold unchanged over a wide variety of models. To be sure, this is partly an illusion, because you would change the inference if you knew the model—not to mention that all classical inferences are an illusion anyway.

Bayesians have made some progress in using randomization (there has been special attention to this in the sample survey literature), but they have not explained its full appeal or exploited it fully.

One remark on robustness having nothing to do with randomization: the model you are using may cause you to estimate a population mean by a sample median or some other nonrobust statistic—nonrobust for estimating the mean. This may happen in both classical and Bayesian analysis, but the tendency is greater in Bayesian analysis since the Bayesian approach almost insists on taking the model literally and making full use of it. In my opinion, a Bayesian is really just as entitled as a classical statistician to use insufficient statistics, but he is less inclined to do so in many situations, because it makes his mathematics harder instead of easier.

I will now make three minor comments and then conclude.

(1) Estimates resulting from preliminary tests are not only biased and nonlinear, they are even *discontinuous* functions of the observations.

(2) People often embarrass me by asking me for references to applications of Bayesian methods in practice, though I have never been so incautious as even to suggest I knew of many. I would be grateful for any references to reports on substantive problems where the primary analysis was Bayesian. (I am not talking about Bayesian re-analysis of data for methodological or illustrative purposes.)

(3) Raiffa's experiments on the difficulty of achieving empirical validity were perhaps not quite as striking as has been suggested by Kaufman, and a similar experiment this year is still less so. The tendency he observed is, however, certainly present.

Finally, lest all these comments on the limitations of the Bayesian approach lead you to think that I am not a Bayesian after all, let me alter de Finetti's famous remark and say that just because the Bayesian house is imperfect and not yet complete, I see no reason to move out of it and into a house with no foundation whatever.

Comments, by Arnold Zellner

I agree with Pratt that a prior pdf is required in a Bayesian approach and this is a reflection of the principle that you rarely, if ever, get something for nothing. On the other hand, non-Bayesians often use strong prior information, e.g. to identify parameters of errors-in-the-variables and "simultaneous equation" models. In such situations, use of prior pdf's may permit the required prior information to be represented more flexibly and suitably.

With respect to "flat" or "noninformative" prior pdf's, these are useful when we know little about parameter values or when we wish to have a posterior pdf reflect mainly just the information in the data. Of course if we have prior information about parameter values and wish to (or have to in order to identify parameters) incorporate it in an analysis, then of course we would not use a "flat" or "noninformative" prior pdf (see the paper by Geisel and myself for results of some analyses using several informative prior pdf's).

On the question of Stein's results on admissibility, Stein himself and Lindley (J.R.S.S., B (1962), 265–287) have indicated the prior pdf's which produce posterior pdf's having means close to or equal to Stein's estimates.[1] The use of such *subjective* prior information to generate an estimator may disturb non-Bayesians. On the other hand, it must be recognized that Stein-like estimators can also be generated by a pre-testing procedure which, however, requires that informal prior information be employed in the selection of a significance level for the pre-test.

With respect to several issues raised by Rothenberg, first it is not the case that the Bayesian approach implies a "charming and peaceful picture" or an "incremental theory" of scientific progress. The Bayesian approach is compatible with different researchers having different priors, that is disagreeing, and with their working to produce new evidence and models to change beliefs in a dramatic and even revolutionary manner. Second, Jeffreys writes, "What is hardly ever mentioned is that induction has often failed in the past and that progress in science is very largely the consequence of direct attention to instances where the inductive method has led to incorrect predictions".[2] In my opinion, Jeffreys' book "Scientific inference", that provides a detailed account

[1] See also B. M. Hill's paper published in the present volume.

[2] H. Jeffreys, Scientific inference, 2nd Ed. Cambridge, University Press, 1957, 14.

and analysis of major developments in physics and related fields and their relation to incorrect predictions and errors, should be considered seriously before a final decision on how science "really" progresses is reached.[3] Last, on the social aspects of science, including the "doctrine of general consent", Jeffreys remarks, "The data and the arguments of an investigator are the essential point; wider acceptance is a by-product. Every individual has to make his own decision. The criterion of general agreement is like the instruction 'do not go into the water until you know how to swim' ".[4] Does all of the above sound like a peaceful, charming, incremental view of scientific progress? I think not.

[3] Also, in chapters I, VII and VIII of his book, Theory of probability, 3rd Rev. Ed. Oxford, Clarendon Press, 1967, Jeffreys presents and analyzes many relevant issues including the relation of decision theory and inference, R. A. Fishers' objections to inverse probability, etc. With respect to Fisher's objections, Jeffreys concludes that, "In spite of his [Fisher's] occasional denunciations of inverse probability I think that he has succeeded better in making use of what it really says than many of its professed users have." (395).

[4] Scientific inference, 194.

PART 3

BAYES' THEOREM AND ECONOMIC THEORY

S. E. Fienberg and A. Zellner (eds.)
Studies in Bayesian Econometrics and Statistics.

3.1 BAYESIAN ANALYSIS AND DUOPOLY THEORY*

Richard M. CYERT and Morris H. DeGROOT

Carnegie-Mellon University

Duopoly theory has a long history in economics, and a distinguished list of names associated with that history. Nevertheless, the problem has proved to be frustrating for economists. The obvious reason for the difficulty is the uncertainty that characterizes the problem. The specific uncertainty revolves around the interrelationship of the two firms and the fact that the decisions of one of them affect the other. Solutions have consisted of finding plausible (or implausible) behavioral assumptions that effectively eliminate the uncertainty. Thus the first solution to the problem, given by Cournot [1, pp. 79–84], was based on the assumption that the two firms viewed themselves as independent of each other. Using this assumption, Cournot showed that an equilibrium output would be achieved in the market. Duopoly and oligopoly theory has developed through the years by attempts at different static assumptions which produce models explaining some regularity believed to exist in duopoly or oligopoly markets.

The difficulty with these models, as a number of economists have noted, is their nonadaptive character. In Cournot's model, for example, the firms continue to follow the same behavior each period despite the fact that their assumption about the behavior of their rival is proved wrong in each decision period.

In practice, however, even casual empiricism leads to the conclusion that firms adapt to their environment. They show a capacity to learn some aspects of the decision-making behavior of their rivals. At a minimum the firm recognizes the information base on which certain

* Reprinted from The Journal of Political Economy 78 (September/October, 1970), 1168–1184, with permission of the University of Chicago. Some of this work was reported to the Second NBER-NSF Seminar on Bayesian Inference in Econometrics, Carnegie-Mellon Univ., May 1971. This research was supported in part by the Guggenheim Foundation and National Science Foundation Grant GP-8824.

decisions of rivals are made. Thus the firm knows that cost changes affect all firms in the market and are highly likely to lead to price changes. Further, the firm has access to information through such sources as trade journals, industry conferences, business reporters in the daily newspapers, industry salesmen, et cetera, that enable it to develop expectations about the kind of behavior rivals may exhibit.

In this paper we intend to discuss briefly the general framework of adaptive models for duopoly and in particular to distinguish between models of simultaneous choice and models of alternating choice. We will then demonstrate how the expectations operator can be used in the analysis of both types of models, and we will examine the process of interfirm learning. Finally we will illustrate the use of formal Bayesian analysis in duopoly theory by two simple adaptive models.

1. Current status of duopoly

The current approach to duopoly owes a large debt to Cournot. The basic approach is one in which a firm makes an assumption about the way it expects a rival will respond to a change in the value of a decision variable (price or quantity) by the firm. This assumption is usually expressed as a derivative, and Frisch [2, p. 31] has given it the name "conjectural variations". Given conjectural variations for each rival, reaction functions for each firm can be derived. These reaction functions show the optimum response for each firm, given the value of a decision variable for the rival. The reaction functions have generally been based on a single-period maximization approach. Only recently, in an important breakthrough, Friedman [3] has shown the existence of equilibrium when the firms use multiperiod horizons for profit-maximizing decisions.

However, in all of these models, including Friedman's, the firm can choose its reaction function only when its rival's reaction function is known. There is no provision in any previous duopoly models in the literature for the firm to learn over time the form of its rival's reaction function and then to adapt its decisions accordingly. Our use of Bayesian analysis is designed to correct this defect.

In addition to the above defect in current models, there has also been a lack of clarity regarding the nature of the decision period in duopoly models. More specifically, there have been two different kinds of decision periods used in duopoly models. It can be assumed either that the firms make their decisions simultaneously or that the firms make their decisions in different periods. The latter approach involves one firm making its decision first while the other firm maintains its position and reacts in the next period. It is, of course, not necessary for the same firm to be the leader each time.

When the decisions are made simultaneously, each firm must attempt to predict the values that the rival will set for the decision variables in the current period. A prediction function might include the actual values for decision variables for recent periods, profit figures, market-share data, changes in the decision variables over time, and other information such as announced plans of the firm.

When the decisions are made in alternating periods, the firm must attempt to predict the values that the rival would choose for the decision variables in the next period as a reaction to any values which the firm might choose for the variables in the current period. In other words, the firm must predict the reaction function of the rival, and it is this prediction that has typically been represented by a conjectural-variations term.

Of course it should be emphasized that in a multiperiod process, even when the decisions are made simultaneously in each period, a firm must consider its rival's reactions in future periods to the choices that are made in the current period. However, in any given period there is a basic difference: When choices are made simultaneously, the firm knows that its rival must choose values for its decision variables without knowledge of the values that are being chosen by the firm in that period. When choices are made in different periods, the firm knows that the rival will have knowledge of the values that the firm has chosen at the time that the rival must choose values for its decision variables. In later sections of this paper we shall consider in greater detail single-period problems in which decisions are made simultaneously and multiperiod problems in which decisions are made alternately by the two firms.

2. *Framework for application of adaptive models*

An adaptive model is one in which a firm can learn and, thereby, modify its previous behavior. This learning can take place with regard to other firms, with regard to the environment, or both. More will be said later about the concept of learning, but at this point we wish to concentrate on the nature of adaptive models. The important point in relation to conventional theory is that an adaptive model is not restricted to a fixed reaction function. In an adaptive model, for example, the firm is able to change its assumption about the way in which its rival will respond to any changes the firm will make in the decision variable. This means that for each decision period the firm can have a different value for the conjectural-variations term and, therefore, a different reaction function *chosen on the basis of the rival's decisions over earlier periods of the process.*

For example, the firm may have a simple rule for shifting its conjectural-variations term, or equivalently, its reaction function, from period to period in accordance with its changing expectations of the rival's behavior. One approach is for the firm to assign subjective probabilities to various modes of behavior of the rival. The firm might then use a given value for the conjectural-variations term over several periods. Then, when the weight of evidence led to a substantial subjective probability in favor of one of the other modes of behavior, the firm might change the value of the conjectural-variations term accordingly.

Another class of adaptive models can be developed by weakening the usual assumption that the precise form of the demand curve is known. This approach has not generally been taken in duopoly theory, but it is possible to derive the interaction between firms through conjectures relating to the rival's perception of the demand curve. We have, in the models that follow, generally taken this approach, partly because of greater ease in exposition but primarily because of the greater simplicity in achieving analytic answers while demonstrating some new approaches to the problems of duopoly. In some cases we have relied on behavioral rules that are different from those usually associated with economic models. Various other behavioral rules might have been presented here. However, the ones we have used serve our basic purpose of introducing the analytic treatment of learning and adaptive behavior in duopoly problems.

3. *An expectations approach to simultaneous-choice models*

One difficulty in dealing with simultaneous-choice models by conventional methods is that the conjectural-variations/reaction-function approach is misleading. The problem arises because the firm must make a prediction about the value for the decision variable that the rival will set. Since both firms make their decision at the same time, the problem cannot be one of conjecturing the reaction of the rival to the firm's decision. We propose in this section to demonstrate a method of approaching the problem that will better describe the process involved in simultaneous decision problems—the use of subjective expectations.

To illustrate the application of subjective expectations, assume that firm 1 must choose a value for its output q, that firm 2 must choose a value for its output r, that the demand function is of the form $p = a-b(q+r)$, and that the firms have no costs. In the examples which we shall consider, it will be assumed that the value of b is fixed and known to both firms, but it will not be necessary to assume that the exact value of a is known to the firms. Indeed, we shall discuss examples in which the firms are uncertain about the value of a, or even have erroneous information about its value, as well as examples in which both firms know the value of a.

The profit of firm 1 is denoted by Q and is specified by the equation $Q = q[a-b(q+r)]$. Hence, the expected profit $E_1(Q)$ of firm 1 is

$$E_1(Q) = qE_1[a-b(q+r)] = qE_1(a)-bq^2-bqE_1(r), \qquad (1)$$

where the expectations in equation (1) are firm 1's subjective expectations. The value of $E_1(Q)$ is maximized when

$$q = \frac{E_1(a)-bE_1(r)}{2b}. \qquad (2)$$

Hence, this value represents the optimal choice for q in a one-period problem. The term $E_1(a)$ represents firm 1's expectation of external conditions. The term $E_1(r)$ represents firm 1's expectation of the value that firm 2 is going to choose. In order to aid firm 1 in evaluating $E_1(r)$, we can analyze this term further.

Firm 1 assumes that firm 2 will be going through the same reasoning as above, and that therefore, just as in equation (2), firm 2 will choose r in accordance with the equation

$$r = \frac{E_2(a)-bE_2(q)}{2b}. \qquad (3)$$

When this value of r is substituted into equation (2), we obtain the following value for q:

$$q = \frac{E_1(a)}{2b}-\frac{E_1E_2(a)}{4b}+\frac{E_1E_2(q)}{4}. \qquad (4)$$

In this equation, a term of the form $E_1E_2(X)$ represents firm 1's expectations as to the value firm 2 expects for the unknown value of some variable X.

Next, if in the final term of (4) we replace q by the value given in (2), we get the following equation:

$$q = \frac{E_1(a)}{2b} - \frac{E_1E_2(a)}{4b} + \frac{E_1E_2E_1(a)}{8b} - \frac{E_1E_2E_1(r)}{8}. \tag{5}$$

If we now replace r in the final term of equation (5) by the value given in (3), and if we continue making substitutions in this fashion, we will obtain the following infinite series:

$$q = \frac{E_1(a)}{2b} - \frac{E_1E_2(a)}{4b} + \frac{E_1E_2E_1(a)}{8b} - \frac{E_1E_2E_1E_2(a)}{16b} + \ldots . \tag{6}$$

A similar expression can be derived for r. These expressions represent a formalization of the difficulties of infinite regress which characterize simultaneous-decision problems. These difficulties have been summarized by referring to this type of model as the "I think that he thinks that I think . . ." model. However, some special cases of interest can be developed in which this chain of regress can be broken and an equilibrium solution obtained.

Example 1

Suppose that both firms know the value of a and that each firm is aware of the other firm's knowledge. Then $a = E_1(a) = E_1E_2(a) = \ldots$, and, therefore,

$$q = \frac{a}{2b} - \frac{a}{4b} + \frac{a}{8b} - \frac{a}{16b} + \ldots = \frac{a}{3b}.$$

The same value will also hold, of course, for r. This result is of particular interest because the value $a/3b$ for the output is precisely the value for the output that would prevail in the case of the Cournot model, as the following argument demonstrates:

Since $Q = aq - bq(q+r)$, then

$$\frac{dQ}{dq} = a - 2bq - br - bq\frac{dr}{dq}. \tag{7}$$

When the value dr/dq is assumed to be 0, the value of q which is determined by the equation $dQ/dq = 0$ is $q = (a-br)/2b$. In a similar fashion it is found that $r = (a-bq)/2b$. Solving these two equations simultaneously, we obtain the values $q = r = a/3b$.

This derivation indicates that the Cournot results can be obtained simply and directly from a model in which the firms make simultaneous decisions, where each firm has accurate expectations of the environment, and where each firm is aware of the other firm's knowledge. This analysis tends to make the equilibrium results of the Cournot model credible, since the assumptions given here are not unlike the assumptions economists frequently make. This is not to argue that the process in the Cournot model as conceived has any particular validity. Rather, this analysis demonstrates that the equilibrium position of the Cournot model can be reached by a model with reasonable assumptions.

Example 2

Suppose that firm 1 thinks that $a = a_1$, and that firm 2 thinks that $a = a_2$. Suppose that each firm is aware of the other's beliefs, but neither firm will change its own beliefs. Then $E_1(a) = a_1$, $E_1E_2(a) = a_2$, $E_1E_2E_1(a) = a_1$, $E_1E_2E_1E_2(a) = a_2, \ldots$. Hence,

$$q = \frac{a_1}{2b} - \frac{a_2}{4b} + \frac{a_1}{8b} - \frac{a_2}{16b} + \ldots = \frac{2a_1 - a_2}{3b}.$$

Example 3

Suppose that firm 1 knows that firm 2's estimate of a is $a = a_2$. Suppose that firm 1 believes also that firm 2 thinks that firm 1 has the same estimate a_2. In fact, however, firm 1's estimate is $a = a_1$. Then $E_1(a) = a_1$, and $E_1E_2(a) = E_1E_2E_1(a) = \ldots = a_2$. Therefore,

$$q = \frac{a_1}{2b} - \frac{a_2}{4b} + \frac{a_2}{8b} - \frac{a_2}{16b} + \ldots = \frac{3a_1 - a_2}{6b}.$$

Example 4

Suppose that both firm 1 and firm 2 know the value of a. Suppose also that firm 1 believes that firm 2 is assuming, for some reason, that firm 1 will produce zero. Then $E_1(a) = E_1E_2(a) = a$, and $E_1E_2(q) = 0$. Therefore, by using equation (4) we obtain the relation

$$q = \frac{a}{2b} - \frac{a}{4b} + 0 = \frac{a}{4b}.$$

This solution is of some special interest because it represents the monopoly output for one firm on the assumption that the firms share equally in the output. Thus, if each firm held the expectations described, they could be led to the monopoly solution even though the expectations were incorrect. This is in contrast with the well-known solution of Chamberlin [4, 46–51]. It is obvious from (4) that a variety of combinations of values of $E_1(a)$, $E_1E_2(a)$, and $E_1E_2(q)$ could lead to the value $a/4b$ for q. An analogous expression exists for r.

This monopoly solution can be derived much more simply in problems in which the firms make their choices alternately rather than simultaneously. Such problems will now be considered.

4. *Expectations and alternating decisions*

In this section we shall again consider a model in which the demand function is of the form $p = a - b(q+r)$ and the profit Q of firm 1 is specified by the equation $Q = q[a - b(q+r)]$. However, we shall now suppose that firm 1 must choose a value of q before firm 2 chooses a value of r, and that firm 2 will be aware of the value of q when it chooses r. In such a problem firm 1 must choose q on the basis of the reaction it expects from firm 2 to each possible value of q. In other words, firm 1's expectation $E_1(r)$ will now be some function $\phi(q)$ of the value it chooses for q. Firm 1 will then choose q to maximize its expected profit $E_1(Q)$ as given by equation (1), with $E_1(r) = \phi(q)$. The function $\phi(q)$ is the expected reaction function, and the derivative $d\phi/dq$ is essentially the conjectural-variations term of the Cournot model.

Example 5

Suppose that the value of a is known, and that firm 1 expects firm 2 to choose its output r to equal the output q of firm 1. Then it is easily found from equation (1) that $E_1(Q)$ is maximized when $q = a/4b$, which is again the monopoly solution.

Example 6

Suppose that both firms know the value of a, and suppose that firm 1 expects firm 2 to choose r after it has learned the value of q in order to maximize its own profit $R = r[a-b(q+r)]$. Thus, firm 1 expects firm 2 to choose r according to the relation $E_1(r) = (a-bq)/2b$. It then follows from equation (1) that the value of q which maximizes $E_1(Q)$ is $q = a/2b$.

Finally, suppose that each firm must choose its output in each period of a multiperiod process; and suppose that the value of a may vary from period to period in accordance with a stochastic process for which the probability distributions are not completely known. Then in each period firm 1 will gain information about this process by observing the value of a for that period, and it will gain information about firm 2's knowledge and behavior by observing the output of firm 2 in the given period. On the basis of this information, firm 1 will revise the relevant expectations for the next period.

Furthermore, when the process involves more than one period, firm 1 will not necessarily choose its output in any given period to maximize its expected profit in that period without regard to expected profits in future periods. In choosing its output in a given period firm 1 must also consider the following three consequences: (1) the information which that choice will reveal to firm 2, (2) the information which that choice will elicit from firm 2, and (3) the expected profit which will be attainable in future periods as a result of this additional information.

Models constructed with such concepts involve learning by firms. The building of models incorporating a learning process is still relatively rare in economics [5]. Therefore, it seems desirable to make a brief diversion to examine the concept of learning as it might apply to firms.

5. *Firm learning*

Learning is a psychological concept, and it is natural to examine the literature in that discipline for an understanding of learning. As might be expected, there is a division of opinion on the exact process by which learning takes place. In recent years a mathematical approach to learning theory has been followed, and the definitions used are compatible with usage in economic theory. For example, Bush and Mosteller, who pioneered in this area with their book on stochastic learning models, define learning as follows [6, p. 3]: "We consider any systematic changes in behavior to be learning whether or not the change is adaptive, desirable for certain purposes, or in accordance with any other such criteria." A probabilistic view of behavior is followed in which the subject (in a psychological experiment) is treated as having some probability p of making a particular response. The effect of the experiment on learning is measured by the change in probability of the subject's making a particular response. The completion of learning would be represented by a stable (obviously needing an operational definition) value for this probability of response. In other words, the learning has ceased when there is no longer any change in the probability of the particular response.

In duopoly theory learning concerns the behavior of rivals. More specifically, it is learning that enables a firm to predict how a rival will respond in the next decision period to an action that the firm takes in this period or to predict the value that a rival will set for its decision variable during a given period of simultaneous decision making. Our concept of learning involves the following three assumptions: (1) The entrepreneur (manager) of the firm starts with an a priori probability that his rival is going to behave in a particular way with respect to his decision making. (2) He observes the actual behavior of his rival in making a decision either in response or simultaneously. (3) He incorporates the results of his experience by modifying his original a priori probability, and the process is then repeated. The utilization of the observation of the rival's behavior to set new a priori probabilities is an act of learning by our definition, and the differences between the original and the new probabilities can be used to define a measure of the amount of learning. We shall not, however, consider that question further.

The specific process by which individuals learn, a learning theory or model, is an area of vigorous activity and disagreement in psychology. The most widely used theory for experimental purposes is the stimulus-response theory. In recent years both mathematical and computer models have been constructed to describe the learning process. It is accurate to say that much is known about the learning process in individuals. At the same time there is no well-established set of propositions that can be immediately transferred to units of analysis other than the individual.

In any event it would be foolish to assume that the learning process in business firms is the same as that observed in individuals. Nevertheless, it is also obvious that firms exhibit adaptive behavior over time, a type of behavior included in the definition of learning given by Bush and Mosteller. As a first approach, it is useful to deal with this adaptive behavior at the aggregate level of the firm without referring to the individual members of the firm. The learning that takes place within the firm can be viewed as falling into two categories. First is the adaptation in the internal processes of the firm with respect to such factors as goals, attention rules, and search processes. The second relates to interfirm learning and is the area directly relevant to the problem of this paper. More specifically, the problem is to exhibit the process by which firms are able to follow the Bayesian process in making decisions in an oligopolistic market. In economic theory the approach followed has been one in which the firm is assumed to know nothing about its rivals and to be unable to learn, or it knows everything and the collusive solution is appropriate.

We are advocating the development of models in which firms make decisions and take actions in the market and then are able to learn from their experience. The learning process takes place in the first instance through observation of the reactions of the firm's rivals to its actions, and the storing of this information. These actions then become the basis for the computation of the new prior probabilities for the firm. These probabilities represent the firm's estimates, on the basis of its learning, of the reactions of its rivals for particular actions of the firm in the next decision period. Thus one aspect of learning as used here is the ability to develop enough knowledge of rivals to enable predictions of their behavior to be made. Increases in learning imply improvements in the accuracy of the predictions.

The learning process in the organization involves the gathering of information from the environment of the form and quantity necessary to make the appropriate inferences. Within the organization it requires the information to be processed and presented in such a manner that judgments can be made about the significance of the information.

The information comes from participants in the organization who are interacting with the environment either in their official organizational roles or as individuals. This information is generally about rival firms' plans and market behavior. Salesmen and managers are an especially fruitful source of information of this kind. Their information comes from interactions with customers who have been in contact with salesmen of other firms, and from direct interactions with rival salesmen. The information is most frequently about price or related selling conditions of the product such as credit terms. Any price changes, including attempts at secret price cuts, will generally be picked up in this information net. This information will not be without bias, since salesmen, because of their payoff function, tend to press for changes in selling conditions that will increase sales. This bias in information is a factor that the organization must learn to evaluate and compensate for in the use of its information system.

It is obvious that the concept of learning in business organizations merits greater analysis than can be given here. Our purpose in this brief discussion has been to motivate some of the assumptions that will be made in the models of the next four sections.

6. *An additive model*

Up to this point we have given some illustrations of the way in which the expectations operator can be utilized in certain problems involving just a single period with either simultaneous choice or alternating choice. Now we will show how the learning described in the preceding section can be incorporated into the models when the decisions are multiperiod and are made in different periods by the two firms.

The basic innovation in these models is the use of reaction functions that involve an unknown parameter and a random component. By using Bayesian analysis the firm is able to adapt its decisions as it gains information (learns) about the unknown parameter from the rival's behavior.

As before, let q denote the quantity to be produced by firm 1, and let r denote the quantity to be produced by firm 2. We shall assume that the demand curve is of the form $p = a-bq-cr$, where a, b, and c are given positive constants. For any values of q and r chosen by the two firms, the resulting profit Q for firm 1 is as follows:

$$Q = aq-bq^2-cqr. \tag{8}$$

We shall make the following assumptions: (1) As in an earlier section, firm 1 is the leader and makes its decision for q before firm 2 chooses the value for r; (2) after learning the value of q, firm 2 chooses r in accordance with the equation

$$r = q+\theta+\varepsilon. \tag{9}$$

Here θ is a constant whose value is chosen and fixed by firm 2 before it learns the value of q, and ε is a random variable with mean 0 whose distribution is completely known to both firms. The exact value of θ that firm 2 is using is unknown to firm 1. This model will be referred to as an additive model because the parameter θ enters additively in equation (9).

The random variable ε represents the random part of firm 2's choice. It may reflect the effect of very recent information about market conditions coming from a market survey, or it may result from firm 2's inability to control output in a completely accurate fashion. In short, the firm may decide on an output of $q+\theta$, but because of random disturbances it actually produces $q+\theta+\varepsilon$. Firm 1 is fully aware of equation (9) and the distribution of ε but does not know the exact value of θ that firm 2 is using. However, firm 1 does have a prior distribution for θ.

By substituting the value of r given in equation (9) into equation (8), firm 1 can compute, for any value of q that he might choose, his expected profit as given by the following equation:

$$E(Q) = aq-(b+c)q^2-cmq. \tag{10}$$

Here $m = E(\theta)$ is the mean of firm 1's prior distribution of θ. The value (10) is maximized when

$$q = \frac{a-cm}{2(b+c)}. \tag{11}$$

Thus, the value given in (11) represents the optimal choice of q for firm 1. The expected profit from this choice of q is

$$E(Q) = \frac{(a-cm)^2}{4(b+c)}. \tag{12}$$

7. *A multiplicative model*

Consider now the same problem as above, but suppose that firm 2's choice of r, instead of being given by equation (9), is now given by the following equation:

$$r = \theta q+\varepsilon. \tag{13}$$

Again, θ is a positive constant chosen and fixed by firm 2, and ε represents the random element in firm 2's choice, with $E(\varepsilon) = 0$. This model will be referred to as a multiplicative model because the parameter θ enters multiplicatively in equation (13).

When (13) is substituted into (8) and expectations are taken, the expected profit of firm 1 is found to be

$$E(Q) = aq-(b+cm)q^2. \tag{14}$$

Here again, $m = E(\theta)$ is computed under firm 1's prior distribution. From (14) we find that firm 1's optimal choice of q is

$$q = \frac{a}{2(b+cm)}, \tag{15}$$

and that the expected profit from this choice is

$$E(Q) = \frac{a^2}{4(b+cm)}. \tag{15a}$$

In these two simple models firm 2 is always restricted to a rather mechanical decision rule. In the additive model, firm 2 tries to produce a fixed amount θ more than firm 1 produces (or less than firm 1 produces if $\theta < 0$). In the multiplicative model, firm 2 always tries to produce a fixed fraction θ of the production of firm 1 (or a fixed multiple if $\theta > 1$). However, these two models will serve to illustrate the results that can be obtained and the difficulties that will be encountered in multiperiod problems in which learning is represented by a Bayesian analysis. As we shall show, the additive model can be handled relatively easily, but the multiplicative model has no optimal solution. A multiplicative model of this type has been introduced in a somewhat different multiperiod problem by Prescott [7].

8. *A multiperiod problem with the additive model*

Consider the additive model, and suppose now that the process continues through n periods. For $j = 1, \ldots, n$, let q_j denote the quantity to be produced by firm 1 in period j, let r_j denote the quantity to be produced by firm 2 in period j, and let Q_j denote the profit of firm 1 in period j. We assume, as before, that

$$Q_j = aq_j - bq_j^2 - cq_jr_j \qquad (j = 1, \ldots, n). \tag{16}$$

We also assume that firm 2 chooses r_j in accordance with the equation

$$r_j = q_j + \theta + \varepsilon_j \qquad (j = 1, \ldots, n). \tag{17}$$

Thus, firm 2 learns the value of q_j before choosing r_j. In equation (17), θ represents the systematic adjustment which firm 2 makes, and the value of θ remains fixed throughout the n periods; and ε_j represents the random component in the value of r_j that is ultimately chosen, and it is assumed that the random variables $\varepsilon_1, \ldots \varepsilon_n$ are independent and identically distributed. Specifically, we make the assumption that each ε_j is normally distributed with mean 0 and precision τ, where τ is a known number. It should be kept in mind that the precision of a normal distribution is equal to the reciprocal of the variance of the distribution; that is, $\tau = 1/\sigma^2$. As we shall see, in many problems involving a Bayesian analysis based on normal distributions, it is simpler to represent these distributions by their mean and precision, rather than by their mean and variance.

It should be noted that by assuming that ε_j has a normal distribution and by allowing q_j, θ, and r_j to vary over all real numbers, we are not adhering to the physical restriction that the amount produced in any period must be nonnegative. This abstraction should not cause the reader any conceptual difficulties. Indeed, this restriction has been ignored throughout all of the earlier sections.

Bayes's theorem can be shown to be directly applicable to firm 1's information about θ. Suppose that, at a given period, firm 1's knowledge about θ can be represented by a normal distribution with mean m and precision h. By a standard application of Bayes's theorem (Pratt et al. [8], sect. 16.4), it can be shown that if we then observe the value x of a random variable whose distribution for any given value of θ is normal with mean θ and precision τ, then the posterior distribution of θ will again be normal with mean m' and precision h' specified as follows:

$$m' = \frac{hm + \tau x}{h + \tau}, \qquad h' = h + \tau. \tag{18}$$

Suppose now that when the prior distribution of θ is as specified above, firm 1 chooses the value q and observes that firm 2 chooses the value r in response. By equation (17), the distribution of $r - q$, for any given value of θ, is a normal distribution with mean θ and precision τ. Therefore, after firm 1 observes the value of r chosen by firm 2, the posterior distribution of θ will be normal with mean m' and precision h' specified as follows:

$$m' = \frac{hm + \tau(r - q)}{h + \tau}, \qquad h' = h + \tau. \tag{19}$$

We assume that firm 1's objective is to maximize its expected total profit over the n periods; that is, firm 1 must find a sequential decision rule which maximizes

$$E(Q_1 + \dots + Q_n). \tag{20}$$

We can now apply the method of "backward induction" to find an optimal sequential rule of this type.

Suppose that the first $n-1$ periods of the process are over and that firm 1 must now make its final choice q_n. Since all profits and losses in the first $n-1$ periods have now been recorded and cannot be changed, firm 1 must choose q_n simply to maximize $E(Q_n)$, its expected profit in the final period. Suppose that, at the beginning of the final period, firm 1's distribution for θ is normal with mean m and precision h. It follows from the results (11) and (12) given earlier that firm 1 should choose

$$q_n = \frac{a-cm}{2(b+c)}, \tag{21}$$

and that the resulting expected profit $E(Q_n)$ on the final stage is

$$E(Q_n) = \frac{(a-cm)^2}{4(b+c)}. \tag{22}$$

We now move back one stage. Suppose that the first $n-2$ periods are over and that firm 1 must now choose q_{n-1}. Suppose again that its distribution for θ at this time is normal with mean m and precision h. Firm 1 must now maximize the sum $E(Q_{n-1})+E(Q_n)$. It follows from (19) and (22) that, for any values of q_{n-1} and r_{n-1}, if it makes an optimal choice of q_n, then $E(Q_n)$ is given by the equation

$$E(Q_n) = \frac{(a-cm')^2}{4(b+c)} = \frac{\left[a-c\,\dfrac{hm+\tau(r_{n-1}-q_{n-1})}{h+\tau}\right]^2}{4(b+c)}. \tag{23}$$

Furthermore, since r_{n-1} is given by (17), we can take the expectation over r_{n-1} in the final part of (23), for any given choice of q_{n-1}. In fact, when the distribution of θ is taken into account, it follows from (17) that the distribution of r_{n-1} is normal with mean $q_{n-1}+m$ and precision $h\tau/(h+\tau)$. It can thus be shown that, for any given choice of q_{n-1},

$$E(Q_n) = (a-cm)^2 + \frac{\tau c^2}{h(\tau+h)}. \tag{24}$$

Since firm 1 must choose q_{n-1} to maximize the sum $E(Q_{n-1})+E(Q_n)$, and since, by (24), the value of $E(Q_n)$ will not depend on the specific value of q_{n-1} which is chosen, firm 1 can choose q_{n-1} simply to maximize

$E(Q_{n-1})$. In other words, he *need not plan ahead.* Again, from (11) and (12) or, equivalently, from (21) and (22), we find that the optimal choice is

$$q_{n-1} = \frac{a-cm}{2(b+c)}, \tag{25}$$

which leads to the value

$$E(Q_{n-1}) = \frac{(a-cm)^2}{4(b+c)}. \tag{26}$$

Hence, by (26) and (24), it is seen that, under the optimal procedure,

$$E(Q_{n-1})+E(Q_n) = \left[1+\frac{1}{4(b+c)}\right](a-cm)^2+\frac{\tau c^2}{h(\tau+h)}, \tag{27}$$

where m is the mean of firm 1's distribution for θ at the beginning of the $(n-1)$st period, and h is the precision of this distribution. This argument can now be carried backward through successively earlier stages by induction.

Summarizing, we can say that when the additive model is appropriate, the optimal sequential decision procedure is the myopic procedure whereby firm 1 makes its choice at each period without regard for future periods: that is, it makes a choice as if each period were the final one, or as if it were in a one-period problem. It does, of course, update the posterior distribution of θ in each period based on the value of r which is observed in that period, as specified in equation (19).

9. *A multiperiod problem with the multiplicative model*[1]

In this section we shall again consider a process with n periods, but we shall now assume that firm 2 chooses its output r in each period in accordance with the multiplicative model (13). We shall show that the optimal procedure for firm 1 in this problem will *not* be the myopic procedure whereby firm 1 chooses q in each period to maximize its expected profit in that period without regard to the effect of this choice on its expected profit in future periods. In fact, the optimal procedure cannot be determined, as will be made clear by attempting to apply the method of backward induction.

[1] This section has been revised for this volume.

However, Bayes' theorem can still be applied easily to update firm 1's information about θ at each period. Again we shall suppose that, at a given time, firm 1's knowledge about θ can be represented by a normal distribution with mean m and precision h. It follows from equation (13) that for any given values of q and θ, the distribution of r/q will be normal with mean θ and precision τq^2. Therefore, by equation (18), if firm 1 chooses q and then observes that firm 2 chooses r, the posterior distribution of θ is again normal with mean m' and precision h' given as follows:

$$m' = \frac{hm+\tau rq}{h+\tau q^2}, \qquad h' = h+\tau q^2. \tag{28}$$

These expressions can be compared with those given in equation (19). The signs which indicate that the optimal sequential procedure for the multiplicative model will be complicated are the appearance of q in the denominator of m' and the appearance of q in the new precision h'. The latter point raises the possibility of an obvious heuristic in approaching the problem. From (28) it can be seen that a large value of q leads to a large value of h', which means that firm 1 will have a very precise posterior distribution for θ. Hence, it might be worthwhile for firm 1 to sacrifice profits at the first stage by choosing a large value of q. It will then know the value of θ very accurately and will be able to exploit this knowledge to its advantage in later periods. In other words, the interrelations among the periods are created by the appearance of the decision variable in the expression for the precision of the posterior distribution.

The optimal value of q for the final period is specified by equation (15), and the expected profit from this period is specified by equation (15a). By making use of this result and equation (14), we find that firm 1 must choose q at period $n-1$ to maximize the following sum, $E(Q_{n-1})+E(Q_n)$:

$$E(Q_{n-1})+E(Q_n) = aq-(b+cm)q^2 + \frac{a^2}{4}E\left[\left(b+c\,\frac{hm+\tau qr}{h+\tau q^2}\right)^{-1}\right], \tag{29}$$

where the random variable r in the final expectation has a normal distribution with mean mq and precision $h\tau/(h+\tau q^2)$. It follows that this expectation will be indeterminate, since the expectation of the reciprocal of a random variable with a normal distribution is indeterminate. Thus, it is impossible to determine an optimum value for q since the value of equation (29) is not well defined.

It should not be inferred that either the concept of the multiplicative model or the method of backward induction is unreasonable. If the normal distribution assumed for r is replaced by a bounded distribution that is concentrated on positive values, then the difficulty of indeterminancy can be eliminated. Another approach would be to formulate the problem as the maximization of a bounded utility function. In this way, risk aversion could also be introduced.

We shall not pursue any of these possibilities here. Our major aim, to reiterate, has been to demonstrate the potential for adaptive models by introducing new tools and concepts.

References

[1] A. Cournot, Researches into the mathematical principles of the theory of wealth. New York, Macmillan, 1897.

[2] R. Frisch, Monopoly-polypoly—the concept of force in the economy, translated by N. Beckerman. In: International economic papers, edited by E. Henderson. New York, Macmillan, 1951.

[3] J. W. Friedman, Reaction functions and the theory of duopoly. Rev. Econ. Studies (July 1968), 257–272.

[4] E. H. Chamberlin, The theory of monopolistic competition. Cambridge, Mass., Harvard Univ. Press, 1946.

[5] Charles E. Ferguson and Ralph W. Pfouts, Learning and expectation in dynamic duopoly behavior. Behavioral Science 7 (April 1962), 223–237.

[6] R. R. Bush and F. Mosteller, Stochastic models for learning. New York, Wiley, 1955.

[7] E. C. Prescott, Adaptive decision rules for macro economic planning. Ph.D. dissertation, Graduate School Indus. Admin., Carnegie-Mellon Univ., 1967.

[8] J. W. Pratt, H. Raiffa, and R. Schlaifer, Introduction to statistical decision theory. New York, McGraw-Hill, 1965.

S. E. Fienberg and A. Zellner (eds.)
Studies in Bayesian Econometrics and Statistics.

3.2 A BAYESIAN APPROACH TO THE THEORY OF EXPECTATIONS*

Stephen J. TURNOVSKY**

Australian National University

1. *Introduction*

When faced with uncertainty, economic decision makers are continually required to make decisions based on present or future values of relevant variables, which are not known at the time of the decision and which must therefore be predicted. Hence, economists have been led to propose a number of schemes which enable the future (and unknown present) values to be predicted from past data (see e.g. [1, 3, 5, 7]). However, these schemes have all been suggested on an ad hoc basis—frequently being justified on the basis of convenience—and are not motivated by any compelling underlying behavioral considerations.[1] In this paper, we therefore attempt to provide a behavioral basis for some of these predictors by showing that they can be derived from a situation where the decision maker forms his predictions by using a Bayesian sampling procedure. Despite the considerable advances that have recently been made in this area of statistical decision theory, it has thus far failed to make any noticeable impact on economic theorizing. In the latter part of the paper, we provide an example of the implications of the Bayesian predictor for a competitive firm faced with uncertainty in the demand for its product.

* Reprinted from the Journal of Economic Theory 1 (1969), 220–227, with permission of the Academic Press, Inc.

** I have benefited from discussions with a number of former colleagues at the University of Pennsylvania and in particular would like to thank Albert Ando, Karl Shell and Deane Terrell. Needless to say, I take sole responsibility for any errors which may remain.

[1] However, it should be pointed out that Nerlove's "adaptive expectations" [7] does have certain optimality properties (see [8]).

2. Bayesian prediction

The Bayesian approach to statistical decision theory emphasizes the accumulation of knowledge by a decision maker in the course of observing the statistical process he is studying. At any point of time he is assumed to have some prior subjective knowledge about this process and this is continually up-dated as more information comes to hand. For the sake of concreteness, we shall assume that the individual is a firm which has to make predictions involving the uncertain actual prices which enter its current decisions. We shall assume the simplest case where the actual prices p_t are normally and independently distributed through time. However, the identical analysis applies to a variety of distributions (see [4, 10]).[2]

The Bayesian approach to forecasting these prices can thus be formulated as follows. Denote the distribution of actual prices of period t by:

$$h(p_t) = N(p_t^*, \sigma_t^2), \tag{1}$$

where p_t^* and σ_t^2 are the mean and variance respectively and N stands for "normal". We shall assume that the firm does not know p_t^*, the expected price, though it does know the variance σ_t^2 about that price. Thus it has perfect knowledge of the variability of the product price. In practice it will probably know neither, but this case is a good deal more complicated to analyze (see [10]).[3] The decision maker does, however, have subjective knowledge of p_t^* in the form of a prior probability distribution which is also assumed to be normal and is denoted by:

$$h'(p_t^*) = N(\bar{p}_t, \varpi_t), \tag{2}$$

where $\bar{p}_t$ and ϖ_t are the subjective mean and variance and reflect the decision maker's feelings about the uncertainty. From (1) and (2) we infer that the firm's *subjective* probability distribution of actual prices

[2] In particular, the lognormal distribution leads to completely analogous results to those we are about to give, with the exception that everything is replaced by its logarithm.

[3] We could similarly study the case where the decision maker knows the mean but not the variance. However, this is considerably less interesting, since in that case we would be unable to interpret any of the existing price expectations schemes in terms of our Bayesian model.

can be written:

$$g(p_t) = N(\bar{p}_t, \theta_t), \tag{3}$$

where

$$\theta_t = \sigma_t^2 + \bar{\omega}_t + 2\,\mathrm{cov}\,(p_t, p_t^*).$$

During period t the firm makes n_t observations on the price prevailing during that period by observing the price in a number of markets. If there is only one market then $n_t = 1$, but it is possible that it may observe prices in other locations too. Letting q_t denote the mean of the prices observed during period t, and applying Bayes' theorem it can be shown that the posterior distribution for p_t^*, after making these observations, is also normal and denoted by:

$$h''(p_t^*) = N(\bar{\bar{p}}_t, \bar{\bar{\omega}}_t), \tag{4}$$

where

$$\bar{\bar{p}}_t = \frac{(1/\bar{\omega}_t)\bar{p}_t + (n_t/\sigma_t^2)q_t}{(1/\bar{\omega}_t) + (n_t/\sigma_t^2)}; \qquad \frac{1}{\bar{\bar{\omega}}_t} = \frac{1}{\bar{\omega}_t} + \frac{n_t}{\sigma_t^2}.$$

The details of this calculation are omitted here but they are worked out in detail in Pratt, Raiffa, and Schlaifer [9].

[In general there is no need for the prior distribution to be normal as well, in which case only the spirit but not the details being presented here apply. However, if the distribution of actual prices is normal, one can give the following argument in favor of prior normals. Suppose originally the firm had no information about the distribution of the expected mean and consequently had a uniform prior (see [9]). In that case its posterior distribution after observing prices during period one will by Bayes' theorem be normal. Thus its prior for period two will be normal and with prices in all periods being normal it follows from repeated applications of Bayes' theorem that all subsequent priors will also be normal.]

If the firm's only source of information is by observing the prevailing prices at each period, then the posterior distribution at time t becomes the prior distribution for period $(t+1)$ so that:

$$h''(p_t^*) = h'(p_{t+1}^*). \tag{5}$$

Thus writing ω_t for $\bar{\omega}_t$, it follows that:[4]

$$\bar{p}_{t+1} = \frac{(1/\omega_t)\bar{p}_t + (n_t/\sigma_t^2)q_t}{(1/\omega_t) + (n_t/\sigma_t^2)}; \qquad \frac{1}{\omega_{t+1}} = \frac{1}{\omega_t} + \frac{n_t}{\sigma_t^2}, \tag{6}$$

[4] Raiffa and Schlaifer [10] define the "precision" to be the inverse of the variance, so that the larger the variance the less the precision. From (6) we see that the precision cannot decrease with sampling.

which in turn enables us to write:

$$\bar{p}_{t+1}-\bar{p}_t = \left(1-\frac{\omega_{t+1}}{\omega_t}\right)(q_t-\bar{p}_t). \tag{7}$$

Equation (7) thus asserts the firm's subjective estimate of the expected price follows an "adaptive" scheme where its previous expectation is revised by some fraction of the previous period's error.[5] This is very similar to the adaptive expectations hypothesis proposed by Nerlove and others [7], the only difference being that in (7) the rate of adaptation is not constant, but varies at a rate which depends upon the relative precisions of the prior and sample means. Equation (7) is a standard first-order linear difference equation with variable coefficients which enables $\bar{p}_t$ to be solved in terms of all previously observed prices.

Equation (7) may be used to justify some of the existing expectational hypotheses that have been advanced in the past. The simplest, known as "static expectations", assumes $\bar{p}_{t+1} = q_t$. In (7) this will occur if and only if $1/\omega_t = 0$—i.e. when the subjective variance at time t is infinite. Hence the decision maker will resort to static forecasting if he feels that there is absolutely no precision to his subjective estimate of the expected price of period t.

In order for the rate of adaptation to remain at a constant rate k, we require that:

$$1-\frac{\omega_{t+1}}{\omega_t} = k,$$

i.e. $$\omega_t = \omega_0(1-k)^t. \tag{8}$$

Thus Nerlove's model of adaptive expectations can be justified in a Bayesian context as occurring in a situation where the decision maker samples in such a way so that his subjective variance decreases geometrically with time. This in turn requires the precision of the sample, n_t/σ_t^2, to increase geometrically with time. Interpreting the precision as being a summary of the amount of knowledge implies that the decision maker must sample in such a way that the ratio of knowledge contained in the sample to that contained in the prior estimate remains constant.

[5] The rather simple adaptive formula describing the change in $\bar{p}_t$ is a direct consequence of the fact that the prior and posterior distributions belong to the same family. Similar formulae apply to certain other distributions, such as the lognormal. However, where the prior and posterior distributions belong to different families, then $\bar{p}_t$ will change in a more complicated fashion.

As a third example let us consider the following particularly important case. Suppose initially $\omega_0 = \infty$, so that at time zero, before he has observed any prices, the decision maker has no faith in his subjective estimate of the mean and assume further that σ_t^2 is constant. If only one price may be observed in each period (as would be the case under perfect competition) so that $n_t = 1$ and $q_t = p_t$, then it can be readily verified that the solution to (7) becomes:

$$\bar{p}_t = \frac{1}{t} \sum_{i=0}^{t} p_i. \tag{9}$$

Thus the expected price is just the unweighted average of all previously observed prices. This expectational scheme was recently proposed by Carlson [1] who found that it always leads to stability in simple market equilibrium models. Thus these three special cases, which have at other times been proposed independently, can all be justified in terms of the above Bayesian procedure.

3. *An application to the theory of the firm*

As an example we shall see the implications of the foregoing for the behavior of a competitive firm which is faced with uncertainty in the demand for its product. Being competitive, the firm can sell as much as it desires in any period at the going price—whatever that may be—so that the uncertainty in demand for its product is effectively an uncertainty in its output price.

The firm is assumed to make its production plans for period t before the start of that period. All inputs required during period t must be purchased at the beginning of the period while revenue from sales is received at the end of the period when the random price becomes known. For the purposes of this example we shall also assume that the firm is unable to carry inventories so that all unsold goods have a zero value to the firm. Hence if the quantity y_t is produced at a cost $c_t(y_t)$ and sold at the prevailing price p_t, then profits in period t, π_t, are given by:

$$\pi_t = p_t y_t - c_t(y_t). \tag{10}$$

We shall assume that the firm chooses its output in each period so as to maximize its subjective expected profits during that period.[6] Accordingly, the firm makes a series of one-period decisions, though in the course of making them it continually incorporates new information obtained from observations. Moreover, by focusing on the accumulation of knowledge this formulation includes some important aspects of dynamic behavior, even though it involves making a series of static decisions. It is also a perfectly plausible behavioral assumption—implying much less sophisticated behavior than does a dynamic programming approach to the problem—and it may approximate reasonably closely the practices of a small, competitive firm. Hence in period t, the firm chooses y_t so as to maximize:

$$\bar{p}_t y_t - c_t(y_t), \tag{11}$$

resulting in the familiar optimality conditions:

$$\bar{p}_t = c_t'(y_t), \tag{12}$$

implying that the marginal cost of production is equated to the firm's subjective estimate of the expected price. Furthermore, provided c is convex then as usual we also have:

$$\frac{\partial y_t}{\partial \bar{p}_t} > 0 \tag{13}$$

so that output (supply) increases with expected price.

Although y_t is directly related to only $\bar{p}_t$, it is also indirectly related via equation (7) to $\bar{p}_{t-1}$ and to all previously observed prices. Thus an increase in $\bar{p}_t$ will increase $\bar{p}_{t+1}$ and hence will lead to an increase in y_{t+1} and so on. Similarly, an increase in q_t, although not affecting y_t, will increase $\bar{p}_{t+1}$ and give rise to subsequent effects. It is therefore of interest to determine the long-run effects of an increase in $\bar{p}_t$, taking into account the repercussions in later periods.[7] The long-run response in output to a change in $\bar{p}_t$ shall be defined as:

$$\frac{\partial y}{\partial \bar{p}_t} = \sum_{i=0}^{\infty} \frac{\partial y_{t+i}}{\partial \bar{p}_t}, \tag{14}$$

[6] This widely used criterion is chosen primarily for illustrative purposes and not because it is necessarily appropriate. Indeed, the appropriate criterion depends upon the type of market structure. In fact Diamond has recently shown [2], that with a stock market present, firms maximizing market value should not necessarily use their own subjective probabilities in their calculations.

[7] The same analysis can be performed for changes in observed price p_t.

which acknowledges the fact that part of the response occurs in subsequent periods. Letting $\partial y_{t+i}/\partial \bar{p}_{t+i} = \varepsilon_{t+i}$, it follows by differentiating (7) that:

$$\frac{\partial y}{\partial \bar{p}_t} = \frac{1}{\omega_t} \sum_{i=0}^{\infty} \omega_{t+i}\varepsilon_{t+i}. \tag{15}$$

For this to be finite, the infinite sum must converge, in which case $\lim_{i\to\infty} \omega_{t+i}\varepsilon_{t+i} = 0$. Since there is no reason to expect the slope of the short-run supply curve in period $t+i$ to tend to zero, a necessary but not sufficient condition for convergence is that $\lim_{i\to\infty} \omega_{t+i} = 0$.[8] As long as n_i/σ_i^2 remains positive so that additional sampling always contributes some information, then (6) ensures that this necessary condition will be satisfied. If this is so, the producer's prior distribution ultimately collapses to the true expected value. Alternatively, if the producer's subjective variance does not tend to zero with time, then the long-run supply curve has an infinite supply elasticity. In this case the firm is at no stage certain about the prior distribution and hence will allow its decision at each period to be influenced by changes that have occurred in previous periods. Therefore, a change in any one period will affect decisions in all subsequent periods and over an infinite time horizon will lead to an infinite response.

Finally, although variances do not explicitly appear in the supply function of any period, nevertheless any period's decision is indirectly dependent on all past uncertainties—both subjective and actual.[9] This dependence also operates through the subjective price expectations which as (6) indicates are influenced by past variances. Thus a change in σ_t^2 will affect y_{t+1} through $\bar{p}_{t+1}$, the precise response being given by:

$$\frac{\partial y_{t+1}}{\partial \sigma_t^2} = \frac{\omega_{t+1}^2}{\omega_t} \cdot \frac{n_t}{\sigma_t^4} (\bar{p}_t - q_t) \frac{\partial y_{t+1}}{\partial \bar{p}_{t+1}}. \tag{16}$$

Just as before this change in $\bar{p}_{t+1}$ will in turn be transmitted to $\bar{p}_{t+2}$ and analogously defining the long-run response to a change in σ_t^2 by:

$$\frac{\partial y}{\partial \sigma_t^2} = \sum_{i=0}^{\infty} \frac{\partial y_{t+i}}{\partial \sigma_t^2}, \tag{17}$$

[8] Notice that Nerlove's version satisfies this condition.

[9] In this respect our model differs quite markedly from the non-Bayesian dynamic programming approach in which the same criterion of expected profit maximization is used. In the latter case, decisions depend only upon the first moments of the distributions and never on the variances (see e.g. Zabel [11], Mills [6, p. 85]).

we can show that:

$$\frac{\partial y}{\partial \sigma_t^2} = \frac{\omega_{t+1}}{\omega_t} \cdot \frac{n_t}{\sigma_t^4} (\bar{p}_t - q_t) \sum_{i=0}^{\infty} \omega_{t+i} \varepsilon_{t+i}. \tag{18}$$

Notice that the sign of the response depends crucially on the relationship of q_t to $\bar{p}_t$. If prices in period t are higher than were expected then an increase in σ_t^2 will tend to reduce $\bar{p}_{t+1}$ and hence all future $\bar{p}_{t+i}$ so that the response will be negative; when $q_t < \bar{p}_t$ the opposite is true.[10]

4. Conclusions

The results of this paper demonstrate that if a decision maker accumulates information by Bayesian sampling, then his subjective expectations are found to change in an adaptive manner, where the rate of adaptation varies at a rate which depends upon the relative precisions of the prior and sample means. By taking special cases we are thus able to justify several existing expectations hypotheses as arising within this Bayesian context. For illustrative purposes it is then shown that if a firm forms its expectations in this way then a change in the prices or uncertainty of one period will indirectly affect all future decisions. Necessary conditions for these long-run effects to imply finite long-run supply elasticities are found to require that the firm's prior subjective distribution of the expected price must ultimately collapse to the true expected price. In general this condition will be satisfied, but in any event any increase in price will always have a positive effect on the output of subsequent periods, while the effect of an increase in uncertainty can be of either sign, depending upon what actual prices turn out to be.

The model has been developed for the case where prices are presumed to be both normally and independently distributed over time. However, both these assumptions may be relaxed without substantial modification to the conclusions. The adaptive formulae can be shown to hold for a number of distributions. In particular, replacing $\bar{p}_t$ by its logarithm in equation (6), enables us to apply this equation to the lognormal distribution. Perhaps more restrictive is the assumption of independence, since it is widely accepted that prices are serially correlated. However,

[10] Similar remarks made in connection with (15) apply to the convergence of the infinite sum appearing in (18). Notice that we could see how ω_t affects future decisions, but results are much the same as these.

if prices are subjected to an appropriate autoregressive transformation the Bayesian procedure described in this paper can be applied to the transformed variables.[11]

References

[1] J. Carlson, An invariably stable cobweb model. Rev. Econ. Studies 35 (July 1968).

[2] P. A. Diamond, The role of a stock market in a general equilibrium model with technological uncertainty. Am. Econ. Rev. 57 (Sept. 1967).

[3] M. Ezekiel, The cobweb theorem. Quart. J. Econ. 52 (Feb. 1938).

[4] G. M. Kaufman, Statistical decision and related techniques in oil and gas. Prentice-Hall, Englewood Cliffs, New Jersey, 1963.

[5] L. A. Metzler, The nature and stability of inventory cycles. Rev. Econ. Stat. 23 (Aug. 1941).

[6] E. S. Mills, Price, output and inventory policy. Wiley, New York, 1962.

[7] M. Nerlove, Adaptive expectations and cobweb phenomena. Quart. J. Econ. 73 (May 1958).

[8] M. Nerlove and S. Wage, On the optimality of adaptive forecasting. Mgt. Science 10 (Jan. 1964).

[9] J. W. Pratt, H. Raiffa, and R. Schlaifer, Introduction to statistical decision theory. McGraw-Hill, New York, 1965.

[10] H. Raiffa and R. Schlaifer, Applied statistical decision theory. Harvard Business School, Boston, 1961.

[11] E. Zabel, A dynamic model of the competitive firm. Internl. Econ. Rev. 8 (June 1967).

[11] Suppose, for example, that the random component of prices follows a first-order autoregressive scheme with autoregressive parameter ρ. Then providing that the decision maker knows ρ, the entire analysis of this paper may be applied to the transformed variables $p_t - \rho p_{t-1}$, etc., which will then be independent and will therefore satisfy the requirements for Bayesian analysis.

PART 4

ASSESSING PRIOR DISTRIBUTIONS

S. E. Fienberg and A. Zellner (eds.)
Studies in Bayesian Econometrics and Statistics.

4.1 ELICITATION OF PERSONAL PROBABILITIES AND EXPECTATIONS*

Leonard J. SAVAGE**

Yale University

1. *Introduction*

1.1. Preface

This article is about a class of devices by means of which an idealized *homo economicus*—and therefore, with some approximation, a real person—can be induced to reveal his opinions as expressed by the probabilities that he associates with events or, more generally, his personal expectations of random quantities. My emphasis here is theoretical, though some experimental considerations will be mentioned. The empirical importance of such studies in many areas is now recognized. It was emphasized for the area of economics in an address by Trygve Haavelmo [28, p. 357]:

> I think most of us feel that if we could use *explicitly* such variables as, e.g., what people *think* prices or incomes are going to be, or variables expressing what people

* Reprinted from the Journal of the American Statistical Association 66 (1971), 783–801, with the permission of the American Statistical Association and executor of the Savage estate. Savage presented this paper at the Third NBER-NSF Seminar on Bayesian Inference in Econometrics, Harvard Univ., October 1971.

** Leonard J. Savage is Eugene Higgins Professor of Statistics and chairman, Department of Statistics, Yale University, New Haven, Conn. 06520. This article was prepared in connection with research supported by the Army, Navy, Air Force and NASA under contracts administered by the Office of Naval Research. Reproduction in whole or in part is permitted for any purpose of the United States Government. Part of the work was supported by the Michigan Institute of Science and Technology. The author wishes to thank Ward Edwards, Allan Murphy, Richard Savage, Carl-Axel Stael von Holstein, Robert Winkler and an excellent referee for their help in improving the article. He is especially grateful to Bruno de Finetti, who provided the impetus for the article.

Editor's Note: Professor Savage died suddenly on November 1, 1971, while this article was in press.

> *think* the effects of their actions are going to be, we would be able to establish relations that could be more accurate and have more explanatory value. But because the statistics on such variables are not very far developed, we do not take the formulation of theories in terms of these variables seriously enough. It is my belief that if we can develop more explicit and a priori convincing economic models in terms of these variables, which are realities in the minds of people even if they are not in the current statistical yearbooks, then ways and means can and will eventually be found to obtain actual measurements of such data.

A special instance of the central general principle of this article was recognized long ago by Brier [5], the general principle itself was briefly but colorfully announced by McCarthy [37], and a considerable literature pertaining to it has grown up, some of which will be cited in context and most of which can be found through the references cited, especially the recent and extensive [52] and others that I call "key references".

Bruno de Finetti and I began to write the present article in the spring of 1960, not yet aware of our predecessors and contemporaries. The impetus was de Finetti's, for he had brought us to rediscover McCarthy's [37] insight about convex functions. We expected to make short work of our "little note", but it grew rapidly in many directions and became inordinately delayed. Now we find that the material in the present article is largely mine and that de Finetti has published on diverse aspects of the same subject elsewhere [12, 13, 14, 17]. De Finetti has therefore withdrawn himself from our joint authorship and encouraged me to publish this article alone, though it owes so much to him at every stage, including the final draft.

The article is written for a diverse audience. Consequently, some will find parts of it mathematically too technical, and others will find parts too elementary. If each skips what puzzles or bores him he will, I hope, find the rest reasonably complete for him.

1.2. Summary

The bare essentials of the economic theory of personal probability and expectation are introduced (section 2). Various difficulties in principle that beset the evaluation of preferences such as those that determine the price at which a man is just willing to sell his car or the probability for him that the car will shortly need a new muffler are discussed (section 3).

A probability is a price, in a manner of speaking. More accurately, it is a marginal rate of substitution. Such rates can be elicited by a general mode of behavioral interrogation rather like one that has been proposed for ordinary prices (section 4). These methods admit mathematically special cases that seem to be of particular interest (section 5).

The rate-eliciting methods, both general and special, are examined with reference to probabilities and expectations (section 6). The methods of eliciting personal probabilities thus arrived at can also be approached through the ideas of statistical decision theory (section 7), and this decision-theoretic method provides a relatively new basis due to de Finetti for defining personal probability (section 8).

The methods are generalized to simultaneous elicitation of several rates, probabilities, or expectations (section 9). This vectorial discussion concludes with a prominent special case in which the subject's income depends only on that event among disjoint and exhaustive possibilities that actually occurs.

Aspects of possible applications are discussed (section 10). These include domains of possible application, impediments to application, and criteria for choice among the methods.

2. *Rates of substitution and probability and expectation*

A man who owns some of a commodity, say wheat, will, under suitable circumstances, be almost indifferent both to buying and selling modest quantities of it at a certain price per bushel, his (marginal) rate of substitution of cash for wheat.

Money payable subject to a contingency, such as the accidental burning of a house or the outcome of a race, can be regarded as a commodity [4]. Such commodities are explicitly dealt in by insurance companies and bookmakers, and we encounter them implicitly wherever we make decisions in the face of uncertainty. There is reason to postulate that an ideally coherent person has a rate of substitution $P(A)$ for money contingent on the event A. When q is not too large, he is indifferent to buying or selling q dollars contingent on A for $qP(A)$ dollars outright, and $P(A)$ is defined as the probability of A for the person [9]. Though the relationship can, because of the nonlinearity of the utility of money, be expected to hold only in the limit for infinitesimal values of q, I shall usually write as though it were exact. This limitation is serious, though a technique for creating utility-free currency and other ways to avoid the effects of nonlinear utility will be mentioned in the next section.

Let U denote the logically certain event and A and B any two logically incompatible events. For an economically coherent person,

$$P(U) = 1, \quad P(A) \geqq 0, \quad P(A \text{ or } B) = P(A)+P(B), \tag{1}$$

as shown in [9]. Thus, such a person's P is a (finitely additive) probability measure.

If a finite sequence $\{A_s\}$ of events is a partition (that is, if every pair of them is incompatible but one or another of them obtains), then

$$\sum P(A_s) = 1, \tag{2}$$

as follows from (1).

To orient the reader in the critical literature on personal probability, the commentary and bibliography of [47] might be useful. Excellent early papers are republished, in English, in [31]. Some more recent ones are [16–18, 51]. An extensive bibliography with special reference to experimental aspects is [19].

More general than the notion of the probability of an event is that of the expectation of a random quantity. Let $\{A_s\}$ be a partition and $\{v_s\}$ a corresponding sequence of numbers; the two together can be regarded as the random quantity V that takes the value v_s if A_s obtains. If a person's probability of A_s is p_s, what ought he to pay for the simultaneous offer of qV dollars, that is, for qv_1 dollars in case A_1 obtains, qv_2 dollars in case A_2 obtains, etc.? Viewing qV as a commodity bundle consisting of qv_1 units of one commodity, qv_2 units of another, etc., the answer (always for moderate q) is clearly

$$qE(V) = q \sum p_s v_s, \tag{3}$$

or q times the expectation of V.

A probability $P(A)$ is itself plainly an expectation, namely $E(I_A)$, where I_A is the random quantity that is 1 if A obtains and 0 if A does not obtain. The event A and its associated random quantity I_A can, to considerable advantage, be rigorously regarded as two aspects of one object, simply denoted by A [15]. Thus it is meaningful and convenient to write $P(A) = E(A)$.

A random quantity V need not be defined in terms of a partition nor need it have only a finite number of possible values; it may be simply any (ordinarily) unknown, empirically determinable number. For the unknown payment qV, the person is presumably willing to exchange some definite payment $qE(V)$. The rate of substitution $E(V)$ is the person's expectation of the unknown, or random, quantity V. (In practice, V would be bounded; mathematically, it can be useful to consider some unbounded V and to exclude others.)

3. *Some difficulties in the experimental elicitation of preferences*

The difficulties mentioned in this section are mainly tangential to the present article. Some of them have little or no bearing on the particular situation to be studied; others are here treated as secondary for the time being. The first type are mentioned only to emphasize the advantage of a certain method of eliciting prices and rates of substitution, including probabilities and expectations; and the others are mentioned to warn of their existence.

Many conceptual experiments on preference will presumably remain conceptual only, some because their financial cost is prohibitive and others because they imply immoral or impractical interference with people's lives. For example, experiments to determine directly what risk of pauperhood a person will take to avoid spending a hundred dollars or to gain a million dollars are literally fantastic. This does not of course preclude learning something indirectly: from the behavior of buyers of insurance and of gamblers and speculators [23, 24]; or by asking subjects to introspect about hypothetical choice; or by observing changes of economic behavior that follow changes in the policies of governments, firms, and other institutions; or in other ways.

Statistical problems and difficulties arise in empirical studies of preference, as they do in all empirical studies. A more special, but perhaps related difficulty, is that all subjects report, or otherwise reveal, that they do not know their own preferences; they experience wavering and indecision that cannot be identified with mere indifference. See, for example, [47, pp. 21, 59, 168–9; 18, section 26; 45; and 22 under "Indifference, intransitive" in the Index].

Another difficulty peculiar to experiments on preference is that once an experimenter satisfies one preference of a subject, he may quite drastically change the subject's pattern of preferences. The thirsty man who now prefers water to wine and wine to whiskey might fail to reveal this in an experiment in which he is first offered his choice between wine and water and then, his thirst quenched with water, is offered his choice between wine and whiskey. A related phenomenon is illustrated by the subject who accepts whiskey (perhaps for later consumption) in preference to wine because he suspects that the experimenter will shortly be offering him an opportunity to obtain water. Such difficulties arise in any attempt to study the gambling (and insuring) preferences of a subject. They are particularly evident and important in experiments to determine the demand price or the offer price of a subject for a specific object or service.

One interesting device for coping with such difficulties was pointed out to me by W. Allen Wallis in 1949 or 1950 and was independently exploited by Allais [2] in experiments conducted in 1952. Let the experimenter put the subject successively in several hypothetical choice situations always with the understanding that, when all the subject's choices have been expressed, one of them will actually be implemented by a chance device that, plain to the subject, has no direct connection with the situations of interest. For instance, a subject asked to rank half a dozen tickets to plays, concerts, and athletic events in the order of his preference will, insofar as he is rational, do so sincerely if he understands that the tickets are to be thoroughly shuffled and that he will be given that one of the top two tickets for which he has indicated preference.

The following application of the same general principle was introduced by Marschak [33, 34]. The experimenter makes a sealed bid for an object in the subject's possession and the subject, before seeing this bid, puts an "asking price" on the object. The sale takes place at the bid price, if and only if the bid is at least as high as the asking price. It is clearly to the advantage of the subject to name his actual offer price, at least if he is not in a position to exclude the possibility of bids in some neighborhood of this price; and even in that exceptional case, there is no advantage in naming any other price. (The procedure can evidently be so modified that the subject is the buyer rather than the seller.)

Actual subjects are of course sometimes blind to their own clear advantage [34; 35, p. 47], failing, for example, to understand that they can only deprive themselves by asking too much. It is no final criticism of such a method to say that subjects do not automatically and instinctively understand it or that, understanding, they have psychological difficulty in doing the rational thing. Such facts do underline the need for education and training prior to, and even during, the application of elicitation devices. Incidentally, such education promises to be of great general benefit to the subject and deserves wide promulgation on its own account.

Marschak's method does not depend on the hypothesis that the utility of money is linear, but throughout most of this article, this hypothesis is relied upon. The hypothesis is presumably a valid approximation if only small transactions are involved. Still better, there would be no approximation at all if payments were made in utiles rather than in money, which may sometimes be roughly feasible.

A certain scheme for effecting payments in utiles is vividly suggested by Smith [50, section 13]. If, to paraphrase Smith, the currency used consists of tickets in a one-prize lottery, which the subject is known to regard as fair and independent of all the uncertainties that are of direct interest for the experiment, then the subject's utility for this currency is linear, though the transactions may be very substantial. Of course, for this scheme to be valid the utility of the prize to the subject must not depend on the outcome of events that are of interest. For example, a lottery in which the prize is a diamond would obviously not provide a valid utility-free currency for exploring a subject's opinions about the future of the diamond market.

In principle, the utility function of a subject for ordinary money can be determined by certain elicitation experiments that give operational meaning to utility. With this function known, the experimenter would know the utility worth to the subject of any proposed payment (positive or negative) and could arrange to make payments in utiles. Practical limitations on this scheme and on the one in the preceding paragraph are severe, but the ideas are at least sufficiently suggestive to be worth bringing out [57, 60].

The linear approximation seems adequate for many applications. Where it is not quite adequate, much can be said for taking as the next approximation a function for the utility of money of the form

$$(1-e^{-\lambda x})/\lambda. \tag{4}$$

(Any function of the form $b-ae^{-\lambda x}$, with a and λ of like sign, would of course amount to the same thing; the choice of the constants a and b made in (4) emphasize that the proper interpretation of $\lambda = 0$ is linear utility, because, for fixed x and small λ, (4) is approximately $[1-(1-\lambda x)]/\lambda = x$.) For almost all applications λ would be positive, which is why a minus sign was introduced in the exponent of (4). A person who is just willing to toss a fair coin for \$10.00 if he receives a side payment of \$0.10 exhibits a λ of about 0.0020 per dollar. To show that a person's utility for money is adequately described by (4) in some range of practical interest and to determine the value applicable to him may in some contexts provide a practical basis for paying him in utiles.

Taken literally, the exponential utilities described what I would call a perfect miser, a person who does not necessarily rank lotteries, and the like, simply by their expected cash value but whose rankings do not fluctuate with his own wealth at the moment. Within sufficiently narrow limits, any person's utilities can be expected to be practically linear. When these limits are somewhat exceeded so that nonlinearity must be taken into account, there should be new limits within which the great experimental simplification of miserly behavior can still be safely assumed as an approximation. Miserly utility, or utility with constant local risk aversion, seems to have been introduced by Pratt [43].

4. The elicitation of a rate of substitution

Some of the difficulties mentioned in the previous section tend to disappear when the investigation is confined to rates of substitution, of which probabilities are for this paper the prime example. First, the transactions envisaged are then moderate, almost by definition, so that typically the experiments that suggest themselves are not grandiose. Second, the satisfaction of one preference here has ideally no effect on the subject's other preferences, though in practice there can well be a conflict between keeping the transactions small enough to avoid important manifestations of saturation (or of nonlinear utility, as we say in connection with probability or expectation) and yet large enough to justify the subject's close attention. Possible remedies for this conflict, at least in probabilistic cases, were mentioned in section 3. Finally, by means of devices related to Marschak's method mentioned there, it is possible to present a subject with a single, relatively simple, economic choice in which it is to his interest to reveal any reasonable number of his rates of substitution, as will now be explained for the case of a single rate.

Suppose the experimenter offers, once and for all, to buy some of a commodity at each possible price—more accurately price rate—so much at each rate. The subject will then have an incentive to satisfy the expressed demands of the experimenter at all rates higher than the subject's rate r but not those at lower rates, thereby revealing r.

In mathematical terms, the experimenter offers, for a certain nonnegative "schedule of demands" f, to buy

$$\int_{x \leqq \rho} f(\rho)\,\mathrm{d}\rho \tag{5}$$

units of the commodity at a total cost of

$$\int_{x \leqq \rho} \rho f(\rho)\,\mathrm{d}\rho \tag{6}$$

for any number x named by the subject. From the subject's own viewpoint, his income from such a transaction, as a function of x, is

$$\begin{aligned} I(x;r) &= -r \int_{x \leqq \rho} f(\rho)\,\mathrm{d}\rho + \int_{x \leqq \rho} \rho f(\rho)\,\mathrm{d}\rho \\ &= \int_{x \leqq \rho} (\rho - r) f(\rho)\,\mathrm{d}\rho. \end{aligned} \tag{7}$$

Plainly, and in correspondence with the verbal argument that preceded this paragraph, the income $I(x;r)$ will be maximized when $x = r$, and for that x alone if $f(\rho)$ is positive near r on both sides. If, for example, $f(\rho) = \rho^{-3}$ for positive ρ, then $I(x;r) = (2x-r)/(2x^2)$, $I(x;r) - I(r;r) = -(x-r)^2/(2rx^2) \leqq 0$ for all positive r.

The mathematical analogy between this device for eliciting a rate and Marschak's device (discussed in paragraph 6 of section 3) for eliciting a price is notable. The possibility that r is negative, in which case the subject regards the commodity as a nuisance, is not necessarily excluded but can often be eliminated from practical consideration. The experimenter might then be content, as in the example above, to define f only for positive prices. Similarly, the experimenter could with impunity attach 0 density to preposterously high prices.

With abbreviations, (7) reads

$$I(x;r) = b(x)r + c(x). \tag{8}$$

The experimenter is therefore in effect offering to give the subject $c(x)$ units of cash and $b(x)$ units of the commodity for any number x chosen by the subject. The vital feature of the functions b and c, which is automatically assured if b and c are of the form implied by (7) with positive f, is that

$$I(x;r) = b(x)r + c(x) \leqq b(r)r + c(r) = J(r), \tag{9}$$

with equality if and only if $x = r$, where $J(r)$ has been introduced as an abbreviation for $I(r, r)$.

The sort of b and c induced by (8) with a positive f satisfies (9), but what can be said of the most general solution of (9)? In this problem, it is to be understood that x and r have a common convex range, such as all real numbers, the nonnegative real numbers, or the real numbers between 0 and 1, inclusive or not. (A reader not familiar with convex sets and convex functions may find [47, Appendix 2] helpful here and elsewhere in the present article.) What (9) requires is that, for each fixed x, the function $I(x; r)$, which is linear in r, lies strictly below the function J except at x, where the linear function and J have the common value $I(x; x) = J(x)$. In short, $I(x; r)$ is, for each x, a linear function of support of J at the point x, and only there. This implies that J is a strictly convex function of r. Such a function J often has only one linear function of support at a given x, its tangent at x, but wherever J has a corner, it has more than one support.

Conversely, let J be any strictly convex function of r, well behaved at the endpoints of the range of r if there are any (that is, neither discontinuous nor vertical there), and let $I(x; r)$ be any support of J at x. This I evidently satisfies (9). (Examples of bad behavior at the endpoint 0 when the range is the nonnegative reals: if $J(0) = 1$ and $J(x) = x^2$ elsewhere, then J is strictly convex but discontinuous and very much without a linear function of support at 0; if $J(x) = -x^{1/2}$, J is strictly convex and continuous but still has no linear function of support at 0.)

Any convex function J has, at each interior x, left and right derivatives $J_L(x)$ and $J_R(x)$. If $x < y$, $J_L(x) \leqq J_R(x) \leqq J_L(y) \leqq J_R(y)$, and the middle inequality is strict if J is strictly convex. The slopes of the supports at x are the numbers between $J_L(x)$ and $J_R(x)$ inclusive. There is just one such slope if and only if this interval degenerates to a single number, which is then the derivative $J'(x)$. Since $b(x)$ is between $J_L(x)$ and $J_R(x)$, the function b is nondecreasing—strictly increasing if J is strictly convex. Consequently, except on an at most denumerable set of points, b is continuous, and except at just those points, b is the derivative of its indefinite integral. The points of continuity of b are the points of differentiability of J, and at those points, $J'(x) = b(x)$.

The function b determines J and c except for an additive constant k; thus

$$J(x) = \int_{y \leqq x} b(y)\,dy + k = b(x)x + c(x) + k. \tag{10}$$

If J is twice differentiable at x, then, as (10) makes clear, b and c are differentiable there, and therefore

$$b(c) = J'(x) = b(x) + b'(x)x + c'(x). \tag{11}$$

Whence

$$b'(x)x = -c'(x). \tag{12}$$

If, therefore, J' is absolutely continuous (that is, an indefinite integral of its own derivative where defined), and f denotes J''; then

$$I(x; r) = -r \int_{x \leqq \rho} f(\rho)\, d\rho + \int_{x \leqq \rho} \rho f(\rho)\, d\rho + dr + e, \tag{13}$$

where d and e are constants. Thus, under a rather mild regularity hypothesis, (7) is the general solution of (9), except for a relatively unimportant linear term. This term expresses the possibility of the experimenter's making the subject outright gifts (possibly negative) of d units of the commodity and e units of cash. Such a gift can have no rational influence on the subject's choice (if his utility is linear in the commodity and in cash), though it might play some practical role, such as insuring that an experimental subject will in net be paid, not penalized, for his cooperation.

Though the integral approach to the problem is of some interest and is not of seriously limited generality for one commodity (especially when extended by Stieljes-like integrals), it does not, over all, seem as useful and informative as (9) and will not be central to this article.

A convenient form for $I(x; r)$ is often

$$\begin{aligned} I(x; r) &= J(x) + b(x)(r-x) \\ &= J(x) + J'(x)(r-x), \end{aligned} \tag{14}$$

where the final line is applicable only if $J'(x)$ exists.

The loss to the subject if he (irrationally) replies with x when his rate is actually r is

$$\begin{aligned} L(x; r) &= I(r; r) - I(x; r) \\ &= J(r) - I(x; r) \\ &= J(r) - J(x) - b(x)(r-x) \\ &= J(r) - J(x) - J'(x)(r-x), \end{aligned} \tag{15}$$

where once more the final line is applicable only where the derivative $J'(x)$ exists.

The function L, and equivalently I, has an easily derived and useful monotonicity [6, p. 43; 30, p. 44], according to which it not only pays to choose x equal to r but to keep any unavoidable discrepancy small. Namely, if x is between r and z, then $L(x; r) \leqq L(z; r)$, with strict inequality if J is strictly convex and $x \neq z$. Since L is nonnegative and b is nondecreasing, the following identity makes this evident.

$$\begin{aligned} L(z; r) - L(x; r) &\\ &= J(x) - J(z) - b(z)(r-z) + J(x)(r-x) \\ &= J(x) - J(z) - b(z)(x-z) + [b(z) - b(x)](x-r) \\ &= L(z; x) + [b(z) - b(x)](x-r). \end{aligned} \tag{16}$$

5. *Some special forms for the eliciting function*

5.1. *Loss as a function of discrepancy; $L(x; r) = H(x-r)$*

One suggestive, and possibly desirable, way of limiting the choice of the convex function J is to require the subject's loss $L(x; r)$, if he replies with x when r is his true rate, to be a function only of his discrepancy $x-r$. Let us then investigate the consequences of supposing that $L(x; r) = H(x-r)$ for some function H, nonnegative and 0 at 0 but not 0 everywhere. As the remaining paragraphs of this subsection are devoted to showing, this condition is so restrictive that $L(x; r)$ must be of the form $k(x-r)^2$ for some positive constant k. (This result has independently been derived with slightly less generality by Brown [6].)

According to (15),

$$H(x-r) = J(r) - J(x) - b(x)(r-x). \tag{17}$$

To see the sort of implication latent in (17), suppose for the moment that b and J are defined for all sufficiently small $|x|$ and $|r|$; as is rather intuitive and as will later be explicitly shown, this entails no real loss of generality. Under the simplifying assumption, an instance of (17) for $|x|$ and $|r|$ small is

$$\begin{aligned} H(x-r) &= H((-r)-(-x)) \\ &= J(-x) - J(-r) - b(-r)(r-x). \end{aligned} \tag{18}$$

Therefore

$$\begin{aligned} J(r) + J(-r) + b(-r)r - b(-r)x = J(x) + J(-x) \\ - b(x)x + b(x)r. \end{aligned} \tag{19}$$

In particular,

$$J(r) + J(-r) + b(-r)r = 2J(0) + b(0)r, \tag{20}$$

$$2J(0) - b(0)x = J(x) + J(-x) - b(x)x. \tag{21}$$

By means of (20) and (21), simplify (19) thus,

$$2J(0)+b(0)r-b(-r)x = 2J(0)-b(0)x+b(x)r; \tag{22}$$

or

$$[b(x)-b(0)]r = [b(0)-b(-r)]x, \tag{23}$$

so $b(x)$ is linear in x, at least for $|x|$ small.

To see that b is linear near every z in the interior of its domain of definition, and therefore linear throughout the interior, let $\hat{x} = x-z$, $\hat{r} = r-z$, $\hat{b}(u) = b(z+u)$, and $\hat{J}(u) = J(z+u)$. Since $(x-r) = (\hat{x}-\hat{r})$, the circumflexed functions and variables also satisfy (17), so $\hat{b}$ is linear for $|\hat{x}|$ small, that is, b is linear near z, as asserted, and consequently linear everywhere, with the (temporarily) possible exception of the endpoints of the domain of definition.

Armed with this information and the fact that H is nonnegative but not identically 0, a reader familiar with the theory of the Cauchy–Hamel equation [1, section 2.1] could easily show (from (17) alone) that $H(x-r) = k(x-r)^2$ for some positive k. But it is more elementary to recall that $b(x) = 2kx+l$ is the slope of a line of support of J at x. Since this slope is continuous in x, J is differentiable, and according to (10),

$$J(x) = kx^2+lx+m, \tag{24}$$

with some positive k, and according to (14),

$$\begin{aligned} I(x; r) &= J(x)+J'(x)(r-x) \\ &= kx^2+lx+m+(r-x)(2kx+l) \\ &= (2kx+l)r-kx^2+m. \end{aligned} \tag{25}$$

And indeed, for any J of the form (24), according to (15),

$$\begin{aligned} L(x; r) &= J(r)-I(x; r) \\ &= kr^2+lr+m-\{[2kx+l]r-kx^2+m\} \\ &= k(x-r)^2, \end{aligned} \tag{26}$$

as anticipated.

(The possibility that the linearity of b might fail at the endpoints of the interval of definition has been left open but can easily be removed by means of (17) and what has now been proved.)

5.2. *Symmetry; $L(x; r) = L(r; x)$*

Another suggestive condition is that the loss for replying with x when r applies should be the same as that for replying with r when x applies; that is, $L(x; r) = L(r; x)$. This condition too leads to (24), (25) and (26).

Briefly, the demonstration is as follows.

$$\begin{aligned} 0 &= L(x; r) - L(r; x) \\ &= [b(r)+b(x)](x-r)+2[J(r)-J(x)]. \end{aligned} \tag{27}$$

Consider the two equations that result from (27) on replacing (x, r) first by (r, z) and then by (z, x), and add all three equations to conclude that

$$b(x)(z-r)+b(r)(x-z)+b(z)(r-x) = 0. \tag{28}$$

Therefore $b(x)$ is linear, and the passage to (24), (25), and (26) can be made as before. Or it can be made thus. When the form $b(x) = 2kx+1/2$ is substituted into (27), $J(x)$ is seen to be of the form kx^2+lx+m.

Quadratic J's are, then, not only the simplest convex functions from the algebraic point of view; they are characterized also by the condition that $L(x; r)$ is a function of $x-r$ and by the condition that $L(x; r) = L(r; x)$.

5.3. *Ratio discrepancy; $L(x; r) = H(r/x)$*

When x ranges over the positive real numbers, the possibility that $L(x; r)$ is of the form $H(r/x)$ might be interesting. This condition too is very restrictive; it implies that J, I, and H are of the compatible forms:

$$J(r) = m+lr-k \log r, \tag{29}$$

$$I(x; r) = (m+k-k \log x)+(l-k/x)r, \tag{30}$$

$$H(u) = k(u-1-\log u), \tag{31}$$

with k positive. The demonstration occupies the rest of this subsection.

To begin with,

$$H(r/x) = J(r)-J(x)-J'(x)(r-x), \tag{32}$$

except for the at most denumerable set of x where $J'(x)$ is not well defined. Therefore, as is seen on differentiating with respect to r,

$$(1/x)H'(r/x) = J'(r)-J'(x), \tag{33}$$

whenever both derivatives on the right exist. Since these two derivatives do exist except on an at most denumerable set, a sequence of conclusions follows one after another:

H is differentiable everywhere; so is J; (32) and (33) hold without exception; (32) can be differentiated on both sides with respect to x;

$$-(r/x^2)H'(r/x) = -(r-x)J''(x); \tag{34}$$

J is twice differentiable everywhere; and

$$rJ'(r)-rJ'(x) = x(r-x)J''(x). \tag{35}$$

Regarded as a differential equation for J in r, (35) easily implies (29); (14) then implies (30); and (15) implies (31). Since J as defined by (31) is strictly convex and results in (30) and (31), the system is indeed compatible.

5.4. *Attempted generalization; $L(x;r) = H(g(r)-g(x))$*

Now consider the seemingly rather general possibility that $L(x;r)$ is of the form $H(g(r)-g(x))$. The special cases $g(x) = -x$ and $g(x) = \log x$ have already been considered. What other functions g are compatible with some strictly convex J? I am content here to examine the question under the simplifying assumption that g is differentiable and has a strictly positive derivative on the domain of J.

The derivation of (35) can be recapitulated to conclude that

$$\frac{J'(r)}{g'(r)} = \frac{(r-x)}{g'(x)}J''(x)+\frac{J'(x)}{g'(r)}. \tag{36}$$

Since the right side of (36) cannot change with x, and since $J'(x)$ is not constant, $1/g'(r)$ must be linear in r.

If this linear function is constant, it can without loss of generality be taken to be 1, with return to (24), (25), and (26). Incidentally, (36) specialized by setting $g' \equiv 1$ provides thus an alternative route to (24), (25) and (26).

If $1/g'(r)$ is linear but not constant, then it is to all intents and purposes of the form $(r-z)$ for some z below the domain of J or of the form $(z-r)$ for some z above the domain of J. In the first case, for example, $g(r) = \log(r-z)+\text{const.}$; and $L(x;r)$ is of the form $G((r-z)/(x-z))$, which is virtually the form that led to (29), (30) and (31). The introduction of g has therefore led to no really new forms of I, J, and L.

For J defined on the unit interval, it might seem interesting to seek

$L(x; r)$ of the form $H(r(1-x)/(1-r)x)$. But that would imply the existence of a $g(r)$ of the form $\log r - \log(1-r)$, which has been shown to be impossible.

6. Probabilistic interpretation

6.1. Application to a probability as a rate of substitution

Suppose that an experimenter, in an effort to elicit the probability p that a subject associates with an event D, invites the subject to choose a number x and promises to pay him $Y(x)$ in case D obtains and $Z(x)$ in case D does not obtain. For what pairs of functions Y and Z will it be to the subject's interest to choose x equal to p?

If p is interpreted as the subject's rate of substitution of dollars for the commodity consisting of dollars that are contingent on D, then the experimenter is in effect offering the subject $Z(x)$ dollars in cash and $Y(x)-Z(x)$ units of the commodity. The worth of such a gift to the subject is

$$\begin{aligned} I(x; p) &= [Y(x)-Z(x)]p+Z(x) \\ &= Y(x)p+Z(x)(1-p), \end{aligned} \tag{37}$$

which is an instance of (8) with p, $Y(x)-Z(x)$, and $Z(x)$ playing the roles of r, $b(x)$, and $c(x)$.

Therefore Y and Z accomplish the objective if and only if

$$\begin{aligned} J(p) &= Y(p)p+Z(p)(1-p) \\ &= Y(p)-(Y(p)-Z(p))(1-p) \\ &= Z(p)+(Y(p)-Z(p))p \end{aligned} \tag{38}$$

is a strictly convex function of p, and $I(x; p)$ is in p a linear function of support of J at x. At values of p where J is differentiable, according to (37),

$$J'(p) = Y(p)-Z(p), \tag{39}$$

so at such values, according to (38) and (39),

$$Y(p) = J(p)+(1-p)J'(p) \tag{40}$$

$$Z(p) = J(p)-pJ'(p). \tag{41}$$

The loss entailed by choice of x when p applies is

$$L(x;p) = [Y(p)-Y(x)]p+[Z(p)-Z(x)](1-p), \tag{42}$$

which is ordinarily—that is, if J is differentiable—

$$L(x;p) = J(p)-J(x)-J'(x)(p-x), \tag{43}$$

as in (15).

The conditions that J is quadratic, that $L(x;p)$ is a function of $p-x$, and that $L(x;p) \equiv L(p;x)$ are, according to section 5, all equivalent. In this special case, (40) and (41) can be put in the suggestive forms

$$Y(p) = m'-k(1-p)^2, \tag{44}$$

$$Z(p) = m-kp^2, \tag{45}$$

which correspond to

$$J(p) = m'p+m(1-p)-kp(1-p). \tag{46}$$

In any real application, p is between 0 and 1, and it might seem natural to confine the range of choice of x to the interval from 0 to 1 inclusive. However, if x is not so confined, a subject who chooses $x < 0$ or $x > 1$ exposes himself to utterly unnecessary loss for any strictly convex J. In fact, according to the monotonicity pointed out in the final paragraph of section 4, for any p in the interval [0, 1], $L(x;p) > L(1;p)$ if $x > 1$, and $L(x;p) > L(0;p)$ if $x < 0$.

6.2. *Application to an expectation as a rate of substitution*

Since the probability of an event D is the expectation of the same D regarded as an indicator, extension of the method of eliciting probabilities just discussed to the elicitation of the personal expectation r of any random quantity V is to be anticipated. And this is indeed straightforward. If the experimenter offers to pay the subject $b(x)V+c(x)$ for any x chosen by the subject, the worth to the subject of choosing x is

$$I(x;r) = b(x)r+c(x), \tag{47}$$

so the loss for choosing x instead of r is

$$L(x;r) = [b(r)-b(x)]r+[c(r)-c(x)]. \tag{48}$$

For this to be positive if and only if x differs from r means, as shown in section 4, that $J(r) = b(r)r + c(r)$ is strictly convex and $b(x)r + c(x)$ is in r a linear function of support of J at x.

If, for example, the subject feels certain that $V \leqq \alpha$ for some constant α, his $r = E(V)$ cannot exceed α. And in fact it is to his advantage to choose α rather than any x larger than α, as can be verified thus. The difference in worth between these two choices is

$$[b(\alpha) - b(x)]V + [c(\alpha) - c(x)] = L(x; V) - L(\alpha; V). \tag{49}$$

In view of the monotonicity mentioned at the end of section 4, (49) is therefore negative if $x > \alpha$. Thus whatever the actual value of V may be—not greater than α—the subject will receive more for choosing α than for choosing any larger number.

6.3. *The most general elicitor of an expectation*

Is $b(x)v + c(x)$ the only possible form for a function $S(x; v)$ for which

$$E(S(x; V)) < E(S(r; V)) \tag{50}$$

for all x different from $r = E(V)$? The answer cannot quite be "yes", because to any such S it is clearly possible to add $f(V)$ with f any function for which $E(f(V))$ is finite for the class of distributions envisaged for V. But little if any further extension is possible if (50) is to hold for a reasonably large class of distributions for V, as the next four paragraphs demonstrate. Even this slight extension is nugatory in case V is the indicator of an event, that is, in case $E(V)$ is a probability, because when V takes only two values, any function f is linear on that pair of values.

To begin with, let $S(x; v)$ be defined for all x and v in an interval, and let (50) hold for every V that is subject to a 2-point distribution in that interval. Specialized to a V that takes the values v and v' with probabilities p and $\bar{p} = 1 - p$, (50) takes the form:

$$pS(x; v) + \bar{p}S(x; v') < pS(pv + \bar{p}v'; v) + \bar{p}S(pv + \bar{p}v'; v'), \tag{51}$$

unless $x = pv + \bar{p}v'$.

For x interior to the interval of definition and S differentiable in x,

$$p \frac{\partial}{\partial x} S(x; v) + \bar{p} \frac{\partial}{\partial x} S(x; v') = 0 \tag{52}$$

if $x = pv + \bar{p}v'$. That is, if $v \leqq x \leqq v'$ and the derivatives exist at x, then

$$(v'-x)\frac{\partial}{\partial x}S(x;v) = (v-x)\frac{\partial}{\partial x}S(x;v'). \tag{53}$$

Therefore, if S is everywhere differentiable with respect to x (more generally, if S is absolutely continuous in x for each v), then $S(x;v)$ is of the form $b(x)v+c(x)+f(v)$ as anticipated. This conclusion obtains even without any differentiability assumption, as will be shown in the next paragraph for those who share an interest in such points.

According to (51), the function

$$(v'-x)S(x;v)+(x-v)S(x;v') \tag{54}$$

is convex in x and has

$$S(x;v')-S(x;v) \tag{55}$$

among its slopes of support at x. Therefore the functions (54) and (55) are of bounded variation in x, whence so is $S(x;v)$ for each v. Let S_0 and S_1 be the singular and the absolutely continuous parts of S with respect to x, rendered unique by the convention that $S_0(x_0;v) = 0$ for some x_0 and for arbitrary v. Since (54), being convex, is absolutely continuous, $S_0(x;v) = (v-x)S_0(x;v')/(v'-x)$ and is therefore of the form $b_0(x)v+c_0(x)$. (Now, the argument employed when S was assumed to be absolutely continuous applies almost unmodified to S_1, the absolutely continuous part of S.)

An important and widely known example is

$$S(x;v) = -(x-v)^2 = -x^2+2xv-v^2.$$

For it is central to the theory of least squares that this S satisfies (50) according to the following familiar calculation.

$$\begin{aligned} E[(x-V)^2] &= E\{[(x-E(V))-(E(V)-V)]^2\} \\ &= (x-E(V))^2+E[(E(V)-V)^2]. \end{aligned} \tag{56}$$

7. *Decision theory*

A different approach to the problem of determining those functions Y and Z that encourage a subject to reveal his true p is implicit in statistical decision theory, as will now be explained.

Imagine a person in an economic situation in which he is free to choose one of several acts and knows that if he chooses the act a he will receive a payment of $Y(a)$ dollars in case the event D obtains and $Z(a)$ dollars in case D does not obtain. If the person's probability for D is p, the worth of the act a to him is

$$I(a;p) = Y(a)p+Z(a)(1-p), \tag{57}$$

a linear function of p. The worth to him of the situation in which he is free to choose among a finite set of acts a is therefore the function

$$J(p) = \max_{a} I(a;p). \tag{58}$$

Indeed, any act a for which the maximum is attained is worth $J(p)$ to him and no available act is worth more. The function $J(p)$ is evidently convex.

When an infinite set of acts is envisaged, little changes, especially if the set is such that the maximum required by (58) is attained. With an unrestricted infinite set of acts, J can be an arbitrary continuous, convex function. (If the end points $p = 0$ and $p = 1$ are to be included and if the maximum is to be attained, J cannot be permitted to be vertical at those points.) If J is strictly convex, a person choosing a support of J will reveal his value of p; for no linear function supports a strictly convex function at more than one point.

The original problem of characterizing functions Y and Z that will elicit the person's true p can now be envisaged thus. For each number x, say in the interval from 0 to 1, an act is made available to the subject that will pay him $Y(x)$ dollars in case D obtains and $Z(x)$ dollars otherwise. For what functions Y and Z will it be optimal for the person to choose x equal to his p, no matter what p may be? As the preceding paragraph shows, it is necessary and sufficient that there be some strictly convex J for which $Y(x)p+Z(x)(1-p)$ is a line of support at x; and this was the main conclusion of section 6.

8. *An alternative definition of probability*

The preceding section suggests an alternative way to define personal probability and to argue that personal probability exists. This avenue has been pointed out and explored by de Finetti, for example, in [13, 17].

Imagine in fact that the very notion of the probability p attached by a person to the event D has not yet been defined, and consider a person confronted with the choice among all those acts a for which the linear functions $I(a;p)$ nowhere exceed a specified convex function $J(p)$

(without vertical endpoints). Suppose that there is one such act a to which he prefers no other. It would be unreasonable for this a not to be one for which $I(a; p)$ is a line of support of $J(p)$ at some point p_0. For otherwise there would be an a' for which $I(a'; p)$ is a line of support of J and is parallel to $I(a; p)$, in which case $Y(a') > Y(a)$ and $Z(a') > Z(a)$, so that a' is clearly superior to a.

If J is strictly convex, the unique p_0 for which $I(a; p)$ is a line of support of J can be defined as the person's personal probability of D. But this raises three important questions: Could a different function J^* lead to a different p_0? Does this definition lead back to the idea of probability as a rate of substitution? Is probability as thus defined a probability measure?

To progress toward answering the first two questions, suppose the person participates simultaneously in two decision problems of the kind under discussion, one determined by J_1 and one determined by J_2, where J_1 and J_2 are convex but not necessarily strictly convex. Assume that he acts in each of the two component problems as he would if he were faced with that problem alone.

(This assumption is not altogether unobjectionable; for it may imply that the person's utility function is linear in money. But such linearity assumptions are made almost throughout the present paper and are presumably tolerable if only moderate sums of money are involved. In the purely mathematical formulation of the decision problems no precautions have been, or need be, taken to keep these sums moderate, but it is fairly clear how such precautions could be taken in applying the theory or how the devices mentioned at the end of section 3 might be invoked. I do not attempt great caution about this point, because my object in this section is only to touch briefly on an approach to personal probability that may be more suggestive and more practical in some respects than such formally more rigorous approaches as those reported on in [47, ch. 3].)

The linear function of support $I_1(a_1; p)$ chosen for J_1 and the linear function of support $I_2(a_2; p)$ chosen for J_2 should support J_1 and J_2 respectively at some common point p^*, as will be argued. The person, in choosing a_1 and a_2, has in effect chosen an act a for which $I(a; p) = I_1(a_1; p)+I_2(a_2; p)$. Some linear function of support $I^*(p)$ of $J = J_1+J_2$ at some point p^* is either the same as, or everywhere higher than, $I(a; p)$. Since, as is not hard to see, I^* can be represented as the sum of linear functions of support of J_1 and J_2 at p^*, I^* represents an act (that is, a choice of a_1 and a_2) that was available to the person. Therefore his choice of a_1 and a_2 is discredited, or seen to be inadmissible, unless $I^*(p) = I(a; p)$ for all p, which is impossible unless $I_1(a_1; p^*) = J_1(p^*)$ and $I_2(a_2; p^*) = J_2(p^*)$ as asserted. This conclusion will yield affirmative answers to the first two questions.

First, as is now evident, two strictly convex functions cannot lead to different values of p if the linear-utility assumption holds and if the person is coherent.

Second, suppose p^* has been determined by means of a strictly convex function J_1, and consider also the decision problem defined by the broken-line convex function J_r with $J_r(p) = \max(rq, pq)$, where r is a positive fraction and q is a positive number. According to what was proved in the paragraph before last, the person will prefer q dollars contingent on D to rq dollars outright if $r > p^*$ and vice versa if $r < p^*$, so p^* is indeed a rate of substitution.

A certain approach to the third question, whether probability as here defined is a probability measure, is best postponed to the next section. But it can be argued now that since personal probabilities defined as rates of substitution constitute a probability measure, the same must be true for the equivalent new definition.

De Finetti [13] has shown how the approach of this section applied to conditional probability.

9. *The simultaneous elicitation of several rates*

9.1. *Vector rates*

When several rates r_s are concerned, each can be independently elicited by means of a function $I_s(x_s; r_s)$ as in section 4, but there are other possibilities.

Generalizing from section 4, let $r = \{r_s\}$ be a finite sequence of rates, which can advantageously be thought of as a vector. (Infinite sequences or even functions on an arbitrary domain might also have a useful interpretation, for instance, as distribution functions or densities. Vectors without any interpretation as functions could also be handled and might have applications. Extensions of this sort are presented by Hendrickson [30].) Let $x = \{x_s\}$ be a sequence of possible numerical responses of a subject, a response vector. The experimenter's aim is to provide the subject with an incentive to choose $x = r$ by offering to give $c(x)$ units of cash and $b_s(x)$ units of the sth commodity for whatever x is chosen by the subject.

The experimenter may already know something about r. For example, in the important case of probabilities, he knows that each probability is between 0 and 1; and when the r_s are probabilities of the elements of a partition, he knows that they add up to 1. In other cases, it will often be known that each $r_s > 0$. For still one more example, the rate for a certain high-quality commodity may be plainly higher than that of certain other commodities.

It may be to the experimenter's advantage to exclude all, or at least some, unreasonable values of x. In particular, he might confine acceptable responses x to some convex set K—a scheme that gives adequate flexibility for the examples just mentioned. For the moment, assume that K has at least one extrinsic interior point z, that is, a point for which $z+\Delta r = \{z_s+\Delta r_s\}$ is in K for all sufficiently small vectors Δr. This temporary simplification excludes at least one important example, that of probabilities p_s so constrained that $\Sigma p_s = 1$.

With the exception of the expression of I by integrals, the whole of section 4 will now easily be paraphrased with sets, or bundles, or commodities (or, more abstractly, vector commodities) playing the role that a single commodity played there.

Though the experimenter does not initially know the vector r, he can in effect offer the subject the income $I(x; r)$ for choosing the vector x, where

$$\begin{aligned} I(x; r) &= \sum_s b_s(x) r_s + c(x) \leqq \sum b_s(r) r_s + c(r) \\ &= I(r; r) = J(r), \end{aligned} \tag{59}$$

with equality if and only if $x = r$.

According to (59), $I(x; r)$ is, for each x, a strict linear function of support at x for the function J. Therefore, J is strictly convex, and virtually all strictly convex functions do lead to at least one I. If K has no boundary points—and it is seldom if ever important to include boundary points in K—then every strictly convex function will serve. If K does have boundary points, then discontinuities at the boundary points and milder sorts of misbehavior incompatible with linear functions of support at the boundary points are to be excluded.

If J is differentiable at the nonboundary point x of K or, equivalently, if the $b_s(x)$ are continuous at x, then

$$b_s(x) = \frac{\partial}{\partial x_s} J(x), \tag{60}$$

and

$$c(x) = J(x) - \sum x_s \frac{\partial}{\partial x_s} J(x). \tag{61}$$

If the $b_s(x)$ are continuous for all such x, they obviously determine J through (60), except for an additive constant. (But continuity is not needed for the conclusion that J is determined except for an additive constant by the b_s.)

The loss incurred by the subject on choosing x when r applies is

$$\begin{aligned} L(x; r) &= [J(r) - J(x)] - \sum b_s(x)(r_s - x_s) \\ &= [J(r) - J(x)] - \sum (r_s - x_s) \frac{\partial}{\partial x_s} J(x), \end{aligned} \tag{62}$$

the last line applying only where J is differentiable.

The condition that $J(x)$ be quadratic, that is, of the form

$$J(x) = \sum_{s,t} k_{s,t} x_s x_t + \sum_s l_s x_s + m \tag{63}$$

is equivalent to the condition that $L(x; r)$ depends only on $x - r = \{x_s - r_s\}$ and to the condition that $L(x; r) = L(r; x)$.

Much the same can be said even without assuming that K has extrinsic interior points, with slight differences because it may no longer be possible to change one x_s without changing other components of x. Equation (59) and the paragraph following it remain in force, but it now may be possible to change the individual $b_s(x)$ without really affecting I. The meaning of (59) and (62) is therefore better conveyed by

$$I(x; r) = b(x)(r) + c(x) \leqq b(r)(r) + c(r) = J(r), \tag{64}$$

$$L(x; r) = J(r) - J(x) - b(x)(r - x), \tag{65}$$

where for each x, $b(x)(r)$ is linear and homogeneous in the vector r and $I(x; r)$ is, as always, a linear function of support of J at x.

9.2. *Vector expectations*

The case of a sequence $V = \{V_s\}$ of random numbers considered as commodities is important. A little more generally, V can be a random vector whose values lie in a finite dimensional convex set K. The random payment to the subject who chooses the vector x is

$$b(x)(V) + c(x), \tag{66}$$

which in V is linear and supports a strictly convex J at x. Where J is differentiable, (66) can be suggestively written as

$$J(x)+J'(x)(V-x). \tag{67}$$

As in subsection 6.3, it can be asked whether there are functions $S(x; v)$ that elicit the expected value of a vector V other than S of the form $I(x; v)+f(v)$, where I is as in (64). The negative answer given in subsection 6.3 is not hard to generalize, at least under generous regularity hypotheses.

The interesting and familiar instance of quadratic S mentioned in subsection 6.3 has a hardly less familiar extension to the present, multidimensional, situation. For if S is a homogeneous, strictly negative-definite, quadratic function of $x-v$, then

$$\begin{aligned} E[S(x-V)] &= S(x-E(V))+E[S(V-E(V))] \\ &> E[S(V-E(V))] \end{aligned} \tag{68}$$

if $x \neq E(V)$.

A subject who feels sure that V is in a specified closed convex subset K^* of K will have $E(V)$ in K^* and therefore ought not choose an x not in K^*. But, under suitable regularity hypotheses, such an x is also discredited in the deeper sense that there is a v in K^* for which $I(v; r) > I(x; r)$ for every r in K^*. This is demonstrated in the next paragraph, which is followed by one showing that some regularity hypothesis is indispensable.

Suppose that J is differentiable in K^* and that the infimum of $L(x; r)$ as a function of r in K^* is attained, say, at $r = v$. Then, for all r in K^*,

$$\begin{aligned} L(x; r) &= J(r)-J(x)-b(x)(r-x) \\ &\geqq J(v)-J(x)-b(x)(v-x) = L(x; v), \end{aligned} \tag{69}$$

$$0 \leqq L(x; r)-L(x; v) = [J'(v)-b(x)](r-v)+o(r-v). \tag{70}$$

Therefore, $[J'(v)-b(x)](r-v) \geqq 0$ for all r in K^* sufficiently close to v; but if such a linear inequality holds for r in K^* close to v, it holds for all r in K^*. Put the facts together, thus.

$$\begin{aligned} L(x; r)-L(v; r) &= J(v)-J(x)-b(x)(r-x)+J'(v)(r-v) \\ &= [J'(v)-b(x)](r-v)+L(x; v) \\ &\geqq 0, \end{aligned} \tag{71}$$

as asserted.

(In one dimension, as has been seen in subsection 6.2, no regularity hypothesis is required for this conclusion, but the following counter-example shows that some such hypothesis is needed in two and more dimensions. Let $J(x, y) = |x|+|y|+\varepsilon(x^2+y^2)$ for some small positive ε. This J is strictly convex over the whole plane, and I can be consistently defined thus.

$$I(x, y; r, s) = J(x, y)+(\operatorname{sgn} x+2\varepsilon x)(r-x) \\ +(\operatorname{sgn} y+2\varepsilon y)(s-y).$$

Let K be the convex set

$$\{x, y\colon 1 \leqq 2x+y; x \leqq 2, y \leqq 3\}.$$

Then $I(0, 0; r, s) = 0$ for all (r, s), but, for each (x, y) in K,

$$\inf_{(r,s)\in K} I(x, y; r, s) \leqq -1+0(\varepsilon).)$$

The significance of the general conclusion in case of regularity is particularly vivid if J is quadratic. For then, in terms of the Euclidean distance associated with the quadratic function J, (71) says that the point v in K^* closest to x is closer to each point of K^* than x is. This is geometrically rather evident and is easy to prove directly or as an instance of the general argument.

To apply the argument about K^*, if the person is sure that a linear equality or inequality is satisfied by several random variables $V_1, \ldots, V_n$, he will expose himself to needless loss if the $x_1, \ldots, x_n$ with which he responds to any regular J do not satisfy this equality or inequality. In particular, if some V_s is sure to be 1, the person should choose $x_s = 1$; if V_s is the indicator of an event, he should choose between 0 and 1; if V_s is the indicator of the union of two disjoint events of which V_j and V_k are the indicators, he should so choose that $x_s = x_j+x_k$. As these remarks show, the program for defining probabilities in section 8 does lead to a probability measure for any person who does not blunder. (More completely, the remarks do lead directly to the anticipated conclusion when the program is applied to any J that is regular. But, as was shown in section 8, all strictly convex J elicit the same probabilities.)

9.3. *Probability distributions as vector expectations*

Consider now the case in which the V_s $(s = 1, \ldots, n)$ are the indicators of the elements of a partition $\{D_s\}$. All previous literature on rate elicitation seems to be confined to this important case. (A few references on the method not mentioned elsewhere in this article are [7, 42, 54].)

Since it is (especially patently for regular J) wasteful in the present case for the person to choose any $x = \{x_s\}$ other than a probability distribution (that is, an x for which $x_s \geqq 0$ and $\Sigma x_t = 1$), K can for many applications be taken to be the simplex of all probability distributions p. If $J(p)$ is a differentiable function on K, the person who chooses x will receive $I(x; V)$, which can, in a somewhat figurative but not unnatural notation, be written,

$$\begin{aligned} I(x; V) &= J(x)+J'(x)(V-x) \\ &= J(x) \sum_t J_t'(x)(V_t - x_t); \end{aligned} \tag{72}$$

this is figurative because x cannot be varied one coordinate at a time and remain in the simplex. In case D_s is the event in the partition that actually obtains, this is,

$$I(x; V) = J(x) - \sum_t J_t'(x)x_t + J_s'(x). \tag{73}$$

The numbers $J_s'(x)$ here are not necessarily derivatives and are determined only up to an additive constant. (They can of course be so chosen that $\Sigma J_t'(x) = 0$, but the symmetry of that particular choice ought not to be invested with much importance.) If J is originally defined not only on the simplex K but, say, for all positive n-tuples of numbers, then the $J_s'(x)$ can be taken to be $\partial J(x)/\partial x_s$, which would not be meaningful on K alone.

One suggestive choice for J, whether on K or on a larger set, is $J(x) = \Sigma x_t^2$. In this case, (73) on K becomes

$$\begin{aligned} I(x; V) &= -\sum_t x_t^2 + 2x_s \\ &= -(1-x_s)^2 - \sum_{t \neq s} x_t^2 + 1 \\ &= -\sum_t (V_t - x_t)^2 + 1, \end{aligned} \tag{74}$$

that is, 1 minus the square of the usual Euclidean distance from x to the vector $(0, 0, \ldots, 1, \ldots, 0, 0)$ representing the D_s that actually obtains. The final constant 1 is of course not very important in (74). Specialized to a single event D (with indicator V) and its complement, (74) without its 1 becomes

$$\begin{aligned} I(x; V) &= -[V-x]^2 - [(1-V)-(1-x)]^2 \\ &= -2(V-x)^2 \\ &= \begin{cases} -2(1-x)^2 & \text{if } D \text{ obtains} \\ -2x^2 & \text{if } D \text{ does not obtain.} \end{cases} \end{aligned} \tag{75}$$

If each of the events D_s is treated separately according to (75), the net effect is practically the same as that of using $J = \Sigma x_t^2$; it is exactly the same as using $2J-2$.

The most general quadratic function of n variables is

$$J(x) = \sum_{s,t} k_{st} x_s x_t + \sum_s l_s x_s + m. \tag{76}$$

To this J corresponds

$$I(x; V) = J(V) - \sum_{s,t} k_{st}(V_s - x_s)(V_t - x_t). \tag{77}$$

If further, J is symmetric in the x_s, (76) specializes to

$$J(x) = k' \sum_s x_s^2 + k''(\sum_s x_s)^2 + l \sum_s x_s + m, \tag{78}$$

which is convex if and only if $k' > 0$ and $k' + nk'' > 0$. For this J,

$$I(x; V) = J(V) - k' \sum_s (V_s - x_s)^2 - k''[\sum_s (V_s - x_s)]^2. \tag{79}$$

If the V_s are the indicators of a partition, $\Sigma V_s^2 = \Sigma V_s = 1$; so $J(V) = k' + k'' + l + m$, which simplifies (79). If further $\Sigma x_s = 1$ as it "ought to", then (79) simplifies to

$$(k' + k'' + l + m) - k' \sum_s (V_s - x_s)^2, \tag{80}$$

which is not essentially different from (74).

9.4. *Separated income for distributions*

It is somewhat attractive to seek functions J on the simplex K such that when D_s obtains the income of the subject will depend on x_s alone. The pertinent facts have been known for some time, a review of them here without any supplementary hypotheses may be useful. (An early reference is [37], which attributes the main fact to Andrew Gleason.)

If $n = 2$, the condition is plainly vacuous, but for $n > 2$ it is very severe—strictly speaking, not quite attainable. What is wanted is n functions f_s on the interval from 0 to 1 (inclusive if possible) such that, for any distribution p,

$$\sum_t f_t(x_t) p_t < \sum_t f_t(p_t) p_t \tag{81}$$

for every distribution x different from p. As will be shown, after preliminary discussion, if this is to hold for all p with each p_s strictly positive, then

$$f_s(z) = k \log z + l_s, \tag{82}$$

for some $k > 0$. Whence

$$J(p) = k \sum_s p_s \log p_s + \sum l_s p_s; \tag{83}$$

$$L(x; p) = k \sum_s p_s \log \frac{p_s}{x_s}. \tag{84}$$

Of course (81) and (82) are not compatible with any assignment of a finite value for $f_s(0)$.

The loss (84) is a constant multiple of what is sometimes called the information of the distribution p with respect to the distribution x. The scheme of elicitation implied by (83) seems to have been suggested first by Good [26, p. 112], who confined himself to two-fold partitions, for which the uniqueness theorem is not relevant. An interesting application and discussion are given by Mosteller and Wallace [38, section 4.9].

It is not hard to prove (82) once it is known that each f_s is sufficiently regular. The next three paragraphs are devoted to proving that regularity.

Apply (81), for $n > 2$, to distributions x and p of the special form $x = \{yw, \bar{y}w, \bar{w}/(n-2), \ldots, \bar{w}/(n-2)\}$ and $p = \{qw, \bar{q}w, \bar{w}/(n-2), \ldots, \bar{w}/(n-2)\}$, where $\bar{y} = 1-y$, $\bar{q} = 1-q$, $\bar{w} = 1-w$, and $0 < y, q, w < 1$.

$$f_1(yw)q + f_2(\bar{y}w)\bar{q} < f_1(qw)q + f_2(\bar{q}w)\bar{q} \tag{85}$$

if $y \neq q$. The left side of (85) is therefore, in q, a strict linear function of support at y of g_w, where

$$g_w(y) = f_1(yw)y + f_2(\bar{y}w)\bar{y}. \tag{86}$$

Therefore, g_w is strictly convex in (0, 1). So g_w is continuous in y and, except possibly on a denumerable set, differentiable in y.

Since $f_1(yw) - f_2(\bar{y}w)$ is a slope of support of g_w at y, it is locally of bounded variation, and the same can therefore be said of $f_1(yw) = g_w(y) + \bar{y}[f_1(yw) - f_2(\bar{y}w)]$ as a function of y. Therefore each f_s has at most a denumerable number of discontinuities and is differentiable almost everywhere.

Returning to (86) and recalling that g_w is continuous in y, it can be seen that if f_1 is discontinuous at z, then f_2 is discontinuous at $w - z$ for all w between z and 1; so f_1, and therefore all f_s, must be continuous everywhere. Therefore, g_w is differentiable at all y. Arguing as before, if f_1 is not differentiable at z, then f_2 is not differentiable at $w-z$. At least, we know that each f_s is differentiable at all z in (0, 1). Let f'_s be the derivative of f_s.

The rest is straightforward:

$$\frac{\mathrm{d}}{\mathrm{d}y} g_w(y) = f_1(yw) - f_2(\bar{y}w)$$

$$= \frac{\mathrm{d}}{\mathrm{d}y} \{f_1(yw)y + f_2(\bar{y}w)\bar{y}\}$$

$$= w\{f'_1(yw)y - f'_2(\bar{y}w)\bar{y}\} + f_1(yw) - f_2(\bar{y}w) \tag{87}$$

$$f'_1(yw)yw = f'_2(\bar{y}w)\bar{y}w. \tag{88}$$

This means that $f'_s(z)z = k$ independently of s or z, which establishes (82).

10. Practical sidelights

10.1. Scope of this section

This article, which is largely about the mathematical aspects of the procedures for eliciting personal probabilities that are now often called proper scoring rules (and also admissible probability measurement procedures) would be misleading and incomplete without some discussion of their actual and potential applications. My preparation for that is inadequate; for I have done no practical work in the area nor even followed the practically oriented literature with energy and care. But it seems incumbent on me to mention, to the best of my knowledge, the sorts of applications that have been envisaged, some difficulties that threaten them, and criteria that might help in selecting among the plethora of proper scoring rules. The subject is a ramified one, and even a cursory survey demands considerable space.

One serious omission is discussion of the experiments that have been done on proper scoring rules, for which some key references are [12; 52, section 3.4 and ch. 10–11; 55; 56].

The applications thus far envisaged have tended to emphasize the elicitation of probabilities over that of other expectations. For this reason and for vividness, I focus here on probabilities, but extension of the ideas to other expectations will often be obvious and ought to be kept in mind.

10.2. The uses of opinions

Strictly proper scoring rules enable us, in principle, to discover people's opinions, so possible fields of application are brought to mind by asking why and when we are interested in opinions.

Often, we want to make use of the opinion of a person whom we regard as an expert. Does the weatherman think that it will rain, the doctor that we shall soon get well, the lawyer that it would be better to settle out of court, or the geologist that there might be lots of oil at the bottom of a deep hole? Most of the following subsections are concerned explicitly with the utilization of experts.

And often, we want to know a person's opinions in order to judge how well informed he is. Every academic examination can be viewed in that light. It is, therefore, interesting to explore the possible usefulness of proper scoring rules in academic examinations, and the final subsection returns to academic testing. This domain of application shades into that of trying to determine which among possible experts are most valuable for a given task by means of the relation between their past opinions and reality.

Since public-opinion polls are ostensibly concerned with finding out the opinions of the public, we might at first expect proper scoring rules to have important applications there, in harmony with the remarks of Haavelmo quoted in section 1. This has, however, apparently never been suggested, partly perhaps because public-opinion polls are seldom so much concerned with the opinions of the public as with the preferences of the public. But investigators do occasionally want to know how firmly the public believes some as yet unresolved matter of fact, and in a few of these cases, proper scoring rules might have some role. Literally paying the participants in a poll on the basis of the accuracy of their predictions would of course ordinarily be infeasible, especially in the common case of casual, one-time participants. Yet scoring rules might be used in training panelists to assess their own probabilities. perhaps partly in paid practice sessions about immediately verifiable predictions. This would seem less far-fetched should scoring rules come to be widely applied in the schools.

One domain of potential application of proper scoring rules is suggested not so much by the problem of obtaining the opinions of others as by the difficulty of obtaining our own. For it is by no means easy to elicit your own probabilities. Vagueness is a major obstacle, and your first reactions are often greatly modified when you reflect upon their implications. Those who have experimented on themselves and on others generally feel that frequent practice with proper scoring rules and with other probability elicitors helps a person to combat vagueness and to arrive more promptly and accurately at his personal probability. This proposition might be difficult to investigate experimentally and even seems difficult to state with precision, but its promise of benefit is great. (See, for example, [12, 59].)

10.3. Yes, no, or maybe is not enough

Traditionally, experts, except for turf experts, have not communicated their opinions in probabilities. The doctor says, "That child will soon be well and then he had better have his tonsils out." The geologist says, "That looks like a good place to drill." And, until recently, most of us were content when the weatherman said simply, "Rain tonight."

The importance of a system of communication in which experts express themselves in terms of genuine personal probabilities and in which those who utilize the opinions of experts—that is, all of us—are trained in understanding and using such probabilities was energetically underlined by Grayson [27]. In recent years, meteorologists have been announcing forecasts in terms of probability on some radio and television services, but not all who offer these probability forecasts think in terms of personal probability, and some seem to be very vague indeed about what they mean by probability in a forecast. The earliest known reference to proper scoring rules is by the meteorological statistician Brier [5], and much of the current literature on proper scoring rules is inspired by meteorology, as in [21, 39–41, 46, 53, 58–60] and works cited in them.

Applying a proper scoring rule to obtain better opinions from an expert might mean, at one extreme, merely using the rule to keep score as a training device to give the expert a mild incentive to understand what probability means and to ask himself whether it is really his personal probabilities that he is reporting. At the opposite extreme, the scoring rule might be implemented by substantial cash payments. For example, an oil geologist willing to buy a $100 interest in a million-dollar drilling investment is saying very clearly that, to him, the expected revenue of the well is greater than the expected cost. More accurately, that is what he is saying if his expression of opinion is not affected by nonlinearities in his utility, possibly reflecting the nonlinearity of income after taxes and possibly associated with trying to explain to his wife how he lost $100 gambling, or with a very small probability of an enormous revenue.

This same example, which suggests so vividly how any proper scoring rule might be applied to elicit from a geologist his opinions about the yield of a well that is to be drilled also serves to illustrate what seems to be an insurmountable obstacle to the application of arbitrary scoring rules, and even of any proper scoring rules to all, to many situations in which opinion is sought as a basis for decision. If whether the well is to be drilled does not depend on this geologist's opinion, then any proper scoring rule concerning its expected yield can be implemented, in particular by offering to sell shares in the well at various prices. But if whether to drill the well depends in part on the opinion of the geologist, then some of the events about which his opinion is wanted may never be tested precisely because of his advice, so the scoring-rule contract cannot be offered literally. The phenomenon is ubiquitous. We can never know what would have happened had surgery been ventured, had a certain product been marketed, or had a certain student been admitted. Business sharing, to be discussed in subsection 10.12, seems to offer some possibility of circumventing this widespread obstacle to the literal application of proper scoring rules.

10.4. Big money

The interrogator has an interest in making the possible fluctuations in the wealth of the expert large. For this motivates the expert to reflect hard and well before answering and yet need not add systematically to his expected fee. The larger the fluctuations are, however, the more the expert is motivated to report not his real probabilities but numbers that reflect in part the nonlinearity of his utility. The theory of these distortions has been somewhat explored by Winkler and Murphy [60], and some possible ways to avoid them or compensate for them were reported on in section 3. My own hope and expectation is that the skillful use of small, or even purely symbolic, scoring-rule payments (in addition to the usual compensations) will enable experts to know and to communicate their opinions much more accurately than has been usual.

The only practical experience with elicitation involving substantial cash prizes and losses thus far apparently consists of observations on gambling behavior that Ward Edwards has made, and expects soon to publish. The use of substantial, or even of token, scoring-rule payments in business would often be a considerable break with tradition likely to involve legal and other administrative problems.

10.5. Tiny probabilities and the expert

Probabilities corresponding to odds of one in a million or even less, such as the probabilities of specific disasters, are sometimes important [32]. Any scheme to give an expert a serious cash incentive to reveal his personal probabilities for very improbable events would seem to court insuperable difficulties with the nonlinearity of utility. And, even if we do not despair of obtaining sincere opinions about such matters by engaging sincerely inclined experts in the right kind of make-believe, the difficulty of such make-believe, and the special training required for it, are in danger of being greatly underrated.

Magnification is sometimes possible. For example, though it would not be practical to engage me in a meaningful bet that the birth awaited by the recently married Smiths will be quadruplets, my expectation for the number of sets of quadruplets among the next million American births (about two, based on a little reading) could be elicited and might be of some use. But this relief is largely illusory. For a real expert on multiple births would be expected to take into account such data as that Mrs. Smith is herself a twin and a very young bride, so there is no practical possibility of counting the quadruplet births in a large number of cases similar in the respects considered pertinent.

10.6. Employing expert opinion

In just what ways can you expect to profit from the opinions of experts in serious matters? According to a very broad model, you have an important decision to make in the light of all sorts of data at your disposal, and these data may include the behavior of experts. You could, in principle, explore all sorts of ways of interrogating an expert—not confining yourself to eliciting his opinions—and study empirically how his responses can profit your business. Conceivably, pain in the weatherman's great toe would better help you plan picnics than would his opinion about the weather. Yet I presume that ordinarily little of importance would be lost if you could obtain only the opinion of the expert, that is, his personal probabilities. What should be done with such an opinion? The simplest thing, and sometimes the appropriate thing, would be to make the expert's opinion your own.

This is by no means mandatory. For example, an expert who always ascribes very small probability to what actually occurs and to nothing else would be as useful as one who is omniscient, but you would of course not make the opinions of this perfect fool your own. Again, you might discover with experience that your expert is optimistic or pessimistic in some respect and therefore temper his judgments. Should he suspect you of this, however, you and he may well be on the escalator to perdition.

10.7. Divergent opinion

We often have access to more than one expert, and what to do when doctors disagree has always been, and will always be, a quandary. One important thing to do, but far outside the scope of this paper, is to encourage the right kind of communication between the experts. Exploration of how to do this is, for example, one of the aims of the Delphi technique [3; 8; 29, part II]. In general, good communication is what makes the experts share factual information and help each other think their opinions through thoroughly, and bad communication is what encourages various vices such as exaggeration and excessive deference. Sooner or later, despite all techniques of communication, divergent expert opinions will have to be faced. Perhaps you will make some composite of one or more expert opinions and your own opinion. An extreme way to do that would be to decide, on the basis of past experience or otherwise, that a particular one of the experts is the only one worth listening to and to make his opinion your own. A more general procedure would be to average the opinions, that is, to average the probability distributions associated with the experts (possibly including yourself), giving each the weight you think appropriate. Thus, rather than simply choose one expert among several, you can choose among the infinite number of synthetic experts that constitute the convex closure of the several. (A few key references bearing on the subject of this paragraph are [10; 11; 47, ch. 10; 52 pp. 65ff; 56].)

10.8. The expert as an instrument

When is one expert, real or synthetic, to be preferred to another? An "expert" in this context is a mechanism, possibly with human components, generating numbers that you contemplate using instead of your own personal probabilities (or, more generally, expectations) in certain contexts. One crude, practical answer sometimes available and appropriate, is this: Employ, until you have further experience, that expert whose past opinions, applied to your affairs, would have yielded you the largest average income.

No rule of this sort can claim absolute or objective validity, and this one has been couched especially roughly for the sake of simplicity. For example, actual past experience with the experts may be extensive, moderate, meager, or absolutely lacking. When past experience is extensive, but not too extensive, the rule often has much to recommend it; when direct past experience with the experts is meager, the rule is silly; and when such experience is altogether lacking, the rule results in a tie and is therefore empty. Actually, if you have little or no past experience with the experts, you will have to ponder them in terms of whatever information it was that brought you to regard them as promising in the first place: this well finder is regarded by the whole neighborhood as infallible with the hazel fork; that one is a professor of geology and the author of an important treatise on subsurface hydrology but has never before tried to help anyone locate a well. In such a context, the subjective aspect of your decision is thrown into prominence, but no matter how much direct past experience you may have with the experts, the ultimate subjectivity of your choice among them never disappears, though its effects may become less agonizing—according to the personalistic Bayesian theory of statistics, as in [48].

When your past experience with the experts is very extensive indeed, it may become profitable for you to refine the original rough rule by dividing up circumstances into categories and confiding in different experts for different categories of decisions. Such discussion could continue indefinitely; for the situations are innumerable and tend to parallel the whole field of decision-theoretic statistics.

An interesting process for coordinating the efforts of experts in different fields known as PIP (for probability information processing) has been vigorously pursued [20].

10.9. Slippery utilities

A different kind of complication in applying, or adapting, the rough rule is that the notion of average income may not be readily applicable to your affairs. For one thing, it may be important to measure your own income in utility rather than in cash. If this involves only determining your utility for cash, you may be able to do that reasonably well with moderate effort. If, however, the consequences of your act are not easily converted into cash but involve values difficult to weigh against each other, such as beauty, justice, and health, your dilemma may be especially severe. Raiffa [44] has recently published a book largely on these subtle problems of pondering the imponderable and evaluating the invaluable that reviews, and contributes to, a considerable literature.

Another difficulty in measuring utility is seen in this example: You are the person responsible for choosing which of several televised meteorological forecasters shall serve your city. There is, I assume for simplicity, abundant evidence of past performance, and the members of the community who use the forecasts will behave in accordance with the probabilities announced in them. Since yours is a public trust, you would like to choose the forecaster that in the past would have maximized the mean income of members of the community.

Subtle welfare-economic decisions about the relative importance of bent-pin anglers and barn painters could complicate your problem, but, even more important, relatively little is really known about the uses to which public weather information is put and what its economic consequences are. Thought has been given to the problem of the economic value of meteorological forecasts, both for the general public and for special purposes, but difficult, important empirical aspects of the question remain to be explored. (Key references are [36, 40].)

10.10. Which scoring rule for the trained respondent?

There are as many proper scoring rules for a trichotomy, for example, as there are convex functions over the baricentric triangle, or two-dimensional simplex. It would therefore seem important to study in what respects one scoring rule is better than another. But this question has thus far proved surprisingly unproductive. Its elusiveness is brought out by the consideration that an ideal subject responds to all proper scoring rules, including those involving extremely small payments, in exactly the same way. Therefore, any criteria for distinguishing among scoring rules must arise out of departures of actual subjects from the ideal.

Since we all do depart markedly from the ideal, it might seem that one proper scoring rule would be much more effective with a real person than another, and this presumably is often so. But if a person is reasonably sophisticated, though far from ideal, the form of a proper scoring rule for eliciting his probabilities, for, say, a trichotomy should—provided its amplitude is sufficient to command attention—have little or no effect on his response. To see this, put yourself in his place. You are offered a contract that will result in certain cash payments to you depending on your choice of three numbers p, q, and r and on whether a certain game ends in win, lose, or tie. If you know what personal probabilities are and understand that the contract is so drawn that it is to your interest to report your personal probabilities, then the details of the contract seem unimportant; for no matter which proper scoring rule it corresponds to, you should ask yourself what your personal probabilities for the three events are and report them.

Yet, the terms of the contract might make a modest difference to you. Suppose, for example, that very little money is to change hands in case of a tie, no matter what your response is. In this case, you have little incentive to ask yourself carefully the probability of a tie and are thus left free to focus on the relative probability of a win given that there is not a tie. In this case, your questioner will be well served if he is mainly interested in that conditional probability, and he will be badly served if he particularly wants to know the probability that you attach to a tie.

Thus, at least a vague criterion applicable even to sophisticated but human respondents emerges. Insofar as responding is hard work, the scoring rule should encourage the respondent to work hardest at what the questioner most wants to know. If this is to be effective, it must not merely be mathematically true but also plain to the respondent that he will be rewarded most for working on the right aspects of his opinion. This is in part a psychological question of human communication, subject to much speculation and experimentation. To illustrate, you are faintly curious to know whether the respondent thinks that a tie is likely and desperate to know whether he believes that the home team will win. You might be best served by entering into two palpably separate contracts with the respondent, a small one hinging only on whether there is a tie and resulting in a rough casual elicitation of the respondent's r, and a larger one, involving no payment in case of a tie, resulting in a well considered evaluation of the ratio $p:q$ for the respondent. Of course the two contracts together amount to a single scoring rule, though presenting them separately might work better psychologically.

The appropriate incentive for you to offer a respondent for his opinion depends not only on the importance for you of obtaining that opinion with a specified accuracy but also on the difficulty for the respondent in obtaining it from himself. This makes the choice of a scoring rule designed to evoke the right degree of effort from the respondent on the various components of his task particularly subtle.

10.11. There are no bargain scoring rules

Since a scoring rule is a scheme of payments, it might seem natural to choose the rule that promises to obtain the required information as cheaply as possible or perhaps the one that obtains the most for the money. There may be something to this line of thinking, but not in first approximation, and I have been unable to make progress with it.

The respondent will presumably work for what he regards as an acceptable wage. In the presence of a scoring rule, he will perceive his wage as random with an expectation that can be adjusted by adding a constant to the scoring rule. So, within the linearity approximation, the respondent does not charge extra for submitting to the scoring rule, and it therefore seems roughly reasonable to reckon that the questioner is not charged for it.

The higher the amplitude of the scoring rule, the more incentive it gives the respondent to reply with care. On this account, the respondent might in principle come to insist on a higher mean wage for facing a highly fluctuating scoring rule, and this could tend to deter the questioner from using high amplitudes. But the important practical limitation on amplitude would seem to be the need to avoid the distortion of response induced by the nonlinearity of the respondent's utility.

10.12. Business sharing

In common sense, we feel without any overt reference to economic models that some responses are not so wrong as others and ought not to be so heavily penalized. If a respondent is pretty sure that the home team will win, and there is in fact a tie, then he is perhaps not so wrong as if it had lost. Scoring rules reflecting this idea have been sought and easily found. See, for example, [21, 53]. One interesting way to adjust the rewards and penalties of the respondent to the interests of the interrogator, which was brought out in a dramatic and more radical form in McCarthy's [37] pioneering note on proper scoring rules, is to give the respondent a fractional interest in the business involved. To illustrate with an overidealized example, an oil prospecting company could give its geologist a small fraction of all profits and losses with the understanding that all decisions in the business would be made using the geologist's personal probabilities about geological uncertainties.

(It would be interesting to consider with some care the respects in which such an example is realistic and unrealistic, but I can only go a step or two in that direction here. Stock in the company would seem to give the geologist an interest in reporting his probabilities honestly if he could be assured that they would be adopted as the personal probabilities of the management for the events concerned. But this incentive may not be fully in harmony with the expert's incentives to appear worthy, as opposed to simply being worthy, of retention and promotion—a complication that affects not only business sharing but all applications of proper scoring rules to a professional expert. Business sharing does not present the expert with an explicit scoring rule in any business complicated enough to provide a more than mechanical role for the managers, in particular in any business in which there are other uncertainties than those about which the expert is consulted, but an implicit rule is as effective in principle as an explicit one.)

Long ago, Gauss [25, section 6] proposed that economic losses (such as those in a game of chance) provide a good model for the incentive to estimate accurately even in the most academic contexts. Decisively to uphold or to overthrow this suggestion does not seem possible. Personally, it appeals to me. Correspondingly, when we say that a tie should not be regarded as so distant from a win as a loss would be, I am inclined to think that that is because we have in mind various uses for sport forecasts in which the penalty is less for one kind of error than another. Of course, the penalty need not be a monetary one; it might involve, for example, loss of social prestige. Fortunately, the elusive question of whether all that is good and bad about a forecast can ultimately be referred to profit and loss in economic decisions, sufficiently, widely interpreted, need not be resolved in order to show the interest and utility of viewing proper scoring rules as a share in a real or a fictional business. For the business model is certainly a mathematically general model for all proper scoring rules and a fertile point of view for the generation of proper scoring rules that penalize some errors more than others.

The technical point that every proper scoring rule can be viewed as a share in a business and that every such share leads to an at least weakly proper scoring rule should be appreciated. Section 7 makes these points clear. Every strictly proper scoring rule amounts to the possibility of choosing among acts, only one of which is appropriate to each system of personal probabilities. Conversely, a person knowing that an act in a specified economic situation is to be chosen for him in accordance with his announced system of personal probabilities will have no incentive to announce a false one. However, if the convex function arising from the family of acts has flat places—technically, is not strictly convex—then the person has no positive incentive to distinguish among certain systems of probabilities, and the scoring rule is only weakly proper.

Insofar as business sharing is a practical method of elicitation, it makes possible the use of a proper scoring rule even in those situations stressed at the end of subsection 10.3 in which arbitrary proper scoring rules cannot be implemented because some of the conditional probabilities to be elicted will not be tested, depending on the opinions expressed by the expert.

The parallelism between implementing a scoring rule by business sharing and the rough rule for rating experts according to how their advice would have affected your business in the past (discussed in subsection 10.8) is evident, but the two things must not be confused. In particular, the rating rule can be used regardless of what scoring rule if any is used to elicit the opinions.

10.13. Some armchair psychology

Consider now a subject quite untrained in personal probability and the theory of scoring rules whose only incentives are provided by the scoring itself. It is questionable whether, in any serious application, this ought to be allowed to happen. Though any strictly proper scoring rule is a sufficient guide for an ideally intelligent Robinson Crusoe no matter how uninstructed, can we expect real people to respond well with no other coaching than is provided by the scoring rule itself or even by extended experience with the scoring rule? And even if investigation should yield a somewhat affirmative reply, is there any point in withholding instruction? Whatever the answers, it does seem stimulating to speculate on what kinds of scoring rules presented in what way would most nearly operate on naive subjects as all strictly proper scoring rules are supposed to operate on sophisticated ones, and this should lead to ideas of practical value for subjects of intermediate sophistication.

In the first place, the subject must understand the scoring rule. If it makes explicit reference to logarithms, or even to squares, most ordinary people will not understand it at all; and even those with mathematical training may not be nearly apt enough at calculation to use the rule effectively. This is an important reason to present the rule through some vivid tabular or graphic device, which could, for example, take the elaborate form of conversational-mode digital computation, or more simply of some slide-rule device such as those of the Shuford-Massengill Corporation (Lexington, Mass.), or perhaps tabulation of the scoring rule itself, possibly very boldly rounded [14].

Perhaps it is helpful to a subject responding about a partition of possible events if the economic consequence of his response is a function only of the element of the partition that happens to obtain. This condition imposes no constraint at all for two-fold partitions, but for n-fold partitions with $n > 2$, it leads to the logarithmic scoring rules of subsection 9.4. The possible advantages of the simplicity might often be outweighed by the inappropriateness of a symmetric scoring rule in asymmetric situations or of a scoring rule that lays emphasis on the correct elicitation of small probabilities. I have sometimes heard the possibility that a subject responding to a logarithmic scoring rule could be subjected to an infinite (or at any rate, unlimited) penalty raised as an overwhelming objection. This possibility does of course imply that the method cannot be applied literally, but approximate applications in which the subject is not allowed to name probabilities less than, say, 10^{-3} suggest themselves. And, as mentioned in subsection 10.5, obtaining probabilities very close to 0 by means of direct incentives does not seem practical by any scoring rule.

10.14. Proper scoring rules in school

Proper scoring rules hold forth promise as more sophisticated ways of administering multiple-choice tests in certain educational situations [14, 49]. The student is invited not merely to choose one item (or possibly none) but to show in some way how his opinion is distributed over the items, subject to a proper scoring rule or a rough facsimile thereof.

Though requiring more student time per item, these methods should result in more discrimination per item than ordinary multiple-choice tests, with a possible net gain. Also they seem to open a wealth of opportunities for the educational experimenter.

Above all, the educational advantage of training people—possibly beginning in early childhood—to assay the strengths of their own opinions and to meet risk with judgment seems inestimable. The usual tests and the language habits of our culture tend to promote confusion between certainty and belief. They encourage both the vice of acting and speaking as though we were certain when we are only fairly sure and that of acting and speaking as though the opinions we do have were worthless when they are not very strong.

Effects of nonlinearity in educational testing deserve some thought, but presumably nonlinearity is not a severe threat when a test consists of a large number of items. One source of nonlinearity that has been pointed out to me is this. A student competing with others for a single prize is motivated to respond so as to maximize the probability that his score will be the highest of all. This need not be consistent with maximizing his expected score, and presumably situations could be devised in which the difference would be important.

References

[1] J. Aczel, Lectures on functional equations and their applications. New York, Academic Press, 1966.

[2] Maurice Allais, Le Comportement de l'homme rationnel devant le risque: critique des postulats et axioms de l'école americaine. Econometrica 21 (October 1953), 503–546.

[3] Robert H. Ament, Comparison of Delphi forecasting studies in 1964 and 1969. Futures 2 (March 1970), 35–44.

[4] Thomas Bayes, Essay towards solving a problem in the doctrine of chances. The Philosophical Transactions 53 (1763), 370–418. Reprinted in: Facsimiles of two papers by Bayes, edited by W. Edwards Deming. Washington, D.C. The Graduate School, The Department of Agriculture, 1940; and in: Thomas Bayes's essay towards solving a problem in the doctrine of chances. Biometrika 45 (December 1958), 293–315.

[5] G. W. Brier, Verification of forecasts expressed in terms of probability. Monthly Weather Review 78 (January 1950), 1–3.

[6] T. A. Brown, Probabilistic forecasts and reproducing scoring systems. Santa Monica, Calif., RAND Corporation, RM-6299-ARPA, June 1970, 65 pp.

[7] Robert J. Buehler, Measuring information and uncertainty. In: Proceedings of the Symposium on Foundations of Statistical Inference. Toronto, Holt, Rinehart, and Winston, 1971.

[8] N. C. Dalkey, An experimental study of group opinion: the Delphi method. Futures 1, 5 (September 1969), 408–426.

[9] Bruno de Finetti, La prévision: ses lois logiques, ses sources subjectives. Annales de l'Institut Henri Poincaré 7 (1937), 1–68. English translation in [31].

[10] Bruno de Finetti, La notion de distribution d'opinion comme base d'un essai d'interpretation de la statistique. Publications de l'Institut de Statistique de l'Université de Paris 1, 2 (1952), 1–19.

[11] Bruno de Finetti, Media di decisioni e media di opinioni. Bulletin de l'Institut International de Statistique, 28th session, 34, 2 (1954), 144–157.

[12] Bruno de Finetti, Does it make sense to speak of "good probability appraiser"? In: The scientist speculates: An anthology of partly-baked ideas. New York, Basic Books, 1962, 357–364.

[13] Bruno de Finetti, Probabilità composte e teoria delle decision. Rendiconti de Matematica 23 (1964), 128–134.

[14] Bruno de Finetti, Methods for discriminating levels of partial knowledge concerning a test item. The British Journal of Mathematical and Statistical Psychology 18 (May 1965), 87–123.

[15] Bruno de Finetti, Quelques conventions qui semble d'être utile. Revue Roumaine de Mathematique Pures et Appliquées 12, 9 (1967), 1227–1233.

[16] Bruno de Finetti, Probability: interpretations, Vol. 12, 496–505, in the International encyclopedia of the social sciences. New York, The Macmillan Company, 1968.

[17] Bruno de Finetti, Logical foundations and measurement of subjective probability. Acta Psychologica 34 (December 1970), 129–145.

[18] Bruno de Finetti and Leonard J. Savage, Sul modo di scegliere le probabilità iniziali. Sui fondamenti della statistica. Biblioteca del Metron, Series C 1 (1962), 81–154 (English summary, 148–151).

[19] Ward Edwards, A bibliography of research on behavioral decision processes to 1968. Memorandum Report No. 7, Human Performance Center, University of Michigan, January 1969, 93.

[20] Ward Edwards, L. D. Phillips, W. L. Hays and B. C. Goodman, Probability information processing systems: design and evaluation. IEEE Transactions on Systems Science and Cybernetics, SSC-4 (1968), 248–265.

[21] Edward S. Epstein, A scoring system for probability forecasts of ranked categories. Journal of Applied Meteorology 8 (December 1969), 985–987.

[22] Peter C. Fishburn, Utility theory for decision making. New York, John Wiley and Sons, Inc., 1970.

[23] Milton Friedman, Choice, chance, and personal distribution of income. Journal of Political Economy 61, 4 (August 1953), 277–290.

[24] Milton Friedman and Leonard J. Savage, The utility analysis of choices involving risk. Journal of Political Economy 56 (August 1948), 279–304. Reprinted in: Readings in price theory. Chicago, Richard D. Irwin, 1952; and in: Landmarks in political economy (edited by Earl J. Hamilton et al.). The University of Chicago Press, 1962.

[25] Carl Friedrich Gauss, Theoria combinationis observationum erroribus minimis obnoxiae. Commentationes Societatis Regiae Scientarum Gottingensis Recentiores 5 (1821), 33–90. In German translation: A. Börsch and P. Simon, Abhandlungen zur Methode der kleinsten Quadrate. Berlin, Westdruckerei Joachim Frickert, 1887 (reprinted Wurzburg, Physica-Verlag, 1964). In French translation: J. Bertrand, Methode des moindres carrés. Paris, Mallet-Bachelier, 1855.

[26] I. J. Good, Rational decisions. Journal of the Royal Statistical Society, Ser. B 14 (1952), 107–114.

[27] Charles Jackson Grayson, Decisions under uncertainty: Drilling decisions by oil and gas operators. Boston, Harvard University, Division of Research, Graduate School of Business Administration, 1960.

[28] Trygve Haavelmo, The role of the econometrician in the advancement of economic theory. Econometrica 26 (July 1958), 351–357. (Presidential address, Econometric Society, Philadelphia, December 29, 1957.)

[29] Olaf Helmer, Social technology. New York, Basic Books, 1966.

[30] Arlo D. Hendrickson, Payoffs to probability forecasters. Technical Report No. 145, University of Minnesota, September 1970.

[31] Henry E. Kyburg, Jr. and Howard E. Smokler, Studies in subjective probability. New York, John Wiley and Sons, 1964.

[32] Clifford J. Maloney, A probability approach to catastrophic threat. ARO-D Report 70–2, U.S. Army Research Office–Durham, Durham, North Carolina, 1970.

[33] Jacob Marschak, Management research and behavioral science. Working Paper No. 2, Western Management Science Institute, University of California, Los Angeles, 1961.

[34] Jacob Marschak, Actual versus consistent decision behavior. Behavioral Science 9 (April 1964), 103–110.

[35] Jacob Marschak, Decision making: economic aspects, Vol. 4, 42–45, in the International encyclopedia of the social sciences. New York, the Macmillan Company, 1968.

[36] W. J. Maunder, The value of the weather. London, Methuen, 1970.

[37] John McCarthy, Measures of the value of information. Proceedings of the National Academy of Sciences (1956), 654–655.

[38] Frederick Mosteller and David L. Wallace, Inference and disputed authorship: the federalist. Reading, Mass., Addison-Wesley, 1964.

[39] Allan H. Murphy, The ranked probability score and the probability score: a comparison. Monthly Weather Review 98 (December 1970), 917–924.

[40] Allan H. Murphy and R. A. Allen, Probabilistic prediction in meteorology: a bibliography. Silver Spring, Md., Environmental Science Services Administration, Technical Memorandum No. WBTM TDL 35, 1970.

[41] Allan H. Murphy and Edward S. Epstein, Verification of probabilistic predictions; a brief review. Journal of Applied Meteorology 6 (October 1967), 748–755.

[42] Allan H. Murphy and Robert L. Winkler, Scoring rules in probability assessment and evaluation. Acta Psychologica 34 (December 1970), 273–286.

[43] John W. Pratt, Risk aversion in the small and in the large. Econometrica 32, 1–2 (January–April 1964), 122–136.

[44] Howard Raiffa, Decision analysis: introductory lectures on choices under uncertainty. Reading, Massachusetts, Addison-Wesley, 1968.

[45] Harry V. Roberts, Risk, ambiguity, and the Savage axioms: comment. Quarterly Journal of Economics 77 (1963), 327–342.

[46] Harry V. Roberts, On the meaning of the probability of rain. In: Proceedings of the First National Conference on Statistical Meteorology. Boston, American Meteorological Society (1968), 133–141.

[47] Leonard J. Savage, The foundations of statistics. New York, John Wiley & Sons, Inc.; and London, Chapman and Hall, 1954.

[48] Leonard J. Savage, The foundations of statistics reconsidered. In: Proceedings of the Fourth Berkeley Symposium. Berkeley, University of California Press, 1961, 575–586. Reprinted in [31].

[49] Emir H. Shuford, Jr., Albert Arthur and H. Edward Massengill, Admissible probability measurement procedures. Psychometrika 31 (June 1966), 125–145.

[50] Cedric A. B. Smith, Consistency in statistical inference and decision. Journal of the Royal Statistical Society, Ser. B 23 (1961), 1–25.

[51] Cedric A. B. Smith, Personal probability and statistical analysis. Journal of the Royal Statistical Society, Ser. A 128 (1965), 469–489.

[52] Carl-Axel S. Staël von Holstein, Assessment and evaluation of subjective probability distributions. Stockholm, The Economic Research Institute at the Stockholm School of Economics, 1970.

[53] Carl-Axel S. Staël von Holstein, A family of strictly proper scoring rules which are sensitive to distance. Journal of Applied Meteorology 9 (1970), 360–364.

[54] Masanao Toda, Measurement of subjective probability distribution. Report No. 3, Division of Mathematical Psychology, Institute for Research, State College, Pennsylvania, 1963.

[55] Robert L. Winkler, The assessment of prior distributions in Bayesian analysis. Journal of the American Statistical Association 62 (September 1967), 776–800.

[56] Robert L. Winkler, The consensus of subjective probability distributions. Management Science 15, 2 (October 1968), B-61–B-75.

[57] Robert L. Winkler, Scoring rules and the evaluation of probability assessors. Journal of the American Statistical Association 64 (September 1969), 1073–1078.

[58] Robert L. Winkler and Allan H. Murphy, Evaluation of subjective precipitation probability forecasts. In: Proceedings of the First National Conference on Statistical Meteorology. Boston, American Meteorological Society, 1968, 148–157.

[59] Robert L. Winkler and Allan H. Murphy, "Good" probability assessors. Journal of Applied Meteorology 7 (October 1968), 751–758.

[60] Robert L. Winkler and Allan H. Murphy, Nonlinear utility and the probability score. Journal of Applied Meteorology 9 (February 1970), 143–148.

PART 5

IDENTIFICATION AND BAYESIAN ANALYSIS

S. E. Fienberg and A. Zellner (eds.)
Studies in Bayesian Econometrics and Statistics.

5.1 BAYESIAN THEORY OF IDENTIFICATION IN SIMULTANEOUS EQUATIONS MODELS*

Jacques H. DRÈZE

University of Louvain

1. *Introduction*

1.1. *Contents*

This paper treats, from a Bayesian viewpoint, the identification problem for parameters of the simultaneous equations model.

That model is introduced in section 1.2. The theory of identification for *parameters* of that model is summarized in section 2. The results in that section are proved, for instance, in [9]. A more detailed treatment can be found in [3].

Section 3 first points out the scope of a Bayesian approach, which permits replacing some of the deterministic constraints of a simultaneous equations model by probabilistic constraints—that is, by a prior density on parameters. This approach may lead the econometrician to analyze, with an informative prior density, models that are not identified. A simple theorem captures the implications of underidentification for the posterior density. The classical conditions for identification are then extended into necessary conditions for the existence of a proper posterior density.

* An earlier version of this paper was presented to the Third NBER-NSF Seminar on Bayesian Inference Econometrics, Harvard Univ., October 1971. I have greatly benefitted from the discussion during this meeting, and in particular from the remarks of L. J. Savage, to whose memory this paper is dedicated. I have also been influenced by the work of Joseph Kadane [5] and Dennis V. Lindley [6]. I am grateful to Anton Barten, Michel Mouchart, Jean-François Richard, Thomas Rothenberg, Tran Van Hoa, Léon Wegge and Arnold Zellner for their comments on an earlier version of this paper.

1.2. The simultaneous equations model

In its simplest form, the simultaneous equations model consists of a system of m linear stochastic equations

$$By' + \Gamma z' = u' \tag{1}$$

where

y' is an $m \times 1$ vector of endogenous variables;

z' is an $n \times 1$ vector of exogenous variables;

B is an $m \times m$ nonsingular matrix of coefficients, with $B_{ii} = 1$, $i = 1 \dots m$ (normalization);

Γ is an $m \times n$ matrix of coefficients;

u' is an $m \times 1$ vector of unobserved disturbances.

Let (Y, Z) denote a $T \times (m+n)$ matrix of observations, and $U = YB' + Z\Gamma'$ be the corresponding $T \times m$ matrix of unobserved disturbances. The data-generating process may be described by the joint density of the disturbances

$$\psi(U|Z)\,\mathrm{d}U = \psi(YB' + Z\Gamma')\,\mathrm{d}U. \tag{2}$$

Let $\mathring{\Sigma}$ denote the parameters of the density ψ—an arbitrary set. The joint density of the observations, or data density, is obtained from (2) by performing the integrand transformation (in Tm space) from U to Y, with Jacobian $\|B\|^T$. Thus:

$$p(Y|Z, B, \Gamma, \mathring{\Sigma})\,\mathrm{d}Y = \psi(YB' + Z\Gamma')\|B\|^T\,\mathrm{d}Y. \tag{3}$$

The set of parameters of this density is $(B, \Gamma, \mathring{\Sigma})$.

Since B is nonsingular, we may solve (1) explicitly for y, obtaining:

$$y' = -B^{-1}\Gamma z' + B^{-1}u' =_{\text{def}} \Pi z' + v'. \tag{4}$$

Equation (4) is called the "reduced form" of the "structural system" (1). The $T \times m$ matrix of unobserved reduced form disturbances is

$$V = Y - Z\Pi' = UB'^{-1}. \tag{5}$$

The joint density of the reduced form disturbances is obtained from (2) by performing the integrand transformation from U to V, with Jacobian $\|B\|^T$. Thus:

$$\phi(V|Z)\,\mathrm{d}V = \psi(VB')\|B\|^T\,\mathrm{d}V. \tag{6}$$

Let $\mathring{\Omega}$ denote the parameters of the density ϕ—a set of functions of the parameters $(B, \mathring{\Sigma})$. From (5) and (6), we obtain an alternative expression for the data density, namely:

$$p(Y|Z, \Pi, \mathring{\Omega})\,\mathrm{d}Y = \phi(Y - Z\Pi')\,\mathrm{d}Y. \tag{7}$$

Notation. If A is a matrix, vec A denotes the row-vector formed from the elements of A, taken one row at a time.

2. *Observational equivalence and identification*

2.1. *Observational equivalence*

A *structure* S is a complete specification of the probability distribution function $P(y)$ of a set of observations y. A set of structures $\mathscr{S}$ is called a *model.* A typical problem of statistical inference consists in making judgments (either deterministic or probabilistic) about structures, given the model $\mathscr{S}$ and the observations y. In most applications, S is a point, and $\mathscr{S}$ is a set of points, in parameter space.

Definition 2.1. *Two structures S and $\bar{S}$ in $\mathscr{S}$ are* "observationally equivalent" *if and only if $P(y|S) = P(y|\bar{S})$ identically in y.*

Given an arbitrary structure $S \in \mathscr{S}$, the set of all structures in $\mathscr{S}$ which are observationally equivalent to S is an *equivalence class*; it will be denoted $\mathscr{E}(S)$.

Application 2.1. Consider the system (1) under the specification that $\psi\,(U|Z)$ is $N(\text{vec } U'|0, \Sigma \otimes I_T)$. The subset $\mathscr{S}$ of $R^{m(2m+n)}$, $\mathscr{S} = \{(B, \Gamma, \Sigma) \in R^{m(2m+n)} \,|\, |B| \neq 0, B_{ii} = 1, i = 1 \ldots m, \Sigma \text{ P.D.S.}\}$ defines a model. A point $S \in \mathscr{S}$ is a structure. Let

$$\begin{aligned} \Pi &=_{\text{def}} -B^{-1}\Gamma, \qquad & B\Pi + \Gamma &= 0; \\ \Omega &=_{\text{def}} B^{-1}\Sigma B'^{-1}, \qquad & B\Omega B' - \Sigma &= 0. \end{aligned} \tag{8}$$

It follows from (6) that $\phi(V|Z)$ is $N(\text{vec } V'|0, \Omega \otimes I_T)$, and from (7) that $p(Y|Z, \Pi, \Omega)$ is $N(\text{vec } Y'|\text{vec } \Pi Z', \Omega \otimes I_T)$. Consequently, the probability density of the data is completely specified by the reduced form parameters (Π, Ω). If S and $\bar{S}$ are two points in $\mathscr{S}$ such that

$$B^{-1}\Gamma = \bar{B}^{-1}\bar{\Gamma}, \quad B^{-1}\Sigma B'^{-1} = \bar{B}^{-1}\bar{\Sigma}\bar{B}'^{-1}, \tag{9}$$

then S and $\bar{S}$ are observationally equivalent.

2.2. Identification

Definition 2.2. (i) *The structure S in $\mathscr{S}$ is* "identified" *if and only if $\mathscr{S}$ does not own any $\bar{S} \neq S$ such that S and $\bar{S}$ are observationally equivalent; i.e. if and only if $\mathscr{E}(S) = \{S\}$.*

(ii) *The model $\mathscr{S}$ is* "identified" *if and only if every $S \in \mathscr{S}$ is identified; i.e., if and only if the partition of $\mathscr{S}$ into equivalence classes is a partition into singletons.*

Application 2.2. Under the normality assumption of application 2.1, the structural model $\mathscr{S} = \{(B, \Gamma, \Sigma) \in R^{m(2m+n)} \mid |B| \neq 0,\ B_{ii} = 1,\ i = 1 \ldots m, \Sigma \text{ P.D.S.}\}$ is not identified; but the reduced form model $\mathscr{S} = \{(\Pi, \Omega) \in R^{m(m+n)} | \Omega \text{ P.D.S.}\}$ is identified. Conditionally on non-singular B, the transformation from the parameter space (B, Γ, Σ) to the parameter space (B, Π, Ω) is one-to-one. In the parameter space (B, Π, Ω), the subspace defined by "$B = \bar{B}, \Omega$ P.D.S." is an identified model; and the intersection of $\mathscr{S}$ with the subspace defined by "$\Pi = \bar{\Pi}, \Omega = \bar{\Omega}$" is an equivalence class.

2.3. Local identification

When $\mathscr{S}$ is a subset of a metric space, one can define the distance between two structures. It is then possible to introduce a weaker, very useful concept.

Definition 2.3. *The structure S is* "locally identified" *if and only if, for some open neighborhood $\mathscr{B}(S)$, $\mathscr{S} \cap \mathscr{B}(S)$ does not own any $\bar{S} \neq S$ such that S and $\bar{S}$ are observationally equivalent, i.e., if and only if $\mathscr{E}(S) \cap \mathscr{B}(S) = \{S\}$.*

Application 2.3. Under the normality assumption of application 2.1, the structural model

$$\mathscr{S} = \{(B, \Gamma, \Sigma) \in R^{m(2m+n)} \mid |B| \neq 0, \quad \Psi_k(B, \Gamma, \Sigma) = 0, \quad k = 1 \ldots K\} \tag{10}$$

is identified whenever the set of relations

$$B\Pi + \Gamma = 0, \quad B\Omega B' - \Sigma = 0, \quad \Psi_k(B, \Gamma, \Sigma) = 0, \quad k = 1 \ldots K, \tag{11}$$

can be solved *uniquely* for (B, Γ, Σ) as functions of (Π, Ω).

In some cases ("just identification"), a solution exists for all $(\Pi, \Omega) \in R^{m(m+n)}$, Ω P.D.S.; in other cases ("overidentification"), a solution exists for some but not all $(\Pi, \Omega) \in R^{m(m+n)}$, Ω P.D.S. The uniqueness problem is raised for values of (Π, Ω) such that a solution exists.

(i) Let (β, γ, σ) be the triplet of $1 \times m^2$, $1 \times mn$ and $1 \times m^2$ vectors (vec B, vec Γ, vec Σ). Let also

$$\Psi_\beta = \left[\frac{\partial \Psi_i}{\partial \beta_j}\right], \quad \Psi_\gamma = \left[\frac{\partial \Psi_i}{\partial \gamma_j}\right], \quad \Psi_\sigma = \left[\frac{\partial \Psi_i}{\partial \sigma_j}\right]$$

be the $K \times m^2$, $K \times mn$ and $K \times m^2$ matrices of partial derivatives of the constraints Ψ_k, $k = 1 \ldots K$, with respect to the structural parameters (β, γ, σ). At points where the $K \times m^2$ matrix

$$[\Psi_\beta \ \ \Psi_\gamma \ \ \Psi_\sigma][I \otimes B \ \ I \otimes \Gamma \ \ I \otimes 2\Sigma]' = \Psi^*$$

is of rank m^2, solutions to (11) are locally unique. This sufficient condition is necessary for local uniqueness, at regular points of the matrix Ψ^*. When the constraints Ψ_k are linear in (B, Γ) and independent of Σ, local uniqueness implies global uniqueness.

(ii) A simple case is obtained when the conditions $\Psi_k = 0$, $k = 1 \ldots K \geqq m^2$, consist in assigning predetermined values to a subset of the parameters (B, Γ).

An arbitrary equation from (1), say the first, may then be written as:

$$\bar{B}_{11} y_1' + B_{12} y_2' + \bar{\Gamma}_{11} z_1' + \Gamma_{12} z_2' = u_1 \tag{12}$$

where $(\bar{B}_{11}, \bar{\Gamma}_{11})$ is the subvector of predetermined parameters. The first row of the first matrix equation in (11) may be written as:

$$(\bar{B}_{11} \ \ B_{12}) \begin{bmatrix} \Pi_{11} & \Pi_{12} \\ \Pi_{21} & \Pi_{22} \end{bmatrix} + (\bar{\Gamma}_{11} \ \ \Gamma_{12}) = 0. \tag{13}$$

Equation (13) may be solved uniquely for B_{12} and Γ_{12} (as functions of Π) if and only if Π_{21} has full row-rank (rank condition). This requires that the dimension of $\bar{\Gamma}_{11}$ should be at least equal to that of B_{12} (order condition); i.e., that $(\bar{B}_{11}, \bar{\Gamma}_{11})$ be of dimension m at least.

Remarks 2.3. (i) Definition 2.2 gives a precise content to the *concept* of identification, for structures and models. The *theory* of identification, for a specific class of parametric models, is concerned with alternative statements of necessary and/or sufficient conditions for a model to be identified. The restatements aim at providing easy checks of the fact that a particular model is, or is not identified. These checks can be applied directly to the (a priori) specifications of the model, without explicit reference to the likelihood. (Of course, analysis of the likelihood underlies the theory.) The order conditions for identification of a simultaneous equations model by means of exclusion restrictions provide a good example.

(ii) The ultimate purpose of a study of identification has been aptly linked to the design of experiments (see the remark by L. J. Savage, quoted by Kadane [5, p. 9]). The statistician who wishes to make inferences about certain parameters will attempt to design an experiment such that these parameters are identified. Because econometric research proceeds usually from nonexperimental data, that remark should be extended. The econometrician who is concerned with inference about parameters that are not identified may try to overcome this difficulty by collecting richer data, or by resorting to a more restrictive theory. For instance, he may switch from annual to quarterly data, and introduce a specific lag structure. A study of identification may then prove useful to guide his strategy for collecting data and developing theoretical specifications.

3. *A Bayesian extension*

3.1. *Scope of a Bayesian analysis*

Whether it be based upon theory or past observation, economic knowledge is typically subject to uncertainties of various kinds. Probabilistic statements would seem more adequate than deterministic statements to reflect such knowledge as we have. In general, model specification should thus be partly stochastic. The prior information would then be described by means of exact constraints on some of the parameters, as well as by means of a probability distribution function defined on some or all of the remaining parameters.

Classical methods are not very well suited to handle such situations. The econometrician relying on these methods will be tempted to disregard probabilistic prior information, or else to translate it into deterministic constraints.[1] It would be more natural to carry out the analysis

[1] I do not mean to deny the possibility of analyzing from a classical viewpoint models with random parameters, or pooled data. However, little use has been made of this possibility, in the econometric literature.

within the Bayesian framework. Statistical methods have been developed to that effect, for the simultaneous equations model [2, 4, 7, 8, 10].[2]

This approach may well lead to models that are not identified, according to the definitions given above, if one considers only the exact prior restrictions. The Bayesian econometrician should thus be prepared to carry out inference within unidentified models—a topic largely ignored (and understandably so) by classical econometric theory.

3.2. *Implications of underidentification*

To say that a model is unidentified is to say that some equivalence classes contain more than one element. These elements are observationally equivalent structures, the respective likelihoods of which are identical. Whatever judgements an econometrician might make about the relative probabilities of these structures must remain unaltered by observation. This common-sense remark is reflected in the theorem below about simultaneous equations models. Its counterpart, from a classical viewpoint, is that no tests can discriminate among alternative specifications of unidentified, or just identified, models.

Let, as in section 1.2, the set of structural parameters be $(B, \Gamma, \mathring{\Sigma})$, where $\mathring{\Sigma}$ is the set of parameters of the density of the disturbances, $\Psi(V|Z)$. The set of parameters of the density $\phi(V|Z)$ of reduced form disturbances is $(\mathring{\Omega})$, where the elements of $(\mathring{\Omega})$ are functions of $(B, \mathring{\Sigma})$. Let $f(B, \Gamma, \mathring{\Sigma})$ be the prior density on the structural parameters, and assume that, for given B, the transformation from $(\mathring{\Sigma})$ to $(\mathring{\Omega})$ is nonsingular. One may then perform an integrand transformation from the parameter space $(B, \Gamma, \mathring{\Sigma})$ to the parameter space $(B, \Pi, \mathring{\Omega})$ to define the alternative prior density $h(B, \Pi, \mathring{\Omega})$. To simplify notation, it is assumed here that the normalization and other exact constraints are implicit in $f(B, \Gamma, \mathring{\Sigma})$; that is, the domain of f is $\mathscr{S}$.

Example 3.2. When $\Psi(V|Z)$ is $N(\text{vec}\, U'|0, \Sigma \otimes I_T)$, then $(\mathring{\Sigma}) = \Sigma$,

[2] "This remark should bring comfort to those who regard underidentification as the natural state of an econometric model: indeed, stochastic prior information, even if imprecise, should enable them to proceed with structural estimation. Turn now to overidentification. In principle, deterministic restrictions are maintained, even if they conflict with the observations: the prior information dominates the sample information. In contrast, stochastic prior information gets combined with sample information through Bayes' theorem; neither type of information dominates the other, but both count in proportion to their respective precisions" [1, p. 9].

$(\mathring{\Omega}) = \Omega = B^{-1}\Sigma B'^{-1}$. The Jacobian of the transformation from (B, Γ, Σ) to (B, Π, Ω) is $\|B\|^{n+m+1}$ and $h(B, \Pi, \Omega)\,\mathrm{d}B\,\mathrm{d}\Pi\,\mathrm{d}\Omega = f(B, -B\Pi, B\Omega B')\cdot\|B\|^{n+m+1}\cdot\mathrm{d}B\,\mathrm{d}\Pi\,\mathrm{d}\Omega$.

The prior density $h(B, \Pi, \mathring{\Omega})$ may always be factorized into a conditional density on B, given $(\Pi, \mathring{\Omega})$, and a marginal density on $(\Pi, \mathring{\Omega})$, that is:

$$h(B, \Pi, \mathring{\Omega})\,\mathrm{d}B\,\mathrm{d}\Pi\,\mathrm{d}\mathring{\Omega} \equiv h_c(B|\Pi, \mathring{\Omega})\,\mathrm{d}B\,h_m(\Pi, \mathring{\Omega})\,\mathrm{d}\Pi\,\mathrm{d}\mathring{\Omega}. \tag{14}$$

Let $h(B, \Pi, \mathring{\Omega}|Y, Z)$ be the posterior density, in parameter space $(B, \Pi, \mathring{\Omega})$, which obtains (from h) after the sample (Y, Z) has been observed.

Theorem 3.2.

$$p(Y|Z, B, \Gamma, \mathring{\Sigma}) \underset{Y}{\equiv} p(Y|Z, \Pi, \mathring{\Omega})$$

implies

$$h(B, \Pi, \mathring{\Omega}\,|\,Y, Z)\,\mathrm{d}B\,\mathrm{d}\Pi\,\mathrm{d}\mathring{\Omega} = h_c(B|\Pi, \mathring{\Omega})\,\mathrm{d}B\,h_m(\Pi, \mathring{\Omega}|\,Y, Z)\,\mathrm{d}\Pi\,\mathrm{d}\mathring{\Omega},$$

whenever the posterior density exists.

Remark. The conclusion in theorem 3.2 could equivalently be written as: $h_c(B|\,Y, Z, \Pi, \mathring{\Omega}) \underset{Y}{\equiv} h_c(B|\Pi, \mathring{\Omega})$. Stated in this way, the conclusion remains valid even when $h_m(\Pi, \mathring{\Omega}|\,Y, Z)$ is not a proper density.

Proof

$$h(B, \Pi, \mathring{\Omega}|\,Y, Z) =_{\mathrm{def}} \frac{h(B, \Pi, \mathring{\Omega})\cdot p(Y|Z, B, \Pi, \mathring{\Omega})}{\int h(B, \Pi, \mathring{\Omega})p(Y|Z, B, \Pi, \mathring{\Omega})\,\mathrm{d}B\,\mathrm{d}\Pi\,\mathrm{d}\mathring{\Omega}}$$

$$= \frac{h_c(B|\Pi, \mathring{\Omega})h_m(\Pi, \mathring{\Omega})p(Y|Z, \Pi, \mathring{\Omega})}{\int h_m(\Pi, \mathring{\Omega})p(Y|Z, \Pi, \mathring{\Omega})\cdot[\int h_c(B|\Pi, \mathring{\Omega})\,\mathrm{d}B]\,\mathrm{d}\Pi\,\mathrm{d}\mathring{\Omega}}$$

$$=_{\mathrm{def}}\ h_c(B|\Pi, \mathring{\Omega})\cdot h_m(\Pi, \mathring{\Omega}|\,Y, Z),$$

where the definitions are given by Bayes' theorem.

Remarks 3.2. (i) A general theorem, of which the foregoing may be viewed as a special case, has been proved by Kadane [5, theorem 5].

Kadane comments upon my result as follows: "Because the map from the structure to the reduced form is identifying, the entire Bayesian analysis can be done on the reduced form" [5, p. 191].

(ii) In the parameter space $(B, \Pi, \mathring{\Omega})$ equivalence classes are obtained by fixing Π and $\mathring{\Omega}$ (see application 2.2). Thus, the conditional density $h_c(B|\Pi, \mathring{\Omega})$ defines, for each equivalence class, a conditional probability measure over the elements of the class.

In lecturing on this subject, I have used the term "identification in probability" to characterize a situation where there exists a proper conditional prior density over the elements of every equivalence class. This choice was motivated by analogy with the use of the term "identification" to characterize a situation where every equivalence class is a singleton.

Reactions to this presentation have led me to recognize that it was misleading to use the word "identification" in defining a property of the prior density for the parameters of unidentified models. I agree with Kadane's view "that identification is a property of the likelihood function, and is the same whether considered classically or from the Bayesian approach" [5, p. 175].[3]

3.3. *Analysis of prior density*

Whenever the prior density function is well defined, and the likelihood is dominated, the posterior density will be well defined as well, whether the model is identified or not. This is the rationale for Lindley's remark [6, p. 46]: "In passing it might be noted that unidentifiability causes no real difficulty in the Bayesian approach." Indeed, the difficulty has been shifted to the level of defining a prior density on *all* the parameters of the model; in the context of simultaneous equations models, which always involve many parameters, elicitation of a proper prior density may raise serious difficulties. In particular, natural-conjugate prior densities will prove inadequate, in the case of unidentified models. Indeed, a natural-conjugate prior density may be expressed in terms of reduced form parameters only, because the likelihood may be so expressed.

[3] The "property" under consideration is that equivalence classes should be singletons. The property should be verified *after* all the relevant constraints have been embodied in the likelihood function. The property can sometimes be verified without full specification of the functional form of the likelihood function—see, e.g., [9, theorem 4].

When the prior density is expressed in terms of the parameters $(B, \Pi, \mathring{\Omega})$, the decomposition appearing in theorem 3.2 is suggestive, because it brings out explicitly a conditional density that will not be altered by the observations, namely $h_c(B|\Pi, \mathring{\Omega})$. Analysis of that conditional density reveals immediately certain characteristics of the posterior density. By contrast, the prior $h_m(\Pi, \mathring{\Omega})$ will be altered by the observations, to an extent that increases (loosely speaking) with the precision of the sample.

Typically, however, the prior information will be expressed more naturally in terms of the structural parameters $(B, \Gamma, \mathring{\Sigma})$. And one may expect the prior density to be truly informative about a subset of these parameters only. One may then wish to know whether the prior density on such a proper subset of structural parameters implies a proper prior density on B, given the reduced form parameters $(\Pi, \mathring{\Omega})$. In that case, one may also wish to evaluate the Jacobian of the transformation. A theory relating the prior information on an arbitrary subset of structural parameters to the implied prior information on B, given $(\Pi, \mathring{\Omega})$, plays, in the Bayesian analysis, a role exactly similar to the classical theory of identification, which relates an arbitrary set of constraints on the structural parameters to identification of the model. The application below shows that the similarity exists at the technical as well as the logical level.

Application 3.3. Consider once more the structural model

$$\mathscr{S} = \{(\beta, \gamma, \sigma) \in R^{m(2m+n)} \big| \, |B| \neq 0, \quad \Psi_k(\beta, \gamma, \sigma) = 0,$$
$$k = 1 \dots K\}.$$

Let $(\tilde{\beta}, \tilde{\gamma}, \tilde{\sigma})$ be a *subvector* of (β, γ, σ), with $\tilde{\beta}$, $\tilde{\gamma}$ and $\tilde{\sigma}$ of dimension $1 \times \mu$, $1 \times \nu$ and $1 \times \rho$ respectively, and define $\lambda = \mu + \nu + \rho$.

We want to investigate the relationship between a prior density on $(\tilde{\beta}, \tilde{\gamma}, \tilde{\sigma})$ and the prior density on B, conditionally on (Π, Ω); and we want to investigate that relationship under the constraints $\Psi_k(\beta, \gamma, \sigma) = 0$, $k = 1 \dots K$.

A major technical simplification in the evaluation of Jacobians is obtained by treating Ψ_k, $k = 1 \dots K$, as a random variable, with the implicit proviso that inference should proceed conditionally on $\Psi_k = 0$. This approach would not be well suited to calculate the posterior density but it is very convenient for our present purposes.

Let then $\psi \doteq (\Psi_1 \dots \Psi_k \dots \Psi_K)$, a $1 \times K$ vector, and choose $(\tilde{\beta}, \tilde{\gamma}, \tilde{\sigma})$ of dimensionality $\lambda = m^2 - K$, by restricting attention to an appropriate subvector of (β, γ, σ). One may then consider the integrand transformation between the m^2-dimensional vector $(\psi, \tilde{\beta}, \tilde{\gamma}, \sigma)$ and the m^2-dimensional vector β. Conditionally on (Π, Ω), that transformation is well defined by means of (8), provided the Jacobian of the transformation has full rank m^2. The transformation relates two alternative parametrizations of the prior density, namely (B, Π, Ω) and $(\psi, \tilde{\beta}, \tilde{\gamma}, \tilde{\sigma}, \Pi, \Omega)$. If the transformation is well defined, a proper prior density on $(\tilde{\beta}, \tilde{\gamma}, \tilde{\sigma})$, conditionally on $\psi(=0)$ and (Π, Ω), implies a proper prior density on B, conditionally on $\psi(=0)$ and (Π, Ω).

(i) *Conditionally on* (Π, Ω), we have from (8):

$$\left[\frac{\partial \gamma'}{\partial \beta}\right] = [I_m \otimes -\Pi'], \qquad \left[\frac{\partial \sigma'}{\partial \beta}\right] = [I_m \otimes 2B\Omega]. \tag{15}$$

Remember (from application 2.3) that

$$\Psi_\beta = \left[\frac{\partial \psi'}{\partial \beta}\right], \qquad \Psi_\gamma = \left[\frac{\partial \psi'}{\partial \gamma}\right], \qquad \Psi_\sigma = \left[\frac{\partial \psi'}{\partial \sigma}\right],$$

where these matrices have dimension $K \times m^2$, $K \times mn$ and $K \times m^2$ respectively.

Define the $\lambda \times m^2$, $\lambda \times mn$ and $\lambda \times m^2$ matrices Φ_β, Φ_γ and Φ_σ as follows:

Let $\tilde{\beta}_i$, the ith element of $\tilde{\beta}$, be the kth element of $\beta (i \leqq \mu, k \leqq m^2)$; row i of Φ_β has an entry equal to 1 in the kth position, entries equal to 0 everywhere else, $i = 1 \ldots \mu$; rows $\mu+1$ to λ of Φ_β have all their entries equal to 0.

Similarly, let $\tilde{\gamma}_i = \gamma_k$; row $\mu+i$ of Φ_γ has an entry equal to 1 in the kth position, entries equal to 0 everywhere else, $i = 1 \ldots v$; rows 1 to μ and $\mu+v+1$ to λ of ϕ_γ have all their entries equal to 0.

Let finally $\tilde{\sigma}_i = \sigma_k$; row $\mu+v+i$ of Φ_σ has an entry equal to 1 in the kth position, entries equal to 0 everywhere else; rows 1 to $\mu+v$ of Φ_σ have all their entries equal to 0.

Conditionally on (Π, Ω), the $(K+\lambda) \times m^2$ Jacobian matrix of the transformation from $(\psi, \tilde{\beta}, \tilde{\gamma}, \sigma)$ to β is defined by:

$$\begin{pmatrix} \Psi_\beta + \Phi_\gamma \left[\dfrac{\partial \gamma'}{\partial \beta}\right] + \Psi_\sigma \left[\dfrac{\partial \sigma'}{\partial \beta}\right] \\ \\ \Phi_\beta + \Phi_\gamma \left[\dfrac{\partial \gamma'}{\partial \beta}\right] + \Phi_\sigma \left[\dfrac{\partial \sigma'}{\partial \beta}\right] \end{pmatrix} =$$

$$= \begin{bmatrix} \Psi_\beta & \Psi_\gamma & \Psi_\sigma \\ & & \\ \Phi_\beta & \Phi_\gamma & \Phi_\sigma \end{bmatrix} \begin{bmatrix} I_m \otimes I_m \\ I_m \otimes -\Pi' \\ I_m \otimes 2B\Omega \end{bmatrix}. \tag{16}$$

That Jacobian must be of full rank m^2 in order for a one-to-one transformation to exist, conditionally on (Π, Ω), between β and

$(\psi, \tilde{\beta}, \tilde{\gamma}, \sigma)$.

Under the assumption $|B| \neq 0$, the rank of the Jacobian matrix in (16) is the same as the rank of the matrix

$$\begin{bmatrix} \Psi_\beta & \Psi_\gamma & \Psi_\sigma \\ & & \\ \Phi_\beta & \Phi_\gamma & \Phi_\sigma \end{bmatrix} \begin{bmatrix} I_m \otimes B' \\ I_m \otimes \Gamma' \\ I_m \otimes 2\Sigma \end{bmatrix} = \begin{bmatrix} \Psi^* \\ \\ \Phi^* \end{bmatrix}. \tag{17}$$

The matrix

$$\begin{bmatrix} \Psi^* \\ \Phi^* \end{bmatrix}$$

is a natural extension of the matrix Ψ^*, whose rank was found relevant (in application 2.3) to ascertain the identification of the model.

(ii) Let us now consider case (ii) of application 2.3, under a mixture of exact and stochastic prior information. More specifically, let us assume that the conditions $\Psi_k = 0$, $k = 1 \ldots K$, consist in assigning predetermined values to a subset of the parameters (B, Γ); and that the prior density factors into $h(\Pi, \Omega)$ and

$$f(\tilde{\beta}, \tilde{\gamma}, \sigma) = \prod_{i=1}^{m} f_i(\tilde{B}_i, \tilde{\Gamma}_i),$$

where $(\tilde{B}_i, \tilde{\Gamma}_i)$ denotes a subvector from the ith row of (B, Γ).

Consider then an arbitrary equation from (1), say the first, and write it as follows (under appropriate relabeling, if necessary):

$$\bar{B}_{11}y_1' + B_{12}y_2' + B_{13}y_3' + B_{14}y_4' + \bar{\Gamma}_{11}z_1' + \Gamma_{12}z_2' + \Gamma_{13}z_3' + \Gamma_{14}z_4' = u_1 \tag{18}$$

where

$(\bar{B}_{11} \quad \bar{\Gamma}_{11})$ is the subvector of predetermined parameters;

$(B_{12} \quad \Gamma_{12} \quad \Gamma_{13})$ is the subvector of parameters appearing as arguments of the prior density $f_1(\tilde{B}_1, \tilde{\Gamma}_1)$;

the dimension of B_{13} does not exceed that of $\bar{\Gamma}_{11}$, and B_{14} has positive dimension only when B_{13} has maximal dimension;

the dimension of Γ_{12} does not exceed that of B_{14}, and Γ_{13} has positive dimension only when Γ_{12} has maximal dimension.

The first row of the matrix equation $B\Pi+\Gamma = 0$ may be written as:

$$(\bar{B}_{11} \quad B_{12} \quad B_{13} \quad B_{14}) \begin{bmatrix} \Pi_{11} & \Pi_{12} & \Pi_{13} & \Pi_{14} \\ \Pi_{21} & \Pi_{22} & \Pi_{23} & \Pi_{24} \\ \Pi_{31} & \Pi_{32} & \Pi_{33} & \Pi_{34} \\ \Pi_{41} & \Pi_{42} & \Pi_{43} & \Pi_{44} \end{bmatrix}$$

$$+(\bar{\Gamma}_{11} \quad \Gamma_{12} \quad \Gamma_{13} \quad \Gamma_{14}) = 0, \tag{19}$$

with Π_{31} and Π_{42} square. Solving explicitly for B_{13} and B_{14} in terms of $\bar{B}_{11}$, B_{12}, $\bar{\Gamma}_{11}$, Γ_{12} and Π, we find

$$B_{13} = -[(\bar{B}_{11}\Pi_{11}+\bar{\Gamma}_{11})+(B_{12}\Pi_{21}+B_{14}\Pi_{41})]\Pi_{31}^{-1} \tag{20}$$

$$B_{14} = -[\bar{B}_{11}\Pi_{12}-(\bar{B}_{11}\Pi_{11}+\bar{\Gamma}_{11})\Pi_{31}^{-1}\Pi_{32}+B_{12} \times(\Pi_{22}-\Pi_{21}\Pi_{31}^{-1}\Pi_{32})+\Gamma_{12}](\Pi_{42}-\Pi_{41}\Pi_{31}^{-1}\Pi_{32})^{-1}. \tag{21}$$

The explicit solution requires that the matrix

$$\begin{pmatrix} \Pi_{31} & \Pi_{32} \\ \Pi_{41} & \Pi_{42} \end{pmatrix}$$

be nonsingular. This condition is equivalent to the rank and order conditions mentioned under application 2.3 (ii).

$(\bar{\Gamma}_{11} \quad \Gamma_{12})$ must have the dimension of $(B_{13} \quad B_{14})$, i.e., the subvector $(\bar{B}_{11} \quad B_{12} \quad \bar{\Gamma}_{11} \quad \Gamma_{12} \quad \Gamma_{13})$ from $(B_1 \quad \Gamma_1)$ on which prior information (exact or stochastic) is available must have dimension at least m (order condition);

the matrix

$$\begin{bmatrix} \Pi_{31} & \Pi_{32} & \Pi_{33} \\ \Pi_{41} & \Pi_{42} & \Pi_{43} \end{bmatrix}$$

must have full row rank (rank condition). This matrix is the counterpart of Π_{21} in (13).

Provided these conditions are fulfilled, one may perform an integrand transformation from Γ_{12} to B_{14}, with Jacobian $\|\Pi_{42}-\Pi_{41}\Pi_{31}^{-1}\Pi_{32}\|$, and write the prior density $f_1(\tilde{B}_1, \tilde{\Gamma}_1)$ as:

$$\begin{aligned} &f(B_{12}, B_{14}(\Pi_{42}-\Pi_{41}\Pi_{31}^{-1}\Pi_{32})+\bar{B}_{11}\Pi_{12} \\ &-(\bar{B}_{11}\Pi_{11}+\bar{\Gamma}_{11})\Pi_{31}^{-1}\Pi_{32}+B_{12}(\Pi_{22}-\Pi_{21}\Pi_{31}^{-1}\Pi_{32}),\Gamma_{13}) \\ &\cdot\|\Pi_{42}-\Pi_{41}\Pi_{31}^{-1}\Pi_{32}\|\cdot \mathrm{d}B_{12}\mathrm{d}B_{14}\mathrm{d}\Gamma_{13}. \end{aligned} \tag{22}$$

Conditionally on Π, the elements of B_1 are then subject to a prior density, with mass concentrated in a subspace having the dimensionality of (B_{12}, B_{14}); that density is defined by (20), (22) and $B_{11} = \bar{B}_{11}$. If similar conditions are satisfied for all other equations as well, prior information (either exact or stochastic) is available on all the elements of B, conditionally on Π and Ω, and $h_c(B|\Pi, \Omega)$ is a well-defined density.

In view of theorem 3.2, the necessary conditions for the existence of a well-defined conditional density $h_c(B|\Pi, \Omega)$ are also necessary for the existence of a well-defined posterior density. The conditions obtained in application 3.3 take the form of generalizations of the conditions for local identification mentioned in application 2.3.

Remark 3.3. The rank of the Jacobians in (16) and (22) depends upon the rank of the matrix Π, or of a submatrix from Π. The true value of Π may be such that the rank condition is not satisfied. However, in a Bayesian analysis, the posterior density of Π will typically be continuous. Because the set of values for Π where the rank condition is *not* met has Lebesgue measure zero in R^{mn}, the posterior probability that the rank condition is fulfilled will then be equal to one, after observing an arbitrary finite sample. G. S. Maddala has recognized this problem, in a recent unpublished note,[4] and shown that it may have serious implications for inference. An adequate test of rank conditions should be incorporated into the existing Bayesian techniques.

[4] Weak priors and sharp posteriors in simultaneous equation models, mimeographed, 1972.

3.4. *Concluding remarks*

By way of conclusion, it may be appropriate to comment again about the relevance of identification, in a Bayesian approach. What is gained, from knowledge that a model is identified? Kadane [5, theorems 3 and 4] presents an interesting characterization of identification in the context of modification of opinion, and in the context of decision-making. I have elsewhere [1, p. 9] noted that identification (resp. local identification) is a necessary condition for uniqueness (resp. local uniqueness) of maximum likelihood estimators.[5] In a Bayesian approach, maximum likelihood estimators correspond (numerically) to posterior modes obtained from a locally uniform prior density.[6] From the viewpoint of making decisions, uniqueness of posterior modes is seldom relevant. Identification is neither necessary nor sufficient for other, more relevant characteristics of the posterior density to obtain. The existence of posterior moments is a good example. In this respect, it should be noted that the "order of overidentification" has been found relevant for the existence of sampling moments, in a classical approach, and of posterior moments, in a Bayesian approach.[7] Again, the "order of overidentification" is a by-product of the theory of identification in simultaneous equations models; it is not part of the general concept of identification.

[5] *Proof*: Let S^* be a maximum likelihood estimator of the true structure. If there exists $\bar{S}$ observationally equivalent to S^*, then $\bar{S}$ is also a maximum likelihood estimator.

[6] In the case of an unidentified model, uniqueness (resp. local uniqueness) of the posterior mode S^* requires uniqueness (resp. local uniqueness) of the prior conditional mode over the set of structures $\mathscr{E}(S^*)$ which are observationally equivalent to S^*.

[7] See [1, p. 11] or [2, theorems 3.1 and 3.2].

References

[1] J. H. Drèze, Econometrics and decision theory. Econometrica 40 (1972), 1–17. (Reprinted in the present volume.)

[2] J. H. Drèze, Bayesian limited information analysis of the simultaneous equations model. CORE DP 7111.

[3] F. M. Fisher, The identification problem in econometrics. New York, McGraw-Hill, 1966.

[4] R. Harkema, Simultaneous equations, a Bayesian approach. Rotterdam, Rotterdam University Press, 1971.

[5] J. Kadane, The role of identification in Bayesian theory. Mimeographed, 1972. (A revised version of this paper appears in the present volume.)
[6] D. V. Lindley, Bayesian statistics: a review. Philadelphia, SIAM, 1971.
[7] J. A. Morales, Bayesian full information structural analysis, Berlin, Springer, 1971.
[8] J. F. Richard, Posterior and predictive densities for simultaneous equations models. Berlin, Springer, 1973.
[9] T. J. Rothenberg, Identification in parametric models. Econometrica 39 (1971), 577–591.
[10] A. Zellner, An introduction to Bayesian inference in econometrics, New York, Wiley, 1971.

S. E. Fienberg and A. Zellner (eds.)
Studies in Bayesian Econometrics and Statistics.

5.2 THE ROLE OF IDENTIFICATION IN BAYESIAN THEORY*

Joseph B. KADANE

Carnegie-Mellon University

1. *Introduction*

Bayesian theory has attracted increasing attention as a coherent theory of statistics. Among the consequences of this attention have been investigations into the relationship between Bayesian theory and various ideas and models in classical statistics [1–3]. Identification figures prominently in classical statistics, and especially in the consideration of the simultaneous equations model in econometrics [4, 5].

One general question unresolved in this literature is, whether Bayesian theory requires a different definition of identification from the classical one. For instance, Rothenberg [6, p. 14] says "we leave unanswered the question of an appropriate Bayesian definition of identification", and Morales [7, p. 20] reports "The concept of identification in a Bayesian context is not altogether clear. We shall adopt the view of considering a structure 'identified' if the posterior density of the parameters of the model is not 'flat' on a subspace of the parameter space. This point of view may not be entirely satisfactory."

At the same time Lindley [2, p. 46, footnote 34] remarks "In passing it might be noted that unidentifiability causes no real difficulty in the Bayesian approach."

The point of view of this paper is that identification is a property of the likelihood function, and is the same whether considered classically or from the Bayesian approach. The usual theory is reviewed and

* Presented to the Winter meeting of the Econometric Society, Toronto, December 1972. This paper is dedicated to the memory of my friend L. J. Savage. In its preparation I have benefited from discussions with several members of the NBER-NSF Seminar on Bayesian Inference in Econometrics, most notably Morris DeGroot. The author acknowledges partial support of the National Science Foundation under Grant GS-38609.

extended in section 2 along the lines of an earlier paper [8], using the concept of identified functions. I characterize real-valued identified functions in section 3, and discuss examples in section 4. After a brief review of Bayesian statistical theory in section 5, in section 6 I characterize identified functions in the context of the modification of opinion, and in section 7 in the context of making decisions. Finally in section 8 I give and discuss a general result on Bayesian analysis of a possibly unidentified model.

2. *Identification in the classical context*

Let $(\mathscr{X}, \mathscr{C})$ be a measurable space, and let $P_\theta, \theta \in \Theta$ be a family of probability distributions on $(\mathscr{X}, \mathscr{C})$. Two parameters θ and θ' are said to be *observationally equivalent* (written $\theta \sim \theta'$) iff

$$P_\theta(A) = P_{\theta'}(A) \quad \text{for all} \quad A \in \mathscr{C}. \tag{1}$$

Θ *is identified* iff

$$\forall \quad \theta, \theta' \in \Theta, \qquad \theta \sim \theta' \Rightarrow \theta = \theta'. \tag{2}$$

This is the concept of identification used in Koopmans and Riersol [4] and in most of the later work on the subject.

However, identification of Θ is insufficient for some purposes. If Θ is not identified, there might be a temptation to discard the model, even though it might be unidentified in uninteresting ways. For this reason Kadane [8] introduced a generalization of (2). Suppose that g is a function with domain Θ and range in some space $\mathscr{S}$. Then g is identified if and only if

$$\forall \quad \theta, \theta' \in \Theta, \qquad \theta \sim \theta' \Rightarrow g(\theta) = g(\theta'). \tag{3}$$

The concept of identification of functions is a generalization of the earlier notion of identification of structures since $\mathscr{S}$ can be taken to be Θ and g can be the identity function. Some consequences of this theory are given in Kadane [8] and need not be repeated here.

3. *Identification of real-valued functions*

An *identifying function* $T:\Theta \to \mathscr{T}$ is a function satisfying

$$\forall \quad \theta, \theta' \in \Theta, \quad \theta \sim \theta' \quad \text{if and only if} \quad T(\theta) = T(\theta'). \tag{4}$$

Intuitively such an identifying function carries with it all information about the identification of functions. Examples of identifying functions are given in the next section.

Lemma 1. *Let* $f:\Theta \to \mathscr{S}$, *and suppose* $T:\Theta \to \mathscr{T}$ *is an identifying function.* f *is identified if and only if there exists a function* $g:\mathscr{T} \to \mathscr{S}$ *satisfying* $f(\theta) = g(T(\theta))$.

Proof. Suppose such a function exists and suppose $\theta \sim \theta'$. Then $T(\theta) = T(\theta')$ so $f(\theta) = g(T(\theta)) = g(T(\theta')) = f(\theta')$. Hence f is identified.

Now suppose no such function exists. Then there are θ_1 and $\theta_2 \ni f(\theta_1) \neq f(\theta_2)$ but $T(\theta_1) = T(\theta_2)$. In this case $\theta_1 \sim \theta_2$ but $f(\theta_1) \neq f(\theta_2)$. Hence f is not identified. Q.E.D.

Now endow Θ with a σ-field $\mathscr{A}$ of subsets. Then the function T generates in $\mathscr{T}$ the σ-field $\mathscr{B}$ of sets B whose inverse images

$$A = T^{-1}(B) = \{\theta:\theta \in \Theta, T(\theta) \in B\}$$

are in $\mathscr{A}$. Thus with respect to $\mathscr{B}$, $T:(\Theta, \mathscr{A}) \to (\mathscr{T}, \mathscr{B})$ is measurable. Furthermore, T induces a sub-σ-field $\mathscr{A}_0$ of $\mathscr{A}$ defined by

$$\mathscr{A}_0 = \{T^{-1}(B):B \in \mathscr{B}\}.$$

Consider the following lemma in Lehmann [9, p. 37]:

Lemma 2. *Let the statistic T from* $(\Theta, \mathscr{A})$ *into* $(\mathscr{T}, \mathscr{B})$ *induce the sub-*σ*-field* $\mathscr{A}_0$*. Then a real-valued* $\mathscr{A}$*-measurable function f is* $\mathscr{A}_0$ *measurable if and only if there exists a* $\mathscr{B}$*-measurable function g such that*

$$f(\theta) = g(T(\theta)) \quad \textit{for all} \quad \theta.$$

In order to use lemma 2 in the context of lemma 1, a third lemma is needed:

Lemma 3. *Let f be a real-valued* $\mathscr{A}$ *measurable function and suppose there a function g satisfying*

$$f(\theta) = g(T(\theta)) \quad \textit{for all} \quad \theta.$$

Then g is $\mathscr{B}$*-measurable.*

Proof. Let P be a Borel set. Let B be the set (in $\mathscr{T}$) defined by

$$B = \{t \in \mathscr{T}:g(t) \in P\}.$$

It must be shown that $B \in \mathscr{B}$. Let

$$A = \{\theta \in \Theta:T(a) \in B\}.$$

Then

$$f^{-1}(P) = T^{-1}(g^{-1}(P)) = T^{-1}(B) = A \in \mathscr{A}$$

by the measureability of f. But then $B \in \mathscr{B}$. Hence g is $\mathscr{B}$-measurable. Q.E.D.

Putting Lemmas 1, 2 and 3 together, we have

Theorem 1. *Let f be a real-valued $\mathscr{A}$-measurable function and suppose an identifying function T exists. Then f is identified if and only if f is $\mathscr{A}_0$-measurable.*

The analogy of the theory of identified functions to the theory of sufficient statistics is now obvious. The reader is referred to Lehmann [9, chapter 2] and Halmos and Savage [10] for the measure-theoretic treatment of sufficiency.

4. *Examples*

This collection of examples is given mainly to allow the reader to gain intuition about the definitions of sections 2 and 3 before proceeding to their consequences in sections 5, 6, and 7. Some of these examples have been discussed by others.

4.1. *The sum of two normals, variance one*

Suppose $X_i \sim \mathcal{N}(\mu_i, 1)$ $(i = 1, 2)$, and only

$$X = X_1 + X_2$$

is observed. The parameter space here is $\Theta = \mathbb{R}^2$ and a typical point in Θ is (μ_1, μ_2). An identifying function is given by

$$T(\mu_1, \mu_2) = \mu_1 + \mu_2.$$

4.2. *The sum of two normals with variances not given*

Here $X_i \sim \mathcal{N}(\mu_i, \sigma_i^2)$ $(i = 1, 2)$, and again only

$$X = X_1 + X_2$$

is observed. Now the parameter space is $\Theta = \mathbb{R}^2 \times (\mathbb{R}^+)^2$ and a typical point in Θ is $(\mu_1, \mu_2, \sigma_1^2, \sigma_2^2)$. An identifying function is

$$T(\mu_1, \mu_2, \sigma_1^2, \sigma_2^2) = (\mu_1 + \mu_2, \sigma_1^2 + \sigma\ \ ).$$

4.3. *Simultaneous equations*

Consider a system of simultaneous equations

$$By_t+\Gamma Z_t = u_t \qquad (t=1,\ldots,T)$$

where B is $m\times m$, Γ is $m\times n$, u_t and y_t are $m\times 1$. Also suppose u_t are normal, with zero mean, zero inter-temporal correlation, and Σ covariance matrix. Then a typical parameter point is (B, Γ, Σ), and an identifying function is

$$T(B,\Gamma,\Sigma) = (B^{-1}\Gamma, B^{-1}\Sigma B^{-1\prime}).$$

This is one way of understanding the import of the theorems of Koopmans *et al.* [11]. See Kadane [8] for a review and a correction.

4.4. *Linear relationship between variables subject to error*

Suppose that a true, unobserved relation

$$\eta = \theta\xi$$

holds, and that x_i is normally distributed, given ξ_i, with mean ξ_i and variance ϕ_1, y_i is normally distributed with mean $\theta\xi_i$ and variance ϕ_2, ξ_i is independently normal with mean 0 and variance t. Then, unconditionally, x_i and y_i are normal with zero mean and variance covariance matrix

$$\begin{bmatrix} t+\phi_1 & \theta t \\ \theta t & \theta^2 t+\phi_2 \end{bmatrix}.$$

Thus the function

$$T(\phi_1, \phi_2, t, \theta) = (t+\phi_1, \theta t, \theta^2 t+\phi_2)$$

is identifying. This problem has a long history; see for example Riersol [12], and, for a Bayesian treatment, Lindley and El-Sayyad [13].

5. *Bayesian statistical theory*

Suppose that the family of probability measures $P_\theta(x)$ is dominated by a σ-finite measure μ, so that they have densities $l(x|\theta)$ with respect to μ, i.e.

$$P_\theta(A) = \int_A l(x|\theta)\,\mathrm{d}\mu(x). \tag{5}$$

Also suppose that Θ has a σ-field $\mathscr{A}$, and that $p(\theta)$ is the density of a probability measure with respect to some dominating σ-finite measure ν. Then more such measures are obtained by Bayes' theorem:

$$p(\theta|x) = l(x|\theta)p(\theta)/\int l(x|\theta)p(\theta)\,\mathrm{d}\nu(\theta). \tag{6}$$

The measure $p(\theta)\,\mathrm{d}\nu(\theta)$ is your opinion about Θ before an experiment, and $p(\theta|x)$ is your opinion after observing x. The elicitation of $p(\theta)$ is described in Savage [14] and the references given there.

Additionally suppose that there is a decision space D and a bounded utility function $U(d, \theta)$ on $D \times \Theta$ which gives the worth, to you, of the decision d if θ obtains. Then a Bayesian chooses $d \in D$ to maximize his expected utility, where the expectation is taken with respect to whatever your current opinion is. For example, if a decision must be made before the data are observed, the optimal decision maximizes

$$\int U(d, \theta)p(\theta)\,\mathrm{d}\nu(\theta), \tag{7}$$

and if the decision is made after x is observed, the optimal decision maximizes

$$\int U(d, \theta)p(\theta|x)\,\mathrm{d}\nu(\theta). \tag{8}$$

In practice, Bayesians often study the likelihood function $l(x|\theta)$ as a function of θ to understand the import of the data. In typical experiments many parameters have the same likelihood (fall on the same contour of the likelihood function). The study of identification permits the exposure of parameters sure to lie on the same contour, regardless of the data. In this sense, identification is properly a part of the study of the design of experiments.[1] This contrasts with the view of Morales [7, quoted above] and with that of Drèze [15, p. 9]: "From the Bayesian viewpoint, classical identification theory is really concerned with local uniqueness of posterior modes."

6. *Identification and the modification of opinions*

Consider a measurable function f: $(\Theta, \mathscr{A}) \to (\mathscr{S}, \mathscr{E})$. An opinion p on θ induces an opinion on f, namely

$$Pr\{E\} = p\{f^{-1}(E)\} \text{ for all } E \in \mathscr{E}.$$

An experiment informative about f is one which might change your

[1] I am indebted to L. J. Savage for this remark.

opinion about f. More formally, suppose the experiment is characterized by the likelihood function $l(x|\theta)$, $\theta \in \Theta$. Your posterior on f is

$$P\{f \in E|C\} = p\{f^{-1}(E)|C\}$$
$$= \iint_{Cf^{-1}(E)} l(x|\theta)p(\theta)\, \mathrm{d}v(\theta)\, \mathrm{d}\mu(x) \Big/ \iint_{C\Theta} l(x|\theta)\, p(\theta)\, \mathrm{d}v(\theta)\, \mathrm{d}\mu(x). \qquad (9)$$

An experiment is *informative about f with respect to the prior p* iff $\exists C \in \mathscr{C}$ with $0 < \mu(C) < 1$ and $\exists E \in \mathscr{E}$ such that

$$p\{f^{-1}(E)|C\} \neq p\{f^{-1}(E)|C^c\}.$$

In the special case in which $(\mathscr{S}, \mathscr{E}) = (\Theta, \mathscr{A})$ and f is the identity function, an experiment is informative with respect to the prior p iff p has positive weight on nonobservationally equivalent θ's.

Clearly whether an experiment is informative for a function f depends on your opinion p as well as on f and on the experiment. For example, if you are opinionated, that is, if you are sure that one particular θ^* obtains, then no experiment is informative for any function for you. This dependence on the prior indicates that the concepts of identification and informativeness are different. Nevertheless identification can be characterized in terms of informativeness, as the following theorem shows.

First, we say that f distinguishes $\Theta_1 \subset \Theta$ if $\theta_1, \theta_2 \in \Theta_1$, $\theta_1 \neq \theta_2$ $\Rightarrow f(\theta_1) \neq f(\theta_2)$.

Theorem 2. *f is identified for an experiment if and only if the experiment is informative for f with respect to every distinguished two-point prior on Θ.*

Proof. Suppose first that f is not identified. Then $\exists \theta_1, \theta_2 \in \Theta$ such that

$$l_{\theta_1}(x) = l_{\theta_2}(x) \quad \text{for all} \quad x \quad \text{and} \quad f(\theta_1) \neq f(\theta_2).$$

Let $s_i = f(\theta_i)$, $i = 1, 2$. Let $0 < p(\theta_1) = 1 - p(\theta_2) < 1$.

Then

$$P\{f(\theta) = s_1|x\} = \frac{p(\theta_1)l_{\theta_1}(x)}{p(\theta_1)l_{\theta_1}(x) + p(\theta_2)l_{\theta_2}(x)} = p(\theta_1).$$

Similarly $P\{f(\theta) = s_2|x\} = p(\theta_2)$.

Then if $C \in \mathscr{C}$

$$P\{f = s_1|C\} = p(\theta_1) = p\{f = s_1|C^c\}$$
$$P\{f = s_2|C\} = p(\theta_2) = p\{f = s_2|C^c\}$$

so the experiment is not informative for f and this distinguished two-point prior.

Suppose now that f is identified. Let θ_1 and θ_2 be arbitrary distinguished points in Θ. Again let $s_i = f(\theta_i)$, $i = 1, 2$. From identification, there is a set $C \in \mathscr{C}$ such that

$$P_{\theta_1}(C) \neq P_{\theta_2}(C). \tag{10}$$

Then let $0 < p(\theta_1) = 1 - p(\theta_2) < 1$, so that

$$P\{f(\theta) = s_1 | C\} = p\{\theta_1 | C\} = \frac{p(\theta_1)P_{\theta_1}(C)}{p(\theta_1)P_{\theta_1}(C) + p(\theta_2)P_{\theta_2}(C)}.$$

Now

$$p\{f(\theta_1) = s_1 | C\} = p\{f(\theta) = s_1 | C^c\} \quad \text{iff}$$

$$\frac{p(\theta_2)P_{\theta_2}(C)}{p(\theta_1)P_{\theta_1}(C)} = \frac{p(\theta_2)P_{\theta_2}(C^c)}{p(\theta_1)P_{\theta_1}(C^c)}, \tag{11}$$

i.e., iff $P_{\theta_1}(C) = P_{\theta_2}(C)$, which contradicts (10). Therefore (11) cannot hold, so the experiment is informative for f with respect to this distinguished two-point prior. Q.E.D.

Notice that even if f is identified, there may be a prior on three points such that your opinion about f is sure not to change, regardless of the outcome of the experiment. Suppose, for example, that the random variable X is binomial, i.e., $Pr\{X = 0\} = \theta_i$, $i = 1, 2, 3$; $Pr\{X = 1\} = 1 - \theta_i$. Also suppose your prior probability on θ_i is p_i. Let $\theta_2 > \theta_3 > \theta_1$ and let $s_1 = f(\theta_1) = f(\theta_2)$; $s_3 = f(\theta_3)$. Thus f distinguishes the middle θ, θ_3, from the other two. Then the posterior probability of s_3 is

$$Pr\{f(\theta) = s_3 | x\} = Pr\{\theta_3 | x\} = \frac{p_3 \theta_3^{1-x}(1-\theta_3)^x}{\sum_{i=1}^{3} p_i \theta_i^{(1-x)}(1-\theta_i)^x}.$$

This, and $Pr\{f(\theta) = s_1 | x\}$ will not depend on the data if p_1, p_2 and p_3 can be specified to satisfy

$$p_1 + p_2 + p_3 = 1$$

$$p_1\theta_1 + p_2\theta_2 = \frac{1}{h} p_3\theta_3$$

$$p_1(1-\theta_1) + p_2(1-\theta_2) = p_3(1-\theta_3)\frac{1}{h} \quad \text{for some} \quad h > 0. \tag{12}$$

These equations are satisfied by

$$p_1 = \frac{1}{h+1}\left(\frac{\theta_2-\theta_3}{\theta_2-\theta_1}\right)$$

$$p_2 = \frac{1}{h+1}\left(\frac{\theta_3-\theta_1}{\theta_2-\theta_1}\right)$$

$$p_3 = \frac{h}{h+1}\,. \tag{13}$$

Intuitively this means that whether $X = 0$ or $X = 1$ is observed the posterior probability of θ_3 will be unchanged. The posterior for θ_1 and θ_2 will each change, θ_2 being more likely and θ_1 less likely if $X = 1$ is observed, and conversely. Thus two such experiments might change your opinion about f, because the prior for the second experiment would not be (13).

Another example of some interest is a single observation x from a normal distribution with mean μ and variance σ^2. The parameter space (μ, σ^2) is identified, and the maximum likelihood estimates are $\hat{\mu} = x$, $\hat{\sigma}^2 = 0$. Intuitively it seems possible that this experiment is not informative about σ^2.

Suppose that the priors on μ and σ^2 are independent, so that

$$p(\mu, \sigma^2) = p_1(\mu)p_2(\sigma^2).$$

Then the posterior on σ^2 is proportional to

$$\int_{-\infty}^{\infty} \frac{1}{\sigma}\exp\left[-\frac{1}{2}\left(\frac{x-\mu}{\sigma}\right)^2\right]p_1(\mu)p_2(\sigma^2)\,d\mu$$

$$= \frac{1}{\sigma}p_2(\sigma^2)\int_{-\infty}^{\infty} \exp\left[-\tfrac{1}{2}\mu^{*2}\right]p_1(x-\sigma\mu^*)d\mu^*\cdot\sigma$$

$$= p_2(\sigma^2)\int_{-\infty}^{\infty} \exp\left[-\tfrac{1}{2}\mu^{*2}\right]p_1(x-\sigma\mu^*)\,d\mu^*.$$

Now if $p_1(\cdot)$ is translation invariant, the above integral is independent of σ, and so the posterior on σ^2 is proportional to the prior and the experiment is noninformative for σ^2. On the other hand if $p_1(\cdot)$ is not translation invariant, the experiment can be seen to give information.

If μ is known to be positive, for instance, a negative value of x large in absolute value would be evidence of a large variance, a positive value of x evidence of a smaller variance.

Finally, consider the following example of Zellner [3, p. 253ff]: suppose there are two models: M_1 with parameter θ_1 and M_2 with parameter θ_2. Putting it together, the entire parameter space is (i, θ_1, θ_2), where $i = 1, 2$. Zellner says M_1 with its prior information on θ_1 is not identified with respect to M_2 with its prior information on θ_2 if

$$p(y|M_1) = p(y|M_2) \quad \text{for all} \quad y,$$

where

$$p(y|M_i) = \int p(y|\theta_i, M_i)p(\theta_i|M_i)\, d\theta_i\,.$$

In terms of the discussion above, the function being discussed is

$$f(i, \theta_1, \theta_2) = i,$$

and Zellner is discussing the informativeness of the experiment for f. Our language differs slightly, but the concepts are identical.

7. Identification and optimal statistical decisions

Suppose now that the decision space D and a utility function $U(d, \theta)$ on $D \times \Theta$ are given. Then the expected value of an experiment e is given by

$$v(e) = \int_X \{\max_d \int U(d, \theta)p(\theta|x, e)\, d\nu(\theta)\}p(x|e)\, d\mu(x)$$
$$-\max_d \int_\Theta U(d, \theta)p(\theta)\, d\nu(\theta). \tag{14}$$

Raiffa and Schlaifer [16] and Lindley [2, p. 25] discuss this concept, and show $v(e) \geqq 0$. The first expression in the difference is the expected value of the situation after the experiment e; the second is the value of the situation before the experiment.

The expected value of sample information is an important concept here because of the widespread belief that an unidentified model is not worth estimating, that is, that $v(e) = 0$ for such an experiment. While this statement is not true without qualification, this section gives results showing the sense in which it is true.

A more convenient expression for (14) can be given using the function

$$h(d, x) = \int_\Theta U(d, \theta)p(\theta|x, e)\, d\nu(\theta).$$

Then

$$v(e) = \int_X \max_d h(d, x)\, \mathrm{d}\mu(x) - \max_d \int_X h(d, x)\, \mathrm{d}\mu(x). \qquad (15)$$

This expression for $v(e)$ shows the advantage of experimentation: you make the decision after seeing x, rather than (conceptually) before. Notice that if the experiment e is not informative h does not depend on x and so the experiment is valueless.

Let d^* maximize $\int_X h(d, x)\, \mathrm{d}\mu x)$.

Let $B = \{x | \exists d \ni h(d, x) > h(d^*, x)\}$.

Theorem 3. (characterization of valuable experiments)

$$v(e) > 0 \quad \textit{iff} \quad \mu(B) > 0.$$

Proof. Suppose $\mu(B) = 0$. Let d be an arbitrary decision

$$h(d, x) \leqq h(d^*, x)$$

except on a set of μ-measure zero. Hence

$$\int h(d, x)\, \mathrm{d}\mu(x) \leqq \int h(d^*, x)\, \mathrm{d}\mu(x),$$

so

$$\max_d \int h(d, x)\, \mathrm{d}\mu(x) \leqq \int h(d^*, x)\, \mathrm{d}\mu(x),$$

and hence $v(e) = 0$.

Now suppose $\mu(B) > 0$. Let

$$B_i = \{x | \exists d \ni \frac{1}{i+1} \leqq h(d, x) - h(d^*, x) < 1/i\}$$

$$B_0 = \{x | \exists d \ni h(d, x) - h(d^*, x) \geqq 1\}$$

$$B \subseteq \bigcup_{i=0}^{\infty} B_i.$$

Hence $\exists i \ni \mu(B_i) > 0$.

Now

$$\int_X \max h(d, x)\, \mathrm{d}\mu(x) = \int_{B_i} \max h(d, x)\, \mathrm{d}\mu(x) + \int_{B_i^c} \max h(d, x)\, \mathrm{d}\mu(x),$$

$$\int_{B_i} \max h(d, x)\, \mathrm{d}\mu(x) \geqq [\mu(B_i)/(i+1)] \int_{B_i} h(d^*, x)\, \mathrm{d}\mu(x)$$

$$> \int_{B_i} h(d^*, x)\, \mathrm{d}\mu(x),$$

$$\int_{B_i^c} \max h(d, x)\, \mathrm{d}\mu(x) \geqq \int_{B_i^c} h(d^*, x)\, \mathrm{d}\mu(x).$$

Thus

$$\int_X \max h(d, x)\, \mathrm{d}\mu(x) > \int_X h(d^*, x)\, \mathrm{d}\mu(x) \quad \text{so} \quad v(e) > 0. \qquad \blacksquare$$

Theorem 3 says that an experiment has strictly positive value if and only if the set of outcomes for which another action is strictly better than what you would have done without the experiment has positive measure. Notice that if there is only one decision available, every experiment has zero value.

To link the decision-making apparatus with the notion of an identified function, I make the following assumption: (A1) there is a function $U^*: D \times Y \to \mathbb{R}$ such that

$$U(d, \theta) = U^*(d, f(\theta)) \quad \text{identically in} \quad d \text{ and } \theta.$$

This assumption means that your utility function depends on θ only through $f(\theta)$. Were this not the case, f should be expanded to include all the aspects of Θ you care about. Such an assumption is not restrictive in that $\mathscr{S}$ can be taken to be Θ and U^* then taken to be U. However, making assumption A1 mitigates the severity of assumption A2 below.

Above I noted that if only one decision is available every experiment has zero value. The same is true if one decision dominates all others, regardless of which θ obtains. Thus in order to have valuable experiments an assumption is needed to insure enough "effective" decisions are available. Below is such an assumption:

A2: $$\forall s \in \mathscr{S}\ \exists d(s) \in D \ni U^*(d(s), s) > U^*(d'(s), s)\ \forall d'(s) \neq d(s)$$

$$\text{and} \quad d(s) \neq d(s') \quad \text{if} \quad s \neq s'.$$

Assumption A2 says that for each range-value s of the function f, there is a strictly best decision $d(s)$ different from the best decision for any s' different from s. Such an assumption is true in many estimation problems. Again notice that if $\mathscr{S}$ is a "small" space, assumption A2 is not as strong as it would be in a "large" space.

Theorem 4. (characterization of identified functions in terms of valuable experiments).

Under assumptions A1 *and* A2, *f is identified if and only if, for each distinguished pair of points* θ_1 and θ_2, *there is a prior on* θ_1 *and* θ_2 *such that* $v(e) > 0$.

Proof. If f is not identified, by theorem 2 if θ_1 and θ_2 are distinguished there is a prior on θ_1 and θ_2 for which the experiment is uninformative, and hence is valueless. (This part of the proof does not depend on assumptions A1 and A2.)

Suppose now that f is identified. Choose θ_1 and θ_2 so that $s_1 = f(\theta_1) \neq s_2 = f(\theta_2)$. Let $d_i = d(\theta_i)$, $i = 1, 2$. By assumption A2, $d_1 \neq d_2$.

Choose any number $\tilde{p}$ such that $0 < \tilde{p} < 1$. Consider the prior p^* putting $\tilde{p}$ on θ_2 and $1-\tilde{p}$ on θ_1. Let $\tilde{d}$ be an optimal action for the prior p^*. Now since $d_1 \neq d_2$, $\tilde{d}$ must be different from at least one of d_1 and d_2. Suppose, without loss of generality that $\tilde{d} \neq d_1$.

Let $\tilde{P} = \{p | \tilde{d}$ is an optimal action if prior probability p is on θ_2 and $1-p$ is on $\theta_1\}$. Note $\tilde{p} \in \tilde{P}$ by construction.

Lemma 1. *P is convex.*

Proof of lemma 1. Suppose to the contrary that $\tilde{d}$ is an optimal action for p_1 and p_2, but not for p_3 where $p_1 < p_3 < p_2$. Then there is some $d_3 \neq \tilde{d}$ strictly better than $\tilde{d}$ at p_3.

Let $R(p, d) = (1-p)U^*(d, s_1)+pU^*(d, s_2)$. By the above assumptions,

$$R(p_1, \tilde{d}) \geqq R(p_1, d_3)$$
$$R(p_3, \tilde{d}) < R(p_3, d_3)$$
$$R(p_2, \tilde{d}) \geqq R(p_2, d_3).$$

But

$$R(p_3, d_3)-R(p_3, \tilde{d}) = \left(\frac{p_2-p_3}{p_2-p_1}\right)[R(p_1, d_3)-R(p_1, \tilde{d})] + \left(\frac{p_3-p_2}{p_2-p_1}\right)[R(p_2,d_3)-R(p_1, \tilde{d})] \leqq 0.$$

Contradiction. Therefore $\tilde{P}$ is convex. Q.E.D. lemma 1.

Therefore $\tilde{P}$ is an interval or $\tilde{P} = \{\tilde{p}\}$. In either case, let $p^* = \text{glb}\,\tilde{P}$.

Lemma 2. $p^* > 0$.

Proof.
$$R(\tilde{p}, \tilde{d}) \geqq R(\tilde{p}, d_1)$$
$$R(0, \tilde{d}) < R(0, d_1)$$
$$R(p, \tilde{d})-R(p,d_1)= \left(\frac{\tilde{p}-p}{\tilde{p}}\right)[R(0, \tilde{d})-R(0, d_1)] + \left(\frac{p}{\tilde{p}}\right)[R(0, \tilde{d})-R(\tilde{p}, d_1)] < 0,$$

if

$$p < \frac{\tilde{p}[R(0, d_1) - R(0, \tilde{d})]}{[R(\tilde{p}, \tilde{d}) - R(\tilde{p}, d_1)] + [R(0, d_1) - R(0, \tilde{d})]}.$$

Thus for such p's, d_1 is definitely better than $\tilde{d}$, so

$$p^* \geqq \frac{\tilde{p}[R(0, d_1) - R(0, \tilde{d})]}{[R(\tilde{p}, \tilde{d}) - R(\tilde{p}, d_1)] + [R(0, d_1) - R(0, \tilde{d})]} > 0.$$

Q.E.D. lemma 2.

Lemma 3. *There is some set $C^* \in \mathscr{C}$ with $\mu(C^*) > 0$ and some $\varepsilon' > 0$ such that*

$$f(x) = \frac{l(x|\theta_1)}{l(x|\theta_2)} > 1 + \varepsilon' \quad \forall x \in C^*.$$

Proof of lemma 3. By identification, there is a set $C \in \mathscr{C}$ such that $P_{\theta_1}(C) \neq P_{\theta_2}(C)$. Without loss of generality, take C or its complement so that

$$0 < P_{\theta_1}(C) - P_{\theta_2}(C) = \int_C [l(x|\theta_1) - l(x|\theta_2)] \, \mathrm{d}\mu(x).$$

If $\mu\{x|l(x|\theta_2) = 0\} > 0$ then the lemma is true for any value of ε'. So suppose $\mu\{x|l(x|\theta_2) = 0\} = 0$, and take $C' = C - \{x|l(x|\theta_2) = 0\}$. Then

$$\begin{aligned} 0 &< \int_C [l(x|\theta_1) - l(x|\theta_2)] \, \mathrm{d}\mu(x) \\ &= \int_{C'} [l(x|\theta_1) - l(x|\theta_2)] \, \mathrm{d}\mu(x) \\ &= \int_{C'} \frac{l(x|\theta_1) - l(x|\theta_2)}{l(x|\theta_2)} \, l(x|\theta_2) \, \mathrm{d}\mu(x) \\ &= \int_{C'} (r(x) - 1) l(x|\theta_2) \, \mathrm{d}\mu(x). \end{aligned}$$

Take

$$C_i = \{x|1/i > r(x) - 1 > 1/(i+1)\} \cap C'$$

$$C_0 = \{x|r(x) - 1 \geqq 1\} \cap C'$$

$$\mu(\bigcup_{i=0}^{\infty} C_i) > 0, \quad \text{so} \quad \exists_i \ni \mu(C_i) > 0.$$

If $i \geqq 1$ take $\varepsilon' = 1/i+1$ and $C^* = C_i$. If $i = 0$ take $\varepsilon' = 1$ and $C^* = C_0$. Q.E.D. lemma 3.

Let $p^* = \text{lub}\,\tilde{p}$, so $p^* \geqq p_*$. Let

$$\varepsilon = \min\left\{\varepsilon', \frac{p^*-p_*}{p_*(1-p^*)}\right\}.$$

Define the prior p^{**} as follows: p^{**} puts weight

$$p = \frac{p_*+\varepsilon p_*}{1+\varepsilon p_*}$$

on θ_2, and $1-p$ on θ_1. Clearly $0 < p < 1$

$$p-p_* = \frac{\varepsilon p_*(1-p_*)}{1+\varepsilon p_*} > 0$$

if $\varepsilon(1-p_*) > 0$, i.e., if $\tilde{p}$ is an interval, and $p-p_* = 0$ if $\tilde{P} = \{\tilde{p}\}$

$$p^*-p = \frac{(p^*-p_*)+\varepsilon P_*(p^*-1)}{1+\varepsilon p_*} \geqq 0.$$

Thus, whether $\tilde{P}$ is an interval or a single point, $\tilde{d}$ is optimal for the prior p^{**}.

However, on C^*, the posterior on θ_2 is

$$\frac{pl(x|\theta_2)}{pl(x|\theta_2)+(1-p)l(x|\theta_1)} = \frac{1}{1+\dfrac{1-p}{p}\,r(x)} < \frac{1}{1+\dfrac{1-p}{p}\,(1+\varepsilon')}$$

$$= \frac{1}{1+\left(\dfrac{1-p_*-\varepsilon p_*}{p_*+\varepsilon p_*}\right)(1+\varepsilon')} = \frac{p_*+\varepsilon p_*}{1+\varepsilon'} \leqq p_*,$$

so $\tilde{d}$ is definitely not optimal on C^*. Application of theorem 3 now proves theorem 4. Q.E.D. theorem 4.

8. *A general Bayesian treatment*

The results of sections 6 and 7 characterize identified functions in terms of the existence or nonexistence of two-point priors with certain properties. While of some theoretical interest, two-point priors are not the general case. Hence the question rises of whether a general treatment of a possibly unidentified model can be given.

Such a treatment can be given, using the framework developed in section 3. However, understanding the result is difficult without added explanation. Recall the general framework of section 3: measurable real-valued function f on $(\Theta, \mathscr{A})$ can be decomposed, if it is identified, into two parts: the mapping from T: $(\Theta, \mathscr{A}) \to (\mathscr{T}, \mathscr{B})$ which is fixed, and the mapping g: $(\mathscr{T}, \mathscr{B}) \to \mathbb{R}$ which depends on f. Thus it would seem reasonable to hope that the whole Bayesian analysis can be done in $(\mathscr{T}, \mathscr{B})$ space. That this can be done is the import of the next theorem.

Theorem 5. *Let $A \in \mathscr{A}$. Then the posterior probability of A can be computed as an integral in $(\mathscr{T}, \mathscr{B})$ space using the formula*

$$P(A|x) = E \frac{l^*(T^*(\theta)|x) Pr\{T^{-1}(T^*) \in A\}}{p(x)}.$$

Proof.

$$\begin{aligned} P(A|x) &= \int_A p(\theta|x)\, \mathrm{d}\, v(\theta) \\ &= \underset{v}{E} I_A(\theta) p(\theta|x) \\ &= E I_A(\theta) \left\{ \frac{l(\theta|x) p(\theta)}{p(x)} \right\} \\ &= E \mathscr{E} \left\{ \frac{I_A(\theta) p(\theta) l(\theta|x)}{p(x)} \middle| T = t^* \right\} \end{aligned}$$

Above

$$I_A(\theta) = \begin{cases} 1 & \text{if} \quad \theta \in A \\ 0 & \text{if} \quad \theta \notin A \end{cases}.$$

Now $l(\theta|x)$ depends on θ only through T so that the likelihood function may be rewritten

$$l(\theta|x) = l^*(T^*(\theta)|x).$$

Then

$$P(A|x) = E \left\{ \frac{l^*(T^*(\theta)(x)}{p(x)} \mathscr{E}[I_A(\theta) p(\theta)|T = T^*] \right\}.$$

Now the inner expectation is the prior probability of A, conditional on $T = T^*$. That is

$$\mathscr{E}\{I_A(\theta) p(\theta)|T = T^*\} = Pr\{\theta : \theta \in A, T(\theta) = T^*\}.$$

This completes the proof of theorem 5.

Theorem 5 says that to know the posterior probability of any set, its prior in $\mathcal{T}$ must be computed and the likelihood function used to compute its posterior in $\mathcal{T}$. For example, in the case of simultaneous equations, any set of structural parameters is mapped into some reduced form set, any prior on the structure implying some prior on the reduced form. Theorem 5 says that because the map from structure to reduced form is identifying, the entire Bayesian analysis can be done on the reduced form. Seen in this light, theorem 5 is a generalization of the theorem of Drèze [15, p. 9].

References

[1] M. DeGroot, Optimal statistical decisions. New York, McGraw-Hill, 1970.

[15] J. Drèze, Econometrics and decision theory. Econometrica 40 (1972), 1–17. (Reprinted in the present volume.)

[5] F. Fisher, The identification problem in econometrics. New York, McGraw-Hill, 1966.

[10] P. R. Halmos and L. J. Savage, Application of the Radon-Nikodym theorem to the theory of sufficient statistics. Ann. Math. Stat. 20 (1949), 225–241.

[8] J. Kadane, Identification of functions of parameters. Unpublished.

[4] T. C. Koopmans and O. Riersol, The Identification of structural characteristics. Ann. Math. Stat. 21 (1950), 165–181.

[11] T. C. Koopmans, Rubin and Leipnik, Measuring the equation systems of dynamic economics, in: Statistical inference in dynamic economic models, edited by T. C. Koopmans. New York, John Wiley, 1950, pp. 53–237. Cowles Commission Monograph no. 10.

[9] E. Lehmann, Testing statistical hypothesis. New York, John Wiley, 1959.

[2] D. V. Lindley, Bayesian statistics: a review. Philadelphia, SIAM, 1971.

[13] D. V. Lindley and G. M. El-Sayyad, The Bayesian estimation of a linear functional relationship. JRSSB 30 (1968), 190–202.

[7] J. A. Morales, Bayesian full information structural analysis. Berlin, Springer-Verlag, 1971.

[16] H. Raiffa and R. Schlaifer, Applied statistical decision theory. Cambridge, M.I.T. Press, 1961.

[12] O. Riersol, Identifiability of a linear relation between variables which are subject to error. Econometrica 18 (1950), 375–389.

[6] T. Rothenberg, The value of structural information: a Bayesian approach. Unpublished.

[14] L. J. Savage, Elicitation of personal probabilities and expectations. Journal of the American Statistical Association 66 (1971), 783–801. (Reprinted in the present volume.)

[3] A. Zellner, An Introduction to Bayesian inference in econometrics. New York, John Wiley, 1971.

PART 6

APPLICATIONS OF BAYESIAN REGRESSION

S. E. Fienberg and A. Zellner (eds.)
Studies in Bayesian Econometrics and Statistics.

6.1 A BAYESIAN APPROACH TO REAL ESTATE ASSESSMENT*

Hal R. VARIAN**

University of California at Berkeley

Real estate assessors are required by law to estimate current market value on all taxable properties in their districts each year. For even moderate-sized districts this is a huge task. In recent years, this burden has been somewhat alleviated by the application of regression techniques to the appraisal process [1–4]. The assessor's office of San Mateo, California, for example, routinely assesses single-family homes by applying a formula estimated from data on previous sales using ordinary least squares procedures (OLSQ).

The purpose of this study is to investigate this problem from the Bayesian viewpoint, emphasizing the ways in which loss minimizing estimators and prior information can lead to more appropriate methods of mechanized appraisal.

1. The loss function approach

In this study, we will be concerned entirely with appraisal of single-family homes. In this case, the goal of the assessor is to estimate the current market value of a house, given the values of certain characteristics of the house such as total living area, number of bedrooms, etc. If the appraiser's estimate is denoted by y_e and the actual value is y_a, certain losses are incurred, denoted by $L(y_e, y_a)$.

* Research performed under the auspices of the Center for Real Estate and Urban Economics, University of California, Berkeley, California. An earlier version of this paper was presented to the Western Economic Association meeting, Santa Clara, California, August 1972.

** I would like to thank Arnold Zellner for his kind help and encouragement in writing this paper. Helpful suggestions were also made by Darius Gaskins. Of course any errors remain the sole responsibility of the author.

In general, these losses could arise from various causes, for example private costs versus social costs. However, for this study, we will consider only losses accruing to the assessor's office. If the assessor's office underestimates the value of a house, the loss is equal to the amount of the underestimate. If the office overestimates the value of the house, the California homeowner has two recourses: he can (1) complain to the assessor's office, confer with them and attempt to convince them that his home was overassessed; or (2) present an appeal to the Assessment Board of Equalization, which evaluates evidence presented by both sides as to the correct assessment. (Other states have similar appeal processes.)

Both of these are lengthy procedures which incur considerable expense for both parties. If the assessor's office loses the case, it not only loses the amount of the overassessment but also the costs involved in the appeal procedure.

If the case is won, the county must still pay the court costs involved. In cases involving substantial assessment error, the losses may amount to a considerable fraction of the tax. Therefore, the assessors would want to choose an estimate for home value that would minimize the total expected loss.

It is well known that if the loss function is quadratic, that is

$$L(y_e, y_a) = a(y_e - y_a)^2,$$

the loss minimizing estimate is just the expected value of y_a. However, in the situation under consideration, a quadratic loss function seems inappropriate in that it assigns the same loss to overestimates as to equal underestimates.

But by virtue of the above considerations it is clear that the assessor's loss function should be asymmetric. The following features may be noted:

(1) The loss function should be linear for large negative errors.

(2) The loss function should be increasing for positive errors at a greater than linear rate.

(3) Even if court costs are constant, the probability of a complaint being submitted increases with the magnitude of the overestimate, and thus the loss function should increase monotonically for positive errors.

For the purposes of exposition I have chosen the following asymmetric loss function:

$$L(y_e, y_a) = b \exp\,[a(y_e - y_a)] - c(y_e - y_a) - b. \qquad (1)$$

This function, which I call the *linex* loss function (for linear-exponential), has many appealing properties: (1) for $c = 1$, $b = a^{-1}$ it has a minimum loss of 0 for $y_e = y_a$; (2) it has exponential losses for large overestimates and linear losses for large underestimates; (3) parameters can be chosen to provide a variety of asymmetric effects; and (4) it combines readily with a normal probability distribution. The shape of L is depicted in fig. 1 for various values of the parameters a and b.

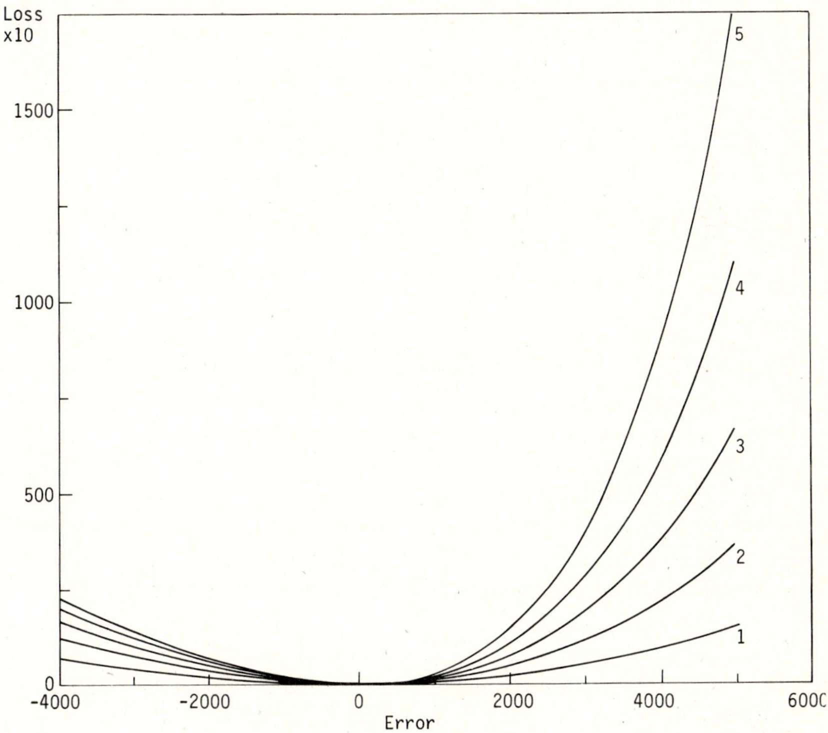

Fig. 1. Plot of loss:

	a	b	c
1	0.0001	10,000	1.0
2	0.0002	5,000	1.0
3	0.0003	3,333	1.0
4	0.0004	2,500	1.0
5	0.0005	2,000	1.0

The estimation of the correct parameters for the assessor's loss function would require considerable information about the probability of a complaint arising, the probability of a complaint winning a reversal, and the magnitude of the court costs involved. However, it can be seen that $a = 0.0004$, $b = 2500$, $c = 1$ gives values that are not unreasonable and will be adequate for purposes of this study.

Given the loss function L, and a probability distribution $p(y_a)$ on y_a, expected losses can be expressed by the following:

$$E[L] = \int_{-\infty}^{\infty} L(y_e, y_a) p(y_a) \, \mathrm{d}y_a. \tag{2}$$

Solving this integral gives us expected loss as a function of y_e:

$$E(L) = f(y_e). \tag{3}$$

Solving (3) for the value of y_e that minimizes expected loss gives us our loss minimizing estimate.

2. *The particular model*

It is commonly assumed that the selling price of a house is a linear function of the characteristics of that house plus an error term. Given a sample of previous sales of houses and their characteristics, we can formulate the following regression model:

$$y = X\beta + \varepsilon \tag{4}$$

where y is a m vector of selling prices, X is a k by m matrix of observed characteristics, β is a k vector of parameters, and ε is an m vector of error terms, assumed to be distributed $N(0, \sigma^2 I)$.

Under these assumptions, if we have diffuse prior information about the value of β and σ, Bayesian methods can be used to show that, given a new observation of a house with unobserved value y_a and characteristics x_a, y_a has a t-distribution with mean $x_a \beta_d$ and variance

$$v_d = \frac{\nu s_d^2}{\nu - 2} [1 + x_a (X'X)^{-1} x_a'],$$

where β_d is the OLSQ estimate of β, s_d^2 is the OLSQ estimate of σ^2 and $\nu = n - k$ [5, pp. 73–74]. (This result is similar to the classical OLSQ result concerning the predictive distribution [6, p. 154].)

With large samples of previous sales, the t-distribution can be well approximated by a normal distribution of mean $x_a \beta_d$ and variance v_d.

Under this assumption of normality and the particular loss function we have chosen, we can derive the loss minimizing estimator of y_a, which is done in the Appendix.

The result is:

$$y_e = x_a\beta_d - \frac{av_d}{2} + a^{-1}\ln\frac{c}{ba}. \tag{5}$$

This estimate has an appealing intuitive interpretation. As can be seen in fig. 1 small values of a correspond to a small degree of asymmetry. Thus for no asymmetry, y_e coincides with the usual estimate; as the degree of asymmetry rises, positive errors will incur higher and higher losses, causing the OLSQ estimator to be biased downwards in compensation. Also, if our regression is highly accurate, s_d^2 will be small; as the accuracy of the regression falls, there is a higher probability of large positive errors, again causing the OLSQ estimate to be compensated downwards. Finally, if a house has "average" characteristics, the quadratic term $x_a(X'X)^{-1}x_a'$ will be small and the correction factor will be low; for "unusual" houses, the equation "gives them the benefit of the doubt" and applies a large correction.

The positive term, on the other hand, tends to counteract the negative correction factor as the ratio c/b rises. This ratio indicates the relative magnitudes of the losses from over and underestimates.

3. *The data*

The data for this study consist of 168 observed sales of single-family homes in a middle-sized California town in 1965. In a previous study using these data, Jancsek [3] found the following nine variables to be good predicators of selling price: (1) lot width, (2) view (1 = average or above average view, 0 = below average view), (3) age, (4) number of bedrooms, (5) quality class (see [7]), (6) total living area, (7) flatwork cost (patios, driveways, etc.), (8) fence cost, (9) a constant term.

The sample was split into two groups: the first, consisting of 125 observations, was used in estimation and the second, consisting of 43 observations, was used for projection and comparison purposes.

4. *Regression results*

The regression coefficients, standard errors, and standard error of the regression are summarized in table 1, column 1. The R^2 for this

TABLE 1
Estimation results

Variable	1 OLSQ	2 Non-data-based prior	3 Non-data-based Bayesian reg.	4 Data-based prior	5 Data-based Bayesian reg.
Lot width	33.11 (20.30)	50.00	41.55 (31.18)	98.08 (31.16)	53.88 (15.97)
View	1879.27 (433.21)	2000.00	1939.64 (665.25)	2405.24 (828.87)	2109.04 (379.04)
Effective years	−175.34 (27.11)	−200.00	−187.67 (41.64)	−139.22 (48.21)	−173.50 (23.02)
Number of bedrooms	−722.48 (32.35)	1000.00	138.76 (496.74)	−12.04 (741.66)	−608.90 (289.63)
Quality class	2000.54 (33.61)	2000.00	2002.73 (516.17)	2454.36 (626.19)	2119.04 (285.23)
Total living area	7.95 (0.78)	10.00	8.98 (1.20)	6.12 (1.56)	7.33 (0.67)
Flatwork cost	5.05 (1.29)	5.00	5.03 (1.98)	4.52 (3.50)	4.48 (1.18)
Fence cost	3.89 (1.80)	2.00	2.94 (2.80)	−0.31 (5.78)	3.51 (1.72)
Constant	1356.11 (1840.49)	1000.00	1178.05 (2826.29)	−4413.44 (4016.91)	161.53 (1641.28)
Standard error	1827.16		3993.47	2081.64	1889.34

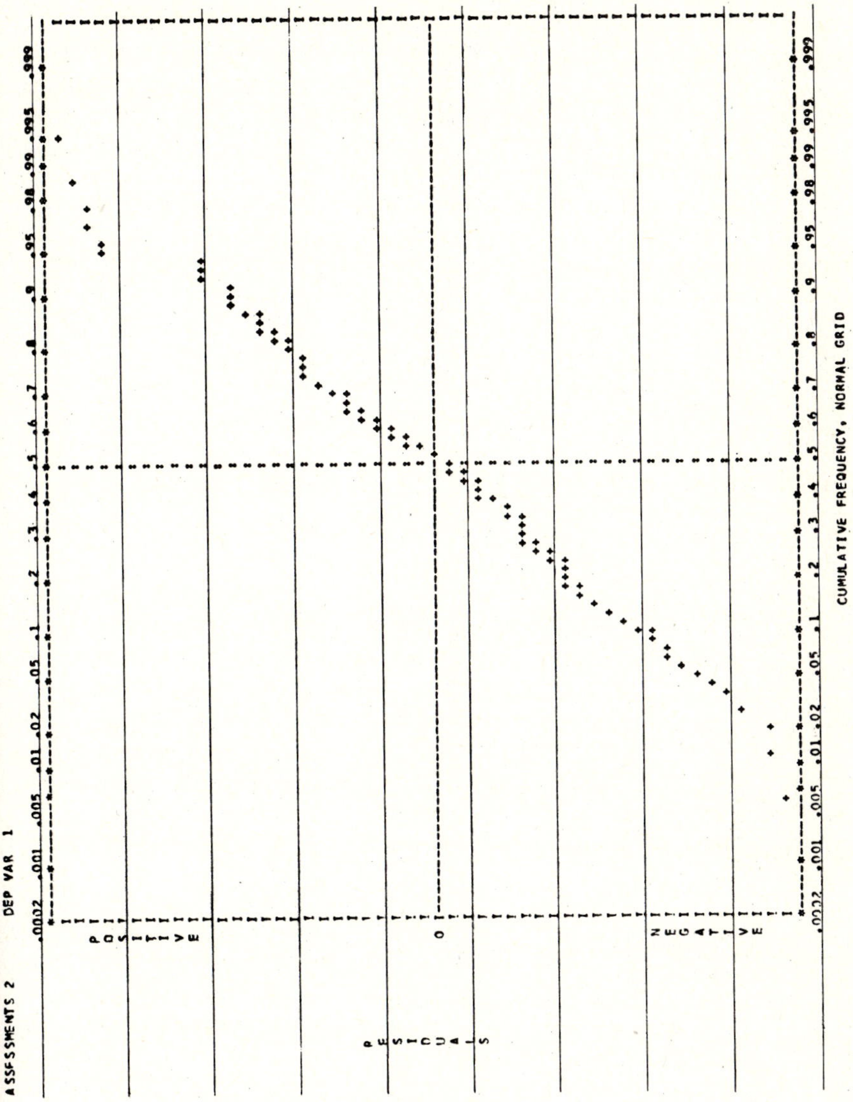

Fig. 2. Cumulative distribution of residuals.

regression was 0.85 which would indicate that the equation would have reasonable predictive ability.

Since we based our derivation of the loss minimizing estimate on the assumption of normality of the errors, it is important to examine the residual behavior to verify this assumption. Fig. 2 is a cumulative normal plot of the residuals; in this plot normally distributed residuals would give rise to a straight line.

Since the scatter of calculated residuals is approximately linear, we can assume the errors to be normally distributed. The comparative results concerning loss minimization will be deferred till a later section.

5. *A prior probability distribution*

When preparing to estimate the model represented by (4), we often have prior information involving the parameter β. For example, such prior information could be of the following types:

(1) Appraisal rules of thumb.—Most appraisers have developed ideas about the relative importance of various factors in determining home value. In our model, for example, we would expect total living area to be more important than fence cost, and the coefficient for total living area ought to be positive. If pressed, an appraiser might even be able to quantify his subjective beliefs about such factors, in so far as he could make statements such as: "I believe that an extra square foot of floor space would, on the average, be worth twice as much as an extra dollar spent on fence costs. However, my estimate is rather imprecise." Suitably formulated, such statements can be used to construct a subjective prior probability distribution on the parameters.

(2) Previous experience.—If similar models have been successfully applied in other communities, their estimations can be used to construct prior probability distributions. Thus the estimates of β for community A can serve as a general guideline for community B while the actual data for community B can serve to pinpoint the estimates more precisely.

(3) Past estimates.—Market conditions continually change and hence it is important to continually revise estimates of β by incorporating new data. However, since market conditions may change relatively slowly, we may believe that the β to be estimated for 1972 will probably be similar to the β estimated in 1971. We can incorporate this belief

into our estimation procedure by using the β estimated in 1971 in the prior probability distribution when estimating β for 1972.

Assuming then, that we do have a prior probability distribution of β representing our beliefs, how do we incorporate it into our analysis?

If $p(\beta, X, y)$ is the joint probability density function for our parameter β and the random variables X and y, we can factor it in two ways:

$$p(\beta, X, y) = p(\beta|X, y)\ p(X, y) = p(X, y|\beta)\ p(\beta).$$

Thus,

$$p(\beta|X, y) \propto p(\beta)\ p(X, y|\beta).$$

The posterior probability density function of β given the data (X, y) is proportional to the product of the prior probability density of β and the likelihood function. This is of course Bayes' formula. Since the integral of $p(\beta|X, y)$ must be 1, this proportionality completely determines the posterior probability distribution.

For the purposes of our analysis we will assume that (β, σ^2) has a Normal-gamma prior probability distribution. That is:

$$p(\beta|\sigma) \propto \frac{1}{\sigma^k} \exp\left[-\frac{1}{2\sigma^2}(\beta-b)'N(\beta-b)\right]$$

$$p(\sigma) \quad \propto \frac{1}{\sigma^{v+1}} \exp\left[-\frac{vs^2}{2\sigma^2}\right] \qquad v = n-p, \quad p = \text{rank}\ (N).$$

Here the parameters of these prior distributions are just (b, N) and (s^2, v). Further, let $p = \text{rank}\ (N)$.

Suppose we are given data (X, y), where X is $k \times n$ and y is $1 \times n$. Then define:

$$\begin{aligned} b_d &= (X'X)^{-1}X'y \\ N_d &= X'X \\ p_d &= \text{rank}\ (N_d) \\ v_d &= n-p_d \\ s_d^2 &= \frac{1}{v_d}(y-Xb_d)'(y-Xb_d). \end{aligned}$$

It is well known (see for example [8, p. 343] or [5, pp. 70–72]) that in this case, the posterior probability distribution is again Normal-gamma with parameters:

$$
\begin{aligned}
N_p &= N_d + N \\
b_p &= N_p^{-1}(N_d b_d + Nb) \\
p_p &= \text{rank } (N_p) \\
v_p &= p_d + v + p - p_p + v_d \\
s_p^2 &= \frac{1}{v_p}\,[(v_d p_d + b_d' N_d b_d) + (y'y - b_p' N_p b_p)].
\end{aligned}
$$

Since we have a normal distribution on β, we will again have a t-distribution on y_a which, exactly as before, we will approximate as a normal distribution. If the loss function is of the linex form, the derivation of the loss minimizing estimate goes through exactly as before yielding:

$$y_e = x_d b_p - \frac{av_p}{2} + a^{-1}\ln \frac{c}{ba}.$$

6. *Construction of the prior and results*

For the purpose of this exposition, I experimented with two prior distributions. One was a data-based prior using the aforementioned 43 cases to represent another "prior" observation of the sample; the other was non-data-based, relying on an intuitive notion of relative importance of the variables in determining the value of a house.

7. *The non-data-based prior*

Examining table 1, column 1, the coefficients appear reasonable, with the exception of the coefficient on bedrooms which is negative, rather than having the positive effect one would expect. This situation arises frequently in this type of analysis; for example, Hinshaw cites an equation with negative signs on the number of bathrooms, fence cost, air conditioning, etc. These effects are probably due to mis-specification of the model; for example, (1) the linear form may be incorrect, (2) there may be left out variables, especially concerning neighborhood effects, and (3) there is often considerable multi-colinearity in the right-hand variables. However, if we want to use the linear functional form and are restricted to this set of data, this negative sign is irrelevant for prediction.

For the purposes of this example, we will deliberately "mis-specify" the prior so as to examine the effect of this on the loss minimizing estimator. We then ask the following question: "What is the best loss minimizing estimate of β given that β has the 'right' signs?"

I chose 1000 as the prior mean for bedrooms in table 1, column 3. I let $s_p = 2460$, and $N = N_d$, the cross-product matrix from the data. The resulting posterior means of β are summarized in table 1, column 4. Notice that the posterior mean for the bedroom coefficient is indeed positive; however, this "right" sign has resulted in a rather large increase in the standard error of the regression. Thus the predictive ability of this regression is undoubtedly poorer than OLSQ.

8. *The data-based prior*

Assume that the 43 cases used for testing purposes in fact come from another sample—perhaps another community, or last year's sample. We can run a regression on these variables which influences our beliefs about the significance of the various parameters. When we encounter the new sample of 125 houses, we wish to use our previous estimations as a prior.

These results are summarized in table 1, columns 4 and 5.

Here the standard error has risen only slightly, while most of the coefficients remain similar, with the exception of the constant. It appears that the posterior distribution may be an improvement over the prior.

9. *Comparative results and conclusions*

Table 2 presents the average linex loss incurred and standard deviation of the loss when the estimation methods discussed are used to assess the aforementioned sample of 43 houses. The linex estimator with data-based prior gives the lowest average loss. One would expect this, since it completely utilizes the information in the entire 168 observations and applies the optimal correction factor. The next best is the diffuse linex estimate (this can also be regarded as an OLSQ regression combined with the correction factor). The three "uncorrected" regressions all give higher average losses than the loss minimizing estimators.

An interesting effect occurs with the non-data-based prior. Since this prior was deliberately misspecified, the Bayesian regression does quite poorly with average loss 13 times worse than with the best method.

However, when the loss minimizing correction term is applied, the estimate's performance improves substantially.

The example worked through in this paper illustrates the feasibility of Bayesian and decision theoretic assessment. The average loss resulting from the loss minimizing Bayesian regression is only 72% of the loss resulting from ordinary least squares. In monetary terms, this could result in a substantial amount of savings.

The next step would be to repeat the type of work done here on a larger scale with an eye towards more realism in the parameter choices. For example: the actual loss function appropriate for the assessor's office could be estimated; other sets of independent variables could be used; and one could experiment with other functional forms.

TABLE 2

Loss minimization results

Method			
Prior	Correction	Average linex loss	Standard deviation
(1) Data-based	Linex	718.0	1367.0
(2) Diffuse (OLSQ)	Linex	790.0	1578.0
(3) Data-based	Quadratic	901.0	1922.0
(4) Diffuse (OLSQ)	Quadratic	999.0	2161.0
(5) Non-data-based	Linex	1278.0	2822.0
(6) Non-data-based	Quadratic	9351.0	14965.0

The estimate is of the form $x_a\beta+$ correction. If correction $=0$ this estimate minimizes quadratic loss, if correction $=-av_a/2$ the estimate minimizes linex loss. The estimate of β depends on the type of prior used. The loss function was:

$$L(x) = 2500 \exp(0.0004x) - x - 2500$$

where x is the error in the estimation; i.e., $x = y_e - y_a$.

Appendix. Derivation of the loss minimizing estimate

We assume the distribution of y_a to be $N(u, v)$. Letting x denote the error, $x = y_e - y_a$, we have a loss function of the form:

$$L(x) = b \exp[ax] - cx - b.$$

The distribution of x is easily seen to be $N(y_e - u, v)$. Letting $d = y_e - u$ and dropping the constant we can express expected loss by:

$$E[L] = \frac{b}{\sqrt{2\pi v}} \int_{-\infty}^{\infty} \exp\left[ax - \frac{(x-d)^2}{2v}\right] dx - \frac{c}{\sqrt{2\pi v}}$$

$$\times \int_{-\infty}^{\infty} x \exp\left[-\frac{(x-d)^2}{2v}\right] dx. \tag{1}$$

The second integral is simply the mean of the distribution, namely d. As for the first we note that:

$$ax - \frac{(x-d)^2}{2v} = \frac{2vax - (x^2 - 2xd + d^2)}{2v}$$

$$= \frac{-x^2 + (2d + 2va)\,x - d^2}{2v}.$$

Completing the square, we have:

$$ax - \frac{(x-d)^2}{2v} = \frac{-[x-(d+va)]^2 + (d+va)^2 - d^2}{2v}$$

$$= \frac{-[x-(d+va)]^2 + 2vad + v^2a^2}{2v}$$

$$= \frac{-[x-(d+va)]^2}{2v} + a\left(d + \frac{va}{2}\right).$$

Substituting this into (1), simplifying and bringing the constant term outside the integral we have:

$$E(L) = \frac{b}{\sqrt{2\pi v}} \exp\left[a\left(d + \frac{va}{2}\right)\right]$$

$$\times \int_{-\infty}^{\infty} \exp\left[-\frac{(x-(d+va))^2}{2v}\right] dx - cd.$$

The integral conveniently cancels out, leaving:

$$E(L) = b \exp\left[a\left(d + \frac{va}{2}\right)\right] - cd. \tag{2}$$

Taking the derivative of (2) with respect to y_e and equating to 0 gives:

$$ba \exp\left[a\left(d+\frac{va}{2}\right)\right]-c = 0$$

$$\exp\left[a\left(d+\frac{va}{2}\right)\right] = \frac{c}{ba}$$

$$a\left(d+\frac{va}{2}\right) = \ln\frac{c}{ba}$$

$$d = a^{-1}\ln\frac{c}{ba}-\frac{va}{2}.$$

Substituting for d and rearranging:

$$y_e = u+a^{-1}\ln\frac{c}{ba}-\frac{va}{2}.$$

In the regression model, u is simply $x_a\beta_d$ and v is v_d, so the optimal estimator is:

$$y_e = x_a\beta_d+a^{-1}\ln\frac{c}{ba}-\frac{v_d a}{2}.$$

References

[1] Robert Gustafson, EDP and the appraiser. Delivered at the 35th Annual Conference of the National Association of Tax Administrators, San Francisco, June 14, 1967.

[2] Andrew J. Hinshaw, The assessor and computerization of data, Assessor's Journal 3, No. 1 (April 1968).

[3] Joel Jancsek, Property tax assessment applications of multiple regression analysis in California. Technical Report No. 4, University of California, Center for Real Estate, 1972.

[4] William C. Pendleton, Statistical inference in appraisal and assessment procedures. The Appraisal Journal (January 1965).

[5] Arnold Zellner, An introduction to Bayesian inference in econometrics. New York, John Wiley, 1971.

[6] J. Johnston, Econometric methods, 2nd Ed. New York, McGraw-Hill, 1972.

[7] Residential building costs. State Board of Equalization, Sacramento, California, 1968.

[8] H. Raiffa and R. Schlaifer, Applied statistical decision theory. Division of Research, Graduate School of Business Administration, Harvard University, 1961.

S. E. Fienberg and A. Zellner (eds.)
Studies in Bayesian Econometrics and Statistics.

6.2. MONOZYGOTIC TWINS REARED APART: A BAYESIAN ANALYSIS

Robin M. HOGARTH*

European Institute of Business Administration

1. *Introduction*

A topic which has intrigued social scientists for many decades is the extent to which mental abilities (and particularly intelligence) are determined by environmental and hereditary factors. Indeed, this issue has recently once again become hotly debated in scientific circles in view of both its considerable educational and social implications (see e.g. [1, 6, 13]). Reducing the question to statistical terms, the problem is to estimate the proportion of variance in a given trait which may be attributable to genetic and environmental forces respectively.

The degree of "heritability" of a given trait is defined by a coefficient which is the ratio of genetic variance to the sum of genetic and environmental variance. The environmental variance in the denominator of the ratio reflects the actual set of environmental conditions, and thus so does the overall ratio. Hence heritability is not a fixed immutable quantity even if the genetic causes of variation remain constant. It can only be applicable to a given population at a specific point in time. In those cases where objective measures of both genetic and environmental forces may be obtained within a controlled experimental framework (e.g., in animal breeding), the determination of the heritability of a given trait is fairly straightforward. However, in human populations such measures are not readily available, nor is it possible to design and implement appropriate experiments. Recourse is there-

* Robin M. Hogarth is Assistant Professor at INSEAD (The European Institute of Business Administration, Fontainebleau, France). He thanks the late Sir Cyril Burt for having provided the data used in the analysis. He is also much indebted to Arnold Zellner for helpful discussions and guidance as well as to Harry V. Roberts who first stimulated interest in the topic and who gave much aid and encouragement.

fore made to postulating a theoretical model linking the unobservable, causal variables, genotype and environment, to the manifestation of the trait examined (the phenotype), and then inferring the effect of the former from such measures of the latter that can be obtained.

This is usually achieved by considering the correlation coefficient for the phenotype studied between pairs of individuals of known degree of relationship. However, note that the correlation coefficient is not used here as a measure of linear relationship, but rather as a ratio of variances, where the numerator expresses the variance that the pairs have in common, and the denominator total variance. To appreciate this, consider the following case: let $\tilde{x}$, $\tilde{y}$ and $\tilde{z}$ be mutually uncorrelated random variables, and let $\tilde{x}$ and $\tilde{y}$ have equal variances. Define

$$\tilde{s} = \tilde{z}+\tilde{x} \tag{1}$$

and

$$\tilde{t} = \tilde{z}+\tilde{y}. \tag{2}$$

This may be viewed as the so-called "common elements" model (see e.g. [12, chapter XV]). Thus it is straightforward that

$$\rho_{s,t} = \frac{\mathrm{Var}(\tilde{z})}{\mathrm{Var}(\tilde{s})} = \frac{\mathrm{Var}(\tilde{z})}{\mathrm{Var}(\tilde{t})}. \tag{3}$$

Consider specifically the case of monozygotic (MZ) twins who, by definition, have identical genotypes. The correlation between pairs of such twins reflects their common variance which may be supposed to be due to the entire effect of genotype plus their shared environmental influences. Of particular interest, therefore, are those studies of MZ twins who have been reared *apart*, for in this case, the correlation coefficient is purported to measure the proportion of variance attributable to genetic effects only, i.e., heritability. Indeed, the three major studies of MZ twins reared apart reported in the literature, have all shown large correlations for IQ,[1] thus suggesting that intelligence (as measured by IQ) is a highly heritable trait.

Heritability estimates based on the comparison of correlation coefficients between the IQ's of MZ and DZ (dizygotic or fraternal) twins reared together have, however, been shown to be inconsistent with those calculated from MZ twins reared apart [10], as well as sensitive to small changes in basic assumptions [9]. In this paper, therefore, the largest sample of MZ twins reared apart reported in

[1] These are Burt, 53 pairs: 0.87 [3]; Shields, 37 pairs: 0.77 [14]; and Newman, Freeman and Holzinger, 19 pairs: 0.77 [11].

the literature [3] is examined with a view to resolving these inconsistencies. The analysis is undertaken in the second section from a Bayesian viewpoint with "vague" or "diffuse" prior information. In the third section, it is noted that the question analyzed is subject to an identification problem (not the sole worry of econometricians!). Finally, in section four, the results of the analysis are discussed.

2. *Analysis of problem*

As the correlation coefficient for a given trait between MZ twins reared apart measures the common influences to which the twins have been subject, it is important that if this statistic is to measure heritability it should not be influenced by common environmental influences. However, it may be argued that MZ twins reared apart could share common environmental influences of two kinds: first, certain *pre*natal influences due to the fact that they shared the same mother; and secondly, similar *post*natal influences. The first of these possible effects is much debated (see, e.g., [4, 15]) and is certainly most difficult to assess (however, see [9]); the second is now investigated by re-analyzing the data of Burt [3].

An interesting feature of Burt's study is that in all 53 pairs, one twin was reared in the family into which he or she was born, whereas the co-twin was reared in a foster home. Hence, there was an ordering of the twins into two groups which shall henceforth be designated as "home" and "away". When a measure of environment is available, it is always possible to correlate this measurement for the separated twins in order to determine the relationship between their postnatal environments. No measures of environment are readily available other than that of S.E.S.—socio-economic status. Therefore, the analysis will proceed with S.E.S. as a surrogate for environmental influences.

Burt has calculated the product moment correlation coefficient between S.E.S.[2] of head of household of the twins' homes to be 0.03 [5].

[2] Burt's classification of S.E.S. levels is based on degree of ability required for occupation pursued by head of household. He uses 6 classes, which he describes as follows: "Class I includes those engaged in the highest type of professional work (university teachers, those of similar standing in law, medicine, education, church, and the top people in commerce, industry or the civil service); class II consists of those engaged in lower professional or technical work (including most teachers, men of business, and executive clerks in the higher grades); class III of those working

With data of this kind, such a correlation is not necessarily very meaningful as it implies that S.E.S. ratings form a cardinal scale. This is probably an unreasonable assumption. It is therefore proposed to proceed in the following manner.

Let $\tilde{w}_1$ be the phenotype of the home twins and denote that of the away co-twins by $\tilde{w}_2$. Further let $\tilde{v}_1$ represent the effects of S.E.S. for the home twins and $\tilde{v}_2$ the effects of S.E.S. for the away twins. The relationship between $\tilde{w}_1$ and $\tilde{v}_1$ on the one hand, and between $\tilde{w}_2$ and $\tilde{v}_2$ on the other, can be represented in a regression framework, that is, let

$$\tilde{w}_1 = \beta_{10}+\beta_{11}\tilde{v}_1+\tilde{u}_1 \tag{4}$$

and

$$\tilde{w}_2 = \beta_{20}+\beta_{21}\tilde{v}_2+\tilde{u}_2 \tag{5}$$

where the β's are coefficients to be estimated, and $\tilde{u}_1$ and $\tilde{u}_2$ are the disturbances of the regressions.

It is illuminating to carry out the regression analyses of (4) and (5) for two reasons. First, if $\tilde{v}_1$ and $\tilde{v}_2$ are represented by *dummy variables* the strength of relationship between IQ and S.E.S. may be estimated in a more flexible manner than by consideration of the simple product moment correlation (see remarks above). Secondly, the correlation between $\tilde{u}_1$ and $\tilde{u}_2$ represents the correlation between the phenotypes of the twins when the effects of S.E.S. are partialled out. That is, it is possible to estimate the relationship between the phenotypes of the twins independently of the effects of their home environments as measured by S.E.S. Clearly, if this correlation ($\rho_{u_1u_2}$) is the same as that when S.E.S. is not partialled out (i.e., $\rho_{w_1w_2}$), it may be concluded (when using the crude S.E.S. measure), that postnatal environmental effects are of little importance in accounting for observed variation in the twins' IQ's. On the other hand, a difference between the correlations indicates the effects of postnatal environments. Specifically, if $\rho_{u_1u_2}$ is smaller than $\rho_{w_1w_2}$, it may be said that $\rho_{w_1w_2}$ has been inflated by common postnatal environmental effects. On the other hand, if $\rho_{u_1u_2}$ is greater than $\rho_{w_1w_2}$, then $\rho_{w_1w_2}$ has been attenuated by differences in postnatal environments.

in intermediate types of clerical, commercial, or technical work; class IV includes those ordinarily classified as skilled workers, but it also contains an appreciable number who are engaged in commercial or industrial work of an equivalent level; class V consists of semi-skilled workers and those holding the poorest type of commercial position; class VI of unskilled labourers, casual labourers, and those employed on course manual work" [2, p. 10].

It is always possible to carry out the regression analyses of equations (4) and (5), and then simply to correlate the least-squares estimates for $\tilde{u}_1$ and $\tilde{u}_2$. However, there is a more efficient procedure.

Equations (4) and (5) may be written as a single (two equation) multivariate regression model, that is,

$$\underset{(2n\times 1)}{\begin{pmatrix} \mathbf{w}_1 \\ \mathbf{w}_2 \end{pmatrix}} = \underset{(2n\times k)}{\begin{pmatrix} V_1 & 0 \\ 0 & V_2 \end{pmatrix}} \underset{(k\times 1)}{\begin{pmatrix} \boldsymbol{\beta}_1 \\ \boldsymbol{\beta}_2 \end{pmatrix}} + \underset{(2n\times 1)}{\begin{pmatrix} \mathbf{u}_1 \\ \mathbf{u}_2 \end{pmatrix}} \tag{6}$$

where

$\underset{(n\times 1)}{\mathbf{w}_i}$ = Vector of measured phenotypes, IQ, of twins in equation i ($i = 1,2$), where, as before, 1 represents the home equation, and 2 the away. (The boldface notation, e.g., $\mathbf{w}_i$ denotes a vector.)

$\underset{(n\times k_i)}{V_i}$ = Matrix of independent variables, the observed S.E.S. ratings. Note $k_1+k_2 = k$; V_1 is of rank k_1; and V_2 of rank k_2. (Capital letters, e.g. V_i, are used to denote matrices.)

$\underset{(k_i\times 1)}{\boldsymbol{\beta}_i}$ = Unobserved vector of coefficients. $\boldsymbol{\beta}_1$ and $\boldsymbol{\beta}_2$ are permitted to be different.

$\underset{(n\times 1)}{\boldsymbol{u}_i}$ = Unobserved disturbance vector.

The above model is a particular example of Zellner's "seemingly unrelated" regression model (see [18]).[3]

In general, when V_1 and V_2 are mutually orthogonal, and the components of $\mathbf{u}_1$ and $\mathbf{u}_2$ are highly correlated, the standard errors of $\boldsymbol{\beta}_1$ and $\boldsymbol{\beta}_2$ are considerably smaller than those calculated from two separate regression analyses. These conditions are suspected to be approximated in the present instance. Thus, there is much to be gained by using the "seemingly unrelated" model. Not only will smaller standard errors for $\boldsymbol{\beta}_1$ and $\boldsymbol{\beta}_2$ increase confidence in the estimates for $\mathbf{u}_1$ and $\mathbf{u}_2$, and thus for $\rho_{u_1u_2}$; but closer insight may be gained as to how the environmental measure, S.E.S., is related to the IQ's of the twins.

[3] In [18], this model is presented from a sampling theory viewpoint. The model has been much used in analyzing economic data, but to the best of my knowledge, this is its first application to a biometric problem.

For simplicity, rewrite (6) as

$$\underset{(2n\times 1)}{\mathbf{w}} = \underset{(2n\times k)}{V}\;\underset{(k\times 1)}{\boldsymbol{\beta}} + \underset{(2n\times 1)}{\mathbf{u}} \tag{7}$$

where

$$\mathbf{w}' = (\mathbf{w}_1', \mathbf{w}_2'), \quad \boldsymbol{\beta}' = (\boldsymbol{\beta}_1', \boldsymbol{\beta}_2'), \quad \mathbf{u}' = (\mathbf{u}_1', \mathbf{u}_2')$$

and

$$V = \begin{pmatrix} V_1 & 0 \\ 0 & V_2 \end{pmatrix}.$$

(Primes, e.g. $\mathbf{w}'$, are used to denote transposes, and below $\otimes$ will be used to denote Kronecker, or direct product, multiplication.) Note $\mathbf{u}$ is assumed normally distributed with mean $\mathbf{0}$ and covariance matrix $\Sigma \otimes I_n$, where

$$\Sigma = \begin{pmatrix} \sigma_{11} & \sigma_{12} \\ \sigma_{21} & \sigma_{22} \end{pmatrix},$$

and σ_{ii} is the variance of equation i, and $\sigma_{12} = \sigma_{21}$ is the covariance of the errors of the two equations. This covariance is assumed nonzero in view of the fact the twins are genetically identical.[4] Note, however, that the covariance of errors *between* different twin pairs is assumed zero.

The analysis is undertaken from a Bayesian viewpoint using a "diffuse" or "vague" prior distribution for the model's parameters. That is, the sample information will be allowed to dominate the posterior distribution.

The diffuse prior distribution used is the one employed in previous analysis of the traditional multivariate regression model,[5] namely

$$p(\boldsymbol{\beta}, \Sigma^{-1}) = p(\boldsymbol{\beta})p(\Sigma^{-1}) \propto |\Sigma^{-1}|^{-3/2}. \tag{8}$$

This implies that $\boldsymbol{\beta}$ and Σ^{-1} are independently distributed, a priori.

The likelihood function is:

$$l(\boldsymbol{\beta}, \Sigma|\mathbf{w}) \propto |\Sigma^{-1}|^{n/2} \exp\{-\tfrac{1}{2}(\mathbf{w} - V\boldsymbol{\beta})'\Sigma^{-1} \otimes I_n(\mathbf{w} - V\boldsymbol{\beta})\} \tag{9}$$

with $n = 53$ for the present problem.

[4] Were this covariance term zero, there would be no point in using the "seemingly unrelated" regression model. As will be illustrated, estimation by this model is more efficient than "equation by equation" estimation in the sense that the parameters of the "seemingly unrelated" model have smaller variances.

[5] The presentation follows closely that given by Zellner [19]. See also Tiao and Zellner [17] and Geisser [7].

Combining the prior and likelihood the joint posterior pdf for the parameters is obtained by Bayes' theorem, that is,

$$p(\boldsymbol{\beta}, \Sigma^{-1}|\mathbf{w}) \propto |\Sigma^{-1}|^{(n-3)/2} \exp\{-\tfrac{1}{2}(\mathbf{w}-V\boldsymbol{\beta})'\Sigma^{-1}\otimes I_n(\mathbf{w}-V\boldsymbol{\beta})\}$$
$$\propto |\Sigma^{-1}|^{(n-3)/2} \exp\{-\tfrac{1}{2}\,\mathrm{tr}\,A\Sigma^{-1}\} \qquad (10)$$

where "tr" denotes the trace operation, and

$$A = \begin{bmatrix} (\mathbf{w}_1 - V_1\boldsymbol{\beta}_1)'(\mathbf{w}_1 - V_1\boldsymbol{\beta}_1) & (\mathbf{w}_1 - V_1\boldsymbol{\beta}_1)'(\mathbf{w}_2 - V_2\boldsymbol{\beta}_2) \\ (\mathbf{w}_2 - V_2\boldsymbol{\beta}_2)'(\mathbf{w}_1 - V_1\boldsymbol{\beta}_1) & (\mathbf{w}_2 - V_2\boldsymbol{\beta}_2)'(\mathbf{w}_2 - V_2\boldsymbol{\beta}_2) \end{bmatrix}.$$

From (10) the conditional posterior pdf for $\boldsymbol{\beta}$ given Σ is the multivariate normal form with mean

$$E(\boldsymbol{\beta}|\Sigma^{-1}, \mathbf{w}) = (V'\Sigma^{-1} \otimes I_n V)^{-1} V'(\Sigma^{-1} \otimes I_n)\mathbf{w} \qquad (11)$$

and covariance matrix

$$\mathrm{Covar}\,(\boldsymbol{\beta}|\Sigma^{-1}, \mathbf{w}) = (V'\Sigma^{-1} \otimes I_n V)^{-1}. \qquad (12)$$

As Σ^{-1} is unknown, for inferential purposes it can be integrated out to obtain the marginal posterior distribution for $\boldsymbol{\beta}$, that is,

$$p(\boldsymbol{\beta}|\mathbf{w}) \propto |A|^{-n/2}. \qquad (13)$$

However, (13) will not be employed since it has not as yet been fully analyzed (see [19, chapter 8]).[6]

Consequently proceed as if $\Sigma = \hat{\Sigma}$, where $\hat{\Sigma} = \{\hat{\sigma}_{ij}\}$, $\hat{\sigma}_{ij}$ being the *i, j*th element of the least-squares estimate of the covariance matrix of the two-equation system. In this case diffuse Bayesian analysis and sampling theory provide identical numerical results.

There were seven S.E.S. classes represented in Burt's sample of twins, including one classification which grouped those children reared in residential institutions. It was decided to effect the analysis in three different ways:

(i) With all seven classes ($n = 53$).

(ii) With all classes except that for residential institutions ($n = 46$). This was done to provide a comparison with a model which assumed that S.E.S. classifications could be measured on a cardinal scale.[7]

(iii) In (ii) 46 pairs of observations had been allocated to 6 classes.

[6] In private conversation, Arnold Zellner has informed me that it is possible to obtain the marginal distribution for $\boldsymbol{\beta}$ in a convenient analytical form by a route which involves a different parameterization of the "seemingly unrelated" model.

[7] That is, where the ranked number assigned to each class is the independent variable in a simple regression model.

This allocation left few observations in some classes. Therefore, some categories were grouped in order to produce 4 classes with more equal numbers of observations.

The classes[8] of (ii) and (iii) were:

Six classes		*Four classes*	
6	higher professional	4	higher and lower professional
5	lower professional	3	clerical and skilled
4	clerical	2	semi-skilled
3	skilled	1	unskilled.
2	semi-skilled		
1	unskilled		

The classes of (i) were those of (ii) plus the 7 pairs of cases where one twin in each pair was brought up in a residential institution.

TABLE 1
Results of regression analyses

Models[c]	Home[a]		Away[a]		Correlation between residuals[b]
	R^2	s^2	R^2	s^2	
(i) 7 Classes, $n = 53$	0.157	182.4	0.017	227.5	0.778
(ii) 6 Classes, $n = 46$	0.196	185.0	0.032	230.5	0.748
(iii) 4 Classes, $n = 46$	0.187	187.1	0.019	233.5	0.758

[a] The R^2's and s^2's are taken from "equation by equation" regressions, and *not* from the "seemingly unrelated" model. These are thus the statistics obtained by treating the problem as two separate regression analyses. The s^2's are also the least-squares estimates of σ_{11} and σ_{22}, the diagonal elements of Σ, the covariance matrix of the two equation system. The R^2's have been adjusted for the number of independent variables according to the formula

$$\left\{1-\frac{n-1}{n-p}(1-R^2)\right\}.$$

[b] The correlation between the dependent variables of the two equations is: 0.877 for $n = 53$; 0.875 for $n = 46$; 0.874 is the intra-class coefficient [3], $n = 53$.

[c] Normal plots of the dependent variables and residuals were made to test the distributional specifications of the models. These suggested that the dependent variables and residuals were approximately normally distributed.

[8] The classes are defined in footnote 2. In the regression models, Burt's numbering of classes was reversed in order to have a positive slope in the comparative model which embodied the cardinal scale assumption.

In summary, three separate dummy variable regressions were carried out of the form:

$$\begin{pmatrix} \mathbf{w}_1 \\ \mathbf{w}_2 \end{pmatrix} = \begin{pmatrix} V_1 & 0 \\ 0 & V_2 \end{pmatrix} \begin{pmatrix} \boldsymbol{\beta}_1 \\ \boldsymbol{\beta}_2 \end{pmatrix} + \begin{pmatrix} \mathbf{u}_1 \\ \mathbf{u}_2 \end{pmatrix}$$

$$(2n \times 1) \qquad (2n \times k) \qquad (k \times 1) \quad (2n \times 1) \tag{14}$$

where for (i) $n = 53$, $k = 13$, where V_1 is (53×6) and V_2 (53×7); for (ii) $n = 46, k = 12$; for (iii) $n = 46, k = 8$.

The regression analyses were effected so that the slope coefficients would represent effects relative to the highest class, six and four as appropriate. This was necessary to avoid multicollinearity problems (see [8, pp. 218–227]).

Results of the three regression analyses are presented in table 1. In addition, table 2 shows details of model (ii), which is further illustrated in figs. 1 and 2. (Model (ii) is chosen for particular examination as it does not contain the residential institution classification which tends to obscure the comparison between the home and away equations.)

One of the principal motivations for the analysis is the examination of the correlation between the residuals of the home and away equations. Estimates are shown in table 1. For all three models, the estimated correlation is some 0.10 to 0.12 smaller than the correlation between the phenotypes of the twins ($r_{w_1w_2} = 0.87$). In accordance with the preceding discussion, this indicates that this latter correlation has been inflated by postnatal environmental similarities.[9] It should be added that (as mentioned above) model (ii) was also effected in a "standard" manner by treating the S.E.S. levels as though they represented a cardinal scale. The correlation between the residuals of this model was essentially that of the phenotypes ($r = 0.88$). The analysis therefore illustrates the dangers of treating data as though they were measured on a cardinal scale, when this is not in fact the case. The cardinal scale assumption leads to a different substantive conclusion from that obtained by the more flexible dummy variables approach used above.

[9] It is noted that from the sampling theory viewpoint, there appears to be no test to determine whether the correlations between the dependent variables and the least-squares estimates of the residuals are significantly different. That is, the problem is to test whether the correlation coefficient between two variables is different from the correlation coefficient between the same variables when other variables have been partialled out. The two coefficients are *not* generated from independent samples and therefore a t-test (applied after using Fisher's z transformation) is not appropriate.

TABLE 2

Regression coefficients—6 class model

		"Efficient" estimates		Equation by equation estimates	
		Home equation	Away equation	Home equation	Away equation
		$(i = 1)$	$(i = 2)$	$(i = 1)$	$(i = 2)$
β_{i0}	Intercept—S.E.S. 6	111.20 (5.6)	99.41 (5.4)	122.0 (7.9)	100.7 (7.6)
β_{i1}	Effect of S.E.S. 1, relative to S.E.S. 6	−16.47 (6.1)	−2.69 (6.1)	−26.91 (8.9)	−8.3 (9.0)
β_{i2}	Effect of S.E.S. 2, relative to S.E.S. 6	−14.81 (5.9)	−0.41 (5.6)	−26.38 (8.7)	1.09 (8.3)
β_{i3}	Effect of S.E.S. 3, relative to S.E.S. 6	−20.16 (6.8)	8.27 (8.0)	−30.6 (9.9)	11.25 (11.6)
β_{i4}	Effect of S.E.S. 4, relative to S.E.S. 6	−11.35 (6.4)	−0.78 (7.3)	−29.86 (9.4)	−3.0 (10.7)
β_{i5}	Effect of S.E.S. 5, relative to S.E.S. 6	−7.65 (6.5)	−4.16 (6.7)	−14.71 (9.4)	−9.92 (9.8)

The figures in parentheses are the standard deviations associated with the normal posterior distributions for the β's. The equation-by-equation estimates in the right of the table have been added to show the gain in "efficiency" in going to the "seemingly unrelated" model.

Another interesting aspect of the analysis is the gain in "efficiency" associated with using the "seemingly unrelated" model. This is shown in table 2 where the first two columns of coefficients represent the "efficient" estimates, and the last two columns show those which would have been obtained by use of equation by equation estimation. The

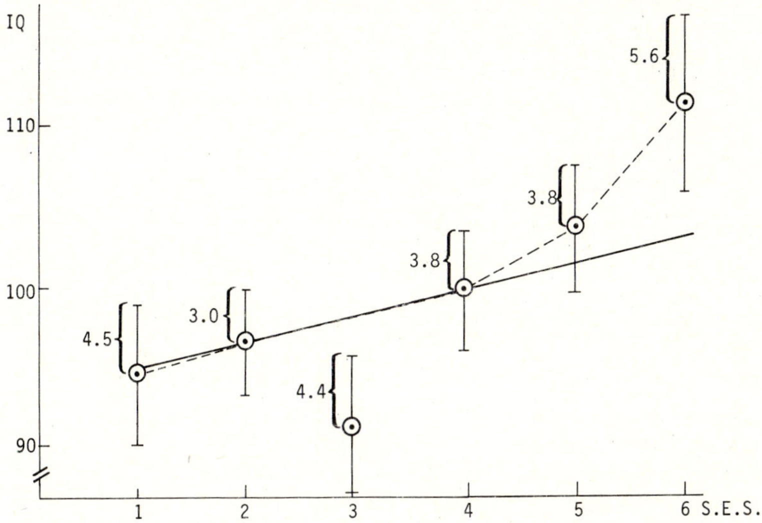

Fig. 1. Relationship of IQ to S.E.S.–home. ⊕ represents point estimates of the effects of S.E.S. levels. These are established by adding the appropriate β coefficients from table 2 to the intercept of the regression, i.e., S.E.S. 6. The lines on either side of the point estimates each represent *one* standard deviation, e.g., the standard deviation of S.E.S. 1 is 4.5. With the exception of that for S.E.S. 6, the standard deviations indicated above are not those of table 2. The variance of the sum of two random variables is by definition the sum of their individual variances and covariances. In this case, the slope coefficients are highly negatively correlated with the intercept, S.E.S. 6. This has the effect that the standard deviations indicated above are slightly smaller than those given in table 2. A straight line has been drawn through the point estimates of S.E.S. 2 and 4. A dotted line connects all means except that of S.E.S. 3.

standard deviations associated with the "efficient" estimates are some two-thirds of the others.

Figs. 1 and 2 provide a direct graphical comparison of the home and away equations. If S.E.S. class 3 is omitted—this only has 5 observations in the home equation and 3 in the away—it can be seen from fig. 1 that the relationship of IQ to S.E.S. tends to be more curvilinear than linear.[10] However, this relationship is only suggestive (consider the indicated standard deviations), and there are not sufficient observations in order to come to any firm conclusions. Fig. 2, illustrating the away equation, shows no apparent relationship between IQ and S.E.S.

In interpreting figs. 1 and 2, as well as the regression statistics shown in table 1,[11] care should be exercised in respect of model misspecifications. S.E.S. is only a crude surrogate for environmental effects—and is thus a misspecification. The result of this is to confuse the relationship between IQ and environmental effects. However, insofar

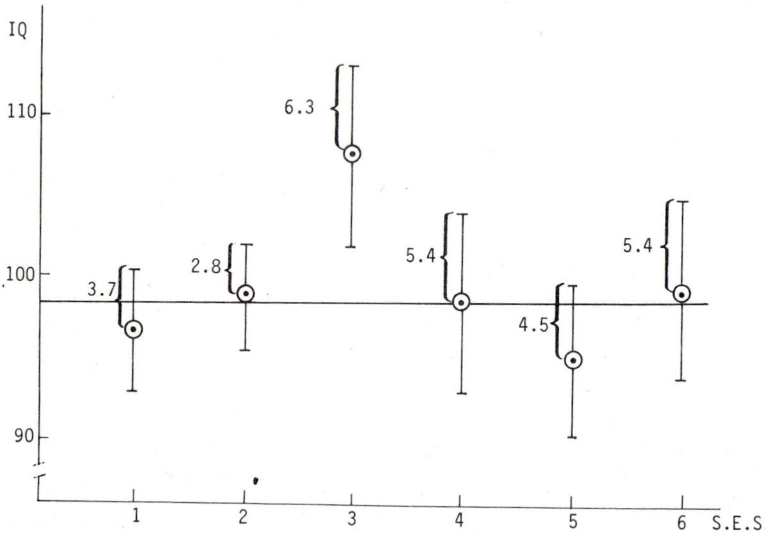

Fig. 2. Relationship of IQ to S.E.S.–away. This graph is to be interpreted in the same manner as fig. 1, i.e., ⊕ represents point estimates of the effects of S.E.S. levels calculated in the same manner as in fig. 1. The vertical lines and standard deviations shown above are also those of fig. 1. The horizontal line is drawn through the mean of the away sample 98.6.

[10] It is interesting to note that the correlation in the linear (not dummy variables) model between IQ and S.E.S. is 0.36 ($n = 46$). This figure is close to the 0.32 reported to hold in the British population [2], thus suggesting that the home sample is fairly representative. Note, however, that the square root of the *adjusted* R^2 for the 6 class dummy variable model is 0.44, table 1. Thus, a stronger relationship exists between IQ and S.E.S. than is revealed by the simple linear correlation.

[11] It should also be noted that there is no small sample test of the linear hypothesis for the "seemingly unrelated" repression model [18].

as such a relationship exists, misspecifications would act to raise the estimated correlation between the residuals of the home and away equations above its true level. This means that the observed 0.10 to 0.12 difference between the correlations of the phenotypes and residuals would be an underestimate. It should be added that this result does *not* permit us to reduce the estimate of heritability based on these data by an amount of, say, 0.10 to 0.12. For this to occur, the common elements model outlined in the first section would require that the residual variances of the home and away equations be equal, which is not the case here (see table 1). What may be said, however, is that since the phenotypes of the home and away twins were affected by some similar postnatal environmental conditions, the use of the correlation coefficient for estimating heritability for these MZ twins reared apart is inappropriate and certainly overestimates the proportion of genetic variance in the trait.

3. *An identification problem*

A further difficulty in interpreting the above results arises because even if one assumes that genotype and environment are uncorrelated in the away equation, this is not necessarily true of the home equation. For if society is highly stratified into different genetic pools, then it follows that an S.E.S. classification is an indication of both genetic *and* environmental factors. In the home equation, we have no certain way of distinguishing these two forces.

It is therefore interesting to extend the analysis in the following way. In the home equation regress IQ on S.E.S., where S.E.S. is considered a measure of indistinguishable environmental and genetic effects. In the away equation, regress IQ on S.E.S. of *both* the class into which each twin is born (as a surrogate for the genetic influence), *and* the S.E.S. class in which the twin is reared—as a rough measure of environmental effects.

Again, use the "unrelated regression" model, which we redefine below as

$$\underset{(2n\times 1)}{\begin{pmatrix} w_1 \\ w_2 \end{pmatrix}} = \underset{(2n\times k)}{\begin{pmatrix} V_1 & 0 \\ 0 & V_2 \end{pmatrix}} \underset{(k\times 1)}{\begin{pmatrix} \beta_1 \\ \beta_2 \end{pmatrix}} + \underset{(2n\times 1)}{\begin{pmatrix} u_1 \\ u_2 \end{pmatrix}} \tag{15}$$

where $n = 53$; $k = k_1 + k_2$, with $k_1 = 6$ and $k_2 = 13$; V_1 is of rank k_1 and V_2 of rank k_2.

The first equation of this model is exactly that of the first equation of (14). In the second (away) equation, the number of dummy variables is extended to 13. That is, there is one dummy variable for each of the seven possible S.E.S. classes in which a twin could have been reared (i.e., the S.E.S. classes 1 through 6 and the residential institution classification), *plus* a dummy variable for each of the six possible classes into which a twin could have been born. In order to avoid problems of multicollinearity the regression analysis was again implemented so that the slope coefficients would be relative to the (environmental) effects of the highest S.E.S. classification. The results of the analysis are presented in table 3.

Examination of table 3 shows that the home equation yields almost exactly the same results as the equation-by-equation estimates of the model presented in table 2. (The small differences, which a comparison between the two tables reveals, are due to the fact that in table 2, $n = 46$, in table 3, $n = 53$.) Apart from the fact that there was no gain in efficiency in going to the "seemingly unrelated" model in the present analysis, for all practical purposes we can still envisage the home equation as in fig. 1.

In the away equation, the efficient estimates of the environmental effects of S.E.S. are very nearly the same as those of table 2. At the same time, the "birth" effects of S.E.S., with the exception of S.E.S. classes 4 and 6, are all centered close to zero, but with large (relative to environmental effects) standard deviations. In addition, although the birth effect of S.E.S. class 4 is low, and that of S.E.S. class 6 is high, the standard deviations associated with these posterior distributions are also large. It is clear that the away equation still shows no regression effect, even when we supplement the environmental influences of S.E.S. by its genetic effects. Lastly, note that the residual correlation is not so different from the correlation between the dependent variables. In this case, however, the same interpretation is not attached to the residual correlation as in the case where only environmental effects were considered.

4. *Discussion*

A number of interesting conclusions may be drawn from the above analysis. First, in the sample examined, the phenotypic correlation between the MZ twins reared apart was inflated by common postnatal

TABLE 3

Regression analysis containing information as to S.E.S. levels at which away twins were born and reared

β-coefficients

		"Efficient" estimates		Equation-by-equation estimates	
		Home	Away	Home	Away
		$(i = 1)$	$(i = 2)$	$(i = 1)$	$(i = 2)$
β_{i0}	Intercept S.E.S. 6	122.0 (7.8)	101.4 (10.8)	122.0 (7.8)	86.4 (17.6)
β_{i1}	Effect of S.E.S. 1, relative to S.E.S. 6	−27.6 (8.6)	−2.1 (5.1)	−27.6 (8.6)	−8.8 (9.0)
β_{i2}	Effect of S.E.S. 2, relative to S.E.S. 6	−25.9 (8.5)	−0.8 (4.7)	−25.9 (8.5)	0.4 (8.2)
β_{i3}	Effect of S.E.S. 3, relative to S.E.S. 6	−30.6 (9.9)	7.2 (6.8)	−30.6 (9.9)	10.7 (12.0)
β_{i4}	Effect of S.E.S. 4, relative to S.E.S. 6	−29.9 (9.3)	−0.1 (6.1)	−29.9 (9.3)	−1.4 (10.7)
β_{i5}	Effect of S.E.S. 5, relative to S.E.S. 6	−18.8 (9.0)	−2.0 (5.7)	−18.8 (9.0)	−7.0 (9.9)
β_{2R}	Effect of S.E.S.—Residential Institution relative to S.E.S. 6	—	−0.5 (5.5)	—	14.1 (16.0)
β_{2B1}	Effect of birth in S.E.S. 1[a]	—	−3.5 (10.5)	—	14.1 (16.0)
β_{2B2}	Effect of birth in S.E.S. 2[a]	—	−1.8 (10.5)	—	14.2 (16.2)
β_{2B3}	Effect of birth in S.E.S. 3[a]	—	−1.9 (12.0)	—	15.1 (17.4)
β_{2B4}	Effect of birth in S.E.S. 4[a]	—	−14.1 (11.5)	—	2.8 (17.1)
β_{2B5}	Effect of birth in S.E.S. 5[a]	—	−0.7 (9.5)	—	20.2 (16.6)
β_{2B6}	Effect of birth in S.E.S. 6[a]	—	14.7 (13.2)	—	27.9 (18.5)

[a] Effect relative to being reared in S.E.S. 6.

(1) As in table 2, the figures in parentheses are the standard deviations associated with the normal posterior distributions for the β's. Again, the equation-by-equation estimates are shown on the right-hand side to illustrate the gain in efficiency by going to the "seemingly unrelated" model. In this instance, there is no gain for the home equation. (2) Statistics from the equation-by-equation regressions are: Home: R^2 (adjusted) $= 0.1546$, $s^2 = 182.44$ (variance of dependent variable $= 211.74$). Away: R^2 (adjusted) $= 0.0731$, $s^2 = 214.53$ (variance of dependent variable $= 227.08$). (3) The correlation between the residuals is 0.8218.

environmental influences. Thus, the use of this statistic as a measure of heritability overstates the degree of genetic determination of the trait examined, IQ. This is also important in the context of reconciling estimates of heritability reached by other means (e.g., comparison of MZ and DZ twins); it goes some way to explaining, for example, the discrepancies noted by Light and Smith ([10], pp. 506–507). In addition, the result is particularly striking in view of the crudity of the S.E.S. measure used as a surrogate for the effects of environment on IQ. It should be noted that the conclusion reached was dependent upon use of the dummy variables regression technique which makes no assumptions concerning the scale upon which S.E.S. classes could be considered to be measured. When the S.E.S. levels were used in the model as a cardinal scale (for which there would seem little or no justification) a different substantive conclusion was reached: there were no common postnatal environmental effects. This situation clearly shows the necessity for paying close attention to scaling problems when constructing indices which might be used in analyses such as the above.

Secondly, it was also observed that in the home equation the relationship between IQ and S.E.S. was more curvilinear than linear. Indeed, the almost logarithmic curve suggested by fig. 1 indicates that the effects of S.E.S. on IQ may be cumulative rather than additive, thus providing some empirical support for the theorizing of Stinchcombe [16].

Thirdly, when in the preceding section the away equation was supplemented with the S.E.S. levels for the classes into which the away twins were born, there was still no regression effect for that equation. This result was perhaps surprising in view of both the large correlation between the phenotypes of home and away twins, and the fact that there was an effect for S.E.S. in the home equation. That is, S.E.S. was found to be a significant variable if a twin was both born and reared in a household with the same S.E.S., but not if he or she were born at one level and reared at another. The immediate significance of this situation is unclear although a possible explanation is that the analysis points to the existence of an interaction effect between genotype and environment. This is clearly a matter which should be explored in future work.

This paper could be described as exploratory in nature. We have investigated the largest sample of MZ twins reared apart reported in the literature from a novel viewpoint and, in view of the crudity of the variables considered, achieved interesting results. That IQ is largely determined by genetic factors is a hypothesis supported by the analysis, although we have failed to answer what the extent of this might be. However, it might be said that to some degree this question is somewhat beside the point. For example, we have noted that there is a crude logarithmic relationship between IQ and S.E.S. which, in future work, could probably be refined by considering other variables. Birth order within a family, for instance, is known to be related to IQ as are probably other environmental factors such as family size, age of parents, etc. In a sense it seems somewhat premature to worry about the heritability of IQ without fully understanding how it is distributed in the population.

References

[1] W. F. Bodmer and L. L. Cavalli-Sforza, Intelligence and race. Scientific American 223, 4 (October 1970), 19–29.

[2] C. Burt, Intelligence and social mobility. British Journal of Statistical Psychology 14 (1961), 3–24.

[3] C. Burt, The genetic determination of differences in intelligence: a study of monozygotic twins reared together and apart. British Journal of Psychology 57 (1966), 137–153.

[4] C. Burt, The genetic determination of intelligence: a reply. British Journal of Psychology 53 (1967), 153–162.

[5] C. Burt, personal communication. February 1970.

[6] Environment, heredity, and intelligence. Harvard Educational Review Reprint Series No. 2, 1969.

[7] S. Geisser, Bayesian estimation in multivariate analysis. Annals of Mathematical Statistics 36 (1965), 150–159.

[8] A. S. Goldberger, Econometric theory. New York, John Wiley, 1964.

[9] R. M. Hogarth, Statistical aspects of the nature-nurture controversy. Unpublished manuscript, University of Chicago, Graduate School of Business, 1972.

[10] R. J. Light and P. V. Smith, Social allocation models of intelligence: a methodological enquiry. Harvard Educational Review 39 (1969), 484–510.

[11] H. Newman, F. Freeman and K. Holzinger, Twins: a study of heredity and environment. Chicago, University of Chicago Press, 1937.

[12] H. V. Roberts, Statistical inference and decision. Chicago, Graduate School of Business, University of Chicago, 1966.

[13] S. Scarr-Salapatek, Race, social class, and IQ. Science 174 (1971), 1285–1295.

[14] J. Shields, Monozygotic twins brought up apart and brought up together. London, Oxford University Press, 1962.

[15] D. H. Stott, Commentary on "The genetic determination of differences in intelligence: a study of monozygotic twins reared together and apart" by Cyril Burt: Congenital influences on the development of twins. British Journal of Psychology 57 (1966), 423–429.

[16] A. L. Stinchcombe, Environment: the cumulation of effects is yet to be understood. Harvard Educational Review 39 (1969), 511–522.
[17] G. C. Tiao and A. Zellner, On the Bayesian estimation of multivariate regression. Journal of the Royal Statistical Society, Series B (1964), 277–285.
[18] A. Zellner, An efficient method for estimating seemingly unrelated regressions and tests for aggregation bias. Journal of the American Statistical Association 57 (June 1962), 348–368.
[19] A. Zellner, An introduction to Bayesian inference in econometrics. New York, John Wiley, 1971.

S. E. Fienberg and A. Zellner (eds.)
Studies in Bayesian Econometrics and Statistics.

6.3 BAYESIAN COMPARISONS OF SIMPLE MACROECONOMIC MODELS*

Martin S. GEISEL

Carnegie-Mellon University

1. *Introduction*

Several years ago Friedman and Meiselman [10] published the results of extensive comparisons of simple Keynesian and Quantity Theory models. Their work generated comment and criticism from many sources including Ando and Modigliani [2], DePrano and Mayer [7] and Hester [16]. While much of this discussion focused on economic issues it also served to point out the inadequacies of existing statistical procedures for comparing models. During the last decade considerable statistical research efforts, both Bayesian and non-Bayesian, have been concerned with development of procedures for comparing models. The Bayesian approach has begun to yield some fruitful results—see, e.g. [4, 13, 25, 26]. Progress, both Bayesian and non-Bayesian, has also been made on the related problem of choosing explanatory variables in multiple regression—see [14; 18; 22; 24, chapter 1].

In the next section of this paper we present a brief discussion of a Bayesian procedure for comparing models. In the third section a brief discussion of the Friedman–Meiselman procedure and earlier comments on it is given. In sections 4 and 5, the procedure outlined in section 2 is

* A preliminary version of this paper was presented at the Winter Meeting of the Econometric Society, Evanston, Illinois, December 1968. Reprinted from the Journal of Banking, Money and Credit, with permission of the Ohio State University Press. This research was undertaken in the author's Ph.D. dissertation, "Comparing and choosing among parametric statistical models: a Bayesian analysis with macro-economic applications", at the University of Chicago. I am indebted to the members of my thesis committee, S. James Press, Harry Roberts, and Arnold Zellner for their guidance and criticism. I have also benefited from the comments of Jacques Drèze, Joseph Kadane, Walter Oi, and George Tiao on earlier drafts of this and related papers. Any remaining errors are the author's responsibility.

applied to several versions of the Friedman–Meiselman comparisons. Concluding remarks are given in section 6.

2. A Bayesian procedure for comparing parametric stochastic models[1]

For purposes of this paper we take a parametric stochastic model of a process to be a family of data densities for the random variable(s) of the process which depend on a (usually finite) number of parameters and a specified set of predetermined variables, and a prior density for the parameters of the data densities. Let us illustrate this definition by reference to the standard linear model

$$y_i = \beta_0 + \beta_1 X_{1i} + \ldots + \beta_k X_{ki} + u_i, \qquad i = 1, 2, \ldots, T$$

where the u_i are assumed to be normally and independently distributed with zero mean and common variance σ^2 and the X's are fixed numbers. Then, for our purposes, the model is completely specified by:

(1) The values X_{ji}, $j = 1, \ldots, k$; $i = 1, \ldots, T$.

(2) A density function for y_i given the parameters and the explanatory variables and the assumption of independence. For example, $p(y_i|\boldsymbol{\beta}, \sigma, X_{ji}, j = 1, \ldots, k)$ is a univariate normal density with mean $\beta_0 + \Sigma_{j=1}^{k} \beta_j X_{ji}$ and variance σ^2 and $p(y_1, \ldots, y_T|\boldsymbol{\beta}, \sigma, X_{ji}, j = 1, \ldots, k$; $i = 1, \ldots, T)$ is the product of the univariate densities above. Here $\boldsymbol{\beta} = (\beta_0, \beta_1, \ldots, \beta_k)'$.

(3) A prior density on $\boldsymbol{\beta}$ and σ.[2]

The first step in the process of comparing models is an enumeration of the set of models to be considered. We call this set the model space. We assume it is a finite set. Since the results of the comparisons to be made depend on the model space chosen this enumeration should be given careful consideration. In particular, the statistical models should

[1] This section is a brief outline of material which is developed more fully in Geisel [13]. An analogous discussion of a problem of comparing forecasters is given in Roberts [21].

[2] If a lagged dependent variable is included among the X's then we must extend our definition to include a specification of the relevant initial conditions. The case of stochastic X's may be included in the definition if we presume to work conditionally on the observed values. If not, we must include a specification of the process generating them in our definition.

be adequate representations of the economic theories they represent. Further, the possibility of the existence of more general models which may encompass apparently separate elements of the space should be examined. Once the model space is specified, the next step is to assess a prior probability mass function on its elements. We denote this pmf by $P(M_i)$ and denote a data density of M_i by $p(y|\theta_i, M_i)$ and a prior density for the parameters, θ_i, of M_i by $\pi(\theta_i|M_i)$. Note that for notational convenience we suppress the dependence of the data density on the values of the predetermined variables.

We may now compute the marginal density of the observations (given the model) by

$$p(y|M_i) = \int_{\Theta_i} p(y|\theta_i, M_i)\pi(\theta_i|M_i)\,\mathrm{d}\theta_i \tag{1}$$

and average over the models to obtain a density which is unconditional of them:

$$p(y) = \sum_i p(y|M_i)P(M_i). \tag{2}$$

When a particular value of the random variable $\tilde{y}$, say y, is observed we may employ Bayes' rule to revise the prior probabilities $P(M_i)$ to become posterior probabilities. That is,

$$P(M_i|y) = \frac{P(M_i)p(y|M_i)}{p(y)}. \tag{3}$$

Further, we may express our knowledge of future sample evidence y_F via the predictive pdf

$$p(y_F|y) = \sum_i P(M_i|y)p(y_F|M_i, y). \tag{4}$$

On observing new sample evidence we may once again revise the probabilistic weight for each model in accord with equation (3) and obtain the density for still another sample via equation (4). Thus, the observed data automatically revise the probabilistic weight for each model; the revision of the weights depends entirely on the predictive density placed on the observed data by each of the models. The computations involved in obtaining $p(y)$ and $p(y_F|y)$ are in principle straightforward if every M_i implies a proper density for y (or y_F). At any stage, the contribution of each model to the predictive pdf is clearly stated.

Note that this procedure does not select one model as "true" or "best" and eliminate the rest. If the probabilistic weight of a model is positive it contributes to our knowledge of the future observations and there is no reason to neglect this contribution. Thus, any decision theoretic procedure which is designed to eliminate some of the models is viewed as an approximation which is used for reasons of simplicity of view or to reduce the cost of computation. Note further that even if some simplification is undertaken, there is no reason that only one model should be retained and all the rest eliminated in all cases. Moreover, once the model space has been specified carefully, the added costs of working with the full space rather than some simplified version may not be great.

We have not discussed procedures for assessing prior distributions. We note, however, that this method of comparing models may be specialized to serve as a method of comparing alternative priors. That is, it applies to the case where all the models under consideration employ the same family of data densities and differ only in the priors on the parameters of the data density.

This discussion of comparing models has made no mention of the details of the statistical analysis of a given model. Nevertheless, such analysis is an essential component of the procedure. That is, conventional procedures, both formal and informal, for checking the appropriateness of the assumed data and prior densities (e.g., serial correlation, non-normality, sign restrictions on parameters) are used as part of and prior to the model comparison procedure. The application of the procedure below will illustrate this.

3. *Discussion of the Friedman–Meiselman procedure*

Friedman and Meiselman [10], hereafter denoted FM, investigate whether the money supply or the level of "autonomous" expenditures is the better explanatory variable for consumption (or income) by estimation of the following two equations:

$$M_1: \quad C_t = \alpha_1 + \beta_1 M_t + u_{1t}$$

$$M_2: \quad C_t = \alpha_2 + \beta_2 A_t + u_{2t} \tag{5}$$

where C_t represents consumption expenditures in period t, M_t is the stock of money in period t, and A_t is autonomous expenditures for period t. All variables are expressed in nominal terms; that is, in current dollars. They also considered equations of the same form as (5) with money income, Y_t, as the dependent variable and equations which contained the price level, P_t, as an additional independent variable. These tests utilized annual data for 1897–1958 and various sub-periods thereof. In addition, relationships involving lags were investigated using post-World War II quarterly data.

The first equation of (5) may be regarded as coming from a simple Quantity Theory model:

$$\begin{aligned} &\text{(1) Demand for money:} && M_t^D = \gamma+\delta Y_t+v_t \\ &\text{(2) Supply of money:} && M_t^S = M_t \text{ (exogenous)} \\ &\text{(3) Equilibrium:} && M_t^D = M_t^S. \end{aligned} \tag{6}$$

Substituting (2) and (3) into (1) we have

$$Y_t = \alpha+\beta M_t+u_t$$

where $\beta = 1/\delta$, $\alpha = \gamma/\delta$, and $u_t = -v_t/\delta$. This equation is in the form of the first equation of (5), except that income is the dependent variable rather than consumption. FM switch to consumption for statistical reasons to be explained below.

The second equation of (5) is the reduced form for the simplest Keynesian model:

$$\begin{aligned} &\text{1) Consumption function:} && C_t^D = a+bY_t^D+w_t \\ &\text{2) Autonomous expenditure:} && A_t^D = A_t \text{ (exogenous)} \\ &\text{3) Aggregate demand:} && Y_t^D = C_t^D+A_t^D \\ &\text{4) Equilibrium:} && Y_t^D = Y_t^S = Y_t. \end{aligned} \tag{7}$$

It is thus assumed that aggregate supply, Y_t^S, and aggregate demand, Y_t^D, equal measured income, Y_t. Substituting 2), 3) and 4) into 1) we have

$$C_t = \alpha_2+\beta_2 A_t+u_{2t}$$

where $\alpha_2 = a/(1-b)$, $\beta_2 = b/(1-b)$, and $u_{2t} = w_t(1-b)$. Since this system could also be solved for Y_t one might wonder why FM used C_t rather than Y_t as their dependent variable. The reason for this is that since $Y_t = C_t+A_t$, the correlation between Y_t and A_t is greater (in absolute value) than that between C_t and A_t and since correlation coefficients were used to compare the money model and the autonomous expenditure model this would introduce a bias in favor of the latter. FM felt that the switch to consumption also might bias the comparison against the money model; however, the correlation between C and M was greater than that between Y and M for many of the periods they considered [10, pp. 176–177].

FM's methodology has been criticized on several grounds by Ando and Modigliani [2], hereafter denoted AM; De Prano and Mayer [7]; and Hester [16]. The essential points of the criticisms may be summarized as follows.

1) The dichotomy (5) is inappropriate and "... cannot be expected to throw any light on such basic issues as how our economic system works, or how it can be more effectively stabilized" [2, p. 693].

2) The functional forms chosen by FM are inappropriate and much too simplistic.

3) The definitions of M_t and especially A_t used by FM are unconventional and make the comparison look more favorable to the money model than it should.

4) Even if these alleged deficiencies are disregarded, the comparison of these models by means of correlation coefficients is inappropriate.

Let's examine these criticisms from the viewpoint of the previous section. The first three deal mainly with the specification of the model space and provide a good illustration of the fact that its enumeration requires a great deal of thought.

The first section of FM's paper contains a thorough discussion of their views on the model space. They consider alternatives to the simple dichotomy (5) and discuss the problem of definition of variables in detail. They feel that the dichotomy is appropriate because "There has long been a striking division among students of economic affairs about the role of money in determining the course of economic events. One view is that the quantity of money matters little; the other, that it is a key factor in understanding, and even more, in controlling economic change." Further, they suggest that the first view is held either because the quantity of money is regarded as a passive element in the economic system which adapts to the "needs of trade" and is determined by and responsive to other factors but is not a source of disturbances or because the ratio of money to other assets is regarded as variable and hence the economic effects of changes in the quantity of money are highly unpredictable. The second view holds that the quantity of money is important because:

(a) it can be fairly accurately controlled by deliberate policy;

(b) changes in the quantity of money can produce substantial changes in the flow of income, prices, and other variables;

(c) the relationships between the stock of money and other assets are relatively stable and dependable [10, pp. 166–167].

Up to this point it seems that the question is whether or not the stock of money should be included in the list (the other variables are not specified) of important variables affecting the economy. The idea that it is either money or something else is an idea generated by the Keynesian revolution. FM point out that since the 1930's the controversy about money has proceeded on two levels. The first is a theoretical issue regarding long-run full-employment equilibrium. The second, with which they are and we shall be concerned, is an empirical question relating to the stability of short-run relations. Here the assertion (of the "Keynesians") was that there is a stable and persistent relationship between "investment" expenditures and total expenditures (income). This view had the effect of propagating the idea that "money does not matter". FM point out that in actual practice attention has been concentrated on the simplest version of the theory, the relationship between "investment" and "consumption" flows with the connection between the two being the main source of changes in income [10, pp. 167–168].

FM consider more complicated models because they recognize the fact that post-World War II experience led to a reincorporation of money into the Keynesian model; the mechanism being that money affects interest rates, which in turn affect investment and thereby consumption and income. Further, a somewhat more sophisticated Quantity Theory argument would allow that investment could affect income by affecting interest rates and thereby the quantity of money demanded and velocity and thus income.

FM point out that the issue is again on this slightly more sophisticated level empirical rather than theoretical. They suggest that it is possible to construct a model including both multiplier relations between investment and consumption and velocity relations between money and income. The central questions are then:

(a) Which set of relations is more critical in the sense of being the primary source of disturbance and change?

(b) Which set is more stable in the sense of expressing empirically consistent relations which don't change greatly from time to time?

FM's objective is to provide evidence on the second of these issues. To do so they chose the simple framework (5) rather than a more complicated approach because far more data were available for the former [10, pp. 168–170]. It should be pointed out, however, that FM's analysis is directed at question (a) above as well as question (b). To be specific, if the first equation of (5) has a higher R^2 than the second, we may say that this implies that variations in the stock of money accounted for more of the variation in income than did variations in autonomous expenditures or that it tells us something about both the importance of money as a source of variation and the stability of the velocity coefficient but it does not give an answer to question (b) alone. Thus, we shall interpret FM's analysis (and ours) as being tests of the adequacy of the simple Quantity Theory model relative to that of the simplest Keynesian model rather than simply as tests of the stability of coefficients. FM's critics have in general taken this view without saying so explicitly and Friedman's other work indicates that he would not be adverse to this interpretation (see, e.g., [11]). Note that if we view the equations (5) as only a part of some larger unspecified model rather than as reduced forms for the simple Quantity Theory model and the simple Keynesian model it is difficult if not impossible to say how the results should be interpreted.

Given that the objective is to learn something about whether money or autonomous expenditures is the more important source of disturbances in the economy and to gain some knowledge about the reliability of the relationships between these variables and income, is it appropriate to seek answers to these questions by means of FM's approach? As we mentioned above, some of FM's critics think not. Since our interest focuses on the statistical methods as well as the economic issues we shall assume that the FM dichotomy is a sensible one and shall continue to employ it and various modifications of it. There are no valid a priori statistical reasons for rejecting such a dichotomy. However, we should be aware that the two dichotomous models are not necessarily exhaustive. In particular, models which include both real and monetary sectors and dichotomize only under special conditions may merit consideration. (For discussion of such models see, e.g. [3].) It must also be recognized that a comparison of more complicated versions of the two models might not yield the same results as comparison of the simpler versions does. To be specific, the results may be sensitive to the particular functional forms assumed.

This leads us directly into the second criticism. Much work of both a theoretical and an empirical nature has been done on both the demand for money function and the consumption function which indicates that the forms used by FM are inappropriate. (On the demand for money see, e.g., [6, 9, 19, 23]. On the consumption function see [1, 8, 20].) This does not necessarily render the analysis of (5) invalid but does serve to point up another factor to which FM's comparisons may be sensitive. It would seem reasonable to extend the model space in order to examine empirically how important this sensitivity is; indeed, AM did this to some extent.

The next criticism regards the definition of the independent variables, A_t and M_t. Unfortunately, theory does not tell us precisely what constitutes money and what makes up autonomous expenditures. There are some fairly widely accepted conventions such as "money is the sum of currency held by the public and demand deposits at commercial banks" but there is no guarantee that these are appropriate. There are several possible approaches for dealing with this problem. First, we could simply state "This is the definition of money which we shall use: . . ." and similarly for autonomous expenditures. With this approach we are not likely to achieve substantial agreement and the results of the comparisons may be sensitive to the definition chosen. Second, we could attempt to determine the appropriate definitions according to some empirical criterion as FM do (see [10, pp. 180–186]). On the definition of money the recent work of Chetty [5] provides a somewhat more sophisticated empirical criterion. His results for a limited sample are supportive of the FM definition. Finally, models which differ only in the definitions of variables may be considered as separate elements of the model space. Haavelmo [15] does something similar to this. He approaches the question of whether or not a certain item is autonomous by considering an expanded model in which the item in question is either autonomous or not accordingly as a certain coefficient is zero or nonzero.

The final criticism hardly merits comment. It is well known that when the number of explanatory variables in each equation is the same and the dependent variable is the same in all equations, then R^2 is a monotonic function of S^2, the estimated residual variance.[3]

[3] In Geisel [13], the problem of choosing one of a set of alternative regression equations is discussed. Sufficient conditions are given under which the procedure of choosing the equation with the smallest value of S^2 is a minimum posterior expected loss procedure.

4. Bayesian comparisons of the simple money and autonomous expenditure models

We now proceed to apply the results of section 2 to the money-autonomous expenditure comparison. We begin by studying four forms of the dichotomy (5). In each case we thus take the model space to consist of the two models in equation (5). The forms differ from each other in definitions of the variables and in the sources of data.

The first comparison uses FM's definitions and annual data for the periods 1929–1941 and 1947–1958.[4] The second comparison retains FM's definition (with a slight modification in M_t) but employs revised data and extends the comparison to 1965.[5] The third and fourth comparisons employ AM's definition of A_t and two alternative definitions of M_t.[6] The periods considered for these comparisons are 1929–1941 and 1947–1965.

[4] The data are given in Friedman and Meiselman [10]. The definitions FM use are - C_t: Personal Consumption Expenditures are reported in U.S., Department of Commerce, Survey of Current Business 39 (July 1959), table I–1. A_t: The sum of Net Private Domestic Investment; Net Foreign Investment; and Government Deficit on Income and Product Account from the same source as above. M_t: The sum oı currency in the hands of the public plus all commercial bank deposits. The data are from unpublished sources. FM's comparisons covered various periods from 1897 to 1958. We shall not consider the pre-1929 comparisons in detail but will give a brief discussion of our results for that period below.

[5] The definitions of C_t and A_t are as in the previous footnote but the data are taken from U.S., Department of Commerce, The National Income and Product Accounts of the U.S., 1929–1965, 1967, tables 1.1, 1.9, and 5.1. These data embody substantial revisions, both definitional and statistical, from the earlier data. These revisions are discussed in the August 1965 issue of the Survey of Current Business. M_t is defined as currency in the hands of the public plus demand and time deposits at commercial banks. The annual numbers are averages of monthly figures reported in U.S., Board of Governors of the Federal Reserve System, Federal Reserve Bulletin, June 1964, and September 1966.

[6] AM's definitions and sources are - C_t: Same as previous footnote. A_t: The sum of: Net Investment in Plant and Equipment, and in Residential Houses; Total Government Purchases of Goods and Services; Exports; Property Tax Portion of Indirect Business Taxes; Net Interest Paid by Government; Government Transfer Payments Except Unemployment Insurance Benefits; Subsidies less Current Surplus of Government Enterprises; and Statistical Discrepancy (minus). The source of these data is The National Income and Product Accounts of the U.S., 1929–1965. M_t: The first definition (M1) is currency in the hands of the public plus demand deposits at commercial banks. The second definition (M2) is M1 plus time deposits at commercial banks. The source of these data is the Federal Reserve Bulletin (see previous footnote).

TABLE 1

Predictive statistics for the simple FM models 1929–1941.[a]
Models 1a and 2a employ original FM data; models 1b and 2b use revised data

Prediction year	Model	R_i^2	Predictive pdf Mean	Predictive pdf Variance	$P(M_i \mid \text{data})$	Prediction error
1934	1a	0.927	48.52	80.17	0.0905	−3.37
	1b	0.937	48.52	64.27	0.1215	−2.82
	2a	0.971	56.81	27.33	0.9095	4.92
	2b	0.971	56.54	25.28	0.8785	5.20
	a		56.06	37.78		4.17
	b		55.56	36.89		4.23
1935	1a	0.924	59.14	37.32	0.2203	2.85
	1b	0.935	58.78	29.94	0.3685	3.09
	2a	0.950	62.52	24.67	0.7797	6.23
	2b	0.946	61.14	25.16	0.6315	5.44
	a		61.77	29.42		5.48
	b		60.27	28.21		4.57
1936	1a	0.917	67.56	29.06	0.5486	4.95
	1b	0.926	66.93	24.59	0.6163	5.01
	2a	0.913	69.21	31.70	0.4514	6.60
	2b	0.915	67.90	28.70	0.3837	5.99
	a		68.31	30.93		5.69
	b		67.30	26.40		5.39
1937	1a	0.895	70.85	29.17	0.6771	3.59
	1b	0.902	69.62	25.58	0.6816	3.11
	2a	0.874	67.16	32.73	0.3229	−0.10
	2b	0.881	66.35	29.02	0.3184	−1.54
	a		69.66	33.29		2.40
	b		68.58	28.99		2.07
1938	1a	0.890	69.72	24.48	0.5897	5.08
	1b	0.899	68.87	21.28	0.6107	4.95
	2a	0.881	62.20	24.36	0.4103	−2.44
	2b	0.888	60.87	21.48	0.3893	−3.05
	a		66.63	38.10		1.99
	b		65.75	36.58		1.84
1939	1a	0.869	75.03	27.34	0.4223	7.45
	1b	0.878	74.18	24.26	0.4699	7.35
	2a	0.877	69.98	22.91	0.5777	2.40

TABLE 1 (*continued*)

Prediction year	Model	R_i^2	Predictive pdf Mean	Variance	$P(M_i \mid \text{data})$	Prediction error
	2b	0.881	68.28	20.72	0.5301	1.45
	a		72.11	31.01		4.53
	b		71.05	31.05		4.22
1940	1a	0.832	81.89	37.37	0.1516	10.00
	1b	0.840	80.90	33.98	0.1449	10.08
	2a	0.877	76.43	23.03	0.8484	4.55
	2b	0.884	74.59	20.20	0.8551	3.76
	a		77.26	29.03		5.38
	b		75.50	27.14		4.68
1941	1a	0.789	87.54	51.68	0.0432	5.66
	1b	0.793	86.17	48.17	0.0294	5.59
	2a	0.874	90.97	32.93	0.9568	9.10
	2b	0.884	88.59	27.79	0.9706	8.01
	a		90.82	34.23		8.94
	b		88.52	28.56		7.94

The posterior probability of model 1a after 1941 is 0.1123. The posterior probability of model 1b after 1941 is 0.0662.

[a] See footnotes 4 and 5.

TABLE 2

Predictive statistics for the simple AM models 1929–1941.
Model 1a employs M1 definition of money; model 1b employs M2[a]

Prediction year	Model	R_i^2	Predictive pdf Mean	Variance	$P(M_i \mid \text{data})$	Prediction error
1934	1a	0.982	54.52	16.36	0.0459	3.18
	1b	0.937	48.52	64.27	0.0021	−2.82
	2a	0.995	54.62	4.85	0.9541	3.28
	2b	0.995	54.62	4.85	0.9979	3.28
	a		54.62	5.38		3.28
	b		54.61	5.06		3.27
1935	1a	0.973	72.44	14.91	0.1711	16.74
	1b	0.935	58.78	29.94	0.0014	3.09
	2a	0.984	59.81	7.33	0.8289	4.11
	2b	0.984	59.81	7.33	0.9986	4.11
	a		61.97	31.25		6.27
	b		59.79	7.67		4.09

TABLE 2 (*continued*)

Prediction year	Model	R_i^2	Predictive pdf Mean	Variance	$P(M_i \mid \text{data})$	Prediction error
1936	1a	0.725	80.72	146.90	0.0006	18.80
	1b	0.926	66.93	24.59	0.0593	5.01
	2a	0.966	65.81	10.92	0.9994	3.90
	2b	0.966	65.81	10.92	0.9407	3.90
	a		65.82	11.15		3.91
	b		65.88	11.80		3.97
1937	1a	0.511	73.96	166.46	0.0001	7.45
	1b	0.902	69.62	25.57	0.0571	3.11
	2a	0.951	71.34	13.12	0.9999	4.84
	2b	0.951	71.34	13.12	0.9429	4.84
	a		71.34	13.13		4.84
	b		71.25	13.99		4.74
1938	1a	0.502	70.46	118.71	0.0001	6.54
	1b	0.899	68.87	21.28	0.1297	4.95
	2a	0.935	67.72	13.62	0.9999	3.80
	2b	0.934	67.72	13.62	0.8703	3.80
	a		67.72	13.63		3.80
	b		67.87	14.76		3.95
1939	1a	0.476	74.76	124.02	0.00008	7.92
	1b	0.878	74.18	24.26	0.1007	7.35
	2a	0.921	69.96	14.09	0.99992	3.13
	2b	0.921	69.96	14.09	0.8993	3.13
	a		69.96	14.10		3.13
	b		70.39	16.72		3.55
1940	1a	0.457	79.33	136.85	0.00003	8.51
	1b	0.840	80.90	33.98	0.0282	10.08
	2a	0.916	76.27	15.21	0.99997	5.45
	2b	0.916	76.27	15.21	0.9718	5.45
	a		76.27	15.22		5.45
	b		74.40	16.33		5.58
1941	1a	0.469	83.09	138.39	0.00004	2.51
	1b	0.793	86.17	48.17	0.0108	5.59
	2a	0.903	96.40	29.76	0.99996	15.83
	2b	0.903	96.40	29.76	0.9892	15.83
	a		96.40	29.77		15.83
	b		96.29	31.08		15.72

The posterior probability of model 1a after 1941 is 0.0017. The posterior probability of model 1b after 1941 is 0.3252.

[a] See footnote 6.

The predictive statistics for the 1929–1941 comparisons are given in tables 1 and 2. Tables 3 and 4 give the predictive statistics for the 1947–1965 comparisons.[7] In all these tables model 1 denotes the money model and model 2 the autonomous expenditure model. Each table gives the following information for two versions of the dichotomy. The coefficient of determination, R_i^2, is given for each model for the period 1929 (1947) to T–1 where T is the prediction year. The mean and variance of the predictive pdf (given the model and the value of the explanatory variable in year T) for year T are given next. The next column gives the posterior probability of model i after year T–1. Finally, the prediction error (mean of predictive pdf minus actual) is given. All numbers are in billions of current dollars. The last lines of each group give characteristics of the predictive pdf which is marginal of the models.

The posterior probabilities are computed under the assumption that $P(\text{model } 1) = P(\text{model } 2) = 0.5$ at the beginning of the sample period.[8] It is also assumed that the prior (i.e., prior to the first observation) pdf on the parameters of each model (α_i, β_i, σ_i) is sufficiently dispersed so that its contribution to the posterior is negligible. Finally, it is assumed that the errors, U_{it}, were distributed normally and independently with mean zero and variance σ_i^2.

Under these assumptions the predictive densities for year T given the model and the value of the explanatory variable in year T are univariate student t densities. The means are given by:

$$E(C_T|\text{model 1, previous data, } M_T) = \hat{\alpha}_{1(T-1)}+\hat{\beta}_{1(T-1)}M_T = \mu_{1(T-1)}$$

and

$$E(C_T|\text{model 2, previous data, } A_T) = \hat{\alpha}_{2(T-1)}+\hat{\beta}_{2(T-1)}A_T = \mu_{2(T-1)}$$

where $\hat{\alpha}_{i(T-1)}$ and $\hat{\beta}_{i(T-1)}$ are the least squares estimates from the regression covering 1929 (1947) to T–1.

The variance of the predictive pdf for C_T given model 1 is given by:

$$V(C_T|\text{model 1, previous data, } M_T)$$

$$= \frac{\nu s_1^2}{\nu-2}\left[1+\frac{1}{T-1}+\frac{(M_T-\overline{M}_{(T-1)})^2}{\sum_{t=1}^{T-1}(M_t-\overline{M}_{(T-1)})^2}\right]$$

[7] Complete regression statistics for all the models are available from the author.

[8] To analyze the sensitivity of the posterior probability to changes in the prior one can simply use the fact that the posterior odds ratio is the product of the prior odds ratio and the (averaged) likelihood ratio. That is,

$$\frac{P(\text{model 1}|\text{data})}{P(\text{model 2}|\text{data})} = \frac{P(\text{model 1})}{P(\text{model 2})}\cdot\frac{P(\text{data}|\text{model 1})}{P(\text{data}|\text{model 2})}.$$

where $v = (T-1)-2$ and

$$\overline{M}_{(T-1)} = \frac{1}{(T-1)} \sum_{t=1}^{T-1} M_t.$$

An analogous expression applies for model 2.

From tables 1 and 2 we observe that for the 1929–1941 period and for all four comparisons the posterior probabilities of the models vary substantially with the length of period considered. This perhaps is not surprising because of the small sample sizes but the magnitude of some of the year-to-year changes is nevertheless dramatic. We also note that the conditional (given next period's M or A) predictions of both models are with few exceptions too high. This could be the result of omitted variables or of instability in the relationships.

Table 1 reveals that the use of the revised data and the M2 definition of money rather than the FM definition does not alter the results substantially.[9] In both cases the posterior probability of the autonomous expenditure model is near 0.9 after 1933, drops to 0.3 in the mid-1930's and then rises to 0.9 or higher by 1940–1941.

Examination of table 2 (the AM comparisons) indicates that the posterior probability of the autonomous expenditure model is never less than 0.67 and is generally greater than 0.9. Its posterior probability is generally lower when the M2 definition of money rather than M1 is used in the money model. The exception to this is in the early part of the period.

The models labeled "b" in each table provide the least ambiguous comparison of the FM and the AM results. The only difference between the two is in the definition of autonomous expenditures. It is immediately clear that the choice of the definition is crucial since use of the FM definition of A_t results in much higher posterior probabilities for the money model (except after 1941).

Before attempting to draw conclusions as to the relative importance of money and autonomous expenditures as determinants of consumption

[9] The posterior probabilities of model 1 using the revised data but using FM's definition of money rather than the M2 definition are:

After 1933:	0.1017	After 1936:	0.6791	After 1939:	0.1645
After 1934:	0.2895	After 1937:	0.6042	After 1940:	0.0349
After 1935:	0.5631	After 1938:	0.4833	After 1941:	0.0775

during the 1929–1941 period we shall make some rather informal checks on the assumptions which were employed in the computations.

Examination of regression results for all the models indicates that the S_i^2 do not vary tremendously over time. The estimated coefficients of A_t and M_t both exhibit a moderate downward trend as the length of time considered increases. Normal probability plots of the least squares residuals (not illustrated due to volume and difficulty of reproduction) do not reveal major departures from normality. However, all the models exhibit substantial positive serial correlation of the residuals. This casts doubt on the above checks for homoscedasticity and normality. Its effect on the posterior probabilities can be sizeable.[10,11] This, combined with the previously noted overpredictions, suggests that the models be revised before conclusions about the relative importance of money and autonomous expenditures are drawn. We shall make some such revisions in the next section. To examine the assumption that M_t or A_t is truly exogenous requires the construction, estimation, and comparison of more complicated models. We have not undertaken that task in this paper.

Before proceeding to the analysis of more complicated models we shall give a brief discussion of the simple comparisons for the period 1947–1965.

Table 3 reveals that use of the revised data and the M2 definition of money rather than the FM definition causes substantial differences in

[10] Under the assumptions employed here the posterior probability depends on S^2 (see [13]). With serially correlated errors S^2 is apt to be biased downward if the explanatory variable also exhibits positive serial correlation. See [17, pp. 189–190] for discussion of this subject. An examination of the sensitivity of the (inappropriately calculated) posterior probabilities has been undertaken in [12]. Their results indicate the size of the error involved in computing the probabilities as if there were no serial correlation is related to the difference between the estimated serial correlation coefficients of the two models.

[11] Bayesian analysis of the regression model with serially correlated errors is relatively straightforward—see [27]. On the assumption of first-order autoregressive correlation (using the revised data) the posterior probabilities of the money model are:

Year	*FM Comparison*	*AM Comparison*
1935	0.807	0.053
1937	0.975	0.025
1939	0.972	0.002
1941	0.928	0.854

the predictive pdf and the posterior probabilities.[12] These changes work to the benefit of the money model.

As in the 1929–1941 period we see in table 4 that use of the M2 definition of money in the 1947–1965 period usually results in a higher posterior probability for the money model than does M1. We also observe that the posterior probabilities and the relative sizes of the prediction errors vary greatly over time. The posterior probability of the money model is low early in the period, rises during the late 1950's and early 1960's and then falls dramatically at the end of the period. Comparison of tables 3 and 4 once again shows that the use of FM's definition of autonomous expenditures results in a higher posterior probability for the money model than the use of AM's definition.

Examination of regression results for these models lead to the same conclusions as in the 1929–1941 period; namely, the serial correlation is sufficient to cast doubt on these results.[13]

We mentioned previously that the FM comparisons covered various periods from 1897–1929 in addition to those we have discussed. We have made comparisons for those periods which we shall now summarize briefly. In general they favor the money model even more impressively than the 1947–1965 FM comparisons. In the period 1897–1913 the posterior probability of the money model is never less than 0.99. For the sample starting in 1913 the posterior probability is in the vicinity of 0.90 in 1916–1918, rises to 0.99 in 1919 and is above 0.999 for the 1920's. The previously mentioned problem of serially correlated residuals is present in these samples also.

[12] The posterior probabilities of model 1 using the M2 definition of money but not the revised data are:

After 1951:	0.5789	After 1955:	0.9997 P (model 2)
After 1952:	0.3348	After 1956:	1.000 (P (model 2) $< 10^{-5}$)
After 1953:	0.8826	After 1957:	1.000 (P (model 2) $< 10^{-6}$)
After 1954:	0.9947	After 1958:	1.000 (P (model 3) $< 10^{-7}$)

This indicates that most of the difference between table 3 and table 4 is due to the use of the revised data rather than the use of the M2 definition of money in table 4.

[13] Allowing for serial correlation as indicated in footnote 11 yields the following results for the posterior probability of the Money Model:

Year	*FM Comparison*	*AM Comparison*
1953	0.613	0.467
1955	0.822	0.583
1957	0.505	0.425
1959	0.881	0.304
1961	0.569	0.028
1963	0.623	0.019
1965	0.793	0.044

TABLE 3

Predictive statistics for the simple FM models 1947–1965[a]. Models 1a and 2a employ original FM data; models 1b and 2b use revised data

Prediction year	Model	R_i^2	Predictive pdf Mean	Predictive pdf Variance	$P(M_i\|\text{data})$	Prediction error
1952	1a	0.893	248.20	612.30	0.6254	28.43
	1b	0.886	244.05	668.19	0.7448	27.37
	2a	0.869	208.57	259.21	0.3746	−11.20
	2b	0.825	216.68	313.15	0.2552	−18.69
	a		233.36	848.02		13.58
	b		232.29	980.71		15.61
1953	1a	0.864	241.45	362.49	0.3262	8.80
	1b	0.868	238.19	367.28	0.7819	8.22
	2a	0.893	211.52	154.91	0.6738	−21.13
	2b	0.798	201.97	290.22	0.2181	−28.00
	a		221.28	419.51		−11.37
	b		230.29	574.23		0.32
1954	1a	0.911	248.83	206.91	0.8775	10.81
	1b	0.915	246.27	209.11	0.9848	9.78
	2a	0.846	207.42	215.21	0.1225	−30.60
	2b	0.719	201.46	413.06	0.0152	−35.03
	a		243.55	398.77		5.52
	b		245.59	242.21		9.10
1955	1a	0.925	258.33	169.39	0.9939	1.39
	1b	0.931	256.80	166.48	0.9991	2.42
	2a	0.731	224.41	422.57	0.0061	−32.53
	2b	0.608	217.73	651.92	0.0009	−36.65
	a		258.12	177.91		1.18
	b		256.77	168.39		2.39
1956	1a	0.950	265.04	117.14	0.9996	−4.88
	1b	0.954	263.03	115.81	0.99994	−3.65
	2a	0.713	216.35	491.05	0.0004	−53.57
	2b	0.606	223.95	758.68	0.00006	−42.72
	a		265.02	118.15		−4.90
	b		263.03	115.95		−3.65

TABLE 3 (*continued*)

Prediction year	Model	R_i^2	Predictive pdf Mean	Variance	$P(M_i \mid \text{data})$	Prediction error
1957	1a	0.964	278.42	97.09	~1.0000	−6.33
	1b	0.967	276.44	94.69	~1.0000	−4.99
	2a	0.569	248.96	1017.87	$< 10^{-5}$	−35.79
	2b	0.589	234.51	933.54	$< 10^{-5}$	−46.92
	a		278.42	97.10		−6.33
	b		276.44	94.70		−4.99
1958	1a	0.973	299.84	90.19	~1.0000	6.88
	1b	0.975	298.32	86.26	~1.0000	8.25
	2a	0.629	232.72	905.95	$< 10^{-6}$	−60.24
	2b	0.587	247.15	1118.42	$< 10^{-6}$	−42.91
	a		299.84	90.19		6.88
	b		298.32	86.26		8.25
1959	1a	0.978			~1.0000	
	1b	0.979	315.08	83.20	~1.0000	3.88
	2a	0.554			$< 10^{-7}$	
	2b	0.608	250.01	1162.43	$< 10^{-7}$	−61.20
	a					
	b		315.08	83.20		3.88
1961	1b	0.986	349.26	71.77	~1.0000	14.11
	2b	0.427	265.80	2336.48	$< 10^{-11}$	−69.36
	b		349.26	71.77		14.11
1963	1b	0.981	412.20	140.84	~1.0000	37.22
	2b	0.533	298.71	2521.51	$< 10^{-11}$	−76.28
	b		412.20	140.84		37.22
1965	1b	0.967	472.31	334.81	1.0000	40.84
	2b	0.650	433.59	3476.95	$< 10^{-9}$	2.13
	b		472.31	334.81		40.84

The posterior probability of model 2b after 1965 is 0.376×10^{-8}.

[a] See footnotes 4 and 5.

TABLE 4

Predictive statistics for the AM simple models 1947–1965[a]

Prediction year	Model	R_i^2	Predictive pdf Mean	Variance	$P(M_i\|\text{data})$	Prediction error
1952	1a	0.787	235.56	1098.83	0.0355	18.89
	1b	0.886	244.05	668.19	0.1492	27.37
	2a	0.943	225.38	198.51	0.9645	8.70
	2b	0.943	225.38	198.51	0.8508	11.48
	a		225.78	234.01		9.06
	b		228.16	312.86		11.48
1953	1a	0.846	230.95	372.49	0.0211	0.98
	1b	0.868	238.19	367.28	0.0331	8.22
	2a	0.957	227.17	91.09	0.9789	−2.80
	2b	0.957	227.17	91.09	0.9669	−2.80
	a		227.25	97.32		−2.72
	b		227.53	104.11		−2.44
1954	1a	0.909	236.46	192.93	0.0129	−0.04
	1b	0.915	246.27	209.11	0.0163	9.78
	2a	0.974	223.20	46.05	0.9871	−13.29
	2b	0.974	223.20	46.05	0.9837	−13.29
	a		223.37	50.19		−13.12
	b		223.58	57.25		−12.92
1955	1a	0.936	251.46	144.16	0.1599	−2.92
	1b	0.931	256.80	166.48	0.1238	2.42
	2a	0.958	237.66	80.64	0.8401	−16.72
	2b	0.958	237.66	80.64	0.8762	−16.72
	a		239.87	116.39		−14.51
	b		240.03	131.03		−14.35
1956	1a	0.957	257.43	102.48	0.7062	−9.25
	1b	0.954	263.03	115.81	0.6383	−3.65
	2a	0.948	261.92	130.23	0.2938	−4.76
	2b	0.948	261.92	130.23	0.3617	−4.76
	a		258.75	114.82		−7.93
	b		262.63	121.31		−4.05

TABLE 4 (*continued*)

Prediction year	Model	R_i^2	Predictive pdf Mean	Variance	$P(M_i\|\text{data})$	Prediction error
1957	1a	0.965	262.94	91.27	0.5922	−18.49
	1b	0.967	276.44	94.69	0.6657	−4.99
	2a	0.962	277.95	110.14	0.4078	−3.48
	2b	0.962	277.95	110.14	0.3343	−3.48
	a		269.06	153.38		−12.37
	b		276.95	100.37		−4.49
1958	1a	0.959	273.55	122.77	0.1036	−16.52
	1b	0.975	298.32	86.27	0.6505	8.25
	2a	0.972	279.65	85.71	0.8964	−10.42
	2b	0.972	279.65	85.71	0.3495	−10.42
	a		279.02	93.01		−11.05
	b		291.80	165.32		1.73
1959	1a	0.958	295.20	146.47	0.0469	−16.01
	1b	0.979	315.08	83.20	0.7377	3.88
	2a	0.975	301.08	91.25	0.9531	−10.13
	2b	0.975	301.08	91.25	0.2623	−10.13
	a		300.80	95.39		−10.40
	b		311.41	123.27		0.20
1961	1a	0.947	311.17	235.49	0.0007	−23.98
	1b	0.986	349.26	71.77	0.8830	14.11
	2a	0.981	332.27	89.25	0.9993	−2.88
	2b	0.981	332.27	89.25	0.1170	−2.88
	a		332.25	89.65		−2.90
	b		347.27	103.63		12.12
1963	1a	0.948	354.13	308.66	0.00001	−20.85
	1b	0.981	412.20	140.84	0.0203	37.22
	2a	0.988	373.24	75.37	0.99999	−1.74
	2b	0.988	373.24	75.37	0.9797	−1.74
	a		373.24	75.38		−1.74
	b		374.03	106.87		−0.95
1965	1a	0.963	429.81	327.97	$< 10^{-6}$	−1.65
	1b	0.967	472.31	334.81	$< 10^{-5}$	40.84
	2a	0.992	428.34	65.75	~1.00	−3.12
	2b	0.992	428.34	65.75	~1.00	−3.12
	a		428.34	65.75		−3.12
	b		428.34	65.76		−3.12

The posterior probability of model 1a after 1965 is less than 10^{-6}. The posterior probability of model 1b after 1965 is less than 10^{-7}.

[a] See footnote 6.

5. Bayesian comparisons of money and autonomous expenditure models which include lagged consumption as an explanatory variable

An ad hoc procedure sometimes used to "remove" serial correlation is to include as another explanatory variable a lagged value of the dependent variable. This was done by AM. It should be noted, however, that this procedure does not by any means guarantee "removal" of serial correlation. If the errors of such equations are serially correlated, the least squares estimators of the coefficients are inconsistent.[14]

We shall present the results of comparisons of the form:

$$M_1: \quad C_t = \beta_{01}+\beta_{11}M_t+\beta_{21}C_{t-1}+u_{1t}$$

$$M_2: \quad C_t = \beta_{02}+\beta_{12}A_t+\beta_{22}C_{t-1}+u_{2t} \qquad (8)$$

where C_t, A_t, and M_t represent the same quantities as in the previous sections. First, however, we note that the second equation of (8) may be regarded as a reduced form equation of

1) Consumption function: $C_t^D = kY_t^* + u_t$
2) Autonomous expenditure: $A_t^D = A_t$ (exogenous)
3) Aggregate demand: $Y_t^D = C_t^D + A_t^D$
4) Equilibrium: $Y_t^D = Y_t^S = Y_t$
5) Mechanism generating expected income (Y_t^*): $(Y_t^* - Y_{t-1}^*) = (1-\lambda)(Y_t - Y_{t-1}^*)$, (9)

where the symbols have the same meaning as in equation (7).[15] This is a simple Keynesian model with a "permanent income" or "expected

[14] For discussion of this subject see, e.g., [17, pp. 214–216].

[15] Solve (5) to yield

$$Y_t^* = (1-\lambda)(Y_t+\lambda Y_{t-1}+\lambda^2 Y_{t-2}+\ldots).$$

Combine this with (1) and (4) to yield

$$C_t^D = \lambda C_{t-1}^D + bY_t + \varepsilon_t$$

where $b = k(1-\lambda)$ and $\varepsilon_t = u_t - \lambda u_{t-1}$. Now use (4) to obtain

$$C_t = \beta A_t + \lambda C_{t-1} + v_t$$

where $\beta = b/(1-b)$, $\gamma = \lambda/(1-b)$, and $v_t = \varepsilon_t/(1-b)$. We have added a constant term in equation (8) for convenience.

income" consumption function. Thus, this model incorporates an important aspect of the revisions in consumption theory which have taken place in the last twenty years. The concept of permanent income has also been used in the study of the demand for money (see [9]). The first equation of (8) does not appear to be a reduced form equation of such a revised Quantity Theory Model, however.[16] Thus, interpretation of the comparisons below as a test of a Quantity Theory Model versus an Autonomous Expenditure Model does not seem possible.

We confine ourselves to two sets of comparisons for the dichotomy (8). Tables 5 and 6 present comparisons using the revised data, the M2 definition of money and both the FM and AM definitions of autonomous expenditure for 1930–1941 and 1948–1965 respectively. The entries in the tables are the same as in tables 1–4.[17]

Examination of table 5 indicates that the posterior probability of the autonomous expenditure model is in the vicinity of 0.99 for both definitions of A_t. On comparing these results with those in tables 1 and 2 we see that the posterior probability of the autonomous model is always higher in the case where lagged consumption is included as an explanatory variable. Also, in this case there is less discrepancy between the results using the AM definition of A_t and those using the FM definition and the posterior probabilities are less variable over time. Finally, the prediction errors are generally smaller here.

Informal checks indicate that the assumption of homoscedasticity is not grossly violated in these equations. Normal probability plots reveal no major departures from normality. The estimated serial correlation coefficients, while appreciably negative early in the period fall to very low levels when the sample is extended to 1938 and beyond. Thus, we may feel moderately confident of our results for this period.

Table 6 presents a similar story for the 1947–1965 period. Once again there is much less discrepancy between the FM and AM results, the posterior probabilities fluctuate less, and the prediction errors are smaller than in (tables 3 and 4) the corresponding models which do not include lagged consumption. For this period neither model dominates

[16] Modification of the simple Quantity Theory Model (6) to include a permanent income demand for money function and proceeding as in the solution of equations (8) and (6) would seem to dictate regression C_t (actually Y_t) on M_t and M_{t-1} rather than M_t and C_{t-1}.

[17] S_i^2 is the estimated residual variance for model i. The assumptions on prior densities and mass functions are the same as in the previous section.

TABLE 5

Predictive statistics for the models with lagged consumption included,[a] 1930–1941. Model 2a employs the FM definition of automous expenditure; model 2b uses the AM definition

Prediction year	Model	S_i^2	Mean of predictive pdf	$P(M_i \| \text{data})$	Prediction error
1935	1a	5.70	61.48	0.0314	5.78
	1b	5.70	61.48	0.0059	5.78
	2a	1.45	57.29	0.9686	1.59
	2b	0.733	56.95	0.9941	1.25
	a		57.42		1.72
	b		56.97		1.27
1936	1a	7.41	65.04	0.0065	3.13
	1b	7.41	65.04	0.0011	3.13
	2a	1.39	63.51	0.9935	1.59
	2b	0.758	62.19	0.9989	0.28
	a		63.52		1.60
	b		62.19		0.28
1937	1a	6.54	66.44	0.0040	−0.07
	1b	6.54	66.44	0.0002	−0.07
	2a	1.35	64.69	0.9960	−1.82
	2b	0.579	68.17	0.9998	1.66
	a		64.69		−1.82
	b		68.17		1.66
1938	1a	5.23	66.67	0.0078	2.75
	1b	5.23	66.67	0.0005	2.75
	2a	1.56	62.61	0.9922	−1.31
	2b	0.785	66.76	0.9995	2.84
	a		62.64		−1.28
	b		66.76		2.84
1939	1a	5.33	71.03	0.0037	4.20
	1b	5.33	71.03	0.0057	4.20
	2a	1.54	67.31	0.9963	0.47
	2b	1.70	67.63	0.9943	0.79
	a		67.32		0.48
	b		67.65		0.81

TABLE 5 (*continued*)

Prediction year	Model	S_i^2	Mean of predictive pdf	$P(M_i \vert \text{data})$	Prediction error
1940	1a	6.12	76.74	0.0005	5.92
	1b	6.12	76.74	0.0009	5.92
	2a	1.34	73.24	0.9995	2.42
	2b	1.52	73.33	0.9991	2.51
	a		73.24		2.42
	b		73.33		2.51
1941	1a	7.64	81.87	0.0002	1.30
	1b	7.64	81.87	0.0004	1.30
	2a	1.65	85.15	0.9998	4.58
	2b	1.85	90.03	0.9996	9.45
	a		85.15		4.58
	b		90.03		9.45

The posterior probability of model 2a after 1941 is 0.9977. The posterior probability of model 2b after 1941 is 0.8618.

[a] For data definitions and sources, see footnotes 4, 5 and 6.

TABLE 6

Predictive statistics for the models with lagged consumption included,[a] 1947–1965

Prediction year	Model	S_i^2	Mean of predictive pdf	$P(M_i \vert \text{data})$	Prediction error
1953	1a	35.63	231.96	0.2733	1.99
	1b	35.63	231.96	0.2790	1.99
	2a	24.10	223.38	0.7267	−6.59
	2b	24.37	227.45	0.7210	−2.52
	a		225.72		−4.25
	b		228.71		−1.26
1954	1a	24.24	243.94	0.3959	7.44
	1b	24.24	243.94	0.2628	7.44
	2a	21.06	237.42	0.6041	0.93
	2b	17.19	232.18	0.7372	−4.31
	a		240.00		3.51
	b		235.27		−1.22

TABLE 6 (*continued*)

Prediction year	Model	S_i^2	Mean of predictive pdf	$P(M_i \mid \text{data})$	Prediction error
1955	1a	25.13	249.34	0.1665	−5.04
	1b	25.13	249.34	0.1096	−5.04
	2a	15.86	246.13	0.8335	−8.25
	2b	13.81	243.80	0.8904	−10.58
	a		246.66		−7.72
	b		244.40		−9.98
1956	1a	22.37	264.72	0.4393	−1.96
	1b	22.37	264.72	0.5682	−1.95
	2a	21.05	265.97	0.5607	−0.71
	2b	23.96	267.27	0.4318	0.60
	a		265.42		−1.26
	b		265.82		−0.85
1957	1a	18.97	277.81	0.4155	−3.63
	1b	18.97	277.81	0.5591	−3.63
	2a	17.59	278.82	0.5845	−2.61
	2b	20.00	280.23	0.4409	−1.20
	a		278.40		−3.03
	b		278.88		−2.56
1958	1a	17.37	296.23	0.3747	6.16
	1b	17.37	296.23	0.4930	6.16
	2a	15.68	295.07	0.6253	5.00
	2b	17.27	292.30	0.5070	2.23
	a		295.51		5.44
	b		294.24		4.17
1959	1a	18.29	306.71	0.3057	−4.50
	1b	18.29	306.71	0.2831	−4.50
	2a	15.76	300.74	0.6943	−10.47
	2b	15.45	302.85	0.7169	−8.36
	a		302.56		−8.65
	b		303.94		−7.27
1961	1a	16.42	343.90	0.8194	8.74
	1b	16.42	343.90	0.5839	8.74
	2a	20.73	340.74	0.1806	5.59
	2b	17.30	339.37	0.4161	4.22
	a		343.33		8.17
	b		342.02		6.86

TABLE 6 (*continued*)

Prediction year	Model	S_i^2	Mean of predictive pdf	$P(M_i \mid \text{data})$	Prediction error
1963	1a	18.27	381.38	0.6285	6.40
	1b	18.27	381.38	0.2736	6.40
	2a	19.59	371.12	0.3715	−3.86
	2b	16.04	372.57	0.7264	−2.41
	a		377.57		2.59
	b		374.98		−0.004
1965	1a	16.65	432.36	0.6495	0.90
	1b	16.65	432.36	0.3549	0.90
	2a	17.90	429.90	0.3505	−1.57
	2b	15.52	425.35	0.6451	−6.11
	a		431.50		0.03
	b		427.84		−3.62

The posterior probability of model 2a after 1965 is 0.3351. The posterior probability of model 2b after 1965 is 0.4024.

[a] See footnotes 4, 5 and 6.

the other. No posterior probability is greater than 0.9 and most are less than 0.7.[18,19]

Once again checks on the validity of the assumptions do not reveal any glaring departures although the estimated serial correlation coefficients for the FM autonomous expenditure model are in the vicinity of 0.5.[20]

[18] For the period 1897–1913, comparisons using FM data and definitions with lagged consumption included favour the money model dramatically (P (model 1) $= 0.99$). For 1913–1929 the posterior probability of the money model ranges from about 0.75 to 0.95.

[19] One possible explanation of this might be that neither M_t nor A_t is very important in a model which includes C_{t-1}. Examination of the regression coefficients of these variables indicates that this is not the case, however. Their t-statistics are almost invariably above 1 and generally in the range 1.5 to 2.0. These are of course much lower than in the models without C_{t-1}, but they are large enough to make substantial reductions in the error variances.

[20] To examine the effect of this "remaining" serial correlation we estimated (using the same data and definitions as in table 6) equations of the form:

$$C_t - \rho_i C_{t-1} = \beta_{0i}(1-\rho_i) + \beta_{1i}(X_{it} - \rho_i X_{i,t-1}) + \beta_{2i} \times (C_{t-1} - \rho_i C_{t-2}) + u_{it}$$

where $X_{1t} = M_t$ and $X_{2t} = A_t$. An iterative procedure (over values of ρ_i) was used to estimate the ρ_i's. Conditional on the likelihood maximizing values (approx.

6. Summary and conclusions

The results for the simple models show great variation with the time period considered and the definitions of the explanatory variables. Even if we agree on the M2 definition of money the two different definitions of autonomous expenditure give vastly different results, the AM definition making the autonomous expenditure model look relatively better. Serial correlation of the errors is serious enough to cast doubt on the appropriateness of these results.

Allowance for possible first-order autocorrelation accentuates the difference between the different definitions of autonomous expenditure for the 1929–1941 period. For the 1947–1965 period wide discrepancies between the two comparisons appear in the 1960's. Neither the money model nor the autonomous expenditure model is overwhelmingly dominant in this period although the autonomous expenditure model performs better when the AM definition is employed.

Models which include C_{t-1} as an explanatory variable seem to have the best statistical characteristics (e.g., smaller prediction errors, lower S^2's) and are not as sensitive to the definition of autonomous expenditures but they also yield widely different results for different time periods. For the pre-1929 periods (FM autonomous expenditure definition only) the posterior probabilities of the money model were quite high. For the 1929–1941 (both definitions) the autonomous expenditure model was dominant. In the 1947–1965 period neither model dominated the other.

This variability over time and sensitivity to definitions of variables would seem to preclude any general conclusions regarding the relative importance of money and autonomous expenditures as determinants of consumption based on comparisons such as those of FM and AM.

$\rho_1 = -0.4$ and $\rho_2 = -0.5$) the posterior probability of the money model is 0.13 after 1958 and 0.67 after 1965. The differences between these numbers and the corresponding ones in table 8 are the result of two changes. First, allowance for possible serial correlation and, second, a change in the effective sample size (two lagged values of C are required so $t = 1$ is now 1949 rather than 1948). If we set $\rho_1 = \rho_2 = 0$ in this model we isolate the second change. The corresponding posterior probabilities of the money model are 0.22 and 0.58 after 1958 and 1965 respectively. Thus, we conclude that the serial correlation present in the models under discussion is not dramatically affecting the posterior probabilities.

Resolution of this issue will probably require application of the methods discussed here and related procedures to simultaneous equations macroeconometric models.

References

[1] A. Ando and F. Modigliani, The "life cycle" hypothesis of saving. American Economic Review LIII (March 1963), 55–84.

[2] A. Ando and F. Modigliani, The velocity and the investment multiplier. American Economic Review LV (September 1965), 693–728.

[3] M. Bailey, National income and the price level. New York, McGraw-Hill, 1962.

[4] G. E. P. Box and W. J. Hill, Discrimination among mechanistic models. Technometrics 9 (February 1967), 57–71.

[5] V. K. Chetty, On measuring the nearness of near moneys. American Economic Review LIX (June 1969), 270–281.

[6] G. Chow, On the long run and short run demand for money. Journal of Political Economy 74 (April 1966), 111–131.

[7] M. DePrano and T. Mayer, Autonomous expenditures and money. American Economic Review LV (September 1965), 729–752.

[8] M. Friedman, A theory of the consumption function. Princeton, Princeton University Press, 1957.

[9] M. Friedman, The demand for money: some theoretical and empirical results. Journal of Political Economy 67 (August 1959), 327–351.

[10] M. Friedman and D. Meiselman, The Relative stability of velocity and the investment multiplier. In: Stabilization policies, Commission on Money and Credit. Englewood Cliffs, N.J., Prentice-Hall, 1963, 165–268.

[11] M. Friedman and A. Schwartz, Money and business cycles. Review of Economics and Statistics XLV (February 1963, Part 2), 32–78.

[12] K. M. Gaver and M. S. Geisel, Comparing linear regression models with serially correlated errors. Carnegie-Mellon University manuscript, 1972.

[13] M. S. Geisel, Comparing and choosing among parametric statistical models: a Bayesian analysis with macroeconomic applications. Unpublished Ph.D. dissertation, University of Chicago, 1970.

[14] J. W. Gorman and R. J. Toman, Selection of variables for fitting equations to data. Technometrics 8 (February 1966), 27–51.

[15] T. Haavelmo, Methods of measuring the marginal propensity to consume. Journal of the American Statistical Association 42 (March 1947), 105–122.

[16] D. Hester, Keynes and the quantity theory: comment on the Friedman and Meiselman CMC paper. Review of Economics and Statistics XLVI (November 1964), 364–377.

[17] J. Johnston, Econometric methods. New York, McGraw-Hill, 1963.

[18] D. V. Lindley, The Choice of variables in multiple regression. Journal of the Royal Statistical Society (B) 30, No. 1 (1968), 31–66.

[19] A. Meltzer, The demand for money: the evidence from the time series. Journal of Political Economy 71 (June 1963), 219–246.

[20] F. Modigliani and R. Brumberg, Utility analysis and the consumption function. In: Post Keynesian economics, edited by K. Kurihara. New Brunswick, N.J., Rutgers University Press, 1954, 388–436.

[21] H. V. Roberts, Probabilistic prediction. Journal of the American Statistical Association 60 (March 1965), 50–62.

[22] S. L. Sclove, Choosing a "best" regression equation for prediction. Carnegie-Mellon University manuscript, 1969.

[23] R. Teigen, Demand and supply functions for money in the U.S.: some structural estimates. Econometrica 32 (October 1964), 476–509.

[24] H. Theil, Principles of econometrics. New York, Wiley, 1971.

[25] E. H. Thornber, Applications of decision theory to econometrics. Unpublished Ph.D. dissertation, University of Chicago, 1966.

[26] A. Zellner and M. Geisel, Analysis of distributed lag models with application to consumption function estimation, Econometrica 38 (1970), 865–887.

[27] A. Zellner and G. C. Tiao, Bayesian analysis of the regression model with autocorrelated errors. Journal of the American Statistical Association 59 (September 1964), 763–778.

PART 7

MISSING OBSERVATION PROBLEMS

S. E. Fienberg and A. Zellner (eds.)
Studies in Bayesian Econometrics and Statistics.

7.1 MISSING VARIABLES IN BAYESIAN REGRESSION*

S. James PRESS

University of Chicago

and

Alastair SCOTT

University of Auckland

1. *Introduction*

This work concerns the problem of parameter estimation in normal regression, with some of the observations missing. A Bayesian approach with vague prior distributions is taken. The missing observations are assumed to be normally distributed and missing in a nonsystematic way. No distributional assumptions are made about the independent variables for which no observations are missing; therefore, these variables may be deterministic, or binary, or whatever. Joint estimators of the parameters are obtained as the joint mode of the posterior distribution.

Interest in the present problem began at least as early as 1932 when Wilks [1] considered simple linear regression, under bivariate normality, with missing observations. His prescription was to replace each missing datum by its mean value and then do the usual regression. In 1933, Yates [2] suggested that the sum of squared residuals be minimized with respect to the incomplete observations (using only complete observations to estimate variable means). It may be shown that such a procedure is equivalent to discarding all observed dependent variables

* An earlier version of this paper was presented to the Fourth NBER-NSF Seminar on Bayesian Inference in Econometrics, Univ. of Chicago, May 1972. An extension of this work to the case where the missing observations, on the average, are linear functions of the non-missing ones, appears in a sequel to this paper, entitled "Missing variables in Bayesian regression, II", which will be published shortly.

corresponding to missing observations on any of the independent variables.

A very instructive Bayesian interpretation of this last result is obtained by adopting a vague prior distribution on the regression parameters and on the missing observations. Thus, the missing observations are assumed to have a uniform distribution over the entire real line, a priori. Under a normality assumption on the dependent variables given the independent variables, it is easy to check that the joint modal value of the posterior distribution given the observed data is the same as the value sought in the Yates' solution. Since only vague prior information is provided for the missing data in this case (all values are equally likely) it is not surprising that the conclusion should be to discard the corresponding dependent variable observations, since they really add nothing to our previous knowledge.[1]

Anderson [3] suggested a maximum likelihood estimation approach for cases in which the observations are missing in a special monotone pattern. Some other basic approaches include those of Glasser [4], Afifi and Elashoff [5], and Woodbury and Orchard [6]. A simulation study of the Yates approach was carried out by Haitovsky [7]. Bayesian approaches include those of Dagenais [8], Kaufman and Krieger [9], and Mehta and Swamy [10]. Some fifty-one references to work on this problem are given in [6].

In this paper, explicit estimators are obtained in the case of one independent variable in section 2. Iterative methods of solution are required for the multiple regression case, developed in section 3. For interval estimation purposes, distributional and asymptotic results are developed in section 4.

2. *Simple linear regression*

2.1. *Joint estimation of regression parameters and missing independent observations*

Suppose $(x_1, y_1), \ldots, (x_N, y_N)$ are mutually independent bivariate random vectors, $(y_j|x_j)$ follows the distribution $N(\alpha+\beta x_j, \sigma^2)$, x_j is distributed as $N(\mu, \delta^2)$, and $j = 1, \ldots, N$. Suppose further that $z \equiv (z_1, \ldots, z_n)' \equiv (x_1, \ldots, x_n)'$ is a vector of observations of x which are missing in a nonsystematic way, and none of the

[1] We are grateful to A. Zellner for suggesting this Bayesian analogue.

y_j's are missing. Let $w \equiv (w_1, \ldots, w_{N-n})' \equiv (x_{n+1}, \ldots, x_N)'$ and $y \equiv (y_1, \ldots, y_N)'$. Adopt the vague prior density[2]

$$p(\alpha, \beta, \sigma, \mu, \delta) \propto \frac{1}{\sigma\delta}, \tag{1}$$

where $\propto$ denotes proportionality. The joint density of all the variables is given by

$$\begin{aligned} p &= p(\alpha, \beta, \sigma, \mu, \delta, y, z, w) \\ &\propto (\sigma\delta)^{-(N+1)} \exp\left(-\tfrac{1}{2}\right)\left\{\frac{1}{\sigma^2}\sum_{j=1}^{N}(y_j-\alpha-\beta x_j)^2 \right. \\ &\quad \left. +\frac{1}{\delta^2}\sum_{j=1}^{N}(x_j-\mu)^2\right\}. \end{aligned} \tag{2}$$

Define the least squares estimators

$$\hat{\beta} \equiv \frac{\sum_1^N (x_j-\bar{x})(y_j-\bar{y})}{\sum_1^N (x_j-\bar{x})^2}, \qquad \hat{\alpha} = \bar{y}-\hat{\beta}\bar{x}, \tag{3}$$

where $N\bar{x} = \Sigma_1^N x_j$, $N\bar{y} = \Sigma_1^N y_j$. Substituting (3) into (2) and simplifying gives

$$\begin{aligned} p &\propto (\sigma\delta)^{-(N+1)} \exp\left(-\tfrac{1}{2}\right)\left\{\frac{1}{\sigma^2}\sum_{j=1}^{N}(y_j-\hat{\alpha}-\hat{\beta}x_j)^2 \right. \\ &\quad +\frac{1}{\sigma^2}(\beta-\hat{\beta})^2\sum_{j=1}^{N}(x_j-\bar{x})^2+\frac{N}{\sigma^2}[(\alpha-\hat{\alpha})+(\beta-\hat{\beta})\bar{x}]^2 \\ &\quad \left. +\frac{1}{\delta^2}\sum_{j=1}^{N}(x_j-\mu)^2\right\}. \end{aligned} \tag{4}$$

Integrating (4) with respect to σ and δ (the integrands are kernels of inverted gamma distributions) gives the marginal posterior density

$$\begin{aligned} f &\equiv f(\alpha, \beta, \mu, y, z, w) \\ &\propto \left\{\left[N(\bar{x}-\mu)^2+\sum_{j=1}^{N}(x_j-\bar{x})^2\right]\left[\sum_{j=1}^{N}(y_j-\hat{\alpha}-\hat{\beta}x_j)^2 \right.\right. \\ &\quad \left.\left. +N[(\alpha-\hat{\alpha})+(\beta-\hat{\beta})\bar{x}]^2+(\beta-\hat{\beta})^2\sum_{j=1}^{N}(x_j-\bar{x})^2\right]\right\}^{-N/2} \end{aligned} \tag{5}$$

[2] This is the invariant prior density of Jeffreys [11].

Note that the conditional density of the regression parameters and the missing observations given the observed data, $(\alpha, \beta, \mu, z|y, w)$ is proportional to the right-hand side of (5).

The joint mode of the posterior distribution of $(\alpha, \beta, \mu, z|y, w)$ is obtained by maximizing the right-hand side of (5) simultaneously with respect to α, β, μ, and z; that is, by minimizing the quantity in braces in (5). It is clear from inspection of (5) that the modal Bayes estimators $(\mu_B, \beta_B, \alpha_B)$, are given by $\mu_B = \bar{x}$, $\beta_B = \hat{\beta}$, and $\alpha_B = \hat{\alpha}$. Note that these estimators still depend upon the missing observations. Thus, the problem remaining is to minimize

$$\Omega \equiv \sum_{j=1}^{N} (x_j - \bar{x})^2 \left[\sum_{j=1}^{N} (y_j - \hat{\alpha} - \hat{\beta} x_j)^2 \right] \tag{6}$$

with respect to the elements of z (the missing observations vector).

Substituting the least-squares estimators from (3) into (6), and simplifying, gives

$$\Omega = a_0 \left[\sum_{j=1}^{N} x_j^2 - N\bar{x}^2 \right] - \left[\sum_{j=1}^{N} a_j x_j \right]^2, \tag{7}$$

where $a_0 \equiv \Sigma_1^N (y_j - \bar{y})^2$ and $a_j \equiv (y_j - \bar{y}), j = 1, \ldots, N$.

Differentiating (7) with respect to $z_j \equiv x_j$, $j = 1, \ldots, n$, setting the result equal to zero, and rearranging, shows that $\hat{z}_k$, the Bayesian modal estimator of z_k, must satisfy the equation

$$(\hat{z}_k - \bar{x}) = (y_k - \bar{y})\hat{b}, \quad k = 1, \ldots, n, \tag{8}$$

where

$$\hat{b} \equiv \frac{\sum_1^N (x_j - \bar{x})(y_j - \bar{y})}{\sum_1^N (y_j - \bar{y})^2}$$

denotes the least squares estimator of the slope coefficient in the backwards regression of x on y. That is, the Bayesian modal estimator of any missing observation is that implied by requiring the missing value to lie on the backwards regression line of x on y, when the line is calculated, assuming no observations are missing. All elements of z are to be estimated simultaneously. Once z has been estimated, α_B, β_B, and μ_B are estimated by substituting $\hat{z}$ for any missing elements in the maximum likelihood estimators of (α, β, μ).

2.2. *Explicit estimators of missing observations*

The Bayesian modal estimators of the missing observations were shown to satisfy the relation given in (8). It is now shown that z may be found in explicit form, obviating the necessity for iterative solutions.

If the backwards regression of x on y is denoted by $E(x_j|y_j) \equiv a+by_j, j = 1, \ldots, N$, the least squares estimators, $(\hat{a}, \hat{b})$, minimize

$$\sum_{j=1}^{N} (x_j-a-by_j)^2 \equiv \sum_{j=1}^{n} (x_j-a-by_j)^2 + \sum_{j=n+1}^{N} (x_j-a-by_j)^2.$$

But by (8), $\Sigma_1^n(x_j-\hat{a}-\hat{b}y_j)^2 = 0$. Therefore, $(\hat{a}, \hat{b})$ also minimize $\Sigma_{n+1}^N(x_j-a-by_j)^2$. That is, $(\hat{a}, \hat{b})$ are identical with the least squares estimators of the regression line determined by regressing x on y, and ignoring all pairs in which an x is missing. Therefore, if

$$\bar{\bar{x}} \equiv \frac{1}{N-n} \sum_{j=n+1}^{N} x_j, \quad \text{and} \quad \bar{\bar{y}} \equiv \frac{1}{N-n} \sum_{j=n+1}^{N} y_j, \tag{9}$$

$$\hat{a} = \bar{\bar{x}}-\hat{b}\bar{\bar{y}}, \qquad \text{and} \quad \hat{b} = \frac{\sum_{j=n+1}^{N} (x_j-\bar{\bar{x}})(y_j-\bar{\bar{y}})}{\sum_{j=n+1}^{N} (y_j-\bar{\bar{y}})^2}. \tag{10}$$

Substituting into (8) shows that the Bayesian modal estimator of the kth missing x is given explicitly by

$$\hat{z}_k = \bar{\bar{x}}+\hat{b}(y_k-\bar{\bar{y}}), \quad k = 1, \ldots, n, \tag{11}$$

where $\bar{\bar{x}}$, $\bar{\bar{y}}$, and $\hat{b}$ are given in (9) and (10), and do not depend upon any missing observations. That is, the missing x is constrained to lie on the backwards regression line obtained after discarding all incomplete pairs of observations.

The complete modal estimated Bayesian solution for the forward regression parameters is now $\hat{\alpha}$ and $\hat{\beta}$, with all missing x's in $\hat{\alpha}$ and $\hat{\beta}$ replaced by their values given by (11). For comparison, note that if the missing x's were given a vague prior distribution (instead of a normal), as discussed in section 1, $\hat{\alpha}$ and $\hat{\beta}$ would be the values computed by ignoring all incomplete pairs of (x, y).

2.3. *Joint estimation of regression parameters and missing dependent and independent observations*

In section 2.1 it was assumed that some of the observations on x were missing but none of the observations on y were missing. Now relax this assumption requiring only that at least one member of the pair (x_j, y_j) is not missing for a given $j, j = 1, \ldots, N$. The joint density of all the variables is still given by (5), and the conditional density of the regression parameters and all missing x's and y's, given the observed data, is still proportional to the right-hand side of (5). Hence, modal Bayes' estimators are obtained as before, but Ω in (6) must now be minimized simultaneously with respect to missing x's and missing y's. Carrying out the minimization with respect to missing x's gives (8). Carrying it out with respect to missing y's gives (by analogy with (8)),

$$(\hat{\ }_i-\bar{y}) = (x_{n+i}-\bar{x})\hat{\beta}, \quad i = 1, \ldots, m, \tag{12}$$

where $\hat{\beta}$ is given in (3), $\eta_i \equiv y_{n+i}$, and $\hat{\eta}_i$ denotes the estimated value of the ith missing y (m is the number of missing y's and $n+m \leqq N$). Bayesian modal estimators of missing x's and y's may be obtained by solving (8) and (12) simultaneously. Note that if there are no missing x's but only missing y's, the indicated Bayesian modal solution is to fill in the missing y's so that they lie on the forward regression line of y on x obtained by ignoring all incomplete pairs of points.

3. *Multiple linear regression*

The "simple" linear regression model of section 2 is now extended to models in which there are p^* independent variables, of which p have at least one observed value missing. The results reduce to those of the simple linear regression model by taking $p^* = p = 1$. Notation used in this section is independent of that in section 2.

Assume $(X_1, y_1), \ldots, (X_N, y_N)$ are mutually independent random vectors with X_j':$p^* \times 1$ and the y_j's are scalars, $j = 1, \ldots, N$. Suppose

$(y_j|X_j, \alpha, \beta, \sigma^2)$ follows the law $N(\alpha+\beta' X_j, \sigma^2)$. Define

$$X_j \equiv \begin{pmatrix} u_j \\ w_j \end{pmatrix},$$

where u_j:$p\times 1$, and for some of the j's, $j = 1, \ldots, N$, at least one element of u_j is missing (no element of any w_j is missing). Assume $(u_j|\mu, \Sigma)$ follows the law $N(\mu, \Sigma)$, Σ positive definite symmetric, $j = 1, \ldots, N$. Adopt the vague, invariant prior density of Jeffreys (see [11, pp. 72–76])

$$p(\alpha, \beta, \sigma, \mu, \Sigma) \propto \frac{1}{\sigma|\Sigma|^{(p+1)/2}}. \tag{13}$$

Denote the matrix of independent variables with missing elements and the vector of dependent variables by $Z \equiv (u_1, \ldots, u_N)$, and $y \equiv (y_1, \ldots, y_N)'$, respectively. Denote the matrix of independent variables with no missing observations by $W \equiv (w_1, \ldots, w_N)$. Then, given W, the joint density of all the remaining variables is given by

$$\begin{aligned} p &\equiv p(\alpha, \beta, \sigma, \mu, \Sigma, Z, y|W) \\ &\propto \sigma^{-(N+1)}|\Sigma|^{-(N+p+1)/2} \exp(-\tfrac{1}{2}) \\ &\quad \times\left\{\frac{1}{\sigma^2}\sum_{j=1}^{N}(y_j-\alpha-\beta' X_j)^2+\sum_{j=1}^{N}(u_j-\mu)'\Sigma^{-1}(u_j-\mu)\right\}. \end{aligned} \tag{14}$$

Define the least squares estimators

$$\hat{\beta} = \sum_{j=1}^{N}(y_j-\bar{y})G^{-1}(X_j-\bar{X}), \quad \hat{\alpha} = \bar{y}-\hat{\beta}'\bar{X}, \tag{15}$$

where $N\bar{y} \equiv \Sigma_1^N y_j$, $N\bar{X} \equiv \Sigma_1^N X_j$, and $G \equiv \Sigma_1^N(X_j-\bar{X})(X_j-\bar{X})'$. Substituting (15) into (14) and simplifying gives

$$\begin{aligned} p \propto \sigma^{-(N+1)}|\Sigma|^{-(N+p+1)/2} \exp(-\tfrac{1}{2})&\left\{\frac{1}{\sigma^2}\sum_1^N(y_j-\hat{\alpha}-\hat{\beta}'X_j)^2\right. \\ &+\frac{1}{\sigma^2}(\beta-\hat{\beta})'G(\beta-\hat{\beta})+\frac{N}{\sigma^2}[(\alpha-\hat{\alpha})+(\beta-\hat{\beta})'\bar{X}]^2 \\ &\left.+N(\bar{u}-\mu)'\Sigma^{-1}(\bar{u}-\mu)+\text{tr}\,\Sigma^{-1}V\right\}, \end{aligned} \tag{16}$$

where $V \equiv \Sigma_1^N(u_j-\bar{u})(u_j-\bar{u})'$, and $N\bar{u} \equiv \Sigma_1^N u_j$.

Now integrate p with respect to σ (the integrand is the kernel of an inverted gamma distribution) and with respect to Σ (the integrand is the kernel of an inverted Wishart distribution, see, e.g. [12, p. 109]) to obtain the marginal density

$$f \equiv f(\alpha, \beta, \mu, Z, y|W)$$

$$\propto \left\{|V|[1+N(\mu-\bar{u})'V^{-1}(\mu-\bar{u})]\left[\sum_{j=1}^{N}(y_j-\hat{\alpha}-\hat{\beta}'X_j)^2 + N[(\alpha-\hat{\alpha})+(\beta-\hat{\beta})'\bar{X}]^2+(\beta-\hat{\beta})'G(\beta-\hat{\beta})\right]\right\}^{-N/2}. \quad (17)$$

Note that even if some of the components of y are missing, the conditional density of the regression parameters and the missing observations given the observed data is proportional to the right-hand side of (17).

As in the simple regression case, the joint mode may be obtained by taking as Bayes' estimators, $\alpha_B = \hat{\alpha}$, $\beta_B = \hat{\beta}$, $\mu_B = \bar{u}$, and minimizing the remaining expression

$$\Omega \equiv |V| \sum_{j=1}^{N} (y_j-\hat{\alpha}-\hat{\beta}'X_j)^2 \quad (18)$$

with respect to the missing elements. In the multiple regression case the differentiation is not as straightforward, but is outlined below.

Define the residual vector

$$\hat{\varepsilon} \equiv y-\hat{\alpha}e-Z_1'\hat{\beta}, \quad (19)$$

where $Z_1' \equiv (Z', W')$, and where e denotes an N-vector of ones. Thus, (18) becomes

$$\Omega = |V|\hat{\varepsilon}'\hat{\varepsilon}. \quad (20)$$

Let $\bar{\mathscr{D}}$ denote the set of all x_{kl} and y_j data points which are missing. Let $Z \equiv (x_{ij})$, and suppose $x_{kl} \in \bar{\mathscr{D}}$. Then,

$$\frac{\partial\Omega}{\partial x_{kl}} = \frac{\partial|V|}{\partial x_{kl}}\hat{\varepsilon}'\hat{\varepsilon}+|V|\frac{\partial\hat{\varepsilon}'\hat{\varepsilon}}{\partial x_{kl}}. \quad (21)$$

The first derivative in (21) may be evaluated as follows.

It is straightforward to check that

$$\frac{\partial|V|}{\partial x_{kl}} = \text{tr}\left[|V|V^{-1}\frac{\partial V}{\partial x_{kl}}\right],$$

and if $A \equiv [I_N-(ee'/N)]$,

$$\frac{\partial V}{\partial x_{kl}} = \frac{\partial}{\partial x_{kl}}(ZAZ') = ZAU'_{kl}+U_{kl}AZ',$$

where U_{kl} denotes a matrix of zeros except for a one in the kl-place. So

$$\frac{\partial|V|}{\partial x_{kl}} = \text{tr}\,[|V|V^{-1}][ZAU'_{kl}+U_{kl}AZ'].$$

Thus, if $H \equiv V^{-1}ZA \equiv (H_{ij})$,

$$\frac{\partial|V|}{\partial x_{kl}} = 2|V|H_{kl}. \tag{22}$$

The second derivative in (21) is evaluated as follows. Note that

$$\frac{\partial}{\partial x_{kl}}(\hat{\varepsilon}'\hat{\varepsilon}) = \left[\frac{\partial\hat{\varepsilon}'\hat{\varepsilon}}{\partial\hat{\varepsilon}}\right]'\left[\frac{\partial\hat{\varepsilon}}{\partial x_{kl}}\right],$$

and since $\partial(\hat{\varepsilon}'\hat{\varepsilon})/\partial\hat{\varepsilon} = 2\hat{\varepsilon}$,

$$\frac{\partial}{\partial x_{kl}}(\hat{\varepsilon}'\hat{\varepsilon}) = 2\hat{\varepsilon}'\left(\frac{\partial\hat{\varepsilon}}{\partial x_{kl}}\right).$$

Substituting (15) into (19) and simplifying shows that the residual vector may be written as

$$\hat{\varepsilon} = (I-t)(y-\bar{y}e)+\frac{e}{N}\,[e't(y-\bar{y}e)], \tag{23}$$

where $t \equiv Z_1'G^{-1}Z_1$. Note that $\hat{\varepsilon}$ depends upon x_{kl} only through t. Thus,

$$\frac{\partial\hat{\varepsilon}}{\partial x_{kl}} = -\frac{\partial t}{\partial x_{kl}}(y-\bar{y}e)+\frac{e}{N}\left[e'\,\frac{\partial t}{\partial x_{kl}}(y-\bar{y}e)\right].$$

But

$$\frac{\partial t}{\partial x_{kl}} = Z_1'G^{-1}\frac{\partial Z_1}{\partial x_{kl}}+Z_1'\frac{\partial G^{-1}}{\partial x_{kl}}Z_1+\frac{\partial Z_1'}{\partial x_{kl}}G^{-1}Z_1,$$

and since

$$\partial Z_1/\partial x_{kl} = U_{kl}, \qquad \partial G^{-1}/\partial x_{kl} = -G^{-1}(\partial G/\partial x_{kl})G^{-1},$$

and

$$\partial G/\partial x_{kl} = Z_1 A U'_{kl} + U_{kl} A Z'_1, \quad \partial t/\partial x_{kl} = J_{kl} + J'_{kl},$$

where

$$J_{kl} \equiv Z'_1 G^{-1} U_{kl}(I - AZ'_1 G^{-1} Z_1).$$

Combining terms gives

$$\frac{\partial}{\partial x_{kl}} = \frac{e}{N}\left[e'(J_{kl}+J'_{kl})(y-\bar{y}e)\right] - (J_{kl}+J'_{kl})(y-\bar{y}e). \tag{24}$$

Substituting (22) and (24) into (21) gives

$$\begin{aligned}\frac{\partial\Omega}{\partial x_{kl}} &= 2|V|H_{kl}\hat{\varepsilon}'\hat{\varepsilon}\\ &\quad + 2|V|\hat{\varepsilon}'\left\{\frac{e}{N}\left[e'(J_{kl}+J'_{kl})(y-\bar{y}e)\right]\right.\\ &\quad \left. -(J_{kl}+J'_{kl})(y-\bar{y}e)\right\}.\end{aligned} \tag{25}$$

By direct computation it may be found that $\hat{\varepsilon}'e = 0$ and $\hat{\varepsilon}'Z'_1 = 0$.

Using these results, setting $\partial\Omega/\partial x_{kl} = 0$, and solving, gives (since $\hat{\beta} = G^{-1}Z_1 Ay$)

$$H_{kl}\hat{\varepsilon}'\hat{\varepsilon} = \hat{\varepsilon}'U'_{kl}\hat{\beta} = \hat{\varepsilon}_l\hat{\beta}_k. \tag{26}$$

It is straightforward to check from (22) that if $V^{-1} \equiv (v^{ij})$, and $\bar{x}_k \equiv N^{-1}\sum_{j=1}^{N} x_{kj}$,

$$H_{kl} = \sum_{i=1}^{p} v^{ki}(x_{il}-\bar{x}_i). \tag{27}$$

Using (27) and the definitions of $\hat{\varepsilon}_l$ and $\hat{\beta}_k$ shows that (26) may be rewritten as

$$\begin{aligned}&\sum_{i=1}^{p} v^{ki}(x_{il}-\bar{x}_i)\\ &= \frac{(y_l-\bar{y})\hat{\beta}_k - \hat{\beta}'(X_l-\bar{X})\hat{\beta}_k}{\sum_{j=1}^{N}(y_j-\bar{y})^2 - \sum_{i=1}^{N}\sum_{j=1}^{N}(y_i-\bar{y})(y_j-\bar{y})(X_i-\bar{X})'G^{-1}(X_j-\bar{X})}.\end{aligned} \tag{28}$$

If none of the dependent variables is missing (28) represents a system of equations (one for each $x_{kl} \in \bar{\mathscr{D}}$) which must be solved simultaneously for all $x_{kl} \in \bar{\mathscr{D}}$. If some of the dependent variables are missing, Ω in (20) must be minimized with respect to the missing y's. Suppose $y_\alpha \in \bar{\mathscr{D}}$. Then, from (20),

$$\frac{\partial \Omega}{\partial y_\alpha} = |V| \frac{\partial(\hat{\varepsilon}'\hat{\varepsilon})}{\partial y_\alpha}.$$

But because

$$\frac{\partial(\hat{\varepsilon}'\hat{\varepsilon})}{\partial y_\alpha} = \left[\frac{\partial(\hat{\varepsilon}'\hat{\varepsilon})}{\partial \hat{\varepsilon}}\right]'\left[\frac{\partial \hat{\varepsilon}}{\partial y_\alpha}\right],$$

and the first derivative is $2\hat{\varepsilon}$,

$$\frac{\partial \Omega}{\partial y_\alpha} = 2|V|\hat{\varepsilon}'\left(\frac{\partial \hat{\varepsilon}}{\partial y_\alpha}\right).$$

Recall from (23) that $\hat{\varepsilon} \equiv L(y-\bar{y}e)$, where $L \equiv (I-AZ_1'G^{-1}Z_1)$ depends only on the X's. So

$$\frac{\partial \hat{\varepsilon}}{\partial y_\alpha} = L\phi_\alpha,$$

where ϕ_α denotes an N-vector whose jth component is $(-1/N)$ for $j \neq \alpha$, and is $(1-1/N)$ for $j = \alpha$. Thus,

$$\frac{\partial \Omega}{\partial y_\alpha} = 2|V|\hat{\varepsilon}'L\,\phi_\alpha = 2|V|(y-\bar{y}e)'L'L\phi_\alpha.$$

Setting $\partial\Omega/\partial y_\alpha = 0$ and solving shows that we must have

$$(y-\bar{y}e)'L'L\phi_\alpha = 0. \tag{29}$$

Since $\hat{\beta} = G^{-1}Z_1(y-\bar{y}e)$, $(y-\bar{y}e)'L' = (y-\bar{y}e)'-\hat{\beta}'(Z_1-\bar{X}e')$.

Substituting in (29) shows we must have

$$(y-\bar{y}e)'L\phi_\alpha = \hat{\beta}'(Z_1-\bar{X}e')L\phi_\alpha. \tag{30}$$

It is straightforward to check that

$$(y-\bar{y}e)'L\phi_\alpha = (y_\alpha-\bar{y})-\hat{\beta}'(X_\alpha-\bar{X}),$$

and that

$$\hat{\beta}'(Z_1-\bar{X}e')L\phi_\alpha = 0.$$

Hence, $\partial\Omega/\partial y_\alpha = 0$ implies

$$(y_\alpha - \bar{y}) = \hat{\beta}'(X_\alpha - \bar{X}). \tag{31}$$

Note that (31) is the multivariate generalization of (12), the simple linear regression result.

If no independent variable observations are missing, the missing y_α's may be estimated simultaneously by solving the system of equations in (31) (there is one such equation for each missing y_α). The system implies that every missing dependent variable observation is constrained to lie on the forwards regression "line" of y on all of the independent variables.

If some y's and some of the observed components of the X's are missing, (28) and (31) must be solved simultaneously for all missing observations.

4. *Interval estimation*

The joint density of the regression parameters and the data, given W, is given in (17). Therefore, the marginal density of all the data, given W, is given by

$$g(Z, y|W) \equiv \int\int\int f(\alpha, \beta, \mu, Z, y|W)\, d\alpha\, d\beta\, d\mu. \tag{32}$$

The integration may be readily carried out by noting that the integrals over μ, α and β all involve kernels of t-distributions. By integrating first with respect to μ, then with respect to α, and finally with respect to β, we easily obtain

$$g(Z, y|W) \propto \frac{1}{|V|^{(N-1)/2}|G|^{1/2}(\hat{\varepsilon}'\hat{\varepsilon})^{(N-p^*-1)/2}}, \tag{33}$$

where $\hat{\varepsilon}$ is defined in (19).

Suppose first that none of the independent variable observations is missing, but that if

$$y' \equiv (\dot{y}'; \ddot{y}') \equiv (y_1, \ldots, y_m; y_{m+1}, \ldots, y_N)',$$

$\dot{y}: m\times 1$ is missing. Next note that V and G do not depend on y, and from (23), $\hat{\varepsilon} \equiv L(y - \bar{y}e)$, where $L \equiv (I - AZ_1'G^{-1}Z_1)$ does not depend upon y, either. Therefore, the density of y given Z_1 is proportional to

$$(\hat{\varepsilon}'\hat{\varepsilon})^{-(N-p^*-1)/2} \equiv [(y-\bar{y}e)'L'L(y-\bar{y}e)]^{-\nu/2},$$

where $\nu \equiv N - p^* - 1$. Since $y - \bar{y}e \equiv Ay$, where $A \equiv (I - ee'/N)$, if

$B = AL'LA = B'$, the density of y given Z_1 is proportional to $(y'By)^{-\nu/2}$. It may be checked that $B = LA$. Partitioning $B \equiv (B_{ij})$, $i, j = 1, 2$ so that $B_{11}: m \times m$, it may be checked that the marginal posterior density of $\dot{y}$, given $\ddot{y}$ and Z_1 is given by

$$\dot{g}(\dot{y}|\ddot{y}, Z_1) \propto \frac{1}{[(N-p^*-m-1)+(\dot{y}-\dot{y}_0)'H^{-1}(\dot{y}-\dot{y}_0)]^{(N-p^*-1)/2}} \tag{34}$$

where

$$H \equiv \tau B_{11}^{-1}/(N-p^*-m-1),$$

$$\tau \equiv \ddot{y}'(B_{22}-B_{21}B_{11}^{-1}B_{12})\ddot{y},$$

and

$$\dot{y}_0 \equiv -B_{11}^{-1}B_{12}\ddot{y}.$$

That is, $(\dot{y}|\ddot{y}, Z_1)$ follows a multivariate Student t-distribution of dimension m, with $(N-p^*-m-1)$ degrees of freedom, mean $\dot{y}_0$, and covariance matrix $\Delta \equiv (N-p^*-m-1)H/(N-p^*-m-3)$. Thus, a single missing y_α, say, y_1, follows a marginal univariate t-distribution of the form

$$\dot{g}(y_1|\ddot{y}, Z_1) \propto \frac{1}{\left[(N-p^*-m-1)+\frac{1}{H_{11}}(y_1-\dot{y}_{10})^2\right]^{(N-p^*-m)/2}}, \tag{35}$$

where $\dot{y}_{10}$ denotes the first element of $\dot{y}_0$, and H_{11} denotes the (1, 1) element of H.

If some of the elements of the X_j's are missing, but none of the y_α's are missing, the posterior distribution of the missing values given the observed ones is obtained from (33) as

$$p(\bar{\mathscr{D}}|\mathscr{D}, y) \propto \frac{1}{|V|^{(N-1)/2}|G|^{1/2}\left[\sum_1^N (y_j-\hat{\alpha}-\hat{\beta}'X_j)^2\right]^{(N-p^*-1)/2}}, \tag{36}$$

where $\bar{\mathscr{D}}$ denotes the collection of missing data points and $\mathscr{D}$ the non-missing ones.

Consider the case of $p = 1$ (one independent variable), and assume N is very large. Then, it is not hard to check from (36) that since

$G = V$ in this case, asymptotically, missing X_α's follow a joint normal distribution; in particular, if $x_N \equiv z_N$ is missing, $(z_N - \hat{z}_N)/\sqrt{c}$ follows a $N(0, 1)$ distribution, asymptotically, where

$$c = \frac{1}{N-2}\left\{\frac{\sum_1^N (y_j - \bar{y})^2 - \left[\sum_1^{N-1} x_j y_j - \bar{y}\sum_1^{N-1} x_j\right]^2}{\sum_1^{N-1}(y_j - \bar{y})^2} - \hat{z}_N^2\right\},$$

and $\hat{z}_N$ is determined from (11); that is, $\hat{z}_N$ is the estimated value of z_N obtained by forcing it to lie on the backwards regression line of the completed observations. Thus, the asymptotic normal distribution is centered at the Bayesian modal estimator.

References

[5] A. A. Afifi and R. M. Elashoff, Missing observations in multivariate statistics I. Review of the literature. Journal of the American Statistical Association 61 (1966), 595–604.

[3] T. W. Anderson, Maximum likelihood estimates for a multivariate normal distribution when some observations are missing. Journal of the American Statistical Association 52 (1957), 200–203.

[8] M. G. Dagenais, Multiple regression analysis with incomplete observations, from a Bayesian viewpoint. Presented at the Fourth NBER-NSF Symposium on Bayesian Inference in Econometrics, 1972. (Reprinted in the present volume.)

[4] M. Glasser, Linear regression analysis with missing observations among the independent variables. Journal of the American Statistical Association 59 (1964), 834–844.

[7] Y. Haitovsky, Missing data in regression analysis. Journal of the Royal Statistical Society, Ser. B 30 (1968), 67–82.

[11] H. Jeffreys, Theory of probability, 3rd ed. Oxford, Clarendon Press, 1961.

[9] G. M. Kaufman and A. Krieger, A Bayesian treatment of bivariate normal sampling with missing observations. Unpublished manuscript circulated through personal communication, 1972.

[10] J. S. Mehta and P. A. V. B. Swamy, Bayesian analysis of bivariate normal distribution when some observations are missing. Presented at the Fourth NBER-NSF Symposium on Bayesian Inference in Econometrics, 1972. (Reprinted in the present volume.)

[12] S. J. Press, Applied multivariate analysis. New York, Holt, Rinehart & Winston, 1972.

[1] S. S. Wilks, Moments and distribution of estimates of population parameters from fragmentary samples. Ann. Math. Statist. 3 (1932), 163–195.

[6] M. A. Woodbury and T. Orchard, A missing information principle: theory and applications. Presented at the Sixth Berkeley Symposium.

[2] F. Yates, The analysis of replicated experiments when the field results are incomplete. The Empire Journal of Experimental Agriculture 1 (1933), 129–142.

S. E. Fienberg and A. Zellner (eds.)
Studies in Bayesian Econometrics and Statistics.

7.2 MULTIPLE REGRESSION ANALYSIS WITH INCOMPLETE OBSERVATIONS, FROM A BAYESIAN VIEWPOINT*

Marcel G. DAGENAIS

University of Montreal

1. *Introduction*

Consider the following standard regression model:

$$Y_t = \sum_{j=1}^{q} \beta_j X_{jt} + u_t \qquad (t = 1, \ldots, N) \tag{1}$$

where Y_t is the dependent variable, the X_{jt}'s are the independent variables and the u_t's are normally distributed with zero mean, zero covariances and common variance σ^2. The sample size is N. Some observations are incomplete: Y_t or one or more of the X_{jt}'s are missing.

Other variables relating to the same N observations but not included in the regression model are also available; in particular it is assumed that X_{1t} and X_{2t} are related to these other variables by the following simultaneous regression equations:

$$\begin{pmatrix} X_{1t} \\ X_{2t} \end{pmatrix} = \begin{pmatrix} Z_{1t} & 0 \\ 0 & Z_{2t} \end{pmatrix} \begin{pmatrix} \theta_1 \\ \theta_2 \end{pmatrix} + \begin{pmatrix} v_{1t} \\ v_{2t} \end{pmatrix}, \tag{2}$$

where Z_{1t} and Z_{2t} are row vectors with respectively r_1 and r_2 elements, θ_1 and θ_2 are column vectors of parameters and each of the pairs v_{it} $(i = 1, 2)$ have joint normal distributions with zero mean and covariance matrix

$$S = \begin{pmatrix} \sigma_1^2 & \sigma_{12} \\ \sigma_{12} & \sigma_2^2 \end{pmatrix}.$$

* This paper was presented at the Fourth NBER-NSF Seminar on Bayesian Inference in Econometrics held at the Univ. of Chicago, May 1972. The research was supported by the Canada Council of Arts. Mr. Tran Cong Liem made a significant contribution as research assistant.

The v's are assumed to be independent between observations and uncorrelated with the u's. The elements of R_t, where $R_t = (X_{3t}, \ldots, X_{qt})$, may also appear in Z_{1t} or Z_{2t}.

No side relationships similar to those contained in equation (2) are considered for the other X_{jt}'s $(j = 3, \ldots, q)$. Our approach could be extended to include a greater number of side regressions, but the algebraic development of the formulas would become excessively cumbersome. Furthermore, it is felt that the bivariate case discussed here should cover satisfactorily most situations encountered in practice.

Missing data problems may occur in a variety of situations. In time series analysis, it frequently happens that the records on one or two of the independent variables do not extend as far back in the past as those of the other variables in the regression. The former variables, which would correspond to X_1 and X_2, may in turn be related to other sets of variables, namely Z_1 and Z_2, for which longer series are available. In econometric applications X_1 might refer to a specific cost index, say the construction cost index for one-family houses, while the Z_1 vector would include related price or cost indices, such as the price index of building materials and the wage index for the construction industry.

Similar situations are encountered with cross-sectional data. In household budget surveys for example, X_1 might refer to the annual operating and maintenance cost of the "first" car while the Z_1 vector might include the age of the car, its list price and the mileage traveled during the year.

2. *The posterior density of the parameters*

2.1. *Preliminary considerations*

Any observable variable, whether Y, the X's or the elements of the Z vectors may be missing in any observation. It will be shown, however, that, under assumptions of vague prior knowledge, the only helpful observations are those which satisfy one of the following criteria:

G1 – Y_t, X_{1t}, X_{2t} and all elements of R_t are present.

G2 – Y_t, X_{1t}, all elements of R_t and Z_{2t} are present, but not X_{2t}.

G3 – Y_t, X_{2t}, all elements of R_t and Z_{1t} are present, but not X_{1t}.

G4 – Y_t, all elements of R_t, Z_{1t} and Z_{2t} are present, but not X_{1t} *and* X_{2t}.

H1 – X_{1t}, X_{2t}, all elements of Z_{1t} and Z_{2t} are present.
H2 – X_{1t} and all elements of Z_{1t} are present, but not X_{2t} *or* one or more elements of Z_{2t}.
H3 – X_{2t} and all elements of Z_{2t} are present, but not X_{1t} *or* one or more elements of Z_{1t}.

The criteria of the G group are mutually exclusive and so are those of the H group. However, the same observation may simultaneously meet both a G and an H criterion.

To derive the posterior probability density function of the parameters, it is useful to express this function in a way that singles out the role played by a given observation t. Assuming, for the moment, that all variables associated with observation t are present, the posterior pdf of the parameters is given by:

$$p(P|\text{data}) \propto p(P)\cdot L^*\cdot L(Y_t, X_{1t}, X_{2t}\,|\,Z_{1t}, Z_{2t}, R_t, P) \qquad (3)$$

where

$$P = (\beta_1, \ldots, \beta_q, \theta_1, \theta_2, \sigma^2, S),$$

$p(P)$ designates the prior pdf of the parameters,
L is the likelihood of observation t, and
L^* is the likelihood of all other observations.

Given the hypotheses made on the conditional distributions of the error terms, equation (3) may also be expressed as follows:

$$p(P|\text{data}) \propto p(P)\cdot L^*\cdot L_{\text{G1}}[Y_t|(\beta_1 X_{1t}+\beta_2 X_{2t}+R_t B), \sigma^2]$$
$$\times L_{\text{H1}}[X_{1t}, X_{2t}|Z_{1t}\theta_1, Z_{2t}\theta_2, S], \qquad (4)$$

where B is a column-vector such that $B' = (\beta_3, \ldots, \beta_q)$.

For all observations meeting criteria G1 and H1, the L_{G1} and L_{H1} functions will be similar to those given in (4). However, the form of the L_{G} function will change for observations corresponding to other categories of the G group, while the L_{H} function will differ for observations associated with other criteria of the H group. The forms of the L_{G} and L_{H} functions are given below, for all seven categories:

$$L_{\text{G1}}[Y_t|(\beta_1 X_{1t}+\beta_2 X_{2t}+R_t B), \sigma^2]$$

$$L_{\text{G2}}[Y_t|(\beta_1 X_{1t}+\beta_2 Z_{2t}\theta_2+R_t B), (\sigma^2+\beta_2^2\sigma_2^2)]$$

$$L_{\text{G3}}[Y_t|(\beta_1 Z_{1t}\theta_1+\beta_2 X_{2t}+R_t B), (\sigma^2+\beta_1^2\sigma_1^2)]$$

$$L_{G4}[Y_t|(\beta_1 Z_{1t}\theta_1+\beta_2 Z_{2t}\theta_2+R_t B), (\sigma^2+\beta_1^2\sigma_1^2+\beta_2^2\sigma_2^2 +2\beta_1\beta_2\sigma_{12})]$$

$$L_{H1}(X_{1t}, X_{2t}|Z_{1t}\theta_1, Z_{2t}\theta_2, S)$$

$$L_{H2}(X_{1t}|Z_{1t}\theta_1, \sigma_1^2)$$

$$L_{H3}(X_{2t}|Z_{2t}\theta_2, \sigma_2^2).$$

L_{G2} is the conditional likelihood of Y_t obtained after having replaced the nonobservable value of X_{2t} in (1) by $Z_{2t}\theta_2+v_{2t}$. Equation (1) then becomes:

$$Y_t = \beta_1 X_{1t}+\beta_2 Z_{2t}\theta_2+R_t B+u_t+\beta_2 v_{2t}. \tag{5}$$

The functions L_{G3} and L_{G4} are obtained by similar transformations. To justify the form of L_{H2}, it will be useful to consider an example. Assuming that all variables of observation t are available except $Z_{21,t}$, the first element of Z_{2t}, the posterior probability of the parameters may be expressed as follows:

$$\begin{aligned} p(P|\text{data}) &= \int p(P, \xi_{2t}|\text{data})\, \mathrm{d}\xi_{2t} \\ &\propto p(P)\cdot L^*\cdot L_{G1}[Y_t|(\beta_1 X_{1t}+\beta_2 X_{2t}+R_t B), \sigma^2] \\ &\quad \times \int L_{H1}[X_{1t}, X_{2t}|Z_{1t}\theta_1, (\xi_{2t}+\sum_{j=2}^{r_2} Z_{2j,t}\theta_{2j}), S] \\ &\quad \times p(\xi_{2t})\, \mathrm{d}\xi_{2t} \end{aligned} \tag{6}$$

where ξ_{2t} stands for $Z_{21,t}\theta_{21}$.

Writing explicitly the algebraic expression for L_{H1}, yields:

$$\begin{aligned} &\int L_{H1}\cdot p(\xi_{2t})\, \mathrm{d}\xi_{2t} \\ &= \int (2\pi)^{-1}|S|^{-\frac{1}{2}} \exp\Bigg\{ -\tfrac{1}{2}\Bigg[(X_{1t}-Z_{1t}\theta_1), \\ &\quad \Big(X_{2t}-\sum_{j=2}^{r_2} Z_{2j,t}\theta_{2j}-\xi_{2t}\Big)\Bigg]S^{-1} \\ &\quad \times \begin{bmatrix} (X_{1t}-Z_{1t}\theta_1) \\ \Big(X_{2t}-\sum_{j=2}^{r_2} Z_{2j,t}\,\theta_{2j}-\xi_{2t}\Big) \end{bmatrix}\Bigg\}\cdot p(\xi_{2t})\, \mathrm{d}\xi_{2t}. \end{aligned} \tag{7}$$

If we assume vague prior knowledge for ξ_{2t} and set $p(P, \xi_{2t}) = p(P) \cdot p(\xi_{2t})$ and $p(\xi_{2t}) \propto$ constant,[1] we obtain:

$$\int L_{\mathrm{H}1} \cdot p(\xi_{2t})\, \mathrm{d}\xi_{2t} \propto (2\pi)^{-\frac{1}{2}} \sigma_1^{-1} \exp\left\{ -\frac{(X_{1t} - Z_{1t}\theta_1)^2}{2\sigma_1^2} \right\}$$

$$\propto L_{\mathrm{H}2}(X_{1t} | Z_{1t}\theta_1, \sigma_1^2). \qquad (8)$$

The derivation of $L_{\mathrm{H}3}$ is similar to that of $L_{\mathrm{H}2}$. Using analogous procedures, it is also possible to prove that:

(1) For observations satisfying one of the H criteria but none of the G criteria, the $L_{\mathrm{G}1}$ function in (4) disappears, while the $L_{\mathrm{H}1}$ function is replaced by $L_{\mathrm{H}j}$, where j corresponds to the number of the H criterion satisfied by the observation.

(2) For observations satisfying one of the G criteria but none of the H criteria, the $L_{\mathrm{H}1}$ function in (4) disappears, while the $L_{\mathrm{G}1}$ function is replaced by $L_{\mathrm{G}j}$, where j corresponds to the number of the G criterion satisfied by the observation.

(3) Observations satisfying none of the G and H criteria do not add any evidence about the posterior distribution of P and must be completely disregarded.

A further example will be given to support these last statements. Assume that all variables are observable except X_{2t} and $Z_{21,t}$. We then have:

$$p(P|\text{data}) = \iint p(P, \psi_{2t}, \xi_{2t}|\text{data})\, \mathrm{d}\psi_{2t}\, \mathrm{d}\xi_{2t}$$

$$\propto p(P) \cdot L^* \cdot \iint L_{\mathrm{G}1}[Y_t | (\beta_1 X_{1t} + \psi_{2t} + R_t B), \sigma^2]$$

$$\times L_{\mathrm{H}1}[X_{1t}, (\psi_{2t}/\beta_2) | Z_{1t}\theta_1, (\xi_{2t} + \sum_{j=2}^{r_2} Z_{2j,t}\theta_{2j}), S]$$

$$\times p(\psi_{2t}) \cdot p(\xi_{2t})\, \mathrm{d}\psi_{2t}\, \mathrm{d}\xi_{2t}$$

$$\propto p(P) \cdot L^* \cdot \int L_{\mathrm{G}1}[Y_t | (\beta_1 X_{1t} + \psi_{2t} + R_t B), \sigma^2]$$

$$\times \{ \int L_{\mathrm{H}1}[X_{1t}, (\psi_{2t}/\beta_2)| Z_{1t}\theta_1, (\xi_{2t} + \sum_{j=2}^{r_2} Z_{2j,t}\theta_{2j}), S]$$

$$\times p(\xi_{2t})\, \mathrm{d}\xi_{2t}\} \cdot p(\psi_{2t})\, \mathrm{d}\psi_{2t} \qquad (9)$$

[1] This approach is very similar to that adopted by Zellner and Tiao, when analyzing the regression model with autocorrelated errors [4]. They also introduce, at one point, a new parameter M which is a function of unobserved quantities and has prior pdf $p(M) \propto$ constant.

where ψ_{2t} stands for $\beta_2 X_{2t}$, ξ_{2t} for $Z_{21,t}\theta_{21}$, and where the joint prior density of P, ψ_{2t} and ξ_{2t} is $p(P, \psi_{2t}, \xi_{2t}) = p(P)\cdot p(\psi_{2t})\cdot p(\xi_{2t})$ and $p(\psi_{2t})\cdot p(\xi_{2t}) \propto$ constant.

Clearly, $\int L_{H1} p(\xi_{2t})\, d\xi_{2t}$ will reduce to L_{H2} given in (8), and then $\int L_{G1} p(\psi_{2t})\, d\psi_{2t}$ will be equal to a constant and will disappear.

2.2. *Derivation of the posterior density*

Let us now assume that there are n_j $(j = 1, \ldots, 4)$ observations satisfying the G_j criterion. Similarly, there are m_j $(j = 1, \ldots, 3)$ observations satisfying the H_j criterion. Clearly,

$$\sum_{j=1}^{4} n_j \leqq N \quad \text{and} \quad \sum_{j=1}^{3} m_j \leqq N.$$

Let us further assume that the prior density of the unknown parameters expresses prior ignorance:

$$p(P) = p(\beta_1, \beta_2, B, \theta_1, \theta_2, \sigma^2, S) \propto (\sigma^2)^{-1} |S|^{-3/2}.$$

The posterior density of the parameters is then given by the following expression:

$$\begin{aligned}
p(P|\text{data}) \propto \sigma^{-(\sum_{j=1}^{4} n_j + \sum_{j=1}^{3} m_j + m_1 + 8)} \\
&\times \exp\left\{-\frac{1}{2\sigma^2}(Y - X_1\beta_1 - X_2\beta_2 - RB)'\right. \\
&\left.(Y - X_1\beta_1 - X_2\beta_2 - RB)_{G1}\right\} \\
&\times w_2^{n_2} \exp\left\{-\frac{w_2^2}{2\sigma^2}(Y - X_1\beta_1 - Z_2\theta_2\beta_2 - RB)'\right. \\
&\left.(Y - X_1\beta_1 - Z_2\theta_2\beta_2 - RB)_{G2}\right\} \\
&\times w_3^{n_3} \exp\left\{-\frac{w_3^2}{2\sigma^2}(Y - Z_1\theta_1\beta_1 - X_2\beta_2 - RB)'\right. \\
&\left.(Y - Z_1\theta_1\beta_1 - X_2\beta_2 - RB)_{G3}\right\}
\end{aligned}$$

$$\times\ w_4^{n_4} \exp\left\{-\frac{w_4^2}{2\sigma^2}(Y-Z_1\theta_1\beta_1-Z_2\theta_2\beta_2-RB)'\right.$$

$$\left.(Y-Z_1\theta_1\beta_1-Z_2\theta_2\beta_2-RB_{\ G4}\right\}$$

$$\times\ \phi^{m_1+3} \exp\left\{-\frac{\phi_1^2}{2\sigma^2}(X_1-Z_1\theta_1)'(X_1-Z_1\theta_1)_{H1}\right.$$

$$+\frac{\phi_{12}}{\sigma^2}(X_1-Z_1\theta_1)'(X_2-Z_2\theta_2)_{H1}$$

$$\left.-\frac{\phi_2^2}{2\sigma^2}(X_2-Z_2\theta_2)'(X_2-Z_2\theta_2)_{H1}\right\}$$

$$\times\ \omega_2^{m_2} \exp\left\{-\frac{\omega_2^2}{2\sigma^2}(X_1-Z_1\theta_1)'(X_1-Z_1\theta_1)_{H2}\right\}$$

$$\times\ \omega_3^{m_3} \exp\left\{-\frac{\omega_3^2}{2\sigma^2}(X_2-Z_2\theta_2)'(X_2-Z_2\theta_2)_{H3}\right\} \tag{10}$$

where

$$w_2^2 = 1/(1+\beta_2^2\lambda_2^2)$$
$$w_3^2 = 1/(1+\beta_1^2\lambda_1^2)$$
$$w_4^2 = 1/(1+\beta_1^2\lambda_1^2+\beta_2^2\lambda_2^2+2\beta_1\beta_2\lambda_{12})$$
$$\lambda_1^2 = \sigma_1^2/\sigma^2$$
$$\lambda_2^2 = \sigma_2^2/\sigma^2$$
$$\lambda_{12} = \sigma_{12}/\sigma^2$$
$$\phi^2 = 1/(\lambda_1^2\lambda_2^2-\lambda_{12}^2);$$
$$\phi_1^2 = \lambda_2^2\phi^2; \qquad \phi_{12} = \lambda_{12}\phi^2; \qquad \phi_2^2 = \lambda_1^2\phi^2;$$
$$\omega_2^2 = 1/\lambda_1^2; \qquad \omega_3^2 = 1/\lambda_2^2.$$

The indices G_j ($j = 1, \ldots, 4$) or H_j ($j = 1, \ldots, 3$) affecting the quadratic forms in the exponentials indicate the set of observations concerned. The t-index previously affecting each variable or set of variables has been omitted to indicate that we are now considering column vectors or matrices where each row represents one observation.

After multiplying exponentials, isolating terms involving θ_1 or θ_2, and forming a quadratic form in the $(r \times 1)$ vector θ, where

$$\theta = \begin{pmatrix} \theta_1 \\ \theta_2 \end{pmatrix} \quad \text{and} \quad r = r_1 + r_2,$$

expression (10) becomes:

$$p(\beta_1, \beta_2, B, \theta, \sigma^2, S|\text{data})$$

$$\propto \Omega \exp\left\{-\frac{K_4}{2\sigma^2}\right\} \exp\left\{-\frac{(\theta-\hat{\theta})' M(\theta-\hat{\theta})}{2\sigma^2}\right\} \tag{11}$$

where

$$\Omega = \sigma^{-(\sum_{j=1}^{4} n_j + \sum_{j=1}^{3} m_j + m_1 + 8)}$$

$$\times w_2^{n_2} w_3^{n_3} w_4^{n_4} \phi^{m_1+3} \omega_2^{m_2} \omega_3^{m_3}$$

and M is a symmetric positive definite matrix of order r:

$$M = \begin{bmatrix} A & C \\ C' & B \end{bmatrix}.$$

The submatrices contained in M are defined as follows:

$$A = w_3^2\beta_1^2(Z_1'Z_1)_{G3} + w_4^2\beta_1^2(Z_1'Z_1)_{G4} + \phi_1^2(Z_1'Z_1)_{H1}$$
$$+ \omega_2^2(Z_1'Z_1)_{H2}$$
$$B = w_2^2\beta_2^2(Z_2'Z_2)_{G2} + w_4^2\beta_2^2(Z_2'Z_2)_{G4} + \phi_2^2(Z_2'Z_2)_{H1}$$
$$+ \omega_3^2(Z_2'Z_2)_{H3}$$
$$C = w_4^2\beta_1\beta_2(Z_1'Z_2)_{G4} - \phi_{12}(Z_1'Z_2)_{H1}.$$

$\hat{\theta}$ and K_4 are given by:

$$\hat{\theta} = M^{-1}F, \qquad K_4 = K_3 - F'M^{-1}F$$

where

$$K_3 = K_1 + K_2$$

and

$$K_1 = (Y - X_1\beta_1 - X_2\beta_2 - RB)'(Y - X_1\beta_1 - X_2\beta_2 - RB)_{G1}$$
$$+ w_2^2(Y - X_1\beta_1 - RB)'(Y - X_1\beta_1 - RB)_{G2}$$

$$+w_3^2(Y-X_2\beta_2-RB)'(Y-X_2\beta_2-RB)_{G3}$$
$$+w_4^2(Y-RB)'(Y-RB)_{G4}$$
$$K_2 = (\phi_1^2X_1'X_1-2\phi_{12}X_1'X_2+\phi_2^2X_2'X_2)_{H1}$$
$$+\omega_2^2(X_1'X_1)_{H2}+\omega_3^2(X_2'X_2)_{H3}$$
$$F = \begin{bmatrix} D \\ E \end{bmatrix}$$
$$D = [w_3^2\beta_1Z_1'(Y-X_2\beta_2-RB)]_{G3}+[w_4^2\beta_1Z_1'(Y-RB)]_{G4}$$
$$+(\phi_1^2Z_1'X_1-\phi_{12}Z_1'X_2)_{H1}+(\omega_2^2Z_1'X_1)_{H2}$$
$$E = [w_2^2\beta_2Z_2'(Y-X_1\beta_1-RB)]_{G2}+[w_4^2\beta_2Z_2'(Y-RB)]_{G4}$$
$$+(\phi_2^2Z_2'X_2-\phi_{12}Z_2'X_1)_{H1}+(\omega_3^2Z_2'X_2)_{H3}.$$

It is now possible to integrate over θ, using the well-known results concerning the multivariate normal distribution:

$$p(\beta_1, \beta_2, B, \sigma^2, S|\text{data}) \propto \Omega\, \sigma^r|M|^{-\frac{1}{2}} \exp\left\{-\frac{K_4}{2\sigma^2}\right\}. \tag{12}$$

To integrate over B we note that only K_4 contains terms involving B. After transforming K_4 we get:

$$p(\beta_1, \beta_2, \sigma^2, S|\text{data})$$
$$\propto \Omega\sigma^r|M|^{-\frac{1}{2}} \exp\left\{-\frac{K_7}{2\sigma^2}\right\}$$
$$\times\int\dots\int \exp\left\{-\frac{(B-\hat{B})'M^*(B-\hat{B})}{2\sigma^2}\right\} \mathrm{d}B \tag{13}$$

where

$$K_7 = K_6-F^{*\prime}(M^*)^{-1}F^*$$
$$K_6 = K_5+K_2-g_1'M^{-1}g_1$$
$$K_5 = (Y-X_1\beta_1-X_2\beta_2)'(Y-X_1\beta_1-X_2\beta_2)_{G1}$$
$$+w_2^2(Y-X_1\beta_1)'(Y-X_1\beta_1)_{G2}$$
$$+w_3^2(Y-X_2\beta_2)'(Y-X_2\beta_2)_{G3}+w_4^2(Y'Y)_{G4}.$$

Furthermore M^* is a symmetric positive definite matrix of order $k = q-2$:

$$M^* = M^{**}-g_2'M^{-1}g_2$$

where

$$M^{**} = (R'R)_{G1} + w_2^2(R'R)_{G2} + w_3^2(R'R)_{G3} + w_4^2(R'R)_{G4}$$

and

$$g_1 = \begin{pmatrix} d_1 \\ e_1 \end{pmatrix}, \qquad g_2 = \begin{pmatrix} d_2 \\ e_2 \end{pmatrix}$$

with

$$\begin{aligned} d_1 &= w_3^2\beta_1[Z_1'(Y - X_2\beta_2)]_{G3} + w_4^2\beta_1(Z_1'Y)_{G4} \\ &\quad + (\phi_1^2 Z_1'X_1 - \phi_{12}Z_1'X_2)_{H1} + \omega_2^2(Z_1'X_1)_{H2} \\ e_1 &= w_2^2\beta_2[Z_2'(Y - X_1\beta_1)]_{G2} + w_4^2\beta_2(Z_2'Y)_{G4} \\ &\quad + (\phi_2^2 Z_2'X_2 - \phi_{12}Z_2'X_1)_{H1} + \omega_3^2(Z_2'X_2)_{H3} \\ d_2 &= w_3^2\beta_1(Z_1'R)_{G3} + w_4^2\beta_1(Z_1'R)_{G4} \\ e_2 &= w_2^2\beta_2(Z_2'R)_{G2} + w_4^2\beta_2(Z_2'R)_{G4}. \end{aligned}$$

Finally:

$$\hat{B} = (M^*)^{-1}F^*$$

and

$$F^* = F^{**} - g_2'M^{-1}g_1$$

with

$$\begin{aligned} F^{**} &= [R'(Y - X_1\beta_1 - X_2\beta_2)]_{G1} + w_2^2[R'(Y - X_1\beta_1)]_{G2} \\ &\quad + w_3^2[R'(Y - X_2\beta_2)]_{G3} + w_4^2(R'Y)_{G4}. \end{aligned}$$

Since the last exponential of (13) corresponds to the kernel of a multivariate normal density, we obtain:

$$\begin{aligned} & p(\beta_1, \beta_2, \sigma^2, S|\text{data}) \\ & \quad \propto \Omega\sigma^r|M|^{-\frac{1}{2}}\sigma^k|M^*|^{-\frac{1}{2}} \exp\left\{-\frac{K_7}{2\sigma^2}\right\}. \end{aligned} \tag{14}$$

To integrate (14) over σ^2 we note that the terms involving σ^2 are:

$$\Omega\sigma^r\sigma^k = \Delta\,\sigma^{-(\sum_{j=1}^{4} n_j + \sum_{j=1}^{3} m_j + m_1 - r - k + 8)} \tag{15}$$

where

$$\Delta = w_2^{n_2} w_3^{n_3} w_4^{n_4} \phi^{m_1+3} \omega_2^{m_2} \omega_3^{m_3}.$$

We change variables from $(\beta_1, \beta_2, \sigma^2, S)$ to $(\beta_1, \beta_2, \sigma^2, \Lambda)$, where

$$\Lambda = \begin{pmatrix} \lambda_1^2 & \lambda_{12} \\ \lambda_{12} & \lambda_2^2 \end{pmatrix}.$$

The Jacobian of the transformation is:

$$J = \left| \frac{\delta(\beta_1, \beta_2, \sigma^2, \sigma_1^2, \sigma_{12}, \sigma_2^2)}{\delta(\beta_1, \beta_2, \sigma^2, \lambda_1^2, \lambda_{12}, \lambda_2^2)} \right| = \sigma^6. \tag{16}$$

Expression (14) can be rewritten as:

$$p(\beta_1, \beta_2, \sigma^2, \Lambda|\text{data}) \propto (\sigma^2)^{-\frac{1}{2}(f+2)} \exp\left\{-\frac{K_7}{2\sigma^2}\right\} \Delta|M|^{-\frac{1}{2}}|M^*|^{-\frac{1}{2}} \tag{17}$$

where

$$f = \sum_{j=1}^{4} n_j + \sum_{j=1}^{3} m_j + m_1 - r - k.$$

Integrating (17) over σ^2, using theorems on the gamma distribution, gives:

$$p(\beta_1, \beta_2, \Lambda|\text{data}) \propto K_7^{-\frac{1}{2}f} \Delta|M|^{-\frac{1}{2}}|M^*|^{-\frac{1}{2}}. \tag{18}$$

The marginal posterior pdf of β_1 or β_2 would then have to be obtained by numerical integration.[2]

If the marginal distribution of one of the elements of B is desired, say β_3, the integration in (13) should be made over all elements of B except β_3. Then integration over σ^2 yields:

$$p(\beta_1, \beta_2, \beta_3, \Lambda|\text{data}) \propto \tilde{K}_7^{-\frac{1}{2}(f+1)} \Delta|M|^{-\frac{1}{2}}|M^*|^{-\frac{1}{2}} h^{-\frac{1}{2}} \tag{19}$$

where $\tilde{K}_7 = K_7 + (\beta_3 - \hat{\beta}_3)^2/h$, h is the element of the first row and first column of $(M^*)^{-1}$, and $\hat{\beta}_3$ is the first element of $\hat{B}$. Numerical integration must now be used to get $p(\beta_3|\text{data})$.[3]

[2] Efficient methods for multivariate numerical integration have been suggested recently by Stroud [2]. If one is prepared to assume that there is no correlation between v_{1t} and v_{2t}, the Λ matrix is now diagonal and the dimension of the integration to be performed is reduced from five to four.

[3] Once the normalizing constant of (18)—say K_{18}—has been evaluated, that of (19) is readily obtained since the conditional distribution of β_3 given β_1, β_2 and Λ, is nonstandardized Student with f degrees of freedom. Hence:

$$K_{19} = K_{18} \frac{\Gamma[(f+1)/2]}{\sqrt{\pi}\,\Gamma(f/2)}.$$

TABLE 1
Sample values

Observations	Y	X_1	X_3	Z_1	Groups	
1	83.09	−0.66	17.81	2.05	G1	H2
2	19.56	−5.55	13.94*	−8.92	—	H2
3	36.26	3.68	−16.96	7.18	G1	H2
4	−35.68	−8.62	−11.16	−0.64	G1	H2
5	115.07	6.77	6.03	1.44	G1	H2
6	−23.42	−2.22	−5.04	−7.87	G1	H2
7	−5.64	−4.24	−13.04	−11.23	G1	H2
8	−10.41	22.91*	−8.16	24.77	G3	—
9	−129.72*	−5.04	−25.01	16.00	—	H2
10	27.53	−2.46	9.65*	−2.71	—	H2
11	−132.60*	−14.27*	−36.08	−11.49	—	—
12	−17.80	10.88	0.96	−1.07	G1	H2
13	−21.69	−4.76*	−6.67	−3.12	G3	—
14	13.11	−15.55	−0.90	−2.70	G1	H2
15	−46.23	−11.46	3.42	−12.88	G1	H2
16	−23.86	−0.41	−14.01	12.25	G1	H2
17	57.47	18.09	10.38	12.68	G1	H2
18	−79.15	7.04	−13.06	25.28*	G1	—
19	−99.99	0.60*	−9.94	−5.01	G3	—
20	−102.12	−11.70	−29.00	−12.85	G1	H2
21	−49.27*	−9.14*	0.21	−15.90	—	—
22	47.71	7.53	2.38	8.13*	G1	—
23	147.30	29.09	17.96	34.64*	G1	—
24	4.38	6.25	10.13	14.35	G1	H2
25	−49.39	−8.12*	−17.99	−15.24	G3	—
26	−121.00	−21.96*	−24.59	−31.33	G3	—
27	57.91	11.90	0.19	13.24	G1	H2
28	−79.05	1.38*	−10.86	3.35	G3	—
29	4.34	−4.26	3.92	−11.74	G1	H2
30	−75.22	−15.79	−15.90	−19.76	G1	H2

The marginal posterior pdf of any element of θ could be obtained by similar techniques. The results obtained in (18) and (19) extend readily to situations where prior information is available on the parameters, provided the form of the joint prior density is appropriate. The following form would be particularly convenient:

$$p(P) = p_v(\theta|S)p(S)p_v(\beta_1, \beta_2, B|\sigma^2)p_\gamma(\sigma^2) \tag{20}$$

where $p_v(\theta \mid S)$ and $p_v(\beta_1, \beta_2, B \mid \sigma^2)$ are multivariate normal and $p_\gamma(\sigma^2)$ is inverted gamma, while $p(S)$ may take any form. Since $p_v(\theta \mid S)$ is a natural conjugate of the L_{H1} function and $[p_v(\beta_1, \beta_2, B \mid \sigma^2) \cdot p_\gamma(\sigma^2)]$ is a natural conjugate of L_{G1} [1, pp. 47–58; 3, pp. 75, 239, 307], the joint posterior density could still be found by the procedure suggested in this paper.

3. *Numerical illustration*

A sample of thirty observations was generated from the following model:

$$Y_t = 2.0X_{1t} + 3.0X_{3t} + u_t \tag{21}$$

$$X_{1t} = 1.0Z_{1t} + v_t. \tag{22}$$

The Z_1, u and v variables were obtained by random drawings from normal distributions centered on zero; the X_{1t}'s were then derived from (22). The values of X_3 were chosen so as to be mildly correlated with X_1 and finally, the Y_t's were calculated from (21). Sample values are reproduced in table 1. The variance of the u_t's as well as that of the v_t's were adjusted to yield the correlation coefficients shown in table 2.

TABLE 2

Sample correlation coefficients, for the thirty observations shown in table 1

	Y	X_1	X_3	Z_1	X_1 and X_3
Y	1.0	0.62	0.80	—	0.84[a]
X_1	0.62	1.0	0.50	0.84	—
X_3	0.80	0.50	1.0	—	—

[a] This value is the square root of the multiple determination coefficient ($R^2 = 0.71$) obtained from the regression of Y on X_1 and X_3.

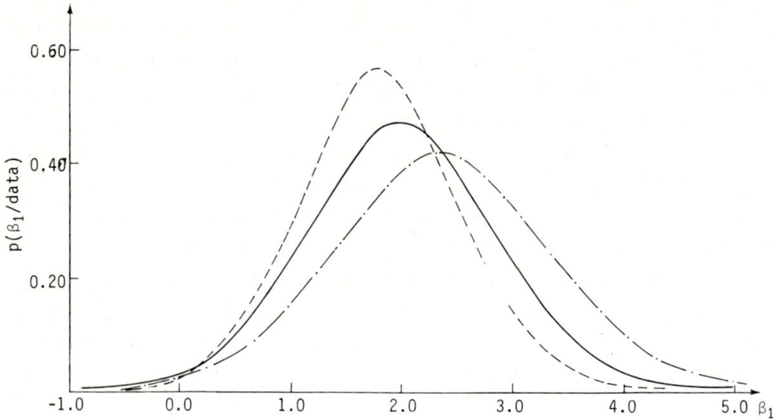

Fig. 1. Posterior densities of β_1. – – –: Posterior pdf obtained from all 30 observations. ———: Posterior pdf obtained from equations (18) and (19). –·–·–: Posterior pdf obtained from the 19 observations of group G1.

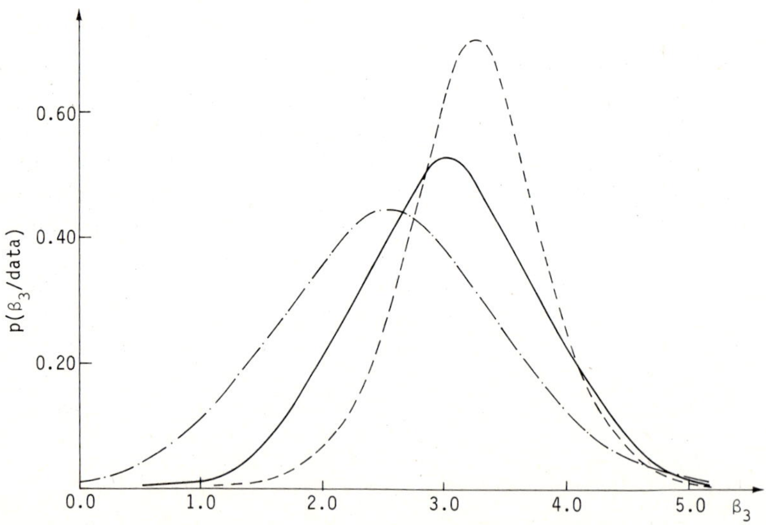

Fig. 2. Posterior densities of β_3. – – –: Posterior pdf obtained from all 30 observations. ———: Posterior pdf obtained from equations (18) and (19). –·–·–: Posterior pdf obtained from the 19 observations of group G1.

"Missing" values of the different variables were also chosen at random with probabilities set at 1/15 for Y, X_3 and Z_1 and 1/4 for X_1. "Missing" data points are identified by asterisks, in table 1. The last columns of the same table indicate the G and H groups to which each observation belongs. In this simplified example, only groups G1, G3 and H2 exist.

Figs. 1 and 2 give the posterior density functions of β_1 and β_3. The posterior densities corresponding to the complete sample as well as the densities obtained from the nineteen observations of group G1 are equally reproduced.

It is readily seen that, as expected, the dispersion of the posterior densities obtained from (18) and (19) is smaller than that of the posterior densities based on the nineteen G1 observations. The modes of the distributions are also closer to the true values of the parameters for the densities derived from the approach suggested in this paper.

References

[1] H. Raiffa and R. S. Schlaifer, Applied statistical decision theory. Boston, Harvard University, 1961.

[2] A. H. Stroud, Approximate calculation of multiple integrals. Englewood Cliffs, N.J., Prentice-Hall, 1971.

[3] A. Zellner, An introduction to Bayesian inference in econometrics. New York, John Wiley and Sons, 1971.

[4] A. Zellner and G. C. Tiao, Bayesian analysis of the regression model with autocorrelated errors, J. Am. Stat. Assoc. 59 (1964), 763–778.

S. E. Fienberg and A. Zellner (eds.)
Studies in Bayesian Econometrics and Statistics.

7.3. BAYESIAN ANALYSIS OF A BIVARIATE NORMAL DISTRIBUTION WHEN SOME OBSERVATIONS ARE MISSING*

J. S. MEHTA

Temple University

and

P. A. V. B. SWAMY

Ohio State University

1. *Introduction*

The problem of estimating parameters from an incomplete sample arises in many empirical studies, cf. Nicholson [7] and Rao [8]. We, in this paper, consider a specific problem of making inferences about the difference of the means in a bivariate normal population when the sample has some missing values corresponding to one or both the variables. Suppose that we have a sample

$$\begin{array}{ll} x_{11}, x_{12}, \ldots, x_{1n}, x_{1,n+1}, \ldots, x_{1,n+n_1}, & \\ x_{21}, x_{22}, \ldots, x_{2n}, & x_{2,n+1}, \ldots, x_{2,n+n_2} \end{array} \tag{1}$$

with n_2 and n_1 missing values of the variables x_1 and x_2 respectively, as indicated. Let the underlying normal population of the 2×1 vector variable $\mathbf{x} = (x_1, x_2)'$ have the mean vector $\boldsymbol{\mu} = (\mu_1, \mu_2)'$ and covariance matrix of the form

$$\begin{bmatrix} \sigma_1^2 & \rho\sigma_1\sigma_2 \\ \rho\sigma_1\sigma_2 & \sigma_2^2 \end{bmatrix}. \tag{2}$$

Using the sample in (1) we wish to draw inferences on the difference of the means

$$\delta = \mu_1 - \mu_2. \tag{3}$$

* An earlier version of this paper was presented to the Fourth NBER-NSF Seminar on Bayesian Inference in Econometrics, Univ. of Chicago, May 1972. Reprinted in revised form from the Journal of the American Statistical Association 68 (1973), 922–926.

Mehta and Gurland [6] approached this problem from the sampling theory point of view when the sample had some missing values corresponding to only one of the variables, and developed a minimum variance unbiased estimator of δ, which involves the nuisance parameters $(\rho, \sigma_1^2, \sigma_2^2)$. The true values of these parameters are usually unknown and inferences about δ can be drawn proceeding conditionally on given sample estimates of nuisance parameters. This procedure is not always optimal as was shown by Mehta and Gurland. In our investigation, we offer a Bayesian solution to the problem. One advantage of the Bayesian approach is that the nuisance parameters can be eliminated by straightforward integration and the effect of extra observations from the populations of x_1 and x_2 on the overall inferences about δ can be easily analyzed.

We derive in section 2 the marginal posterior pdf's for $\boldsymbol{\mu}$ and δ utilizing a specific "non-informative" prior distribution of parameters $(\mu_1, \mu_2, \sigma_1^2, \sigma_2^2, \rho)'$ when some observations are missing on only one of the variables, say x_2 (i.e., $n_2 = 0$). In section 3 we modify these derivations to incorporate certain knowledge in the form of prior distributions which are proper in the nuisance parameters. Finally, in section 4 we draw posterior inferences about δ when some observations on both x_1 and x_2 are missing (i.e., $n_1 > 0$, $n_2 > 0$) and the prior distributions on the parameters are non-informative.

2. *Prior and posterior distributions of parameters*

From (1) and (2) the likelihood function can be written as

$$\begin{aligned} L(\boldsymbol{\theta}|\text{data}) = {} & (2\pi)^{-\frac{1}{2}(2n+n_1+n_2)}(\sigma_1^2)^{-\frac{1}{2}(2n+n_1+n_2)} \\ & \times (1-\rho^2)^{-n/2}(k)^{-\frac{1}{2}(n+n_2)} \\ & \times \exp\left\{-\frac{1}{2\sigma_1^2(1-\rho^2)k}\left[k\sum_{i=1}^{n}(x_{1i}-\mu_1)^2\right.\right. \\ & -2\rho\sqrt{(k)}\sum_{i=1}^{n}(x_{1i}-\mu_1)(x_{2i}-\mu_2) \\ & +\sum_{i=1}^{n}(x_{2i}-\mu_2)^2+k(1-\rho^2)\sum_{i=n+1}^{n+n_1}(x_{1i}-\mu_1)^2 \\ & \left.\left.+(1-\rho^2)\sum_{i=n+1}^{n+n_2}(x_{2i}-\mu_2)^2\right]\right\} \end{aligned} \tag{4}$$

where $k = \sigma_2^2/\sigma_1^2$ and $\boldsymbol{\theta} = (\mu_1, \mu_2, \sigma_1^2, k, \rho)'$.

We consider the problem under two cases. In this section and the next we consider the case where the extra observations $(x_{2,n+1}, \ldots, x_{2,n+n_2})$ are not available.

Case I: $n_2 = 0$. For the prior distributions of $\boldsymbol{\theta}$, we assume that the experimental situation is such that "little" is known about $\boldsymbol{\theta}$. Following Geisser and Cornfield [4], and Tiao and Zellner [10] we take

$$p(\boldsymbol{\theta}) = p(\boldsymbol{\mu})p(\rho, \sigma_1^2, k) \tag{5}$$

with

$$p(\mu) = \text{constant} \tag{6}$$

$$p(\rho, \sigma_1^2, k) \propto \frac{d\sigma_1^2 dk\, d\rho}{\sigma_1^2 k(1-\rho^2)^{3/2}} \tag{7}$$

$$-\infty < \mu_i < \infty\ (i = 1, 2), \quad 0 < \sigma_1^2 < \infty, \quad 0 < k < \infty,$$

$$-1 < \rho < 1.$$

Using Bayes' theorem we can combine the prior pdf's in (6) and (7) with the likelihood function in (4) after eliminating the terms involving the extra observations from the population of x_2 and obtain the joint posterior pdf for $\boldsymbol{\theta}$ as

$$p(\boldsymbol{\theta}|\text{data}) \propto (\sigma_1^2)^{-\frac{1}{2}(2n+n_1+2)}(1-\rho^2)^{-\frac{1}{2}(n+3)}(k)^{-\frac{1}{2}(n+2)}$$
$$\times \exp\left\{-\frac{1}{2\sigma_1^2 k(1-\rho^2)}\left[k\sum_{i=1}^{n}(x_{1i}-\mu_1)^2\right.\right.$$
$$-2\rho\sqrt{(k)}\sum_{i=1}^{n}(x_{1i}-\mu_1)(x_{2i}-\mu_2) \tag{8}$$
$$\left.\left.+\sum_{i=1}^{n}(x_{2i}-\mu_2)^2+k(1-\rho^2)\sum_{i=n+1}^{n+n_1}(x_{1i}-\mu_1)^2\right]\right\}.$$

It is shown in section A of the Appendix to the paper that by applying the transformation $\phi_1 = \mu_1 - \bar{\mu}$, where $\bar{\mu} = (\mu_1+\mu_2)/2$, and by integrating out $\bar{\mu}$ and σ_1^2, the conditional posterior distribution of δ given ρ and k, can be written in the form of Student's t-distribution with $2n+n_1-2$ df and with mean

$$\hat{\bar{\delta}} = A\bar{x}_1 + (1-A)\hat{\mu}_1 - \bar{x}_2 \tag{9}$$

where

$$A = \lambda + \rho\sqrt{(k)}(1-\lambda),\ \lambda = n/(n+n_1),\ \bar{x}_i = \frac{1}{n}\sum_{l=1}^{n} x_{il} \qquad (i = 1, 2)$$

and

$$\hat{\mu}_1 = \frac{1}{n_1}\sum_{l=n+1}^{n+n_1} x_{1l}.$$

It is interesting to note that $\hat{\hat{\delta}}$ is the same as the estimator of δ developed by Mehta and Gurland [6] and the conditional posterior distribution of δ given ρ and k, corresponding to prior distributions (5) is centered at the sampling theory quantity $\hat{\hat{\delta}}$. Notice also that $\hat{\hat{\delta}}$ is the same as the estimator of δ implied by the maximum likelihood estimators of μ_1 and μ_2 derived by Anderson [1]. It is clear from (9) that the effect of extra observations from one of the two populations on the inferences about δ depends upon A whose value is usually unknown. Consequently, in the sampling theory framework the effect of extra observations cannot be easily analyzed. It can, however, be analyzed in a straightforward manner using Bayesian techniques. From the pdf in (8) we can make the transformation from (σ_1^2, k, ρ) to $(\sigma_1^2, k\sigma_1^2(1-\rho^2), \rho\sqrt{k})$.[1] The latter set of parameters can be easily integrated out to yield the unnormalized posterior pdf for $(\mu_1, \mu_2)'$. This can be compared with the posterior pdf for $(\mu_1, \mu_2)'$ when the extra observations are ignored. In the same way the posterior pdf for δ can be obtained from the marginal posterior pdf for $(\mu_1, \mu_2)'$ and a similar comparison can be made.

Let

$$v_1 = \sigma_1^2, \qquad v_2 = k\sigma_1^2(1-\rho^2) \quad \text{and} \quad v_3 = \rho\sqrt{(k)}. \tag{10}$$

The Jacobian of the transformation is $(v_1 v_2 + v_1^2 v_3^2)^{-1/2}$. The transformation (10) when applied to (8), yields

$$p(\mu_1, \mu_2, v_1, v_2, v_3 | \text{data}) \propto (v_1)^{-\frac{1}{2}(n+n_1+1)}(v_2)^{-\frac{1}{2}(n+3)} \times \exp\left\{-\frac{1}{2v_1}\left[\left(1+\frac{v_1 v_3^2}{v_2}\right)\sum_{i=1}^{n}(x_{1i}-\mu_1)^2 - 2\frac{v_3 v_1}{v_2}\sum_{i=1}^{n}(x_{1i}-\mu_1)(x_{2i}-\mu_2) + \frac{v_1}{v_2}\sum_{i=1}^{n}(x_{2i}-\mu_2)^2 + \sum_{i=n+1}^{n+n_1}(x_{1i}-\mu_1)^2\right]\right\}. \tag{11}$$

[1] This transformation was kindly suggested to us by a referee of the Journal of the American Statistical Association.

We now integrate (11) over v_1 and v_2 to obtain

$$p(\mu_1, \mu_2, v_3|\text{data})$$

$$\propto \left\{ \sum_{i=1}^{n} (x_{1i}-\mu_1)^2 + \sum_{i=n+1}^{n+n_1} (x_{1i}-\mu_1)^2 \right\}^{-\frac{1}{2}(n+n_1-1)}$$

$$\times \left\{ v_3^2 \sum_{i=1}^{n} (x_{1i}-\mu_1)^2 - 2v_3 \sum_{i=1}^{n} (x_{1i}-\mu_1)(x_{2i}-\mu_2) \right.$$

$$\left. + \sum_{i=1}^{n} (x_{2i}-\mu_2)^2 \right\}^{-\frac{1}{2}(n+1)} \tag{12}$$

We can complete the square on v_3 and integrate (12) from $-\infty$ to ∞ with respect to v_3, an operation which yields the marginal posterior pdf for $(\mu_1, \mu_2)'$.

$$p(\mu_1, \mu_2|\text{data})$$

$$\propto \left[\sum_{i=1}^{n+n_1} (x_{1i}-\mu_1)^2 \right]^{-\frac{1}{2}(n+n_1-1)} \left[\sum_{i=1}^{n} (x_{1i}-\mu_1)^2 \right]^{-\frac{1}{2}}$$

$$\times \left\{ \sum_{i=1}^{n} (x_{2i}-\mu_2)^2 - \frac{\left[\sum_{i=1}^{n} (x_{1i}-\mu_1)(x_{2i}-\mu_2) \right]^2}{\sum_{i=1}^{n} (x_{1i}-\mu_1)^2} \right\}^{-n/2} . \tag{13}$$

It is instructive to write (13) in the following form,

$$p(\mu_1, \mu_2|\text{data})$$

$$\propto \left\{ 1 + \frac{(\mu_1 - \bar{x}_1^{(G)})^2}{s_{11}^{(G)}} \right\}^{-\frac{1}{2}(n+n_1-1)} \left\{ 1 + \frac{(\mu_1 - \bar{x}_1)^2}{s_{11}} \right\}^{\frac{1}{2}(n-1)}$$

$$\times \{1 + (\boldsymbol{\mu} - \bar{\boldsymbol{x}})' S^{-1} (\boldsymbol{\mu} - \bar{\boldsymbol{x}})\}^{-n/2} \tag{14}$$

where

$$\bar{x}_1^{(G)} = \frac{1}{n+n_1} \sum_{l=1}^{n+n_1} x_{1l}, \quad s_{11}^{(G)} = \frac{1}{n+n_1} \sum_{l=1}^{n+n_1} (x_{1l} - \bar{x}_1^{(G)})^2,$$

$$s_{ij} = \frac{1}{n} \sum_{l=1}^{n} (x_{il} - \bar{x}_i)(x_{jl} - \bar{x}_j) \quad (i, j = 1, 2), \quad \bar{\boldsymbol{x}} = (\boldsymbol{x}_1, \bar{x}_2)',$$

$$S = \begin{bmatrix} s_{11} & s_{12} \\ & s_{22} \end{bmatrix}$$

and $\bar{x}_i$ $(i = 1, 2)$ are as defined in (9).

If we ignore the extra observations $(x_{1n+1}, \ldots, x_{1n+n_1})$ the product of the first two factors on the rhs of (14) is equal to unity and $p(\mu_1, \mu_2|$ data) is proportional to

$$\{1 + (\boldsymbol{\mu} - \bar{\boldsymbol{x}})' S^{-1} (\boldsymbol{\mu} - \bar{\boldsymbol{x}})\}^{-n/2}. \tag{15}$$

The marginal posterior pdf in (14) is the product of three factors. The first factor is in the form of a scaled t-distribution with mean $\bar{x}_1^{(G)}$ and scale parameter $s_{11}^{(G)}$ and with $(n+n_1-2)$ df. The reciprocal of the second factor is in the form of a scaled t-distribution with mean $\bar{x}_1$ and scale parameter s_{11} and with $(n-2)$ df, and the third factor is in the form of a scaled bivariate t-distribution with mean $\bar{\boldsymbol{x}}$, scaling matrix S and $(n-2)$ df. The first two factors arise as a direct consequence of extra observations on x_1.

If interest centers on μ_1, its posterior pdf can be obtained from (14) by integration. This can be done easily by completing the square on μ_2 in the third factor of (14) and using properties of the univariate Student's t pdf to yield

$$p(\mu_1|\text{data}) \propto \left\{1 + \frac{(\mu_1 - \bar{x}_1^{(G)})^2}{s_{11}^{(G)}}\right\}^{-\frac{1}{2}(n+n_1-1)} \tag{16}$$

which is in the form of univariate Student's t pdf with $(n+n_1-2)$ df. It can be seen from (14) that the conditional posterior distribution of μ_2 given μ_1, is also in the form of univariate Student's t-distribution with $n-1$ df. The marginal posterior distribution of μ_2 can be obtained from (14) by numerical integration. Thus, in a Bayesian framework to draw inferences about $\boldsymbol{\mu}$ we do not have to proceed conditionally on given sample estimates of nuisance parameters.

Since our interest centers on δ, we transform the variables in (14) from (μ_1, μ_2) to $(\phi_1+\bar{\mu}, \bar{\mu}-\phi_1)$ to obtain

$$p(\phi_1, \bar{\mu}|\text{data}) \propto \left\{1+\frac{(\phi_1+\bar{\mu}-\bar{x}_1^{(G)})^2}{s_{11}^{(G)}}\right\}^{-\frac{1}{2}(n+n_1-1)}$$
$$\times\left\{1+\frac{(\phi_1+\bar{\mu}-\bar{x}_1)^2}{s_{11}}\right\}^{\frac{1}{2}(n-1)} \{1+s^{11}(\phi_1+\bar{\mu}-\bar{x}_1)^2$$
$$+2s^{12}(\phi_1+\bar{\mu}-\bar{x}_1)(\bar{\mu}-\phi_1-\bar{x}_2)+s^{22}(\bar{\mu}-\phi_1-\bar{x}_2)^2\}^{-n/2} \tag{17}$$

where

$$S^{-1} = \begin{bmatrix} s^{11} & s^{12} \\ & s^{22} \end{bmatrix}$$

is the inverse of S.

If we ignore the extra observations on x_1, $p(\phi_1, \bar{\mu}|\text{data})$ is proportioned to the third factor on the rhs of (17).

We can utilize readily available bivariate numerical integration programs to integrate (17) with respect to $\bar{\mu}$ and obtain the marginal posterior pdf for $2\phi_1 = \delta$.

We study the behavior of the marginal posterior pdf for δ with the help of a generated example. A random sample of size 20 was drawn from each of 12 bivariate normal populations of the vector variable $(x_1, x_2)'$ with the same mean vector $(10, 5)'$ and with different covariance matrices of the form

$$\sigma_1^2\begin{bmatrix} 1 & \rho\sqrt{k} \\ & k \end{bmatrix}.$$

Twelve different covariance matrices were formed by combining the values -0.8, -0.3, 0.3, 0.8 of ρ with each of the values 0.2, 1.0, 5.0 of k. The parameter σ_1^2 was set equal to unity in all cases. By deleting 2, 6, 10 observations on a variable, say x_2, from each sample of size 20 we formed three sets of 12 samples each so that each sample of the first set contained 2 extra observations on x_1, each sample of the second set contained 6 extra observations on x_1 and each sample of the third set contained 10 extra observations on x_1. In each of these 36 samples the number of complete observations on the vector (x_1, x_2) was 10. We evaluated (17) using these 36 samples.

We present the posterior measures like mean and variance of the marginal posterior distribution of δ in table 1. Corresponding to each set of samples we present two rows of values. In the first row we present the mean while in the second we present the variance.

In order to evaluate the effects of extra observations on x_1 we present in the first two rows of table 1 the mean and variance of the marginal posterior distribution of δ when the extra observations are ignored. The following conclusions emerge from the values in table 1.

(1) The posterior mean when the extra observations from the population of x_1 are utilized is not markedly different from the posterior mean when the extra observations are ignored.

(2) The significant effect of extra observations is to reduce the variance of the posterior distribution of δ. That is, the effect of the first two factors in (14) is to increase the sharpness of the posterior distribution of δ. The use of extra observations brings about larger reductions in the posterior variance if the "true" value of ρ is negative than if it is positive depending on the "true" value of k. The reduction in the posterior variance as a result of utilizing the extra observations is greater for smaller values of k than for higher values of k when $\rho > 0$, and smaller for smaller values of k than for higher values of k when $\rho < 0$.

3. *Posterior distribution of δ corresponding to a proper prior distribution of (σ_1^2, ρ, k)*

Our analysis in the previous section is based on the assumption that, initially, we know "little" about $\boldsymbol{\theta}$. This may not always be true. We may have some knowledge about at least some elements of $\boldsymbol{\theta}$. Suppose that we have prior information about (σ_1^2, k, ρ) which can be expressed as

$$p(\sigma_1^2, \rho, k|s^*, v) \propto (k)^{-\frac{1}{2}(v-1)}(\sigma_1^2)^{-(v-2)}(1-\rho^2)^{-v/2}$$

$$\times \exp\left\{-\frac{1}{2k\sigma_1^2(1-\rho^2)}\left[ks_{11}^*-2\rho\surd(k)s_{12}^*+s_{22}^*\right]\right\}. \tag{18}$$

Let $p(\boldsymbol{\theta}) = p(\boldsymbol{\mu})p(\sigma_1^2, \rho, k)$ and $p(\boldsymbol{\mu}) \propto$ constant.

Savage [9] has shown that prior and posterior probabilities can be interpreted as subjective probabilities. It may be verified that (18) is a proper pdf if $v \geqq 5$ and $\sigma_1^2\sigma_2^2(1-\rho^2) > 0$. The constants s_{11}^*, s_{12}^* and s_{22}^* are known. We can easily see that (18) is the same as (7) when

TABLE 1

The mean and variance of the marginal posterior distribution of δ for different values of n_1 and the nuisance parameters when $n = 10$ and $n_2 = 0$

n_1	$\rho = 0.8$			$\rho = 0.3$			$\rho = -0.3$			$\rho = -0.8$		
	$k = 5.0$	$k = 1.0$	$k = 0.2$	$k = 5.0$	$k = 1.0$	$k = 0.2$	$k = 5.0$	$k = 1.0$	$k = 0.2$	$k = 5.0$	$k = 1.0$	$k = 0.2$
0	5.447	5.326	5.270	5.918	5.610	5.471	6.277	5.834	5.623	6.463	5.983	5.721
	0.255	0.128	0.196	0.780	0.446	0.388	1.526	0.819	0.568	2.032	1.118	0.674
2	5.436	5.382	5.357	5.973	5.671	5.533	6.302	5.848	5.636	6.411	5.908	5.664
	0.250	0.109	0.154	0.677	0.337	0.275	1.177	0.567	0.377	1.541	0.735	0.433
6	5.439	5.371	5.339	5.963	5.659	5.521	6.306	5.845	5.634	6.465	5.924	5.677
	0.246	0.079	0.086	0.559	0.218	0.152	0.814	0.332	0.203	0.959	0.401	0.230
10	5.424	5.438	5.444	6.030	5.731	5.596	6.330	5.863	5.650	6.340	5.836	5.608
	0.240	0.076	0.078	0.543	0.200	0.132	0.737	0.286	0.168	0.795	0.318	0.180

$\nu = 3$, $s^*_{11} = s^*_{12} = s^*_{22} = 0$. We again consider Case I: $n_2 = 0$ in this section.

If we combine (18) with the likelihood function (4) using Bayes' theorem, we obtain the joint posterior pdf for $\boldsymbol{\theta}$ as

$$p(\boldsymbol{\theta}|\text{data}) \propto (\sigma_1^2)^{-\frac{1}{2}(2n+n_1+2\nu-4)}(1-\rho^2)^{-\frac{1}{2}(n+\nu)}(k)^{-\frac{1}{2}(n+\nu-1)}$$

$$\times \exp\left\{-\frac{1}{2\sigma_1^2 k(1-\rho^2)}\left[k\sum_{i=1}^{n}(x_{1i}-\mu_1)^2\right.\right.$$

$$-2\rho\surd(k)\sum_{i=1}^{n}(x_{1i}-\mu_1)(x_{2i}-\mu_2)$$

$$+\sum_{i=1}^{n}(x_{2i}-\mu_2)^2+k(1-\rho^2)\sum_{i=n+1}^{n+n_1}(x_{1i}-\mu_1)^2$$

$$\left.\left.+ks^*_{11}-2\rho\surd(k)s^*_{12}+s^*_{22}\right]\right\}. \tag{19}$$

Changing the variables in (19) from $(\sigma_1^2, k\sigma_1^2(1-\rho^2), \rho\surd k)$ to (v_1, v_2, v_3) and integrating (19) with respect to v_1, v_2 and v_3, the marginal posterior distribution of $\boldsymbol{\mu}$ is

$$p(\boldsymbol{\mu}|\text{data}) \propto \left\{1+\frac{(\mu_1-\bar{x}_1^{(G)})^2}{s_{11}^{(G)}+s^*_{11}(n+n_1)^{-1}}\right\}^{-\frac{1}{2}(n+n_1+\nu-4)}$$

$$\times\left\{1+\frac{(\mu_1-\bar{x}_1)^2}{s_{11}+s^*_{11}n^{-1}}\right\}^{\frac{1}{2}(n+\nu-4)}$$

$$\times\left\{1+(\boldsymbol{\mu}-\bar{\mathbf{x}})'\left(S+\frac{S^*}{n}\right)^{-1}(\boldsymbol{\mu}-\bar{\mathbf{x}})\right\}^{-\frac{1}{2}(n+\nu-3)} \tag{20}$$

where

$$S^* = \begin{bmatrix} s^*_{11} & s^*_{12} \\ & s^*_{22} \end{bmatrix}.$$

Numerical methods may be utilized to calculate the marginal posterior pdf for δ from (20). Choosing a set of values for s^*_{ij} $(i, j = 1, 2)$ and using one of the 36 samples referred to in the previous section we evaluated the marginal posterior pdf for δ when $\nu = 4, 6, 10$. Plots of these distributions are given in fig. 1. These plots indicate that if we fix the elements of S^* and increase ν, the variance of the posterior distribution of δ decreases but its mean is invariant. The decrease in the variance is due to the increase in the information contained in (18).

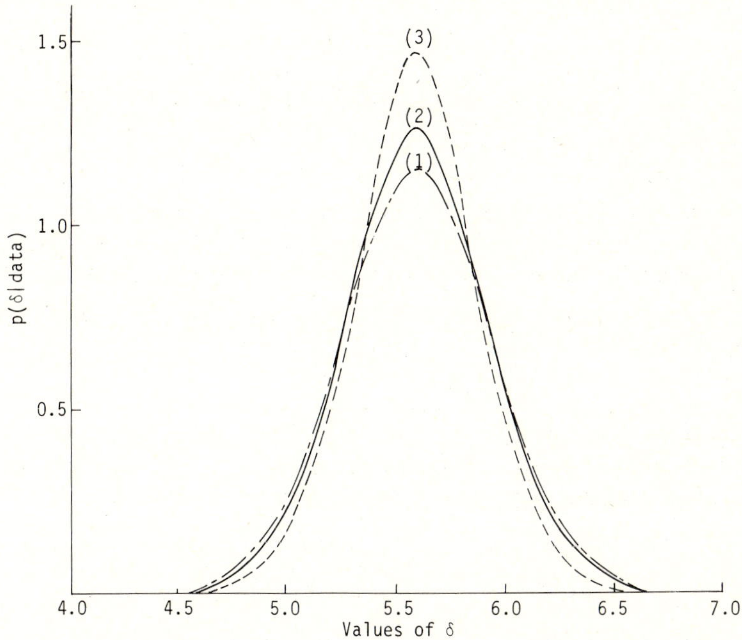

Fig. 1. Marginal posterior pdf for δ from equation (20) for $n_1 = 10$, $n = 10$, $\rho = 0.3$ and $k = 0.2$. (1) $\nu = 4$, mean $= 5.596$, variance $= 0.132$. (2) $\nu = 6$, mean $= 5.596$, variance $= 0.107$. (3) $\nu = 10$, mean $= 5.596$, variance $= 0.079$.

4. *Inferences about δ when some observations on both x_1 and x_2 are missing*

In this section we consider the second case where all the observations indicated in (1) are available.

Case II: $n_1 > 0$, $n_2 > 0$. The likelihood function which is appropriate for this case is given in (4). We consider the same prior pdf's as in (6) and (7). We combine these prior pdf's with the likelihood function (4) via Bayes' theorem to obtain

$$p(\boldsymbol{\theta}|\text{data}) \propto (\sigma_1^2)^{-\frac{1}{2}(2n+n_1+n_2+2)}(k)^{-\frac{1}{2}(n+n_2+2)}(1-\rho^2)^{-\frac{1}{2}(n+3)}$$

$$\times \exp\left\{-\frac{1}{2k(1-\rho^2)\sigma_1^2}[ka_{11}-2\rho\sqrt{(k)}a_{12}+a_{22}\right.$$

$$\left.+k(1-\rho^2)b_1+(1-\rho^2)b_2]\right\} \tag{21}$$

where $a_{ij} = \sum_{l=1}^{n}(x_{il}-\mu_i)(x_{jl}-\mu_j)$ $(i, j = 1, 2)$, $b_1 = \sum_{l=n+1}^{n+n_1}(x_{1l}-\mu_1)^2$ and $b_2 = \sum_{l=n+1}^{n+n_2}(x_{2l}-\mu_2)^2$.

It is shown in section B of the Appendix to the paper that the conditional posterior distribution of δ given ρ and k, is in the form of Student's t-distribution with $2n+n_1+n_2-2$ df and with mean

$$\tilde{\delta} = A_1\bar{x}_1+(1-A_1)\hat{\mu}_1-A_2\bar{x}_2-(1-A_2)\hat{\mu}_2 \tag{22}$$

where

$$A_1 = \frac{1+\lambda_2+\rho\lambda_1\sqrt{(k)}}{(1+\lambda_1)(1+\lambda_2)-\rho^2\lambda_1\lambda_2},$$

$$A_2 = \frac{\sqrt{(k)}(1+\lambda_1)+\rho\lambda_2}{\sqrt{(k)}[(1+\lambda_1)(1+\lambda_2)-\rho^2\lambda_1\lambda_2]},$$

$$\lambda_1 = \frac{n^1}{n}, \qquad \lambda_2 = \frac{n_2}{n}, \qquad \hat{\mu}_2 = \frac{1}{n_2}\sum_{l=n+1}^{n+n_2} x_{2l}$$

and the remaining symbols are as explained in section 2.

Following the same argument as in Mehta and Gurland [6] we can show that $\tilde{\delta}$ is the minimum variance unbiased estimator of δ. Notice that $\tilde{\delta}$ is not a feasible estimator of δ because it depends upon unknown parameters. Let

$$v_1 = \sigma_1^2, \qquad w_1 = (1-\rho^2)/\rho^2 \quad\text{and}\quad w_2 = 1/\rho\sqrt{(k)}. \tag{23}$$

In (21) we make the transformation (23) to obtain

$$p(\boldsymbol{\mu}, v_1, w_1, w_2|\text{data}) \propto (v_1)^{-\frac{1}{2}(2n+n_1+n_2+2)}(w_1)^{-\frac{1}{2}(n+3)}$$

$$\times(1+w_1)^{-n_2/2}(w_2)^{n+n_2-1}\exp\left\{-\frac{1}{2v_1}\left[\left(1+\frac{1}{w_1}\right)a_{11}+b_1\right.\right.$$

$$\left.\left.+\left(\frac{a_{22}}{w_1}+\frac{b_2}{1+w_1}\right)w_2^2-\frac{2a_{12}w_2}{w_1}\right]\right\}. \tag{24}$$

Now the following identity is useful. For $p > 0$

$$\int_{-\infty}^{\infty} x^n e^{-px^2+2qx}\,dx = e^{q^2/p}\frac{\sqrt{\pi}}{p}\left(\frac{q}{p}\right)^n \sum_{l=0}^{E[n/2]} \frac{n!}{(n-2l)!l!}\left(\frac{p}{4q^2}\right)^l \tag{25}$$

where $E[n/2]$ is the integer part of $n/2$.

Using the identity (25) and integrating (24) with respect to w_2, we obtain

$$\begin{aligned} p(\boldsymbol{\mu}, v_1, w_1|\text{data}) \propto & \sum_{l=0}^{E[(n+n_2-1)/2]} \left\{\frac{v_1^{-(2n+n_1+n_2-2l+1)/2}}{(n+n_2-1-2l)!l!2^l}\right\} \\ & \times \exp\left\{-\frac{1}{2v_1}\left[b_1+a_{11}\left(1+\frac{1}{w_1}\right)\frac{a_{12}^2}{w_1}\left(a_{22}+\frac{b_1w_1}{1+w_1}\right)^{-1}\right]\right\} \\ & \times w_1^{-\frac{1}{2}(n+2-2l)}(1+w_1)^{\frac{1}{2}(2n+n_2-2l-1)}(a_{12})^{n+n_2-1-2l} \\ & \times [a_{22}(1+w_1)+b_2w_1]^{-(n+n_2-\frac{1}{2}-l)}. \end{aligned} \tag{26}$$

On integrating (26) with respect to v_1 we obtain the joint posterior pdf for $\boldsymbol{\mu}$ and w_1

$$\begin{aligned} p(\boldsymbol{\mu}, w_1|\text{data}) \propto & \sum_{l=0}^{E[(n+n_2-1)/2]} \frac{2^{[(2n+n_1+n_2-1)/2]-2l}}{(n+n_2-1-2l)!l!} \\ & \times w_1^{(n+n_1+n_2-3)/2}[a_{22}(1+w_1)+b_2w_1]^{(n_1-n_2)/2} \\ & \times \frac{\Gamma\left(n+\dfrac{n_1+n_2-1}{2}-l\right)a_{12}^{(n+n_2-1-2l)}(1+w_1)^{[(2n+n_2-1)/2]-l}}{[\{a_{11}(1+w_1)+b_1w_1\}\{a_{22}(1+w_1)+b_2w_1\} - a_{12}^2(1+w_1)]^{[(2n+n_1+n_2-1)/2]-l}} \end{aligned} \tag{27}$$

where $-\infty < \mu_i < \infty$ $(i = 1, 2)$ and $0 < w_1 < \infty$.

If we ignore the extra observations on x_1 and x_2, we can easily derive the marginal posterior distribution of $\boldsymbol{\mu}$ from (21) and it is the same as (15). We can obtain the marginal posterior distribution of δ from (27) utilizing a trivariate numerical integration program.

We have drawn a sample of size 10 from a bivariate normal population with mean vector $(10, 5)'$ and with variances and covariances given by $\sigma_1^2 = 1$, $\rho = 0.3$ and $k = 0.2$. We deleted 2 observations on x_1 and 3 observations on x_2 so that $n = 5$, $n_1 = 3$ and $n_2 = 2$ for this sample. We evaluated (27) using this sample and obtained the posterior measures like mean and variance of the marginal posterior distribution of δ, which were 5.64 and 0.73 respectively. The mean and variance of the marginal posterior distribution of δ when the extra observations on x_1 and x_2 are ignored are 6.24 and 0.81 respectively. In this case both the mean and variance of the posterior distribution of δ are affected by the extra observations.

Now it is obvious from (27) that in order to obtain the posterior density of δ we have to resort to trivariate integration which even on a fast electronic computer is time consuming and hence expensive. It will be especially so if we consider large values of n and n_2. Therefore there is a need for a technique which could possibly reduce this cost. We present here an approximation which effectively reduces the problem of trivariate integration to one of bivariate integration. For example, let us consider the conditional posterior density of δ given ρ and k, i.e., $p(\delta|\rho, k, \text{data})$ which we know is Student's t-distribution with mean given in (22) and $(2n+n_1+n_2-2)$ df. If we estimate ρ and k by $\hat{\rho}$ and $\hat{k}$ obtained from the following estimates suggested by Rao [8, pp. 161–163]

$$\hat{\rho} = \sum_1^n (x_{1i}-\bar{x}_1)(x_{2i}-\bar{x}_2) \Big/ \left[\sum_1^n (x_{1i}-\bar{x}_1)^2 \sum_1^n (x_{2i}-\bar{x}_2)^2\right]^{\frac{1}{2}} \tag{28}$$

$$\hat{\sigma}_1^2 = \frac{1}{n+n_1-1}\left[\sum_1^{n+n_1} x_{1i}^2 - \frac{1}{n+n_1}\left(\sum_1^{n+n_1} x_{1i}\right)^2\right] \tag{29}$$

$$\sigma_2^2 = \frac{1}{n+n_2-1}\left[\sum_1^{n+n_2} x_{2i}^2 - \frac{1}{n+n_2}\left(\sum_1^{n+n_2} x_{2i}\right)^2\right] \tag{30}$$

$$\hat{k} = \hat{\sigma}_2^2/\hat{\sigma}_1^2,$$

then we can approximate $p(\delta|\text{data})$ by $p(\delta|\rho = \hat{\rho}, k = \hat{k}, \text{data})$. Notice that if we calculate posterior measures like the mean and variance from $p(\delta|\rho = \hat{\rho}, k = \hat{k}, \text{data})$, then the procedure involves evaluation of only a bivariate integral. When this was done using our sample with $n = 5$, $n_1 = 3$ and $n_2 = 2$ we obtained 5.605 and 0.494 as the mean and variance of $p(\delta|\rho = \hat{\rho}, k = \hat{k}, \text{data})$. Although this mean is not substantially different from the mean 5.640 of $p(\delta|\text{data})$, this variance is quite different from the variance 0.73 of $p(\delta|\text{data})$. This shows that in small samples the conditional posterior distribution of δ given $\rho = \hat{\rho}$ and $k = \hat{k}$, does not provide a good approximation to the marginal posterior distribution of δ. However, $p(\delta|\rho = \hat{\rho}, k = \hat{k}, \text{data})$ provides a better approximation to $p(\delta|\text{data})$ if we reduce its df by the number

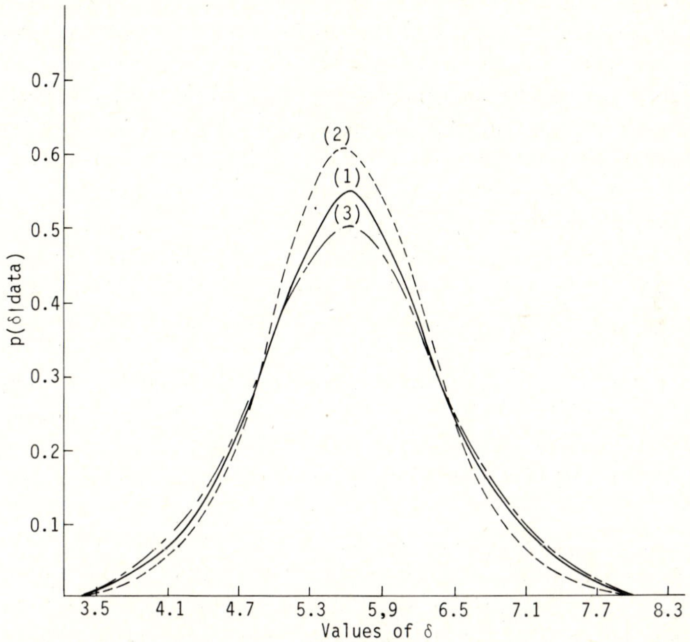

Fig. 2. (1) Marginal posterior distribution of δ from equation (27). (2) Conditional posterior distribution of δ given ρ and k, from equation (B.4). (3) Conditional posterior distribution of δ given ρ and k, adjusted for loss of df from equation (B.4).

of parameters estimated from the data.[2] The mean and variance of Student's t-distribution of the form $p(\delta|\rho = \hat{\rho}, k = \hat{k}, \text{data})$ with $(2n+n_1+n_2-4)$ df are 5.608 and 0.766 respectively. These measures are quite close to the corresponding measures of $p(\delta|\text{data})$. Plots of the marginal posterior distribution of δ, the conditional posterior distribution of δ given $\rho = \hat{\rho}$ and $k = \hat{k}$, and the Student's t-distri-

[2] This procedure was suggested by Box and Cox [2]. For an alternative procedure of approximating a marginal distribution by a conditional distribution, see Tiao and Draper [10].

bution of the form $p(\delta|\rho = \hat{\rho}, k = \hat{k}, \text{data})$ with $(2n+n_1+n_2-4)$ df are given in fig. 2.

It may be pointed out that this technique of approximation can be utilized even in Case I to reduce the problem of bivariate integration to one of univariate integration. In this case we have considered two situations, one with $n = 10$, $n_1 = 2$, $\rho = 0.3$, $k = 0.2$, and the other with $n = 10$, $n_1 = 10$, $\rho = 0.3$, $k = 0.2$. The exact posterior density of δ and $p(\delta|\rho = \hat{\rho}, k = \hat{k}, \text{data})$ with reduced df have been calculated for these two situations and their closeness have been verified by considering the 90% and 95% highest posterior density (HPD) intervals which are provided below.

	$n = 10, n_1 = 2, \rho = 0.3, k = 0.2$		$n = 10, n_1 = 10, \rho = 0.3, k = 0.2$	
	Approximate	*Exact*	*Approximate*	*Exact*
90% HPD	(4.66, 6.41)	(4.68, 6.38)	(4.98, 6.21)	(5.00, 6.19)
95% HPD	(4.47, 6.60)	(4.49, 6.58)	(4.85, 6.34)	(4.88, 6.32)

Acknowledgements

We wish to thank Professor Bradley Efron and the referees of JASA for their constructive suggestions which have substantially improved this paper. This work was reported at the Fourth NBER-NSF Seminar on Bayesian Inference in Econometrics. After this paper was submitted for publication we learned of the independent work by G. M. Kaufman and A. Krieger [5] who established the results contained in our paper through a different technique.

Appendix[3]

A. Conditional posterior pdf for δ given ρ and k, when $n_2 = 0$

Let $\phi_1 = \mu_1 - \bar{\mu}$ where $\bar{\mu} = (\mu_1 + \mu_2)/2$. Transformation of the variables μ_1 and μ_2 in (8) to $\phi_1 + \bar{\mu}$ and $\bar{\mu} - \phi_1$ respectively, yields

$$p(\phi_1, \bar{\mu}, \rho, \sigma_1^2, k|\text{data}) \propto (\sigma_1^2)^{-\frac{1}{2}(2n+n_1+2)}(1-\rho^2)^{-\frac{1}{2}(n+3)}$$

[3] The Appendix was not included in the original version of this paper.

$$\times k^{-\frac{1}{2}(n+2)} \exp\Bigg[-\frac{n}{2\sigma_1^2 k(1-\rho^2)} \{ks_{11}+k(\phi_1+\bar{\mu}-\bar{x}_1)^2$$

$$-2\rho\surd(k)s_{12}-2\rho\surd(k)(\phi_1+\bar{\mu}-\bar{x}_1)(\bar{\mu}-\phi_1-\bar{x}_2)+s_{22}$$

$$+(\bar{\mu}-\phi_1-\bar{x}_2)^2+\lambda_1 k s_1^2(1-\rho^2)$$

$$+\lambda_1 k(1-\rho^2)(\phi_1+\bar{\mu}-\hat{\mu}_1)^2\}\Bigg], \tag{A.1}$$

where $ns_{ij} = \sum_{l=1}^{n} (x_{il}-\bar{x}_i)(x_{jl}-\bar{x}_j)$ $(i, j = 1, 2)$, $n\bar{x}_i = \sum_{l=1}^{n} x_{il}$ $(i = 1, 2)$, $n_1\hat{\mu}_1 = \sum_{l=n+1}^{n+n_1} x_{1l}$, $n_1 s_1^2 = \sum_{l=n+1}^{n+n_1} (x_{1l}-\hat{\mu}_1)^2$ and $\lambda_1 = n_1/n$.

Now we can write

$$k(\bar{\mu}+\phi_1-\bar{x}_1)^2-2\rho\surd(k)(\bar{\mu}+\phi_1-\bar{x}_1)(\bar{\mu}-\phi_1-\bar{x}_2)$$

$$+(\bar{\mu}-\phi_1-\bar{x}_2)^2+\lambda_1 k(1-\rho^2)(\phi_1+\bar{\mu}-\hat{\mu}_1)^2$$

$$= \left\{\bar{\mu}-\left[\frac{(\phi_1+\bar{x}_2)-\rho\surd(k)(\bar{x}_1+\bar{x}_2)-k(\phi_1-\bar{x}_1) \atop +(1-\lambda)\lambda^{-1}k(1-\rho^2)(\hat{\mu}_1-\phi_1)}{k-2\rho\surd(k)+1+\lambda_1 k(1-\rho^2)}\right]\right\}^2$$

$$\times[k-2\rho\surd(k)+1+k\lambda^{-1}(1-\rho^2)(1-\lambda)]$$

$$+[2\phi_1(1-\rho\surd k)+(1-\rho\surd k)\bar{x}_2+(k-\rho\surd k)\bar{x}_1$$

$$-(k-2\rho\surd(k)+1)\hat{\mu}_1]^2$$

$$\times\frac{k(1-\lambda)(1-\rho^2)(k-2\rho\surd(k)+1)^{-1}}{\{\lambda(k-2\rho\surd(k)+1)+k(1-\lambda)(1-\rho^2)\}}$$

$$+\frac{k(1-\rho^2)(2\phi_1-\bar{x}_1+\bar{x}_2)^2}{(k-2\rho\surd(k)+1)} \tag{A.2}$$

where $\lambda = n/(n+n_1) = 1/(1+\lambda_1)$.

On substituting (A.2) in (A.1) and carrying out the integration wrt $\bar{\mu}$ we obtain the joint posterior distribution of ϕ_1, ρ, k and σ_1^2

$$\begin{aligned} p(\phi_1, \rho, k, \sigma_1^2|\text{data}) &\propto (\sigma_1^2)^{-\frac{1}{2}(2n+n_1+1)}(1-\rho^2)^{-\frac{1}{2}(n+2)} \\ &\times (k)^{-\frac{1}{2}(n+1)}[k-2\rho\sqrt{(k)}+1+k\lambda^{-1}(1-\lambda)(1-\rho^2)]^{-\frac{1}{2}} \\ &\times \exp\Bigg[-\frac{n}{2\sigma_1^2 k(1-\rho^2)}\Bigg\{ks_{11}-2\rho\sqrt{(k)}s_{12}+s_{22} \\ &+ks_1^2\lambda^{-1}(1-\lambda)(1-\rho^2)+\frac{k(1-\rho^2)(2\phi_1-\bar{x}_1+\bar{x}_2)^2}{(k-2\rho\sqrt{(k)}+1)} \\ &+[(1-\rho\sqrt{k})(2\phi_1+\bar{x}_2)+(k-\rho\sqrt{k})\bar{x}_1 \\ &-(k-2\rho\sqrt{(k)}+1)\hat{\mu}_1]^2 \\ &\times\frac{k(1-\lambda)(1-\rho^2)(k-2\rho\sqrt{(k)}+1)^{-1}}{[\lambda(k-2\rho\sqrt{(k)}+1)+k(1-\lambda)(1-\rho^2)]}\Bigg\}\Bigg]. \end{aligned} \tag{A.3}$$

Now we can integrate (A.3) with respect to σ_1^2 and from the resulting expression obtain the following conditional posterior density of $\delta(=2\phi_1)$ given ρ and k

$$p(\delta|\rho, k, \text{data}) \propto \left[1+\frac{(\omega_1+\omega_2)}{a_1}(\delta-\hat{\delta})^2\right]^{-\frac{1}{2}(2n+n_1-1)} \tag{A.4}$$

where

$$\begin{aligned} \delta &= 2\phi_1 = \mu_1-\mu_2, \quad \omega_1 = n(k-2\rho\sqrt{(k)}+1)^{-1}, \\ \omega_2 &= \frac{n(1-\lambda)(1-\rho\sqrt{k})^2}{(k-2\rho\sqrt{(k)}+1)[\lambda(k-2\rho\sqrt{(k)}+1)+k(1-\lambda)(1-\rho^2)]}, \\ \eta_1 &= (\bar{x}_1-\bar{x}_2), \\ \eta_2 &= [\hat{\mu}_1(k-2\rho\sqrt{(k)}+1)-\bar{x}_1(k-\rho\sqrt{k})-\bar{x}_2(1-\rho\sqrt{k})] \\ &\quad /(1-\rho\sqrt{k}), \\ \hat{\delta} &= (\omega_1\eta_1+\omega_2\eta_2)(\omega_1+\omega_2)^{-1} = A\bar{x}_1+(1-A)\hat{\mu}_1-\bar{x}_2, \\ A &= \lambda+\rho\sqrt{(k)}(1-\lambda) \end{aligned}$$

and

$$\begin{aligned} a_1 &= [ns_{11}-2\rho ns_{12}/\sqrt{(k)}+ns_{22}/k](1-\rho^2)^{-1}+n_1 s_1^2 \\ &\quad + (\omega_1^{-1}+\omega_2^{-1})^{-1}(\eta_1-\eta_2)^2. \end{aligned}$$

B. Conditional posterior distribution of δ given ρ and k, when $n_1 > 0$, $n_2 > 0$.

Transforming (μ_1, μ_2) in (21) to $(\phi_1+\bar{\mu}, \bar{\mu}-\phi_1)$, we obtain the joint posterior distribution of $\phi_1, \bar{\mu}, \rho, k, \sigma_1^2$ as follows:

$$p(\phi_1, \bar{\mu}, \rho, k, \sigma_1^2|\text{data}) \propto (\sigma_1^2)^{-\frac{1}{2}(2n+n_1+n_2+2)} k^{-\frac{1}{2}(n+n_2+2)}$$
$$\times (1-\rho^2)^{-\frac{1}{2}(n+3)} \exp\left[-\frac{n}{2k\sigma_1^2(1-\rho^2)}\{ks_{11}\right.$$
$$+k(\phi_1+\bar{\mu}-\bar{x}_1)^2-2\rho\sqrt{(k)}s_{12}-2\rho\sqrt{(k)}(\phi_1+\bar{\mu}-\bar{x}_1)$$
$$\times(\bar{\mu}-\phi_1-\bar{x}_2)+s_{22}+(\bar{\mu}-\phi_1-\bar{x}_2)^2+\lambda_1 k(1-\rho^2)s_1^2$$
$$+\lambda_1 k(1-\rho^2)(\phi_1+\bar{\mu}-\hat{\mu}_1)^2+\lambda_2(1-\rho^2)s_2^2$$
$$\left.+\lambda_2(1-\rho^2)(\bar{\mu}-\phi_1-\hat{\mu}_2)^2\}\right] \tag{B.1}$$

where $n_2\hat{\mu}_2 = \sum_{i=n+1}^{n+n_2} x_{2i}$, $n_2 s_2^2 = \sum_{i=n+1}^{n+n_2} (x_{2i}-\hat{\mu}_2)^2$ and $\lambda_2 = n_2/n$. Now we can write

$$k(\phi_1+\bar{\mu}-\bar{x}_1)^2-2\rho\sqrt{(k)}(\phi_1+\bar{\mu}-\bar{x}_1)(\bar{\mu}-\phi_1-\bar{x}_2)$$
$$+(\bar{\mu}-\phi_1-\bar{x}_2)^2+\lambda_1 k(1-\rho^2)(\phi_1+\bar{\mu}-\hat{\mu}_1)^2$$
$$+\lambda_2(1-\rho^2)(\bar{\mu}-\phi_1-\hat{\mu}_2)^2$$
$$=\left[\bar{\mu}-\left\{\frac{\begin{array}{c}k(\bar{x}_1-\phi_1)-\rho\sqrt{(k)}(\bar{x}_1+\bar{x}_2)+(\phi_1+\bar{x}_2)\\ +\lambda_1 k(1-\rho^2)(\hat{\mu}_1-\phi_1)+\lambda_2(1-\rho^2)(\phi_1+\hat{\mu}_2)\end{array}}{k-2\rho\sqrt{(k)}+1+\lambda_1 k(1-\rho^2)+\lambda_2(1-\rho^2)}\right\}\right]^2$$
$$\times[k-2\rho\sqrt{(k)}+1+(\lambda_1 k+\lambda_2)(1-\rho^2)]$$
$$-[k(\bar{x}_1-\phi_1)-\rho\sqrt{(k)}(\bar{x}_1+\bar{x}_2)+(\phi_1+\bar{x}_2)$$
$$+\{\lambda_1 k(\hat{\mu}_1-\phi_1)+\lambda_2(\phi_1+\hat{\mu}_2)\}(1-\rho^2)]^2[k-2\rho\sqrt{(k)}+1$$
$$+(\lambda_1 k+\lambda_2)(1-\rho^2)]^{-1}+k(\bar{x}_1-\phi_1)^2+(\phi_1+\bar{x}_2)^2$$
$$+\lambda_1 k(1-\rho^2)(\hat{\mu}_1-\phi_1)^2+\lambda_2(1-\rho^2)(\phi_1+\hat{\mu}_2)^2$$
$$-2\rho\sqrt{(k)}(\phi_1+\bar{x}_2)(\bar{x}_1-\phi_1), \tag{B.2}$$

where use is made of an identity in Dickey [3, p. 1616]. We can now integrate wrt $\bar{\mu}$ to obtain

$$\begin{aligned} &p(\phi_1, \rho, k, \sigma_1^2|\text{data}) \\ &\quad \propto (\sigma_1^2)^{-\frac{1}{2}(2n+n_1+n_2+1)}(1-\rho^2)^{-\frac{1}{2}(n+2)}k^{-\frac{1}{2}(n+n_2+1)} \\ &\quad \times [k-2\rho\sqrt{(k)}+1+(\lambda_1 k+\lambda_2)(1-\rho^2)]^{-\frac{1}{2}} \\ &\quad \times \exp\Bigg[-\frac{n}{2\sigma_1^2 k(1-\rho^2)}\{ks_{11}-2\rho\sqrt{(k)}s_{12}+s_{22} \\ &\quad +\lambda_1 k(1-\rho^2)s_1^2+\lambda_2(1-\rho^2)s_2^2+k(\bar{x}_1-\phi_1)^2+(\phi_1+\bar{x}_2)^2 \\ &\quad +\lambda_1 k(1-\rho^2)(\hat{\mu}_1-\phi_1)^2+\lambda_2(1-\rho^2)(\phi_1+\hat{\mu}_2)^2 \\ &\quad -2\rho\sqrt{(k)}(\phi_1+\bar{x}_2)(\bar{x}_1-\phi_1)-[k(\bar{x}_1-\phi_1)-\rho\sqrt{(k)}(\bar{x}_1+\bar{x}_2) \\ &\quad +(\phi_1+\bar{x}_2)+\lambda_1 k(1-\rho^2)(\hat{\mu}_1-\phi_1)+\lambda_2(1-\rho^2) \\ &\quad \times(\phi_1+\hat{\mu}_2)]^2[(k)-2\rho\sqrt{(k)}+1+(\lambda_1 k+\lambda_2)(1-\rho^2)]^{-1}\}\Bigg]. \end{aligned} \tag{B.3}$$

Integration of (B.3) with respect to σ_1^2 is straightforward and from the resulting expression we immediately obtain the conditional posterior of δ given ρ and k, as

$$p(\delta|\rho, k, \text{data}) \propto \left[1+\frac{(\delta-\tilde{\delta})^2}{h_2}\right]^{-\frac{1}{2}(2n+n_1+n_2-1)} \tag{B.4}$$

where

$$\begin{aligned} h_2 = {} & [k-2\rho\sqrt{(k)}+1+(\lambda_1 k+\lambda_2)(1-\rho^2)][ks_{11}-2\rho\sqrt{(k)}s_{12} \\ & +s_{22}+(1-\rho^2)\{k\lambda_1 s_1^2+\lambda_2 s_2^2+k(\bar{\mathbf{x}}-\hat{\boldsymbol{\mu}})'\Lambda(\bar{\mathbf{x}}-\hat{\boldsymbol{\mu}})\}] \\ & \times[k(1-\rho^2)\{(1+\lambda_1)(1+\lambda_2)-\rho^2\lambda_1\lambda_2\}]^{-1} \end{aligned}$$

$$\Lambda = \begin{bmatrix} \dfrac{(1+\lambda_2)\lambda_1}{(1+\lambda_1)(1+\lambda_2)-\rho^2\lambda_1\lambda_2} & \dfrac{-\rho\lambda_1\lambda_2}{\sqrt{(k)}[(1+\lambda_1)(1+\lambda_2)-\rho^2\lambda_1\lambda_2]} \\ & \dfrac{(1+\lambda_1)\lambda_2}{k[(1+\lambda_1)(1+\lambda_2)-\rho^2\lambda_1\lambda_2]} \end{bmatrix}$$

$$\tilde{\delta} = A_1\bar{x}_1+(1-A_1)\hat{\mu}_1-A_2\bar{x}_2-(1-A_2)\hat{\mu}_2$$

$$A_1 = \frac{1+\lambda_2+\rho\lambda_1\sqrt{k}}{(1+\lambda_1)(1+\lambda_2)-\rho^2\lambda_1\lambda_2}$$

$$A_2 = \frac{\sqrt{(k)}(1+\lambda_1)+\rho\lambda_2}{\sqrt{(k)}[(1+\lambda_1)(1+\lambda_2)-\rho^2\lambda_1\lambda_2]}$$

$$\bar{\mathbf{x}} = (\bar{x}_1\ \bar{x}_2)'$$

$$\hat{\boldsymbol{\mu}} = (\hat{\mu}_1\ \hat{\mu}_2)'.$$

References

[1] T. W. Anderson, Maximum likelihood estimates for a multivariate normal distribution when some observations are missing. Journal of the American Statistical Association 52 (1957), 200–204.

[2] G. E. P. Box and D. R. Cox, An analysis of transformations. Journal of the Royal Statistical Society (Series B) 26 (1964), 211–252.

[3] J. H. Dickey, Three multidimensional-integral identities with Bayesian applications. Annals of Mathematical Statistics 39 (1968), 1615–1627.

[4] S. Geisser and J. Cornfield, Posterior distributions for multivariate normal parameters, Journal of the Royal Statistical Society (Series B) 25 (1963), 368–376.

[5] G. M. Kaufman and A. Krieger, A Bayesian treatment of bivariate normal sampling with missing observations. Unpublished manuscript circulated through personal communication.

[6] J. S. Mehta and J. Gurland, Some properties and an application of a statistics arising in testing correlation. Annals of Mathematical Statistics 40 (1969), 1736–1745.

[7] G. E. Nicholson, Estimation of parameters from incomplete multivariate samples. Journal of the American Statistical Association 52 (1957), 523–526.

[8] C. R. Rao, Advanced statistical methods in biometric research. New York, John Wiley, 1952.

[9] L. J. Savage, The foundations of statistical inference. London, Methuen, 1962.

[10] G. C. Tiao and A. Zeller, On the Bayesian estimation of multivariate regression. Journal of the Royal Statistical Society (Series B) 26 (1964), 277–285.

[11] G. C. Tiao and N. R. Draper, Bayesian analysis of linear models with two random components with special reference to the balanced incomplete block design. Biometrika 55 (1968), 101–117.

PART 8

TIME SERIES PROBLEMS

S. E. Fienberg and A. Zellner (eds.)
Studies in Bayesian Econometrics and Statistics.

8.1 MULTIPERIOD PREDICTIONS FROM STOCHASTIC DIFFERENCE EQUATIONS BY BAYESIAN METHODS*

Gregory C. CHOW

Princeton University

1. *Introduction*

Existing methods of parameter estimation, be they versions of least squares, maximum likelihood, or Bayes, that have been applied to systems of dynamic econometric equations to produce forecasts were not designed for the purpose of forecasting. In this paper, it is argued that these estimation methods may be inadequate if the resulting estimates are to be used to make ex ante predictions for more than one period ahead, and if the accuracy of the predictions is measured, as it usually is, by the mean squared errors. A formulation of the multiperiod prediction problem is presented. It will then become clear that the same set of parameter estimates cannot be optimal in making predictions for different time periods into the future, when optimality is defined by minimum mean squared errors in small samples.

Recently there has been much interest in comparing different econometric models, or different versions of the same econometric model obtained by applying different estimation techniques, in terms of how well they would have forecasted the dependent variables during the sample period, given the true values of the exogenous variables and given the values of the dependent variables lagged one or more periods. It has now become clear that, as judged by ex post forecasting for the sample period, models or techniques that perform better for one-period predictions may do worse for multiperiod predictions. For example, purely autoregressive models could do better than models

* Reprinted from Econometrica 41 (1973), 109–118, with the permission of the Econometric Society. An earlier version of this paper was presented to the Second NBER-NSF Seminar on Bayesian Inference in Econometrics, Carnegie-Mellon Univ., May 1971. Research supported by NSF Grant GS-2799.

based on structural equations in ex post forecasting for one quarter ahead, but were worse in forecasting three or four quarters ahead, as documented in Hickman [4]. Klein [9] and Fair [3] have compared multiperiod predictions of the sample data by different estimation techniques applied to the same econometric model. A related, though different, question naturally arises as to whether different parameter estimators should be used to produce ex ante forecasts for different periods into the future. The former topic is one of fitting equations to a set of data. The latter topic is one of statistical decision, and is the subject of this paper.[1]

Section 2 treats the case of multiperiod predictions for a univariate time series which satisfies a first-order autoregression. A completely analytical solution to the prediction problem is given by the Bayesian method. This analytical result could be considered complementary to the Monte Carlo results reported in the recent work of Orcutt and Winokur [10] on estimation of, and predictions from, first-order autoregression. Section 3 generalizes the result of section 2 to the case of a higher order system of linear stochastic difference equations with exogenous variables. It provides an alternative approach to multiperiod forecasting by the use of the reduced form of a system of linear econometric equations.

2. *Multiperiod predictions from a first-order autoregression*

Consider the simplest case of a univariate first-order autoregression.

$$y_t = ay_{t-1}+c+u_t \qquad (Eu_t = 0;\ Eu_t u_s = \delta_{ts}h^{-1}). \tag{1}$$

The initial value y_0 is treated as fixed, and u_t is assumed to be normally distributed. The parameters are the coefficients $\pi' = (a, c)$ and the variance h^{-1} of the residuals. The problem is that, having observed $y' = (y_1, \ldots, y_n)$, one wishes to predict y_{n+k}. For the predictor

[1] I should like to acknowledge that my interest in the subject of this paper has partly stemmed from conversations with Ray C. Fair in connection with his work on the former subject [3]. I would also like to thank Harry H. Kelejian, Richard E. Quandt, Arnold Zellner, and a referee for having read an early draft and made helpful comments.

$\hat{y}_{n+k} = \hat{y}_{n+k}(y)$ of y_{n+k}, assume the risk function to be

$$R(\pi, h) = E(\hat{y}_{n+k}-y_{n+k})^2$$
$$= \int (\hat{y}_{n+k}-y_{n+k})^2 p(y, y_{n+k}|\pi, h)\, \mathrm{d}y\, \mathrm{d}y_{n+k} \quad (2)$$

where, as elsewhere in this paper, the symbol p denotes probability density function which is assumed to exist. We will take the Bayesian approach of choosing the prediction function $\hat{y}_{n+k}(y)$ to minimize expected risk, given a prior density $p(\pi, h)$ of the parameters.

It will simplify the derivations to rewrite the risk (2) as an expectation taken over the future $u_{n+1}, \ldots, u_{n+k}$, rather than the future y_{n+k}. By the model (1), and on repeated substitutions of u's for y's, we have

$$y_{n+k} = a^k y_n + (a^{k-1}c + a^{k-2}c + \ldots + c) + (a^{k-1}u_{n+1}$$
$$+ a^{k-2}u_{n+2} + \ldots + u_{n+k}). \quad (3)$$

Then, noting the fact that the future u's are uncorrelated with the observed y, or with the parameters, we can write the risk (2) as

$$E(\hat{y}_{n+k}-y_{n+k})^2$$
$$= E[\hat{y}_{n+k} - a^k y_n - (a^{k-1}c + \ldots + c)]^2$$
$$+ E[a^{k-1}u_{n+1} + \ldots + u_{n+k}]^2. \quad (4)$$

Since the prediction function $\hat{y}_{n+k}(y)$ affects only the first term on the right side of (4), the Bayesian approach amounts to minimizing, with respect to $\hat{y}_{n+k}$, the expectation of this first term, taken over the prior density $p(\pi, h)$:

$$\int E[\hat{y}_{n+k} - a^k y_n - (a^{k-1}c + \ldots + c)]^2 p(\pi, h)\, \mathrm{d}\pi\, \mathrm{d}h$$
$$= \iint [\hat{y}_{n+k} - a^k y_n - (a^{k-1}c + \ldots + c)]^2 p(y|\pi, h)\, \mathrm{d}y$$
$$\times p(\pi, h)\, \mathrm{d}\pi\, \mathrm{d}h. \quad (5)$$

The above expression, treated as an integral over y, will be minimized by minimizing the integrand

$$\int [\hat{y}_{n+k} - a^k y_n - (a^{k-1}c + \ldots + c)]^2 p(y|\pi, h) \cdot p(\pi, h)\, \mathrm{d}\pi\, \mathrm{d}h. \quad (6)$$

Setting the derivative of (6) with respect to $\hat{y}_{n+k}$ equal to zero, one obtains

$$\hat{y}_{n+k} = \frac{1}{\int p(y|\pi, h)p(\pi, h)\,\mathrm{d}\pi\,\mathrm{d}h} \int [a^k y_n + (a^{k-1}c + \ldots + c)]$$

$$\times p(y|\pi, h)p(\pi, h)\,\mathrm{d}\pi\,\mathrm{d}h$$

$$= [\int a^k p(\pi|y)\,\mathrm{d}\pi] y_n + \int (a^{k-1}c + \ldots + c)p(\pi|y)\,\mathrm{d}\pi. \quad (7)$$

The predictor (7) could have been obtained by appealing directly to the well-known result in Bayesian statistics that, for a quadratic loss function, the mean of y_{n+k} given y_n, i.e., $a^k y_n + (a^{k-1}c + a^{k-2}c + \ldots + c)$, evaluated by the posterior density $p(a, c|y) = p(\pi|y)$ is optimal. Thus the optimal predictor $\hat{y}_{n+k}$ consists of two parts, the first being the mean of a^k times the initial value y_n, and the second being the mean of $(a^{k-1}c + a^{k-2}c + \ldots + c)$. If one had formulated a similar Bayesian problem of estimating a and c, and not of predicting y_{n+k}, his point estimates $\hat{a}$ and $\hat{c}$ would be the mean of the posterior density $p(a, c|y)$. A predictor naively constructed from these estimates would be $\hat{a}^k y_n + (\hat{a}^{k-1}\hat{c} + \ldots + \hat{c})$, and would not in general be optimal by the criterion of minimum mean squared error. The two predictors, in small samples, would coincide only for the prediction of y_{n+1}. Note, however, that as the sample size n approaches infinity, the difference between the two predictors is $0(1/n)$.

It is well known that if the density $p(y|\pi, h, y_0)$ is normal, and the prior density $p(\pi, h)$ is either diffuse, being proportional to h^{-1} as suggested by Jeffreys [5], or is normal-gamma as suggested by Raiffa and Schlaifer [11], the posterior density is normal-gamma. Under the assumption of the diffuse prior, the posterior density is

$$p(\pi, h|y) \propto h^{\frac{1}{2}n-1} \exp\{-\tfrac{1}{2}h(y - Z\pi)'(y - Z\pi)\} \quad (8)$$

where

$$y' = [y_1 \ldots y_n]; \qquad Z' = \begin{bmatrix} y_0 \ldots y_{n-1} \\ 1 \ \ldots \ 1 \end{bmatrix}; \qquad \pi' = [a, c]. \quad (9)$$

On integrating over h, the marginal posterior density of π is well-known to be the bivariate t density

$$
\begin{aligned}
p(\pi|y) &\propto [(y-Z\pi)'(y-Z\pi)]^{-\frac{1}{2}n} \\
&\propto [(y-Z\hat{\pi})'(y-Z\hat{\pi})+(\hat{\pi}-\pi)'Z'Z(\hat{\pi}-\pi)]^{-\frac{1}{2}n} \\
&\propto \left[(n-2)+(\pi-\hat{\pi})'\left(\frac{1}{s^2}Z'Z\right)(\pi-\hat{\pi})\right]^{-\frac{1}{2}n} \qquad (10)
\end{aligned}
$$

where

$$
\begin{aligned}
\hat{\pi} &= \begin{bmatrix}\hat{a}\\ \hat{c}\end{bmatrix} = (Z'Z)^{-1}Z'y; \\
(n-2)s^2 &= (y-Z\hat{\pi})'(y-Z\hat{\pi}).
\end{aligned} \qquad (11)
$$

The posterior density (10) can be applied to equation (7) to obtain optimal predictions $\hat{y}_{n+k}$. For $k = 1$,

$$
\begin{aligned}
\hat{y}_{n+1} &= [\textstyle\int ap(\pi|y)\,\mathrm{d}\pi]y_n+\int cp(\pi|y)\,\mathrm{d}\pi \\
&= (Ea)y_n+(Ec) = \hat{a}y_n+\hat{c}, \qquad (12)
\end{aligned}
$$

where the expectation E is taken over the posterior density (10). For $k = 2$,

$$\hat{y}_{n+2} = (Ea^2)y_n+(Eac)+(Ec). \qquad (13)$$

To evaluate the second moments, we note

$$
\begin{aligned}
E\begin{bmatrix}a^2 & ac\\ ac & c^2\end{bmatrix} &= E\begin{bmatrix}a-\hat{a}\\ c-\hat{c}\end{bmatrix}[a-\hat{a} \quad c-\hat{c}]+\begin{bmatrix}\hat{a}\\ \hat{c}\end{bmatrix}[\hat{a} \quad \hat{c}] \\
&= \frac{n-2}{n-4}s^2(Z'Z)^{-1}+\begin{bmatrix}\hat{a}^2 & \hat{a}\hat{c}\\ \hat{c}\hat{a} & \hat{c}^2\end{bmatrix} \qquad (14)
\end{aligned}
$$

where the second moments around the means for the multivariate t density can be found in Raiffa and Schlaifer [11, p. 257]. Thus the predictor (13) differs from the predictor $\hat{a}^2y_n+\hat{a}\hat{c}+\hat{c}$ which is often in use [10] by the amount

$$[Ea^2-\hat{a}^2]y_n+[Eac-\hat{a}\hat{c}] = \frac{(n-2)s^2}{n-4}[g_{11}y_n+g_{12}] \qquad (15)$$

where g_{ij} is the $i-j$ element of $(Z'Z)^{-1}$.

In general, the predictor $\hat{y}_{n+k}$ involves Ea^k and terms of the form Ea^ic. To evaluate Ea^k, one uses the marginal density of a, which is univariate t given by [11, p. 258]

$$p(a|y) \propto [(n-2)+(a-\hat{a})^2/s^2 g_{11}]^{-\frac{1}{2}(n-1)}. \tag{16}$$

Let $\mu_j = E(a-\hat{a})^j$. Then $\mu_j = 0$ for j odd; and, for j even, μ_j is given by [8, p. 60]

$$\mu_j = \frac{\Gamma(\frac{1}{2}j+\frac{1}{2})\Gamma(\frac{1}{2}n-\frac{1}{2}j-1)}{\Gamma(\frac{1}{2})\Gamma(\frac{1}{2}n-1)}\,\omega^j \qquad (n > j+2). \tag{17}$$

where $\omega^2 = (n-2)s^2 g_{11}$. Ea^k can then be calculated by the formula [8, p. 56]

$$Ea^k = \sum_{j=0}^{k} \binom{k}{j} \mu_{k-j}\hat{a}^j. \tag{18}$$

To evaluate $Ea^i c$, one can use the conditional mean of c [11, p. 258],

$$\begin{aligned} \int cp(c|a, y)\,\mathrm{d}c &= \hat{c} - g_{21}g_{11}^{-1}\hat{a} + g_{21}g_{11}^{-1}a \\ &= \bar{y} - a\bar{y}_{-1} \end{aligned} \tag{19}$$

where $-g_{21}g_{11}^{-1}$, from definitions, can easily be shown to be $\bar{y}_{-1} = (\Sigma_{t=0}^{n-1} y_t)/n$, and where use has been made of the relationship $\bar{y} = (\Sigma_{t=1}^{n} y_t)/n = \hat{c} + \hat{a}\bar{y}_{-1}$ from the method of least squares. One can then evaluate

$$\begin{aligned} Ea^i c &= \int a^i [\int cp(c|a, y)\,\mathrm{d}c] p(a|y)\,\mathrm{d}a \\ &= \bar{y}Ea^i - \bar{y}_{-1}Ea^{i+1}. \end{aligned} \tag{20}$$

The optimal predictor (7) can now be written as

$$\begin{aligned} \hat{y}_{n+k} &= (Ea^k)y_n + \sum_{i=0}^{k-1} Ea^i c \\ &= (Ea^k)y_n + \bar{y}\sum_{i=0}^{k-1} Ea^i - \bar{y}_{-1}\sum_{i=0}^{k-1} Ea^{i+1} \end{aligned} \tag{21}$$

where Ea^i can be calculated by (17) and (18). In contrast, the usual predictor can be written as, using $\hat{c} = \bar{y} - \hat{a}\bar{y}_{-1}$,

$$\begin{aligned} y^*_{n+k} &= \hat{a}^k y_n + \sum_{i=0}^{k-1} \hat{a}^i \hat{c} \\ &= \hat{a}^k y_n + \bar{y}\sum_{i=0}^{k-1} \hat{a}^i - \bar{y}_{-1}\sum_{i=0}^{k-1} \hat{a}^{i+1}. \end{aligned} \tag{22}$$

Equations (17) and (18) imply that Ea^k exists if and only if $k < n-2$. Since the predictor (21) is a linear combination of Ea^i ($i = 1, 2, \ldots, k$), it exists if and only if $k < n-2$. Note that if predictions for y_{n+1}, $y_{n+2}, \ldots, y_{n+K}$ are required simultaneously, each can be constructed by the method of equation (7). This procedure can be justified by reformulating the risk function (2) to be $\Sigma_{k=1}^{K} E(\hat{y}_{n+k}-y_{n+k})^2$, and equation (7) for each value of k will follow from the same derivations as given above.

The result of this section complements those of Orcutt and Winokur [10] on first-order autoregression, where results from Monte Carlo experiments on the properties of some estimators and predictors were reported. It might be of interest to compare the multiperiod predictors (21) with those examined by Orcutt and Winokur through further Monte Carlo experiments.

3. *Generalization to systems of stochastic difference equations*

Consider a $(p \times 1)$ vector y_t obeying

$$y_t = A_1 y_{t-1} + \ldots + A_q y_{t-q} + C_0 x_t + \ldots + C_r x_{t-r} + u_t$$
$$(E u_t u_s' = \delta_{t,s} H^{-1}). \tag{23}$$

Again, $y_0, y_{-1}, \ldots, y_{-q+1}$ are treated as fixed, and u_t is assumed to be normal. The parameters are $\Pi' = (A_1 \ldots A_q;\ C_0 \ldots C_r)$ and H. Having observed $Y' = (y_1 \ldots y_n)$ with $n > q+r+1$, one wishes to predict y_{n+k} conditionally on all future values of the exogenous variables up to x_{n+k}. For the predictor $\hat{y}_{n+k}(Y)$ of y_{n+k}, assume the risk to be, with D symmetric and positive definite,

$$R(\Pi, H) = E(\hat{y}_{n+k}-y_{n+k})' D(\hat{y}_{n+k}-y_{n+k}). \tag{24}$$

To achieve a generalization of the result of section 2, we rewrite the model (23) as a first-order system

$$\begin{bmatrix} y_t \\ y_{t-1} \\ \vdots \\ y_{t-q+1} \end{bmatrix} = \begin{bmatrix} A_1 & A_2 \ldots & A_q \\ I & 0 \ \ldots & 0 \\ & \ddots & \\ 0 & \ldots\ I & 0 \end{bmatrix} \begin{bmatrix} y_{t-1} \\ y_{t-2} \\ \vdots \\ y_{t-q} \end{bmatrix} + \begin{bmatrix} C_0 \ldots C_r \\ 0 \ \ldots 0 \\ \ddots \\ 0 \ \ldots 0 \end{bmatrix} \begin{bmatrix} x_t \\ \vdots \\ x_{t-r} \end{bmatrix} + \begin{bmatrix} u_t \\ 0 \\ \vdots \\ 0 \end{bmatrix} \tag{25}$$

or, in new symbols,

$$y_t^* = Ay_{t-1}^* + Cx_t^* + u_t^* . \tag{26}$$

If $\hat{y}_{n+k}^*$ denotes the predictor of y_{n+k}^*, we are interested only in its uppermost subvector $\hat{y}_{n+k}$, but let us first use model (26) to express y_{n+k}^* in terms of the future u^*'s as in (3),

$$y_{n+k}^* = A^k y_n^* + (A^{k-1}Cx_{n+1}^* + \ldots + Cx_{n+k}^*)$$
$$+ (A^{k-1}u_{n+1}^* + \ldots + u_{n+k}^*). \tag{27}$$

The relevant subvector y_{n+k} is obtained as My_{n+k}^*, where the matrix $M = (I_p \quad 0 \ldots 0)$ selects the first p rows of y_{n+k}^*. Using (27), the definition of M, and the independence of the future u^*,s, we write the risk function (24) as

$$E\|\hat{y}_{n+k} - y_{n+k}\|_D^2 = E\|\hat{y}_{n+k} - My_{n+k}^*\|_D^2$$
$$= E\|\hat{y}_{n+k} - MA^k y_n^* - M(A^{k-1}Cx_{n+1}^*$$
$$+ A^{k-2}Cx_{n+2}^* + \ldots + Cx_{n+k}^*)\|_D^2$$
$$+ E\|M(A^{k-1}u_{n+1}^* + \ldots + u_{n+k}^*)\|_D^2 \tag{28}$$

where, the term involving future u^*'s is not affected by the predictor $\hat{y}_{n+k}$.

The development from (5) to (7) can be followed exactly to yield the result

$$\hat{y}_{n+k} = [\int MA^k p(\Pi|Y)\,\mathrm{d}\Pi]y_n^* + [\int MA^{k-1}Cp(\Pi|Y)\,\mathrm{d}\Pi]x_{n+1}^*$$
$$+ \ldots + [\int MCp(\Pi|Y)\,\mathrm{d}\Pi]x_{n+k}^* . \tag{29}$$

The interpretation of (29) is similar to that of (7). Rather than first obtaining the estimate $\hat{\Pi}' = (\hat{A}_1, \ldots, \hat{A}_q; \hat{C}_0 \ldots \hat{C}_r)$ of Π' as the mean of the posterior density and then form the predictor of y_{n+k} as $M\hat{A}^k y_n^* + M\hat{A}^{k-1}\hat{C}x_{n+1}^* + \ldots + M\hat{C}x_{n+k}^*$, one should compute the expectations of MA^k, $MA^{k-1}C, \ldots, MC$ by the posterior density, and then apply them to y_n^*, $x_{n+1}^*, \ldots, x_{n+k}^*$ respectively. Again, the two methods would give the same result only for the prediction of y_{n+1}.

It is well-known that if the density $p(y|\Pi, H)$ is normal, and the prior density $p(\Pi, H)$ is either diffuse, being proportional to $|H|^{-\frac{1}{2}(p+1)}$, or is normal-Wishart, the posterior density is normal-Wishart. Under the assumption of the above diffused prior, the posterior density is

$$p(\Pi, H|Y) \propto |H|^{\frac{1}{2}(n-p-1)} \exp\{-\tfrac{1}{2}\operatorname{tr} H(Y-Z\Pi)'(Y-Z\Pi)\} \tag{30}$$

where Y is $n \times p$, Z is $n \times s$ ($s = q+r+1$), and Π is $s \times p$, as defined by

$$Y' = [y_1 \ldots y_n];$$

$$Z' = \begin{bmatrix} y_0 & & y_{n-1} \\ \vdots & & \vdots \\ y_{q-1} & \vdots & y_{n-q} \\ x_1 & & x_n \\ \vdots & & \vdots \\ x_r & & x_{n-r} \end{bmatrix}; \Pi' = [A_1 \ldots C_r]. \tag{31}$$

On integrating over H, using the integral of the Wishart distribution [1, p. 154], one obtains the marginal posterior density of Π to be the matricvariate t density[2]

$$\begin{aligned} p(\Pi|Y) &\propto |(Y-Z\Pi)'(Y-Z\Pi)|^{-\frac{1}{2}n} \\ &\propto |(Y-Z\hat{\Pi})'(Y-Z\hat{\Pi})+(\Pi-\hat{\Pi})'Z'Z(\Pi-\hat{\Pi})|^{-\frac{1}{2}n} \end{aligned} \tag{32}$$

where $\hat{\Pi}$ is $(Z'Z)^{-1}Z'Y$. The posterior density (32) can be applied to equation (29) for obtaining the optimal prediction $\hat{y}_{n+k}$.

While the problem of numerical integration of (29) remains to be further investigated, the following analytical results may be used at least for evaluating predictions when k is small. To evaluate $\hat{y}_{n+k}$ by (29), it is required to calculate expectations of products of at most k elements of Π. The predictor $\hat{y}_{n+k}$ exists only if these expectations exist, Π having a matricvariate t density given by (32). Let the p columns of Π be arranged into a column vector $\pi = (\pi_1 \ldots \pi_{ps})'$ of ps elements. To

[2] Recent applications of the matricvariate t density to econometric problems include papers by Zellner and Chetty [13], Sewell [12], Kaufman [7], and Drèze and Morales [2], among others. None of these studies deals with the multiperiod prediction problem of the present paper, however.

calculate the expectation of $\pi_i\pi_j \ldots \pi_m$, one can use the posterior density $p(\Pi, H|Y)$ and integrate first over Π:

$$\begin{aligned}\int (\pi_i\pi_j \ldots \pi_m) p(\Pi, H|Y)\, d\Pi\, dH \\ \propto \int |H|^{\frac{1}{2}(n-p-1-s)} [\int (\pi_i \ldots \pi_m) |H|^{\frac{1}{2}s} \\ \times \exp\{-\tfrac{1}{2}\operatorname{tr} H(\Pi-\hat{\Pi})'Z'Z(\Pi-\hat{\Pi})\}\, d\Pi] \\ \times \exp\{-\tfrac{1}{2}\operatorname{tr} H(Y-Z\hat{\Pi})'(Y-Z\hat{\Pi})\}\, dH \end{aligned} \tag{33}$$

where the density of Π, given H, is seen to be multivariate normal with mean $\hat{\Pi}$ and the covariance matrix of the vector π is $H^{-1} \otimes (Z'Z)^{-1} \equiv V = (v_{ij})$. The expectation of the product $\pi_i\pi_j \ldots \pi_m$ of k π's is known, as it can easily be derived from the characteristic function of the multivariate normal density, to be $\hat{\pi}_i\hat{\pi}_j \ldots \hat{\pi}_m$, plus products of one v_{ij} and $(k-2)$ $\hat{\pi}$'s, plus products of two v_{ij} and $(k-4)$ $\hat{\pi}$'s, etc., the k subscripts in each product being a permutation of $(i, j, \ldots, m)$ but with v_{ij} and v_{ji} counted only once.[3] Thus for $k = 2, 3$ at most one v_{ij} is involved in each term; for $k = 4, 5$ the product of at most two v_{ij} is involved in each term. The former involves one element of H^{-1}, and the latter involves the product of at most two elements of H^{-1}.

Having found the expectation of the product $\pi_i \ldots \pi_m$ given H, one has to perform the remaining integration of (33) over H. The remaining density of H in (33) is the Wishart density with parameter set $[(Y-Z\hat{\Pi})'(Y-Z\hat{\Pi}), (n-s)]$. It is thus required to find the expectation of an element (for $k = 2, 3$), or of the product of two elements (for $k = 4, 5$), of H^{-1}, as the $\hat{\pi}$'s and the elements of the matrix $(Z'Z)^{-1}$ are given for the purpose of this integration. Let h^{ij} be the i–j element of the $p \times p$ matrix H^{-1}, s_{ij} be the i–j element of $(Y-Z\hat{\Pi})'(Y-Z\hat{\Pi})$, and let $v = (n-s)-p+1$. Kaufman [6, p. 14] has obtained

$$\begin{aligned} Eh^{ij} &= \frac{1}{v-2} s_{ij}\,; \\ E(h^{ij}h^{kl}) &= \frac{1}{(v-1)(v-2)(v-4)} [(v-3)s_{ij}s_{kl} + s_{ik}s_{jl} + s_{il}s_{kj}]. \end{aligned} \tag{34}$$

[3] Thus for $k = 5$, we have

$$E\pi_i\pi_j\pi_k\pi_l\pi_m = \hat{\pi}_i\hat{\pi}_j\hat{\pi}_k\hat{\pi}_l\hat{\pi}_m + \binom{5}{2} \text{ terms of the form } v_{ij}\hat{\pi}_k\hat{\pi}_l\hat{\pi}_m + \frac{1}{2}\binom{5}{2}\binom{3}{2}$$

terms of the form $v_{ij}v_{kl}\hat{\pi}_m$.

These results complete the analytical solutions to the prediction of y_{n+k} for k up to 5.

Insofar as the reduced form equations in linear econometric models are systems of stochastic difference equations, the result of this section applies to multiperiod predictions by econometric models. While this paper has brought out the dynamic aspect of forecasting, it has ignored the non-linear restrictions on reduced-form coefficients induced by overidentification of the structural equations, an aspect emphasized in the existing literature on the estimation of simultaneous equation systems. Which aspect of the problem will turn out to be more important for forecasting purposes remains an open question.[4]

[4] Of course, both aspects could be incorporated, at least in principle. One way to do this would be to utilize the posterior density $p(B, \Gamma | Y)$ of Drèze and Morales [2] for the coefficients B and Γ in the structure $B'y_t = \Gamma' z_t +$residuals, where a priori restrictions on the elements of B and Γ have been imposed. Our solution would require evaluating elements of the integral

$$\int \Pi^k p(\Pi | Y) \mathrm{d}\Pi = \int (\Gamma B^{-1})^k p(B, \Gamma | Y) \mathrm{d}(B, \Gamma).$$

I have not examined the feasibility of this numerical problem.

References

[1] T. W. Anderson, Introduction to multivariate statistical analysis. New York, John Wiley, 1958.

[2] J. H. Drèze and J. A. Morales, Bayesian full information analysis of the simultaneous equations model. Center for Operations Research and Econometrics, Discussion Paper No. 7031, Catholic University of Louvain (August 1970).

[3] R. C. Fair, A comparison of alternative estimators of macroeconomic models. Econometric Research Program Research Memorandum No. 121 (March 1971).

[4] B. G. Hickman (Ed.), Econometric models of cyclical behavior. National Bureau of Economic Research, New York (to be published).

[5] H. Jeffreys, Theory of probability. Oxford, Clarendon Press, 1948.

[6] G. M. Kaufman, Some Bayesian moment formulae. Center for Operations Research and Econometrics Discussion Paper No. 6710, Catholic University of Louvain (August 1967).

[7] G. M. Kaufman, Conditional prediction and unbiasedness in structural equations. Econometrica 37, No. 1 (January 1969), 44–49.

[8] M. G. Kendall and A. Stuart, The advanced theory of statistics, Vol. I. London, Charles Griffin, 1958.

[9] L. R. Klein, Estimation of interdependent systems in macro-econometrics. Econometrica 37, No. 2 (April 1969), 171–192.

[10] G. H. Orcutt and H. S. Winokur, Jr., First order autoregression: inference, estimation, and prediction. Econometrica 37, No. 1 (January 1969), 1–14.

[11] H. Raiffa and R. Schlaifer, Applied statistical decision theory. Boston, Harvard University Graduate School of Business Administration, 1961.

[12] W. P. Sewell, Least squares, conditional predictions, and estimator properties. Econometrica 37, No. 1 (January 1969), 39–43.

[13] A. Zellner and V. K. Chetty, Prediction and decision problems in regression models from the Bayesian point of view. Journal of American Statistical Association 60, No. 310 (June 1965), 608–616.

S. E. Fienberg and A. Zellner (eds.)
Studies in Bayesian Econometrics and Statistics.

8.2 ESTIMATION OF PARAMETERS IN DISTRIBUTED LAG ECONOMETRIC MODELS*

William S. CLEVELAND

University of North Carolina and Bell Laboratories

1. *Introduction and summary*

One very common occurrence in fitting distributed lag models to economic data is that the least squares estimates of the parameters do not make economic sense, since in general the exogenous variables and their lagged values are so highly correlated and the number of observations so small that there is little information in the data about the true values of the parameters. One common remedy is to eliminate variables from the model. This is an adequate procedure if prediction is the only goal, but it does not help the person interested in understanding the economic system generating the data or the person who wants to make policy decisions.

In this paper, a method of estimating parameters in distributed lag models is presented. A subset of the parameter space is generated which includes the usual maximum likelihood value, a very smooth value, and other values whose smoothness ranges between these two extremes. The subset is the set of posterior means corresponding to a class of prior distributions which range from those with large prior weight on smooth values to those with small prior weight on smooth values. The choice of the prior distribution is aided by considering how much the prior information overrides the information in the sample.

Two classes of prior distributions are presented. The first assumes the parameters are generated by an autoregression. The second breaks up the parameters into two parts: one part satisfies an Almon polynomial assumption and the other is noise. The use of each class is illustrated

* An earlier version of this paper was presented to the Fourth NBER-NSF Seminar on Bayesian Inference in Econometrics, Univ. of Chicago, May 1972.

by its application to estimating the parameters of an equation from the St. Louis Federal Reserve Bank model of the U.S. economy.

Other authors who have dealt with the problem of adding smoothness assumptions to distributed lag models are Almon [1], Leamer [6], and Shiller [9]. Also, Hoerl and Kennard [4, 5] have devised the ridge trace to help study regression models with high correlation among independent variables.

2. *The model and the problem*

To keep notation from getting out of hand, it will be supposed that there are two exogenous variables a_t^* and b_t^*. The discussion that follows holds with obvious changes in the details if there are more than two. The distributed lag model relating a_t^* and b_t^* to the endogenous variable y_t is

$$y_t = \mu^* + \sum_{j=0}^{a} \alpha_j^* a_{t-j}^* + \sum_{j=0}^{b} \beta_j^* b_{t-j}^* + z_t,$$

for $t = 1, \ldots, T$ where z_t is a sequence of independent normal random variables with mean 0 and variance $\sigma^2 = e^{-\varphi}$.

It is convenient for computational reasons and certain statistical reasons to be seen later to relocate each exogenous variable so that its mean is 0 and rescale it so that its sum of squares is 1. Let *ma* and *sa* be the sample mean and standard deviation of a_t^* and *mb* and *sb* be the sample mean and standard deviation of b_t^*. The model can be rewritten in the form

$$y_t = \mu T^{-\frac{1}{2}} + \sum_{j=0}^{a} \alpha_j a_{t-j} + \sum_{j=0}^{b} \beta_j b_{t-j} + z_t,$$

for $t = 1, \ldots, T$,

where

$$a_t = \frac{a_t^* - ma}{sa}$$

$$b_t = \frac{b_t^* - mb}{sb}$$

are the normalized forms of the exogenous variables,

$$\alpha_j^* = \alpha_j \div sa,$$
$$\beta_j^* = \beta_j \div sa,$$

and

$$\mu^* = \mu T^{-\frac{1}{2}} - ma \sum_{j=0}^{a} \alpha_j^* - mb \sum_{j=0}^{b} \beta_j^*.$$

The second set of T equations, which will be referred to as the normalized form of the model, will now be written in matrix form. Let

$$Y = (y_1, \ldots, y_T)'$$
$$Z = (z_1, \ldots, z_T)'$$

and

$$\theta = (\mu, \alpha_0, \ldots, \alpha_a, \beta_0, \ldots, \beta_b)'.$$

Let

$$X = \begin{bmatrix} T^{-\frac{1}{2}} & a_1 & a_0 \ldots a_{1-a} & b_1 & b_0 \ldots b_{1-b} \\ \vdots & \vdots & \vdots \quad\quad \vdots & \vdots & \vdots \quad\quad \vdots \\ \vdots & \vdots & \vdots \quad\quad \vdots & \vdots & \vdots \quad\quad \vdots \\ T^{-\frac{1}{2}} & a_T & a_{T-1} \ldots a_{T-a} & b_T & b_{T-1} \ldots b_{T-b} \end{bmatrix}$$

then the distributed lag model is

$$Y = X\theta + Z.$$

It will be supposed that the columns of X are linearly independent.

The likelihood function of the data given θ and φ is

$$L(\theta, \varphi) \propto e^{T\varphi/2} \exp -\tfrac{1}{2} e^{\varphi} SS(\theta)$$

where $SS(\theta) = (Y - X\theta)'(Y - X\theta)$.

The maximum likelihood estimate of θ is

$$\theta_{MLE} = (X'X)^{-1}X'Y.$$

Since we want to focus on the problem of estimating θ we will study the marginal likelihood function of θ

$$\begin{aligned} L(\theta) &\propto \int L(\theta, \varphi)\, d\varphi \\ &\propto [SS(\theta)]^{-T/2} \\ &= [SS(\theta_{MLE}) + (\theta - \theta_{MLE})'X'X(\theta - \theta_{MLE})]^{-T/2}. \end{aligned}$$

The marginal likelihood L is proportional to a multivariate Student density function [8, p. 256]. The maximum occurs at $\theta = \theta_{MLE}$ and the function is constant on each ellipsoid of the form

$$(\theta-\theta_{MLE})'X'X(\theta-\theta_{MLE}) = \text{constant}.$$

That is, this family of homothetic ellipsoids is the set of contours of L.

The maximum likelihood estimate θ_{MLE} is, on the basis of the data alone, a good candidate as a point estimate of θ since it is in some sense the most likely value. But a value of θ with high relative likelihood (i.e., for which $L(\theta) \div L(\theta_{MLE})$ is not far from 1) is nearly as good a candidate as θ_{MLE}. Economic data very often have the two properties that the columns of X are highly correlated and the value of T, the number of observations, is small. Both properties tend to produce a likelihood function which cannot distinguish with much precision between values of θ which have very different economic meanings. That is, they tend to produce a large region of good candidates containing values of θ which have very different economic meanings. The smaller the value of T, the slower the relative likelihood

$$RL(\theta) = L(\theta) \div L(\theta_{MLE})$$

tends to decrease as θ moves in any direction away from θ_{MLE}. The more the amount of correlation between the columns of X, the more elongated the ellipsoidal contours of L tend to be. Thus as θ moves away from θ_{MLE} in the direction of the major axis of these contours, $RL(\theta)$ tends to decrease slowly. Thus the data alone do not provide enough information to enable one to make a precise specification of the parameters in the model.

3. *A general description of the method*

The one thing that is clear from the last section is that the problem cannot be resolved by statistics alone. There is not sufficient precision in the data to be able to do this. A remedy that works must involve bringing in additional knowledge about the economic system generating the data. In the typical situation it is sensible to suppose there is some smoothness in the coefficients $\alpha_0, \ldots, \alpha_a$ and $\beta_0, \ldots, \beta_b$. That is, there is a tendency to believe that α_j does not differ radically from α_{j-1} or α_{j+1}, nor β_j from β_{j-1} or β_{j+1}. The following procedure allows consideration of values of the parameters which are smoother than the maximum likelihood estimate.

A subset of the parameter space is generated by considering the posterior means corresponding to a class of prior distributions on the parameters. The choice of the particular prior distribution for a given set of data is, as in most applications of Bayesian inference, a difficult and complex problem. To aid in this choice the amount by which the information in the sample is overridden by the information in the prior in determining the posterior mean is studied for the priors in the class. This is done by studying *RL* for a number of values in the subset of posterior means. An estimate $\hat{\theta}$ is taken to be a value which has an amount of smoothness that makes economic sense but for which $RL(\hat{\theta})$ is as large as possible subject to the smoothness constraint.

It would seem that there is nothing in Bayesian theory which dictates that one must completely specify a prior before seeing the data. Indeed for the distributed lag problem, leaving some aspects of the prior undetermined until after the data is gathered is more likely to bring one closer to the ideal of perfect coherence than picking a particular prior beforehand and sticking with it no matter how unhappy one is with the posterior.

4. *The autoregressive subset*

4.1. *Autoregressive priors*

Suppose the prior distribution on θ given σ^2 satisfies the following assumptions:

(i) The distributions on the three sets of variables $\{\mu\}$, $\{\alpha_0, \ldots, \alpha_a\}$, and $\{\beta_0, \ldots, \beta_b\}$ are independent.

(ii) $\alpha_0, \ldots, \alpha_a$ is a mean 0, variance v_α, stationary, normal, first-order autoregression with first-order autocorrelation coefficient r_α.

(iii) $\beta_0, \ldots, \beta_b$ is a mean 0, variance v_β, stationary, normal, first-order autoregression with first-order autocorrelation coefficient r_β.

(iv) μ is uniform on the real line.

While the first-order autoregression assumption has been made, it is clear that higher order autoregressions are possible. However, the class of prior distributions specified in (i)–(iv) allow for a wide range of prior opinion about smoothness. Typically, of course, r_α and r_β will be nonnegative. The prior correlation between α_j and α_k is $r_\alpha^{|j-k|}$ which decreases with increasing $|j-k|$, which would seem to be a desirable property. As r_α increases toward 1 and v_α stays constant the prior distribution on $\alpha_0, \ldots, \alpha_a$ concentrates more weight on values of the parameters which are smooth. As v_α decreases and r_α stays constant the prior distribution concentrates more weight near the origin. A similar statement holds for $\beta_0, \ldots, \beta_b$.

4.2. *The posterior mean of θ given σ^2*

A little more notation will, unfortunately, be necessary in order to describe the posterior mean of θ given σ^2. Let g be a positive integer and let $G_g(r)$ be a $g \times g$ tri-diagonal matrix with the following elements: the first and last elements of the main diagonal are $(1-r^2)^{-1}$; all other diagonal elements are $(1+r^2)(1-r^2)^{-1}$; all elements on the two diagonals parallel and adjacent to the main diagonal are $-r(1-r^2)^{-1}$; and all other elements are zero. For example

$$G_3(r) = (1-r^2)^{-1} \begin{bmatrix} 1 & -r & 0 \\ -r & 1+r^2 & -r \\ 0 & -r & 1 \end{bmatrix}.$$

Let $q_\alpha = \sigma^2 \div v_\alpha$, $q_\beta = \sigma^2 \div v_\beta$, $p = (r_\alpha, q_\alpha, r_\beta, q_\beta)$, and

$$H(p) = \begin{bmatrix} q_\alpha G_{a+1}(r_\alpha) & 0 \\ 0 & q_\beta G_{b+1}(r_\beta) \end{bmatrix}.$$

Then the posterior mean of θ given σ^2 using the autoregressive prior is easily seen by straightforward calculations to be

$$\theta(p) = (X'X + H(r_\alpha, q_\alpha, r_\beta, q_\beta))^{-1}X'Y.$$

When $q_\alpha = q_\beta = 0$, $\theta(p)$ is just the maximum likelihood estimate θ_{MLE}. For $q_\alpha > 0$, as r_α tends to 1 the posterior means of $\alpha_0, \ldots, \alpha_a$ tend to the same value.

4.3. *The methodology and an example*

In practice an estimate of θ can be chosen by calculating $\theta(p)$ for various values of p and selecting an estimate $\hat{\theta}$ which has an amount of smoothness that makes sense with respect to the economic mechanism, but for which $RL(\hat{\theta})$ is not so small as to render $\hat{\theta}$ unlikely in light of the data. It is of interest to study not only $RL(\theta(p))$, but also

$$RSS(\theta(p)) = SS(\theta_{MLE}) \div SS(\theta(p)),$$

the relative residual sum of squares of $\theta(p)$. *RSS* contains, of course, the same information as *RL*, but displays it in a different way which is useful. *RSS* gives us some feeling for how the predictive power of the equation is changing as p changes.

In making displays (tables and graphs) we will not give $\theta(p)$ directly but rather the corresponding values

$$\theta^*(p) = (\mu^*(p), \alpha_0^*(p), \ldots, \alpha_a^*(p), \beta_0^*(p), \ldots, \beta_b^*(p))'$$

of the unnormalized form of the model. The transformation from $\theta(p)$ to $\theta^*(p)$ is described by the equations in the second paragraph of section 2.

The use of the subset $\theta^*(p)$ will be illustrated by its application to the price equation from the economic model in [2]. The equation (before normalization) is

$$y_t = \mu^* + \sum_{j=0}^{5} \alpha_j^* a_{t-j}^* + \beta_0^* b_t^* + z_t$$

where y_t is the change in the price level from quarter $t-1$ to t, a_t^* is the demand pressure in quarter t, and b_t^* is the change from quarter $t-1$ to t in the anticipated price level. $t = 1$ corresponds to the first quarter of 1955 and $T = 60$.

Since $b = 0$ the value of r_β has no influence on the prior. In addition the value of v_β was taken to be infinity; that is, the prior on β_0 is uniform on the real line.

Table 1 shows the results of calculating $\theta^*(p)$ for a number of values of p. The maximum likelihood estimate $\theta_{MLE}^* = \theta^*(p)$ where $q_\alpha = r_\alpha = 0$, does not appear to make very much sense, for the estimates of α_2^* and α_5^* are both negative, which would imply prices increase by a smaller amount during a given quarter if the demand was large 3 quarters earlier or 5 quarters earlier.

$\theta^*(p)$ for $q_\alpha = 0.01$ and $r_\alpha = 0$ is quite different from θ_{MLE}^* even though $RL(\theta^*(p))$ is as high as 0.90. This is exactly a feature of the lack-of-precision phenomenon resulting from high correlations and a small sample size that was described in section 2. With $q_\alpha = 0.01$ the estimates of α_2^* and α_5^* increase as r_α increases, crossing 0 for r_α around 0.7.

TABLE 1

$\theta^*(p)$ for the price equation

q_α	r_α	$RL(\theta(p))$	$RSS(\theta(p))$	$\mu^*(p)$	$\alpha_0^*(p)$	$\alpha_1^*(p)$	$\alpha_2^*(p)$	$\alpha_3^*(p)$	$\alpha_4^*(p)$	$\alpha_5^*(p)$	$\beta_0^*(p)$
0	0	1	1	2.66	0.029	0.038	−0.026	0.021	0.043	−0.017	0.869
0.01	0	0.90	0.997	2.68	0.029	0.028	−0.010	0.017	0.032	−0.009	0.866
0.01	0.1	0.89	0.996	2.68	0.029	0.027	−0.009	0.017	0.032	−0.008	0.866
0.01	0.3	0.86	0.995	2.68	0.030	0.026	−0.006	0.017	0.030	−0.007	0.865
0.01	0.5	0.80	0.993	2.69	0.030	0.024	−0.003	0.016	0.026	−0.005	0.864
0.01	0.7	0.71	0.999	2.70	0.029	0.021	0.003	0.015	0.021	−0.001	0.862
0.01	0.9	0.56	0.981	2.71	0.027	0.020	0.010	0 013	0.014	0.004	0.857
0.01	0.95	0.50	0.977	2.72	0.025	0.020	0.013	0.013	0.012	0.006	0.855
0.01	0.99	0.42	0.971	2.76	0.022	0.019	0.016	0.013	0.011	0.009	0.847
0.05	0.5	0.56	0.981	2.70	0.025	0.021	0.010	0.014	0.015	0.003	0.859

Also, as r_α increases the estimates of $\alpha_0^*, \ldots, \alpha_5^*$ get smoother. With $r_\alpha = 0.5$, the effect of increasing q_α from 0.01 to 0.05 is to pull the estimates of $\alpha_0^*, \ldots, \alpha_5^*$ closer to the origin.

The choice of a particular $\theta^*(p)$ as an estimate of θ would, of course, depend upon one's knowledge about the economic mechanism. $\theta^*(p)$ for $q_\alpha = 0.01$ and $r_\alpha = 0.95$ might make sense to many. The estimates of α_j^*, for $j = 0, \ldots, 5$ are reasonably smooth and decrease as j increases, and $RL(\theta(p))$ and $RSS(\theta(p))$ are not so small as to discredit $\theta^*(p)$ in light of the data. Fig. 1 illustrates graphically the erratic behavior of θ_{MLE}^* and the smoothness of $\theta^*(p)$ for $q_\alpha = 0.01$ and $r_\alpha = 0.95$.

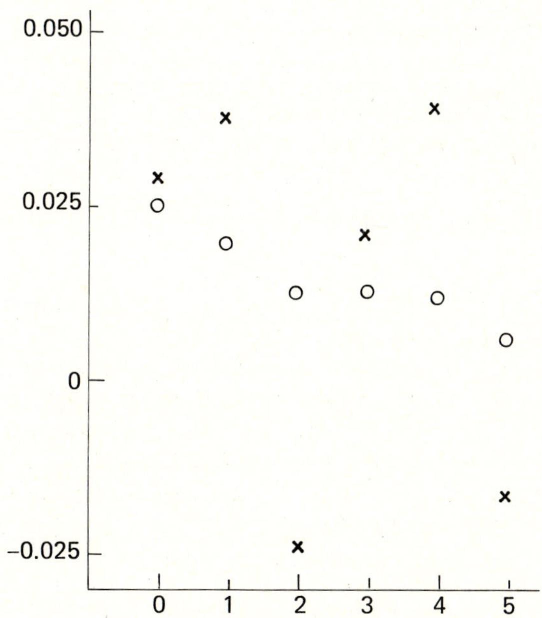

Fig. 1. $\alpha_{MLE}^* = \times$; $\alpha^*(p) = \bigcirc$.

5. The Almon polynomial plus noise subset

5.1. Almon polynomial plus noise priors

One way to incorporate a smoothness assumption into the model is to follow Almon [1] and suppose that both α_j and β_j are generated by polynomials of certain prespecified degrees. The Almon procedure, while very sensible, tends to be somewhat inflexible. Increasing the degree of one of the polynomials by 1 can often have a substantial effect on the estimates, leaving one wishing that a middle ground could be achieved. The subset described in this section provides this added flexibility.

Let

$$\alpha_j = \alpha_j^{(0)} + \alpha_j^{(1)} \quad \text{for} \quad j = 0, \ldots, a$$

and

$$\beta_j = \beta_j^{(0)} + \beta_j^{(1)} \quad \text{for} \quad j = 0, \ldots, b$$

where

$$\alpha_j^{(0)} = \sum_{k=0}^{d_\alpha} \alpha_k^{(2)} j^k$$

and

$$\beta_j^{(0)} = \sum_{k=0}^{d_\beta} \beta_k^{(2)} j^k.$$

Let the prior distributions on the unknown parameters μ, $\alpha_j^{(1)}$, $\beta_j^{(1)}$, $\alpha_k^{(2)}$ and $\beta_k^{(2)}$ given σ^2 be jointly independent and described by the following:

(i) μ, $\alpha_k^{(2)}$, $\beta_k^{(2)}$ are uniform on the real line.

(ii) $\alpha_j^{(1)}$ is normal with mean 0 and variance v_α.

(iii) $\beta_j^{(1)}$ is normal with mean 0 and variance v_β.

Thus α_j and β_j are broken into two components, one satisfying a polynomial assumption and the other a noise whose influence is measured by v_α and v_β. If $v_\alpha = v_\beta = 0$ then we have exactly the Almon model. As v_α and v_β increase there is less prior restriction on the smoothness of the α_j and β_j until, when v_α and v_β are very large, there is virtually no prior restriction on the smoothness.

5.2. The posterior mean of θ given σ^2

The distributed lag model can now be written as

$$\begin{aligned} y_t &= \mu + \sum_{j=0}^{a} \left(\alpha_j^{(1)} + \sum_{k=0}^{d_\alpha} \alpha_k^{(2)} j^k \right) a_{t-j} \\ &\quad + \sum_{j=0}^{b} \left(\beta_j^{(1)} + \sum_{k=0}^{d_\beta} \beta_k^{(2)} j^k \right) b_{t-j} + z_t \\ &= \mu + \sum_{j=0}^{a} \alpha_j^{(1)} a_{t-j} + \sum_{j=0}^{b} \beta_j^{(1)} b_{t-j} + \sum_{k=0}^{d_\alpha} \alpha_k^{(2)} \left(\sum_{j=0}^{a} a_{t-j} j^k \right) \\ &\quad + \sum_{k=0}^{d_\beta} \beta_k^{(2)} \left(\sum_{j=0}^{b} b_{t-j} j^k \right) + z_t, \end{aligned}$$

for $t = 1, \ldots, T$. In matrix form these T equations are

$$Y = W\psi + Z$$

where

$$\psi = (\mu, \alpha_0^{(1)}, \ldots, \alpha_a^{(1)}, \beta_0^{(1)}, \ldots, \beta_b^{(1)}, \alpha_0^{(2)}, \ldots, \alpha_{d_\alpha}^{(2)}, \beta_0^{(2)}, \ldots, \beta_{d_\beta}^{(2)})'.$$

Let $p_\alpha = \sigma^2 \div v_\alpha$ and $p_\beta = \sigma^2 \div v_\beta$. Let $D(p_\alpha, p_\beta)$ be a square diagonal matrix whose main diagonal is given by the following:

(i) The first element is 0.

(ii) The next $a+1$ elements are p_α.

(iii) The next $b+1$ elements are p_β.

(iv) The next $d_\alpha + d_\beta + 2$ elements are 0.

From [7, p. 6] the posterior mean of ψ given σ^2 is

$$\psi(p_\alpha, p_\beta) = (W'W + D(p_\alpha, p_\beta))^{-1} W'Y.$$

Thus the posterior means of α_j and β_j are

$$\alpha_j(p_\alpha, p_\beta) = \alpha_j^{(1)}(p_\alpha, p_\beta) + \sum_{k=0}^{d_\alpha} \alpha_k^{(2)} (p_\alpha, p_\beta) j^k$$

and

$$\beta_j(p_\alpha, p_\beta) = \beta_j^{(1)}(p_\alpha, p_\beta) + \sum_{k=0}^{d_\beta} \beta_k^{(2)} (p_\alpha, p_\beta) j^k.$$

The subset we shall study in this section is this class of posterior means

$$\begin{aligned} \theta(p_\alpha, p_\beta) = (&\mu(p_\alpha, p_\beta), \alpha_0(p_\alpha, p_\beta), \ldots, \alpha_a(p_\alpha, p_\beta), \\ &\beta_0(p_\alpha, p_\beta), \ldots, \beta_b(p_\alpha, p_\beta)), \end{aligned}$$

for $p_\alpha > 0$ and $p_\beta > 0$. If p_α is very close to zero, the prior information specifies that $\alpha_j^{(1)}$ is very likely to be near zero, the posterior mean of

$\alpha_j^{(1)}$ is near zero and thus the posterior mean of α_j is nearly the polynomial part,

$$\sum_{k=0}^{d_\alpha} \alpha_k^{(2)} j^k.$$

A similar statement holds for β_j. If p_α and p_β are both near zero then $\theta(p_\alpha, p_\beta)$ is very nearly the standard Almon estimate.

If p_α is large there is little prior information about the value of $\alpha_j^{(1)}$, the posterior mean of $\alpha_j^{(1)}$ will typically be nonnegligible, and therefore the posterior mean of α_j for $j = 0, \ldots, a$ will typically not look like a polynomial of degree d_α. If p_α and p_β are both very large then $\theta(p_\alpha, p_\beta)$ is very nearly the standard maximum likelihood estimate, θ_{MLE}.

5.3. *The methodology and an example*

As in the previous section we shall not study $\theta(p_\alpha, p_\beta)$ directly but rather the corresponding parameters $\theta^*(p_\alpha, p_\beta)$ of the unnormalized form of the model. $\theta^*(p_\alpha, p_\beta)$ is studied by calculating it for various values p_α and p_β. An estimate $\hat{\theta}$ is selected which has an amount of smoothness that makes economic sense, but for which $RL(\hat{\theta})$ is not so small as to render θ unlikely in light of the data. As in the previous section we shall also study

$$RSS(\theta(p_\alpha, p_\beta)) = SS(\theta_{MLE}) \div SS(\theta(p_\alpha, p_\beta)).$$

In some cases it is quite reasonable to consider values (p_α, p_β) for which $p_\alpha = p_\beta$. This would arise in a situation where there is no a priori reason to believe that the smoothness of $\alpha_0, \ldots, \alpha_\beta$ differs from that of $\beta_0, \ldots, \beta_b$.

To illustrate the use of $\theta(p_\alpha, p_\beta)$, the total spending equation from the economic model in [2] will be analyzed. This equation in unnormalized form is

$$y_t = \mu^* + \sum_{j=0}^{4} \alpha_j^* a_{t-j}^* + \sum_{j=0}^{5} \beta_j^* b_{t-j}^* + z_t,$$

where y_t, a_t^*, and b_t^* are the changes from quarter $t-1$ to quarter t of, respectively, the nominal GNP, the money stock, and high-employment Federal expenditures. $t = 1$ corresponds to the first quarter of 1953 and $T = 68$.

The maximum likelihood estimate θ_{MLE}^* of the unnormalized parameters (i.e., corresponding to θ_{MLE}) is given in table 2. It does not appear

TABLE 2

θ^*_{MLE} and $\theta^*(p_\alpha, p_\beta)$ for the total spending equation

		$RL(\theta_{MLE})$	$RSS(\theta_{MLE})$	μ^*_{MLE}	α^*_{MLE}					β^*_{MLE}					
		1	1	2.61	2.10	0.40	2.31	1.18	−0.52	0.45	0.49	0.15	−0.69	−0.56	−0.31
p_α	p_β	$RL(\theta(p_\alpha, p_\beta))$	$RSS(\theta(p_\alpha, p_\beta))$	$\mu^*(p_\alpha, p_\beta)$	$\alpha^*(p_\alpha, p_\beta)$					$\beta^*(p_\alpha, p_\beta)$					
10^{-5}	10^{-5}	1.00	1.00	2.61	2.10	0.40	2.31	1.81	−0.52	0.45	0.49	0.16	−0.69	−0.56	−0.31
10^{-5}	10^{5}	0.04	0.91	2.56	2.13	0.45	2.31	0.87	−0.27	0.76	0.17	−0.20	−0.35	−0.26	0.05
10^{5}	10^{-5}	0.08	0.93	2.47	1.89	1.51	1.14	0.77	0.40	0.49	0.41	0.10	−0.63	−0.57	0.29
0.05	0.05	0.85	1.00	2.60	1.97	0.73	2.05	1.12	−0.38	0.50	0.44	0.11	−0.63	−0.53	0.27
0.1	0.1	0.63	0.99	2.58	1.91	0.91	1.89	1.07	−0.28	0.53	0.40	0.07	−0.59	−0.51	0.25
0.2	0.1	0.47	0.98	2.57	1.85	1.10	1.72	1.01	−0.15	0.54	0.40	0.06	−0.59	−0.51	0.25
0.3	0.1	0.37	0.97	2.56	1.83	1.20	1.61	0.97	−0.06	0.54	0.39	0.06	−0.58	−0.51	0.25
0.4	0.1	0.31	0.97	2.55	1.82	1.26	1.54	0.94	0.00	0.54	0.39	0.05	−0.58	−0.51	0.24
0.5	0.1	0.26	0.96	2.54	1.82	1.30	1.49	0.92	0.05	0.55	0.38	0.05	−0.58	−0.51	0.24
0.1	0.5	0.24	0.96	2.57	1.92	0.92	1.88	1.00	−0.22	0.64	0.29	−0.06	−0.47	−0.40	0.16
0.2	0.5	0.18	0.95	2.55	1.86	1.11	1.70	0.96	−0.11	0.65	0.29	−0.07	−0.47	−0.40	0.16
0.3	0.5	0.15	0.94	2.54	1.84	1.20	1.60	1.93	−0.03	0.65	0.28	−0.07	−0.47	−0.40	0.16
0.4	0.5	0.13	0.94	2.53	1.83	1.26	1.53	0.91	0.02	0.65	0.28	−0.07	−0.47	−0.40	0.16
0.5	0.5	0.11	0.94	2.53	1.83	1.31	1.48	0.89	0.07	0.65	0.28	−0.07	−0.47	−0.40	0.15

grossly inadequate but the small value of the estimate of α_1^* and the negative value of the estimate of α_5^* might be regarded as somewhat unrealistic.

Fig. 2 shows a plot of the maximum likelihood estimates of α_j^* vs. j. The "trend" in these values appears linear so d_α was taken to be 1. The corresponding plot for β_j^* shown in fig. 3 has a noticeable quadratic effect so that d_β was taken to be 2.

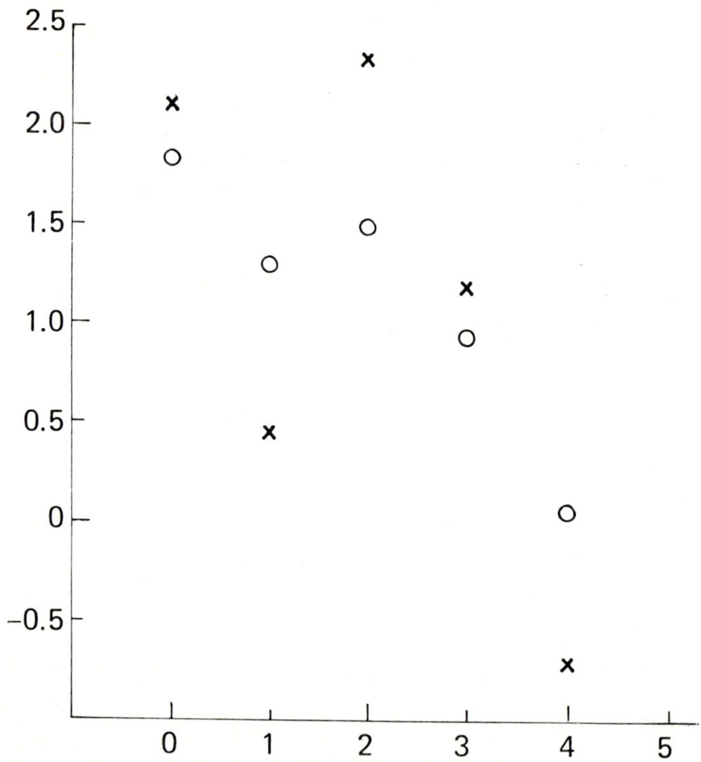

Fig. 2. $\alpha_{MLE}^* = \times$; $\alpha^*(0.5, 0.1) = \bigcirc$.

Table 2 shows the results of calculating $\theta^*(p_\alpha, p_\beta)$ for a number of values of p_α and p_β. The following statements illustrate some of the properties of $\theta^*(p_\alpha, p_\beta)$ described in the previous section: $\theta^*(10^{-5}, 10^{-5})$ is close to θ_{MLE}^*; $\alpha_j^*(10^5, 10^{-5})$ for $j = 0, \ldots, 4$ lie very nearly along a straight line (the first differences are nearly constant); $\beta_j^*(10^{-5}, 10^5)$ for $j = 0, \ldots, 5$ lie very nearly along a quadratic curve (the second differences are very nearly constant). As p_α increases the tendency is to increase $\alpha_1^*(p_\alpha, p_\beta)$ and $\alpha_5^*(p_\alpha, p_\beta)$. $\theta^*(0.5, 0.1)$ appears not to be an unreasonable estimate. Parts of it are plotted in table 1 and fig. 1 along with parts of the maximum likelihood estimate. $\alpha_1^*(0.5, 0.1) = 1.30$ is more nearly in line with $\alpha_0^*(0.5, 0.1) = 1.82$ and $\alpha_2^*(0.5, 0.1) = 1.49$, and $\alpha_5^*(0.5, 0.1) = 0.05$ is positive. $RSS(\theta(0.5, 0.1)) = 0.96$ and

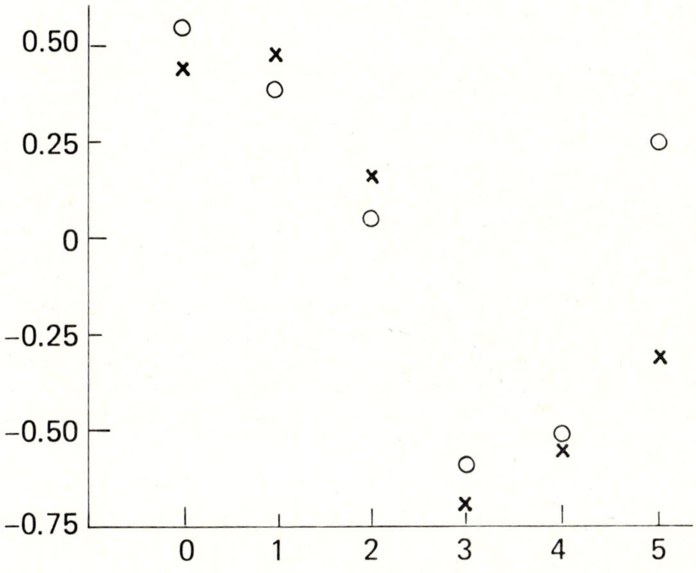

Fig. 3. $\beta_{MLE}^* = \times$; $\beta^*(0.5, 0.1) = \bigcirc$.

$RL(\theta(0.5, 0.1)) = 0.26$ are not so low as to discredit $\theta^*(0.5, 0.1)$ in light of the data. It should be emphasized that others may find another $\theta^*(p_\alpha, p_\beta)$ more sensible, for the choice will (and must since there is not enough information in the data to specify the model with great precision) depend upon one's prior knowledge about the economic mechanism.

Acknowledgements

This paper benefited greatly from the comments of my colleagues in the Fourth NBER-NSF Seminar on Bayesian Inference in Econometrics. In addition the comments of Bruce Hill and Colin Mallows were very helpful.

References

[1] Shirley Almon, The distributed lag between capital appropriations and expenditures. Econometrica 33 (1965), 178–196.

[2] Leonall C. Andersen and Keith M. Carlson, A monetarist model for economic stabilization. Review—Federal Reserve Bank of St. Louis, April 1970, 7–25.

[3] Harris Hancock, Theory of maxima and minima. New York, Dover, 1960.

[4] Arthur E. Hoerl and Robert W. Kennard, Ridge regression: biased estimation for nonorthogonal problems. Technometrics 12 (1970), 55–67.

[5] Arthur E. Hoerl and Robert W. Kennard, Ridge regression: application to nonorthogonal problems. Technometrics 12 (1970), 69–78.

[6] Edward Leamer, Distributed lag analysis with informative prior distributions Harvard University Discussion Paper No. 147, November 1970.

[7] D. V. Lindley and A. F. M. Smith, Bayes estimates for the linear model. Journal of the Royal Statistical Society B-34 (1972), 1–41.

[8] Howard Raiffa and Robert Schlaifer, Applied statistical decision theory. Boston, Division of Research, Harvard Business School, 1961.

[9] R. J. Shiller, A distributed lag estimator derived from smoothness priors. Presented to the Fourth NBER-NSF Seminar on Bayesian Inference in Econometrics, University of Chicago, May 1972. [Editors' note: Shiller's paper is reprinted in the present volume.]

S. E. Fienberg and A. Zellner (eds.)
Studies in Bayesian Econometrics and Statistics.

8.3 A DISTRIBUTED LAG ESTIMATOR DERIVED FROM SMOOTHNESS PRIORS*

Robert J. SHILLER**

University of Minnesota

In most applications of the linear distributed lag model for estimation, the researcher will feel that he has prior knowledge that the lag coefficients should trace out a "smooth" or "simple" curve. The kind of erratic or jagged shape typically provided, when data are poor, by an ordinary least squares estimate of a finite lag is regarded as unreasonable, that is, it is considered implausible that a stable relationship should take such a form. Unless the researcher is prepared, however, to state that the coefficients lie on a function known except for a finite number of parameters, such prior knowledge can only be expressed in probabilistic terms. The classical sampling theory approach to distributed lag estimation is thus ill-equipped to handle the most common or general form of the estimation problem.

Most literature has, however, approached the distributed lag estimation problem from a sampling theory approach, by suggesting parametrizations for the lag curve. One is impressed by the arbitrary and unlikely form of prior knowledge that some of these parametrizations presuppose. It has been assumed that the lag coefficients are known to lie on a polynomial of known degree, over a known interval, but with unknown coefficients [1]. It has alternatively been assumed that the lag coefficients are known to be proportional to a geometric distribution with an unknown decay rate [2, 8]; that they are proportional to a

* Reprinted from Econometrica 41 (1973), 775–788, with the permission of the Econometric Society. An earlier version of this paper was presented to the Fourth NBER-NSF Seminar on Bayesian Inference in Econometrics, Univ. of Chicago, May 1972, and to the December 1971 meeting of the Econometric Society in New Orleans.

** The author wishes to thank Robert E. Hall, Gordon M. Kaufman, Edward E. Leamer and Franco Modigliani for helpful comments. This paper was prepared while the author was supported by a National Science Foundation Graduate Fellowship.

Pascal (negative binomial) distribution with unknown degree r (i.e., the rth convolution of the geometric distribution) [15]; or that the coefficients have a rational generating function whose numerator and denominator have known degree but unknown coefficients [7]. It is indeed difficult to think of circumstances where just such prior knowledge would be forthcoming, or where one would be led to approximate his prior knowledge in such terms.

Apparently the parametrizations are chosen merely so that one or more of the shapes that the parametrization admits has a reasonable appearance. As such, the parametrizations are not efficient representations of prior knowledge. If, as in the most general problem, prior knowledge consists of information that the lag curve is "simple" or "smooth", then although it is true that these parametrizations may exclude many "jagged" shapes, if the number of parameters is kept small, the prior knowledge is incorporated at the expense of constraining the lag coefficients to lie on a small class of elementary shapes. On the other hand, if enough parameters are allowed so that the lag function can well approximate many "smooth" shapes, then the function will also admit many irregular shapes as well. It would be far better to attempt to state, in the appropriate probabilistic terms, what our prior knowledge actually is, and then use this directly to derive an estimate with Bayesian estimation procedures.

1. Distributed lags and prior knowledge

A time series y_t is said to follow a linear distributed lag on time series x_t if:

$$y_t = \sum_{i=0}^{\lambda-1} \beta_i x_{t-i} + \varepsilon_t \tag{1}$$

where λ, the lag length, is a (known) constant, the β_i are the (unknown) coefficients and ε_t is a stationary stochastic process with zero mean.

The problem is, given n observations of y, $Y = \{y_{T-n+1}, y_{T-n+2}, \ldots, y_T\}$ and given a matrix X of current and lagged values of x_t:

$$X = \begin{bmatrix} x_{T-n+1} & x_{T-n} & \cdot & x_{T-n-\lambda+2} \\ x_{T-n+2} & & & \cdot \\ \cdot & & & \cdot \\ x_T & x_{T-1} & \cdot & x_{T-\lambda+1} \end{bmatrix} \tag{2}$$

and given some characterization of the joint distribution of the error term ε_t, to obtain an estimate of $\beta = \{\beta_0, \beta_1, \ldots, \beta_{\lambda-1}\}$. We assume that the ε_t are independent of X and spherically normal:

$$f(\varepsilon_t) \propto h^{\frac{1}{2}} \exp\left[-\tfrac{1}{2}h\varepsilon_t^2\right] \tag{3}$$

where $h \equiv$ process precision $\equiv 1/\sigma^2$. In the sampling theory context in which most of the literature on distributed lags has been cast, it is clear that the ordinary least squares estimate b_0 of β, $b_0 = (X'X)^{-1}X'Y$ is best unbiased. In the Bayesian context, we know that the ordinary least squares estimate is the mean of the posterior distribution based on uninformative (locally uniform) priors (and with an independent locally uniform prior on the log of σ, Jeffreys [6], Savage [13]). The peculiar form of the X matrix (2) is not what distinguishes the problem from the standard regression problem, it must be instead a prior distribution on the β_i.

Expression (1) already involves one aspect of prior information which is not always assumed to be available: a finite length of lag λ. The parameter λ is commonly set at infinity, in the literature, but then estimation could be handled only by assuming a functional constraint on the β_i which is no less arbitrary than the assumption of a finite λ. It is supposed here that one will find it easiest to express prior knowledge about the tail of the lag curve (i.e., the β_i where i is large) in the form of an upper bound on the lag length.

To define dth degree smoothness priors, we first define u as the $d+1$ order differences of the β_i:

$$u = R_d\beta \tag{4}$$

where R_d is a $(\lambda\text{–}d\text{–}1 \times \lambda)$ matrix of $d+1$ differences with rank $\lambda\text{–}d\text{–}1 \equiv p$. For example, if $d = 1$ and $\lambda = 5$,

$$R_1 = \begin{bmatrix} 1 & -2 & 1 & 0 & 0 \\ 0 & 1 & -2 & 1 & 0 \\ 0 & 0 & 1 & -2 & 1 \end{bmatrix}. \tag{5}$$

The prior distribution of u is then assumed to be spherically normal with prior variance ξ^2 and zero mean:

$$f(u|\xi) \propto \left(\frac{1}{\xi}\right)^p \exp\left[-\frac{1}{2}\frac{u'u}{\xi^2}\right]. \tag{6}$$

With uninformative priors on $d+1$ of the β_i and a change of variables we get a prior distribution of β which is uninformative regarding any

β_i considered separately, or regarding the lag sum. The priors are less restrictive than the Almon polynomial constraint of the same degree, which could be represented as a limiting case of the smoothness priors, as ξ goes to zero.

If $d = 1$ and if ξ is small, then the priors attach high probability to all shapes in which the slope changes slowly, highest of all to a straight line, and small probability to jagged shapes. A jagged shape which lies within a narrow band around a straight line is still accorded low probability, while a shape which deviates markedly from a straight line but which does so gradually is given high probability. If $d = 0$ then the priors assert that the lag is unlikely to involve large jumps, it will probably proceed in small steps.

First degree smoothness priors can be likened to a draftsman's flexible curve. The favored shape is a straight line, moreover, the shapes that one can draw with a flexible curve will have small "second differences" unless the curve is bent very hard. In this sense a flexible curve makes relatively jagged shapes improbable, i.e., difficult to obtain.[1]

The likelihood function is:

$$p(Y|\beta, h) \propto h^{\frac{1}{2}n} \exp\left[-\tfrac{1}{2}h(Y - X\beta)'(Y - X\beta)\right]. \tag{7}$$

Thus the posterior distribution (assuming for the moment that h is given) has kernel, by Bayes' law:

$$\begin{aligned} f(\beta | Y, h) &\propto \exp\left[-\frac{1}{2\xi^2}\beta' R_d' R_d \beta\right] \\ &\quad \times \exp\left[-\tfrac{1}{2}h(Y - X\beta)'(Y - X\beta)\right] \\ &\propto \exp\left[-\tfrac{1}{2}h(\beta - b)'\tilde{X}'\tilde{X}(\beta - b)\right] \end{aligned} \tag{8}$$

where

$$\tilde{X} = \begin{bmatrix} X \\ kR_d \end{bmatrix} \qquad k = h^{-\frac{1}{2}}/\xi = \sigma/\xi$$

$$\tilde{Y} = \begin{bmatrix} Y \\ 0 \end{bmatrix} \qquad b = (\tilde{X}'\tilde{X})^{-1}\tilde{X}'\tilde{Y}$$

so that the posterior distribution is multivariate normal with mean b and variance matrix $\sigma^2(\tilde{X}'\tilde{X})^{-1}$. The simplicity of the estimation procedure is thus immediately apparent: to obtain an estimate of β one need only run an ordinary least squares regression using data matrices

[1] Of course, the analogy is not entirely accurate, since the second differences that a flexible curve will allow depend on its slope.

which are augmented with dummy observations, no extra calculations are needed. Implementation of other distributed lag estimators, in contrast, is not so straightforward.

It may be observed that:

(1) $\tilde{X}'\tilde{X}$ is singular if and only if $A_d'X'XA_d$ is singular, where A_d is an Almon transformation matrix (no endpoint constraints) with $d+1$ columns consisting of d degree Lagrangian interpolators (described in [1]). Thus the smoothness priors break collinearity with the same success as the Almon transformation of the same degree. The same number of observations are required for an estimate as in the Almon case of the same degree. Since a degree of one or zero will usually be considered too low for an Almon estimate, the smoothness priors are if anything *less* likely to fail to deal with multicollinearity than is the Almon procedure.

(2) $\lim_{\xi\to 0} (\tilde{X}'\tilde{X})^{-1} = A_d(A_d'X'XA_d)^{-1}A_d'$. Thus, $\lim_{\xi\to 0} b = b_a$ (the Almon estimate of β). $\lim_{\xi\to 0} \sigma^2(\tilde{X}'\tilde{X})^{-1} = \text{var}\,(b_a|X, \beta, h)$ (the improper variance matrix of the Almon estimate).

(3) If $X'X$ is nonsingular, $\lim_{\xi\to\infty} (\tilde{X}'\tilde{X})^{-1} = (X'X)^{-1}$. Then $\lim_{\xi\to\infty} b = b_0$ (the ordinary least squares estimate of β). $\lim_{\xi\to\infty} \sigma^2(\tilde{X}'\tilde{X})^{-1} = \text{var}\,(b_0|X, \beta, h)$ (variance matrix of the ordinary least squares estimate, b_0).

(4) If errors are not independent but follow a first-order autocorrelated scheme, $\varepsilon_t = \rho\varepsilon_{t-1}+\eta_t$, where η_t are themselves independently normally distributed with zero mean and variance σ_η, and given a prior on ρ (which we assume here to be locally uniform), then it is possible to develop along conventional lines a posterior distribution for β and ρ, but this will not yield a tractable marginal distribution for β. Tiao and Zellner [18] have suggested (in the context of strictly uninformative priors on β and $\log \sigma_\eta$ as well as ρ) a simple approximation for the distribution (in the form of a multivariate student), which could be extended to this case. If, however, one is content with a modal estimate of β, then it is only necessary to minimize the expression

$$(Y-\rho Y_{-1}-(X-\rho X_{-1})\beta)'(Y-\rho Y_{-1}-(X-\rho X_{-1})\beta)+k_\eta^2(\beta' R_d' R_d \beta)$$

with respect to β and ρ (here $Y_{-1} = \{y_{T-n}, \ldots, y_{T-1}\}$, X_{-1} is the $n\times\lambda$ matrix whose ijth element is $x_{T-n+i-j}$ and $k_\eta = \sigma_\eta/\xi$). It is easily seen that procedures analogous to those developed by Hildreth and Lu [5] and Cochrane and Orcutt [3] would be applicable here. The limit of such a modal estimate as ξ goes to zero or infinity are Hildreth-Lu (or Cochrane-Orcutt) estimates of β by the Almon and ordinary least squares procedures respectively.

To use smoothness priors one must first choose a value of ξ to represent his prior knowledge. A value of 1 for d will probably do for most applications, but there is no way to suggest in advance a "reasonable" value of ξ since it is not independent of units of measurement. To determine ξ, one must decide in each application how large a $d+1$ order difference is acceptable. This choice, incidentally, has its counterpart in the flexible curve analogy. One must choose among flexible curves on the basis of their stiffness or "minimum radius", a number which is again not independent of the units of measurement.

Once it is understood what ξ means, then there is nothing more that can be said in general about how one might obtain its value; it must be supposed that the researcher has such prior knowledge, specific to his problem. Mention might be made here, however, of a rule of thumb which has seemed useful in some instances, when $d = 1$ and one has a prior expected value for the sum S of the lag coefficients (if one has prior knowledge regarding the sum, one may also wish to use this in the estimation, of course). The rule of thumb is to set:

$$\xi = 8S/\lambda^2. \tag{9}$$

This is just the second difference across the peak of an inverted V, 0 at $i = 0$ and $i = \lambda$, whose integral is S.

The units problem would not exist if one could express smoothness priors in terms of the differences of the logs of the coefficients, or in terms of the differences of the coefficients as a percent of the lag sum. The dth degree log smoothness prior:

$$f(v) \propto \left(\frac{1}{\xi}\right)^p \exp\left[-\frac{1}{2\xi^2} v'v\right] \tag{10}$$

where $v = R_d \log(\beta)$ when combined with the same likelihood function (7) and with a change of variables yields a posterior distribution of the log of β whose mode, $\log(\bar{\beta})$ is given by:

$$X'X\bar{\beta} + \begin{bmatrix} 1/\bar{\beta}_0 & & & & 0 \\ & \cdot & & & \\ & & \cdot & & \\ & & & \cdot & \\ 0 & & & & 1/\bar{\beta}_{\lambda-1} \end{bmatrix} \frac{\sigma^2}{\xi^2} R'_d R_d \log(\bar{\beta}) = X'Y \tag{11}$$

which is not a system which can be solved for $\bar{\beta}$ without an iterative procedure. Still, the priors have some intuitive appeal, and the limiting behavior of the estimator (11) is familiar. It will of course constrain all lag coefficients to be positive, which may be considered desirable in many instances. If $b_0 > 0$, then as $\xi \to \infty$ the estimate $\bar{\beta}$ approaches the ordinary least squares estimate. If $d = 1$ then as $\xi \to 0$ the estimated lag curve must approach the familiar geometric lag. If $d = 2$, then as $\xi \to 0$ the estimated lag approaches a function whose log is a quadratic, a special case of which is, of course, the normal density function. Such priors are not however, considered further here, because of the difficulties posed by the nonlinearities in (11).

It may also seem, incidentally, that there may be an advantage in assuming for u a prior variance matrix ψ which is not diagonal. It should be realized, however, that the equivalent can be achieved by assuming a prior distribution on a different linear combination of the β_i with an identity correlation matrix. As a matter of fact, if we assume a Markovian correlation matrix with a high ρ, then we are in effect increasing the degree of the smoothness priors by one, i.e., if:

$$\psi = \xi^2 M = \xi^2 \begin{bmatrix} 1 & \rho & \rho^2 & \rho^3 & & & \\ \rho & 1 & \rho & & & & \\ \rho^2 & \rho & 1 & & & & \\ \rho^3 & & & & & & \\ & & & & & & \\ & & & & & & \rho \\ & & & & & \rho & 1 \end{bmatrix} \tag{12}$$

then the same assumption could be represented with $\psi' = \tilde{\xi}^2 I$ by changing R_d to $R_{d,\rho}$ by:

$$R_{d,\rho} = mR_d = \frac{1}{\sqrt{1-\rho^2}} \begin{bmatrix} 1 & -\rho & 0 & & & \\ 0 & 1 & -\rho & & & \\ & & & & & 0 \\ & & & & 1 & -\rho \\ & & & & 0 & \sqrt{1-\rho^2} \end{bmatrix} R_d \tag{13}$$

since $m'm = M^{-1}$. Thus, if ρ approaches 1 and ξ^2 is at the same time increased keeping $(1-\rho^2)\xi^2$ constant, then our priors approach the $d+1$ degree smoothness priors with $\psi' = (1-\rho^2)\xi^2 I = \tilde{\xi}^2 I$. The final row of $R_{d,\rho}$ contains an endpoint prior which, however, vanishes as ρ approaches one.

The procedure, where smoothness priors alone are used, may be contrasted with the full-scale Bayesian distributed lag estimation procedure of Leamer [9] which begins with a prior expected value for each lag coefficient. As has already been noted, there is no expected value for β_i associated with smoothness priors. Since Leamer assumed a prior variance-covariance matrix of the coefficients with positive off-diagonal elements, deviations from his expected curve must also be "smooth". The smoothness priors procedure may be regarded as a limiting case of a full-scale Bayesian procedure such as Leamer's as the prior variances of the coefficients are increased to infinity, and, at the same time, the covariances are increased, approaching perfect correlation. That is, such a full-scale procedure is approximately similar to an estimator based on zero degree smoothness priors if the prior variance matrix is large and Markovian with high ρ, *and* if the first differences of the prior expected values of the coefficients are zero. The marginal distribution of the first differences of the coefficients will, in any event, have, in the limit, the variance $\tilde{\xi}^2$ around the first differences of the original prior expected values, whether zero or not.

Instances will undoubtedly arise when it is desired also to use far endpoint priors in estimation, that is, priors which imply that the final coefficients are themselves small. In these circumstances, it will no longer be true that the priors are uninformative regarding the individual coefficients. We may conceive of an Almon-type far endpoint prior with a normal prior distribution with zero mean for the final coefficient in the lag (a prior represented by appending a single row to the matrix R_d which is zero except in the final column). Alternatively, one may wish to impose smoothness priors connecting the final coefficients in the lag with the zero "coefficients" beyond, e.g., when $\lambda = 5$ and $d = 1$:

$$u^* = R_d^*\beta,\ R_d^* = \begin{bmatrix} 1 & -2 & 1 & 0 & 0 \\ 0 & 1 & -2 & 1 & 0 \\ 0 & 0 & 1 & -2 & 1 \\ 0 & 0 & 0 & 1 & -2 \\ 0 & 0 & 0 & 0 & 1 \end{bmatrix},\quad \rho(R_d^*) = \lambda. \tag{14}$$

If u^* is now supposed to be spherically normally distributed with zero mean and variance ξ^2, and if $d = 1$, the marginal prior distribution of

β_i has zero mean with variance $\xi^2\Sigma_{j=1}^{\lambda-i}j^2$. Thus, the prior variance of β_i for small i may be quite wide even though the smoothness priors are very tight.

2. *Posterior distributions: precision unknown*

In most applications, the assumption that h is known will be unrealistic. The assumption of known h may be replaced by the assumption of a prior distribution on h, which we will take for convenience to be gamma:

$$f(h|s^2, v) \propto h^{\frac{1}{2}v-1} \exp[-\tfrac{1}{2}hvs^2] \tag{15}$$

with location parameter s^2 and scale parameter v. If v is zero, then the distribution reduces to the (improper) uninformative prior distribution suggested by Jeffreys [6].

If h is assumed independent of β, in the priors, then, using (6), (7), (15) and Bayes' law, the marginal posterior distribution of β is seen to be:

$$f(\beta) \propto \left[n+v-\rho(X)+\frac{1}{\bar{\sigma}}(\beta-b_0)'X'X(\beta-b_0)\right]^{-(n+v)/2}$$

$$\times \exp[-\tfrac{1}{2}\beta'R_d'R_d\beta/\xi^2] \tag{16}$$

$$\bar{\sigma}^2 \equiv \frac{vs^2+(Y-Xb_0)'(Y-Xb_0)}{n+v-\rho(X)} \tag{17}$$

where b_0 is a solution of $X'Xb_0 = X'Y$. To derive moments of the posterior distribution one may represent the distribution as a normal multiplied by a correction factor which may be expanded in powers of $(n+v-\rho(X))^{-1}$ (following Tiao and Zellner [19]). The normal distribution is obtained by replacing the Student part of (16) with a normal so that we have:

$$f(\beta) \propto \exp[-\tfrac{1}{2}(\beta-b_0)'X'X(\beta-b_0)/\bar{\sigma}^2-\tfrac{1}{2}\beta'R_d'R_d\beta/\xi^2]$$

$$\propto \exp[-\tfrac{1}{2}(\beta-b)'\tilde{X}'\tilde{X}(\beta-b)/\bar{\sigma}^2] \tag{18}$$

where $\tilde{X}$, $\tilde{Y}$, and b are defined as above, except that k is now $\bar{\sigma}/\xi$. If $n+v-\rho(X)$ is large we may ignore all elements in the expansion except the first, and take (18) as itself an approximation to the posterior distribution. If $v = 0$, then this is the approximation used by Leamer [9], and the estimation procedure reduces to one suggested by Theil [17] (derived from a sampling theory approach). A two-stage procedure is thus used. A preliminary ordinary least squares regression (on the largest subset of independent columns of X) is run to obtain $\bar{\sigma}$, and k is obtained as the ratio $\bar{\sigma}/\xi$. In the second stage $\tilde{Y}$ is regressed on $\tilde{X}$, as before.

The situation is simpler if one is prepared to begin with a value of k rather than ξ, i.e., if the conjugate prior is assumed, as in Raiffa and Schlaifer [12]:

$$f^{p}_{N_{\gamma}}(u, h|0, s^2, k^2 I, \nu) \propto h^{\frac{1}{2}p} \exp[-\tfrac{1}{2}hk^2u'u]h^{\frac{1}{2}\nu-1} \times \exp[-\tfrac{1}{2}h\nu s^2] \tag{19}$$

which, when applied as a prior to (7) using Bayes' law yields a Student marginal posterior for β:

$$f(\beta) \propto \left[n+p+\nu-\lambda+\frac{1}{\bar{\sigma}^2}(\beta-b)'\tilde{X}'\tilde{X}(\beta-b)\right]^{-(n+p+\nu)/2} \tag{20}$$

where $\tilde{X}$, $\tilde{Y}$, and b are defined as before and $\bar{\sigma}^2 \equiv [\nu s^2+(\tilde{Y}-\tilde{X}b)'(\tilde{Y}-\tilde{X}b)]/(n+p+\nu-\lambda)$, which is the reciprocal of the mean of h. If $\nu = 0$, then $\bar{\sigma}^2$ is the estimate of σ^2 printed out by the regression program (when $\tilde{Y}$ is regressed on $\tilde{X}$) and $\bar{\sigma}^2(\tilde{X}'\tilde{X})^{-1}$ is the estimate of the variance-covariance matrix of estimated coefficients printed out by the program. In this circumstance, $\lim_{k\to\infty} \bar{\sigma}^2(\tilde{X}'\tilde{X})^{-1}$ is the Almon estimate of the variance-covariance matrix of estimated coefficients, and $\lim_{k\to\infty} \bar{\sigma}^2$ is the Almon estimate of σ^2. Thus, if it weren't for the practical problems of inverting near-singular matrices, one would find that the easiest way (in terms of the amount of programming required) to get an mth degree Almon estimate of all of these quantities would be to run a simple regression of $\tilde{Y}$ on $\tilde{X}$ with $d = m$ and k very large.

3. *Illustrative examples*

To illustrate the practical application of the estimator, two examples[2] of distributed lags were estimated, the first using actual data in a term structure equation relating the corporate bond yield average to the commercial paper rate, the second on artificial data with a known lag

[2] The estimator has also been used to estimate investment equations (for producer durables and producer structures) and a price equation [14].

shape. In both cases the estimator used first-degree smoothness priors, but did not use endpoint priors. All estimation was done with the ordinary least squares command of the Econometric Software Package. The estimate $\bar{\sigma}$ of the standard error of the regression assumes $v = 0$ with the conjugate prior, i.e., these are just the standard errors printed by the regression program when $\tilde{Y}$ was regressed on $\tilde{X}$.

The theory behind the hypothesized relation between the long rate and the short rate has been developed elsewhere [10, 11, 16], and will not be entered into here. The lag coefficients are supposed to reflect an expectational mechanism and thus we would expect that the lag shapes should be, for the most part, smooth. It has been suggested [4, 10], that if expectations are of the extrapolative variety, or a combination extrapolative-regressive variety, then β_0 may not, however, be related to the other coefficients, so that it should be freed from the rest of the lag.

Such a distributed lag (quarterly, $\lambda = 16$) was estimated with a constant term over the period 1955 third quarter to 1969 fourth quarter, with first-degree smoothness priors excluding the first coefficient and with $k = 8$. The dependent variable was the Federal Reserve Board Aaa new issue yield series, and the independent variable was the 4–6 month prime commercial paper rate. Using an ordinary least squares estimate of $\sigma = 0.25$, we note that this value of k corresponds to the value that one would arrive at in the independent priors case with $v = 0$, $\xi = 1/32$ (or that given by expression (9) based on the assumption that $S = 1$). The conjugate estimate $\bar{\sigma}$ of σ is then 0.23, so that the standard errors printed by the program would also correspond quite closely to the posterior standard deviation of β in the independent priors case, expression (18). The estimated shape (shown except for the first coefficient in fig. 1A) suggests that the lag becomes more or less asymptotic after $t-12$. A third-degree Almon (fig. 1B) gives a similar shape, but turns negative and then positive at the end, an unexpected result. Of course, it was constrained to turn up at the end (if it was to match the hump at the beginning) since a third-degree polynomial is assumed. When the estimator is given more freedom to approximate the curve in fig. 1 by increasing the degree of the polynomial, it does not do so. A 6th degree Almon estimate gives rather erratic results, reflecting the weakness of the priors.

In the second example, the estimator is used on artificial data with a known lag shape. A (forty-element) vector ε of standardized normal random numbers was generated with the Lehmer pseudo-random number generator. Then, a (forty-element) vector Y was defined according to:

$$Y = XN + 0.05\varepsilon \tag{21}$$

where N is a (twenty-element) vector of ordinates of the normal density with mean at the 10th element ($t-9$) and a standard deviation of two periods. The sum of the normal coefficients is thus 2.000. To illustrate

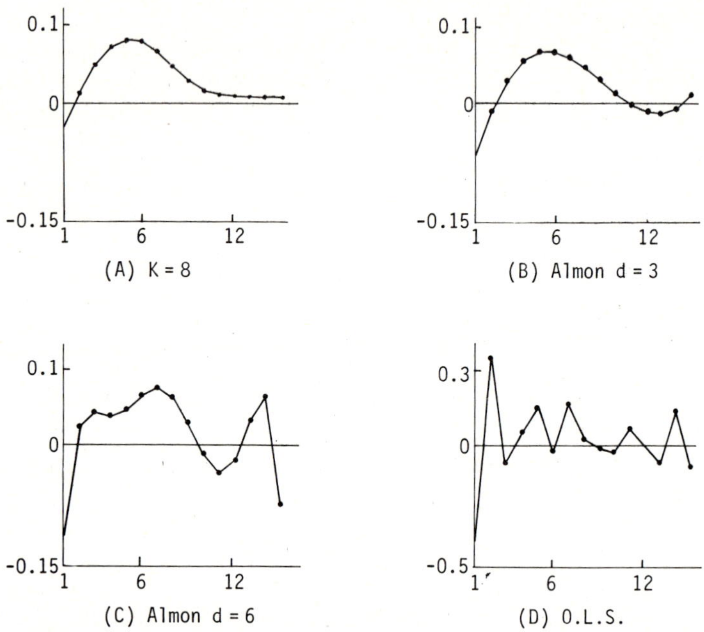

Fig. 1. Estimates of distributed lag coefficients in term structure equation. (A) Estimate with first-degree smoothness priors, $k = 8$, no endpoint priors. Sum of coefficients is 0.8743, $\bar{\sigma}$ is 0.2343. (B) Third-degree Almon polynominal estimate. Sum of coefficients is 0.7333, $\hat{\sigma}$ is 0.2556. (C) Sixth-degree Almon estimate. Sum of coefficients is 0.7327, $\hat{\sigma}$ is 0.2583. (D) Ordinary least squares estimate. Sum of coefficients is 0.8640, $\hat{\sigma}$ is 0.2520. Note different scaling of y axis in (D).

a typical, nearly collinear, X matrix, the same data (the commercial paper rate) that was used in the first example is used here to generate a (40×20) matrix of current and lagged x_t.

A number of values of k are tried, the first corresponding to the rule of thumb mentioned. Since the sum of weights is 2.000, the ξ dictated by expression (9) would be 0.04 so that, with the known value of

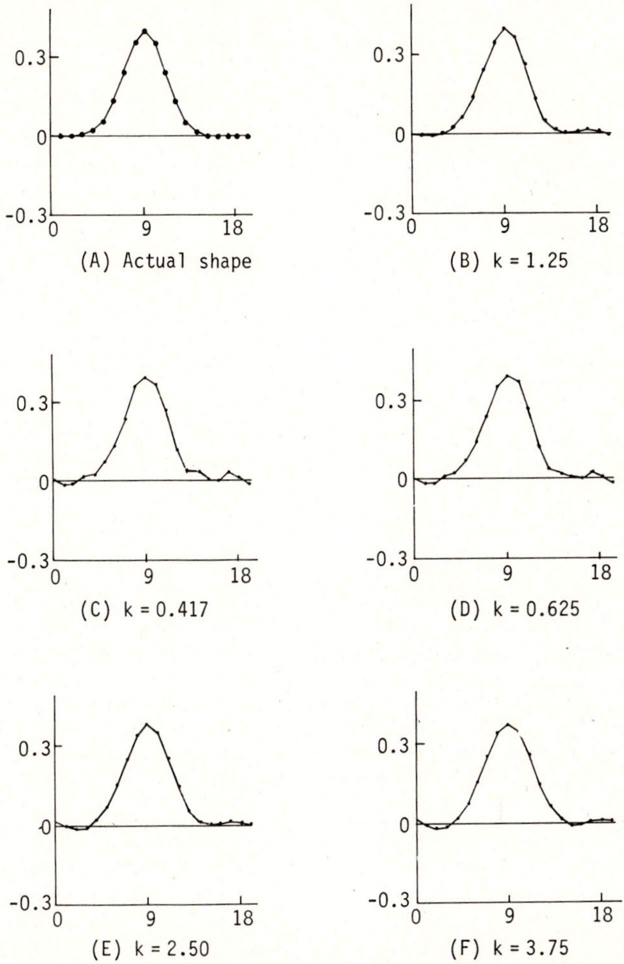

Fig. 2. Known lag shape in artificial data and estimates with first-degree smoothness priors, no endpoint priors. (A) Actual shape, normal density. Maximum second difference is 0.094, sum of coefficients is 2.000, σ is 0.050. (B) Estimate using $\xi = 0.04$, i.e., $k = 1.25$. Sum of coefficients is 2.003, $\bar{\sigma}$ is 0.053. (C) Estimate with very loose priors, $k = 0.417$. Sum of coefficients is 2.003, $\bar{\sigma}$ is 0.045. (D) Estimate with loose priors, $k = 0.625$. Sum of coefficients is 2.003, $\bar{\sigma}$ is 0.047. (E) Estimate with tight priors, $k = 2.5$. Sum of coefficients is 2.004, $\bar{\sigma}$ is 0.070. (F) Estimate with very tight priors, $k = 3.75$. These priors put the actual second difference at the peak 7 standard deviations out on the prior distribution. Sum of coefficients is 2.004, $\bar{\sigma}$ is 0.088.

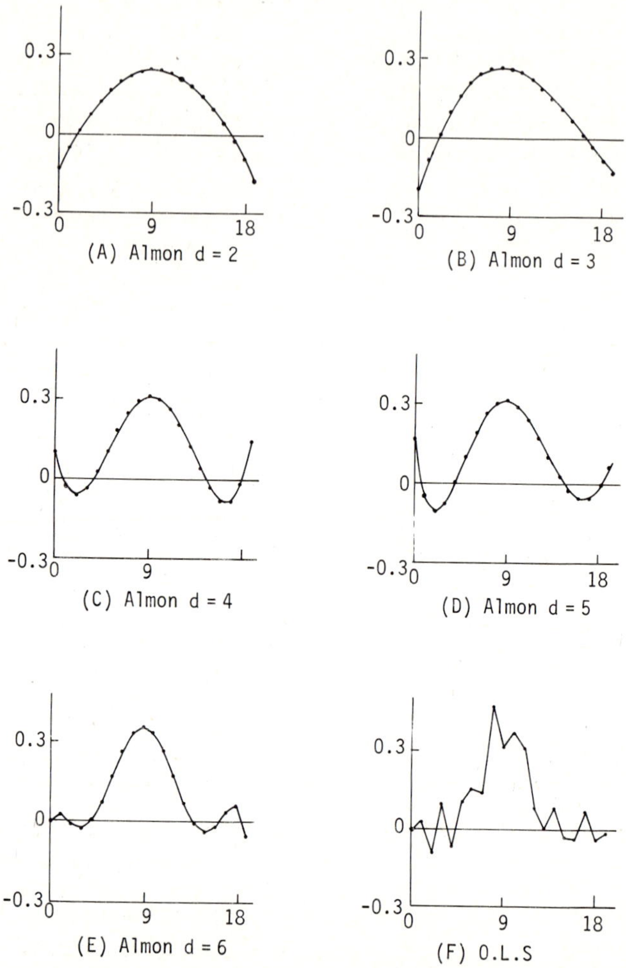

Fig. 3. Conventional estimators applied to artificial data used for fig. 2. (A) Second-degree Almon polynominal estimate. Sum of coefficients is 2.002, $\hat{\sigma}$ is 0.501. (B) Third-degree Almon estimate. Sum of coefficients is 2.004, $\hat{\sigma}$ is 0.480. (C) Fourth-degree Almon estimate. Sum of coefficients is 2.011, $\hat{\sigma}$ is 0.164. (D) Fifth-degree Almon estimate. Sum of coefficients is 2.004, $\hat{\sigma}$ is 0.131. (E) Sixth-degree Almon estimate. Sum of coefficients is 2.003, $\hat{\sigma}$ is 0.080. (F) Ordinary least squares estimate. Sum of coefficients is 2.003, $\hat{\sigma}$ is 0.062.

$\sigma = 0.05$, k will be 1.25. An estimate was made with this value of k, and with a k of 1/3, 1/2, 2 and 3 times this. These estimates appear in fig. 2. Here, it appears that if data are good (i.e. if σ is small), within a wide range the value of k is not crucial, since all of the Bayesian estimates are quite satisfactory.

Also shown are the second-degree through sixth-degree Almon estimates, and an ordinary least squares estimate (fig. 3). The estimates illustrate the restrictions that the polynomial parametrizations place on the estimates. The strong upturns at both ends of the fourth- and fifth-degree polynomial reflect the inability of the polynomial to provide a horizontal asymptote. A higher degree polynomial can approximate an asymptote better by oscillating around it (as the sixth degree begins to do in this example) but may also give erratic results. The ordinary least squares estimate is of course a 19th degree polynomial. Had Almon estimates in the range 7–18 been attempted, it is clear that they would share some of the jagged appearance of the ordinary least squares estimate.

References

[1] Shirley Almon, The distributed lag between capital appropriations and expenditures. Econometrica 33 (1966), 178–196.

[2] Phillip Cagan, The monetary dynamics of hyperinflation. In: Studies in the quantity theory of money, edited by M. Friedman. Chicago, University of Chicago Press, 1956, 25–117.

[3] D. Cochrane and G. H. Orcutt, Application of least squares regressions to relationships containing auto-correlated error terms. Journal of the American Statistical Association 44 (1949), 32–61.

[4] F. de Leeuw, A model of financial behavior. In: The Brookings quarterly econometric model of the United States economy, edited by J. Duesenberry et al. Chicago, 1965.

[5] Clifford Hildreth and John Y. Lu, Demand relationships with autocorrelated disturbances. Technical Bulletin 276, Michigan State University Agricultural Experiment Station, East Lansing, Michigan, 1960.

[6] H. Jeffreys, Theory of probability. London, Oxford University Press, 1961.

[7] Dale Jorgenson, Rational distributed lag functions. Econometrica 34 (1966), 135–149.

[8] L. M. Koyck, Distributed lags and investment analysis. Amsterdam, North-Holland, 1954.

(9) Edward E. Leamer, A class of informative priors and distributed lag analysis. Econometrica 40 (1972), 1059–1081.

[10] Franco Modigliani and Richard Sutch, Innovations in interest rate policy. American Economic Review Papers and Proceedings 56 (1966), 178–197.

[11] Franco Modigliani and Richard Sutch, Debt management and the term structure of interest rates. Journal of Political Economy 75 (1967), 569–589.

[12] H. Raiffa and R. Schlaifer, Applied statistical decision theory. Boston, Harvard University Press, 1961.

[13] L. J. Savage, The subjective basis of statistical inference. Manuscript, University of Michigan, 1961.

[14] R. J. Shiller, Estimation of the investment and price equations of a macroeconometric model. Federal Reserve Board Staff Economic Study No. 61, Washington, 1971.

[15] Robert M. Solow, On a family of lag distributions. Econometrica 28 (1960), 393–406.

[16] Richard Sutch, Expectations, risk and the term structure of interest rates. Unpublished Ph.D. thesis, M.I.T., 1968.

[17] H. Theil, On the use of incomplete prior information in regression analysis. Journal of the American Statistical Association 58 (1963), 401–414.

[18] George C. Tiao and Arnold Zellner, Bayesian analysis of the regression model with autocorrelated errors. Journal of the American Statistical Association 59 (1964), 763–778.

[19] George C. Tiao and Arnold Zellner, Bayes theorem and the use of prior knowledge in regression analysis. Biometrika 51, 1 and 2 (1964), 219–230.

PART 9

ANALYSES OF SIMULTANEOUS EQUATION MODELS

S. E. Fienberg and A. Zellner (eds.)
Studies in Bayesian Econometrics and Statistics.

9.1 BAYESIAN ANALYSIS OF HAAVELMO'S MODELS*

V. K. CHETTY

Indian Statistical Institute

1. Introduction

In recent years, the Bayesian approach has been used to analyze the robustness of specifying assumptions in stochastic models. Box and Tiao [3] used this approach to assess the effects of a departure from normality in the comparison of variances. Zellner and Tiao [18], using the Bayesian approach, analyzed the effects of departures from serial independence of the error terms in a multiple regression model on inferences about the model's parameters.

In this paper, we make use of Bayesian methods similar to those used in the cited studies to analyze the robustness of assumptions in two simple simultaneous equation econometric models. The exact finite sample posterior distributions of the parameters are derived for given prior distributions of the parameters for the two models. Compared with some models of actual economies, the models that we analyze are very simple. The two models are chosen for convenience in experimenting with the Bayesian techniques, rather than for their inherent economic merits.

The plan of the paper is as follows. In section 2 we state Haavelmo's first model and its specifying assumptions, and review the sampling theory analysis of the model. In section 3 we analyze the model from the Bayesian point of view and get the exact posterior distribution of the parameters. In section 4 the predictive density of the model is

* Reprinted with minor changes from Econometrica 36 (1968), 582–602, with permission of the Econometric Society. This is a revised version of a chapter of the author's thesis submitted to the Univ. of Wisconsin [4]. The author wishes to acknowledge Professors A. Zellner and Jacques Drèze and the referees for their valuable comments. Research financed in part by NSF Grant GS-151.

derived. Specification analysis of the model is done in section 5. In section 6 Haavelmo's second model is analyzed from the Bayesian point of view. In section 7 informative prior distributions for a parameter are introduced for the second model in order to study how the sample modifies the investigator's prior belief.

2. First model and specifying assumptions

Haavelmo's first model [9] consists of a consumption equation and an identity, namely,

$$c_t = \alpha y_t + \beta + u_t, \tag{1}$$

$$y_t = c_t + z_t \qquad (t = 1, 2, \ldots, T), \tag{2}$$

where c_t represents consumers' expenditures in constant dollars per capita, y_t is disposable income in constant dollars per capita, and z_t is investment expenditures in constant dollars per capita. In this model, investment is assumed to be exogenous. The u's are structural disturbance terms and α and β are structural coefficients. Regarding the disturbance terms, the following assumptions are made:

(a) the u's are normally distributed;
(b) $E(u_t) = 0$ and $E(u_t^2) = \sigma^2$ for every value of t;
(c) $E(u_t u_{t-\tau}) = 0$ for $\tau \neq 0$;
(d) the time series z_t $(t = 1, 2, \ldots, T)$ is exogenous in relation to c_t and y_t.

Condition (d) is fulfilled if either (d.1) the sequence z_t is a sequence of given numbers, or (d.2) each z_t is a random variable that is stochastically independent of u_t.

The reduced form equation for c_t is

$$c_t = \frac{\alpha}{1-\alpha} z_t + \frac{\beta}{1-\alpha} + \frac{u_t}{1-\alpha}. \tag{3}$$

Haavelmo [9] has shown that a consistent estimate of α can be obtained from the least squares estimate of $\alpha/(1-\alpha)$ derived from equation (3). Since this is a just-identified model, all sampling theory estimation procedures including maximum likelihood, two-stage least squares, three-stage least squares, etc., will lead to the same consistent estimate of α, namely, $\hat{\alpha} = \hat{k}/(\hat{k}+1)$, where

$$\hat{k} = \frac{\sum_{t=1}^{T} (c_t - \bar{c})(z_t - \bar{z})}{\sum_{t=1}^{T} (z_t - \bar{z})^2}, \tag{4}$$

$\bar{c} = (1/T)\Sigma_{t=1}^{T} c_t$ and $\bar{z} = (1/T)\Sigma_{t=1}^{T} z_t$. The exact sampling distributions of the least squares and the maximum likelihood estimators of α for this model have been derived by Bergstrom [2] for the special case where the z's are assumed to be nonstochastic.

3. *Bayesian analysis of the model*

The likelihood function for the model's parameters α, β, and σ is given by

$$l(\alpha, \beta, \sigma|\text{data}) \propto \frac{(1-\alpha)^T}{\sigma^T} \exp\left\{-\frac{1}{2\sigma^2} \sum_{t=1}^{T} (c_t - \alpha y_t - \beta)^2\right\}. \tag{5}$$

Regarding the prior probability density functions for the parameters α, β, and σ, we assume (following Jeffreys [10], Savage [12], Geisser and Cornfield [8], and Tiao and Zellner [14]) that $p(\alpha, \beta, \sigma) \propto \sigma^{-1}$, $0 < \alpha < 1$.[1]

We believe that our prior beliefs are adequately represented by these pdf's. The analysis of the model, however, can be easily done using different prior distributions. Now, the posterior distribution of the parameters is given by

$$p(\alpha, \beta, \sigma|\text{data}) \propto \frac{(1-\alpha)^T}{\sigma^{T+1}} \exp\left\{\frac{-1}{2\sigma^2} \sum_{t=1}^{T} (c_t - \alpha y_t - \beta)^2\right\}. \tag{6}$$

Since we are interested in making inferences about α, we need its marginal distribution. On integrating out β from (6), we have,

$$p(\alpha, \sigma|\text{data}) \propto \frac{(1-\alpha)^T}{\sigma^T} \exp\left\{\frac{-1}{2\sigma^2} \sum_{t=1}^{T} (w_t^2 - T\bar{w}^2)\right\} \tag{7}$$

[1] Here and below we use the symbol p to denote pdf's generally and not one specific pdf. The argument of the function p identifies the particular distribution being considered.

TABLE 1

U.S. data used in study

Year	Disposable income (dollars per capita), deflated y_t	Consumer expenditures (dollars per capita), deflated c_t	$z_t = y_t - c_t$	Gross investment (dollars per capita), deflated x_t
1929	543	474	60	128
1930	478	439	39	97
1931	440	399	41	81
1932	372	350	22	45
1933	381	364	17	45
1934	419	392	27	66
1935	449	416	33	78
1936	511	463	48	103
1937	520	469	51	98
1938	477	444	33	83
1939	517	471	46	105
1940	548	494	54	122
1941	629	529	100	165

Source of data: Haavelmo [9].

where $w_t = c_t - \alpha y_t$ and $\bar{w} = \Sigma_{t=1}^{T} w_t/T$. Now integrating out σ, we have

$$p(\alpha|\text{data}) \propto \frac{(1-\alpha)^T}{\{Ts^2 + (\alpha-\hat{\alpha})^2 \cdot M\}^{(T-1)/2}},$$

where $Ts^2 = \Sigma_{t=1}^{T} (c_t - \hat{\alpha} y_t - \hat{\beta})^2$, $M = \Sigma y_t^2 - ((\Sigma y_t)^2/T)$, and $\hat{\alpha}$ and $\hat{\beta}$ are the least squares coefficients computed by regressing c_t on y_t.

Rothenberg [11] obtained this posterior distribution of α when analyzing this model. In order to evaluate the normalizing factor, we have to perform the following integration:

$$\int_0^1 \frac{(1-\alpha)^T}{\{(1+(\alpha-\hat{\alpha})^2 k)\}^{(T-1)/2}} \, d\alpha \tag{8}$$

where $k = M/Ts^2$. In actual analysis, the normalized density function can be easily obtained using the numerical integration methods described in the Appendix.

The posterior distribution of α is derived for U.S. data for the period 1929–41. The data used are given in table 1. Numerical integration methods have been used to evaluate the normalized density and its moments. The mean value of α is found to be 0.6597. This can be used as an estimate of α. The mean of the posterior distribution is an optimal point estimate, if the quadratic loss function $L(\alpha, \hat{\alpha}) = (\alpha - \hat{\alpha})^2$ is used. The marginal posterior distributions of α and β are shown in fig. 1. The variance of the distribution of α is found to be 0.0043. Since we have the complete distribution of the parameters, we can compute

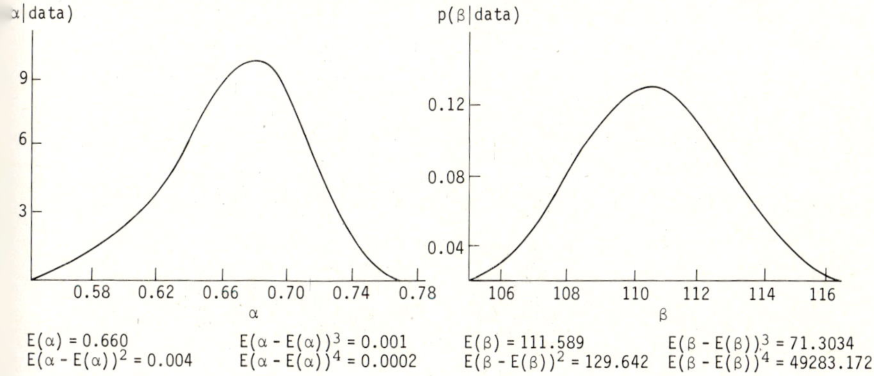

Fig. 1. Marginal posterior distributions of α and β for Model I.

the probability that the parameter lies in any particular interval. Such an interval is called the Bayesian confidence interval.

The sampling theory estimate of α for the same data is found to be 0.674, with a variance of 0.005.

4. *Predictive density*

On many occasions we will be interested in making inferences about observations that are as yet unobserved, that is, future observations. In the Bayesian approach, the probability density function for the as yet unobserved observations can be derived and is known as the predictive pdf. Here we derive the predictive pdf for c_{T+1}—the consumption expenditures in period $T+1$—given the investment expenditures z_{T+1}. The joint posterior distribution of c_{T+1}, α, β, and σ is

$$P(c_{T+1},\alpha, \beta, \sigma|\text{data}) = p(c_{T+1}|\alpha, \beta, \sigma, \text{data})\, p(\alpha, \beta, \sigma|\text{data}).$$

Now

$$p(c_{T+1}|\alpha, \beta, \sigma, \text{data}) \propto \frac{1-\alpha}{\sigma} \times \exp\left[-\frac{(1-\alpha)^2}{2\sigma^2}\left(c_{T+1}-\frac{\alpha}{1-\alpha}z_{T+1}-\frac{\beta}{1-\alpha}\right)^2\right]$$

and

$$p(\alpha, \beta, \sigma|\text{data}) \propto \frac{(1-\alpha)^T}{\sigma^{T+1}} \times \exp\left[-\frac{(1-\alpha)^2}{2\sigma^2}\left\{\sum_{t=1}^{T}\left(c_t-\frac{\alpha}{1-\alpha}z_t-\frac{\beta}{1-\alpha}\right)^2\right\}\right].$$

Hence

$$p(c_{T+1}, \alpha, \beta, \sigma|\text{data}) \propto \frac{(1-\alpha)^{T+1}}{\sigma^{T+2}} \times \exp\left\{-\frac{(1-\alpha)^2}{2\sigma^2}\sum_{t=1}^{T+1}\left(c_t-\frac{\alpha}{1-\alpha}z_t+\frac{\beta}{1-\alpha}\right)^2\right\}.$$

Let $w_t = c_t-\alpha/(1-\alpha)z_t$. Then

$$p(c_{T+1}, \alpha, \beta, \sigma|\text{data}) \propto \frac{(1-\alpha)^{T+1}}{\sigma^{T+2}} \times \exp\left\{-\frac{(1-\alpha)^2}{2\sigma^2}\left(\sum_{t=1}^{T+1}\left(w_t-\frac{\beta}{1-\alpha}\right)^2\right)\right\}.$$

Now

$$\sum_{t=1}^{T+1}\left(w_t-\frac{\beta}{1-\alpha}\right)^2 = \sum_{t=1}^{T+1} w_t^2+\frac{(T+1)\beta^2}{(1-\alpha)^2}-\frac{2\beta}{(1-\alpha)}\sum_{t=1}^{T+1} w_t.$$

Hence

$$p(c_{T+1},\alpha,\beta,\sigma|\text{data}) \propto \frac{(1-\alpha)^{T+1}}{\sigma^{T+2}}$$

$$\times \exp\left[-\frac{(1-\alpha)^2}{2\sigma^2}\left\{\sum_{t=1}^{T+1} w_t^2-\frac{1}{T+1}\left(\sum_{t=1}^{T+1} w_t\right)^2\right.\right.$$

$$\left.\left.+\frac{T+1}{(1-\alpha)^2}\left[\beta-\frac{(1-\alpha)}{T+1}\left(\sum_{t=1}^{T+1} w_t\right)\right]^2\right\}\right].$$

Integrating out β, we get

$$p(c_{T+1},\alpha,\sigma|\text{data}) \propto \frac{(1-\alpha)^{T+1}}{\sigma^{T+1}}$$

$$\times \exp\left[-\frac{(1-\alpha)^2}{2\sigma^2}\left\{\sum_{t=1}^{T+1} w_t^2-\frac{1}{T+1}\left(\sum_{t=1}^{T+1} w_t\right)^2\right\}\right].$$

Integrating out σ, we get

$$p(c_{T+1},\alpha|\text{data}) \propto \frac{1-\alpha}{\left\{\sum_{t=1}^{T+1} w_t^2-\frac{1}{T+1}\left(\sum_{t=1}^{T+1} w_t\right)^2\right\}^{T/2}}.$$

Substituting for w_t,

$$p(c_{T+1},\alpha|\text{data}) \propto (1-\alpha)\bigg/\left[\sum_{t=1}^{T}\left(c_t-\frac{\alpha}{1-\alpha}z_t\right)^2\right.$$

$$-\frac{1}{T}\left(\bar{c}-\frac{\alpha}{1-\alpha}\bar{z}\right)^2$$

$$\left.+\frac{T}{T+1}\left(c_{T+1}-\frac{\alpha}{1-\alpha}z_{T+1}-\bar{c}+\frac{\alpha}{1-\alpha}\bar{z}\right)^2\right]^{T/2}, \qquad (9)$$

where $\bar{c} = (1/T)\Sigma_{t=1}^{T} c_t$ and $\bar{z} = (1/T)\Sigma_{t=1}^{T} z_t$. From (9) it is apparent that the predictive density for c_{T+1}, given α, is a univariate Student t distribution. The predictive pdf for c_{T+1} can be obtained by integrating out α using numerical methods. The predictive pdf for c_{T+1} has been computed assuming $z_{T+1} = 110$ and is shown in fig. 2. As an estimate of the consumption expenditure in period $T+1$, the mean of the predictive pdf, namely 561.49, can be used.

For a loss function of the type $L(c_{T+1}, \hat{c}_{T+1}) = (c_{T+1} - \hat{c}_{T+1})^2$, where $\hat{c}_{T+1}$ is the predictive value of c_{T+1}, the mean of the predictive pdf will be optimal in the sense of minimizing the expected loss. Also we note that the variance of the predictive pdf is 600.314 and hence the distribution has a wide spread. This indicates that c_{T+1} cannot be predicted with a great degree of precision.

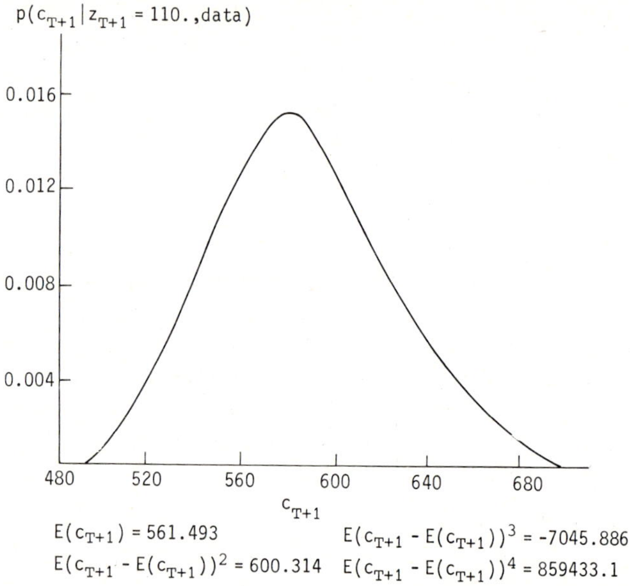

Fig. 2. Predictive density for c_{t+1} for Model I.

This distribution can be used for analyzing decision and control problems. For example, if $\bar{y}_{T+1}$ represents a target value, as defined by Tinbergen [15], for y_{T+1}, we can determine the distribution of $d_{T+1} = y_{T+1} - \bar{y}_{T+1}$ from $p(c_{T+1}, \alpha | \text{data})$ in (9), since $d_{T+1} = \bar{y}_{T+1} = c_{T+1} + z_{T+1} - \bar{y}_{T+1}$.

The joint distribution of d_{T+1} and α is obtained by substituting $d_{T+1} - z_{T+1} + \bar{y}_{T+1}$ for c_{T+1} in (9). The marginal distribution of d_{T+1} can then be obtained by integrating out α using numerical methods. Further, if we have a loss function which depends upon $y_{T+1} - \bar{y}_{T+1}$, it seems possible to determine the z_{T+1} that minimizes the mathematical expectation of the loss function.

It is interesting to compare this approach to prediction with the sampling theory approach. In the sampling theory framework, the reduced form equation for c_t, namely (3), can be used to predict the future consumption, say c_{T+1}, for a given value of z_{T+1}. The forecast which minimizes the expected value of the loss function of the type $L(c_{T+1}, \hat{c}_{T+1}) = (c_{T+1} - \hat{c}_{T+1})^2$, is $\hat{c}_{T+1} = z_{T+1}\hat{\alpha}/(1-\hat{\alpha})$, where $\hat{\alpha}$ is the least squares estimator of α in the regression (3). The sampling theory prediction for c_{T+1}, when $z_{T+1} = 110$, is 565.865 with a variance of forecast 998.56. It can be seen that in this approach, the prior information about α, namely $0 < \alpha < 1$, is not incorporated. This prior information accounts for the substantial reduction of the variance (from 999 to 600) when the Bayesian approach is used.

5. *Specification analysis of the model*

One of the assumptions we have made about the error terms is that they are serially independent. In time series econometric models, however, the error terms are often found to be serially dependent. In order to assess the effect of departures from specifying assumptions about the error terms on inferences made about the model's coefficients, we analyze the model by assuming that the error terms are generated by the first-order autoregressive process, namely,

$$u_t = \rho u_{t-1} + \varepsilon_t, \qquad -1 < \rho < 1 \qquad (t = 1, \ldots, T). \qquad (10)$$

The approach used here is similar to the one used by Zellner and Tiao [18]. It is assumed that the ε_t's are normally and independently distributed with zero means and common variance σ^2. From (1) and (10), we obtain

$$c_t = \rho c_{t-1} + \alpha(y_t - \rho y_{t-1}) + (1-\rho)\beta + \varepsilon_t \qquad (t = 1, \ldots, T). \qquad (11)$$

We assume that y_0 is fixed and known. We assume the following prior distributions: $p(\alpha, \beta, \rho, \sigma) \propto \sigma^{-1}$, $0 < \alpha < 1$, and $-1 < \rho < 1$. The likelihood function for the model is

$$l(\alpha, \beta, \rho, \sigma|\text{data}) \propto \frac{(1-\alpha)^T}{\sigma^T} \exp\left\{\frac{-1}{2\sigma^2} \times \sum_{t=1}^{T} (c_t - \rho c_{t-1} - \alpha(y_t - \rho y_{t-1}) - \beta(1-\rho))^2\right\}. \qquad (12)$$

Combining the likelihood function with the prior distribution, we get the posterior distribution of the parameters:

$$p(\alpha, \beta, \rho, \sigma | \text{data}) \propto \frac{(1-\alpha)^T}{\sigma^{T+1}}$$

$$\times \exp\left[\frac{-1}{2\sigma^2} \sum_{t=1}^{T} (c_t - \rho c_{t-1} - \alpha(y_t - \rho y_{t-1}) - \beta(1-\rho))^2\right].$$

Let $w_t = c_t - \rho c_{t-1} - \alpha(y_t - \rho y_{t-1})$ and $\bar{w} = \Sigma_{t=1}^{T} w_t/T$. Then

$$p(\alpha, \beta, \rho, \sigma | \text{data}) \propto \frac{(1-\alpha)^T}{\sigma^{T+1}} \exp\left\{\frac{-1}{2\sigma^2} \sum_{t=1}^{T} (w_t - \beta(1-\rho))^2\right\}.$$

Integrating out β,

$$p(\alpha, \rho, \sigma | \text{data}) \propto \frac{(1-\alpha)^T}{\sigma^T (1-\rho)} \exp\left[\frac{-1}{2\sigma^2}\left\{\sum_{t=1}^{T} w_t^2 - T\bar{w}^2\right\}\right].$$

Integrating out σ,

$$p(\alpha, \rho | \text{data}) \propto \frac{(1-\alpha)^T}{(1-\rho)\cdot\left[\sum_{t=1}^{T} w_t^2 - T\bar{w}\right]^{(T-1)/2}}. \tag{13}$$

The normalizing constant can be evaluated by bivariate numerical integration over α and ρ. The marginal distribution of α can be obtained by integrating out ρ numerically. The conditional distribution $p(\alpha|\rho$, data) provides inferences about α for an assumed value of ρ. If the conditional distribution is sensitive to changes in the value of ρ, then it is clear that assuming that ρ equals some fixed value—say $\rho = 0$, corresponding to the assumption that the errors are independent—could lead to a posterior distribution of α far different from the one that can be obtained from (13).

The posterior distributions of α and ρ for the data given in table 1 are shown in fig. 3. The mean of the posterior distribution of α is 0.5273 as compared with the mean of α, namely 0.6597, obtained under the assumption of zero correlation among the error terms. This difference in $E(\alpha|\text{data})$ reflects the fact that $E(\rho|\text{data})$ is very different from 0. The difference seems to be considerable in terms of the multiplier $1/(1-\alpha)$, viz., 2.118 versus 2.924. It is also seen from figs. 1 and 3 that the spread of the marginal distribution of α is sensitive to the specification about ρ. If ρ is assumed to be zero, the marginal posterior distribution of α has a variance of 0.004, as compared with a variance of 0.0105 when ρ is not assumed to be zero. The variance in the latter case is about 2.6 times the variance in the former case.

To assess the effect of departures from the specifying assumptions about ρ, we have also computed conditional distributions of α for various values of ρ which are plotted in fig. 4. The conditional posterior distribution of α, given $\rho = 0$, is similar to $p(\alpha)$ shown in fig. 1. These two are not exactly the same since $p(\alpha|\rho = 0)$ is computed using the sample of size 12 while the other is based on the sample of size 13. This figure shows that the mean and the spread of the distribution

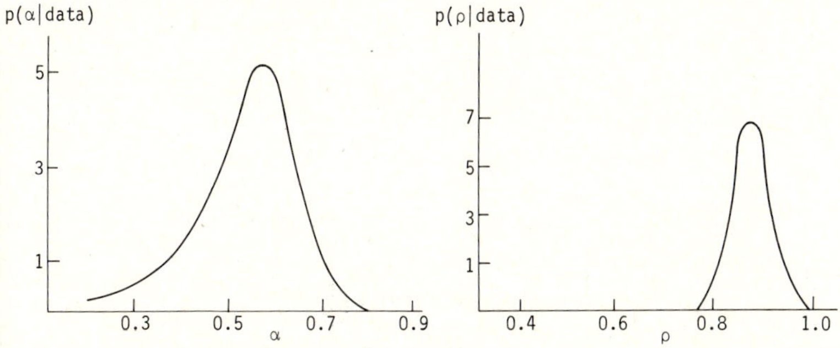

$E(\alpha) + 0.5273$ $E(\alpha - E(\alpha))^3 = -0.0014$ $E(\rho) = 0.898$ $E(\rho - E(\rho))^3 = 0$

$E(\alpha - E(\alpha))^2 = 0.0105$ $E(\alpha - E(\alpha))^4 = 0.0006$ $E(\rho - E(\rho))^2 = 0.003$ $E(\rho - E(\rho))^4 = 0.000001$

Fig. 3. Marginal posterior distributions of α and ρ for Model I.

are quite sensitive to such changes. Thus an inappropriate assumption about ρ can vitally affect the mean and the spread of the distribution.

From fig. 3, it is seen that the marginal posterior distribution of ρ is highly concentrated in the range 0.8 to 1. This clearly shows that ρ cannot be assumed to be zero. In this case, assuming that $\rho = 1$ should lead to a posterior distribution of α which is very similar to that of the marginal posterior distribution of α in fig. 3, because the posterior distribution of ρ is highly concentrated in the range 0.8 to 1. This is shown by the fact that the conditional distribution of α when $\rho = 1$ in fig. 4 is very similar to the marginal distribution of α in fig. 3. Thus, in this case, using first differences of the observations, which corresponds to the assumption $\rho = 1$, will lead to a posterior distribution of α that is similar to the marginal posterior distribution of α in fig. 3.

Another assumption that should be examined is that investment is an exogenous variable. In some theories (e.g., Schumpeter's theory of innovations), investment is considered as an exogenous variable. In other theories, the notion of induced investment is introduced; investment is assumed to be partly autonomous—its main determinants being such external factors as growth of population, new inventions,

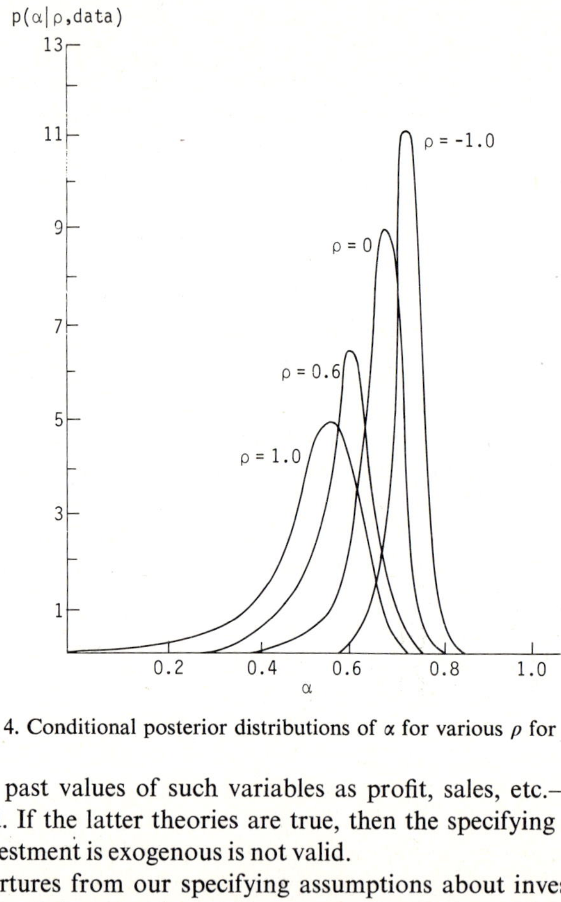

Fig. 4. Conditional posterior distributions of α for various ρ for Model I.

etc., or past values of such variables as profit, sales, etc.—and partly induced. If the latter theories are true, then the specifying assumption that investment is exogenous is not valid.

Departures from our specifying assumptions about investment may affect our inference about the parameters of the model to a great extent. This can be seen from the controversy between Friedman and Meiselman [6], Ando and Modigliani [1], and others. Friedman and Meiselman, using simple macro-models which are very similar to Haavelmo's model I, come to the conclusion that the Keynesian income-expenditures approach to the prediction of national income is almost completely useless [6, p. 187] when compared to the quantity theory approach. Ando and Modigliani and De Prano and Mayer [5] point out that one of the shortcomings of Friedman and Meiselman's approach is that investment is specified to be autonomous, whereas, in fact, it is partly induced; they contend that such shortcomings in the Friedman and Meiselman procedure make the results of their elaborate battery of tests essentially worthless.

In order to see the effects of departures from the specifying assumption about investment on our inference about the model's parameters, we turn to Haavelmo's expanded model [9]. It may be noted that the following analysis is not intended to shed any particular light on the Friedman–Meiselman controversy, but merely to illustrate the general problem.

6. *The second model*

Haavelmo [9] assumes that investment, z_t, is the difference between gross investment, x_t, and gross business savings, r_t. Gross investment here is defined as gross private capital formation plus government net deficit. This part of investment is exogenous. Gross business savings are defined as the sum of depreciation and depletion charges, capital outlay charged to current expenses, income credited to other business reserves, and corporate savings minus revaluation of business inventories. Hence $z_t = x_t - r_t$, $t = 1 \ldots T$. The gross business savings is assumed to be a function of what might be called gross disposable income, $c_t + x_t$. Hence the expanded model is:

$$c_t = \alpha y_t + \beta + u_t, \tag{14}$$

$$r_t = \mu(c_t + x_t) + \nu + v_t, \tag{15}$$

$$y_t = c_t + x_t - r_t,$$

$$x_t \text{ is an autonomous variable} \qquad (t = 1, 2, \ldots, T). \tag{16}$$

The variables c, y, r, x are all as defined before, u_t and v_t are structural disturbance terms, and α, β, μ, ν are structural coefficients. We note that if $\mu = 0$ and $\sigma_{uv} = Eu_t v_t = 0$, the model in (14)–(16) reduces to the first model (1)–(2). When $\mu = 1$ the model reduces to one in which income is measured with error and has constant mean. We shall investigate how sensitive our inferences are to departures from our assumptions, namely $\mu = 0$ and $\sigma_{uv} = 0$.

Regarding the error terms, we assume that $E(u_t) = E(v_t) = 0$, $E(u_t^2) = \sigma_u^2$, $E(v_t^2) = \sigma_v^2$, $E(v_t v_{t-\tau}) = E(u_t u_{t-\tau}) = 0$ for $\tau \neq 0$, and $E(u_t v_t) = \sigma_{uv}$. We do not assume that u_t and v_t are independent.

The likelihood function for the model's parameters is:

$$l(\alpha, \beta, \mu, \nu, \sigma_v^2, \sigma_u^2, \sigma_{uv}) \propto |\Sigma^{-1}|^{T/2} |1-\alpha(1-\mu)|^T \times \exp\{-\tfrac{1}{2}\operatorname{tr} V'V\Sigma^{-1}\} \tag{17}$$

where $V = (u, v)$, a $T \times 2$ matrix of disturbance terms and

$$\Sigma = \begin{bmatrix} \sigma_u^2 & \sigma_{uv} \\ \sigma_{uv} & \sigma_v^2 \end{bmatrix}.$$

Regarding the prior distributions for the parameters, we assume that

$$p(\alpha, \beta, \mu, \nu, \Sigma^{-1}) \propto |\Sigma^{-1}|^{3/2}, \quad 0 < \alpha < 1, \quad 0 < \mu < 1.$$ [2]

On combining the prior pdf's with the likelihood function in (17), we obtain the following pdf for the model's parameters:

$$p(\alpha, \beta, \mu, \nu, \Sigma^{-1}|\text{data}) \propto |\Sigma^{-1}|^{(T-3)/2}(1-\alpha(1-\mu))^T \times \exp\{-\tfrac{1}{2}\operatorname{tr} V'V\Sigma^{-1}\}. \tag{18}$$

Integrating with respect to Σ^{-1}, using the properties of Wishart's distribution, we obtain the posterior pdf for the structural coefficients:

$$p(\alpha, \beta, \mu, \nu|\text{data}) \propto \frac{(1-\alpha(1-\mu))^T}{|V'V|^{T/2}}. \tag{19}$$

Except for the factor $\{1-\alpha(1-\mu)\}^T$, the form of (19) is completely analogous to the posterior distribution of coefficients in a multivariate regression model analysed by Tiao and Zellner [13]. On integrating out β and ν, the posterior pdf for α and μ is found to be

$$p(\alpha, \mu|\text{data}) \propto \frac{\{1-\alpha(1-\mu)\}^T}{|\tilde{V}'\tilde{V}|^{(T-1)/2}} \tag{20}$$

where $\tilde{V} = (\tilde{v}_1, \tilde{v}_2)$ is a $T \times 2$ matrix in which $\tilde{v}_1$ and $\tilde{v}_2$ have typical elements given by

$$\tilde{v}_{1t} = c_t - \bar{c} - \alpha[c_t - \bar{c} + x_t - \bar{x} - (r_t - \bar{r})]$$

[2] Geisser [7] has pointed out that this prior distribution for Σ^{-1} may be justified by various rules, e.g., invariance, conjugate families, stable estimation, etc.

and

$$\tilde{v}_{2t} = r_t - \bar{r} - \mu[c_t - \bar{c} + x_t - \bar{x}],$$

respectively.

It can be seen that if the covariance between the contemporaneous error terms, namely, $\sigma_{uv} = E(u_t v_t)$, $t = 1, \ldots, T$, is assumed to be zero, the posterior distribution of α and μ reduces to

$$p(\alpha, \mu | \sigma_{uv} = 0, \text{data}) \propto \frac{[1-\alpha(1-\mu)]^T}{[Ts_2^2 + (\alpha-\hat{\alpha})^2 M_1]^{(T-1)/2} [Ts_1^2 + (\mu-\hat{\mu})^2 M_2]^{(T-1)/2}} \quad (21)$$

where

$$Ts_2^2 = \sum_{t=1}^{T} (c_t - \hat{\alpha} y_t)^2 - \frac{1}{T}\left\{\sum_{t=1}^{T} (c_t - \hat{\alpha} y_t)\right\}^2,$$

$$Ts_1^2 = \sum_{t=1}^{T} (r_t - \hat{\mu}(c_t + x_t))^2 - \frac{1}{T}\left\{\sum_{t=1}^{T} (r_t - \hat{\mu}(c_t + x_t))\right\}^2,$$

$$M_1 = \sum_{t=1}^{T} (c_t + x_t)^2 - \frac{1}{T}\left\{\sum_{t=1}^{T} (c_t + x_t)\right\}^2,$$

$$M_2 = \sum_{t=1}^{T} y_t^2 - \frac{1}{T}\left(\sum_{t=1}^{T} y_t\right)^2,$$

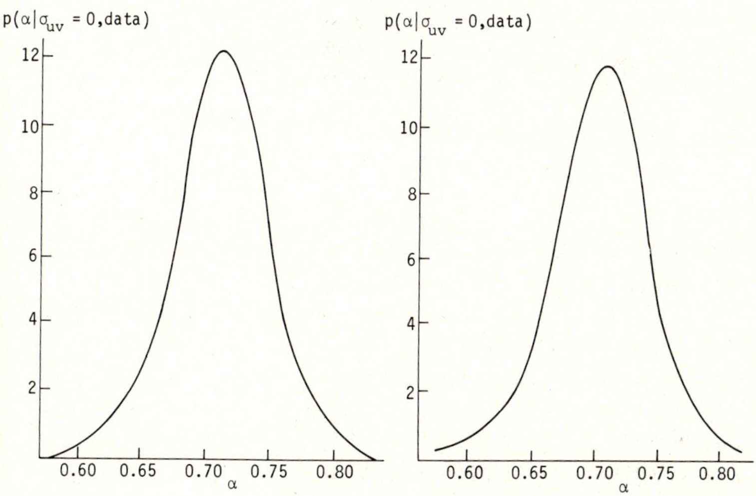

$E(\alpha) = 0.7123$ $E(\alpha - E(\alpha))^3 = 0$ $E(\alpha) = 0.7016$ $E(\alpha - E(\alpha))^3 = 0$

$E(\alpha - E(\alpha))^2 = 0.0011$ $E(\alpha - E(\alpha))^4 = 0.000014$ $E(\alpha - E(\alpha))^2 = 0.0013$ $E(\alpha - E(\alpha))^4 = 0.000016$

Fig. 5. Marginal posterior distributions of α for Model II.

and $\hat{\alpha}$, $\hat{\mu}$ are as defined before. From the posterior joint distributions of α and μ, we can compute the conditional posterior pdf for α, the marginal propensity to consume, for given values of μ, and these pdf's can be used to find out how sensitive our inferences about α are to what is assumed about investment's dependence on income in (15).

Using the data for the variables given in table 1, the posterior distributions in (20) and (21) are analyzed using bivariate numerical

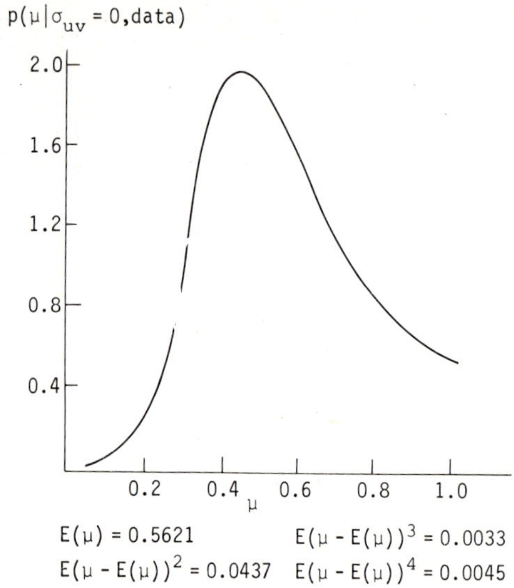

Fig. 6. Marginal posterior distribution of μ for Model II.

integration techniques. The marginal distributions of α and μ are shown in figs. 5, 6, and 7 for the two cases $\sigma_{uv} \neq 0$ and $\sigma_{uv} = 0$.

It is interesting to compare the marginal posterior distribution of α obtained from the second model with that of the first model. As can be observed from figs. 1 and 5, the mean of the distribution α for the first model is a little less than that obtained for the second model. There is considerable difference among the dispersions of the distributions of α obtained for the two models. The variance of the marginal posterior distribution of α obtained from the first model is about three times the variance of the distributions of α that is obtained from the second model.

As can be seen from figs. 6 and 7, the posterior distribution of μ is very sensitive to the specification of σ_{uv}. In the case when $\sigma_{uv} \neq 0$, distribution of μ is highly concentrated around $\mu = 16$. The distribution of α does not appear to be very sensitive to the specification of σ_{uv}. In fig. 8, the conditional distributions of α for various values of μ

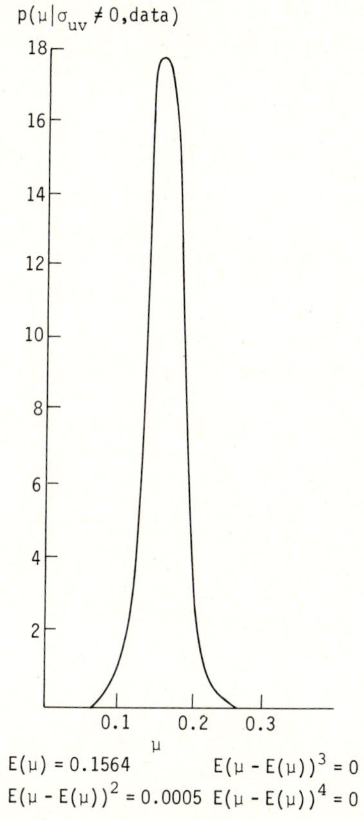

Fig. 7. Marginal posterior distribution of μ for Model II.

are shown, when σ_{uv} is assumed to be zero. It can be seen that these conditional distributions are somewhat sensitive to the specification of μ. The conditional distributions of α for various values of μ, in the case when σ_{uv} is not assumed to be zero, are shown in fig. 9. It can be easily seen that both the mean and the spread of the distribution

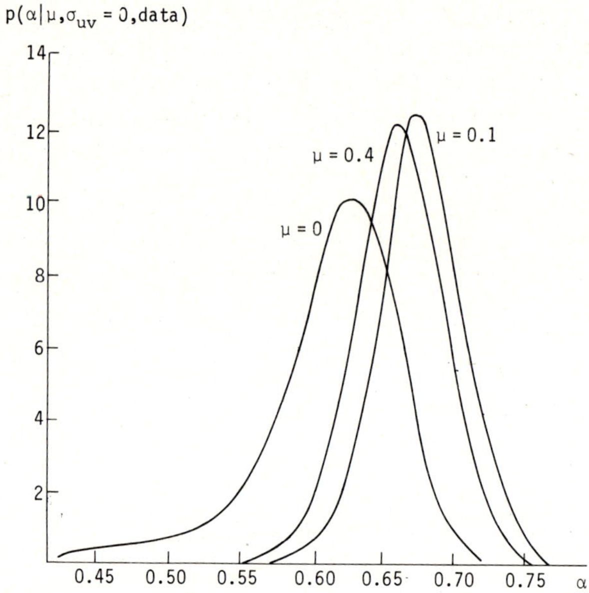

Fig. 8. Conditional posterior distributions of α for various μ for Model II when $\sigma_{uv} = 0$.

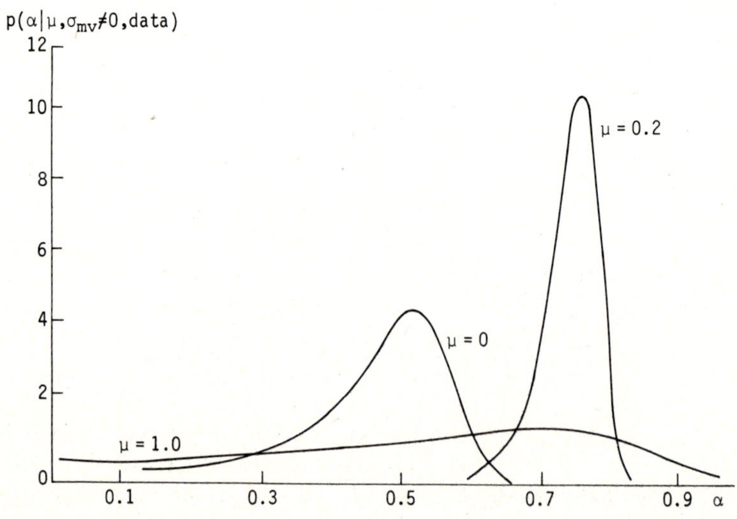

Fig. 9. Conditional posterior distributions of α for various μ for Model II when $\sigma_{uv} \neq 0$.

change remarkably and hence the posterior distribution of α is highly sensitive to the specification of μ. This means that our inference about α, the marginal propensity to consume, is highly dependent upon our assumption about the investment variable. The contours of the joint density function for the cases $\sigma_{uv} = 0$ and $\sigma_{uv} \neq 0$ are shown in figs. 10 and 11, respectively.

Our analysis clearly shows that specifications about the nature of the variables (namely endogenous or exogenous) and about the correlation between the contemporaneous error terms condition our inferences about the parameters of the model to a remarkable extent. Our analysis also shows how Bayesian methods can be conveniently used to assess the effects of departures from specifying assumptions.

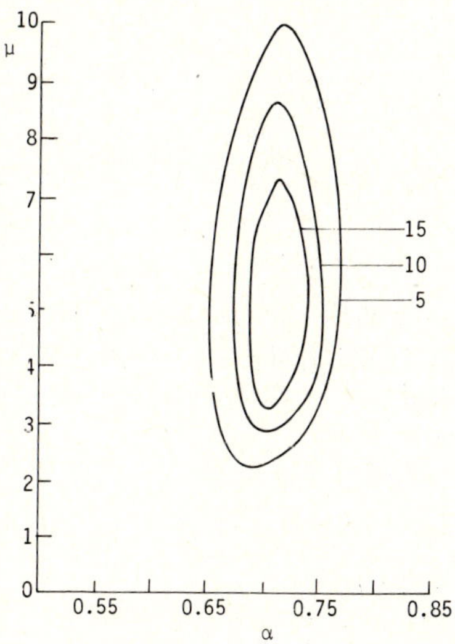

Fig. 10. Contours of the joint posterior distribution of α and μ for Model II when $\sigma_{uv} = 0$.

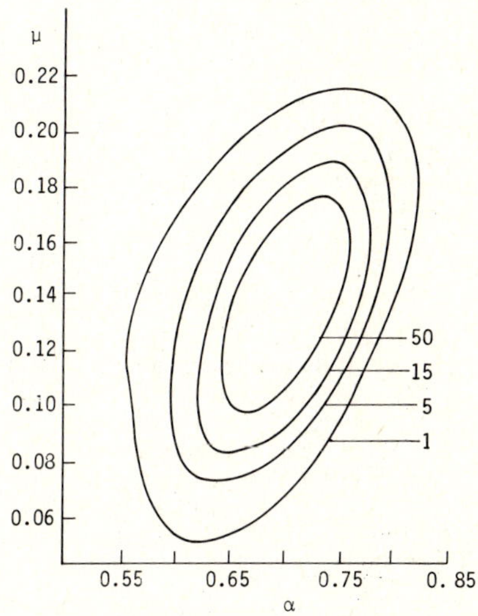

Fig. 11. Contours of the joint posterior distribution of α and μ for Model II when $\sigma_{uv} \neq 0$.

7. *Informative prior distributions*

The choice between the two models we discussed depends upon the investigator's belief about the value of μ, that is, whether he regards investment to be exogenous or not. In order to investigate how the prior information about μ is modified by the sample information, different prior distributions are used for μ and the posterior distributions of μ are obtained. One prior distribution used for μ is $p(\mu) \propto \mu^{0.5}(1-\mu)^{27.5}$, $0 < \mu < 1$, which has mean 0.05 and has very small

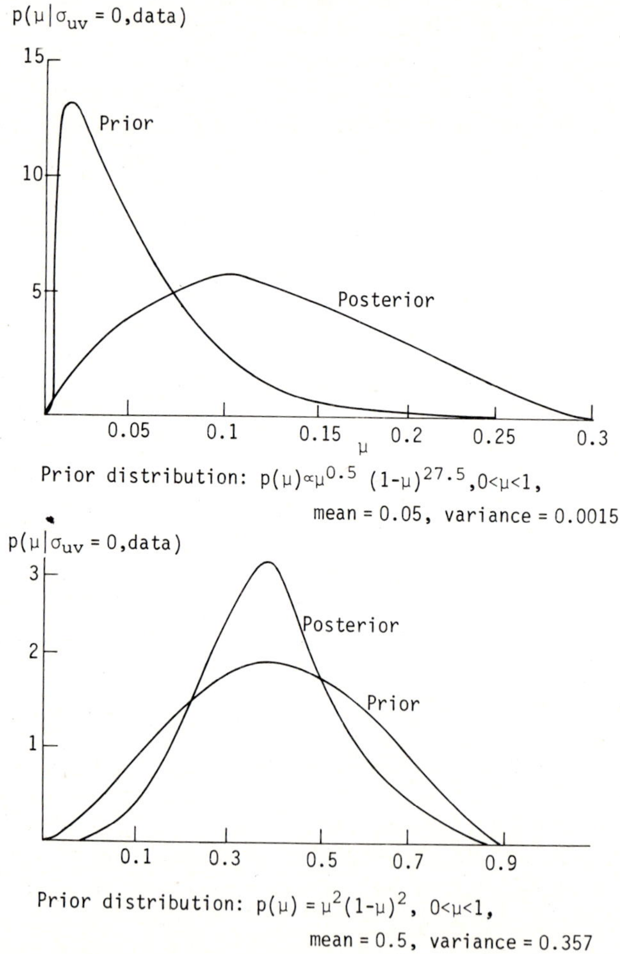

Fig. 12. Posterior distributions of μ for different priors when $\sigma_{uv} = 0$.

probabilities beyond $\mu = 0.15$. This can represent the prior knowledge of an investigator who strongly believes that investment is exogenous. Another prior distribution used for μ is $p(\mu) \propto \mu^2(1-\mu)^2$, $0 < \mu < 1$, which has mean 0.5 and is symmetrically distributed around the mean in the interval (0,1). The posterior distributions of μ are shown in figs. 12 and 13 for the cases $\sigma_{uv} = 0$ and $\sigma_{uv} \neq 0$ respectively. Also it is to be noted that the marginal posterior distributions of μ, when μ is assumed a priori to be distributed uniformly in the interval (0,1), are shown in figs. 6 and 7, for the two specifications about σ_{uv}.

In the case when σ_{uv} is assumed to be zero, it is seen from figs. 6 and 12 that the investigator's belief that μ is nearly zero is modified by the sample to some extent. In the case when σ_{uv} is not assumed to be zero, it can be seen from figs. 7 and 13 that the sample modifies the prior

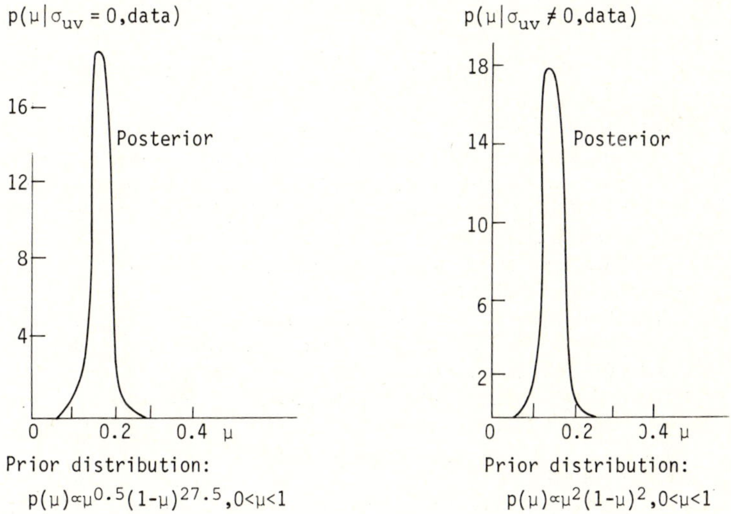

Fig. 13. Posterior distributions of μ for different priors when $\sigma_{uv} \neq 0$.

belief very markedly. The posterior distribution of μ is very sharp and is concentrated around $\mu = 0.156$, irrespective of the nature of the prior distribution.

Appendix

In order to integrate a function, say $y = f(x)$ numerically, Simpson's rule was used. To use this rule, the interval of integration was subdivided into 100 subintervals, of equal length, h, using 101 points.

The value of the function at each one of the points was evaluated. If these values are $y_0, y_1, \ldots, y_{100}$, the value of the integral is given by

$$A = (\tfrac{1}{3})h\{y_0 + y_{100} + 4(y_1 + y_3 + \ldots y_{99}) \\ + 2(y_2 + y_4 + \ldots y_{98})\}.$$

When the function to be integrated, the range of integration, and the number of subintervals are specified, the program evaluates the y_i and computes A. Also the program prints out y_i/A, the ordinates of the normalized density function, and x_i and its first four moments about the mean. For evaluating a double integral, this rule was used twice.

References

[1] Albert Ando and Franco Modigliani, The relative stability of monetary velocity and the investment multiplier. American Economic Review 55 (1965), 693–728.

[2] A. R. Bergstrom, The exact sampling distributions of least squares and maximum likelihood estimators of the marginal propensity to consume. Econometrica 30 (1962), 480–490.

[3] G. E. P. Box and G. C. Tiao, A further look at robustness via Bayes' theorem. Biometrika 49 (1962), 419–433.

[4] V. K. Chetty, Bayesian analysis of some simultaneous equation models and specification errors. Unpublished doctoral dissertation submitted to the University of Wisconsin, 1966.

[5] Michael De Prano and Thomas Mayer, Tests of the relative importance of autonomous expenditures and money. American Economic Review 55 (1965), 729–752.

[6] Milton Friedman and D. Meiselman, The relative stability of monetary velocity and the investment multiplier in the United States, 1897–1958. In: Commission on Money and Credit, Stabilization Policies. Englewood Cliffs, 1963, 165–288.

[7] S. Geisser, Bayesian estimation in multivariate analysis. University of Wisconsin Mathematics Research Center Report No. 503, published in the Annals of Mathematics and Statistics 60 (1965), 150–159.

[8] S. Geisser and J. A. Cornfield, Posterior distributions for multivariate normal parameters. Journal of the Royal Statistical Society, Series B, 25 (1963), 368–376.

[9] T. Haavelmo, Methods of measuring the marginal propensity to consume. Journal of American Statistical Association 42 (1947), 105–122. Reprinted as Ch. 4 in: Wm. C. Hood and T. C. Koopman's Studies in econometric method. New York, John Wiley, 1953.

[10] H. Jeffreys, Theory of Probability, 3rd ed. Oxford, Clarendon Press, 1961.

[11] T. J. Rothenberg, A Bayesian analysis of simultaneous equation systems. Report 6315, Econometric Institute, Netherlands School of Economics, Rotterdam, 1963.

[12] L. J. Savage, The subjective basis of statistical practice. Manuscript, University of Michigan, 1961.

[13] G. C. Tiao and A. Zellner, On the Bayesian estimation of multivariate regression. Journal of the Royal Statistical Society, Series B, 26 (1964), 277–285.

[14] G. C. Tiao and A. Zellner, Bayes' theorem and the use of prior knowledge in regression analysis. Biometrika 51 (1964), 219–230.

[15] J. Tinbergen, Economic policy: principles and design. Amsterdam, North-Holland, 1956.

[16] A Zellner, Bayesian inference and simultaneous equation econometric models. Paper presented to the First World Congress of the Econometric Society, Rome, September 9–14, 1965.

[17] A. Zellner and V. K. Chetty, Prediction and decision problems in regression models from the Bayesian point of view. Systems Formulation and Methodology Workshop Paper 6403, University of Wisconsin, 1964. Published in Journal of the American Statistical Association 60 (1965), 608–616.

[18] A. Zellner and G. C. Tiao, Bayesian analysis of the regression model with autocorrelated errors. Journal of the American Statistical Association 59 (1964), 763–778.

S. E. Fienberg and A. Zellner (eds.)
Studies in Bayesian Econometrics and Statistics.

9.2 POSTERIOR INFERENCE FOR STRUCTURAL PARAMETERS USING CROSS-SECTION AND TIME SERIES DATA*

G. M. KAUFMAN**

Massachusetts Institute of Technology

1. Introduction

Our purpose is to discuss features of an explicit Bayesian analysis of a simultaneous equation system when both time series and cross-section data are available. The process generating time series data we shall assume to be the usual set of stochastic equations

$$\boldsymbol{B}\tilde{\mathbf{y}}^{(j)}+\boldsymbol{\Gamma}\mathbf{z}^{(j)} = \tilde{\mathbf{u}}^{(j)} \qquad j = 1, 2, \ldots, \tag{1}$$

where $\boldsymbol{B}(m \times m)$ non-singular and $\boldsymbol{\Gamma}(m \times r)$ are coefficient matrices, fixed for all j, $\mathbf{z}^{(j)}$ $(r \times 1)$ is a vector of predetermined observable variables and $\tilde{\mathbf{y}}^{(j)}$ and $\tilde{\mathbf{u}}^{(j)}$ $(m \times 1)$ are random vectors. In particular we assume that $\{\tilde{\mathbf{u}}^{(j)}, j = 1, 2, \ldots\}$ is a sequence of mutually independent identically distributed normal random vectors with mean $\mathbf{0}$ and variance matrices $\boldsymbol{\Sigma}$, and that neither $\boldsymbol{B}$, $\boldsymbol{\Gamma}$, nor $\boldsymbol{\Sigma}$ are known with certainty. The system may be re-expressed in *reduced form*

$$\tilde{\mathbf{y}}^{(j)} = \boldsymbol{\Pi}\mathbf{z}^{(j)}+\tilde{\varepsilon}^{(j)} \qquad j = 1, 2, \ldots, \tag{2}$$

when $\boldsymbol{B}$ is non-singular, where $\boldsymbol{\Pi} \equiv -\boldsymbol{B}^{-1}\boldsymbol{\Gamma}$ and $\tilde{\varepsilon}^{(j)} = \boldsymbol{B}^{-1}\tilde{\mathbf{u}}^{(j)}$. Defining $\boldsymbol{Y} = [\mathbf{y}^{(1)}, \ldots, \mathbf{y}^{(N)}]$, $\boldsymbol{Z} = [\mathbf{z}^{(1)}, \ldots, \mathbf{z}^{(N)}]$, and

* Paper written for the Session on Bayesian Analysis of Simultaneous Equations Models, World Congress of the Econometric Society, Cambridge, England, September 1970.

** I wish to express my thanks to Abba Krieger for doing the numerical calculations in section 2. This research was supported in part by a grant from the Cambridge Project.

$$M = \begin{bmatrix} YY^t & YZ^t \\ ZY^t & ZZ^t \end{bmatrix},$$

the likelihood function for B, Γ, and Σ is

$$|BB^t|^{\frac{1}{2}N} \exp\{-\tfrac{1}{2} \operatorname{tr} \Sigma^{-1}[B\,\Gamma]M[B\,\Gamma]^t\}|\Sigma|^{-\frac{1}{2}N}. \tag{3}$$

The identification problem is essentially one of determining structural parameters from those of the reduced form. Its Bayesian counterpart is that time series observations generated according to (1) do not alter a posteriori the prior conditional density of $\tilde{B}$ given reduced form parameters. Cross-section data bearing directly on elements of B and Γ clearly will.

Here we shall explore properties of $\tilde{B}$ and $\tilde{\Gamma}$ posterior to observing cross-section data bearing on individual rows of $(\tilde{B}\,\tilde{\Gamma})$ under two differing sets of assumptions. The first is: observations bearing on row i of (B, Γ), $(\mathbf{B}_i\,\mathbf{\Gamma}_i)$, are generated according to an independent normal regression process with residual error variance v_i, v_i a nuisance parameter unrelated to structural parameters. This process is independent of the process generating time series data; in addition, the processes generating observations bearing on individual rows $(B\,\Gamma)$ are mutually independent. The second assumption varies from the first only in that the regression process generating cross-section data bearing on $(\mathbf{B}_i\,\mathbf{\Gamma}_i)$ has residual error variance proportional to σ_{ii}, the ith diagonal element of Σ.

While we shall usually assume that the prior assigned to structural parameters $\tilde{B}$, $\tilde{\Gamma}$, and $\tilde{\Sigma}$ or alternately to reduced form parameters $\tilde{\Omega} = \tilde{B}^{-1}\tilde{\Sigma}(\tilde{B}^t)^{-1}$ and $\tilde{\Pi}$, is non-informative, the functional form of the posterior density of $\tilde{B}$, $\tilde{\Gamma}$ and $\tilde{\Sigma}$ will be precisely the same as if we had assigned a prior natural conjugate to the time series process, to the cross-section process, or to both, the only difference appearing in the values of parameters of the posterior. We leave the modifications so induced as an exercise for the reader.

Under the assumption that cross-section residual error variances are not related to the time series variance matrix Σ, the marginal density of $(\tilde{B}\,\tilde{\Gamma})$ has components in the form of a product of (multivariate) Student kernels—*poly-t* kernels. Posterior poly-*t* densities have been studied by Tiao and Zellner [7], Dickey [3, 4], and appear also in Chetty [2]. We show that under certain circumstances the marginal posterior density of the reduced form parameter $\tilde{\Pi}$ is proportional to a ratio of *matric* poly-*t* densities. This ratio bears an interesting relationship to the posterior density for a single row of $\tilde{B}$ that Drèze [5] uses to derive a Bayesian version of a LIMLE estimator when cross-section residual error variances are proportional to corresponding time series error variances.

Using the same data as Drèze [5], we compute the posterior mode of parameters in the demand equation of Tintner's model of supply and demand for meat, poultry, and fish. While purely illustrative in nature, the computations indicate that the procedure is surprisingly robust given that there is a substantive difference in specification of the model for cross-section data used by Drèze and that used in our calculations.

The assumption that cross-section residual error variances are proportional to corresponding σ_{ii}'s leads to a posterior density for $(\tilde{B}\,\tilde{\Gamma})$ unconditional as regards $\tilde{\Sigma}$ that is extremely complicated. We present some results for the special case $m = 2$, and give an asymptotic expansion of this density.

2. *Time series and cross-section variances unrelated*

When time series and cross-section variances, σ_{ii} and υ_i respectively, are unrelated, observation of cross-section data under the assumptions outlined earlier leads to a likelihood function for B, Γ, and the υ_i's of the form

$$\prod_{i=1}^{m} \upsilon_i^{-\frac{1}{2}n_i} \exp\{-\tfrac{1}{2}\{[(\mathbf{B}_i\mathbf{\Gamma}_i)-(\overline{\mathbf{B}}_i\overline{\mathbf{\Gamma}}_i)]\boldsymbol{P}_i \\ [(\mathbf{B}_i\mathbf{\Gamma}_i)-(\overline{\mathbf{B}}_i\overline{\mathbf{\Gamma}}_i)]^t + S_i\}/\upsilon_i\}. \tag{4}$$

Here υ_i is a nuisance parameter and $\{n_i, S_i, \boldsymbol{P}_i, (\overline{\mathbf{B}}_i\overline{\mathbf{\Gamma}}_i)\}$ a set of statistics sufficient for inference about $(\mathbf{B}_i\mathbf{\Gamma}_i)$ and υ_i.

If in addition, time-series data generated according to (1) are observed, the likelihood function for B, Γ, Σ and the υ_i's is of the form (4) times (3).

Assigning a non-informative prior $|BB^t|^{-\frac{1}{2}m}|\Omega|^{-\frac{1}{2}(m+1)}\,\mathrm{d}\Omega\,\mathrm{d}B\,\mathrm{d}\Pi$ to $\tilde{B}$ and the reduced form parameters $\tilde{\Pi}$ and $\tilde{\Omega}$, induces a prior

$$|BB^t|^{-\frac{1}{2}(m+r)}|\Sigma|^{-\frac{1}{2}(m+1)}\,\mathrm{d}B\,\mathrm{d}\Sigma\,\mathrm{d}\Gamma \tag{5}$$

for $\tilde{B}$, $\tilde{\Gamma}$ and $\tilde{\Sigma}$. Assignment of a uniform prior $\mathrm{d}B$ to $\tilde{B}$ in place of a non-informative prior leads to a substantial computational simplification: the determinant in $\boldsymbol{B}$ "disappears" under certain conditions and we shall examine the implications of this modification. With no increase in computational complexity, we can as asserted earlier assign independent normal-gamma priors to the $(\tilde{\mathbf{B}}_i\tilde{\mathbf{\Gamma}}_i)$, $\tilde{v}_i$'s and/or a (degenerate) prior of the form

$$|\boldsymbol{BB}^t|^{-\frac{1}{2}(m+r)}|\boldsymbol{\Sigma}|^{-\frac{1}{2}(m+1)} \exp\{-\tfrac{1}{2}\operatorname{tr}\boldsymbol{\Sigma}^{-1}[\boldsymbol{B\Gamma}]\boldsymbol{M}'[\boldsymbol{B\Gamma}]^t\}. \quad (6)$$

Upon assigning a prior proportional to

$$|\boldsymbol{BB}^t|^{-\frac{1}{2}(m+r)}|\boldsymbol{\Sigma}|^{-\frac{1}{2}(m+1)}v_1^{-1}\ldots v_m^{-1} \quad (7)$$

to $\tilde{\boldsymbol{B}}$, $\tilde{\boldsymbol{\Sigma}}$, $\tilde{\boldsymbol{\Gamma}}$ and the $\tilde{v}_i$'s, and then integrating the product of (7) and the likelihood (4) times (3) over[1] $\boldsymbol{\Sigma} > \boldsymbol{0}$ and $v_i \in (0, \infty)$, $i = 1, 2, \ldots m$ yields a posterior density for $\tilde{\boldsymbol{B}}$ and $\tilde{\boldsymbol{\Gamma}}$ unconditional as regards $\tilde{\boldsymbol{\Sigma}}$ and the $\tilde{v}_i$'s proportional to:

$$|\boldsymbol{BB}^t|^{\frac{1}{2}(N-m-r)}|[\boldsymbol{B\Gamma}]\boldsymbol{M}[\boldsymbol{B\Gamma}]^t|^{-\frac{1}{2}N}\prod_{i=1}^{m}\{[(\mathbf{B}_i\mathbf{\Gamma}_i)-(\bar{\mathbf{B}}_i\bar{\mathbf{\Gamma}}_i)]\boldsymbol{P}_i \times[(\mathbf{B}_i\mathbf{\Gamma}_i)-(\bar{\mathbf{B}}_i\bar{\mathbf{\Gamma}}_i)]+S_i\}^{-\frac{1}{2}n_i}. \quad (8)$$

From a Bayesian standpoint, a "full information" analysis bearing on, say, the first row of $(\tilde{\boldsymbol{B}}\tilde{\boldsymbol{\Gamma}})$ requires use of *all* of (8); i.e. we must compute the marginal density of $(\tilde{\mathbf{B}}_1\tilde{\mathbf{\Gamma}}_1)$ by integrating (8) over the range set of all parameters in $(\boldsymbol{B\,\Gamma})$ save $(\mathbf{B}_1\mathbf{\Gamma}_1)$. The resulting posterior density takes into account all information generated by the prior as well as all time series and cross-section data. This integration appears difficult, and a useful representation of the resulting density remains to be found. However, partitioning

$$[\boldsymbol{B\,\Gamma}] = \begin{bmatrix}\mathbf{B}_1 & \mathbf{\Gamma}_1\\ \boldsymbol{B}_2 & \boldsymbol{\Gamma}_2\end{bmatrix},$$

$$|[\boldsymbol{B\,\Gamma}]\boldsymbol{M}[\boldsymbol{B\,\Gamma}]^t| = (\mathbf{B}_1\mathbf{\Gamma}_1)\boldsymbol{M}(\mathbf{B}_1\mathbf{\Gamma}_1)^t\cdot|[\boldsymbol{B}_2\boldsymbol{\Gamma}_2]\boldsymbol{E}[\boldsymbol{B}_2\boldsymbol{\Gamma}_2]^t|$$

where $\boldsymbol{E} = \boldsymbol{M}-\boldsymbol{M}(\mathbf{B}_1\mathbf{\Gamma}_1)^t\{(\mathbf{B}_1\mathbf{\Gamma}_1)\boldsymbol{M}(\mathbf{B}_1\mathbf{\Gamma}_1)^t\}^{-1}(\mathbf{B}_1\mathbf{\Gamma}_1)\boldsymbol{M}$. Upon substituting, we see that the density is

$$\{(\mathbf{B}_1\mathbf{\Gamma}_1)\boldsymbol{M}(\mathbf{B}_1\mathbf{\Gamma}_1)\}^{-\frac{1}{2}N}\{[(\mathbf{B}_1\mathbf{\Gamma}_1)-(\bar{\mathbf{B}}_1\bar{\mathbf{\Gamma}}_1)]\boldsymbol{P}_1 \times[(\mathbf{B}_1\mathbf{\Gamma}_1)-(\bar{\mathbf{B}}_1\bar{\mathbf{\Gamma}}_1)]^t+S_1\}^{-\frac{1}{2}n_1} \quad (9a)$$

[1] By $\boldsymbol{\Sigma} > \boldsymbol{0}$ we mean the set $\{\boldsymbol{\Sigma}|\boldsymbol{\Sigma}$ is positive definite symmetric$\}$ or alternately that $\boldsymbol{\Sigma}$ is positive definite symmetric.

times

$$|BB^t|^{\frac{1}{2}(N-m-r)}|[B_2\Gamma_2]E[B_2\Gamma_2]^t|^{-\frac{1}{2}N}$$

$$\times \prod_{i=2}^{m} \{[(\mathbf{B}_i\mathbf{\Gamma}_i)-(\bar{\mathbf{B}}_i\bar{\mathbf{\Gamma}}_i)]P_i[(\mathbf{B}_i\mathbf{\Gamma}_i)-(\bar{\mathbf{B}}_i\bar{\mathbf{\Gamma}}_i)]^t+S_i\}^{-\frac{1}{2}n_i}. \tag{9b}$$

Thus the posterior density of $(\tilde{B}_2\tilde{\Gamma}_2)$ given $(\tilde{\mathbf{B}}_1\tilde{\mathbf{\Gamma}}_1) = (\mathbf{B}_1\mathbf{\Gamma}_1)$ is, aside from the complication introduced by the first row of B being fixed, is of the same functional form as that of $(\tilde{B}\,\tilde{\Gamma})$, while (9a) is the kernel[2] of a (degenerate) *poly-t* density (cf. Dickey [3], Tiao and Zellner [7]). Replication of this decomposition $m-2$ times allows us to write the kernel (9) as $|BB^t|^{\frac{1}{2}(N-m-r)}$ times a marginal poly-t kernel in $(\mathbf{B}_1\mathbf{\Gamma}_1)$ and $(m-1)$ conditional poly-t kernels in rows $(\mathbf{B}_i\mathbf{\Gamma}_i)$, $i = 2, \ldots, m$.
If we transform from (B, Γ) to (B, Π), we find that $(\tilde{B}, \tilde{\Pi})$ has a posterior kernel

$$|BB^t|^{-\frac{1}{2}m}|[\Pi-M_{12}M_{22}^{-1}]M_{22}[\Pi-M_{12}M_{22}^{-1}]+M_{11.2}|^{-\frac{1}{2}N}$$

$$\times \prod_{i=1}^{m} \{[\mathbf{B}_i(I(-\Pi))-(\bar{\mathbf{B}}_i\bar{\mathbf{\Gamma}}_i)]P_i[\mathbf{B}_i(I(-\Pi))$$

$$-(\bar{\mathbf{B}}_i\bar{\mathbf{\Gamma}}_i)]^t+S_i\}^{-\frac{1}{2}n_i}. \tag{10}$$

We may complete the square in the $\mathbf{B}_i$ and write terms in them as

$$\{[\mathbf{B}_i-\bar{\boldsymbol{\beta}}_i(\Pi)]V_i(\Pi)[\mathbf{B}_i-\bar{\boldsymbol{\beta}}_i(\Pi)]^t+S_i^0(\Pi)\}^{-\frac{1}{2}n_i} \tag{11}$$

where

$$V_i(\Pi) \equiv [I(-\Pi)]P_i[I(-\Pi)]^t$$

$$\bar{\boldsymbol{\beta}}_i(\Pi) \equiv [\bar{\mathbf{B}}_i\bar{\mathbf{\Gamma}}_i]P_i[I(-\Pi)]^tV_i^{-1}(\Pi)$$

and

$$S_i^0(\Pi) = S_i+(\bar{\mathbf{B}}_i\bar{\mathbf{\Gamma}}_i)P_i(\bar{\mathbf{B}}_i\bar{\mathbf{\Gamma}}_i)^t-\bar{\boldsymbol{\beta}}_i(\Pi)V_i(\Pi)\bar{\boldsymbol{\beta}}_i^t(\Pi).$$

Aside from the term $|BB^t|^{-\frac{1}{2}m}$ arising via the non-informative prior assigned to $\tilde{B}$, (11) shows that the distribution of $\tilde{B}$ given $\tilde{\Pi} = \Pi$ is proportional to a product of (conditionally) independent Student densities in the rows $\mathbf{B}_i$ of B.

A slight alteration of the prior (7) will in fact lead to $\tilde{B}$ given $\tilde{\Pi} = \Pi$ having rows that *are* conditionally independent and Student:

[2] We use "kernel" in the sense defined by Raiffa and Schlaifer [6, ch. 2]; i.e., that portion of a density or likelihood depending on a parameter not known with certainty.

assign a uniform prior $\mathrm{d}\boldsymbol{B}$ to $\tilde{\boldsymbol{B}}$ in place of a non-informative prior or alternatively assign a prior $|\Sigma|^{-\frac{1}{2}(2m+1)}$ to $\tilde{\Sigma}$ in place of $|\Sigma|^{-\frac{1}{2}(m+1)}$. (This latter assignment changes N in (8) to $N-m$.) Then given $\tilde{\boldsymbol{\Pi}} = \boldsymbol{\Pi}$, the rows $\tilde{\mathbf{B}}_i$ of $\tilde{\boldsymbol{B}}$ are conditionally independent and Student with $\tilde{\mathbf{B}}_i$ having mean $\bar{\boldsymbol{\beta}}_i(\boldsymbol{\Pi})$ and variance $\{S_i^0(\boldsymbol{\Pi})/(n_i-m-2)\}V_i^{-1}(\boldsymbol{\Pi})$ when $n_i-m > 2$.

Adopting a uniform prior $\mathrm{d}\boldsymbol{B}$ for $\tilde{\boldsymbol{B}}$ in place of $|\boldsymbol{B}\boldsymbol{B}^t|^{-\frac{1}{2}m}\,\mathrm{d}\boldsymbol{B}$, an integration over $\boldsymbol{B} \in \mathscr{R}^{m^2}$ gives the kernel of the marginal posterior density of $\tilde{\boldsymbol{\Pi}}$.[3] To this end observe that the reciprocal of the normalizing constant of (11) is proportional to $|V_i(\boldsymbol{\Pi})|^{-\frac{1}{2}}(S_i^0(\boldsymbol{\Pi}))^{-\frac{1}{2}(n_i-m)}$. Letting $c_i = S_i + (\bar{\mathbf{B}}_i\bar{\boldsymbol{\Gamma}}_i)\boldsymbol{P}_i(\bar{\mathbf{B}}_i\bar{\boldsymbol{\Gamma}}_i)^t$ and using the well-known fact that $(\mathbf{x}V\mathbf{x}^t + c) = c|V||c^{-1}\mathbf{x}^t\mathbf{x} + V^{-1}|$, we may write $|V_i(\boldsymbol{\Pi})|^{-\frac{1}{2}}$ $(S_i^0(\boldsymbol{\Pi}))^{-\frac{1}{2}(n_i-m)}$ as proportional to

$$|V_i(\boldsymbol{\Pi})|^{\frac{1}{2}(n_i-m-1)}|[\boldsymbol{I}(-\boldsymbol{\Pi})][\boldsymbol{P}_i - c_i^{-1}\boldsymbol{P}_i(\bar{\mathbf{B}}_i\bar{\boldsymbol{\Gamma}}_i)^t(\bar{\mathbf{B}}_i\bar{\boldsymbol{\Gamma}}_i)\boldsymbol{P}_i] \times [\boldsymbol{I}(-\boldsymbol{\Pi})]^t|^{-\frac{1}{2}(n_i-m)}. \tag{12}$$

Partitioning

$$\boldsymbol{P}_i = \begin{bmatrix} \boldsymbol{P}_{11}^{(i)} & \boldsymbol{P}_{12}^{(i)} \\ \boldsymbol{P}_{21}^{(i)} & \boldsymbol{P}_{22}^{(i)} \end{bmatrix} \qquad \boldsymbol{P}_{11}^{(i)}(m\times m),$$

and

$$\boldsymbol{H}_i \equiv \boldsymbol{P}_i - c_i^{-1}\boldsymbol{P}_i(\bar{\mathbf{B}}_1\bar{\boldsymbol{\Gamma}}_1)^t(\bar{\mathbf{B}}_1\bar{\boldsymbol{\Gamma}}_1)\boldsymbol{P}_i = \begin{bmatrix} \boldsymbol{H}_{11}^{(i)} & \boldsymbol{H}_{12}^{(i)} \\ \boldsymbol{H}_{21}^{(i)} & \boldsymbol{H}_{22}^{(i)} \end{bmatrix} \qquad \boldsymbol{H}_{11}^{(i)}(m\times m),$$

and completing the square in (12), it becomes

$$|[\boldsymbol{\Pi} - \boldsymbol{P}_{12}^{(i)}\boldsymbol{P}_{22}^{(i)-1}]\boldsymbol{P}_{22}^{(i)}[\boldsymbol{\Pi} - \boldsymbol{P}_{12}^{(i)}\boldsymbol{P}_{22}^{(i)-1}]^t + \boldsymbol{P}_{11.2}^{(i)}|^{\frac{1}{2}(n_i-m-1)} \times |[\boldsymbol{\Pi} - \boldsymbol{H}_{12}^{(i)}\boldsymbol{H}_{22}^{(i)-1}]\boldsymbol{H}_{22}^{(i)}[\boldsymbol{\Pi} - \boldsymbol{H}_{12}^{(i)}\boldsymbol{H}_{22}^{(i)-1}]^t + \boldsymbol{H}_{11.2}^{(i)}|^{-\frac{1}{2}(n_i-m)}.$$

The complete kernel of the posterior density of $\tilde{\boldsymbol{\Pi}}$ is then:

$$|[\boldsymbol{\Pi} - \boldsymbol{M}_{12}\boldsymbol{M}_{22}^{-1}]\boldsymbol{M}_{22}[\boldsymbol{\Pi} - \boldsymbol{M}_{12}\boldsymbol{M}_{22}^{-1}]^t + \boldsymbol{M}_{11.2}|^{-\frac{1}{2}N} \times \prod_{i=1}^{m} \frac{|[\boldsymbol{\Pi} - \boldsymbol{P}_{12}^{(i)}\boldsymbol{P}_{22}^{(i)-1}]\boldsymbol{P}_{22}^{(i)}[\boldsymbol{\Pi} - \boldsymbol{P}_{12}^{(i)}\boldsymbol{P}_{22}^{(i)-1}]^t + \boldsymbol{P}_{11.2}^{(i)}|^{\frac{1}{2}(n_i-m-1)}}{|[\boldsymbol{\Pi} - \boldsymbol{H}_{12}^{(i)}\boldsymbol{H}_{22}^{(i)-1}]\boldsymbol{H}_{22}^{(i)}[\boldsymbol{\Pi} - \boldsymbol{H}_{12}^{(i)}\boldsymbol{H}_{22}^{(i)-1}]^t + \boldsymbol{H}_{11.2}^{(i)}|^{\frac{1}{2}(n_i-m)}}; \tag{13}$$

[3] The range set of $\boldsymbol{B}$ is in fact the ordinary matrix representation of the real linear group of order m^2. Since the set of values of $\boldsymbol{B}$ such that $|\boldsymbol{B}| = 0$ is of measure zero in $\mathscr{R}^{m^2} \equiv$ euclidean m^2-space, we can integrate over $\mathscr{R}^{m^2}$.

namely, a *ratio* of matric-poly-t kernels. The denominator is a matric version of the poly-t kernels that have appeared thus far.

When cross-section data is available only on the first row of $(\boldsymbol{B\Gamma})$, and *no* time series data is available, the resulting kernel suggests an interesting rough analogy: Drèze [5] has invented a Bayesian analogue of the usual LIMLE; he finds a posterior density of a single row of $(\tilde{\boldsymbol{B}}\,\tilde{\boldsymbol{\Gamma}})$ consisting of a ratio of quadratic forms and chooses its mode as a point estimator. Here we have a matric-version of this ratio for the reduced form . . . a reversal of roles!

If we continue to assume that cross-section data are available only on the first row, and assign a prior of the form (7), then the posterior kernel of $(\tilde{\boldsymbol{B}}\,\tilde{\boldsymbol{\Gamma}})$ is expressible as $|\boldsymbol{BB}^t|^{\frac{1}{2}(N-m-r)}$ times a poly-t kernel:

$$|\boldsymbol{BB}^t|^{\frac{1}{2}(N-m-r)}|[\boldsymbol{B}_2\boldsymbol{\Gamma}_2]\boldsymbol{E}[\boldsymbol{B}_2\boldsymbol{\Gamma}_2]^t|^{-\frac{1}{2}N}\{(\mathbf{B}_1\mathbf{\Gamma}_1)\boldsymbol{M}(\mathbf{B}_1\mathbf{\Gamma}_1)^t\}^{-\frac{1}{2}N}$$
$$\times\{[(\mathbf{B}_1\mathbf{\Gamma}_1)-(\overline{\mathbf{B}}_1\overline{\mathbf{\Gamma}}_1)]\boldsymbol{P}_1[(\mathbf{B}_1\mathbf{\Gamma}_1)-(\overline{\mathbf{B}}_1\overline{\mathbf{\Gamma}}_1)]^t+S_1\}^{-\frac{1}{2}n_1} \qquad (14)$$

where $(\boldsymbol{B}_2\boldsymbol{\Gamma}_2)$ are rows 2 through m of $(\boldsymbol{B\Gamma})$.

Re-assembling the quadratic forms in (14) generated by the time series data and writing them as $|\boldsymbol{BM}_{11.2}\boldsymbol{B}^t+[\boldsymbol{\Gamma}+\boldsymbol{BM}_{12}\boldsymbol{M}_{22}^{-1}]\boldsymbol{M}_{22}\times[\boldsymbol{\Gamma}+\boldsymbol{BM}_{12}\boldsymbol{M}_{22}^{-1}]^t|^{-\frac{1}{2}N}$, if we use a fundamental decomposition from the general linear hypothesis, and write this determinant as a product of a Student kernel for $\tilde{\mathbf{\Gamma}}_1$ given $\tilde{\boldsymbol{B}}=\boldsymbol{B}$ and a matric Student kernel for $\tilde{\boldsymbol{\Gamma}}_2$ given $\tilde{\boldsymbol{B}}=\boldsymbol{B}$ and $\tilde{\mathbf{\Gamma}}_1=\mathbf{\Gamma}_1$, upon integrating out $\boldsymbol{\Gamma}_2$ the kernel of $\tilde{\boldsymbol{B}}$ and $\tilde{\mathbf{\Gamma}}_1$ is found to be

$$|\boldsymbol{BB}^t|^{-\frac{1}{2}m}(\mathbf{B}_1\boldsymbol{M}_{11.2}\mathbf{B}_1^t)^{\frac{1}{2}(N-m-r+1)}$$
$$\times\{(\mathbf{B}_1\boldsymbol{M}_{11.2}\mathbf{B}_1^t)+[\mathbf{\Gamma}_1+\mathbf{B}_1\boldsymbol{M}_{12}\boldsymbol{M}_{22}^{-1}]\boldsymbol{M}_{22}$$
$$\times[\mathbf{\Gamma}_1+\mathbf{B}_1\boldsymbol{M}_{12}\boldsymbol{M}_{22}^{-1}]^t\}^{-\frac{1}{2}(N-m+1)}$$
$$\times\{(\mathbf{B}_1-\overline{\mathbf{B}}_1)\boldsymbol{P}_{11.2}^{(1)}(\mathbf{B}_1-\overline{\mathbf{B}}_1)^t+S_1$$
$$+[\mathbf{\Gamma}_1-\overline{\mathbf{\Gamma}}_1+(\mathbf{B}_1-\overline{\mathbf{B}}_1)\boldsymbol{P}_{12}^{(1)}\boldsymbol{P}_{22}^{(1)-1}]$$
$$\times\boldsymbol{P}_{22}^{(1)}[\mathbf{\Gamma}_1-\overline{\mathbf{\Gamma}}_1+(\mathbf{B}_1-\overline{\mathbf{B}}_1)\boldsymbol{P}_{12}^{(1)}\boldsymbol{P}_{22}^{(1)-1}]^t\}^{-\frac{1}{2}n_1}. \qquad (15)$$

Here we have partitioned

$$\boldsymbol{P}_1=\begin{bmatrix}\boldsymbol{P}_{11}^{(1)} & \boldsymbol{P}_{12}^{(1)}\\ \boldsymbol{P}_{21}^{(1)} & \boldsymbol{P}_{22}^{(1)}\end{bmatrix},$$

$\boldsymbol{P}_{11}^{(1)}(m\times m)$ so as to show that $\tilde{\mathbf{\Gamma}}_1$ given $\tilde{\boldsymbol{B}}=\boldsymbol{B}$ is poly-t.

Again, a slight modification of the prior assigned to $\tilde{\boldsymbol{B}}$ simplifies analysis. A uniform prior in place of a non-informative prior drops the determinant $|\boldsymbol{B}\boldsymbol{B}^t|^{-\frac{1}{2}m}$ from (15) and it may be written as the product of a poly-t kernel in $\boldsymbol{\Gamma}_1$ given $\mathbf{B}_1$ and a poly-t kernel in $\mathbf{B}_1$. To this end adopt the normalization rule, set the first column of $\boldsymbol{B}$ equal to a column of 1's, and similarly set $\overline{\mathbf{B}}_{i1} = 1$, $i = 1, \ldots, m$. Define for notational simplicity

$$\boldsymbol{M}_{11.2} = \begin{bmatrix} M_{11} & \mathbf{M}_{12} \\ \mathbf{M}_{21} & \boldsymbol{M}_{22} \end{bmatrix}$$

M_{11} scalar,

$$\boldsymbol{P}^{(1)}_{11.2} = \begin{bmatrix} P_{11} & \mathbf{P}_{12} \\ \mathbf{P}_{21} & \boldsymbol{P}_{22} \end{bmatrix},$$

P_{11} scalar,

$$\begin{aligned} \mathbf{B}_1\boldsymbol{M}_{11.2}\mathbf{B}_1^t &= M_{11.2} + (\mathbf{B}_{12} - \mathbf{M}_{12}\boldsymbol{M}_{22}^{-1})\boldsymbol{M}_{22} \\ &\quad \times (\mathbf{B}_{12} - \mathbf{M}_{12}\boldsymbol{M}_{22}^{-1})^t \equiv Q_1(\mathbf{B}_{12}) \end{aligned} \tag{16}$$

and

$$\begin{aligned} & S_1 + (\mathbf{B}_1 - \overline{\mathbf{B}}_1)\boldsymbol{P}^{(1)}_{11.2}(\mathbf{B}_1 - \overline{\mathbf{B}}_1)^t \\ &\quad = S_1 + (\mathbf{B}_{12} - \overline{\mathbf{B}}_{12})\mathbf{P}_{22}(\mathbf{B}_{12} - \overline{\mathbf{B}}_{12})^t \equiv Q_2(\mathbf{B}_{12}). \end{aligned}$$

Then letting

$$\begin{aligned} \nu_1 &= N - m - r + 1, \qquad \nu_2 = n_1 - r, \\ \boldsymbol{M}^*_{22} &= \boldsymbol{M}_{22}/\nu_1 Q_1(\mathbf{B}_{12}), \\ \boldsymbol{P}^*_{22} &= \mathbf{P}_{22}/\nu_2 Q_2(\mathbf{B}_{12}), \\ \boldsymbol{\Gamma}^*_B &= -\mathbf{B}_1\boldsymbol{M}_{12}\boldsymbol{M}_{22}^{-1}, \\ \boldsymbol{\Gamma}^{**}_B &= \overline{\boldsymbol{\Gamma}}_1 - (\mathbf{B}_{12} - \overline{\mathbf{B}}_{12})\mathbf{M}_{12}\boldsymbol{M}_{22}^{-1}, \end{aligned}$$

the kernel (15) becomes $\nu_1^{\frac{1}{2}(N-m-1)}\nu_2^{\frac{1}{2}n_1}$ times

$$\begin{aligned} & [Q_1(\mathbf{B}_{12})]^{-\frac{1}{2}r}[Q_2(\mathbf{B}_{12})]^{-\frac{1}{2}n_1}\{1 + [\boldsymbol{\Gamma}_1 - \boldsymbol{\Gamma}^*_B]\boldsymbol{M}^*_{22} \\ & \times [\boldsymbol{\Gamma}_1 - \boldsymbol{\Gamma}^*_B]^t\}^{-\frac{1}{2}(N-m+1)}\{1 + [\boldsymbol{\Gamma}_1 - \boldsymbol{\Gamma}^{**}_B]\boldsymbol{P}^*_{22} \\ & \times [\boldsymbol{\Gamma}_1 - \boldsymbol{\Gamma}^{**}_B]^t\}^{-\frac{1}{2}n_1}. \end{aligned} \tag{17}$$

The mode of the posterior kernel (17) of $(\tilde{\mathbf{B}}_1\tilde{\boldsymbol{\Gamma}}_1)$ is a Bayesian version of a maximum likelihood estimator, and when the only cross-section information available bears solely on the first row of $(\boldsymbol{B\Gamma})$, given that the prior we assigned doesn't do too much violence to a priori beliefs, we can label this a *full* information estimator, since we are using all available information. If (17) ignores additional prior information or cross-section data bearing on rows 2, . . ., m of $(\boldsymbol{B\Gamma})$ then it is in a loose sense a *limited* information procedure.

Finding a maximizer $(\mathbf{B}_{12}^0, \mathbf{\Gamma}_1^0)$ of (17) is in general quite involved. When both N and n_1 become large, however, a reasonable approximation to $(\mathbf{B}_{12}^0, \mathbf{\Gamma}_1^0)$ can be found by using an expansion of two-factor poly-t kernels given by Tiao and Zellner [7]. The leading term in their expansion is a normal kernel, which for the poly-t kernel of $\mathbf{\Gamma}_1$ given $\tilde{\mathbf{B}}_1 = \mathbf{B}_1$ in (17), is centered at $\mathbf{\Gamma}_1^a \equiv [\mathbf{\Gamma}_B^* \boldsymbol{M}_{22}^* + \mathbf{\Gamma}_B^{**} \boldsymbol{P}_{22}^*][\boldsymbol{M}_{22}^* + \boldsymbol{P}_{22}^*]^{-1}$. If we adopt this approximation, then we may compute an (approximate) maximizer of (17) in two steps: choose as an approximation to $\mathbf{B}_{12}^0$ that value of $\mathbf{B}_{12}$ which maximizes (17) *subject to* $\mathbf{\Gamma}_1$ fixed at $\mathbf{\Gamma}_1^a$ as defined above; i.e., for a generic $\mathbf{B}_{12}$ compute the value of (17) at $\mathbf{B}_{12}$ and $\mathbf{\Gamma}_1^a$, the latter a function of $\mathbf{B}_{12}$. Search over values of $\mathbf{B}_{12}$ for a maximizer of (17).

2.1. *An example*

To illustrate the application of the preceding calculations, we shall analyze the example treated by Drèze in [5] under a different specification as to how cross-section data come into play. He examines Tintner's model of supply and demand for meat, poultry, and fish, using as "cross-section" data estimates of price and income elasticities of demand for meat products in Sweden computed by Wold and Jureen [8]. Use of this example has an advantage: we can adopt it as a vehicle for studying the effect of differing assumptions about the role of cross-section data in a Bayesian context.

Since this is intended solely as a numerical example to illustrate the methodology, we do not attempt to justify the particular numbers we use. (See Drèze [5] for a discussion.) Drèze uses the cross-section data to generate a normal-gamma prior for $(\bar{\mathbf{B}}_{12}\bar{\mathbf{\Gamma}}_1)$ ($\mathbf{B}_{12}$ is (1×1) in this example) of the form

$$\sigma_{11}^{-1} \exp \{-\tfrac{1}{2}[(B_{12}\mathbf{\Gamma}_1)-(\bar{B}_{12}\bar{\mathbf{\Gamma}}_1)]\boldsymbol{P}_1^{\square}[(B_{12}\mathbf{\Gamma}_1)-(\bar{B}_{12}\bar{\mathbf{\Gamma}}_1)]^t/\sigma_{11}\} \quad (18)$$

where σ_{11} is the (1,1)th element of $\boldsymbol{\Sigma}$, thus assuming that the (residual error) variance of the cross-section data is (proportional to) that of the residuals in the first time series equation. This prior is improper, but informative. He then computes a maximizer of

$$\frac{\{(1\ \mathbf{B}_{12})\boldsymbol{M}_{11.2}(1\ \mathbf{B}_{12})^t\}^{\frac{1}{2}(N-m-r+1)}}{\{(1\ \mathbf{B}_{12})\boldsymbol{W}^*(1\ \mathbf{B}_{12})^t\}^{-\frac{1}{2}(N+\nu-2)}}$$

where in our notation $\boldsymbol{W}^* = \boldsymbol{M}_{22}+\boldsymbol{P}^{(1)}_{11.2}$ and ν is a parameter Drèze manipulates in his numerical calculations (see [5, p. 28] for the rationale). Upon finding a maximizer of this ratio, he computes an estimator of $\mathbf{\Gamma}_1$, by observing that since $\mathbf{\Gamma}_1$ given $\tilde{\mathbf{B}}_1 = \mathbf{B}_1$ is Student, a maximizer of the density of $\tilde{\mathbf{\Gamma}}_1$ given $\tilde{\mathbf{B}}_1 = \mathbf{B}_1$ is the mean of $\tilde{\mathbf{\Gamma}}_1$ given $\tilde{\mathbf{B}}_1 = \mathbf{B}_1$.

From (15), we see that the factor

$$\{(1\ \mathbf{B}_{12})\boldsymbol{M}_{11.2}(1\ \mathbf{B}_{12})^t\}^{\frac{1}{2}(N-m-r+1)}$$

also appears under the assumption that cross-section residual error variances and time series error variances are unrelated. However, in place of a *single* Student kernel for $(\tilde{\mathbf{B}}_{12}\tilde{\mathbf{\Gamma}}_1)$ whose parameter set blends time series data $\{\boldsymbol{M}, N\}$ and cross-section data $\{\boldsymbol{P}^{\square}_{1}, \overline{\mathbf{B}}_{12}, \overline{\mathbf{\Gamma}}_1, n_1\}$, we have a product of Student kernels in $(\mathbf{B}_{12}\mathbf{\Gamma}_1)$.

In terms of deviations from means, the demand and supply equations in Tintner's model are of the form[4]

$$\tilde{y}_1+B_{12}\tilde{y}_2+\Gamma_{11}z_1+\Gamma_{12}z_2+\Gamma_{13}z_3 = \tilde{u}_1$$

and

$$\tilde{y}_1+B_{22}\tilde{y}_2+\Gamma_{21}z_1+\Gamma_{22}z_2+\Gamma_{23}z_3 = \tilde{u}_2.$$

We do not distinguish between specifications but rather leave all exogenous variables in the model and assume that, in a classical sense, it is under-identified in each equation. Tintner examined two models with different a priori (exact) specifications. Under-identification in this particular instance corresponds reasonably well with a priori judgments, since there is some question as to which exogenous variables really belong in the model. Furthermore, under-identification is more natural in a Bayesian context since it avoids the necessity of assuming that we know the value of a particular parameter with certainty when in fact we don't.

An exact comparison of results is not possible, but by some jiggling of the numbers, we can match up the prior in [5] and a Student prior

[4] See [5] for a detailed description of the model.

for $(\tilde{\mathbf{B}}_{12}\tilde{\boldsymbol{\Gamma}}_1)$, then use the latter as the cross-section "component" of (15). (In particular, the prior (18) unconditional as regards $\tilde{\sigma}_{11}$ is improper and has infinite variance.) Drèze [5] sets

$$\boldsymbol{P}_1^{\square} = \begin{bmatrix} 83.000 & & & \boldsymbol{0} \\ & 3.333 & & \\ & & 333.300 & \\ \boldsymbol{0}^t & & & 333.300 \end{bmatrix}$$

and

$$(\overline{\mathbf{B}}_{12}\overline{\boldsymbol{\Gamma}}_1) = (0.54,\ -0.17, 0, 0).$$

The matrix $\boldsymbol{P}_1^{\square}$ is expressed in units of the computed variance of the indirect least squares residuals (in the just-identified case): 33.33. Consequently, we scale the raw time series data $\boldsymbol{M}$ to correspond with $\boldsymbol{P}_1^{\square}$ and use $\{N, \boldsymbol{M}\}$, $N = 23$ and

$$\boldsymbol{M} = \begin{bmatrix} 1.36954 & -0.35244 & 3.67191 & -0.53648 & 0.98386 \\ & 1.58149 & 8.35459 & 0.85033 & 1.23576 \\ & & 83.43365 & 3.61172 & 12.20477 \\ & & & 2.53480 & 0.73078 \\ & & & & 2.62699 \end{bmatrix}$$

Under the specifications for cross-section data we have been exploring, we must (see (15)) assign values to elements of the parameter set $\{\boldsymbol{P}_1^{\square}, S_1, n_1, (\overline{\mathbf{B}}_1, \overline{\boldsymbol{\Gamma}}_1)\}$. As previously stated, an exact match with (18) is not possible[5] because of the difference in models; but we can set $\boldsymbol{P}_1^{\square}$ and $(\overline{\mathbf{B}}_1, \overline{\boldsymbol{\Gamma}}_1)$ equal to the values displayed above and then manipulate S_1 and n_1 to explore features of (15).

To this end we used the method for approximating $(\boldsymbol{B}_{12}^0\boldsymbol{\Gamma}_1^0)$ suggested earlier and calculated table 1a for ($S_1 = 14$, $n_1 = 3, 5, 6, 10, 20$) and ($S_1 = 2000$, $n_1 = 7, 8, 10$). While in practice n_1 and S_1 will usually be determined by the process generating cross-section data, here our choice of values is ad hoc. However, when a prior (18) is

[5] The "cross-section" factor in (15) may be written as $\{[(\mathbf{B}_1\boldsymbol{\Gamma}_1)-(\overline{\mathbf{B}}_1\overline{\boldsymbol{\Gamma}}_1)]\boldsymbol{P}_1[\cdot]^t+S_1\}^{-\frac{1}{2}n_1}$; setting $S_1 = 0$ duplicates the exponent of (18) in the synthetic representation

$$\frac{1}{\Gamma(\frac{1}{2}n_1)} \int_0^\infty \exp(-\tfrac{1}{2}h\{\cdot\})h^{\frac{1}{2}n_1-1}\,\mathrm{d}h,$$

but then (15) "blows up" at $(B_{12}\boldsymbol{\Gamma}_1) = (\bar{B}_{12}\overline{\boldsymbol{\Gamma}}_1)$.

TABLE 1a

Posterior modal values for (15)

N	s_1	n_1	B_{12}	Γ_{11}	Γ_{12}	Γ_{13}
23	14	3	0.380	−0.10205	0.0002533	−0.0001052
			1.611	−0.1948	−0.00003659	−0.0001684
		5	0.370	−0.084180	0.006892	−0.006214
			1.483	−0.1884	−0.001285	−0.02747
		6	1.23	−0.15910	0.008316	−0.03899
			0.449	−0.08845	−0.009888	−0.009568
		10	0.494	−0.09123	−0.02148	−0.02296
		20	0.514	−0.08955	−0.04546	−0.05446
23	2000	7	0.874	−0.06674	0.1443	−0.4856
		8	0.874	−0.06430	0.1477	−0.5026
		10	0.890	0.0640	0.1462	−0.5226

TABLE 1b

Modal values from cases B.1 and B.2 of [5]

	B_{12}	Γ_{11}	Γ_{12}	Γ_{13}
Case B.1: ($\nu = 2$)	1.22	−0.166	0.0003	−0.0014
Case B.2: ($\nu = r+m+1$)	1.32	−0.174	0.00018	−0.0014

assigned to $(\tilde{\mathbf{B}}_{12}\tilde{\mathbf{\Gamma}}_1)$ the *conditional* variance of $(\tilde{\mathbf{B}}_{12}\tilde{\mathbf{\Gamma}}_1)$ given $\tilde{\sigma}_{11} = \sigma_{11}$ is $(\boldsymbol{P}_1^{\square})^{-1}$. Under our specification the cross-section component of (15) is a Student kernel; the density generated by this kernel has variance $S_1(\boldsymbol{P}_1^{\square})^{-1}/(n_1-6)$ provided that $n_1 > 6$. Hence at $S_1 = 14$ and $n_1 = 20$, this variance is also $(\boldsymbol{P}_1^{\square})^{-1}$. Of course the same is true for all S_1 and n_1 such that $S_1/(n_1-6) = 1$. For fixed $\boldsymbol{P}_1^{\square}$, the ratio[6] $S_1/(n_1-6)$ determines the "peakedness" of this component of (15); hence we expect that increasing S_1 with n_1 held fixed affects the estimates in the same way as decreasing n_1 with S_1 held fixed.

When $n_1 \leqq 4$, a prior of the form $c\{[(\mathbf{B}_{12}\mathbf{\Gamma}_1)-(\bar{\mathbf{B}}_{12}\bar{\mathbf{\Gamma}}_1)]\boldsymbol{P}_1^{\square}[(\mathbf{B}_{12}\mathbf{\Gamma}_1)-(\bar{\mathbf{B}}_{12}\bar{\mathbf{\Gamma}}_1)]^t+S_1\}^{-\frac{1}{2}n_1}$ is improper even when $\boldsymbol{P}_1^{\square} > \boldsymbol{0}$ and $S_1 > 0$. For our specification this corresponds to the improper prior (18) in the specification of [5].

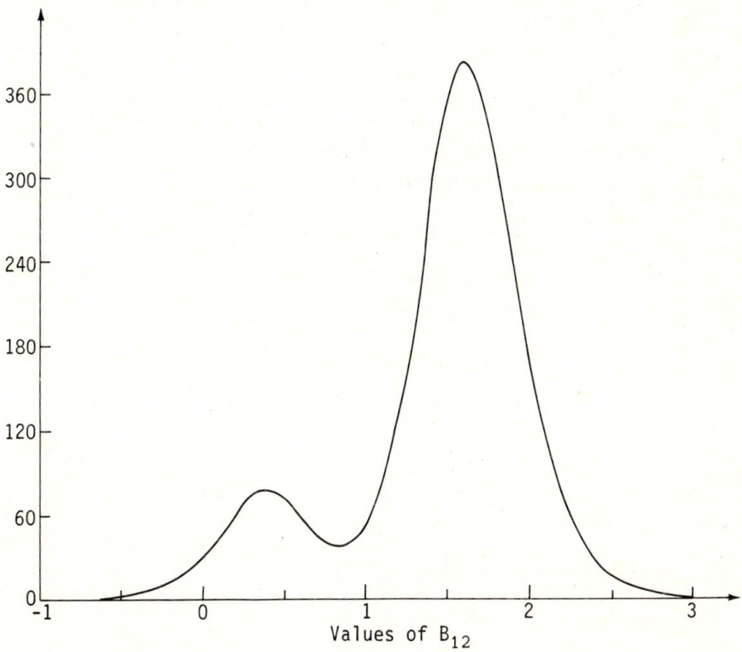

Fig. 1. Graph of kernel of $\tilde{\mathbf{B}}_{12}|\mathbf{\Gamma}_1$ as a function of B_{12} with $\mathbf{\Gamma}_1 = [\mathbf{\Gamma}_B{}^*\boldsymbol{M}_{22}{}^* + \mathbf{\Gamma}_B{}^{**}\boldsymbol{P}_{22}{}^*][\boldsymbol{M}_{22}{}^* + \boldsymbol{P}_{22}{}^*]^{-1}$ for $s_1 = 14$, $N = 23$, $n_1 = 3$.

[6] For $n_1 > 6$.

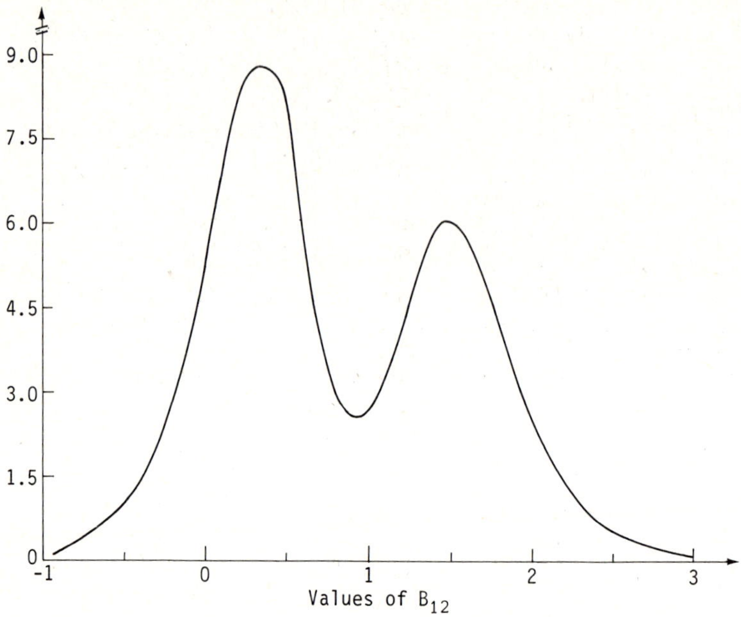

Fig. 2. Graph of kernel of $\tilde{\mathbf{B}}_{12}|\mathbf{\Gamma}_1$ as a function of B_{12} with $\mathbf{\Gamma}_1 = [\mathbf{\Gamma}_B{}^*\boldsymbol{M}_{22}{}^* + \mathbf{\Gamma}_B{}^{**}\boldsymbol{P}_{22}{}^*][\boldsymbol{M}_{22}{}^* + \boldsymbol{P}_{22}{}^*]^{-1}$ for $s_1 = 14, N = 23, n_1 = 5$.

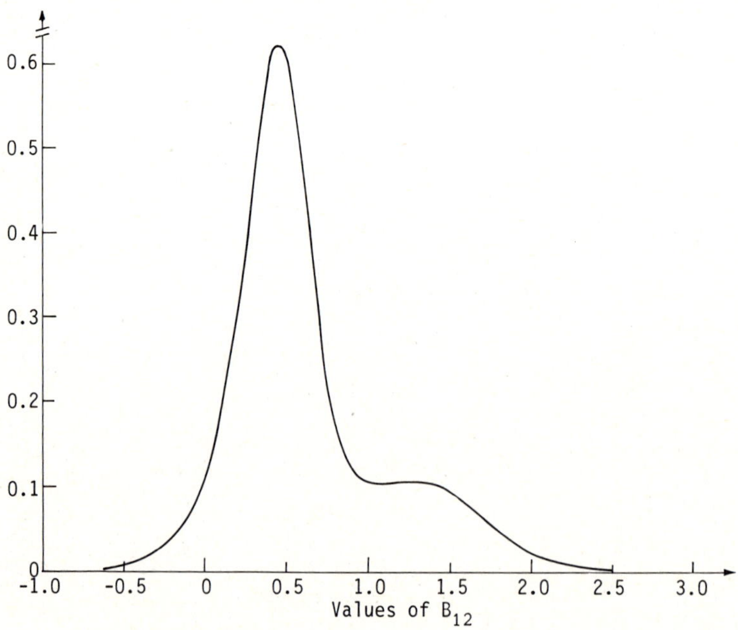

Fig. 3. Graph of kernel of $\tilde{\mathbf{B}}_{12}|\mathbf{\Gamma}_1$ as a function of B_{12} with $\mathbf{\Gamma}_1 = [\mathbf{\Gamma}_B{}^*\boldsymbol{M}_{22}{}^* + \mathbf{\Gamma}_B{}^{**}\boldsymbol{P}_{22}{}^*][\boldsymbol{M}_{22}{}^* + \boldsymbol{P}_{22}{}^*]^{-1}$ for $s_1 = 14, N = 23, n_1 = 6$.

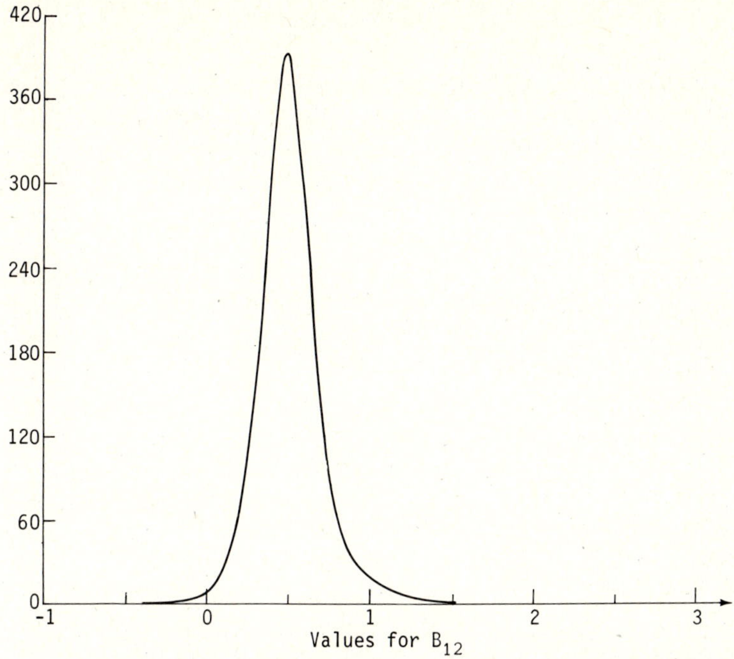

Fig. 4. Graph of kernel of $\tilde{\mathbf{B}}_{12}|\mathbf{\Gamma}_1$ as a function of B_{12} with $\mathbf{\Gamma}_1 = [\mathbf{\Gamma}_B{}^*\boldsymbol{M}_{22}{}^* + \mathbf{\Gamma}_B{}^{**}\boldsymbol{P}_{22}{}^*][\boldsymbol{M}_{22}{}^* + \boldsymbol{P}_{22}{}^*]^{-1}$ for $s_1 = 14, N = 23, n_1 = 10$.

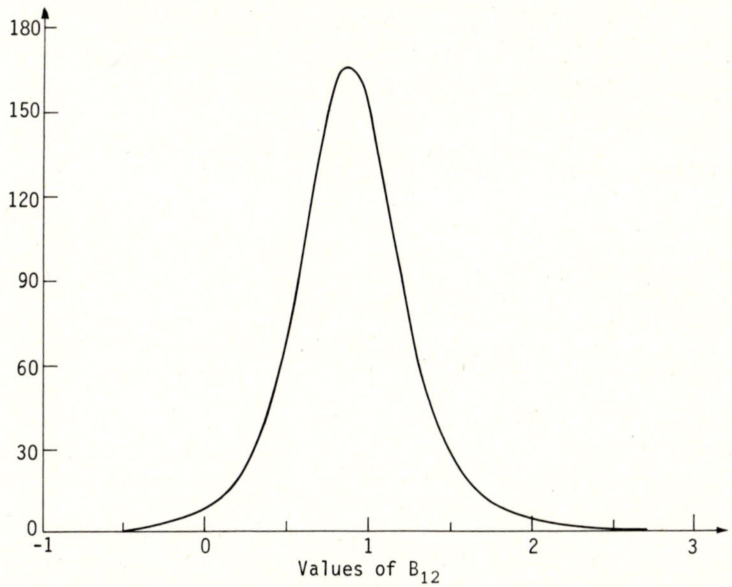

Fig. 5. Graph of kernel of $\tilde{\mathbf{B}}_{12}|\mathbf{\Gamma}_1$ as a function of B_{12} with $\mathbf{\Gamma}_1 = [\mathbf{\Gamma}_B{}^*\boldsymbol{M}_{22}{}^* + \mathbf{\Gamma}_B{}^{**}\boldsymbol{P}_{22}{}^*][\boldsymbol{M}_{22}{}^* + \boldsymbol{P}_{22}{}^*]^{-1}$ for $s_1 = 2000, N = 23, n_1 = 7$.

Poly-t densities need not be unimodal. Graphs[7] of (15) (with $|\boldsymbol{B}\boldsymbol{B}^t|^{-\frac{1}{2}m}$ deleted) for $\boldsymbol{\Gamma}_1$ fixed at $[\boldsymbol{\Gamma}_B^* \boldsymbol{M}_{22}^* + \boldsymbol{\Gamma}_B^{**} \boldsymbol{P}_{22}^*][\boldsymbol{M}_{22}^* + \boldsymbol{P}_{22}^*]^{-1}$ and ($N = 23$, $S_1 = 14$, $n_1 = 3, 5, 6$) show that in fact it is *bimodal* (figs. 1-5). As can be seen, for $n_1 = 3$, the maximizer of (15) is the larger of the two modal values of (15), while for $n_1 = 5, 6$, it is the smaller. The smaller modal values of $\mathbf{B}_{12}$ are close to $\bar{\mathbf{B}}_{12}$ while the larger ones are in the range of estimates derived in [5] by Drèze using his analogue of LIMLE estimation. For $n \geqq 10$, the larger modal value disappears and the graph becomes unimodal, with the modal value of $\mathbf{B}_{12}$ not far from $\bar{\mathbf{B}}_{12}$. We conclude that as n_1 increases with S_1 and N held fixed, the cross-section component of (15) quickly begins to play a dominant role. The behavior for $S_1 = 2000$ is somewhat different, as can be seen from the graphs. Tables 2a and 2b display variance matrices for $\tilde{\boldsymbol{\Gamma}}_1$ corresponding to the estimates in tables 1a and 1b. Those in table 2a were computed using a large (time series) sample approximation to the exact variance of $\tilde{\boldsymbol{\Gamma}}_1$ given $\tilde{\mathbf{B}}_{12} = \mathbf{B}_{12}^0$; i.e., $[\boldsymbol{M}_{22}^* + \boldsymbol{P}_{22}^*]^{-1}$. While having substantially the same flavor as those computed by Drèze, the conditional variances of the $\tilde{\boldsymbol{\Gamma}}_{1i}$'s are much larger, and increase as n_1 increases relative to S_1.

TABLE 2a

Large sample estimate of variance matrix of $\tilde{\boldsymbol{\Gamma}}_1$ given $\tilde{\mathbf{B}}_{12} = \mathbf{B}_{12}^0$

$N = 23$, $s_1 = 14$, $n_1 = 3$

	Γ_{11}	Γ_{12}	Γ_{13}
Γ_{11}	0.08971	−0.00009	−0.00030
Γ_{12}		0.00299	~0
Γ_{13}			0.00299

$N = 23$, $s_1 = 14$, $n_1 = 5$

	Γ_{11}	Γ_{12}	Γ_{13}
Γ_{11}	0.11335	−0.01904	−0.07287
Γ_{12}		0.45937	−0.00549
Γ_{13}			0.50078

$N = 23$, $s_1 = 14$, $n_1 = 6$

	Γ_{11}	Γ_{12}	Γ_{13}
Γ_{11}	0.07565	0.00533	−0.01886
Γ_{12}		0.12571	−0.00054
Γ_{13}			0.12980

[7] The kernels displayed in these graphs have *not* been normalized.

TABLE 2b
Variance matrix, case B.2 of Drèze [5]

	Γ_{11}	Γ_{12}	Γ_{13}
Γ_{11}	$(0.019)^2$	$-(0.002)^2$	$-(0.003)^2$
Γ_{12}		$(0.010)^2$	~ 0
Γ_{13}			$(0.010)^2$

3. *Cross-section and time series variances related*

In the sequel we shall assume that the variance of the error term in each cross-section equation is proportional to the marginal variance of the error term in the corresponding structural equation. Namely, with $(\mathbf{B}_i\,\mathbf{\Gamma}_i)$, S_i, and n_i defined as before, the likelihood function generated by cross-section data $\{(\overline{\mathbf{B}}_i\overline{\mathbf{\Gamma}}_i), S_i, \boldsymbol{P}_i, n_i\}$ bearing on individual rows of $(\boldsymbol{B}\,\boldsymbol{\Gamma})$ is[8]

$$\prod_{i=1}^{m} \sigma_{ii}^{-\frac{1}{2}n_i} \exp\{-\tfrac{1}{2}\{[(\mathbf{B}_i\mathbf{\Gamma}_i)-(\overline{\mathbf{B}}_i\overline{\mathbf{\Gamma}}_i)]\boldsymbol{P}_i \times [(\mathbf{B}_i\mathbf{\Gamma}_i)-(\overline{\mathbf{B}}_i\overline{\mathbf{\Gamma}}_i)]^t + S_i\}/k_i\sigma_{ii}\} \tag{19}$$

where σ_{ii} is the ith component of Σ and the scale factors k_i, $i = 1, 2, \ldots, m$ are known with certainty. Provided that the k_i are known with certainty we may set them each equal to one with no loss in generality and do so.

If we assign a non-informative prior to $\tilde{\mathbf{B}}$, $\tilde{\mathbf{\Pi}}$, and $\tilde{\mathbf{\Omega}}$, posterior to observing cross-section data leading to (19) and N time series observations generated according to (1) and summarized by the sufficient set $\{\boldsymbol{M}, N\}$, the kernel of $(\tilde{\boldsymbol{B}}\,\tilde{\boldsymbol{\Gamma}})$ unconditional as regards $\tilde{\boldsymbol{\Sigma}}$ is, letting $\boldsymbol{A} = [\boldsymbol{B}\,\boldsymbol{\Gamma}]\boldsymbol{M}[\boldsymbol{B}\,\boldsymbol{\Gamma}]^t$ and $q_i = [(\mathbf{B}_i\mathbf{\Gamma}_i)-(\overline{\mathbf{B}}_i\overline{\mathbf{\Gamma}}_i)]\boldsymbol{P}_i[(\mathbf{B}_i\mathbf{\Gamma}_i)-(\overline{\mathbf{B}}_i\overline{\mathbf{\Gamma}}_i)]^t + S_i$,

$$|\boldsymbol{B}\boldsymbol{B}^t|^{\frac{1}{2}(N-m-r)} \int_{\Sigma>0} \exp\{-\tfrac{1}{2}\operatorname{tr}\boldsymbol{\Sigma}^{-1}\boldsymbol{A}\}|\boldsymbol{\Sigma}|^{-\frac{1}{2}(N+m+1)} \times \prod_{i=1}^{m} \exp\{-\tfrac{1}{2}q_i/\sigma_{ii}\}\,\sigma_{ii}^{-\frac{1}{2}n_i}\,\mathrm{d}\boldsymbol{\Sigma}, \tag{20}$$

where $q_i \equiv 0$ when $n_i = 0$ and $q_i \geqq 0$ when $n_i > 0$.

The integral has not to the writer's knowledge been explored before, and appears difficult to evaluate in general. We show below that when $m = 2$, the posterior kernel of $(\tilde{\boldsymbol{B}}\,\tilde{\boldsymbol{\Gamma}})$ may be expressed as a product of

[8] By definition, when $n_i = 0$, $\boldsymbol{P}_i = \boldsymbol{0}$.

(1) $|BB^t|^{\frac{1}{2}(N-m-r)}$ times independent Student kernels in $(\mathbf{B}_1\mathbf{\Gamma}_1)$ and $(\mathbf{B}_2\mathbf{\Gamma}_2)$; and

(2) an infinite sum with generic term expressible in terms of a hypergeometric function F_1 of two arguments (cf. Appell and Kampé de Feriet [1]), the arguments being ratios $q_1/(A_{11}+q_1)$ and $q_2/(A_{22}+q_2)$ of quadratic forms.

3.1. *Evaluation of* (20) *when* $m = 2$

To evaluate the integral

$$\int_{\Sigma>0} \exp\{-\tfrac{1}{2}\operatorname{tr}\Sigma^{-1}A\}|\Sigma|^{-\frac{1}{2}(N+3)}\sigma_{11}^{-\frac{1}{2}n_1}\sigma_{22}^{-\frac{1}{2}n_2} \times \exp\{-\tfrac{1}{2}([q_1/\sigma_{11}]+[q_2/\sigma_{22}])\}\,\mathrm{d}\Sigma, \tag{21}$$

first transform from Σ to $H = \Sigma^{-1}$ and then from H to (H_{11}, H_{22}, ρ), $\rho = H_{12}/\sqrt{(H_{11}H_{22})}$. Expanding the term $\exp\{-\rho A_{12}\sqrt{(H_{11}H_{22})}\}$ appearing in the integrand in a Taylor series and then integrating term by term over $H_{11} > 0$ and $H_{22} > 0$ yields an infinite sum each of whose terms involves an integral with integrand ρ ranging over $[-1, 1]$. A transformation from ρ to $u = \rho^2$ shows that each of these integrals is an integral representation of a hypergeometric function F_1 of two variables[9] $R_1 = q_1/(A_{11}+q_1)$ and $R_2 = q_2/(A_{22}+q_2)$. Namely (21) is equal to

$$2^{N+\frac{1}{2}(n_1+n_2)}\Gamma(\tfrac{1}{2}(N+n_1+n_2-1))(A_{11}+q_1)^{-\frac{1}{2}(N+n_1)} \times (A_{22}+q_2)^{-\frac{1}{2}(N+n_2)}\sum_{j=0}^{\infty}\left\{\frac{c_j}{(2j!)}([A_{12}^2/(A_{11}A_{22})] \times(1-R_1)(1-R_2))^j F_1(j+\tfrac{1}{2}, \tfrac{1}{2}N+\tfrac{1}{2}n_1+j, \tfrac{1}{2}N+\tfrac{1}{2}n_2+j, \tfrac{1}{2}N+\tfrac{1}{2}n_1+\tfrac{1}{2}n_2+j; R_1, R_2)\right\} \tag{22a}$$

where

$$c_j = 2^{2j}\Gamma(j+\tfrac{1}{2})\Gamma(\tfrac{1}{2}N+\tfrac{1}{2}n_1+j)\Gamma(\tfrac{1}{2}N+\tfrac{1}{2}n_2+j)/\Gamma(\tfrac{1}{2}N+\tfrac{1}{2}n_1+\tfrac{1}{2}n_2+j), \tag{22b}$$

and

$$F_1(j+\tfrac{1}{2}, \tfrac{1}{2}N+\tfrac{1}{2}n_1+j, \tfrac{1}{2}N+\tfrac{1}{2}n_2+j, \tfrac{1}{2}N+\tfrac{1}{2}n_1+\tfrac{1}{2}n_2+j; R_1, R_2) \equiv \sum_{m=0}^{\infty}\sum_{k=0}^{\infty}\frac{[j+\frac{1}{2}]_{m+k}[\frac{1}{2}N+\frac{1}{2}n_1+j]_m[\frac{1}{2}N+\frac{1}{2}n_2+j]_k}{m!k![\frac{1}{2}N+\frac{1}{2}n_1+n_2+j]_{m+k}}R_1^m R_2^k. \tag{22c}$$

[9] Cf. Appell and Kampé de Feriet [1].

Then (22) implies that given $\tilde{\mathbf{B}}^{(1)} = \mathbf{1} = \bar{\mathbf{B}}^{(1)}$, the posterior density of $\tilde{\mathbf{B}}^{(2)}$ and $\tilde{\Gamma}$ unconditional as regards $\tilde{\Sigma}$ has a kernel proportional to $|\boldsymbol{BB}^t|^{\frac{1}{2}(N-r-2)}$ times

(1) a product of independent Student kernels, one involving $(B_{12}\mathbf{\Gamma}_1)$, the other $(B_{22}\mathbf{\Gamma}_2)$; and

(2) a complicated function expressible in terms of powers of (a) ratios of quadratic forms appearing in the cross-section and time series likelihoods and (b) a cross-product term $(1B_{12}\mathbf{\Gamma}_1)\boldsymbol{M}(1B_{22}\mathbf{\Gamma}_2)^t$; namely,

$$\begin{aligned} &|\boldsymbol{BB}^t|^{\frac{1}{2}(N-r-2)}\{S_1'' + [(B_{12}\mathbf{\Gamma}_1) - (\bar{B}_{12}''\bar{\mathbf{\Gamma}}_1'')]\boldsymbol{n}_1 \\ &\times [(B_{12}\mathbf{\Gamma}_1) - (\bar{B}_{12}''\bar{\mathbf{\Gamma}}_1'')]\}^{-\frac{1}{2}(N+n_1-1)}\{S_2'' + [(B_{22}\mathbf{\Gamma}_2) \\ &-(\bar{B}_{22}''\bar{\mathbf{\Gamma}}_2'')]\boldsymbol{n}_2[(B_{22}\mathbf{\Gamma}_2) - (\bar{B}_{22}''\bar{\mathbf{\Gamma}}_2'')]\}^{-\frac{1}{2}(N+n_2-1)} \\ &\times \psi(N, n_1, n_2; \boldsymbol{A}, q_1, q_2) \end{aligned} \tag{23}$$

where the function ψ is implicitly defined by (22) and S_i'', $\bar{B}_{i2}''$,$\bar{\mathbf{\Gamma}}_i''$, and $\boldsymbol{n}_i$ for $i = 1, 2$, are found by completing the square in $A_{ii} + q_i = (1B_{i2}\mathbf{\Gamma}_i) \times \boldsymbol{M}(1B_{i2}\mathbf{\Gamma}_i)^t + S_1 + [(1B_{i2}\mathbf{\Gamma}_i) - (1\bar{B}_{i2}\bar{\mathbf{\Gamma}}_i)]\boldsymbol{P}^{(i)}[(1B_{i2}\mathbf{\Gamma}_i) - (1\bar{B}_{i2}\bar{\mathbf{\Gamma}}_i)]^t$.

In the special case where cross-section data are available only on, say, the first row of $(\boldsymbol{B}\boldsymbol{\Gamma})$, (22) leads directly to the densities (2.23) and (2.24) derived by Drèze in [5]. For example, when $q_2 = n_2 = 0$, terms in (22a) reduce as follows: $c_j = 2^{2j}\Gamma(\frac{1}{2}N+j)$ and $F_1(j+\frac{1}{2}, \frac{1}{2}(N+n_1)+j, \frac{1}{2}N+j, \frac{1}{2}(N+n_1)+j; R_1, 0) = (1-R_1)^{-j-\frac{1}{2}}$. Then use of the relation $\Gamma(j+\frac{1}{2})/\Gamma(2j+1) = \sqrt{\pi}/2^{2j}\Gamma(j+1)$ together with some algebraic manipulation reduces (22a) to

$$\begin{aligned} &2^{N+\frac{1}{2}n_1}\sqrt{\pi}\Gamma(\tfrac{1}{2}N)\Gamma(\tfrac{1}{2}(N+n_1-1))A_{11}^{\frac{1}{2}(N-1)} \\ &\times (A_{11}+q_1)^{-\frac{1}{2}(N+n_1-1)}|\boldsymbol{A}|^{-\frac{1}{2}N}. \end{aligned} \tag{24}$$

Aside from a constant, (24) is expressible as

$$(A_{11}+q_1)^{-\frac{1}{2}(N+n_1-1)}A_{11}^{-\frac{1}{2}}A_{22.1}^{-\frac{1}{2}N}.$$

The first term is the kernel of a Student density in $(B_{12}\mathbf{\Gamma}_1)$ centered at $(\bar{B}_{12}''\bar{\mathbf{\Gamma}}_1'')$ with $N+n_1-r-1$ degrees of freedom and variance matrix $S_1''\boldsymbol{n}_1^{-1}/(N+n_1-r-3)$ provided that $N+n_1-r-1 > 2$. The second term $A_{11}^{-\frac{1}{2}}A_{22.1}^{-\frac{1}{2}N}$ is the kernel of a Student density in $(B_{22}\mathbf{\Gamma}_2)$ conditional on $(\tilde{B}_{12}\tilde{\mathbf{\Gamma}}_1) = (B_{12}\mathbf{\Gamma}_1)$. The marginal posterior kernel of $\tilde{\mathbf{B}}^{(2)}$ may be found by integrating $|\boldsymbol{BB}^t|^{\frac{1}{2}(N-r-2)}$ times (24) first over

$\mathbf{\Gamma}_2 \in \mathscr{R}^r$ and then over $\mathbf{\Gamma}_1 \in \mathscr{R}^r$. Although derived somewhat differently, the resulting posterior kernel is given in Drèze [5].[10]

3.2. *Asymptotic expansion of* (21)

A density with kernel (23) is clearly very unwieldy. However, for large N, an approximation to (21) of order $1/N$ is available; i.e., (21) is asymptotically proportional to

$$|A|^{-\frac{1}{2}N} \frac{A_{11}^{\frac{1}{2}(N-1)} A_{22}^{\frac{1}{2}(N-1)}}{(A_{11}+q_1)^{\frac{1}{2}(N+n_1-1)}(A_{22}+q_2)^{\frac{1}{2}(N+n_2-1)}}. \tag{25a}$$

Writing out A and q_i, $i = 1, 2$, in terms of B and Γ and multiplying by $|BB^t|^{\frac{1}{2}(N-r-2)}$ we have an asymptotic approximation to (21), namely

$$|BB^t|^{\frac{1}{2}(N-r-2)}|[B\Gamma]M[B\Gamma]^t|^{-\frac{1}{2}N} \times \prod_{i=1}^{2} \frac{\{(\mathbf{B}_i\mathbf{\Gamma}_i)M(\mathbf{B}_i\mathbf{\Gamma}_i)^t\}^{\frac{1}{2}(N-1)}}{\{S_i+[(\mathbf{B}_i\mathbf{\Gamma}_i)-(\overline{\mathbf{B}}_i\overline{\mathbf{\Gamma}}_i)]P^{(i)}[(\mathbf{B}_i\mathbf{\Gamma}_i) - (\overline{\mathbf{B}}_i\overline{\mathbf{\Gamma}}_i)]^t+(\mathbf{B}_i\mathbf{\Gamma}_i)M(\mathbf{B}_i\mathbf{\Gamma}_i)^t\}^{\frac{1}{2}(N+n_i-1)}}. \tag{25b}$$

Given $\tilde{\mathbf{B}}^{(1)} = \mathbf{1} = \overline{\mathbf{B}}^{(1)}$, the denominator of (25b) can, upon completing the square in $(B_{12}\mathbf{\Gamma}_1)$ and in $(B_{22}\mathbf{\Gamma}_2)$, be expressed as shown in (23). When say, $n_2 = q_2 = 0$, (25) reduces in form to $|BB^t|^{\frac{1}{2}(N-r-2)}$ times the synthetic representation (24)—the exact posterior kernel of $(\tilde{\mathbf{B}}^{(2)}\tilde{\Gamma})$ when cross section observations bearing on $(B_{12}\mathbf{\Gamma}_1)$ alone are available.

To demonstrate (25a) we examine the behavior of (22a) as $N \to \infty$ with n_1 and n_2 fixed.

(22a) is, aside from $2^{N+\frac{1}{2}(n_1+n_2)}\Gamma(\frac{1}{2}(N+n_1+n_2-1))(A_{11}+q_1)^{-\frac{1}{2}(N+n_1)} \times (A_{22}+q_2)^{-\frac{1}{2}(N+n_2)}$,

$$\sum_{j=0}^{\infty} \frac{2^{2j}\Gamma(j+\frac{1}{2})\Gamma(\frac{1}{2}N+\frac{1}{2}n_1+j)\Gamma(\frac{1}{2}N+\frac{1}{2}n_2+j)}{2j!\Gamma(\frac{1}{2}N+\frac{1}{2}n_1+\frac{1}{2}n_2+j)} \times \left\{\frac{A_{12}^2(1-R_1)(1-R_2)}{A_{11}A_{22}}\right\}^j$$

[10] This result is easily verified directly from (21) by transforming from Σ to $\boldsymbol{H}$, $\boldsymbol{H}$ to $(H_{11.2}, H_{12}, H_{22})$, $H_{11.2} = H_{11}-(H_{12}^2/H_{22})$ and then integrating first over $H_{12} \in (-\infty, +\infty)$ and then over $H_{22} > 0$ and $H_{11.2} > 0$.

$$\times \sum_{m=0}^{\infty} \sum_{k=0}^{\infty} \frac{[j+\frac{1}{2}]_{m+k}[\frac{1}{2}N+\frac{1}{2}n_1+j]_m[\frac{1}{2}N+\frac{1}{2}n_2+j]_k R_1^m R_2^k}{m!k![\frac{1}{2}N+\frac{1}{2}n_1+\frac{1}{2}n_2+j\,]_{m+k}}. \tag{26}$$

Using the relations $\Gamma(\frac{1}{2}N+x)/\Gamma(\frac{1}{2}N) = (\frac{1}{2}N)^x(1+o_1(N^{-1})) = [\frac{1}{2}N]_x \times(1+o_2(N^{-1}))$ for integral x and $\Gamma(x+\frac{1}{2})/\Gamma(2x) = \sqrt{\pi}/2^{2x-1}\Gamma(x)$, a generic coefficient in the triple sum (26) is to order $1/N$,

$$\sqrt{\pi}\Gamma(\tfrac{1}{2}N)\times\frac{[\frac{1}{2}N]_j[j+\frac{1}{2}]_{m+k}}{m!k!j!}. \tag{27}$$

Writing $[j+\frac{1}{2}]_{m+k} = [j+\frac{1}{2}]_m[j+m+\frac{1}{2}]_k$ we have

$$\sum_{m=0}^{\infty} \sum_{k=0}^{\infty} \frac{[j+\frac{1}{2}]_m[j+m+\frac{1}{2}]_k R_1^m R_2^k}{m!k!} = (1-R_1-R_2)^{-(j+\frac{1}{2})}$$

provided that N is large enough[11] so that $R_1+R_2 < 1$; and so substitution of (27) for coefficients in (26) yields the following approximation of (26) to order $1/N$:

$$\sqrt{\pi}\Gamma(\tfrac{1}{2}N)(1-R_1-R_2)^{-\frac{1}{2}} \sum_{j=0}^{\infty} \frac{[\frac{1}{2}N]_j}{j\,!}$$

$$\times\left\{\frac{A_{12}^2(1-R_1)(1-R_2)}{A_{11}A_{22}(1-R_1-R_2)}\right\}^j$$

$$= \sqrt{\pi}\Gamma(\tfrac{1}{2}N)(1-R_1-R_2)^{-\frac{1}{2}}\left(1-\frac{A_{12}^2(1-R_1)(1-R_2)}{A_{11}A_{22}(1-R_1-R_2)}\right)^{-\frac{1}{2}N}. \tag{28}$$

Using (28) and the definitions $R_i = q_i/(A_{ii}+q_i)$ we have

$$\{\sqrt{\pi}2^{N+\frac{1}{2}(n_1+n_2)}\Gamma(\tfrac{1}{2}(N+n_1+n_2-1))\Gamma(\tfrac{1}{2}N)\}|A|^{-\frac{1}{2}N}$$

$$\times\frac{A_{11}^{\frac{1}{2}(N-1)}A_{22}^{\frac{1}{2}(N-1)}}{(A_{11}+q_1)^{\frac{1}{2}(N+n_1-1)}(A_{22}+q_2)^{\frac{1}{2}(N+n_2-1)}}$$

$$\times\left(1-\frac{q_1q_2}{A_{11}A_{22}}\right)^{\frac{1}{2}(N-1)}\left(1-\frac{q_1q_2}{|A|}\right)^{-\frac{1}{2}N} \tag{29}$$

as an asymptotic approximation to (22a). Ignoring terms of order $1/N^2$, (29) is proportional to

[11] We assume that components of $\boldsymbol{M}$ are of order N as $N\to\infty$.

$$|A|^{-\frac{1}{2}N} \frac{A_{11}^{\frac{1}{2}(N-1)} A_{22}^{\frac{1}{2}(N-1)}}{(A_{11}+q_1)^{\frac{1}{2}(N+n_1-1)}(A_{22}+q_2)^{\frac{1}{2}(N+n_2-1)}}, \tag{30}$$

which upon writing out $\boldsymbol{A}$ and $A_{ii}+q_i$, $i=1, 2$ in terms of $\boldsymbol{B}$ and $\boldsymbol{\Gamma}$ yields (25).

We conclude by noting the obvious: the approximation (25) is an order of magnitude simpler than the exact posterior kernel (22a) and appears amenable to numerical analysis in a straightforward way.

References

[1] P. Appell and J. Kampé de Feriet, Fonctions hypergeométriques. Paris, Gauthiers-Villars, 1926.

[2] V. K. Chetty, On pooling of time series and cross-section data. Econometrica 38 (1968), 279–290.

[3] J. M. Dickey, Matricvariate generalizations of the multivariate *t* distribution and inverted multivariate *t* distribution. Annals of Mathematical Statistics 38 (1967), 503–511.

[4] J. M. Dickey, Expansions of *t* densities and related complete integrals, Annals of Mathematical Statistics 38 (1967), 511–518.

[5] J. Drèze, Limited information estimation from a Bayesian viewpoint. CORE Discussion Paper No. 6816, University of Louvain, 1968.

[6] H. Raiffa and R. Schlaifer, Applied statistical decision theory. Boston, Division of Research, Harvard Business School, 1961.

[7] G. Tiao and A. Zellner, Bayes theorem and the use of prior knowledge in regression analysis. Biometrika 51 (1964), 219–230.

[8] H. Wold and L. Jureen, Demand analysis. New York, John Wiley, 1953.

S. E. Fienberg and A. Zellner (eds.)
Studies in Bayesian Econometrics and Statistics.

9.3 BAYESIAN ANALYSIS OF SIMULTANEOUS EQUATIONS MODELS

Thomas J. ROTHENBERG*

1. *The logic of simultaneous equations systems*

In a series of important theoretical studies [8, 9, 11, 12], researchers associated with the Cowles Commission in the 1940s developed what has become known as the "simultaneous equations model" for the analysis of economic data. This statistical model has proved to be extraordinarily useful and has dominated much of the subsequent development in applied and theoretical econometrics. In recent years there has been considerable research activity concerned with recasting the simultaneous equations model using the methods of Bayesian inference. The present paper will survey some of this recent research and indicate the current status of the Bayesian approach to econometric model building.

The fundamental concept underlying any economic model is the *experiment* which is assumed to generate the observed data. An experiment consists of a set of "conditions" which are imposed on a situation and a set of "outcomes". Most economic theories specify that the outcome of an economic experiment is a random variable with a probability distribution which is a smooth function of the various variables describing the conditions of the experiment. The task of the econometrician is to measure as precisely as possible this functional relationship between the distribution of outcomes and the important economic conditions.

If the available economic data were in fact generated from simple controlled experiments, econometrics could use the standard statistical methods that have been developed for analyzing experimental data. Unfortunately, economic data come from the very complicated process

* The author is Associate Professor at the University of California, Berkeley.
Research for this paper was supported by the National Science Foundation.

of everyday economic life, not from the isolated experimentation of the social scientist. More often than not, the "explanatory" variables of an economic relationship are themselves the result of a complex stochastic mechanism. Although economic theory discusses a number of isolated "thought" experiments, the world consists of a large number of experiments performed all at once. Since an economic relationship is simply the description of some hypothetical experiment, we see that a model of the world which involves the interplay of many experiments naturally gives rise to the study of systems of simultaneous stochastic equations.

The simultaneous equations model can be applied to a variety of economic situations. However, in considering the logic of econometric model building and inference, it is important to distinguish between two general classes of applications. On the one hand, there are applications where the basic economic question involves only one hypothetical experiment and the problem is to draw inferences about the parameters of a single structural equation. Other experiments and other relationships are considered only as a means for learning about the given equation. These applications are typically studied using so-called "single-equation" or "limited-information" techniques. On the other hand, although somewhat less common, there are also applications where the basic economic question being asked depends in an essential way on an interdependent system of experiments. Inference about the parameters of a system of equations is required and "full information" methods of analysis are typically employed.

This distinction is not simply one of mathematical technique, but is fundamental to the entire process of model building. An example may clarify the matter. Consider the problem of estimating the production function for some economic unit. The question can be easily stated in terms of a stimulus-response experiment: give the firm a certain amount of inputs (e.g., labor, tools, buildings) and find out the maximum level of output attainable. In principle one could perform this experiment a number of times with different input levels and one could observe the output. The maintained hypothesis is that the observed output has a probability distribution such that its mean is a smooth function of the input levels. With sufficient data—and perhaps some assumed parametric form for the functional dependency—it would be possible to estimate the production function. Unfortunately, such an experiment is difficult and expensive to actually perform. But one does have information about various firms at various points of time—and this information includes data on inputs purchased and outputs produced. The levels of inputs, however, were not determined by some econometrician in an experimental setting; rather, they were determined (at least, in part) by the very managers of the firm who control the production process. In order to use these data to learn about the production technology, some assumptions have to be made about the mechanism which determines the input levels. The usual linear-model assumption that the errors in the production relationship are distributed independently of the input variables is not always plausible when production and input decisions are made by the same people. Although our question concerns the production function (defined in terms of an isolated controlled experiment), our analysis must also take into account other relationships and other experiments.

This example captures the essence of many econometric problems: we want to learn about a relationship defined in terms of a hypothetical experiment but the data we have available were in fact generated from a more complex experiment (typically describable by a system of interrelated experiments). We are not particularly interested in studying the process that actually generated the data, except insofar as it helps us to learn about the process we *wish* had generated the data.

Occasionally, however, the economic problem is of a different sort. There is no distinction between the hypothetical experiment and the actual one; the question concerns only the process that really did generate the data. That process, however, may be a very complicated one and the best way of learning about the process is to break it up into a set of subprocesses. For example, we may be interested in understanding the relationship between aggregate factor supplies and factor prices. One way of studying this is to examine a "reduced-form" relation between factor prices and the important exogenous variables (including aggregate factor supplies). However, a more efficient procedure might involve estimating a set of structural equations (including a number of production functions) and solving for the reduced form. That is, the best way to learn about the process determining factor prices might be to study a complete set of structural equations that describe the behavior of all the important groups of agents who constitute the market. No one structural equation is sufficient to answer the basic economic question.

The distinction between these two cases can be succinctly stated: in one case we are interested in a structural equation for its own sake; in the other case we estimate the entire structure to get better reduced-form estimates. Now I think it is universally agreed that our understanding of complicated economic processes is very limited. Models of actual economic mechanisms involving many people and many sectors are only very crude approximations to the models we would like to have. Thus two simple principles of model building are widely accepted: first, assumptions should be made only if actually needed to answer the question at hand; second, models should use efficiently the little information one has about the phenomena being studied. These principles suggest that the specification of econometric models must be closely tailored to the data that is available and the economic question being asked. Just because the world "really" consists of a large system of interdependent relationships does not mean that one should necessarily use large-scale models to answer economic questions. The task of the econometrician is to develop models that relate as efficiently as possible the available information and the economic questions to be answered. If assumptions about the world can be expressed in terms of subjective probability distributions, then a very flexible and powerful mode of analysis is provided by Bayesian statistical theory. With the use of subjective probabilities from the beginning, a wealth of possible models is available to help the researcher capture the crucial aspects of a very complicated stochastic process.

2. *A simple example*

Before we turn to general simultaneous equations models, it will be convenient to examine a special case. Nearly all of the interesting features of the simultaneous equations problem can be illustrated here without the need of complicated matrix manipulation. We consider the simple model

$$y_{1t} = \alpha y_{2t} + u_t \qquad (t = 1, \ldots, T), \tag{1}$$

where y_{1t} and y_{2t} are observations on two economic variables and u_t is an unobserved error. The unknown parameter α is given a causal interpretation in terms of a hypothetical experiment where y_{2t} is controlled and u_t is random noise. Unfortunately, the sample data were generated by a more complicated process where y_{2t} is uncontrolled and possibly related to the error u_t.

The Bayesian approach to the problem of learning about α involves two steps. First, the econometrician must specify a joint probability distribution for the observed variables and the parameter α. Second, on the basis of this joint distribution, the conditional distribution of α given the sample observations must be determined. This conditional (or posterior) distribution is then the basis for any inference statement about α. Now the second step involves purely mathematical manipulations; there are no conceptual problems involved. The difficult problem for the econometrician is at the first stage: to decide on a convenient specification that approximates his subjective beliefs and is at the same time computationally feasible.

It is necessary to specify a distribution on the $2T+1$ random variables $y_{11}, \ldots, y_{1T}, y_{21}, \ldots, y_{2T}$ and α. However, by relation (1), this is equivalent to specifying a distribution on the $2T+1$ variables $u_1, \ldots, u_T, y_{21}, \ldots, y_{2T}$ and α. Letting u be the T-dimensional vector of the u_t and y_2 be the T-dimensional vector of the y_{2t}, we can represent the basic $2T+1$ random variables by the triple (u, y_2, α). Before we specify a distribution on (u, y_2, α) it is useful to spell out more carefully the story which underlies the formal model. Presumably the error vector u represents some unobserved factors which influence the outcome of our hypothetical experiment. Since u is thought to be correlated with the independent variable y_2, there must be some mechanism which gives rise to this correlation. Our choice of a probability distribution on (u, y_2, α) must ultimately be based on our beliefs about this mechanism.

It will be convenient to begin the discussion by assuming that very little is known about the process generating the errors and the independent variables. Consider the T pairs (u_t, y_{2t}) as seen by the econometrician. Weak knowledge of the process suggests that the pairs are exchangeable; that is, the joint distribution of the T pairs is the same under any permutation of the order in which they are listed. This implies that the marginal density function for (u, y_2) can be written in the form

$$f(u, y_2) = \int \prod_t g(u_t, y_{2t}; \theta) h(\theta)\, \mathrm{d}\theta, \tag{2}$$

where g is a bivariate density function depending on a parameter vector θ and $h(\theta)$ is a density function for θ.[1] Of course (2) is not very restrictive since g and h can be *any* density functions. However, since

[1] For a further discussion of exchangeability see Lindley and Smith [13]. Throughout this paper we shall assume that all probability distributions can be represented by continuous density functions.

normality is so often a reasonable approximation for economic processes, it seems natural to assume that g is a bivariate normal density with mean vector zero[2] and covariance matrix $\Sigma = [\sigma_{ij}]$. If the process generating u and y_2 is not closely related to the causal process described by (1), it would seem plausible to assume that α and Σ are independently distributed. Further, if little is known about α, we might wish to use a prior on that parameter which is relatively uninformative compared to the sample evidence.

The above discussion suggests the following set of assumptions which define the inference problem:

(A1) The econometrician possesses a probability distribution over the $2T+1$ variables (u, y_2, α).

(A2) The distribution on the vectors u and y_2 reflects the exchangeability hypothesis; it is a mixture of the product of independent, identical bivariate distributions like (2). Conditional on α and Σ, each pair (u_t, y_{2t}) is normally distributed with mean zero and covariance matrix Σ.

(A3) The pair (α, Σ) has a joint distribution with α independent of Σ. The distribution of α is uniform over the entire real line.

Under these assumptions it is easy to derive the result that, conditional on Σ, the posterior distribution of α is normal with mean and variance given by

$$E(\alpha|y_1, y_2, \Sigma) = a - \frac{\sigma_{12}}{\sigma_{22}}$$

$$\operatorname{Var}(\alpha|y_1, y_2, \Sigma) = \frac{1}{y_2'y_2}\left(\sigma_{11} - \frac{\sigma_{12}^2}{\sigma_{22}}\right), \tag{3}$$

where a is the least-squares estimate $y_1'y_2/y_2'y_2$. The unconditional posterior distribution depends on the distribution of Σ which we have not yet specified. Two possibilities are worth exploring. First, suppose that the distribution is degenerate so that Σ is a fixed known matrix. In that case we are assuming that the T pairs (u_t, y_{2t}) are independent normal random variables, each with mean vector zero and covariance

[2] Since (1) contains no constant term, it is natural to center y_2 at zero. The simplest interpretation is that all variables are deviations from sample means. The general case of regression with constant term is covered in section 3.

matrix Σ. If $\sigma_{12} = 0$, then u_t and y_{2t} are independent and hence, with weak prior information on α, the posterior distribution is centered at the least-squares estimate. If $\sigma_{12} \neq 0$, then u_t and y_{2t} are correlated and the posterior distribution of α is shifted away from the least-squares estimate by an amount that depends on the degree of correlation.

A second possibility is to assume that Σ is a random variable with an inverted Wishart distribution. That is, Σ^{-1} has a density function proportional to

$$|S|^{\eta/2}|\Sigma|^{-(\eta-3)/2} \exp\{-\tfrac{1}{2}\operatorname{tr} S\Sigma^{-1}\}, \tag{4}$$

where the 2×2 positive definite matrix S is proportional to the mean of Σ; the scalar η is some positive integer. The Wishart distribution combines nicely with the normal distribution already assumed and it is possible to find a closed-form expression for the posterior density on α. In fact, that posterior is a Student distribution with mean given by

$$E(\alpha) = a - \frac{s_{12}}{s_{22}}.$$

Thus, the posterior mean is obtained by subtracting from the least-squares estimate a measure of the prior expectation of the regression coefficient of u_t on y_{2t}. The assumption of a known Σ is just the limiting case where the Wishart distribution is concentrated at one point.

Note that our assumptions do not distinguish between a likelihood function which gives "objective" probabilities and a prior distribution which gives "subjective" probabilities. The problem is to specify a joint distribution on u, y_2, and α such that it captures the uncertainties felt by the economist. The variable Σ is introduced simply as a device to help make the specification easier. There is never a need to assume that u_t and y_{2t} are "in fact" generated by a physical experiment that produces normal random variables with a covariance matrix Σ. All we are doing is suggesting that a family of plausible subjective distributions on u and y_2 can be represented by the mixture of normal distributions given in assumptions (A1–A3).

Now it may be plausible in situations of great ignorance to make assumptions (A1–A3) and to specify, say, a Wishart mixture. However, it is unlikely that these assumptions will be satisfactory for many economic problems; a crucial aspect is missing. The mechanism which causes the omitted variables constituting u_t to move systematically with the y_t is more complicated than that described by simple bivariate correlation. In practice, the relationship between the explanatory variable and the error involves other observed variables which enter in a predictable way. To use the usual language, there exist instrumental variables that are available to help make inferences about α.

Suppose, for example, there are K exogenous variables which are thought to influence y_{2t} but are unrelated to u_t. Then it seems natural to rewrite (1) as

$$y_{1t} = \alpha y_{2t} + u_t$$

$$y_{2t} = \pi_1 x_{1t} + \ldots + \pi_K x_{Kt} + v_t, \tag{5}$$

where the second equation reflects some causal mechanism determining y_{2t}. The errors u_t and v_t represent omitted variables and may contain common influences. Our model then might be defined in terms of the following assumptions:

(B1) The observed variables y_1 and y_2 are generated by two interrelated experiments described by the two equations of (5). The econometrician possesses a subjective joint probability distribution on the $K+1$ parameters $(\alpha, \pi_1, \ldots, \pi_K)$ and the T error pairs (u_t, v_t).

(B2) The distribution on the $2T$ errors reflects the exchangeability hypothesis; it is a mixture of the product of T independent, identical bivariate distributions like (2) with v_t replacing y_{2t}. Conditional on $\alpha, \pi_1, \ldots, \pi_K$, and Σ, each pair (u_t, v_t) is normally distributed with mean zero and covariance matrix Σ.

(B3) The distribution on α and the π's is independent of the distribution on Σ. The distribution on α and the π's is uniform in $K+1$ dimensions.[3]

Again, it is easy to derive the posterior distribution of α conditional on Σ. Introducing some simplifying notation, we find that the distribution is normal with

$$E(\alpha | y_1, y_2, \Sigma) = \frac{a - (\sigma_{12}/\sigma_{22}) - \rho^2 R^2 a^* + R^2(\sigma_{12}/\sigma_{22})}{1 - \rho^2 R^2}$$

$$\text{Var}\,(\alpha | y_1, y_2, \Sigma) = \frac{\sigma_{11}(1-\rho^2)}{y_2' y_2 (1 - \rho^2 R^2)}, \tag{6}$$

where $\rho^2 = \sigma_{12}^2/\sigma_{11}\sigma_{22}$ is the squared correlation coefficient between

[3] The assumption of a uniform prior density over an infinite range causes problems, especially when the dimensionality is large. Cf. Lindley and Smith [13]. We retain the assumption since attractive alternative assumptions have not yet been explored in the present context.

u_t and v_t; $R^2 = y_2'X(X'X)^{-1}X'y_2/y_2'y_2$ is the squared multiple correlation coefficient from the regression of y_2 on all the X's; $a^* = y_2'X(X'X)^{-1}X'y_1/y_2'X(X'X)^{-1}X'y_2$ is the two-stage least-squares estimate of α. Again, if $\sigma_{12} = 0$ (so that u_t and v_t are independent), the posterior mean is equal to the least-squares estimate; if $R^2 = 0$ (and hence there are *no* relevant instrumental variables), the mean is the same as given in (3) for the earlier model.

To obtain the actual posterior density, we must specify the distribution on Σ and integrate it out of the conditional posterior density given above. If the distribution is degenerate so that Σ is a known matrix, then the posterior on α is normal with moments given by (6). If Σ has the inverted Wishart distribution given in equation (4), the posterior is no longer normal but has a density proportional to

$$\left[\kappa_1 + s_{22}\left(\alpha - \bar{a} + \frac{s_{12}}{s_{22}}\right)^2\right]^{-(T^*-K+1)/2}$$
$$\times [s_{11} + (y_1 - \alpha y_2)'M(y_1 - \alpha y_2)]^{(T^*-K)/2}$$
$$\times [s_{11} + (y_1 - \alpha y_2)'(y_1 - \alpha y_2)]^{-(T^*/2)}, \tag{7}$$

where $T^* = T + \eta - 1$ and $M = I - X(X'X)^{-1}X'$; κ_1 is a constant depending on the data and $\bar{a} = y_1'My_2/y_2'My_2$.

If S is the zero matrix and η is a small positive number, the Wishart density (4) becomes the "non-informative" prior suggested by Drèze [4] and Zellner [18, p. 267]. In that case the posterior density for α becomes

$$\kappa_2[(y_1 - \alpha y_2)'M(y_1 - \alpha y_2)]^{(T^*-K)/2}$$
$$\times [(y_1 - \alpha y_2)'(y_1 - \alpha y_2)]^{-(T^*/2)}, \tag{8}$$

which is the "limited information" Bayesian result reported by both Drèze and Zellner.[4] The density (8) is closely related to the concentrated likelihood function of the limited-information maximum likelihood approach to the simultaneous equations problem. Moreover, for large sample size T, (8) is well approximated by the exponential

$$\kappa_3 \exp\{-\tfrac{1}{2}(y_1 - \alpha y_2)'X(X'X)^{-1}X'(y_1 - \alpha y_2)/\hat{\sigma}^2\}, \tag{9}$$

which is the kernel of a normal density centered at the two-stage least-squares estimate a^*. Thus, for large samples, Bayesian inference

[4] Drèze and Zellner use different values for η and hence their results end up with slightly different values for T^*. Also, it appears that Zellner has made an error in footnote 33 on page 269; his equation (9.76) is correct and equivalent to our equation (8), but (9.78) and (9.79) are incorrect.

procedures based on weak prior information yield essentially the classical limited information estimates.

3. General single-equation approaches

The results given for the one-regressor model generalize quite easily. Consider the linear model

$$y = Z\delta + u, \tag{10}$$

where y is a T-dimensional (column) vector of observations on a random variable, Z is a $T\times n$ matrix of observations on n explanatory variables, and u is a T-dimensional vector of unobserved errors; the n-dimensional vector of parameters δ is unknown and must be estimated from the sample. If Z were nonrandom and u had zero mean and scalar covariance matrix, we would have the traditional linear model. Here, however, we assume that some of the columns of Z are random and, along with u, possess a joint probability distribution. Specifically, we assume that Z can be partitioned as $[Y_1 X_1]$, where Y_1 is the $T\times g$ matrix consisting of those columns of Z which are random and X_1 is the $T\times k$ matrix consisting of the remaining non-random columns. Partitioning δ into corresponding subvectors, we can rewrite (10) as

$$y = Y_1\beta + X_1\gamma + u. \tag{11}$$

Bayesian inference on β and γ involves the calculation of a conditional probability distribution on the parameters given the observed sample. This distribution can be derived once we have specified the joint probability distribution of the parameters and the data. Using equation (11), this joint distribution is determined by specifying the joint distribution of u, Y_1, and δ. The error term u consists of unmeasured economic variables that are thought to be correlated with the variables in Y_1. Thus the econometrician must have in mind some economic mechanism which gives rise to this correlation. Just as in the simple case examined in section 2, we have two possible ways to proceed. On the one hand, we can suppose that the econometrician has some prior information about that correlation and that this information is not modified by the available data. On the other hand, we can make the alternative assumption that the economic mechanism relating u and Y_1 also involves other observed economic variables and can be described by a "reduced-form" equation. However, since the former case can, under certain assumptions, be subsumed as a special case of the latter, we shall not discuss it separately.

Let X be the $T \times K$ matrix of all the important exogenous variables that influence Y_1. Then we write

$$Y_1 = X\Pi' + V, \tag{12}$$

where Π is a $g \times K$ matrix of "reduced-form" coefficients and V is a matrix of reduced-form errors. The arguments given in section 2 suggest the following specifications:

(C1) The observed random variables y and Y_1 are generated by two interrelated experiments represented by equations (11) and (12). The econometrician possesses a subjective joint probability distribution on the $gK+n$ parameters (δ, Π) and the $T(g+1)$ components of the error matrix $[u \;\; V]$.

(C2) The error distribution is a mixture of the product of independent, identical $(g+1)$-variate distributions. Conditional on δ, Π, and the mixing parameter Σ, each row of $[u \;\; V]$ has an independent normal distribution with mean vector zero and $(g+1) \times (g+1)$ covariance matrix

$$\Sigma = \begin{pmatrix} \sigma_{11} & \omega' \\ \omega & \Omega \end{pmatrix},$$

where σ_{11} is the variance of each u_t, ω is the g-dimensional column vector of covariances between u_t and the v_{ti}. The $g \times g$ matrix Ω is the covariance matrix for the $(v_{t1}, \ldots, v_{tg})$.

(C3) The distribution on (δ, Π) is independent of the distribution on Σ. The distribution on (δ, Π) is uniform in $gK+n$ dimensions.

Conditional on Σ, the posterior distribution on δ is easily shown to be normal with mean vector and covariance matrix given by

$$\begin{aligned} E(\delta|y, Y, \Sigma) &= [Z'Z - \rho^2 Z'X(X'X)^{-1}X'Z]^{-1} \\ &\quad \times [Z'y - \rho^2 Z'X(X'X)^{-1}X'y \\ &\quad - Z'Y_1\phi + Z'X(X'X)^{-1}X'Y_1\phi] \\ &= [I_n - \rho^2 R^2]^{-1}[d - \rho^2 R^2 d^* - (I_n - R^2)\phi]; \\ \operatorname{Cov}(\delta|y, Y, \Sigma) &= \sigma_{11}(1-\rho^2)[Z'Z - \rho^2 Z'X(X'X)^{-1}X'Z]^{-1} \\ &= \sigma_{11}(1-\rho^2)[I_n - \rho^2 R^2]^{-1}(Z'Z)^{-1}, \end{aligned} \tag{13}$$

where $d = (Z'Z)^{-1}Z'y$ is the least-squares estimate of δ and $d^* =$

$[Z'X(X'X)^{-1}X'Z]^{-1}Z'X(X'X)^{-1}X'y$ is the two-stage least-squares estimate; R^2 (the matrix generalization of the reduced-form squared correlation coefficient), ρ^2 (the generalized squared correlation coefficient between u and V), and ϕ (the regression coefficient of u on V) are defined as follows:

$$R^2 = (Z'Z)^{-1}Z'X(X'X)^{-1}X'Z,$$

$$\rho = \frac{\omega'\Omega^{-1}\omega}{\sigma_{11}},$$

$$\phi = \Omega^{-1}\omega.$$

Although more complicated to write, these formulas for the mean and covariance matrix are obvious generalizations of those given in (6) for the single-regressor model.

Again, we must specify a probability distribution for Σ in order to obtain the marginal posterior density for δ. If the distribution is degenerate so that Σ is a fixed, known matrix, then δ is normal with moments given by (13). If Σ has an inverted Wishart distribution with density proportional to

$$|S|^{\eta/2}|\Sigma^{-1}|^{(\eta-g-2)/2}\exp\{-\tfrac{1}{2}\operatorname{tr} S\Sigma^{-1}\}, \tag{14}$$

then the posterior density on δ is given by

$$|A+S|^{-(T^*+K+1)/2}[s_{11}+(y-Z\delta)'M(y-Z\delta)]^{(T^*-K)/2}$$
$$\times[s_{11}+(y-Z\delta)'(y-Z\delta)]^{-(T^*/2)} \tag{15}$$

where $T^* = T+\eta-g$, s_{11} is the upper-left element of S, and A is the square matrix of order $g+1$

$$A = \begin{bmatrix} (y-Z\delta)'M(y-Z\delta) & (y-Z\delta)'MY \\ Y'M(y-Z\delta) & Y'MY \end{bmatrix}.$$

The special case of a "non-informative" prior proportional to $|\Sigma|^{-(\eta-g-2)/2}$ with a small value for η leads to the posterior density on δ proportional to

$$[(y-Z\delta)'M(y-Z\delta)]^{(T^*-K)/2}[(y-Z\delta)'(y-Z\delta)]^{-(T^*/2)} \tag{16}$$

where M again is the idempotent matrix $I-X(X'X)^{-1}X'$. As before, if T^* is very large, the posterior density is normal with mean given by the two-stage least-squares estimate. However, if K is large compared to T^*, the density given in (16) may depart significantly from normality and its mean may be far from the $2SLS$ estimate. Drèze [5] has investigated the density (16) and has shown that it is a proper density if $n \leqq K$ and that the rth moment of δ exists if $r \leqq K-n$. He has also applied this model to an economic problem and graphs the density for a particular data set.

All of the analysis so far assumes that there is no information about δ independent of the data. However, it is not at all difficult to relax that assumption and assume that δ has a prior normal density instead of a uniform one. As long as Σ is known, the posterior on δ remains normal, but the formulas in (6) and (13) must be modified to incorporate the prior information. If Σ^{-1} is Wishart distributed, a new factor representing the prior information multiplies the three factors making up (7) and (15), and the two factors making up (16). If the prior on δ takes the conjugate normal-gamma form, then considerable simplification occurs; for example, the posterior (16) remains valid under a simple redefinition of the symbols. For details, see Drèze [5].

In summary, under weak information on Π,[5] and a Wishart mixture of normals to represent the distribution of the errors, u and V, reasonably simple expressions can be found for the posterior density on δ. For small values of g this density can be plotted and moments calculated. Thus, for equations with few endogenous explanatory variables, computational Bayesian methods are available.

4. *Full-information methods*

The strategy employed in sections 2 and 3 was to make the simplest assumptions possible for the process generating the independent variables and the errors. As argued in section 1, this seems to be a reasonable procedure since our understanding of the complex mechanism which actually produces our data must be very small in most practical problems. However, sometimes there will be cases where information on a complete system of structural equations is available. The problem is no longer one of finding the simplest form for expressing our prior

[5] There should be no particular difficulty in generalizing these results for a natural conjugate prior on Π. However, as discussed in section 4, this is not a very satisfactory generalization when g is greater than one.

information, but one of finding a rich enough model for expressing this information.

A convenient starting point is the complete system of structural equations

$$By_t + \Gamma x_t = u_t, \tag{17}$$

where y_t is a G-dimensional column vector of endogenous variables, x_t is a K-dimensional column vector of exogenous variables, and u_t is a G-dimensional column vector of errors; B and Γ are parameter matrices of dimensions $G \times G$ and $G \times K$, respectively. Each equation in (17) represents some hypothetical controlled experiment and hence each parameter element of B and Γ has a causal interpretation. However, the data consist of T observations on the vectors (y_t, x_t) which were generated by the complete system of equations (17). Assuming B is nonsingular, we can write the reduced-form system

$$\begin{aligned} y_t &= -B^{-1}\Gamma x_t + B^{-1}u_t \\ &= \Pi^* x_t + v_t, \qquad (t = 1, \ldots, T), \end{aligned} \tag{18}$$

where Π^* is, by definition, the $G \times K$ matrix $-B^{-1}\Gamma$. Since y_t is a linear function of u_t, it is clear that the elements of y_t are in general correlated with the elements of u_t. Thus the individual equations constituting (17) cannot be studied using standard regression techniques.

The models discussed in sections 2 and 3 are easily seen to be special cases of the model presented here. Equation (11), for example, represents (for T sample points) one of the structural equations in (17); a normalization rule which sets one element of each row of B equal to unity can trivially be imposed. The mechanism which relates the other endogenous variables in a given equation to the errors is summarized by the reduced form. Thus the equations in (12) are a subset of the equations in (18).

In the previous sections we assumed very little about the mechanism relating the "right-hand" explanatory variables and the errors. Here we specify that the mechanism consists of the set of structural equations (17), one of which is the equation of particular interest. Indeed, here we will no longer distinguish between one equation of interest and the others; our task is to learn about all the elements of B and Γ. That is, the problem for the econometrician is to make inferences about the entire system of relations on the basis of a sample. If Y is defined as the $T \times G$ matrix of observations on the endogenous variables and X is defined as the $T \times K$ matrix of observations on the exogenous variables, equation (17) may be written as

$$BY' + \Gamma X' = U', \tag{19}$$

where U is a $T \times G$ error matrix. The model can be specified by the following set of assumptions:

(D1) The observed random variable Y is generated by a system of G interrelated experiments represented by the G equations in (17). The econometrician possesses a subjective probability distribution on the $G(G+K)$ parameters constituting (B, Γ) and the TG components of the error matrix U.

(D2) The distribution of the errors is a mixture based on T independent, identical G-dimensional distributions. Conditional on B, Γ, and the mixing parameter Σ^*, each row of U has an independent normal distribution with mean vector zero and $G \times G$ covariance matrix Σ^*.

(D3) The distribution on B, Γ, and Σ^* is informative and, in general, does *not* factor into a uniform prior on (B, Γ) and an independent distribution on Σ^*.

Assumptions (D1) and (D2) look very similar to assumptions (C1) and (C2) given earlier. This resemblance, however, is a bit deceptive and is based in part on a pun—a play on notation. The matrix Σ used in section 3 refers to the joint distribution of one structural error and g reduced-form errors. The matrix Σ^* concerns the joint distribution of G structural errors. It is by no means obvious that assumptions appropriate for the earlier model should carry over to the present one. Nevertheless, it seems fruitful to pursue the analogy and to note that the crucial difference between the full-information approach and the limited-information approach is in the third assumption. Here we assume that definite information about the whole structure is available; otherwise, it would be useless to introduce the structural mechanism into the analysis.

But what is a plausible family of distribution functions to describe the econometrician's information about B, Γ, and Σ^*? Unfortunately, this is an unanswered question at the present stage of theoretical Bayesian research. Distributions which combine conveniently with the normal distribution already assumed have very restrictive properties. In particular, they require that the $K \times K$ matrix of all the variances of the elements of Γ have rank one.[6] This is clearly an unacceptable restriction on subjective beliefs. Less restrictive prior distributions, however, do not combine easily with the normal distribution and hence lead to difficult computations.

Up to now, two approaches have been developed to cope with these difficulties. Harkema [10] uses the natural conjugate prior (which combines nicely with the normal distribution) and simply ignores the restrictive assumptions it imposes on prior beliefs. He argues that, despite its restrictiveness, this prior can approximate plausible information about simultaneous equations models. Drèze and Morales [7], Morales [14], and Richard [15], on the other hand, use an "extended" natural conjugate prior distribution which is somewhat richer in the type of information that can be incorporated. Even these extended prior distributions, however, are rather restrictive; and they yield computationally feasible results only for very small models. It appears that extensive numerical integrations will be necessary if flexible prior distributions on the structural parameters are to be employed.

Assumptions (D1–D3) express the econometrician's information about the process generating the data in terms of the structural parameters (B, Γ) and the structural errors u_t. In practice, however, it is often the case that the reduced-form matrix Π^* is the relevant parameter for economic decision making. For such cases, the final inference should be in terms of a posterior distribution on Π^* given the data and the prior information about the process. One possible approach would be to begin with the reduced-form equations (18) and replace (D1–D3) with assumptions that express the econometrician's information directly in terms of Π^* and the v_t. However, it is usually assumed that economists have better information about the individual structural equations (which describe hypothetical isolated experiments concerning autonomous subprocesses) than about the reduced form (which summarizes the entire economic mechanism). Thus an important part

[6] If Σ^* is permitted to have a degenerate distribution concentrated at one point, no difficulties arise; a general normal distribution on B and Γ combines nicely with the conditional normal distribution on U. If, however, we make the more usual assumption that Σ^* has a continuous density, then only the normal-Wishart family of distributions on B, Γ, and Σ^* seems to give rise to easy computations (see, for example, Ando and Kaufman [1] for a discussion of this "natural conjugate" family). The normal-Wishart family of distributions, unfortunately, forces the covariance matrices for each row of Γ to be proportional to each other. For details, see Morales [14, pp. 26–28] and Rothenberg [17, pp. 140–144].

of econometric theory is concerned with the implications for the reduced-form coefficient matrix Π^* and errors v_t of prior information on B, Γ, and the structural errors u_t. In traditional approaches to econometrics this involves the study of overidentifying restrictions on the reduced form that result from exact knowledge of some of the elements of B and Γ. In Bayesian terms, we want to know the relationship between probability distributions on (B, Γ, Σ^*) and the induced probability distributions on (Π^*, Ω^*), where $\Omega^* = B^{-1}\Sigma^* B'^{-1}$ is the mixing parameter for the reduced-form errors v_t. Unfortunately, the nonlinear equations that relate the structure and the reduced form make this study very difficult. Again, only very restrictive and unrealistic distributions on the structural parameters (B, Γ, Σ^*) give rise to simple and computationally convenient distributions on the reduced-form parameters (Π^*, Ω^*).

The distribution theory for systems of simultaneous equations appears to be very difficult. Plausible distributions that can capture our prior information about the economic process do not combine easily with the data to yield useful posterior distributions for either structural or reduced-form parameters. For small models, numerical integration techniques are available and, as Chetty, Morales, and Richard have shown, can be used to analyze actual data. Larger models cannot be analyzed using current methods without putting severe restrictions on the structural information permitted.

The full-information methods for analyzing simultaneous equations models assume that substantial information about B, Γ, and the structural errors is available. The technical problem of combining this information with the data is the major stumbling block. In practice, however, it is by no means obvious that considerable structural information is available. It may well be the case that our major information concerns specific structural equations and indicates at most only the degree of correlation between the explanatory variables and the errors. That is, we may in practice have better knowledge about the matrix Σ of sections 2 and 3 than about the matrices B, Γ, and Σ^*. If this is true, then limited information approaches may be more fruitful for econometric model building than the full information approaches discussed here.

The appropriate starting point for the analysis of economic data must depend crucially on the particular questions being asked and on the particular types of information available. If the entire mechanism which generates the data is well understood and considerable information about the various subprocesses is available, then full-information methods may be very useful. It is more likely, however, that the economic mechanism is not well understood and any system of equations is only a crude approximation to the actual mechanism. The problem of the econometrician is to find convenient ways to capture the little information we have. Bayesian methods must be applied in a flexible and imaginative way if they are to be useful for the difficult and vaguely specified problems that the practicing economist typically faces.

5. *Historical survey*

The presentation of results given in the past sections is organized according to the logical nature of the subject matter. Various contributions of different researchers have been merged and reinterpreted in order to tell a reasonably unified story. It will be useful, perhaps, to pause here and indicate the historical development of these results.

Two of the earliest attempts to apply Bayesian methods to the simultaneous equations model were the mimeographed papers by Drèze [4] and Rothenberg [16] that appeared in 1962 and 1963. Both papers take a full-information point of view and examine a complete system of structural equations. The Drèze paper is particularly concerned with the manner in which data can modify prior information on the structure. One of its major theorems is the identifiability result: conditional on the reduced-form parameters, the prior distribution on the structure is unaffected by the observations. This paper, although never published, has had a great impact on later work on the subject. Many of the results reported in Drèze [5, 6], Morales [14], Richard [15], and Rothenberg [16, 17] are hinted at in Drèze's earlier paper.

The problem with using natural conjugate priors for systems of equations was first emphasized in Rothenberg [16]. Also in that paper is a very simple example of Bayesian structural analysis where weak prior information is imposed on Haavelmo's consumption function model. This same example is extended in many directions in the 1966 doctoral thesis by Chetty; some of the major results are reported in his article [2]. Further case studies are made by Harkema [10], Morales [14], and Richard [15].

In a series of papers beginning in 1962, Zellner has produced a striking number of results on the application of Bayesian methods to econometrics. Most of these results are summarized in his book [18].

The development given in sections 2 and 3 for single-equation models is essentially a reformulation of models given in chapters 5 and 9 of that book. Very similar results were obtained by Drèze [5, 6] using a somewhat different starting point.

Research on the simultaneous problem continues and many important contributions have undoubtedly been omitted. Alas, a survey of a rapidly growing young subject must necessarily be out of date before it is in print. That must surely be a sign of the healthy state of Bayesian econometric research.

References

[1] A. Ando and G. Kaufman, Bayesian analysis of the independent multinormal process—neither mean nor precision known. Journal of the American Statistical Association 60 (1965), 347–358.

[2] V. K. Chetty, Bayesian analysis of Haavelmo's models. Econometrica 36 (1968), 582–602. (Reprinted in the present volume.)

[3] J. Dickey, Matric-variate generalizations of the multivariate *t* distribution and the inverted multivariate *t* distribution. Annals of Mathematical Statistics 38 (1967), 511–518.

[4] J. Drèze, The Bayesian approach to simultaneous equation estimation. O.N.R. Research Memorandum 67, Northwestern University, 1962.

[5] J. Drèze, Bayesian limited information analysis of the simultaneous equations model. CORE Discussion Paper 7111, 1971.

[6] J. Drèze, Econometrics and decision theory. Econometrica 40 (1972), 1–17.

[7] J. Drèze and J. Morales, Bayesian full information analysis of the simultaneous equations model. CORE Discussion Paper 7031, 1970.

[8] T. Haavelmo, The statistical implications of a system of simultaneous equations. Econometrica 11 (1943), 1–12.

[9] T. Haavelmo, The probability approach to econometrics. Supplement to Econometrica 12 (1944).

[10] R. Harkema, Simultaneous equations: A Bayesian approach. Rotterdam University Press, 1971.

[11] W. Hood and T. C. Koopmans (eds.), Studies in econometric method. Cowles Commission Monograph 14. New York, Wiley, 1953.

[12] T. C. Koopmans (ed.), Statistical inference in dynamic economic models. Cowles Commission Monograph 10. New York, Wiley, 1950.

[13] D. V. Lindley and A. F. M. Smith, Bayes estimates for the linear model. Journal of the Royal Statistical Society, Series B, 34 (1972), 1–18.

[14] J. A. Morales, Bayesian full information structural analysis. Berlin, Springer-Verlag, 1971.

[15] J. F. Richard, Posterior and predictive densities for simultaneous equations models. Doctoral thesis, Université Catholique de Louvain, 1973.

[16] T. J. Rothenberg, A Bayesian analysis of simultaneous equations systems. Report 6315, Econometric Institute, Rotterdam, 1963.
[17] T. J. Rothenberg, Efficient estimation with a priori information. Cowles Foundation Monograph 23, Yale University Press, 1973.
[18] A. Zellner, Introduction to Bayesian inference in econometrics. New York, Wiley, 1971.

PART 10

CONTROL PROBLEMS

S. E. Fienberg and A. Zellner (eds.)
Studies in Bayesian Econometrics and Statistics.

10.1 ADAPTIVE DECISION RULES FOR MACROECONOMIC PLANNING*

Edward C. PRESCOTT**

Carnegie-Mellon University

Leading economists have proposed the use of linear decision rules to formulate macroeconomic policy.[1] Provided the loss function is quadratic and the econometric model linear with known values of coefficients, this method provides a simple and elegant means of finding optimal policy. But, in any macroeconomic application, the coefficients' values are not known and must be estimated using the limited number of available and relevant historical observations and consequently, the coefficients' values will not be known with certainty. Proponents [13, p. 77] of the use of linear decision rules recognize this saying "their specification is a problem of statistical inference". For the single period problem Fisher [5] has shown Bayesian estimators do in fact exist, but no analogous result exists for the multiperiod problem.

This analysis also assumes linear constraints and quadratic preferences, but does not assume the values of the coefficients of the constraints are known with certainty. Instead of estimating the coefficients, the Bayesian approach of using a distribution to represent one's knowledge of the unknown parameters is utilized. The coefficient uncertainty introduces two additional complications. First, the variances of the performance variables now depend upon the levels of the government instruments of control. In addition, there is the element of experimentation for the decisions selected today affect what is known to-

* Reprinted with minor changes from the Western Economic Journal 9 (1971), 369–378, with permission of the Western Economic Association.

** The author acknowledges helpful comments of Morris H. DeGroot, Michael C. Lovell, Henri Theil, and Arnold Zellner. He retains full responsibility for the views presented.

[1] The techniques commonly referred to as linear decision rules were developed by Simon [11] and Theil [12]. Bogaard and Theil [2], Holt [6], and Theil [13] have proposed the use of this technique for economic planning.

morrow (actually affect the distribution of posterior distribution). For example, by trying a tax cut, a more precise estimate of the tax multiplier could be obtained, useful information on selecting future decisions. The adaptive decision rules procedure is an approximate solution to this complicated problem for the case when experimentation is not an important consideration. It selects in each period the decision that would be optimal if no further information about the coefficients would become available. Like linear decision rules, these rules can be computed for complex systems involving large numbers of variables and complex quadratic loss functions.

Section 1 explores a simple structure for which computation of the optimal decisions is possible. The performances of linear and adaptive decisions rules are compared to the optimal procedure. Except when uncertainty about the unknown coefficients is great, exceeding those levels generally encountered by econometricians, adaptive decision rules are nearly optimal. Linear decision rules, however, perform well only when uncertainty is small. Section 2 presents the problem and develops the adaptive decision rules. Section 3 summarizes the results of tests comparing the performance of linear and adaptive decisions rules, using econometric models to simulate the American economy. For these simulated environments, adaptive decision rules performed significantly better, reducing average loss by between 30 and 50%.

1. *An example*

The following simple example illustrates the concept of experimentation and indicates when it is important. Suppose national income $y(t)$ is linearly related to a policy instrument $x(t)$ plus an error term $u(t)$:

$$y(t) = \beta x(t) + u(t). \tag{1}$$

The multiplier β is unknown and the $u(t)$ are independent and normally distributed with mean 0 and known variance, say 1. The loss function is the sum of the squared deviation of national income from the full employment level g,

$$\text{Loss} = [y(1) - g]^2 + [y(2) - g]^2 \tag{2}$$

and the policymaker wishes to minimize its expected value.

Suppose also the policy maker has no prior information about β except that provided by past observations. It is well-known [10, p. 337] that the distribution on β at time t will be normal with mean, $m(t)$, equal the least squares estimator and variance, $\sigma(t)^2$, equal the variance of the least squares estimator:

$$m(t) = \sum_{-N}^{t-1} x(i)y(i) / \sum_{-N}^{t-1} x(i)^2 \quad \text{and} \quad \sigma(t)^2 = 1 / \sum_{-N}^{t-1} x(i)^2. \tag{3}$$

From (3) the knowledge of β at time two will depend upon $x(1)^2$, the larger its value, the smaller the variance of the posteriori distribution. The selection of a large $x(1)^2$ results in a very precise knowledge of the unknown multiple β in the second period.

The conventional approach to this decision problem is to estimate the parameter β and select the decisions that would be optimal if the estimate b were without error; that is, select

$$x(t) = g/b. \tag{4}$$

This is my interpretation of the procedure proposed by the proponents of linear decision rules for macroeconomic planning as applied to this simple problem. The performance of the rule will depend upon the method of estimation. Traditional econometricians would surely agree that the least squares estimator is appropriate, for it has all the "desirable" statistical properties. This implies

$$x(t) = g/m(t). \tag{5}$$

Since the expected loss in period t is

$$E[y(t)-g]^2 = [m(t)x(t)-g]^2 + x(t)^2\sigma(t)^2 + 1 \tag{6}$$

substituting for $x(t)$ yields

$$[\sigma(t)/m(t)]^2 g^2 + 1 \tag{7}$$

as the expected loss for decision rule (5). When uncertainty in β is great, namely when $[m(t)/\sigma(t)]^2$ is small, the conventional approach has large expected loss. But, in such situations the estimate is not statistically significant, and traditional econometricians would recommend a passive policy. If the variables are measured as deviations from their initial values, the decision $x(t) = 0$ corresponds to no change in policy and is the logical passive policy. It has smaller expected loss than (5) whenever $|\sigma(t)/m(t)| \leqq 1$, which suggests the following linear decision rule procedure:

$$x(t) = g/m(t), \quad \text{if} \quad |m(t)/\sigma(t)| \geqq 1, \quad \text{and} \qquad (8)$$
$$x(t) = 0, \qquad \text{otherwise.}$$

An alternative procedure is to select the decision that minimizes expected loss if learning were not a relevant consideration. For this simple problem,[2] this adaptive decision rule is obtained by minimizing (6) with respect to $x(t)$ to obtain

$$x(t) = gm(t)/\{m(t)^2 + \sigma(t)^2\}. \qquad (9)$$

The third procedure is the optimal decision rule which selects the best trade-off between stabilization and experimentation. The larger $x(1)^2$, the more informative is the experiment and the smaller the minimal obtainable expected loss in period two. For this very simple example, the optimal solution can be computed using backward inductions and numerical methods.[3]

The expected losses for each of the three decision procedures depend upon $m(1)/\sigma(1)$ and the target income. Table 1 presents the expected losses for the decision rules for selected $m(1)/\sigma(1)$ and the corresponding first-period decisions when $g = 8$. The table reveals experimentation is important only when uncertainty is large, say $|m(1)/\sigma(1)| < 0.6$, and the greater the uncertainty, the greater the return for experimentation. If the t-statistic (actually the ratio of the least squares estimator to its standard deviation) is large, both the linear and adaptive rules are nearly optimal. In the intermediate range that characterizes many macroeconomic situations, only the adaptive decision rule is nearly optimal.

2. *The example*

Suppose the decision environment is characterized by stochastic linear constraints with unknown coefficients and the objective is to minimize

[2] This rule corresponds to the conventional approach if Fisher's [5] certainty equivalence estimator is used. If, however, dynamic complications resulting from lagged variables or inter-period terms in the loss function are present, Fisher's concept of a certainty equivalence estimator is not applicable.

[3] See [9] for a more complete analysis of the multiperiod control problem for this simple regression model.

TABLE 1

Comparison of alternative decision rules, $T = 2$ and $g = 8$

	Expected losses			First period decision $[\sigma(1) = 1]$[a]		
$m(1)/\sigma(1)$	Certainty equivalent	Adaptive	Optimal	Certainty equivalent	Adaptive	Optimal
0.0	130.0	130.0	94.4	0.0	0.0	2.78
0.3	130.0	84.8	79.2	0.0	2.20	3.60
0.6	130.0	64.1	62.5	0.0	3.53	4.44
1.0	130.0	44.4	43.9	0.0	4.00	4.47
1.4	41.0	31.1	30.8	5.71	3.78	4.06
2.0	22.1	19.0	18.9	4.00	3.20	3.33
4.0	7.0	6.8	6.8	2.00	1.88	1.91
8.0	3.5	3.5	3.5	1.00	0.98	0.99

[a] To obtain decision values when $\sigma(1) \neq 1$, divide by $\sigma(1)$.

the expected value of a quadratic loss function. In vector equation notation the model is

$$y(t) = Ay(t-1) + Bx(t) + Cz(t) + u(t) \qquad t = 1, \ldots, T \quad (10)$$

where $y(t)$ is a column vector of endogenous variables, $x(t)$ the decision instruments, $z(t)$ observable exogenous elements, and $u(t)$ unobservable disturbance. The elements A, B, and C are matrices of coefficients, some of which are unknown. In a macroeconomic application, (10) would be the reduced form of the econometric model approximating the economy. The loss function is assumed to be of the form

$$\text{Loss} = \sum_{t=1}^{T} q_t[y(t)] \quad (11)$$

where the q_t are quadratic and nonnegative. The error vectors are identically and independently distributed normal variates with mean 0. The vectors $z(t)$ are distributed independently of each other and of the disturbances, and have known distribution with finite second moment.

The representation, (10) and (11), is far more general than it may appear. By expanding the dimensionality of $y(t)$, all finite lag structures can be put into this form. For example, if the model were

$$y(t) = A_1 y(t-1) + A_2 y(t-2) + B_1 x(t) + B_2 x(t-1) + u(t) \quad (12)$$

it could be represented as

$$\begin{bmatrix} y(t) \\ y(t-1) \\ x(t) \end{bmatrix} = \begin{bmatrix} A_1 & A_2 & B_2 \\ I & 0 & 0 \\ 0 & 0 & 0 \end{bmatrix} \begin{bmatrix} y(t-1) \\ y(t-2) \\ x(t-1) \end{bmatrix} + \begin{bmatrix} B_1 \\ 0 \\ I \end{bmatrix} x(t) + \begin{bmatrix} u(t) \\ 0 \\ 0 \end{bmatrix} \tag{13}$$

putting it into the form of (10). This also demonstrates that the arguments of the loss function can include instruments, exogenous, and lagged variables. For example, if change in an instrument were an argument in the loss function then $x(t-1)$ as well as $x(t)$ must be included in the left-hand side vector of (13).

The unknown parameters in the model are the unknown coefficients of the linear constraints and parameters specifying the covariance matrix of the error vector $u(t)$. Bayesian analysis can be used to form an initial prior using past observations and other information provided by economic theory.[4] Let $I(t)$ be a set of sufficient statistics, specifying the distribution at time t. The distribution at time $t+1$ will be a function of $I(t)$ and the most recent observation:

$$I(t+1) = I[I(t), y(t), x(t), z(t), y(t-1)]. \tag{14}$$

This relationship describes the evolution of the information pattern during the decision process.

A discrete dynamic programming formulation will be used for this problem. The state of the system is described by the information vector $I(t)$ and $y(t-1)$. The distribution of future outcomes depends only upon these variables and future decisions. Letting f_t be the minimal obtainable expected loss for periods t through T, this Markov property implies

$$\begin{aligned} f_t[I(t), y(t-1)] = \min_{x(t)} E\{q_t[y(t)] \\ + f_{t+1}[I(t), y(t)] \,|\, I(t), y(t-1)\} \end{aligned} \tag{15}$$

for $t = 1, \ldots, T$ and with $f_{T+1} = 0$. The first term in the squared brackets gives the expected effect of $x(t)$ upon current loss and the second, the expected effect of $x(t)$ upon future losses, given the best policies for $t > T$. The $x(t)$ that minimizes the expression on the right

[4] In recent years a considerable literature has appeared concerning the problem of developing the posterior or approximate posterior distribution while incorporating structural form information. A good summary is provided by Zellner [15].

of (15) depends upon $I(t)$ and $y(t-1)$, so the optimal decision rule is of the form

$$x(t)^{\circ} = d_t[I(t), y(t-1)]. \tag{16}$$

The above system can be solved in theory by determining each f_t beginning with the last and using it to determine the one before. But, except for simple cases, as in section 1, solution does not appear to be computationally feasible.

When experimentation is not an important consideration, however, an approximate solution can be obtained even for complex systems involving tens or even hundreds of variables. With this approximate solution, called adaptive decision rules, the decision selected in period t is the one that would be best if the distribution on the unknown parameters were not updated subsequently. Before selecting the next decision, however, the most recent observation is used to update the distributions. Zellner [15] has called this procedure the sequential updating rule. The results of section 1 indicate that this approach is likely to be far better than the linear decision rules if there is moderate uncertainty in the values of the coefficients.

Computationally, the initial information vector $I(1)$ is substituted for $I(t)$ in the functional equation system (15) to obtain

$$h_t[y(t-1)] = \min_{x(t)} E\{q_t[y(t)] + h_{t+1}[y(t)] \mid I(1), y(t-1)\} \tag{17}$$

for $t = 1, \ldots, T$ and with $h_{T+1} = 0$. This system of functional equations is easily solved because h_{t+1} being quadratic implies h_t is quadratic, so h_{T+1} being quadratic implies by mathematical induction all the h_t are quadratic. Thus, only a finite number of parameters are needed to characterize the h_t functions.[5] An examination of (17) reveals only the first two moments of the prior distribution on the unknown coefficients is needed to determine these functions. Once h_2 has been determined, the approximately optimal $x(1)$ is implied. At the time of decision t, one must solve a $T-t+1$ period problem given $I(t)$ for this is the number of periods remaining in the planning horizon.

[5] Computationally, adaptive decision rules and the random coefficients model of the control engineers [1, pp. 37–53] are equivalent. Chow [3] has an excellent exposition of the computational procedure and has developed a general computer program which does the computations.

3. Tests

Economic policy games, simulating certain important elements of the decision environment facing President Johnson in the 1965–1968 period, were used to compare the performance of linear and adaptive decision rules in macroeconomic planning.[6] A computer program, using econometric models with stochastic shocks, was written to simulate the American economy. Chow's [4] econometric model formed the basis for the two models utilized in the study. It is a small five-equation model which is basically Keynesian with distributed lags to describe the equilibrium adjustment process, but it has the money supply as an argument of the consumption function. The Chow model was designated as Economy-1. Whether money did in fact affect consumption directly is uncertain, since the standard error of the estimated effect was as large as the estimate itself. Therefore, Economy-2, which was like Economy-1 except that money was dropped from the consumption function, was also considered. For both models the government instruments of control were the money supply and government spending. In the latter, money had a smaller effect, and more precise information was needed to use it effectively.

The historical period used to draw initial inference about the coefficients of the econometric model was 1951–1964 inclusive. Actual observations for this period could not be used, because they were part of the data used by Chow to estimate his model. If the sample period and estimation procedure are the same as the ones used by Chow, the estimated parameters will be equal to those of Economy-1. If the estimates are equal to the true values, linear decision rules would be optimal. For this reason, stochastic simulations were used to obtain the historical data.[7]

3.1. Loss function

The loss function was

$$\sum_{t=1965}^{1968} \{2[Y(t)-Y(t)^D]^2+[C(t)-C(t)^D]^2 + [G(t)-G(t)^D]^2+[I(t)-I(t)^D]^2\} \tag{18}$$

[6] A more complete description of the simulation experiments can be found in [8].

[7] A rule was needed to specify how decisions were selected during this period. Regression analysis was used to explore relationships between lagged variables and

where $Y(t)$ denotes GNP, $C(t)$ consumption, $G(t)$ government spending, and $I(t)$ gross private investment, and where the superscript D indicates desired levels. Okun's [7] estimates for capacity gross national product were used for the $Y(t)^D$; $I(t)^D$; $G(t)^D$ and $G(t)^D$ were set at 16, 20 and 64% of capacity respectively. The first term in the loss function reflects the trade-off between unemployment and inflation, while the remaining terms reflect a desired distribution of product among private and public consumption and investment.

3.2. *Decision procedures*

The reduced form for the models was estimated by regression analysis. Linear decision rules were then used to determine the decision that would be optimal if the estimates were without error. After each observation the model was re-estimated before computing the next decision. This simulated a policymaker using linear decision rules (LDR). Experiments were made using two-stage least squares to estimate the structural form imposing certain sign constraints. For these examples the simple method of estimating the reduced form directly performed much better when LDR were used to select decisions.

To use adaptive decision rules (ADR), a method for finding the first two moments of the posterior on the unknown coefficients is needed. Proceeding in the simplest way, using the reduced form, diffuse priors, and ignoring structural constraints, the posterior distribution was derived. Using a result of [14], the mean of the distribution is just the least squares estimators of the reduced form coefficients, while the covariance matrix equals the usual unbiased estimator of the covariance matrix of the least squares estimators.

In practice, there would be a considerable amount of prior information about the magnitude of many of the coefficients in the model. By incorporating such information using Bayesian methods, the performance of adaptive decision rules would surely have been better. This was not done because we were testing whether Bayesian decision procedures which recognize parameter uncertainty would improve economic planning and not whether Bayesian estimation methods can be used to reduce coefficient uncertainty.

the instruments. Since no significant relationships were found, changes in the controls were set equal to a constant plus a random term having variance equal to the maximum likelihood estimator for the period.

3.3. *Summary of results*

Both of the decision procedures were tested a number of times to compare their performance in achieving the economic objectives with Economy-1 and with Economy-2 simulating the American economy. As a benchmark, the loss for perfect information decision rules (PIDR) was also computed. With this technique, the decision selected was the one that would be optimal if the true coefficient values, but not the size of the stochastic shocks, were known. The tests were performed for three uncertainty levels that were defined by the variance of the error terms in the structural equations simulating the American economy. The high level used the maximum likelihood estimates of the variances, with actual 1953–1964 data and Chow's model. The level of uncertainty appeared excessive, so a medium and low level were also used. The mid-level was obtained by reducing the variances by approximately a factor of 4. The resulting reduction in the coefficient uncertainty was equivalent to that which would have occurred if the sample period had been doubled. The variances were divided by another factor of 4 to obtain the low uncertainty level.

In order to increase the power of the test, both decision procedures were evaluated for the same set of random disturbances, each set constituting a simulation. Table 2 lists the losses and their averages. For both economies at each level of uncertainty, adaptive decision rules had the smaller average losses. For Economy-2, LDR had average loss nearly double that of ADR. The difference in performance is statistically significant in every case except Economy-1 low uncertainty.[8] Apparently in this case, the degree of uncertainty in the coefficients was sufficiently low that LDR was nearly optimal, and little was gained by using the adaptive procedure. This was not true for Economy-2 with low uncertainty, because for that model money had a smaller effect and a more precise estimate of its effect was needed to use that instrument effectively.

In summary, the results of these tests clearly demonstrate the superiority of the adaptive Bayesian approach in the context of these simulated economies. Before using the techniques as a guide to policy formulation,

[8] Actually I tested whether the median of the differences in performance was less than zero. Given that the distribution is skewed to the left, the mean will be less than median, making this a conservative test. Because the distribution of the differences did not approach normality and sample sizes were small, the t-test was not appropriate.

TABLE 2

Losses for the alternative decision procedures[a]

	Low uncertainty					
	Economy–1			Economy–2		
Simulation	LDR	ADR	PIDR	LDR	ADR	PIDR
1	2.27	1.96	1.27	20.7	11.4	3.1
2	1.56	1.45	1.26	16.1	12.5	2.8
3	1.36	1.59	1.11	17.5	9.9	3.0
4	1.96	1.81	1.29	37.4	14.5	5.5
5	2.00	2.05	1.25	15.8	12.4	2.6
6	2.91	1.99	1.02	26.0	12.9	3.0
7	1.50	1.61	1.45	13.2	6.4	2.8
8	1.44	1.55	1.23	31.7	11.2	3.1
9	1.58	1.23	0.98	29.9	9.4	5.7
10	1.21	1.38	1.20	20.7	5.8	3.7
11	3.00	2.55	1.33	20.2	14.6	4.4
12	1.34	1.34	1.28	19.5	12.4	3.0
Average	1.84	1.71	1.22	22.4	11.1	3.6

	Medium uncertainty					
	Economy–1			Economy–2		
Simulation	LDR	ADR	PIDR	LDR	ADR	PIDR
1	3.62	4.11	1.85	18.4	10.2	2.2
2	2.15	1.96	1.59	5.7	6.6	3.1
3	5.75	3.38	2.27	51.8	20.0	3.3
4	3.63	1.69	1.54	40.3	15.7	4.0
5	2.54	2.45	2.34	11.8	12.2	4.6
6	12.21	4.81	1.40	47.8	18.1	4.0
7	5.80	2.13	1.26	29.2	15.0	3.0
8	1.72	2.31	1.41	21.1	14.2	2.8
9	1.48	1.58	1.36	47.9	13.8	2.3
10	1.94	1.72	1.72	20.6	18.9	4.9
Average	4.08	2.61	1.67	29.5	14.5	3.4

TABLE 2 (*continued*)

	High uncertainty[b]			
	Economy–1		Economy–2	
Simulation	LDR	ADR	LDR	ADR
1	5.5	5.2	24.8	9.9
2	8.7	6.0	12.4	8.7
3	8.7	6.5	7.4	8.9
4	10.9	5.9	6.1	5.9
5	10.6	10.0	18.0	4.0
6	5.9	4.2	14.3	11.0
7	9.5	4.4	24.4	11.6
8	5.7	3.9	8.6	3.9
9	6.5	6.0	23.4	40.8
10	9.0	6.6	50.7	22.7
Average	8.1	5.9	19.0	12.7

[a] Actual loss for the period was approximately 5 given the units in which loss were measured.

[b] With high uncertainty and linear decision rules, the implied values for the instruments were frequently excessively large or small. In these cases, G_t was set at its desired level and the money supply grew by 5 per cent. For high uncertainty, the capacity output was assumed to grow at 3 rather than 3.5 per cent.

however, additional tests using more complex models and realistic loss functions are clearly warranted. Additional research is needed to determine whether linear constraints can adequately approximate economic systems and how best to approximate a policymaker's preference ordering using a quadratic function.

References

[1] M. Aoki, Optimization of stochastic systems. New York, John Wiley, 1967.

[2] P. J. M. vandem Bogaard and H. Theil, Macro dynamic policymaking: an application of strategy and certainty equivalence concepts to the economy of the United States, 1933–1936. Metroeconomica 11 (1959), 149–167.

[3] G. C. Chow, Effect of uncertainty on optimal control policies. Econometric Research Program Research Memorandum No. 145, September 1972.

[4] G. C. Chow, Multiplier, accelerator, and liquidity preference in the determination of national income in the United States. Rev. Econ. Stat. 49 (1967), 1–15.

[5] W. Fisher, Estimation in the linear decision model. Internat. Econ. Rev. 3 (1962), 1–29.

[6] C. C. Holt, Linear decision rules for economic stabilization and growth. Quart. Jour. Econ. 76 (1962), 20–45.

[7] A. Okun, The battle against unemployment. New York, W. W. Norton, 1965.

[8] E. C. Prescott, Adaptive decision rules for macro-economic planning. Unpublished doctoral dissertation, Carnegie-Mellon University, 1967.

[9] E. C. Prescott, The multi-period control problem under uncertainty. Econometrica 40 (1972), 1043–1058.

[10] H. Raiffa and R. Schlaifer, Applied statistical decision theory. Boston, Harvard Business School, 1961.

[11] H. A. Simon, Dynamic programming under uncertainty with a quadratic criterion. Econometrica 24 (1956), 74–81.

[12] H. Theil, A note on certainty equivalence in dynamic planning. Econometrica 25 (1957), 346–349.

[13] H. Theil, Optimal decision rules for government and industry. Amsterdam, North-Holland, 1964.

[14] G. C. Tiao and A. Zellner, On the Bayesian estimation of multivariate regression. Jour. Royal Stat. Soc., B, 27 (1964), 277–285.

[15] A. Zellner, Introduction to Bayesian inference in econometrics. New York, John Wiley, 1971.

[16] A. Zellner and V. K. Chetty, Prediction and decision problems in regression models from the Bayesian point of view. Jour. Am. Stat. Assoc. 60 (1965), 608–616.

[17] A. Zellner and M. S. Geisel, Sensitivity of control to uncertainty and form of the criterion function. In: The future of statistics, edited by D. G. Watts. New York, Academic Press, 1968, 269–289.

S. E. Fienberg and A. Zellner (eds.)
Studies in Bayesian Econometrics and Statistics.

10.2 STOCHASTIC OPTIMIZATION IN RECURSIVE EQUATION SYSTEMS WITH RANDOM PARAMETERS WITH AN APPLICATION TO CONTROL OF THE MONEY SUPPLY*

H. Woods BOWMAN

University of Illinois, Chicago

and

Anne Marie LAPORTE**

Federal Reserve Bank of Chicago

A promising approach to decision making with econometric models has been developed by Holt and Theil who postulate a quadratic utility (or loss) function in the criteria variables.[1] Provided the model is linear with *known* coefficients, the optimal policy is found to be one for which the criterion function is an extremum.[2] Since, in practice, the coefficients of a model are not known the technique utilizes the mean values of the coefficient estimators and for this reason it is known as the *certainty equivalence* approach.

* Reprinted from Annals of Economic and Social Measurement 1 (1972), 419–435, with permission of the National Bureau of Economic Research, Inc. Some of the work contained in the paper was presented to the First NBER-NSF Seminar on Bayesian Inference in Econometrics, Univ. of Chicago, December 1970. Most of the work on this paper was done while Dr. Bowman was affiliated with the Federal Reserve Bank of Chicago. The authors would like to express their appreciation to Karl A. Scheld of the Federal Reserve Bank of Chicago and Arnold Zellner of the University of Chicago for their guidance and encouragement throughout the project. Calculations were performed using computer facilities at the Federal Reserve Bank of Chicago.

** Ms. Laporte is primarily reponsible for the final section which treats the application. Dr. Bowman is primarily responsible for the remainder of the paper.

[1] C. C. Holt, Linear decision rules for economic stabilization and growth, Quarterly Journal of Economics 67 (1962), and Henri Theil, Optimal decision rules for government and industry (Amsterdam, North-Holland, 1964), ch. 6.

[2] Among their other virtues, quadratic functions possess unique extrema, so that a single "best" policy will always exist.

Unfortunately this approach does not take account of the variance of the population parameters. An alternative, not yet popular, utilizes an explicit criterion function, but *presumes* that the population parameters are *unknown*. In this case, the criterion function is a function of random variables and, hence, an extremum does not exist. If the function is quadratic, however, its mathematical expectation is sometimes tractable and an extremum can be located for the expectation of the function.

This technique, known as the Bayesian or expected loss technique, has been explored by Fisher and Zellner, among others, and recently an application was made to a money multiplier model at the Federal Reserve Bank of St. Louis.[3] Fisher and Zellner derive exact results for single equation models and indicate that the difference in control settings and expected loss arising from using the certainty equivalence solution instead of the Bayesian solution is rather small in some cases.

In multi-equation models, the difference can be magnified by the particular shape of the quadratic criterion function employed and by the relationships among the endogenous variables. Research in this area had been held up because exact solutions are difficult to find for multi-equation systems. One type of multi-equation system for which exact solutions can be easily found is the recursive system. This paper develops the exact form for the one period solution in the case of (1) a quadratic criterion function, (2) a single policy instrument, (3) any number of equations which are linear in the parameters and recursive in structure, and (4) any number of criteria variables.

An application of the solution is then made to a model of the U.S. economy developed at the Federal Reserve Bank of St. Louis.[4] The results indicate that the difference in expected loss can be extremely large. The magnitude of the difference is significantly affected by information contained in the data (for a given model) over which the policy-maker has no control. However, if more than one criterion variable is

[3] Walter Fisher, Estimation in the linear decision model, International Economic Review 3 (1962); Arnold Zellner, An introduction to Bayesian inference in econometrics (New York, John Wiley, 1971), ch. 11; Albert E. Burger, Lionel Kalish III and Christopher T. Babb, Money stock control and its implications for monetary policy, Federal Reserve Bank of St. Louis Review 53 (October 1971), 6–22.

[4] Leonall C. Andersen and Keith M. Carlson, A monetarist model for economic stabilization, Federal Reserve Bank of St. Louis Review 52 (April 1970), 7–25. Hereafter referred to as the St. Louis model.

used, the policymaker employing a certainty equivalence solution can minimize the difference by emphasizing in the criterion function the variable about which he or she is more certain.

1. Control in general recursive systems

The problem is to

$$\text{minimize: } EL = E(\mathbf{y}-\mathbf{a})'Q(\mathbf{y}-\mathbf{a}), \text{ expected loss} \tag{1}$$

$$\text{subject to: } Y = Y\Gamma + XB + U \tag{2}$$

where these quantities are defined as follows:

$\mathbf{y}$ is an m-element column vector of "future" observations (that is, the $T+1$ observations) on the endogenous variables of the system, having unconditional mean value $\bar{\mathbf{y}}$;

$\mathbf{a}$ is an m-element column vector of targets corresponding to these variables;

Q is an $m \times m$ positive definite symmetric matrix of constants—the parameters of the loss function;[5]

Y is a $T \times m$ matrix of observations on the endogenous variables in the system;

Γ is an $m \times m$ upper triangular matrix of random coefficient parameters with zeros along the diagonal;

X is a $T \times n$ matrix of observations on the predetermined variables of the system;

B is an $n \times m$ matrix of random coefficient parameters;

U is a $T \times m$ matrix of unobserved random error terms.

Furthermore, it is assumed that the columns of U are normally and independently distributed, such that letting

$$\mathbf{v}' \equiv (\mathbf{u}_1', \mathbf{u}_2', \ldots, \mathbf{u}_m'),$$

[5] Q may be of rank $k < m$, where k is the number of criteria variables in the control problem.

then

$$E\mathbf{v}' = \mathbf{0}$$

$$E\mathbf{v}\mathbf{v}' = D(\sigma_i^2) \otimes I_T \qquad i = 1, 2, \ldots, m$$

where $\mathbf{u}_i$ represents the ith column of U and $D(\sigma_i^2)$ is a diagonal matrix with diagonal elements given by σ_i^2. The recursive nature of the system is evident from two conditions: (1) the Γ matrix is upper triangular and (2) the error terms of different equations are independently distributed.

The same model is assumed to generate "future" observations, that is

$$\mathbf{y}' = \mathbf{y}'\Gamma + \mathbf{x}'B + \mathbf{u}' \tag{3}$$

where $\mathbf{y}'$ is an m-element row vector and $\mathbf{x}'$ is an n-element row vector representing observations on the endogenous and predetermined variables respectively in the $T+1$ period. The m elements of the row vector $\mathbf{u}'$ are unobserved error terms in the $T+1$ period, independently and normally distributed such that

$$E\mathbf{u} = \mathbf{0},$$

$$E\mathbf{u}\mathbf{u}' = D(\sigma_i^2) \qquad i = 1, 2, \ldots, m,$$

$$E\mathbf{u}\mathbf{u}_k' = [0] \qquad k = 1, 2, \ldots, T,$$

where $\mathbf{u}_k$ is the kth row of U.

Having defined these new quantities and placed the appropriate restrictions on the error terms, we are now in a position to consider the loss function[6]

$$EL = E(\mathbf{y}-\mathbf{a})'Q(\mathbf{y}-\mathbf{a}) = E(\mathbf{y}-\bar{\mathbf{y}})'Q(\mathbf{y}-\bar{\mathbf{y}})+(\bar{\mathbf{y}}-\mathbf{a})'Q(\bar{\mathbf{y}}-\mathbf{a}). \tag{4}$$

Note that the second term is the loss function associated with the certainty equivalence solution (CEL) to the control problem. It is simply a weighted sum of the deviations of the mean values of the criteria variables about their respective targets. The Bayesian approach, which is the sum of the two terms, takes into account additional loss arising from the possible failure to predict the mean values of the criteria variables accurately. The first term clearly expands into a linear

[6] Alternatively the problem could be expressed in a utility maximizing format. If so, completing the square on $\mathbf{y}$ yields a declining monotonic function of (4). Maximizing utility and minimizing loss are therefore equivalent statements of the problem.

combination of variances and covariances of the multivariate predictive probability density function (pdf) for $\mathbf{y}$, which are easily derivable because the form of the distribution is known.

Consider a single element of the vector $\mathbf{y}$, say y_i. The predictive pdf for this element conditional upon the other elements of $\mathbf{y}$ can be set up using a noninformative prior on the appropriate elements of B, Γ and $D(\sigma_i)$, and then integrating over these unknown quantities, thusly

$$\begin{aligned} p(y_i|\mathbf{y}_{i-1}, \#) &= \int p(y_i, \gamma_i, \boldsymbol{\beta}_i, \sigma_i|\mathbf{y}_{i-1}, \#)\, \mathrm{d}\boldsymbol{\beta}_i\, \mathrm{d}\gamma_i\, \mathrm{d}\sigma_i \\ &= \int p(y_i|\boldsymbol{\beta}_i, \gamma_i, \sigma_i, \mathbf{y}_{i-1}, \#) \\ &\quad \times p(\boldsymbol{\beta}_i, \gamma_i, \sigma_i)\, \mathrm{d}\boldsymbol{\beta}_i\, \mathrm{d}\gamma_i\, \mathrm{d}\sigma_i \end{aligned} \tag{5}$$

where $\mathbf{y}_{i-1}$ represents the vector of endogenous variables in the ith equation of the system, $\boldsymbol{\beta}_i$ is the ith column of B (the coefficients of the predetermined variables in the ith equation), γ_i represents the ith column of Γ and # represents observed data (X, Y, $\mathbf{x}$). The first pdf in the second line above is univariate normal because of the normality assumption concerning the error terms and the second pdf is proportional to $1/\sigma_i$ because we are using noninformative prior pdf's. When the indicated integration is performed the resulting conditional predictive pdf for y_i is Student-t with mean and variance given by[7]

$$\bar{y}_i^c = \mathbf{y}'\bar{\gamma}_i + \mathbf{x}'\bar{\boldsymbol{\beta}}_i, \tag{6}$$

and

$$v_i^c = \bar{s}_i^2(1 + \mathbf{z}_i' M_i \mathbf{z}_i), \tag{7}$$

respectively, where the bars in (6) represent the means of the corresponding quantities and

$$\mathbf{z}_i \equiv (\mathbf{y}_{i-1}, \mathbf{x}_i),$$

$$\bar{s}_i^2 \equiv [\nu_i/(\nu_i - 2)]s_i^2,$$

$$M_i \equiv \begin{bmatrix} Y_{i-1}'Y_{i-1} & Y_{i-1}'X_i \\ X_i'Y_{i-1} & X_i'X_i \end{bmatrix},$$

ν_i being the degrees of freedom in the ith equation, s_i^2 being the estimated OLS residual variance for the ith equation, and where the subscripts on the X and Y quantities indicate that only the variables in the ith equation are to be considered.

[7] The integration is performed in Zellner, op. cit., pp. 72–74.

We may now use these quantities to find the mean and variance of the corresponding marginal distribution of y_i.

First, the mean

$$\begin{aligned}\bar{y}_i &\equiv \int y_i p(y_i|\#)\,\mathrm{d}y_i \\ &= \int y_i p(y_i|\mathbf{y}_{i-1}, \#)p(\mathbf{y}_{i-1}, \#)\,\mathrm{d}\mathbf{y}_i \\ &= \int \bar{y}_i^c p(\mathbf{y}_{i-1}, \#)\,\mathrm{d}\mathbf{y}_{i-1} \\ &= \bar{\mathbf{y}}'\bar{\gamma}_i + \mathbf{x}'\bar{\boldsymbol{\beta}}_i. \end{aligned} \tag{8}$$

By induction, therefore,

$$\bar{\mathbf{y}} = \bar{\mathbf{y}}'\,\bar{\mathbf{\Gamma}} + \mathbf{x}'\bar{B}, \tag{9}$$

and since it is easily seen that $(I_m - \bar{\Gamma})$ is non-singular,

$$\bar{\mathbf{y}} = \mathbf{x}'\bar{B}(I_m - \bar{\Gamma})^{-1} \equiv \mathbf{x}'\bar{\Pi}. \tag{10}$$

The mean of the marginal predictive pdf for an arbitrary element of $\mathbf{y}$ is therefore given by

$$\bar{y}_i = \mathbf{x}'\bar{\pi}_i \tag{10a}$$

where $\bar{\pi}_i$ is the ith column of $\bar{\Pi}$.

The variance of the marginal pdf is more difficult to find but the same technique of integrating unwanted variables out of the joint distribution can still be used:

$$\begin{aligned} v_{ii} &\equiv \int (y_i - \bar{y}_i)^2 p(y_i|\#)\,\mathrm{d}y_i \\ &= \int [(y_i - \bar{y}_i^c)^2 + 2y_i(\bar{y}_i^c - \bar{y}_i) + \bar{y}_i^2 - \bar{y}_i^{c2}] \\ &\qquad \times p(\mathbf{y}_i|\mathbf{y}_{i-1}, \#)p(\mathbf{y}_{i-1}|\#)\,\mathrm{d}\mathbf{y}_i. \end{aligned} \tag{11}$$

All terms in the brackets will cancel except for $(y_i - \bar{y}_i)^2$. The others were added in order to make use of the expression for the variance of the conditional predictive pdf, thusly

$$\begin{aligned} v_{ii} &= \int [v_i^c + (\bar{y}_i^c - \bar{y}_i)^2] p(\mathbf{y}_{i-1}|\#)\,\mathrm{d}\mathbf{y}_{i-1}, \\ &= \int \{\bar{s}_i^2[1 + (\mathbf{y}_{i-1}, \mathbf{x}_i)' M_i(\mathbf{y}_{i-1}, \mathbf{x}_i)] \\ &\qquad + [(\mathbf{y}_{i-1} - \bar{\mathbf{y}}_{i-1})'\bar{\gamma}_i]^2\} p(\mathbf{y}_{i-1}|\#)\,\mathrm{d}\mathbf{y}_{i-1}. \end{aligned}$$

The first term in brackets is a quadratic form in the elements of $\mathbf{y}_{i-1}$. The indicated integration transforms these random variables into their respective means. Using the well-known relation $Ey_k^2 = v_{kk} + \bar{y}_k^2$, this term becomes a quadratic form in the elements of $\bar{y}_{i-1}$ plus a weighted sum of variances, that is

$$1+(\bar{\mathbf{y}}_{i-1},\mathbf{x}_i)'M_i(\bar{\mathbf{y}}_{i-1},\mathbf{x}_i)+\sum_{j=1}^{i-1} M_i^{jj}v_{jj}$$

or

$$1+\bar{\mathbf{z}}_i'M_i\bar{\mathbf{z}}_i+\sum_{j=1}^{i-1} M_i^{jj}v_{jj}$$

where M_i^{jj} is the jth diagonal element of M_i.

The second term in brackets is also a quadratic function and performing the indicated integration transforms it into a weighted sum of variances and covariances thusly

$$\sum_{j=1}^{i-1}\bar{\gamma}_{ij}^2 v_{jj}+2\sum_{j=2}^{i-1}\sum_{k=1}^{j-1}\bar{\gamma}_{ij}\bar{\gamma}_{ki}v_{jk}.$$

The covariance terms can be evaluated as follows for $j > k$

$$\begin{aligned}
v_{jk} &\equiv \int (y_j-\bar{y}_j)(y_k-\bar{y}_k)p(\mathbf{y}_j|\#)\,\mathrm{d}\mathbf{y}_j,\\
&= \int (y_j-\bar{y}_j)(y_k-\bar{y}_k)p(y_j|\mathbf{y}_{j-1},\#)p(\mathbf{y}_{j-1}|\#)\,\mathrm{d}\mathbf{y}_j,\\
&= \int (y_k-\bar{y}_k)(\mathbf{y}'\bar{\gamma}_j+\mathbf{x}'\bar{\beta}_j-\bar{y}_j)p(\mathbf{y}_{j-1}|\#)\,\mathrm{d}\mathbf{y}_{j-1},\\
&= \int [(y_k-\bar{y}_k)(\mathbf{y}'\bar{\gamma}_j-\bar{\mathbf{y}}'\bar{\gamma}_j)+(y_k-\bar{y}_k)(\bar{\mathbf{y}}'\bar{\gamma}_j+\mathbf{x}'\bar{\beta}_j-\bar{y}_j)]\\
&\quad\times p(\mathbf{y}_{j-1}|\#)\,\mathrm{d}\mathbf{y}_{j-1}.
\end{aligned} \tag{12}$$

The first term in the brackets becomes a weighted sum of other covariance terms while the second vanishes when integration is carried out over y_k. Therefore we can simplify the above to an element of V, an upper triangular $m\times m$ matrix of covariances with variances along the diagonal. If $\mathbf{v}_i'$ is a row vector from this matrix and $\mathbf{i}'$ is the ith row vector of I_m, then

$$\mathbf{v}_i' = \mathbf{v}_i'\bar{\Gamma}+v_{ii}\mathbf{i}'. \tag{13}$$

And further letting D be an $m\times m$ diagonal matrix of variances, this generalizes to

$$V = V\bar{\Gamma}+D. \tag{14}$$

Since $(I_m-\bar{\Gamma})$ is non-singular,

$$V = D(I_m-\bar{\Gamma})^{-1}. \tag{15}$$

Thus, a covariance relation between any two endogenous variables can be expressed as a multiple of the variance associated with the variables having the lower index number. For example, v_{ij} is some multiple of v_{ii} if $i < j$, and v_{jj} if $j < i$.

Returning to the second term in brackets in the expression for the variance, (11), we can now see that it will be a linear combination of variances of endogenous variables occurring in preceding equations. In particular, if we consider the complete variance-covariance matrix $(V+V'-D)$, then the second term in brackets becomes, after integration,

$$\bar{\gamma}_i'[D(I_m-\bar{\Gamma})^{-1}+(I_m-\bar{\Gamma}')^{-1}D-D]\bar{\gamma}_i.$$

This cumbersome expression is of the form

$$\sum_{j=1}^{i-1} g_{ij}v_{jj}.$$

Finally, the complete expression for the variance of the endogenous variable of the ith equation is

$$v_{ii} = \bar{s}_i^2(1+\bar{\mathbf{z}}_i'M_i\bar{\mathbf{z}}_i)+\sum_{j=1}^{i-1}(\bar{s}_i^2M_i^{jj}+g_{ij})v_{jj}. \tag{16}$$

Before proceeding to evaluate the expected loss function, it is instructive to pause and study this expression. The first part is the variance of the ith endogenous variable conditional upon the other variables in the system being equal to their mean values. This will tend to understate the true variance, however, insofar as the mean of the *conditional* distribution of the ith endogenous variable is not equal to the mean of its marginal distribution. Any difference between these quantities will serve to increase the variance. Moreover, the effect upon the variance will be magnified by the variance of the other endogenous variables appearing in the ith equation, hence the terms in g_{ij}. The remaining terms $\bar{s}_i^2M_i^{jj}$ will be recognized as the variance of the parameter estimates of the jth endogenous variable appearing in the ith equation. This indicates that any uncertainty introduced by random variables appearing in the ith equation is also magnified by the uncertainty associated with the corresponding coefficients.

To adapt this expression for a matrix format we let $\mathbf{d}$ be an m-element column vector consisting of the diagonal elements of D, the ith element of which is v_{ii}. Then,

$$\mathbf{d}' = \bar{\mathbf{v}}'+\mathbf{d}'\bar{G} \tag{17}$$

where we have defined the ith element of $\bar{\mathbf{v}}$ to be $\bar{s}_i^2[1+\bar{\mathbf{z}}_i'M_i\bar{\mathbf{z}}_i]$ and the i,

jth element of the $m \times m$ matrix $\bar{G}$ to be $(M_i^{ij} + g_{ij})$. Therefore,

$$\mathbf{d}' = \bar{\mathbf{v}}'(I_m - \bar{G})^{-1}. \tag{18}$$

Finally, we are ready to evaluate the loss function and to find its extremum. Recall that the loss function was given by

$$EL = E(\mathbf{y} - \bar{\mathbf{y}})'Q(\mathbf{y} - \bar{\mathbf{y}}) + (\bar{\mathbf{y}} - \mathbf{a})'Q(\bar{\mathbf{y}} - \mathbf{a}). \tag{4}$$

The expectation operator applied to the first term merely yields a weighted sum of variances and covariances. That is,

$$\text{trace } [Q(V + V' - D)]. \tag{19}$$

But the covariances are themselves functions of the variances,

$$\begin{aligned} &\text{trace } \{Q[D(I_m - \bar{\Gamma})^{-1} + (I_m - \bar{\Gamma}')^{-1}D - D]\} \\ &\quad = \text{trace } \{DQ[(I_m - \bar{\Gamma})^{-1} + (I_m - \bar{\Gamma}')^{-1} - I_m]\} \\ &\quad \equiv \mathbf{d}'\mathbf{q}. \end{aligned}$$

Upon substituting for $\mathbf{d}$ we have

$$\mathbf{d}'\mathbf{q} = \bar{\mathbf{v}}'(I_m - \bar{G})^{-1}\mathbf{q} \equiv \bar{\mathbf{v}}'\mathbf{w}, \quad \text{where} \quad \mathbf{w} \equiv (I_m - \bar{G})^{-1}\mathbf{q}. \tag{20}$$

So we see that the first term of the loss function is a weighted sum of variances of conditional pdf's given that other variables are equal to their mean values. The weights are given by the corresponding elements of the vector $(I_m - \bar{G})^{-1}\mathbf{q}$, that is w_i.

From this point it is a straightforward, but tedious, task to differentiate the loss function with respect to the policy instrument (control variable) and then solve for the loss minimizing setting for that variable, so we simply state the result without proof. However, some new quantities need to be defined:

- x_1 is the control variable;
- n_i is the number of predetermined variables in the ith equation;
- $\mathbf{p}_i'$ is an $n_i + i - 1$ element vector equal to $\boldsymbol{\iota}'(\bar{\Pi}_{i-1}, P_i)$ where
- $\boldsymbol{\iota}$ is an n-element vector consisting of zeros and ones, such that $\boldsymbol{\iota} = \mathrm{d}\mathbf{x}/\mathrm{d}x_1$ and
- $\bar{\Pi}_{i-1}$ is a matrix consisting of the first $i-1$ columns of $\bar{\Pi}$ and

P_i is an $n \times n_i$ matrix consisting of zeros and ones such that $\iota' P_i = \mathrm{d}\mathbf{x}_i/\mathrm{d}x$, where $\mathbf{x}_i$ is a vector of predetermined variables entering in the ith equation;

$\mathbf{x}_0$ is an n-element vector of "future" values of all predetermined variables in the system with the exception of x_1, the value of which is set equal to zero;

$\mathbf{x}_{0i}$ is an n_i-element vector of all predetermined variables in the ith equation with the exception of x_1, the value of which is set equal to zero;

P_i^* is an $n \times n_i$ matrix consisting of zeros and ones such that $\mathbf{x}'_{0i} = \mathbf{x}'_0 P_i^*$;

$\mathbf{p}_i^{*\prime}$ is an $n_i + i - 1$ element vector equal to $\mathbf{x}'_0(\Pi_{i-1}, P_i^*)$.

Using these quantities, the optimal setting for the control variable is given by

$$x_1^* = \frac{-\sum_{i=1}^{m} w_i \bar{s}_i^2 \mathbf{p}'_i M_i \mathbf{p}_i^* - \iota' \Pi Q(\Pi' \mathbf{x}_0 - \mathbf{a})}{\sum_{i=1}^{m} w_i \bar{s}_i^2 \mathbf{p}'_i M_i \mathbf{p}_i + \iota' \Pi Q \Pi' \iota} . \tag{21}$$

The other quantities, such as w_i, $\bar{s}_i^2$, M_i, Q and $\mathbf{a}$, are defined elsewhere above. We are assured that x_1^* minimizes the expected loss function because the matrix Q was assumed to be positive definite symmetric.

Had we elected to merely minimize certainty equivalence loss the terms involving sums of squares and cross-products (M_i) would not appear in the solution:

$$x_1^{ce} = -\frac{\iota' \Pi Q(\Pi' \mathbf{x}_0 - \mathbf{a})}{\iota' \Pi Q \Pi' \iota} . \tag{22}$$

Clearly, x_1^{ce} is a linear function of x_1^*. A sufficient condition for x_1^{ce} to exceed x_1^* is that the indicated summation in the numerator of (21) be non-negative, which is always the case when there are no other predetermined variables in the system except the control variable.

2. *The loss function*

In the experimental example which follows, we employ two criteria variables—price changes and the unemployment rate—so it is useful at this point to consider some of the properties of a two variable loss function. Specializing (4) we have:

$$CEL = q_1(\bar{y}-a_y)^2+2q_2(\bar{y}-a_y)(\bar{z}-a_z)+q_3(\bar{z}-a_z)^2, \tag{23}$$

$$EL = q_1 V(y)+2q_2 \operatorname{Cov}(y, z)+q_3 V(z)+CEL. \tag{24}$$

A regular minimum exists for these functions if and only if $q_1 > 0$ and $q_1q_3-q_2^2 > 0$.

These second-order conditions imply elliptical iso-loss contours for both (23) and (24). One senses immediately, however, that only a portion of the ellipse would be economically relevant. To make the discussion concrete, consider fig. 1 and suppose that z measures the rate of change in prices and y measures the unemployment rate.

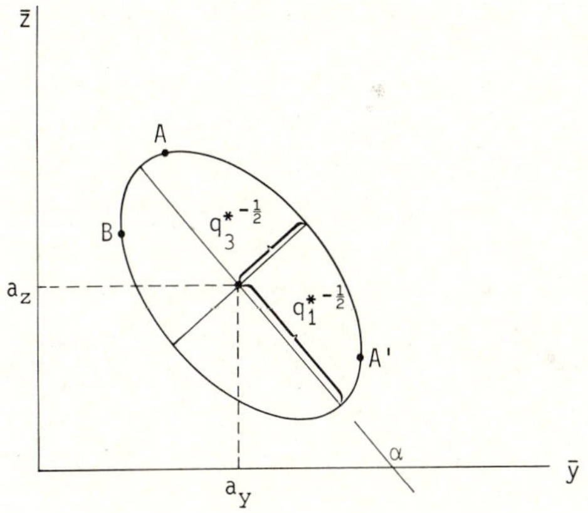

Fig. 1. Elliptical iso-loss function.

If one were to begin at point B and move clockwise, one would be trading more inflation for *more* unemployment. This bizarre implication is, of course, due to the symmetric nature of the quadratic function: undershooting a target is as undesirable as overshooting by the same amount. Those who find this symmetry unappealing may be tempted to abandon the quadratic function and search for a nonsymmetric one. But one may retain the quadratic function if one is satisfied with being confined to a limited region of the ellipse between A and A'. Within this region one always trades more inflation for less unemployment, or vice versa. Moreover, one experiences an *increasing* marginal rate of substitution—which is as it should be since deviations from either target are *undesirable*.

A set of necessary and sufficient conditions for operating in the relevant range of the ellipse are (1) the implied Phillips' curve of the system must be negatively sloped, and (2) the targets must be chosen to be below and to the left of the implied Phillips' curve. The reason for these conditions is clear when one considers that the system of regression equations acts as a constraint to minimizing expected loss. The optimal solution x^* is that one for which the implied Phillips' curve is tangent to the lowest possible iso-loss contour projected onto its plane. When operating with an optimal policy the policymaker's marginal rate of substitution will be equal to the slope of the implied Phillips' curve of the system.

Bearing these things in mind, we seek to reparameterize the loss function in terms of easily interpretable quantities with geometric significance. Consider first the certainty equivalence loss (23). The center of each elliptical iso-loss curve will be at the point (a_y, a_z). From analytic geometry we recognize that the loss function parameters q_1 and q_3 represent simple transformations of the projections of the elliptical axes upon the axes of the reference system, and if we rotate the ellipse about the point (a_y, a_z) until its axes are parallel to those of the reference system, q_2 vanishes and q_1 becomes q_1^*, q_3 becomes q_3^*, $\bar{y}$ becomes $\bar{y}^*$ and $\bar{z}$ becomes $\bar{z}^*$.

A rotation of the elliptical axes through an angle α can be expressed as

$$\bar{y}^* = \bar{y} \cos \alpha - \bar{z} \sin \alpha, \tag{25}$$

$$\bar{z}^* = \bar{y} \sin \alpha + \bar{z} \cos \alpha. \tag{26}$$

Substituting these transformations into the certainty equivalence loss function and using the following trigonometric identities:

$$\cos^2 \alpha = \frac{1}{\tan^2 \alpha + 1}, \tag{27}$$

$$\sin^2 \alpha = \frac{\tan^2 \alpha}{\tan^2 \alpha + 1}, \tag{28}$$

$$\sin \alpha \cos \alpha = \frac{\tan \alpha}{\tan^2 \alpha + 1}, \tag{29}$$

we have

$$CEL = \left[\frac{r+\tan^2\alpha}{\tan^2\alpha+1}\right](\bar{y}-a_y)^2+2\left[\frac{(1-r)\tan\alpha}{\tan^2\alpha+1}\right]$$

$$(\bar{y}-a_y)(\bar{z}-a_z)+\left[\frac{r\tan^2\alpha+1}{\tan^2\alpha+1}\right](\bar{z}-a_z)^2 \qquad (30)$$

where $r = q_1^*/q_3^*$. This quantity is also by definition equal to $1-e^2$ where e is the eccentricity of the ellipse, and measures the departure of the ellipse from circularity. For a circle $e = 0$ and as the ellipse becomes "tighter" $e \to 1$. Notice, however, that $e = 1$ is not possible if the loss function is to remain an ellipse.

The coefficient of $(\bar{y}-a_y)^2$ is evidently q_1, the coefficient of $(\bar{y}-a_y)(\bar{z}-a_z)$ is $2q_2$, and the coefficient of $(z-a_z)^2$ is q_3. All q's are expressed in terms of two simple quantities—the slope of the major axis of the ellipse and the ratio of the elliptical axes. Although $\tan \pi/2$ is infinite, the elements of the loss function possess finite limits as $\tan\alpha \to \infty$. Systematic variation of the parameters is facilitated by this particular parameterization because $\tan\alpha$ is a periodic function of α, and the eccentricity is bounded by zero and one.

We are now in a position to consider the effect of taking the mathematical expectation of the loss function instead of simply replacing the random variables with their respective means. To make the discussion concrete, let us consider a simple two equation system:

$$y_i = bx_i+u_i \qquad i = 1, 2, \ldots, T \qquad u_i \text{ are } NID(0, \sigma_u^2) \qquad (31)$$

$$z_i = cy_i+v_i \qquad i = 1, 2, \ldots, T \qquad v_i \text{ are } NID(0, \sigma_v^2)$$

$$\text{and } E(u_i v_i) = 0. \qquad (32)$$

For simplicity, consider a circular loss function where $q_2 = 0$. In this case

$$EL = [V(y)+(\bar{y}-a_y)^2]+[V(z)+(\bar{z}-a_z)^2]. \qquad (33)$$

Note that the difference between this result and the certainty equivalence loss is the addition of a weighted sum of variances which are themselves expressible in terms of the variable means $\bar{y}$ and $\bar{z}$ and other quantities calculated from the data:

$$V(z) = \bar{s}_z^2+[V(c)+\bar{c}^2][V(y)+\bar{y}^2]+\bar{z}(\bar{z}-2\bar{c}\bar{y}), \quad (34)$$

$$V(y) = \bar{s}_y^2+\bar{y}^2 V(b)/\bar{b}^2. \quad (35)$$

However, we do not make use of the system constraint $\bar{z} = \bar{c}\bar{y}$.

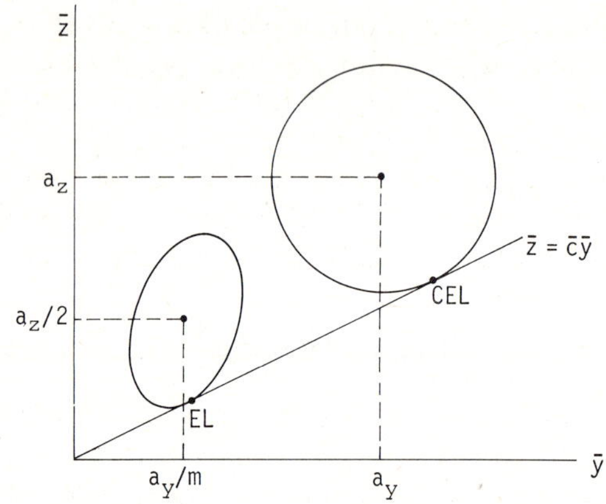

Fig. 2. Comparison of certainty equivalence loss and expected loss functions.

Upon inserting these quantities into (33) and completing the square on $\bar{y}$ and $\bar{z}$ we have:

$$EL = m(\bar{y}-a_y^*)-2\bar{c}(\bar{y}-a_y^*)(\bar{z}-a_z^*)+2(\bar{z}-a_z^*)+\text{constants} \quad (36)$$

where $m = (1+\bar{c}^2)/t_b^2+\bar{c}^2/t_b^2t_c^2+\bar{c}^2/t_c^2+\bar{c}^2+1$ and the new, or virtual, targets are given by $a_y^* = a_y/m$ and $a_z^* = a_z/2$. Thus, the new iso-loss contour will be elliptically shaped, rotated with respect to the coordinate axes and centered closer to the origin. The angle of rotation and eccentricity are given by:

$$\cot 2\alpha = (2-m)/2\bar{c} \quad \text{and} \quad (37)$$

$$e^2 = 1-(2-\bar{c}^2)/(m+\bar{c}^2), \quad \text{if } m+\bar{c}^2 > 2-\bar{c}^2, \quad \text{otherwise} \quad (38a)$$

$$e^2 = 1-(m+\bar{c}^2)/(2-\bar{c}^2). \quad (38b)$$

If (38a) holds, the major axis of the ellipse corresponds to the $\bar{z}$ dimension in an unrotated system, while if (38b) holds, the major axis of the ellipse corresponds to the $\bar{y}$ dimension in an unrotated system.

To illustrate this example, consider fig. 2. The system constraint $\bar{z} = \bar{c}\bar{y}$ is a straight line passing through the origin which is associated with a zero setting for the control variable. The certainty equivalence loss function is a circle with center at (a_y, a_z). The certainty equivalence solution is the point of tangency of the certainty equivalence loss function and the system constraint, and is denoted by *CEL*. The expected loss function, on the other hand, is an ellipse centered at $(a_y/m, a_z/2)$. Its tangency with the system constraint, denoted by *EL*, represents the solution to the Bayesian control problem.

The *EL* solution is a more conservative solution than the *CEL* solution because it moves the policymaker in the direction of a zero setting for the policy instrument (i.e., towards the origin).[8] The conservativeness of the policy is a function of the certainty with which the coefficients b and c are known. For example, suppose that there were a great deal of uncertainty about either b or c or both, such that $m \gg 2$, then the major axis of the ellipse would be the $\bar{z}$ axis and the angle of rotation would be close to zero or π. As uncertainty increases, m increases and the ellipse becomes "tighter" as the control solution assigns increasing weight to deviations of $\bar{y}$ about its virtual target.

3. *Application of control solution*

In the final section of this paper we apply the Bayesian control solution to variations of the St. Louis model of the U.S. economy. The model's structure can be easily seen from the flow chart on the following page. The model is fully recursive—unemployment is influenced by prices and nominal spending through a real output identity; prices, in turn, are influenced by nominal spending; nominal spending depends only upon exogenous variables, including a money variable.[9]

[8] This conclusion may not be true if there are exogenous variables in the system which are not under control.

[9] In addition, the necessary assumption about the error terms are made. The original St. Louis model contained additional equations for long- and short-term interest rates, but these are not considered here since the criteria variables of interest are prices and unemployment and the recursive nature of the system does not require their inclusion for either estimation or computation of the optimal solution.

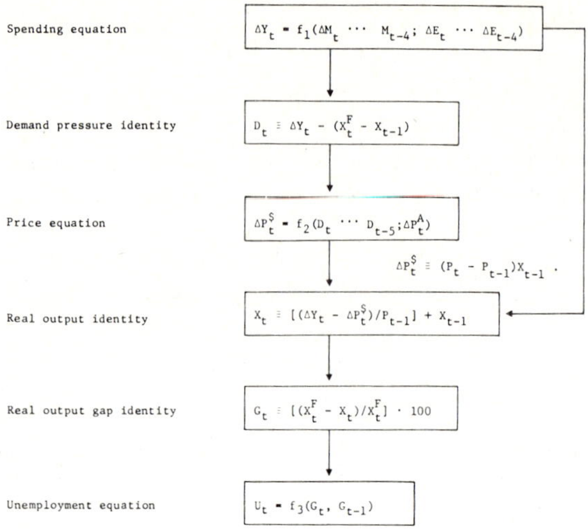

Exogenous variables:

ΔM_t, change in currency plus demand deposits;

ΔE_t, change in high employment government expenditures;

X_t^F, full employment level of real output;

Endogeneous variables:

ΔY_t, change in nominal spending;

D_t, demand pressure;

$\Delta P_t^{\$}$, dollar change in nominal spending due to change in price deflator;

ΔP_t^A, anticipated change in prices;

X_t, real output;

G_t, gap in real output;

U_t, unemployment rate;

Fig. 3. Flow Chart of the St. Louis Model.

While this form of the model was satisfactory for estimating parameters and performing simulation experiments, it could not be used for computing the optimal value of the instrument variable. A problem arises because the identities operate on the endogenous variables of the system before they enter subsequent equations. The unemployment equation was respecified in terms of a new dependent variable, $U_t X_t^F P_{t-1}$, because the real income identity and the gap identity caused ΔY_t and $\Delta P_t^{\$}$ to be multiplied by $(X_t^F P_{t-1})^{-1}$ in the original unemployment equation. This new unemployment variable is an approximate value of foregone production (in period $t-1$ prices) due to underutilized resources. The measure is not exact, however, because the elasticity of output with respect to labor input may not be unity.

TABLE 1

Coefficient estimates

A. *Spending Equation*

$$\Delta Y_t = \underset{(3.56)}{2.74} + \sum_{i=0}^{4} m_i \Delta M_{t-i} + \sum_{i=0}^{4} e_i \Delta E_{t-i}$$

$m_0 =$	1.259	(2.86)	$e_0 =$	0.569	(2.66)
$m_1 =$	1.755	(7.21)	$e_1 =$	0.504	(3.91)
$m_2 =$	1.485	(3.90)	$e_2 =$	0.061	(0.32)
$m_3 =$	0.711	(3.02)	$e_3 =$	−0.440	(−3.31)
$m_4 =$	−0.045	(−0.10)	$e_4 =$	−0.611	(−2.79)

Sample Period: 1953/I to 1969/IV
Constraints: 4th degree polynomial ($m_{-1} = m_5 = 0$; $e_{-1} = e_5 = 0$)
$\bar{R}^2 = 0.656$
S.E. = 3.844

B. *Price Equation*

$$\Delta P_t^{\$} = \underset{(6.62)}{2.60} + \underset{(8.78)}{0.941}\, \Delta P_t^A + \sum_{i=0}^{5} d_i D_{t-i}$$

$d_0 =$	0.0222 (2.32)	$d_3 =$	0.0146 (3.12)
$d_1 =$	0.0205 (5.95)	$d_4 =$	0.0105 (2.06)
$d_2 =$	0.0179 (6.82)	$d_5 =$	0.0056 (1.59)

Sample Period: 1955/I to 1969/IV
Constraints: 2nd degree polynomial ($d_{-1} \neq 0$; $d_6 = 0$)
$\bar{R}^2 = 0.865$
S.E. = 1.134

C. *Unemployment Equation*

1. Lagged gap term included

Sample Period: 1955/I to 1969/IV

$$\frac{U_t X_t^F P_{t-1}}{100} = \underset{(1.05)}{4.554}\left[\frac{X_t^F P_{t-1}}{100}\right] \underset{(-0.15)}{-0.0066}(\Delta Y_t - \Delta P_t^{\$} + Y_{t-1})$$

$$+\underset{(7.34)}{0.315} G_{t-1}\left[\frac{X_t^F P_{t-1}}{100}\right]$$

$\bar{R}^2 = 0.915$
S.E. = 1.817

2. No lagged gap term

Sample Period: 1955/I to 1969/IV
$\bar{R}^2 = 0.838$
S.E. = 2.511

$$\frac{U_t X_t^F P_{t-1}}{100} = \underset{(22.1)}{35.3}\left[\frac{X_t^F P_{t-1}}{100}\right] \underset{(-19.2)}{-0.314}(\Delta Y_t - \Delta P_t^{\$} + Y_{t-1})$$

(t-values in parentheses)

Since the Bayesian formulation of the control problem differs from the certainty equivalence formulation by taking account of the uncertainty in the model's parameters, information can be gained by comparing models with differing degrees of uncertainty in the parameters. To this end an alternative unemployment equation was selected. It is the same as the revised unemployment equation without the term

TABLE 2

Ratio of expected certainty equivalence loss to expected Bayesian loss in 1970/I for alternative loss function parameterizations assuming zero targets for rate of price change and unemployment rate

A. *Model-I*

Angle of rotation (radians)	Eccentricity				
	0.0 (circle)	0.75	0.87	0.97	0.99
$\pi/12$	1.001	1.000	1.000	1.000	1.000
$\pi/6$	1.001	1.092	1.188	1.512	2.258
$\pi/4$	1.001	1.354	1.815	2.925	3.759
$\pi/3$	1.001	1.672	2.752	5.288	6.837
$5\pi/12$	1.001	1.769	3.515	9.652	14.857
$\pi/2$	1.001	1.404	2.802	15.705	52.060
$7\pi/12$	1.001	1.007	1.024	1.402	859.029
$2\pi/3$	1.001	1.210	2.087	12.304	54.893
$3\pi/4$	1.001	1.469	2.569	6.599	10.292
$5\pi/6$	1.001	1.400	1.946	2.993	3.590
$11\pi/12$	1.001	1.206	1.389	1.609	1.772
$\pi(\equiv 0)$	1.001	1.055	1.100	1.150	1.328

B. *Model-II*

Angle of rotation (radians)	Eccentricity				
	0.0 (circle)	0.75	0.87	0.97	0.99
$\pi/12$	1.067	1.035	1.018	1.002	1.000
$\pi/6$	1.067	1.040	1.028	1.019	1.041
$\pi/4$	1.067	1.052	1.049	1.056	1.073
$\pi/3$	1.067	1.067	1.071	1.085	1.093
$5\pi/12$	1.067	1.081	1.090	1.104	1.110
$\pi/2$	1.067	1.091	1.104	1.120	1.126
$7\pi/12$	1.067	1.098	1.115	1.135	1.143
$2\pi/3$	1.067	1.101	1.122	1.151	1.164
$3\pi/4$	1.067	1.096	1.121	1.169	1.197
$5\pi/6$	1.067	1.082	1.104	1.182	1.266
$11\pi/12$	1.067	1.061	1.063	1.125	1.573
$\pi(\equiv 0)$	1.067	1.042	1.027	1.005	1.327

introduced by the lagged real output gap. As can be seen in table 1, however, the properties of the two resulting models differ in that the model with the lagged gap term (Model-I) has much less precise coefficient estimates for the contemporaneous endogenous variables in the unemployment equation than does the model without the lagged gap term (Model-II).

To study the differences between the solutions, the targets in the criteria function were set at zero price change and zero unemployment while the shape and orientation of the quadratic criteria function were systematically varied. The two solutions were compared (1) in terms of the ratio of the expected losses generated for each particular choice of criteria function parameters and (2) in terms of the ratio of the corresponding settings for the control variable. Although the choice of targets is arbitrary, zero values represent the most extreme values which satisfy the necessary and sufficient conditions for a relevant economic solution.[10]

Because the variances of the coefficients associated with contemporaneous endogenous variables are much larger in Model-I than in Model-II, we expect that the relative differences in expected loss between the certainty equivalence solution and the Bayesian solution for Model-I would be greater than the relative difference for Model-II. Further, since the control setting in the Bayesian solution tends toward zero as one becomes more uncertain, we expect the relative difference in the control settings between the certainty equivalence solution and the Bayesian solution to be greater for Model-I than for Model-II.

The results of the experiments in terms of expected loss are given in table 2. In general, they are in accord with expectations. In the case of Model-II the relative difference between the two solutions is rarely very great. In the case of Model-I the relative difference between the solutions is more sensitive to the parameters of the loss function, often increasing by an order of magnitude or more as the eccentricity of the function approaches unity. The greatest differences occur when the angle of rotation is in the vicinity of $\pi/2$ and the eccentricity approaches unity. This particular parameterization of the criteria function puts the maximum weight on the unemployment target. Since we are very uncertain about the unemployment equation parameters in Model-I (relative to Model-II) the loss associated with using the certainty equivalence approach is much larger than the Bayesian approach which takes this

[10] Friedman has suggested that the optimal rate of price change is actually negative, but this raises other issues which are beyond the scope of this paper.

TABLE 3

Ratio of certainty equivalance control variable setting to Bayesian control variable setting in 1970/I for alternative loss function parameterizations assuming zero targets for rate of price change and unemployment rate

A. *Model-I*

Angle of rotation (radians)	Eccentricity				
	0.0 (circle)	0.75	0.87	0.97	0.99
$\pi/12$	2.705	1.670	1.312	0.979	0.876
$\pi/6$	2.705	3.356	2.914	2.413	2.181
$\pi/4$	2.705	4.267	4.059	3.785	3.680
$\pi/3$	2.705	5.537	5.964	6.449	6.639
$5\pi/6$	2.705	7.105	9.012	12.373	14.239
$\pi/2$	2.705	8.274	12.785	28.088	48.233
$7\pi/12$	2.705	5.045	8.317	30.653	4328.440
$2\pi/3$	2.705	10.377	14.138	30.686	59.371
$3\pi/4$	2.705	6.970	7.733	9.503	10.616
$5\pi/6$	2.705	4.934	4.432	3.873	3.650
$11\pi/12$	2.705	3.826	2.959	2.096	1.792
$\pi(\equiv 0)$	2.705	3.506	2.497	1.634	1.354

B. *Model-II*

Angle of rotation (radians)	Eccentricity				
	0.0 (circle)	0.75	0.87	0.97	0.99
$\pi/12$	1.130	1.114	1.098	1.048	0.987
$\pi/6$	1.130	1.107	1.091	1.063	1.049
$\pi/4$	1.130	1.108	1.097	1.083	1.078
$\pi/3$	1.130	1.114	1.107	1.100	1.097
$5\pi/6$	1.130	1.121	1.118	1.115	1.114
$\pi/2$	1.130	1.129	1.129	1.129	1.129
$7\pi/12$	1.130	1.138	1.142	1.145	1.146
$2\pi/3$	1.130	1.147	1.155	1.164	1.167
$3\pi/4$	1.130	1.154	1.169	1.191	1.200
$5\pi/6$	1.130	1.157	1.180	1.233	1.269
$11\pi/12$	1.130	1.149	1.173	1.285	1.577
$\pi(\equiv 0)$	1.130	1.131	1.133	1.111	1.354

uncertainty into account. However, if the loss function is circular, the certainty equivalence solution and the Bayesian solution generate very similar expected losses, regardless of the model employed.

On the basis of these results one may be tempted to conclude that the certainty equivalence solution could be used with impunity as a suitable approximation to the Bayesian solution if the loss function did not depart significantly from circularity.[11] But, according to table 3 which reports the relative difference in control settings, the certainty equivalence solution typically calls for much larger settings for the control variable than the Bayesian solution in Model-I. Again, the largest differences occur for the loss function parameterizations emphasizing the unemployment target. Only in Model-II does there seem to be little difference in the control settings. Thus, if there is a cost to changing the control variable which has not been taken into account in formulating the control problem, use of the certainty equivalence solution instead of the Bayesian solution may impose a severe penalty upon the policymaker.

4. *Concluding remarks*

In this paper we have found the exact setting for a single control variable in the case where the control variable is linked to several criteria variables by a stochastic linear recursive equation system and the criterion function is quadratic. By taking the uncertainty in system parameters into account the solution differs from the certainty equivalence approach developed by other authors. By being applicable to recursive systems the solution represents an extension of the single equation solution already known. In the application developed here, the optimal setting provided a rule for the conduct of monetary policy one period at a time.

The results of the application were sensitive to the manner in which the equations were specified in each of two models. The two models, which had identical specifications of the price equation but different specifications of the unemployment rate equation, had vastly different control properties. The relative differences between the certainty equivalence approach and the Bayesian approach were generally greater

[11] This conclusion is, however, conditional upon the particular set of criteria variable targets chosen.

for that model specification having the less precise contemporaneous coefficient estimates but better overall predictive properties. For judicious choices of loss function parameters the relative differences for expected losses were modest, but the relative differences in control settings remained large.

PART 11

REPORTING RESULTS

S. E. Fienberg and A. Zellner (eds.)
Studies in Bayesian Econometrics and Statistics.

11.1 REPORTING OF BAYESIAN STUDIES*

Harry V. ROBERTS

University of Chicago

1. Introduction

With a few exceptions, some of which will be mentioned later, reporting of statistical results, Bayesian or otherwise, has seldom received more than superficial attention from writers on statistical methodology. Yet current practice in reporting is often poor: relatively few statistical studies are reported in sufficient detail for a critical reader to be able to reach an informed judgment about the appropriateness of the statistical analysis and therefore about the believability of the results. There is no shortage of possible explanations for inadequate reporting: editorial pressure for brevity; the emphasis of much statistical teaching on formalistic analysis and stylized conclusions—such as the ritual of "p-values"; the easy accessibility of packaged computer programs to those who understand little about statistics; and a climate of opinion in which statistics is seldom taken more seriously than any other mechanical prerequisite for publication, such as correct spelling or inclusion of references. A documentation of some of these complaints, together with suggestions in the spirit of this paper although in a non-Bayesian framework, has been given by Thomas [1].

It therefore seems worthwhile to survey the problem of statistical reporting with the aim of formulating useful guidelines. Although Bayesian studies are not yet so numerous as to warrant the negative pronouncement of the previous paragraph, there is no reason to believe that those who wear the Bayesian armor will be immune to the danger of reporting lapses. Many shortcomings can arise in applications regardless of the theoretical framework espoused by the statistical analyst, as Professor William Kruskal has rightly stressed from the beginning of the current Bayesian revival. Nonetheless the Bayesian orientation

* An earlier version of this paper was presented to the Econometrics and Statistics Colloquium, Univ. of Chicago, February 1973.

both encounters new problems of reporting and offers fresh insights for their resolution.

Leonard J. Savage often said, "do research for your enemies, not your friends". This injunction conveys a good state of mind for approaching the task of reporting: imagine an adversary skilled both in statistics and the tactics of a trial lawyer who has been given the widest latitude to cross-examine you on any aspect of your study. Then try to draft your report in such a way that his every question can be answered convincingly by a reference to the report itself. The aim is to make the conclusions of the analysis completely believable, not simply to go through the motions to achieve a report superficially like those typically produced by others in the field.

For some the metaphor of the malevolent adversary may require modification. The trial lawyer, whose tactics are not shaped by a desire to get at the truth but only by the aim of appearing to have the better of the argument, may be replaced by a scientific skeptic, well respected but inclined toward a different view from that of the author of the report. My own observation is that scientific controversy is often less principled than courtroom strife between skilled attorneys, and I should want to anticipate not only honest, penetrating criticism, but also the possibility of deliberate distortion by my enemies.

Concentration on reporting to your enemies risks some disservice to the friendly reader who simply wants to know what you think you have found out. The very diversity of audience of most reports poses familiar problems of communication that are easy to mention and hard to solve. To serve both friend and foe, for example, you may have to put much of the detailed documentation in an appendix, in a separate paper for a more technical journal, or even in an archival document available by special request. Similarly there are many possibilities for attractive statistical packaging that can make the report interesting as well as informative and convincing. I shall only offer passing comments on such matters in order to concentrate my focus upon the key problem of achieving clarity and believability. For this purpose I shall have in mind the sometime unrealistic picture of a monolithic report.

2. *The investigator's posterior distribution is not enough—and is often inappropriate*

It is useful to distinguish the private and public phases of any statistical analysis. In the private phase, the investigator is interested mainly in clarifying his own state of mind about the subject of the study; in the public phase, he wishes to communicate to others. The full Bayesian apparatus is appropriate to the private phase. The investigator combines his prior distribution with the likelihood function of his data to arrive at a posterior distribution by application of Bayes' theorem. Typically the distributions, prior and posterior, concern constructs called parameters. More generally, however, they pertain to characteristics of finite samples of observations not yet made, and the parameters are only useful fictions for reaching predictive distributions of such observations. Posterior distributions or predictive distributions are objectives of the private phase of Bayesian analyses. If the investigator is concerned with the formal resolution of a decision problem, one of these distributions serves as an input to a Bayesian decision analysis in which utilities must be assessed and an action chosen so as to maximize expected utility.

Although the posterior distribution is the objective of a private analysis, it rarely serves as the appropriate focus of a public report. The leaders of the current Bayesian revival have been very clear about this. In a letter to Egon S. Pearson on November 1, 1959, which was widely circulated at the time, Bruno de Finetti wrote, "Except in special cases sanctioned by the NP [Neyman–Pearson] theory, statisticians should not assign numerical values to prior distributions and utilities". He went on to add, "they should . . . be aware that such evaluations are unavoidable in a full picture of the problem, instead of rejecting them as a scandal or sin of past centuries". L. J. Savage once said, "Sometimes listeners to an exposition of Bayesian statistics get the misimpression that they are being urged to publish their own opinions as their analysis of an empirical study" [2, pp. 98–99]. Bayesian writers have agreed that the public phase analysis entails the reporting of the import of the data, not of the investigator's private opinions. For the posterior distribution is a blend of the import of the data as reflected in the likelihood function and of the investigator's opinions as reflected in the prior distribution. If only the posterior distribution is reported, there is in general no way in which the reader can decompose this information into its two constituents. The reader cannot therefore make his own private analysis, which is for him the most important use of the information.

In the early days of the Bayesian revival, an extremely capable statistician, who unfortunately regarded anything Bayesian as heresy, felt that the Bayesian approach gave limitless opportunities for fakery and deception because an investigator could simply rig the prior distribution to achieve any desired posterior distribution, regardless of the data. I think that I never convinced this statistician that Bayesian principles did not condone indiscriminate reporting of posterior distributions; that these principles, in fact, provided some safeguard against dishonest reporting by their clear delineation of the steps involved in making up one's mind in a coherent way. (Yet I can still sympathize with his difficulty in understanding this point, because he, more than most statisticians, had devoted a great deal of effort to the uncovering of statistical quackery, so that it was natural for him to react emotionally to any device that appeared to provide an easy loophole for statistical crime.)

There appear to be four exceptions to the general injunction against reporting of posterior distributions:

(1) Several prior distributions might be formulated to represent common positions taken by experts on the subject matter under investigation. The posterior distributions appropriate to these priors might then be reported. Even when appropriate, however, this approach would not fully discharge the reporting obligation, as will be explained in later sections.

(2) It may be possible to collect data that would serve as a basis for a prior distribution acceptable to almost anyone. The terms *data-based* or *data-benched* are sometimes used in this context. In an application of Bayesian methods to medical diagnosis, or auditing, or safety studies, for example, there may be extensive past frequency evidence on the incidence of a disease, of clerical errors, or of accidents under circumstances indistinguishable from the occasion of the present application. Indeed, studies using the discriminant function have often been regarded by sampling theorists as legitimate exceptions to the general presumption against Bayesian methods, since then Bayes' theorem may be given a frequentistic interpretation. If data-based priors are used, the sources should be fully reported, as Zellner has stressed [3, pp. 40–41].

So far, few attempts to assess data-based priors have appeared. The most impressive and convincing effort along these lines known to me is to be found in the Mosteller–Wallace study of the Federalist Papers [4]. Even in this study, the effort was focused on a prior distribution for the usage of words, not for the central question of the prior probability of the authorship of the disputed papers. It remains to be seen whether it will be fruitful to exert similar effort in other studies; the Mosteller–

Wallace study, however, points to the care that is needed if data-based priors are to be attempted.

(3) A diffuse prior distribution is one that is extremely sensitive to revision by even a small sample. Various lines of reasoning have led to identical or nearly identical mathematical expressions of "diffuseness" for most of the standard statistical models. Such a distribution can serve as an approximation to the judgments of an open-minded person for whom the data being reported will have a preponderate influence on the conclusion of the analysis; hence the happy expression of Box and Tiao, "the jury principle". The posterior distribution implied by the likelihood function in the presence of a diffuse prior distribution therefore offers an attractive solution to the reporting problem, a solution that has been frequently embraced in reports of Bayesian studies.

Besides the jury principle, there are two other practical arguments in favor of this form of reporting. Since this posterior distribution is at least approximately proportional to the likelihood function (when no nuisance parameters have been integrated out), it can be interpreted as an implementation of likelihood reporting, which will be discussed in section 4. Further, this form of reporting often leads to probability intervals that are identical or nearly identical to the corresponding confidence intervals of sampling theory, a fact that may be reassuring to a non-Bayesian reader.

Although the reporting of the posterior distributions that follow from diffuse prior distributions has the advantages just discussed, it also entails dangers. It can lead the reader to the false conclusion that evidence other than that of the sample has no legitimate place in the "full picture of the problem", to use de Finetti's phrase. It can lull the reader to overlook the fact that even for diffuse priors, there are statistical models for which the conclusions expressed in posterior distributions and confidence intervals can diverge, as in the Behrens–Fisher problem, or in drawing inferences about variance components, or in the treatment of autoregressive models of time series. Finally, it may happen that the improper prior distributions used to represent diffuse prior distributions lead to paradoxes that collide with the judgments of any rational person, however open-minded about the non-sample evidence. One important example is that of problems of the k-means type in which the number of parameters is large. In an early discussion of this problem, Anscombe showed that even the natural-conjugate multivariate normal prior that follows from an improper "prior–prior" combined with fictitious prior sample evidence can imply a predictive distribution for the sample to be analyzed that is often contradicted by the sample observations themselves when the number of parameters is large [5]. Lindley has provided a recent summary of multiparameter problems [6, pp. 49–56], and Zellner and Vandaele treat these problems in an explicitly econometric context [7]. It may be, as Anscombe's example suggests, that the root of the difficulty lies in the a priori assessment of independence of parameters rather than in improper distributions as such. In any event, blind reliance on diffuse priors as an automatic solution of the reporting problem may thus divert the reader's attention from matters of genuine importance. (The same caution extends to interpretation of the outputs of sampling-theory investigations. Concern with the k-means problem, in fact, can be traced to the pioneering work of Charles Stein in the context of sampling theory [8].)

(4) Finally, the investigator may choose to publish his private analysis, including a full discussion of how he assessed his prior distribution. Although this procedure runs the risk of violating the injunction against reporting opinions, it need not preclude adequate reporting of the data, and it may sometimes enrich the reader's grasp of the situation. So far there has been little tendency to report private analyses in explicitly Bayesian form, but this may reflect the ease of using diffuse prior distributions rather than a conscious rejection of this option. (Curiously, sampling-theorists often do discuss aspects of what a Bayesian would call the private prior distribution. Many econometric studies, for example, include a discussion of the expected signs of regression coefficients.) An interesting assessment of a non-diffuse prior distribution in a large econometric model has been given by Theil [9, Ch. 6].

3. *On reporting the data*

In introducing what they call the "vexed question of the proper way to report a 'scientific' experiment", Raiffa and Schlaifer begin by saying that "an obvious 'solution' to this problem is simply to report the data" [10, pp. 68–69]. Zellner comments further: "With respect to reporting the data employed in an analysis, it is good procedure to describe in detail how they were obtained and have them available for any interested party by including them in the report or by making it known that they can be obtained on request" [3, p. 41]. Full reporting in this sense would permit any reader to make his own analysis of the data, starting from scratch if he wished. Full disclosure also adds to the credibility of a study, even to a casual reader who has not the slightest interest in even glancing at the data.

Savage has expressed the ideal of full reporting as follows:

> Now, when we Bayesians emphasize that all opinions are but opinion, we do not mean that a scientist publishing the results of his investigation has said the last word when he tells the world what his opinion is. Quite the contrary, the first thing that he ought to tell is what he has observed. In principle, he should do this so well that his peers will know what happened as well as if they had done the experiment themselves. This is an idealization quite unachievable in practice, but approximation to it is the core of a serious research report. In particular, numerical data should be reported as fully as is practical [2, p. 99].

Reporting of the data for many investigations in econometrics can often be accomplished by physical inclusion in the main body or an appendix of a study. This is possible because the total numbers of observations and variables are often modest. It is useful even when the data are drawn from generally-available sources, although references to these sources are still desirable if only to permit the most avid reader to check for accuracy of the data transcription, a problem that is far from negligible. The sources also may include more detailed descriptions of the methods of collection of the data and warnings about their limitations than are warranted in the report. Methods of seasonal adjustment provide one example of special importance in econometrics. If the investigation employs seasonally adjusted data, it is desirable to provide a reference to the source in which the original unadjusted data can be found and the methods of seasonal adjustment are given.

Other bodies of data, especially those generated by sample surveys, may be far too extensive for literal publication in any compact form. A solution here is to announce the availability of computer tapes or cards, or even of the original questionnaires, to the serious student at some reasonable charge to cover marginal costs. Establishment of data banks could go far to make this solution feasible. Current attempts to set up data banks appear to be motivated largely by the possibility of making further analyses of the same data, such as new cross tabulations or the fitting of alternative statistical models. For example, a great deal of the work by academic researchers in the area of marketing has comprised sophisticated analyses of data collected by consumer panels. But there is room also for private analyses that cover exactly the same ground of the published study, with perhaps slight variations on the statistical model used. Among the many benefits of such data availability is the opportunity for students to attack realistic problems and cases with the incentive of better statistical training and also of the chance to make useful new discoveries. In my own experience with such student assignments, I have found that at least some of the students in the class achieve improvements over the previous analysis of the same data.

However desirable full reporting of data may be, this step does not exhaust the reporting obligation of the investigator. Only rarely will readers be sufficiently motivated to do their own private analyses of the data. Some digestion of the original data is needed to serve the interests of the audience.

4. *Reporting of the likelihood function and the sufficient statistics*

Unlike sampling theory, the Bayesian approach fully endorses the likelihood principle. In analysis, the Bayesian must keep his eye on the likelihood function, so it is natural for Bayesians to look to the likelihood function for guidance in the summarization necessary for meaningful reporting to a broad audience. If non-trivial sufficient statistics exist—that is, if there is a smaller set of sufficient statistics than the entire body of data—these are the objects to be reported. If not, the likelihood principle brings us back to the rule of reporting the data. This latter situation is likely to be pertinent in many serious investigations in econometrics.

Even when there are non-trivial sufficient statistics, their presentation does not fully replace data-reporting, although they are more appropriate in the main body of the report than in an appendix reserved for the full set of data. After the quotation of section 3, Savage went on to say, "The most excusable abbreviations are perhaps attempts to condense the data by means of sufficient statistics, but even these are often detrimental, because sufficient statistics are sufficient only for some model that is nominally accepted, but that might justifiably be rejected in view of some details lost in condensing the data to a sufficient statistic" [2, p. 99].

But even if non-trivial sufficient statistics existed for a model that all readers were willing to accept, the sufficient statistics would not suffice for full reporting. The author must furnish further computation that the reader cannot ordinarily be expected to do. For example, the likelihood function can be computed and displayed—at least for the case of one or two unknown parameters. Here, however, as discussed in section 2, reporting of posterior distributions for diffuse priors may be a more feasible way of reporting since this posterior distribution has interest in its own right. Further, many readers may be unfamiliar with the meaning of the likelihood function, and marginal posterior distributions of parameters are a more natural form of reporting than marginal likelihoods.

Likelihood functions and sufficient statistics are designed to give the reader what he needs from the data if he should want to do his own analysis; the only proviso is that he be willing to accept the author's specification of the model. This objective is often overlooked in current statistical practice and is seldom stressed in orthodox textbooks. Raiffa and Schlaifer remind us:

> What is completely intolerable is to report in such a way that the sufficient statistics cannot be recovered by the user; and while it is probably unnecessary to point out that it is inexcusable to report a level of significance or a confidence interval without also reporting sample size, it does seem necessary to point out that even if sample size *is* reported the user cannot in general recover the information he needs from a statement that the data was or was not significant at some predetermined significance level or from an asymmetric or randomized confidence interval. And even if a report *is* full enough to make it possible to recover the sufficient statistics, the user is not well served if he is forced to calculate these numbers which might have been given to him directly [10, pp. 68–69].

Even worse sins can be found in current reporting, such as publication of significance levels without further information: "The treated group showed superior weight gain ($p = 0.05$)." More exasperating still: "The two groups did not differ significantly."

5. *Reporting on the model specification*

Bayesians emphasize that any statistical analysis must be founded upon a specification of the model—of the likelihood function; and that this specification should be taken seriously with full reliance in its formulation on all available evidence, including that of the sample to be analyzed. Model specification is just another aspect of probability assessment. (Unlike assessment of the prior distribution, when "knowing little" can often be represented by some type of diffuse distribution, there is no "robust" short cut when information bearing on the appropriate likelihood function is meager. In spite of a number of attempts, the Bayesian analogue of non-parametric inference has not yet taken definitive shape.)

The importance of model specification relative to the rest of an analysis can hardly be overstressed. The right perspective is suggested by the emphasis given by Box and Jenkins in their recent book, Time series analysis [11]. The great bulk of the book is taken up with a discussion of the models used, the choice of an appropriate model for a particular application, and the checking of appropriateness of a tentatively chosen model. Statistical inference in the traditional sense appears for the first time in chapter 7, "Model estimation".

There are two broad sources of evidence bearing upon model specification: prior information, including the experience of earlier statistical studies of similar contexts, and information coming from the data of the actual study. Good reporting should bring out this evidence so that the reader can judge its cogency. Of special importance here is the reporting of residuals from fitted models. Although Zellner has recently opened this area of statistics to Bayesian scrutiny [12], and although certain general sampling-theory tests are available, this aspect of statistics still lies largely outside the framework of formal schemes of statistical analysis. The following statement by Tukey and Wilk provides an admirable overview:

> One of the truly incisive tools of exposure is the computation and adequate examination of residuals. Perhaps because of the blinding effects of unrealistic optimism about assumptions, perhaps because of the difficulties of computational practice during the era before modern computing evolved, residuals have not received the attention and use they richly deserve.
>
> There is no substitute for examining the collection of detailed individual residuals in diverse ways. It is almost always a sad inadequacy (though far better than nothing) to try to summarize the exposing information in a body of residuals by a mean square error. Even the computation of the individual residuals and their examination in an unstructured mass is not enough. Several graphs of residuals are usually in order. Calculation of suitable numerical summaries, which answer specific questions more stringently but more general questions hardly at all, can also be useful [13].

Bayesian studies can draw upon two additional strategies for the examination of model specification that are difficult to implement within the sampling-theory approach because of the problems faced by the latter in the elimination of nuisance parameters. The first is the use of the Jeffreys' testing approach to assign probabilities to competing models in proportion to the predictive probabilities assigned by each to the data being analyzed. An example from econometrics is Geisel's comparison of the quantity theory and the income–expenditure theory [14]. The second strategy is illustrated by a pioneering paper by Box and Tiao [15]. In order to deal with the question of possible non-normality, they enlarged the model specification from the normal distribution to the family of symmetrical power distributions in which a non-normality parameter defines a range of distributions varying in shape from extreme blunt-tailed to moderately fat-tailed; when this parameter is zero, the model reduces to the normal distribution. The Bayesian analysis then proceeds in the standard way to reach a joint posterior distribution of all the parameters, including the non-normality parameter. The non-normality parameter can then be treated as a nuisance parameter and integrated out of the joint posterior distribution. In principle this strategy can be enlarged by the inclusion of additional parameters for additional types of departure from the initially-entertained model, such as skewness or time dependence of disturbances. It is safe to assert that the approach complements rather than displaces analysis of residuals. More recent work in this direction can be found in studies by Chetty [16] and Blattberg and Gonedes [17].

The process of model specification often involves a good deal of trial-and-error "data-dredging", both in Bayesian and non-Bayesian studies. Like examination of residuals from explicitly formulated models, data-dredging lies largely outside any formal statistical framework. The reader of a report is well served by a description of the data-dredging that led to simplification, elaboration, or discard of models originally entertained.

Whatever the treatment of the model specification problem, full reporting is essential to make the analysis believable.

6. *Reporting of approximations*

Mosteller and Wallace remark: "The moment he leaves the simplest application of Bayes' theorem, the applier finds himself in a welter of makeshifts and approximations" [4, p. 266]. (A parallel observation would apply to sampling theory; it is not clear that either sampling theory or Bayesian inference has a net advantage of tractability. In econometrics, for example, exact Bayesian results are available for applications in which only large sample approximations are available in sampling theory.)

Bayesian approximations may apply either to the likelihood function or the posterior distribution, or both. For example, as discussed by Pratt [18], it may be sensible to treat non-sufficient statistics as if they were sufficient as a means of approximating the likelihood function. In some applications, the mode may be the only characteristic of the posterior distribution that can be easily ascertained. Often the posterior distribution must be reached by numerical integration, which involves approximation of a computational kind; a good deal of progress on technical aspects of numerical integration has been achieved.

The most serious reporting problems are posed by applications in which the number of parameters is large, especially when the structure of the sampling design is complex, and problems of measurement error and non-response are substantial. All these ingredients are present in a large-scale sample survey. Such a survey might be envisaged as a highly multidimensional contingency table where simple models are inappropriate, where there are thousands of cells of interest, many of them with zero observations, and where the report must serve both the casual and the intensely interested reader. Since no analysis is likely to exhaust the information of such bodies of data, the suggestions about data banks and similar devices in section 3 are appropriate. As to the problems of analysis, there are various beginnings of Bayesian assaults on the components of the problem: realistic diffuse priors for large numbers of parameters, theory for complex designs, treatment of measurement error and non-response. Some of the sampling-theory developments, such as sophisticated treatment of the log-linear model for contingency tables, may be interpretable in a Bayesian way. Finally, some of the remarks about general-purpose estimation in section 7 are pertinent.

In the personalistic view of Bayesian methodology, even the diffuse prior distribution itself is to be interpreted as an approximation. Hence the conventional posterior distribution is regarded as an approximation even if the likelihood function can be treated by exact methods. Savage was interested in ways of evaluating the adequacy of approximation entailed by diffuse priors, but there appears to have been little follow-up of his early work [19, ch. 4]. An alternative approach known as sensitivity analysis has been more often exploited. Here the idea is to try out a number of alternative expressions of diffuseness to see to what extent the resulting posterior distributions differ from each other [10, pp. 66–68].

At whatever stage the approximation comes in—prior distribution, likelihood function, evaluation of the posterior distribution—the report should document the approximations used and give whatever support is possible for the adequacy thereof. The book of Mosteller and Wallace provides an excellent model [4].

In closing this discussion of approximations, I shall mention briefly an aspect of approximation that often goes unmentioned in polite discussion: the numerical accuracy of computations. My own conjecture is that sheer miscalculation often distorts, even beyond recognition, the results of statistical analyses. Reporting of the data, which may then possibly lead to recomputation by some avid reader, provides some protection. Documentation of computational procedures and checks, including especially the computer programs that may have been used, since some widely used programs may provide very poor numerical accuracy, is also desirable.

7. *Reporting for the casual reader: general-purpose estimation*

The emphasis of this paper has been on the highly-interested reader, preferably a hostile one. Most readers of most studies, however, are likely to have only mild interests in the subject being reported. These readers may seek simply to tuck away a few of what Hildreth has called "parcels of information" [20].

Robert Schlaifer has set the following scenario: "In a great many situations a businessman wants additional information on the value of some unknown quantity, not because he is on the point of choosing among a number of definite, well-defined terminal acts whose costs depend on this quantity, but simply to guide his thinking in some general problem area" [21, pp. 587–589]. In Schlaifer's example, the solution is a Bayesian point estimate, which often will be the mean or other central value of the conventional prior distribution following a diffuse prior distribution. The estimate may simply be stored in a person's memory until needed as an input to a later decision analysis, possibly very complicated and not necessarily formalized. Schlaifer used a marketing illustration: ". . . a soap manufacturer who is thinking about developing and marketing a new detergent designed especially for use in automatic dishwashers may want to get some idea of the number of homes equipped with automatic dishwashers long before his thinking reaches the stage where he is ready even to specify his possible terminal acts, let alone to say exactly how the profitability of each act depends on the number in question."

Schlaifer goes on to say, "Until the businessman has completely specified his terminal acts and their conditional costs, he will think primarily in terms of an *estimate* of the quantity in which he is interested rather than in terms of a *probability distribution* of this quantity. He will make rough calculations in which he treats the estimated number of homes equipped with dishwashers as if this estimate were a true value even though he knows that the estimate is almost certainly in error . . .." Since most decisions are made with benefit of rough calculations at best, it follows that the precision of point estimates often will not have the importance to users that is stressed in much statistical writing. This clarifies a point that troubled me for many years: in the period in which probability samples were being introduced to users in the social sciences and business, one of the main arguments in favor of probability sampling was the availability of estimates of sampling error. Yet I noticed that reports of surveys using probability sampling rarely included sampling errors and when they did, gave them small prominence.

A good example is the monthly estimate of percentage unemployment from the Current Population Survey. This is used routinely by many readers—even by econometricians as an input to an econometric model—without serious consideration either of the cross-sectional sampling error or of the error in the estimated change from the previous month's estimate. I do not mean to denigrate the probability calculations of statistical inference but rather to emphasize that very simple and condensed results may be the objects of highest priority in reporting to casual readers.

In many survey applications, the self-weighted sample means or proportions are approximate central values of conventional posterior distributions. The display of these point estimates in simple tables and charts is just as appropriate in Bayesian reporting as it has traditionally been in survey practice.

Although it is natural to presume that symmetrical loss functions are appropriate in the derivation of general-purpose point estimates, one area of application may pose an exception. Auditors feel that estimates should be slanted in the direction of "conservatism", and this implies the use of an asymmetrical loss function. This exception appears to stem from the fear of legal action for errors in one particular direction. Perhaps the treatment of reporting problems in this area could be resolved by the explicit use of posterior distributions to supplement the point estimates.

Up to the present time there has been little explicit attention to appropriate loss functions in the reporting of point estimates, although the auditing example suggests that, in some contexts, such attention would be appropriate. There is a danger worth noting, however. Unbounded loss functions, like improper prior distributions, can lead to paradoxes. An early discussion was provided by Thornber [22].

Although this section has emphasized the usefulness of point estimates, standing alone, for casual readers, I do not mean to suggest that standard errors should be suppressed. David Wallace has reminded me that in the early decades of the century, it was felt that point estimates and accompanying standard errors go a long way toward discharging the reporting obligation. For Bayesians and non-Bayesians alike, these are clearly important numbers. To the extent that conventional posterior distributions are well approximated by the normal distribution, for example, a point estimate and its standard error characterize the distribution. Physical science reporting still emphasizes this mode by statements of the form $x \pm y$, although it is not always made clear whether the y is a standard error, probable error, two standard errors, or what, and the y may not always be calculated from the data alone but may allow judgmentally for systematic error as well as random error. In econometrics and social science, regression studies are often reported with the standard errors accompanying the estimated regression coefficients. (Sometimes the standard errors are reported only indirectly by giving the ratios of estimated coefficients to standard errors, which has the unfortunate effect of suggesting that a test of a null hypothesis is the appropriate thing to do with a standard error.)

I certainly feel that standard errors are appropriate. If we think of a division of a report into a section intended for the casual reader and sections of a more technical nature, I would suggest that the standard errors are often more appropriately included in the latter, where they serve various purposes suggested in earlier sections of this paper.

The brief mention above of systematic error suggests another opportunity for Bayesian studies to give more insightful reporting by exploitation of Bayesian ideas. In a chapter of his 1959 book that has attracted regrettably little follow-up, Robert Schlaifer outlined simple Bayesian procedures for assessing the impact of systematic error on posterior distributions (conventional or other); the chapter heading was, "Biased measurement and biased selection". Schlaifer's approach requires the assessment of a probability distribution on such things as the bias of non-response. In reporting, it could be shown how the conventional posterior distribution for the object of central interest is affected by various assessments ("assumptions") about the bias of non-response. Important work on the Bayesian treatment of the problem of bias has also been done by Harold Jeffreys [23].

8. Summary and discussion

Statistical reporting, Bayesian or otherwise, is usually given scant attention in technical statistical writing, yet current practice leaves much to be desired in achieving the full disclosure needed to make a study convincing to a critical reader and the palatability needed to reach a casual reader. The problems of reporting Bayesian studies are not new and esoteric, but there are differences of emphasis and new technical tools that impinge on reporting as well as analysis. For example, the Bayesian approach makes it easier to report on model specification since nuisance parameters describing certain possible departures can be introduced and then integrated out. Again, Bayesian reasoning, unlike that of sampling theory, suggests that readers need the inputs to make up their own minds, not formal outputs.

The first requirement of reporting is to report the data so that the serious reader not only will know in detail how the study was conducted but so he can, if he wishes, make calculations of his own to check or to extend the analysis. This injunction is widely ignored in current practice. The data sometimes can be included in the main body of the published report or an appendix thereto; sometimes a data-bank arrangement is needed when the data are extensive. Even when the data are taken from published sources, it is a service to the reader to include it in the report; a full reference to the original source is of course appropriate in addition. Adaptation to statistics of the "lab notebook" might facilitate later statistical reporting.

Although the posterior distribution is the main output of a private Bayesian analysis, it is not a full answer to the Bayesian reporting problem, and is often inappropriate when the prior distribution is non-diffuse. If the posterior distribution is reported under these conditions, the report should give the non-diffuse prior distribution explicitly and should explain the reasoning and evidence that led to it.

The posterior distribution following a diffuse prior distribution can be called the conventional posterior distribution. The conventional posterior distribution is often a useful reporting device. It can be thought of as a means of conveying the likelihood function, to which it is proportional, or nearly so, in the unknown parameters. According to the likelihood principle, which is fundamental in Bayesian inference, the likelihood function conveys the entire import of the data. Sometimes further condensation is possible by the reporting of non-trivial sufficient statistics.

Reporting of the likelihood function and the sufficient statistics constitutes full reporting only under the tentative specification of a particular statistical model. The reader may not entertain the same model as the investigator, and the investigator owes the reader a full report on the theory and data that support the choice of model employed in the study. In particular, incisive analysis of residuals is just as important in Bayesian studies as in sampling-theory studies.

Bayesian studies, like sampling-theory investigations, often must rely upon various types of approximations; in the Bayesian approach these approximations may come in the assessment of the prior distribution or the likelihood function, or in the numerical evaluation of the posterior distribution. Documentation and justification of these approximations are an essential part of the report.

Although statistical reports are framed with the most vitally-interested reader, and most acerbic critic, in mind, the casual reader is also important. Casual readers are often interested only in general-purpose point estimates, and central values of the conventional posterior distribution often serve this need. Casual readers may profit also from non-technical, intuitive explanations of the statistical methods that have been used. Further, many readers may have only a sketchy or erroneous conception of the rationale of the Bayesian approach; a brief exposition of that rationale may enlighten readers and help to dispel the impression that Bayesian analysis is occult and non-scientific. It is worth emphasizing that the reader is free to articulate his own prior distribution with the evidence of the study in order to reach his own posterior distribution.

Bayesian reporting has thus far emphasized conventional posterior distributions and point estimates derived therefrom. Further study of easy methods of assessing non-diffuse prior distributions could yield novel reporting devices. For example, the report might include a simple conversational computer program that would permit the reader to answer a few simple questions about his own judgments about the prior distribution, and then to compute the posterior distribution implied by these judgments and by the data of the study. Technical work on assessment and elicitation, as represented by the paper by Savage in this volume, might yield unexpected bonuses for reporting.

Acknowledgements

For many valuable suggestions, I am very much indebted to the editors of this volume, Stephen Fienberg and Arnold Zellner, and to members of the Econometrics and Statistics Colloquium at the University of Chicago, especially Philip Cooper, Walter Vandaele, and Arnold Zellner.

References

[5] F. J. Anscombe, Bayesian inference concerning many parameters, with reference to supersaturated designs. Bulletin of the I.S.I., 34th Session, Ottawa, 1963.

[17] Robert C. Blattberg and Nicholas J. Gonedes, A comparison of the stable and Student distributions as statistical models for stock prices. Manuscript, Graduate School of Business, University of Chicago, Chicago, 1972.

[11] George E. P. Box and Gwilym M. Jenkins, Time series analysis. San Francisco, Holden-Day, 1970.

[15] G. E. P. Box and G. C. Tiao, A further look at robustness via Bayes' theorem. Biometrika 49 (1962), 419–432.

[16] V. K. Chetty, Bayesian analysis of some simultaneous equation models and specification errors. Unpublished dissertation, University of Wisconsin, Madison, 1966.

[14] Martin S. Geisel, Comparing and choosing among parametric statistical models: a Bayesian analysis with macroeconomic applications. Unpublished dissertation, Graduate School of Business, University of Chicago, Chicago, 1970.

[20] Clifford Hildreth, Bayesian statisticians and remote clients. Econometrica 31 (1963), 422–438.

[23] Harold Jeffreys, Theory of probability. 3rd Ed. London, Oxford University Press, 1961.

[6] D. V. Lindley, Bayesian statistics, a review. Philadelphia, Society for Industrial and Applied Mathematics, 1971.

[4] Frederick Mosteller and David L. Wallace, Inference and disputed authorship: the federalist. Reading, Mass., Addison-Wesley, 1964.

[18] John W. Pratt, Bayesian interpretation of standard inference statements. Journal of the Royal Statistical Society (Series B) 59 (1965), 169–203.

[10] Howard Raiffa and Robert Schlaifer, Applied statistical decision theory. Division of Research, Graduate School of Business, Harvard University, Boston, 1961.

[2] L. J. Savage, and other contributors, The foundations of statistical inference. New York, John Wiley, 1962.

[19] Leonard J. Savage, The subjective basis of statistical practice. Manuscript, The University of Michigan, Ann Arbor, Michigan, 1961.

[21] Robert Schlaifer, Probability and statistics for business decisions. New York, McGraw-Hill, 1959.

[8] Charles Stein, Inadmissibility of the usual estimator for the mean of a multivariate normal distribution. In: Proceedings of the Third Berkeley Symposium on Mathematical Statistics and Probability, Vol. I. Berkeley, University of California Press, 1956.

[9] Henri Theil, Theory and measurement of consumer demand. Manuscript, Graduate School of Business, University of Chicago, Chicago, 1973.

[1] J. J. Thomas, The reporting of empirical work in economics. Applied Statistics 16 (1967), 172–176.

[22] Hodson Thornber, Applications of decision theory to econometrics. Unpublished dissertation, Department of Economics, University of Chicago, Chicago, 1966.

[13] J. W. Tukey and M. B. Wilk, Data analysis and statistics: techniques and approaches. In: Proceedings of the Symposium on Information Processing in Sight Sensory Systems, November 1–3, 1964, California Institute of Technology. Reprinted in: The quantitative analysis of social problems, edited by Edward R. Tufte. Reading, Mass., 1970.

[3] Arnold Zellner, An introduction to Bayesian inference in econometrics. New York, John Wiley, 1971.

[12] Arnold Zellner, Bayesian analysis of regression error terms. Manuscript, Graduate School of Business, University of Chicago, Chicago, 1972.

[7] Arnold Zellner and Walter Vandaele, Bayes-Stein estimators for *k*-means, regression and simultaneous equation models. Manuscript, Graduate School of Business, University of Chicago, Chicago, 1972. (Reprinted in the present volume.)

S. E. Fienberg and A. Zellner (eds.)
Studies in Bayesian Econometrics and Statistics.

11.2 SCIENTIFIC REPORTING AND PERSONAL PROBABILITIES: STUDENT'S HYPOTHESIS*

James DICKEY

State University of New York at Buffalo

1. Introduction

In the presence of uncertainty, coherent personal behavior (that is, behavior subject to a few intuitively compelling consistency or rationality principles) is known to require that a person choose acts to maximize some personal expected utility ([1–4], and other references). In this normative theory of rational behavior, personal probabilities are defined in terms of personal behavior in betting contexts. Coherent choice of a statistical decision procedure requires decisions that maximize personal expected utility posterior to sampling, with the posterior probabilities determined by Bayes' theorem from the chooser's prior personal probabilities and the data.

It has been said (Fisher [5], for example) that decision-theoretic formulations may be inappropriate for problems of scientific inference and reporting, where the experimenter–reporter is not concerned with making decisions. However, personal probability can be viewed as a description of personal knowledge, and knowledge ought, as a normative standard, to be coherently act-on-able, even if specific acts seem remote. Even in purely scientific reports, Bayes' theorem should be applied to personal probabilities to show how they would be changed coherently by the data.

But isn't personal knowledge, or opinion, like superstititon, non-objective and unscientific, and therefore to be avoided in science? Who cares to read about a scientific reporter's opinion as described by his prior and posterior probabilities? That may be true about the reporter's opinion: but communicating a single opinion ought not be the purpose

* Reprinted with minor changes from the Journal of the Royal Statistical Society (B) 35 (1973), 285–305, with the permission of the Royal Statistical Society. This work was reported in an invited talk to the Eastern Regional Meeting of the Institute of Mathematical Statistics, May 1970, and to the Second NBER-NSF Seminar on Bayesian Inference in Econometrics, Carnegie-Mellon Univ., May 1971.

of a scientific report; but rather, to let the data speak for itself, by giving the effect of the data on the wide diversity of real prior opinions. For example, instead of merely quoting an irrelevant and misleading "significance level", an experimenter can summarize the application of Bayes' theorem to whole ranges of prior distributions, devised to include the opinions of his readers. Scientific reports should objectively exhibit as much as possible of the *inferential content of the data*, the data-specific prior-to- posterior *transformation* of the collection of all personal probability distributions on the parameters of a realistically rich statistical model. (Of course, the complete likelihood function is itself a representation of the inferential content of the data. The mental gymnastics necessary to visualize the likelihood function as a transformation on the space of personal probability distributions can be aided in scientific reporting.)

Is this not an impractical task, especially for a busy experimenter? I think not. As this paper demonstrates, research statisticians can prefabricate the reporters' tools. We have powerful computers, graphical techniques, and an old interest in descriptive statistics. Low dimensional parametric problems are a hopeful starting point.

This paper proposes the use of GRIP's, Graphical Reports of Inference by Probabilities, graphical displays of the functional dependence of the inference on the prior distribution. For example, the weighted likelihood ratio or Bayes factor for one hypothesis versus another, the coherent ratio of the posterior odds to the prior odds, can be plotted against the parameters of the prior conditional distributions given the two hypotheses.

Such scientific reporting tools are developed for the case of inference about a point null hypothesis versus a vague alternative for a univariate Normal mean. The Jeffreys–Lindley–Savage view is taken that interesting actual prior opinions can be *approximated* by attaching positive probability to the null hypothesis. (Jeffreys, of course, did not use this language.)

Bounds, also, are given for the weighted likelihood ratio.

Many of the ideas here have their roots, and some boughs, in Edwards et al. [6].

2. *Use of the Bayes factor for a dichotomy*

Suppose only two distinct given constants are possible for an unknown quantity θ, either

$$H: \theta = \theta_H, \quad \text{or} \quad H': \theta = \theta_{H'}. \tag{1}$$

If a person's probability for H' is $P(H')$, then his *odds* (*ratio*) *for* H' is defined as

$$O(H') = P(H')/P(H), \qquad P(H) = 1-P(H'). \tag{2}$$

Conversely, given his odds $O(H')$ $(= 1/O(H))$,

$$P(H') = O(H')/[1+O(H')]. \tag{3}$$

Suppose an experiment yields data $\mathbf{x}$,

$$\mathbf{x} = (x_1, \ldots, x_n)^T, \tag{4}$$

with conditional personal probability or probability density $\varphi(\mathbf{x}|\theta)$ parametrized by θ. Sometimes, there are no obvious unknown parameters other than θ (that is, no nuisance parameters), and many persons will agree on the common conditional probability φ for the data, at least as an interesting approximation. Then we speak of the *public* φ, the *statistical sampling model.*

By an easy application of Bayes' theorem, a person's coherent posterior odds for H', $O(H'|\mathbf{x}) = P(H'|\mathbf{x})/P(H|\mathbf{x})$, is given by

$$O(H'|\mathbf{x}) = O(H')\cdot L_{\mathbf{x}}(H', H), \tag{5}$$

where, for the hypotheses of (1),

$$L_{\mathbf{x}}(H', H) = L_{\mathbf{x}}(\theta_{H'}, \theta_H) = \varphi(\mathbf{x}|\theta_{H'})/\varphi(\mathbf{x}|\theta_H). \tag{6}$$

This *Bayes factor*, or *likelihood ratio for* $\theta_{H'}$ *versus* θ_H, depends on only the public probabilities $\varphi(\mathbf{x}|\theta)$. Hence, the ratio of a person's posterior odds to his prior odds is determined by only the data $\mathbf{x}$, whatever his prior odds. The conclusion must be that data bearing on a simple dichotomy should be reported as a likelihood ratio, by which the reader will multiply his prior odds.

Our first GRIP is given by fig. 1. The solid curve is a plot of $O(H')$ on a logarithmic scale as a function of $P(H')$. The dashed curve, $O(H'|\mathbf{x})$ versus $P(H')$, is obtained by a vertical shift of $\log O(H')$ by $\log L_{\mathbf{x}}(H', H)$. Then the horizontal distances from the dashed curve to the solid curve give the changes from $P(H')$ to $P(H'|\mathbf{x})$ wrought by the data $\mathbf{x}$.

The particular dashed curve plotted in fig. 1 reports inference on Fisher's [5, p. 19] problem of whether a given black mouse is homo-

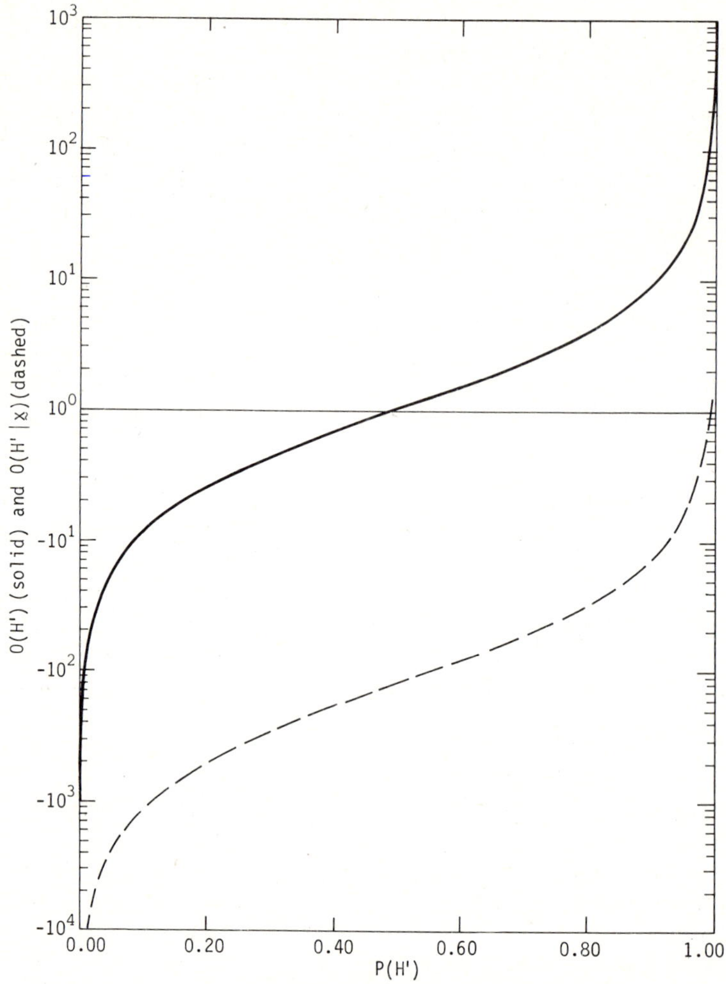

Fig. 1. Personal use of $L_{\mathbf{x}}(H', H)$.

zygous (H) with respect to dominant blackness or heterozygous (H'), in light of the mouse's seven offspring with a brown mate all being black. If the given mouse is homozygous, the probability of the data is $\varphi(\mathbf{x}|H) = 1$; if heterozygous, $\varphi(\mathbf{x}|H') = \frac{1}{2}^7$. Hence $L_{\mathbf{x}}(H', H) = 2^{-7} \doteq 0.00781$.

Fisher obtains, from information about the given mouse's parents, the public prior probability $P(H') = \frac{2}{3}$, from which $O(H') = 2$,

$O(H'|\mathbf{x}) = 2^{-6}$, and hence $P(H'|\mathbf{x}) = \frac{1}{65} \doteq 0.0154$, which fig. 1 gives directly, though crudely, from $P(H')$.

Of course, fig. 1 is proposed for use in reporting data for which the prior probability $P(H')$ is not public. In more general cases, to follow, where not even L itself is public, a whole family of dashed curves could be labeled by the prior opinions to which they pertain. I shall choose, though, to give special GRIP's of L, rather than of P.

3. *A vague alternative to a point null hypothesis*

Commonly, an experimenter proposes no special alternative value $\theta_{H'}$, but considers a continuous range of θ values; thus redenote

$$H' : \theta \neq \theta_H. \tag{7}$$

3.1. *Reporting in terms of simple dichotomies*

One possibility with a vague alternative hypothesis is to report the likelihood ratio for a dichotomy $L_{\mathbf{x}}(\theta, \theta_H)$ as a function of the alternative value $\theta_{H'} = \theta$. Notice that, as such, $L_{\mathbf{x}}(\theta, \theta_H)$ is proportional to the *likelihood function*, the probability of the data, $\varphi(\mathbf{x}|\theta)$ considered as a function of the parameter θ,

$$L_{\mathbf{x}}(\theta, \theta_H) \propto \varphi(\mathbf{x}|\theta). \tag{8}$$

(Barnard [7] has suggested the reporting of a multiple of the whole likelihood function.)

Consider the Normal-sample model with unknown mean $\theta = \mu$ and known variance σ^2,

$$\varphi(\mathbf{x}|\mu) = \varphi(\mathbf{x}|\mu, \sigma^2) = \prod_{i=1}^{n} h_\infty(x_i - \mu; \sigma^2)$$
$$= (2\pi\sigma^2)^{-\frac{1}{2}n} \exp\{-\tfrac{1}{2}[n(\bar{x}-\mu)^2/\sigma^2 + (n-1)s^2/\sigma^2]\}, \tag{9}$$

where $h_\infty(x_i - \mu; \sigma^2)$ denotes the Normal density with variance σ^2 and mean μ, evaluated at x_i (or with mean 0, evaluated at $x_i - \mu$), and where $\bar{x}$ and s^2 are the usual sample mean and unbiased sample variance. (Box and Tiao [8] give a Bayesian analysis for the question of whether given data follow the Normal model.)

For a simple dichotomy μ_H, $\mu_{H'}$, we have

$$L_{\mathbf{x}}(\mu_{H'}, \mu_H) = h_\infty(\bar{x}-\mu_{H'}; \sigma^2/n)/h_\infty(\bar{x}-\mu_H; \sigma^2/n)$$
$$= \exp[\tfrac{1}{2}(Z_H^2 - Z_{H'}^2)], \tag{10}$$

in terms of the standardized deviates,

$$Z_{H'} = (\bar{x}-\mu_{H'})/(\sigma/n^{\frac{1}{2}}), \quad \text{and} \quad Z_H = (\bar{x}-\mu_H)/(\sigma/n^{\frac{1}{2}}). \tag{11}$$

Fig. 2 is a log–log plot of contours of $L_{\mathbf{x}}(\mu_{H'}, \mu_H)$ as a function of $|C|$ and $|D|$, where

$$C = Z_{H'}/Z_H = (\bar{x}-\mu_{H'})/(\bar{x}-\mu_H), \tag{12}$$

$$D = 1/Z_H = \sigma/[n^{\frac{1}{2}}(\bar{x}-\mu_H)]. \tag{13}$$

This master GRIP for L itself can be data-specified by translating the origin to $(-\log_{10}|\bar{x}-\mu_H|, -\log_{10}|\bar{x}-\mu_H| - \frac{1}{2}\log_{10}n)$. Then the new axes are $|\mu_{H'}-\bar{x}|$ and σ.

The GRIP is data-specified in fig. 3 to Darwin's data on the difference in heights of self- and cross-fertilized plants, as discussed in Fisher [9, p. 37]. The special value for the null hypothesis is $\mu_H = 0$. The statistics are $n = 15$, $\bar{x} = 20.933$, $s = 37.75$, $(s/n^{\frac{1}{2}} = 9.746$, $t = 2.148)$. Hence the origin for fig. 3 relative to fig. 2 is $(-\log_{10}20.9, -\log_{10}81.1)$. (An experimenter can easily perform such translations with an overlay sheet.)

For a given value of the abscissa $|\mu_{H'}-\bar{x}|$, the cross-section plot of $L_{\mathbf{x}}(\mu_{H'}, \mu_H)$ against the ordinates σ is a GRIP for the simple dichotomy $\mu_{H'}$, μ_H. The value $\sigma = s$ is, of course, of special interest.

(By taking square–square, rather than log–log, axes in fig. 2, one can obtain concurrent straight-line contours, but then a simple translation will not data-specify the GRIP. Any projective dual of concurrent straight-line contours will yield a nomogram [10].)

As shown in Edwards et al. [6], to report a significance level is to mislead. Note from equation (10) that the traditionally "significant" two standard deviations, $Z_H = 2$ $(\alpha < 0.05)$, yields a maximum e^2 of the dichotomy likelihood ratio, less than 7.4, attained for the alternative value $\mu_{H'}$ exactly equal to the observed value $\bar{x}$, $Z_{H'} = 0$. For the most improbable value under $\mu = \mu_{H'}$, $Z_{H'} = 1.15$ and $Z_H = 2$(two-tailed power $\beta \doteq 1-0.25$), "significance" means a ratio of only 3.8. A prior probability of $\frac{1}{2}$ for H, for example, would be diminished to not less

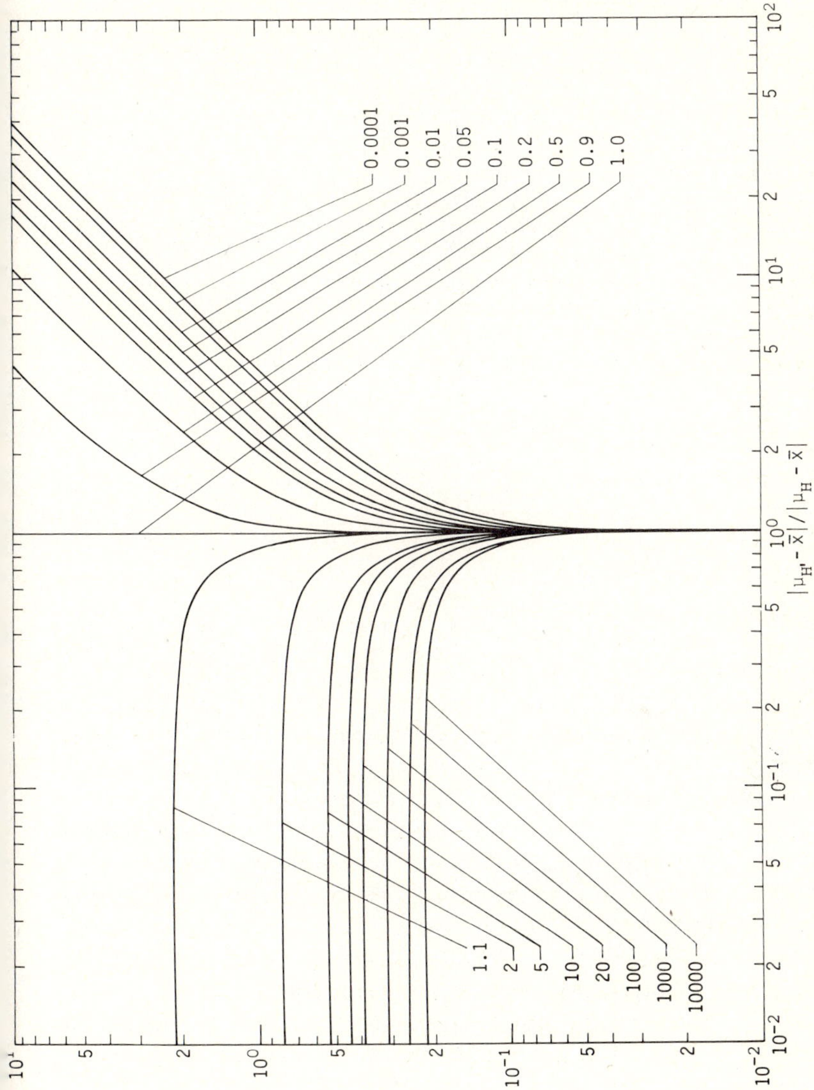

Fig. 2. **GRIP** by contours of $L_{\mathbf{x}}(\mu_{H'}, \mu_H)$, simple dichotomy, simple Normal sample, variance assumed known.

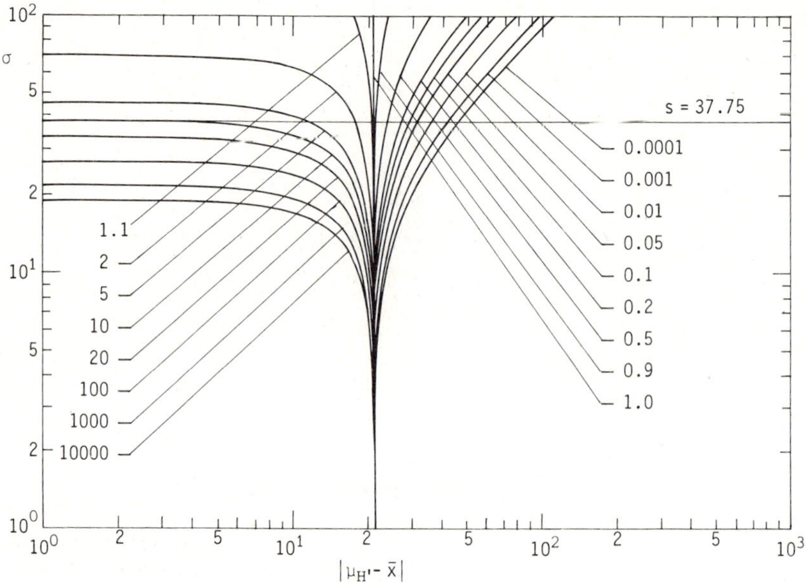

Fig. 3. Darwin's data; specific GRIP by contours of $L_{\mathbf{x}}(\mu_{H'}, \mu_H)$; derived from fig. 2.

than 0.11 and 0.20, respectively, for these two cases; whereas the frequency probability of wrongly rejecting H is less than 0.05 for the two-tailed test, and less than 0.025 for the one-tailed test.

The maximum of the Normal-model simple dichotomy likelihood ratio (10), attained for $Z_{H'} = 0$ ($\mu_{H'} = \bar{x}$), is given as a function of the standard deviate Z_H in column 3 of table 1. This is the first of our Public Upper Bounds,

$$\begin{aligned}\text{PUB1} &= \max_{\mu} L_{\mathbf{x}}(\mu, \mu_H) = h_{\infty}(0; 1)/h_{\infty}(Z_H; 1) \\ &= \exp\left(\tfrac{1}{2}Z_H^2\right). \end{aligned} \tag{14}$$

3.2. *Use of quantified vague prior knowledge*

What if the readers of a scientific report have no specific alternative values $\theta_{H'}$ in mind; rather, conditional on $H': \theta \neq \theta_H$, they have opinions which can be described by personal prior densities $p(\theta|H')$? In this case, the experimenter can report the *weighted likelihood ratio* $L_{\mathbf{x}}(H', \theta_H)$, the ratio of posterior odds to prior odds in (5), as a function of the readers' weight functions $p(\theta|H')$,

$$L_{\mathbf{x}}(H', \theta_H) = \Phi(\mathbf{x}|H')/\varphi(\mathbf{x}|\theta_H), \tag{15}$$

where

$$\Phi(\mathbf{x}|H') = \int \varphi(\mathbf{x}|\theta)p(\theta|H')\,\mathrm{d}\theta. \tag{16}$$

(For extensive studies of L with a vague alternative, see [12–14], and references cited therein.)

We have the immediate *theorem*:

$$L_{\mathbf{x}}(H', \theta_H) = \int L_{\mathbf{x}}(\theta, \theta_H)p(\theta|H')\,\mathrm{d}\theta. \tag{17}$$

Hence, a reader of a scientific report can obtain his personal weighted likelihood ratio $L_{\mathbf{x}}(H', \theta_H)$ as a personal weighted average of reported simple-dichotomy likelihood ratios, such as given by fig. 3. This equation (17) explains why odds for H', against H, are treated here, instead of odds for H, against H'.

Such a prior distribution which gives positive probability to a zero-volume set H and has a conditional density under the complement H' is intended as an approximation to actual continuous-type prior densities which concentrate probability in a neighborhood of θ_H.

Generally speaking, the Bayesian L tends not to reject the null hypothesis as readily as do classical tests. Even so, if $\theta = \theta_H + \varepsilon$ both tests will eventually reject θ_H for increasingly large samples, when we would be willing in practice to say that the true value θ differs so little from θ_H that we do not distinguish them. But one can, of course, wiggle the null value to see if nearby values would also be rejected by L. One can look at the posterior density under H' or at the likelihood function and the maximum likelihood value $\hat{\theta}$, to see just which alternative point hypothesis dominates. (The real rub is from classical goodness-of-fit tests in which the alternative hypotheses may be unclear.)

If a classical test does not reject a null hypothesis, then the Bayesian L cannot *strongly* reject it, since if the null value does not lie far out in the tail of the likelihood function, then the likelihood-function ordinate at the null value cannot be much smaller than even the maximum ordinate. But, on the other hand, the Bayesian L can conceivably strongly accept a classically rejected null value.

A more accurate, if less convenient, Bayesian analysis would start with a continuous-type prior density only, a mixture of overlapping

TABLE 1

Public upper bounds and public lower and upper markers, vague alternative to a point null hypothesis, simple normal sample, known variance

Z_H	Significance level (2 tailed)	PUB1 = $\max_\mu L_{\mathbf{x}}(\mu,\mu_H)$	PUB2 = $\max_{\tau_0} L_{\mathbf{x}}(H',\mu_H)$	at $\tau_0/(\sigma/n^{\frac{1}{2}})$	$\alpha = 0.05$[a] PLUM1	$\alpha = 0.05$[a] PLUM2	$\alpha = 0.01$[b] PLUM1	$\alpha = 0.01$[b] PLUM2
0.0	1.0000	1.000	1.000	—	0.6075	0.1465	0.4819	0.0363
0.1	0.9204	1.005	1.000	—	0.6105	0.1472	0.4843	0.0365
0.2	0.8414	1.020	1.000	—	0.6197	0.1494	0.4916	0.0371
0.3	0.7642	1.046	1.000	—	0.6354	0.1532	0.5040	0.0380
0.4	0.6892	1.083	1.000	—	0.6581	0.1587	0.5220	0.0393
0.5	0.6170	1.133	1.000	—	0.6884	0.1660	0.5460	0.0412
0.6	0.5486	1.197	1.000	—	0.7273	0.1754	0.5769	0.0435
0.7	0.4840	1.278	1.000	—	0.7761	0.1872	0.6156	0.0464
0.8	0.4238	1.377	1.000	—	0.8366	0.2017	0.6636	0.0500
0.9	0.3682	1.499	1.000	—	0.9108	0.2196	0.7224	0.0545
1.0	0.3174	1.649	1.000	—	1.002	0.2415	0.7944	0.0599
1.1	0.2714	1.831	1.010	0.4583	1.112	0.2683	0.8824	0.0665
1.2	0.2302	2.054	1.038	0.6633	1.248	0.3010	0.9899	0.0746
1.3	0.1936	2.328	1.086	0.8307	1.414	0.3410	1.122	0.0846
1.4	0.1616	2.664	1.154	0.9798	1.619	0.3903	1.284	0.0968
1.5	0.1336	3.080	1.245	1.118	1.871	0.4512	1.484	0.1119
1.6	0.1096	3.597	1.363	1.249	2.185	0.5269	1.733	0.1306
1.7	0.0892	4.242	1.513	1.375	2.577	0.6214	2.044	0.1541
1.8	0.0718	5.053	1.703	1.497	3.070	0.7402	2.435	0.1835
1.9	0.0574	6.080	1.941	1.616	3.693	0.8907	2.930	0.2208

2.0	0.0454	7.389	2.241	1.732	4.489	1.082	3.560	0.2684
2.1	0.0358	9.070	2.620	1.847	5.510	1.329	4.371	0.3295
2.2	0.0278	11.25	3.100	1.960	6.832	1.647	5.419	0.4085
2.3	0.0214	14.08	3.714	2.071	8.555	2.063	6.786	0.5115
2.4	0.0164	17.81	4.502	2.182	10.82	2.610	8.584	0.6471
2.5	0.0124	22.76	5.522	2.291	13.83	3.334	10.97	0.8267
2.6	0.0094	29.37	6.852	2.400	17.84	4.303	14.15	1.067
2.7	0.0070	38.28	8.600	2.508	23.26	5.608	18.45	1.391
2.8	0.0052	50.40	10.02	2.615	30.62	7.383	24.29	1.831
2.9	0.0038	67.02	14.02	2.722	40.71	9.818	32.29	2.434
3.0	0.0026	90.02	18.20	2.828	54.68	13.19	43.38	3.270
3.1	0.0020	122.1	23.89	2.934	74.18	17.89	58.84	4.436
3.2	0.0014	167.3	31.72	3.040	101.7	24.51	80.63	6.078
3.3	0.0010	231.6	42.57	3.145	140.7	33.93	111.6	8.412
3.4	0.0006	323.8	57.76	3.250	196.7	47.43	156.0	11.76
3.5	0.0004	457.1	79.22	3.354	277.7	66.97	220.3	16.60
3.6	0.0004	652.0	109.8	3.458	396.1	95.51	314.2	23.68
3.7	0.0002	939.2	154.0	3.562	570.5	137.6	452.5	34.11
3.8	0.0002	1366.0	218.1	3.666	830.1	200.2	658.5	49.63
3.9	—	2008.0	312.3	3.770	1220.0	294.2	967.7	72.94
4.0	—	2981.0	452.0	3.873	1811.0	436.7	1436.0	108.3

[a] $\frac{\alpha}{1-\alpha} = 0.0526$; $Z_{0.975} = 1.96$.

[b] $\frac{\alpha}{1-\alpha} = 0.0101$; $Z_{0.995} = 2.575$.

tractable prior densities, one of which gives high probability to a small interval serving as the null hypothesis, $H = (\theta_0 - \varepsilon, \theta_0 + \varepsilon)$. Then $P(H)$ and $P(H\|\mathbf{x})$ would be mixtures of probability integrals. Such an approach was suggested to the author in 1962 by Leonard J. Savage. See also Hildreth [15], in which some relevant inequalities can be found.

Returning to the Normal-model case, consider the handy class of approximate personal weight functions, the "conjugate" class of Normal densities. If the prior mean and variance at μ under H' are denoted $\bar{x}_0$ and τ_0^2, then the posterior distribution of μ under H' is Normal with mean $\bar{x}_1 = (\tau_0^{-2}\bar{x}_0 + n\sigma^{-2}\bar{x})/\tau_1^{-2}$, a weighted average of $\bar{x}$ and $\bar{x}_0$; the variance is $\tau_1^2 = 1/(\tau_0^{-2} + n\sigma^{-2})$, half the harmonic mean of σ^2/n and τ_0^2.

In practice, the Normal prior densities can be inadequate for the approximation of real partial knowledge. But then a mixture of two Normal densities is nearly as tractable a prior. (The integral-mixture Student-t densities are even possible.)

In an estimation problem, plots of $\bar{x}_1$ and $\bar{x}_1 \pm 2\tau_1$ against $\bar{x}_0$ for various values of τ_0, or against τ_0 for various $\bar{x}_0$, could serve as a GRIP. But, I suppose, an experimenter with an estimation problem would usually take large enough n that $\bar{x}_1$ and τ_1 are approximated by $\bar{x}$ and $\sigma/n^{\frac{1}{2}}$ for his "vague-alternative" readers. For example, if τ_0 is merely five times the standard deviation of the mean $\sigma/n^{\frac{1}{2}}$, then the effect of $\bar{x}_0$ on $\bar{x}_1$ is one-twentyfifth the effect of $\bar{x}$.

The Bayes factor, or weighted likelihood ratio, for a vague alternative (15) becomes

$$L_{\mathbf{x}}(H', \mu_H) = h_\infty(\bar{x} - \bar{x}_0\,;\, \sigma^2/n + \tau_0^2)/h_\infty(\bar{x} - \mu_H\,;\, \sigma^2/n)$$

$$= h_\infty(\mu_H - \bar{x}_0;\, \tau_0^2)/h_\infty(\mu_H - \bar{x}_1\,;\, \tau_1^2)\,. \tag{18}$$

The second of these expressions for L, a ratio of prior and posterior densities, holds in more generality as Savage's density ratio [12].

Of natural interest are the prior densities under H' centered at the null value,

$$\bar{x}_0 = \mu_H\,. \tag{19}$$

With such a Normal (μ_H, σ_0^2) prior density $p(\mu|H')$, we have

$$L_{\mathbf{x}}(H', \mu_H) = (1 + Z_*^2)^{-\frac{1}{2}} \exp\left[\tfrac{1}{2} Z_*^2 Z_H^2/(1 + Z_*^2)\right], \tag{20}$$

where

$$Z_* = \tau_0/(\sigma/n^{\frac{1}{2}}), \quad \text{and} \quad Z_H = (\bar{x} - \mu_H)/(\sigma/n^{\frac{1}{2}})\,. \tag{21}$$

Proceeding again as in (12) and (13) to obtain data-translatable contours of L as a function of τ_0, the personal standard deviation under H', and of σ, the presumed-known experimental-process standard deviation, write

$$C_* = Z_*/Z_H = \tau_0/(\bar{x}-\mu_H) \quad \text{and}$$

$$D = 1/Z_H = \sigma/[n^{\frac{1}{2}}(\bar{x}-\mu_H)]. \tag{22}$$

Fig. 4 is a horizontally weighted average of fig. 2, a log–log plot of contours of $L_{\mathbf{x}}(H', \mu_H)$ as a function of $|C_*|$ and $|D|$. Fig. 5 is the analogue of fig. 3 for Darwin's data.

For a given Z_H, the maximum of this centered-Normal-density-weighted likelihood ratio (20) is attained at $Z_* = (Z_H^2-1)^{\frac{1}{2}}$ if $|Z_H| \geqq 1$, and $Z_* = 0$ if $|Z_H| \leqq 1$:

$$\begin{aligned} \text{PUB2} &= \max_{\tau_0} L_{\mathbf{x}}(H', \mu_H), \\ &= \begin{cases} Z_H^{-1} \exp[\tfrac{1}{2}(Z_H^2-1)], & \text{if } |Z_H| \geqq 1, \\ 1, & \text{if } |Z_H| \leqq 1, \end{cases} \end{aligned} \tag{23}$$

This second Public Upper Bound is given as a function of Z_H in column 4 of table 1. The maximizing Z_* is given in column 5. Note that for the traditional $Z_H = 2$, $L < 2.3$, at $Z_* = \sqrt{3}$.

More generally, the numerator in the first expression in (18) is maximized by $\tau_0 = (\bar{x}-\bar{x}_0)^2-\sigma^2/n$ if $\geqq 0$, and otherwise by $\tau_0 = 0$ for which $L_{\mathbf{x}}(H', \mu_H) = L_{\mathbf{x}}(\mu_{H'} = \bar{x}_0, \mu_H)$.

3.3. *The one-sided test*

The above weighted likelihood ratio (20), in which opinions under the alternative H' are approximated by Normal densities centered at the null value μ_H, might be thought of as giving a Bayesian "two-sided test". To obtain a "one-sided test", consider the class of prior densities, Normal $(\bar{x}_0, \tau_0^2)$, but truncated at c_0, say,

$$p(\mu|H') = \begin{cases} h_{\infty}(\mu-\bar{x}_0\,;\,\tau_0^2)/\mathcal{N}[(\bar{x}_0-c_0)/\tau_0] & \text{for } \mu \geqq c_0, \\ 0 \quad \text{for } \mu < c_0, \end{cases} \tag{24}$$

where $\mathcal{N}(x)$ is the standard Normal cumulative $P[Z \leqq x]$. The corresponding posterior densities are of the same form, with $(\bar{x}_0, \tau_0^2)$ replaced by the conjugate values $(\bar{x}_1, \tau_1^2)$, as previously, and with the

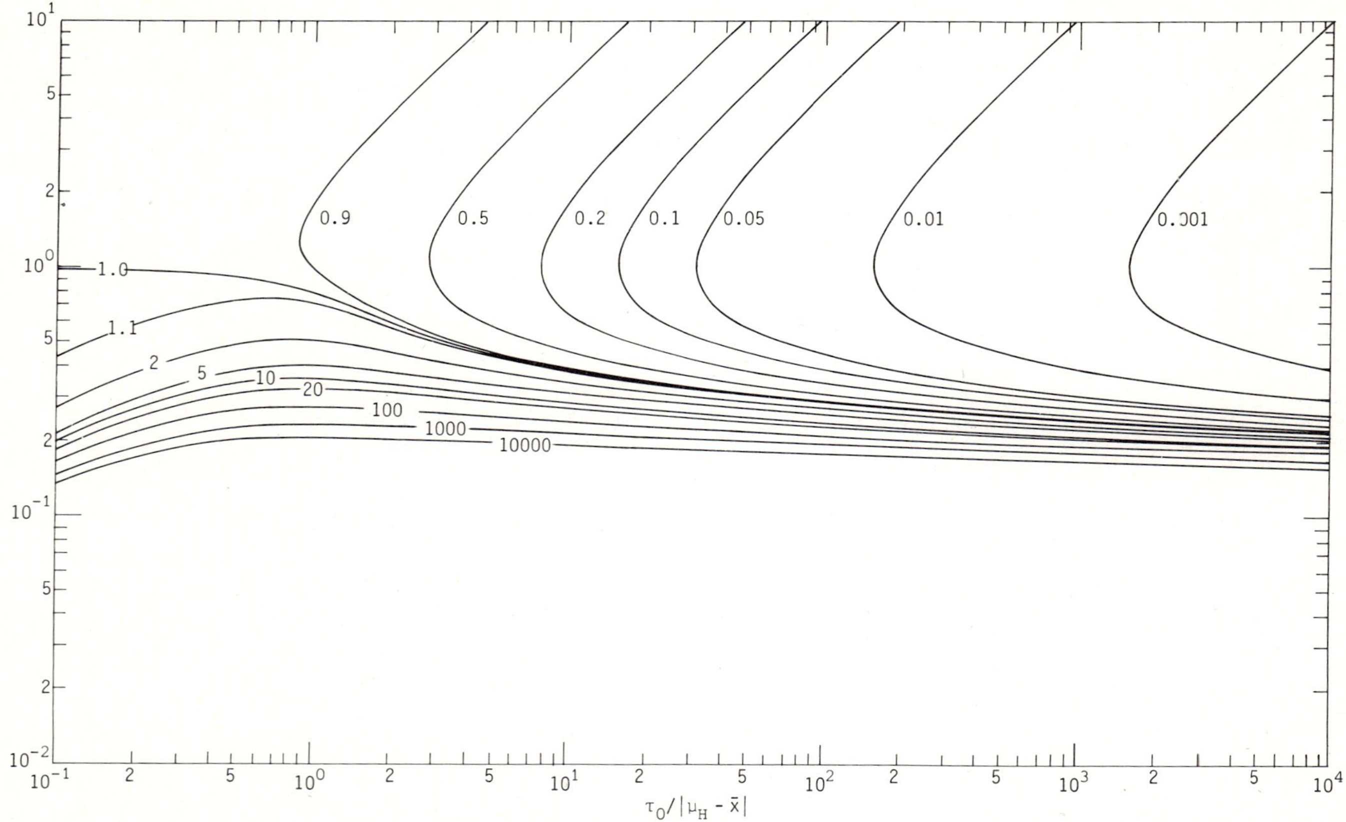

Fig. 4. GRIP by contours of $L_{\mathbf{x}}(H', \mu_H)$, normally weighted alternative to a point null hypothesis, simple Normal sample, variance assumed known.

prior truncation point, $c_1 = c_0$. Under the *right-hand* alternative

$$H'_+ : \mu > \mu_H, \tag{25}$$

take the half-Normal prior density truncated on the left at the null value,

$$\bar{x}_0 = \mu_H, \qquad c_0 = \mu_H. \tag{26}$$

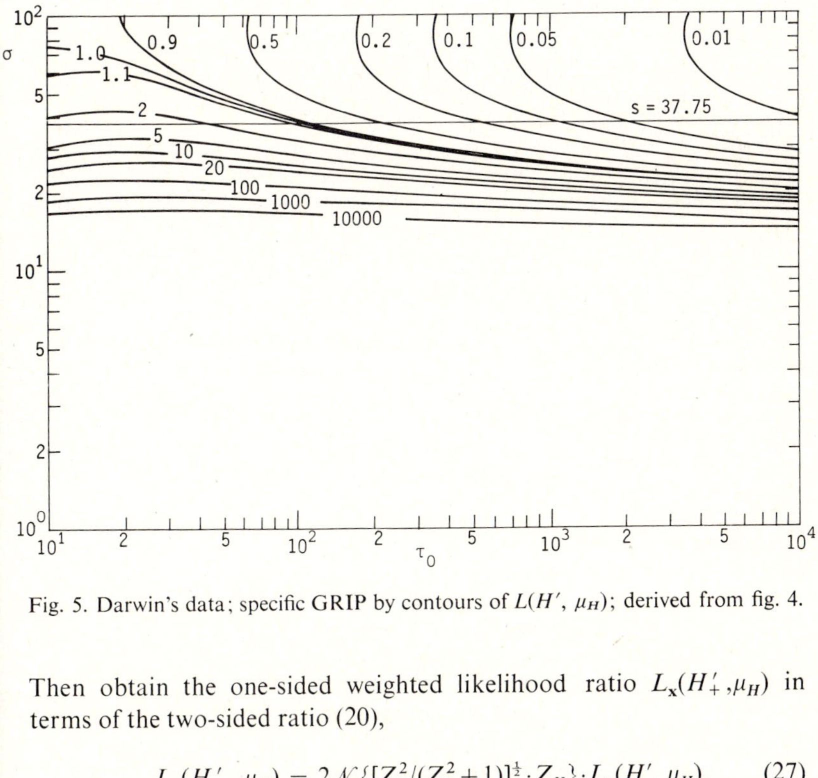

Fig. 5. Darwin's data; specific GRIP by contours of $L(H', \mu_H)$; derived from fig. 4.

Then obtain the one-sided weighted likelihood ratio $L_{\mathbf{x}}(H'_+, \mu_H)$ in terms of the two-sided ratio (20),

$$L_{\mathbf{x}}(H'_+, \mu_H) = 2\mathcal{N}\{[Z_*^2/(Z_*^2+1)]^{\frac{1}{2}} \cdot Z_H\} \cdot L_{\mathbf{x}}(H', \mu_H). \tag{27}$$

Two separate GRIP's, interms of C_* and D (22), can be made for the two cases of the sign of $x - \mu_H$. I imagine the GRIP for the negative sign would receive little use in practice.

For the Normal prior density truncated on the *right*, the ratio $L_{\mathbf{x}}(H'_-, \mu_H)$ follows from $L_{\mathbf{x}}(H'_+, \mu_H)$ by a change of sign in the argument of the Normal cumulative.

3.4. The test based on stable estimation

In Savage's theory of "stable estimation" [6], a prior density of θ is assumed to be approximately locally constant in the vicinity where the likelihood function is other than negligible, and also to be not outlandishly greater elsewhere. Then the posterior density is approximately equal to the normalized likelihood function, here denoted

$$\psi_{\mathbf{x}}(\theta) = \varphi(\mathbf{x}|\theta)/\int \varphi(\mathbf{x}|\tilde{\theta})\, d\tilde{\theta}. \tag{28}$$

To apply this theory to test a sharp (a point) null hypothesis against a vague alternative (as by Savage and by Lindley [16]), note that equation (15) for $L_{\mathbf{x}}(H', \theta_H)$ holds with $\varphi(\mathbf{x}|\theta)$ replaced by $\psi_{\mathbf{x}}(\theta)$,

$$L_{\mathbf{x}}(H', \theta_H) = \int \psi_{\mathbf{x}}(\theta) p(\theta|H')\, d\theta/\psi_{\mathbf{x}}(\theta_H). \tag{29}$$

Denote the maximum likelihood value of θ by $\hat{\theta} = \hat{\theta}(\mathbf{x})$. Then we have the approximation,

$$L_{\mathbf{x}}(H', \theta_H) \doteq p(\theta|H')_{\hat{\theta}}/\psi_{\mathbf{x}}(\theta_H), \tag{30}$$

the personal constant $p(\theta/H')_{\hat{\theta}}$ multiplied by the reciprocal of the "flat-prior" approximate posterior density $\psi_{\mathbf{x}}(\theta_H) \doteq p(\theta|H', \mathbf{x})_{\theta_H}$, a public function of the data $\mathbf{x}$. The next section presents bounds for L harmonious with this stable-estimation form.

3.5. Bounds from likelihood-function intervals

Given an interval about the value of the statistic $\hat{\theta}$,

$$I = [\hat{\theta} - d_1, \hat{\theta} + d_2], \tag{31}$$

the integral in the numerator of (29) equals the sum of the integrals over I and over its complement I'. Hence, we have the lower bound,

$$P_{\min I}(I) \cdot [\text{PLUM1}] \leqq L_{\mathbf{x}}(H', \theta_H), \tag{32}$$

with PLUM1, our first Public Lower and Upper Marker, the ratio of the average over I of the approximate-posterior-density ordinate to the null-value ordinate,

$$\text{PLUM1} = \bar{\psi}(I)/\bar{\psi}_{\mathbf{x}}(\theta_H), \tag{33}$$

$$\bar{\psi}(I) = (1-\alpha)/(d_1 + d_2), \qquad 1-\alpha = \int_I \psi_{\mathbf{x}}(\theta)\, d\theta, \tag{34}$$

and where $P_{\min I}(I)$ is the probability that would be assigned to I by the minimum of the prior density over that interval,

$$P_{\min I}(I) = (d_1 + d_2) \cdot \min_I p(\theta | H'). \tag{35}$$

Note that PLUM 1 can be interpreted as $L_{\mathbf{x}}(I, \theta_H)$ for a prior distribution in which the conditional distribution of θ given H' is uniform over I.

We have the three upper bounds

$$\begin{aligned} L_{\mathbf{x}}(H', \theta_H) &\leqq [P_{\max I}(I) + P_{\mathrm{MAX}}(I)\alpha/(1-\alpha)] \cdot [\mathrm{PLUM1}] \\ &\leqq P_{\mathrm{MAX}}(I)(1-\alpha)^{-1} \cdot [\mathrm{PLUM1}] \\ &= \max p(\theta | H')/\psi_{\mathbf{x}}(\theta_H) \\ &\leqq \mathrm{PLUM1}/(1-\alpha), \quad \text{if} \quad P_{\mathrm{MAX}}(I) \leqq 1, \end{aligned} \tag{36}$$

where $P_{\max I}$ and P_{MAX} follow as in (35) from $\max_I p(\theta | H')$ and the overall $\max\ p(\theta | H')$, respectively. Note that $P_{\max I}(I)$ and $P_{\mathrm{MAX}}(I)$ need not be less than unity. Note that the second upper bound in (36) is constant in I.

If $\psi_{\mathbf{x}}(\theta)$ is decreasing in $|\theta - \hat{\theta}| \geqq \min d_i$, then we have further upper bounds,

$$\begin{aligned} L_{\mathbf{x}}(H', \theta_H) &\leqq P_{\max I}(I) \cdot [\mathrm{PLUM1}] \\ &\quad + P[\theta \in I' | H'] \cdot [\mathrm{PLUM2}] \\ &\leqq [\mathrm{PLUM1}] + [\mathrm{PLUM2}] \quad \text{if} \quad P_{\max I}(I) \leqq 1, \end{aligned} \tag{37}$$

where

$$\begin{aligned} \mathrm{PLUM2} &= \max_i \psi_{\mathbf{x}}(\hat{\theta} \mp d_i)/\psi_{\mathbf{x}}(\theta_H) \\ &= \max_i L_{\mathbf{x}}(\hat{\theta} \mp d_i, \theta_H). \end{aligned} \tag{38}$$

The bounds on L are easily determined by multiplying PLUM's by assessed personal probabilities. The experimenter need only report the PLUM's together with some illustrative calculations.

In general, d_1, d_2, and the shape of the normalized likelihood function ψ could depend on the data, as in section 4.

Turning to the Normal-model case, we have $\theta = \mu$, $\hat{\theta} = \bar{x}$, $\psi_{\mathbf{x}}(\mu) = h_{\infty}(\mu - \bar{x}; \sigma^2/n)$; and hence if

$$I = [\bar{x} - d, \bar{x} + d], \tag{39}$$

we have

$$d = d_1 = d_2 = (\sigma/n^{\frac{1}{2}})Z_{1-\frac{1}{2}\alpha}, \tag{40}$$

$$\text{PLUM1} = [(1-\alpha)/(2Z_{1-\frac{1}{2}\alpha})]/h_{\infty}(Z_H\,;\,1), \tag{41a}$$

$$\begin{aligned}\text{PLUM2} &= h_{\infty}(Z_{1-\frac{1}{2}\alpha}\,;\,1)/h_{\infty}(Z_H\,;\,1)\\ &= \exp\,[\tfrac{1}{2}(Z_H^2 - Z_{1-\frac{1}{2}\alpha}^2)],\end{aligned} \tag{41b}$$

in which

$$1/h_{\infty}(Z_H\,;\,1) = [1/h_{\infty}(0;\,1)]\cdot[\text{PUB 1}], \tag{42}$$

where Z_β is the standard Normal (100β)th percentile, and Z_H is the standardized deviate for $\mu = \mu_H$ (11). The two public quantities are given as functions of Z_H for $\alpha = 0.05$ and 0.01 in columns 6–9 of table 1.

For Darwin's data, if, for example, $\sigma/n^{\frac{1}{2}} = s/n^{\frac{1}{2}}$ ($= 9.75$), then $Z_H = T_H$ ($= 2.15$), $I = [1.8, 40.0]$ or $[-4.2, 46.0]$, PLUM1 $= 6.1$ or 4.9, and PLUM2 $= 1.5$ or 0.37, for $\alpha = 0.05$ or 0.01, respectively. Considering $\alpha = 0.05$, we infer that one's prior personal odds against the null hypothesis H of equal-population-mean-heights for the self- and cross-fertilized plants is multiplied by at least $6.1\cdot P_{\min\,I}(I)$ where $P_{\min\,I}(I)$ is one's minimum prior probability, under the alternative hypothesis H', for the interval $I = [1.8,\ 40.0]$. If the analogous "probability" from one's maximum density on I, $P_{\max\,I}(I)$, is not greater than unity, then the prior odds are multiplied by, at most, $6.1+1.5 = 7.6$. Other upper bounds are PUB1 $= 10.1$, for the point alternative $\mu_{H'} = \bar{x} = 20.9$, and PUB2 $= 2.9$, for the Normally distributed alternative $\mu|H'$ centered at 0 with standard deviation 18.6.

4. *Normal-model case, variance unknown*

Commonly, the variance σ^2 is unknown too (as is the mean μ). Then the foregoing general theory applies again with $\theta = \mu$, but with the conditional experimental probabilities $\varphi(\mathbf{x}|\theta) = \varphi(\mathbf{x}|\mu, \sigma^2)$ replaced by $\Phi(\mathbf{x}|\mu)$,

$$\Phi(\mathbf{x}|\mu) = \int \varphi(\mathbf{x}|\mu, \sigma^2)p(\sigma|\mu)\,\mathrm{d}\sigma, \tag{43}$$

a subjective-weighted average of the statistical-model probabilities. The probability density Φ is less public than φ.

Consider the important common case in which μ and σ are independent a priori, $p(\sigma|\mu) \equiv p(\sigma)$. For example, such might be the case when a calibrated measuring instrument is used to measure a quantity, which the instrument has never before measured.

We apply to σ the rich and tractable two-parameter family of inverted-gamma-2 densities

$$p(\sigma|s_0, \nu_0) = \frac{2(\frac{1}{2}\nu_0 s_0^2/\sigma^2)^{\frac{1}{2}(\nu_0+1)} \exp(-\frac{1}{2}\nu_0 s_0^2/\sigma^2)}{(\frac{1}{2}\nu_0 - 1)!(\frac{1}{2}\nu_0 s_0^2)^{\frac{1}{2}}}, \tag{44}$$

for σ, s_0, $\nu_0 > 0$ [17, sect. 7.22]. The "precision" $1/\sigma^2$ is distributed like a multiple of a chi-squared random variable,

$$\nu_0 s_0^2/\sigma^2 \sim \chi^2_{\nu_0}. \tag{45}$$

The moments of σ easily follow from (45); the mode of σ is $s_0[\nu_0/(\nu_0+1)]^{\frac{1}{2}}$.

Equations (9), (43), and (44) yield the proportionality

$$\begin{aligned}\Phi(\mathbf{x}|\mu) &\propto [1+\nu_1^{-1}(\bar{x}-\mu)^2/(s_1^2/n)]^{-\frac{1}{2}(\nu_1+1)} \\ &\propto h_{\nu_1}(\bar{x}-\mu; s_1^2/n),\end{aligned} \tag{46}$$

the density at $\bar{x}-\mu$ of a scaled Student-t, $(s_1/n^{\frac{1}{2}})\cdot t_{\nu_1}$, where

$$\nu_1 = \nu_0 + \nu, \qquad s_1^2 = \nu_1^{-1}[\nu_0 s_0^2 + \nu s^2], \tag{47}$$

with $\nu = n-1$, the sample degrees of freedom. The proportionality constants in (46) are independent of $\bar{x}$ and μ. The weighted likelihood function, for our given person, is just as if $\bar{x}$ were a single observation by him from a t distribution with unknown center μ, and known scale $s_1/n^{\frac{1}{2}}$ and known degrees of freedom ν_1.

Hence, for a point alternative $\mu_{H'}$,

$$\begin{aligned}L_{\mathbf{x}}(\mu_{H'}, \mu_H) &= h_{\nu_1}(T_{H'}; 1)/h_{\nu_1}(T_H; 1) \\ &= [(1+\nu_1^{-1}T_H^2)/(1+\nu_1^{-1}T_{H'}^2)]^{\frac{1}{2}(\nu_1+1)},\end{aligned} \tag{48}$$

where

$$T_{H'} = (\bar{x}-\mu_{H'})/(s_1/n^{\frac{1}{2}}), \quad \text{and} \quad T_H = (\bar{x}-\mu_H)/(s_1/n^{\frac{1}{2}}), \tag{49}$$

since $h_{\nu_1}(\bar{x}-\mu; s_1^2/n) = h_{\nu_1}(T; 1)/(s_1/n^{\frac{1}{2}})$.

Just as figs. 2 and 3 follow from equation (10) in the known-variance case, master GRIP's can be developed from equation (48) with s_1 playing the role of σ. But a special plot is needed for each value of ν_1. A few selected s_1-cross-sections for selected ν_1's should suffice as a report on specific data. The business of choosing ν_0 and s_0, and hence ν_1 and s_1, is discussed later (section 4.2).

The unknown-variance analogue of PUB1 (14) is our first Partly Public Upper Bound,

$$\begin{aligned}\text{PPUB1} &= \max_{\mu|s_0,\nu_0} L_{\mathbf{x}}(\mu, \mu_H) = h_{\nu_1}(0;1)/h_{\nu_1}(T_H;1)\\ &= [1+\nu_1^{-1}T_H^2]^{\frac{1}{2}(\nu_1+1)}.\end{aligned}$$

where

$$h_{\nu_1}(0;1) = \nu_1^{-\frac{1}{2}}\Gamma[\tfrac{1}{2}(\nu_1+1)]/[\Gamma(\tfrac{1}{2}\nu_1)\pi^{\frac{1}{2}}]. \tag{50}$$

Table 2 gives (column 2) the numerator of PPUB1. Table 3 reports PPUB1 for Darwin's data.

We have ([12, 13], see also Anscombe [18] and Stone [19]) the realistic family of joint prior densities for μ and σ with the above independence assumption and prior densities for σ, and with the prior densities of μ under H' Normal with mean $\bar{x}_0$ and variance τ_0^2. Then the joint (nonindependent) posterior density of μ and σ is given by

$$[n(\mu-\bar{x})^2+\nu_1 s_1^2]/\sigma^2\big|\mu, \mathbf{x} \sim \chi^2_{\nu_1+1}, \tag{51}$$

and

$$\begin{aligned}p(\mu\big|H', \mathbf{x}) &\propto [1+\nu_1^{-1}(\mu-\bar{x})^2/(s_1^2/n)]^{-\frac{1}{2}(\nu_1+1)}\\ &\quad\times \exp[-\tfrac{1}{2}(\mu-\bar{x}_0)^2/\tau_0^2];\end{aligned} \tag{52}$$

or by

$$\mu\big|H', \sigma, \mathbf{x} \sim N(\bar{x}_1, \tau_1^2), \tag{53}$$

and

$$\begin{aligned}p(\sigma\big|H', \mathbf{x}) &\propto (1/\sigma^2)^{\frac{1}{2}(\nu_1+1)}[(1/\sigma^2)/(n/\sigma^2+1/\tau_0^2)]^{\frac{1}{2}}\\ &\quad\times \exp\{-\tfrac{1}{2}[\nu_1 s_1^2/\sigma^2+(\bar{x}-x_0)^2/(\sigma^2/n+\tau_0^2)]\}.\end{aligned} \tag{54}$$

The posterior probability of H' is determined by the weighted likelihood ratio,

$$\begin{aligned}L_{\mathbf{x}}(H', \mu_H) &= \int_{-\infty}^{\infty} h_{\nu_1}(\bar{x}-\mu; s_1^2/n)h_\infty(\mu-\bar{x}_0; \tau_0^2)\,d\mu\\ &\quad /h_{\nu_1}(\bar{x}-\mu_H; s_1^2/n).\end{aligned} \tag{55}$$

As in (17), this weighted likelihood ratio is an average, by the prior density of μ under H', of the dichotomy likelihood ratio (48) with $\mu_{H'} = \mu$. The integral is essentially a Behrens–Fisher density, as are the normalizing constants for (52) and (54) [13, 14, 20].

If, as before, $\bar{x}_0 = \mu_H$, we have

$$L_{\mathbf{x}}(H', \mu_H) = [1 + \nu_1^{-1} T_H^2]^{\frac{1}{2}(\nu_1+1)} \times \int_{-\infty}^{\infty} [1 + \nu_1^{-1}(T_H - z)^2]^{-\frac{1}{2}(\nu_1+1)} (2\pi T_*^2)^{-\frac{1}{2}} \times \exp[-\tfrac{1}{2}z^2/T_*^2]\,\mathrm{d}z, \tag{56}$$

where

$$T_* = \tau_0/(s_1/n^{\frac{1}{2}}), \qquad T_H = (\bar{x} - \mu_H)/(s_1/n^{\frac{1}{2}}). \tag{57}$$

Again, master GRIP's can be prepared for various values of ν_1.

The weighted likelihood ratio (55) is maximized in $\tau_0 \geqq 0$ by either the solution to $\tau_0^2 = E[(\mu - \bar{x}_0)^2 \mid H', \mathbf{x}]$, or $\tau_0 = 0$ for which $L_{\mathbf{x}}(H', \mu_H) = L_{\mathbf{x}}(\mu_{H'} = \bar{x}_0, \mu_H)$. For $\bar{x}_0 = \mu_H$, PPUB 2 would require either the solution $T_* > 0$ to

$$0 = \int_{-\infty}^{\infty} (z^2 - T_*^2)[1 + \nu_1^{-1}(T_H - z)^2]^{-\frac{1}{2}(\nu_1+1)}(2\pi T_*^2)^{-\frac{1}{2}} \times \exp[-\tfrac{1}{2}z^2/T_*^2]\,\mathrm{d}z, \tag{58}$$

or $T_* = 0$ for which (56) is unity.

4.1. *Bounds from weighted-likelihood-function intervals*

To use a weighted-likelihood-function interval about $\bar{x}$, $I = [\bar{x} - d, \bar{x} + d]$, to obtain again bounds on the weighted likelihood ratio, take in place of ψ, $\Psi_{\mathbf{x}}(\mu) = h_{\nu_1}(\mu - \bar{x}; s_1^2/n)$, and hence

$$d = (s_1/n^{\frac{1}{2}}) T_{\nu_1, 1-\frac{1}{2}\alpha}, \tag{59}$$

where $T_{\nu_1,\beta}$ is the (100β)th percentile of Student's t_{ν_1}. Then we have the Partly Public analogues of PLUM1 (41a), and PLUM2 (41b),

$$\begin{aligned} \text{PPLUM1} &= \bar{\Psi}_{\mathbf{x}}(I)/\Psi_{\mathbf{x}}(\mu_H) \\ &= [(1-\alpha)/(2T_{\nu_1,1-\frac{1}{2}\alpha})]/h_{\nu_1}(T_H; 1) \end{aligned} \tag{60}$$

$$\begin{aligned} \text{PPLUM2} &= \Psi_{\mathbf{x}}(d)/\Psi_{\mathbf{x}}(\mu_H - \bar{x}) \\ &= h_{\nu_1}(T_{\nu_1,1-\frac{1}{2}\alpha}; 1)/h_{\nu_1}(T_H; 1). \end{aligned} \tag{61}$$

Note that the interval I, as well as the PPLUM's depends on the prior parameters ν_0 and s_0 through ν_1 and s_1.

Table 2 gives the quotients of PPUB1 divided into PPLUM1 and

TABLE 2

Normalization factor of Student-t densities; and useful data-free factors of Public Lower and Upper Markers, vague alternative to a point null hypothesis, simple Normal sample, unknown variance. The further data-specific factor, PPUB1 $= (1+\nu^{-1}T_H^2)^{\frac{1}{2}(\nu_1+1)}$

		$\alpha = 0.05$			$\alpha = 0.01$		
ν_1	$h_{\nu_1}(0;1)$	$T_{\nu_1,0.975}$	[PPLUM1]/[PPUB1]	[PPLUM2]/[PPUB1]	$T_{\nu_1,0.995}$	[PPLUM1]/[PPUB1]	[PPLUM2]/[PPUB1]
1	0.3183	12.706	0.1174	0.006145	63.657	0.02443	0.0003108
2	0.3536	4.303	0.3122	0.03043	9.925	0.1410	0.001007
3	0.3676	3.182	0.4061	0.05223	5.841	0.2305	0.006532
4	0.3750	2.776	0.4563	0.06824	4.604	0.2867	0.01004
5	0.3796	2.571	0.4868	0.07987	4.032	0.3235	0.01301
6	0.3827	2.447	0.5072	0.08871	3.707	0.3488	0.01548
7	0.3850	2.365	0.5216	0.09545	3.499	0.3675	0.01751
8	0.3867	2.306	0.5327	0.1013	3.355	0.3814	0.01920
9	0.3880	2.262	0.5412	0.1053	3.250	0.3925	0.02061
10	0.3891	2.228	0.5479	0.1090	3.169	0.4014	0.02184
12	0.3907	2.179	0.5580	0.1145	3.055	0.4146	0.02377
15	0.3924	2.131	0.5683	0.1205	2.947	0.4281	0.02587
20	0.3940	2.086	0.5779	0.1266	2.845	0.4416	0.02822
24	0.3948	2.064	0.5828	0.1297	2.797	0.4483	0.02941
30	0.3956	2.042	0.5880	0.1330	2.750	0.4550	0.03066
40	0.3965	2.021	0.5927	0.1362	2.704	0.4618	0.03201
60	0.3973	2.000	0.5978	0.1397	2.660	0.4684	0.03338
120	0.3981	1.980	0.6026	0.1430	2.617	0.4750	0.03482
∞	0.3989	1.960	0.6074	0.1465	2.576	0.4818	0.03625

TABLE 3

Darwin's data, report of bounds for the weighted likelihood ratio, Student's hypothesis (Normal model, variance not known)

ν_0		0			1			6			∞	
s_0/s	½	1	2	½	1	2	½	1	2	½	1	2
$\sigma_{0.25}/s$		∞		0.44	0.87	1.74	0.44	0.88	1.76			
$\sigma_{0.50}/s$		∞		0.74	1.48	3.96	0.54	1.08	2.16			
$\sigma_{0.75}/s$		∞		1.44	2.88	5.76	0.66	1.32	2.64			
ν_1		14			15			20			∞	
s_1/s		1		0.98	1	1.08	0.88	1	1.38	½	1	2
T_H		2.15		2.20	2.15	1.82	2.44	2.15	1.56	4.30	2.15	1.07
PPUB1		8.5		9.4	8.6	4.9	15.4	8.9	3.3	10353	10.1	1.8
d		20.9		20.4	20.8	22.4	17.9	20.4	28.1	9.6	19.1	38.2
PPLUM1		4.8		5.3	4.9	2.8	8.9	5.1	1.9	6288	6.1	1.1
PPLUM2		1.01		1.1	1.03	0.59	1.9	1.1	0.42	1517	1.5	0.26
($\alpha = 0.05$)												

PPLUM2 ($\alpha = 0.05$ and 0.01) as functions of ν_1. Linear interpolation on ν_1 suffices, since great accuracy is not needed. Table 3 reports PPLUM1 and PPLUM2 for Darwin's data.

4.2. *Prior parameter values for an unknown variance*

The six-parameter family of personal joint densities for μ and σ under H, defined by (51)–(54), is closed under three kinds of sampling: (i) from a Normal population with the unknown mean μ and the unknown variance σ^2; (ii) from a different Normal population with the same unknown mean μ but its own known variance; and (iii) from a different Normal population with its own known mean but the same

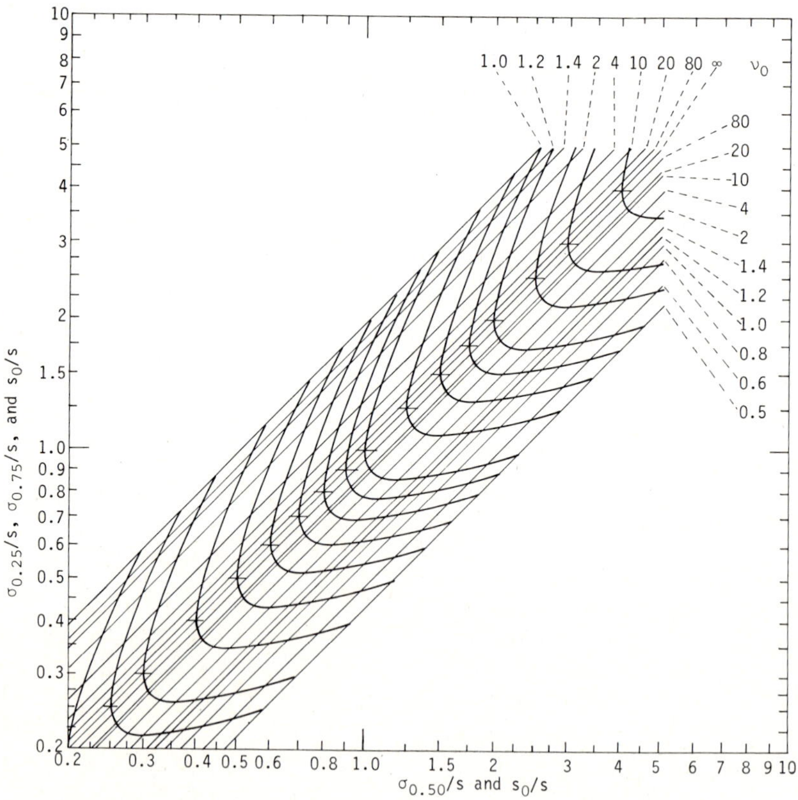

Fig. 6. Parameters and quartiles of the inverted-gamma-2 distributions.

unknown variance σ^2. Only under the sampling types (ii) and (iii) is a prior independence of μ and σ preserved in the transition to the posterior density. In particular, under sampling of type (iii), the prior density for σ (44) is transformed into the posterior density given by the same expression with ν_0 and s_0 replaced by ν_1 and s_1 (47) but with $\nu = n$ and $\nu s^2 = \Sigma_1^n (x_i - \text{pop. mean})^2$. Hence for such inference about an unknown variance, as well as to use our GRIP's, PPUB's, and PPLUM's

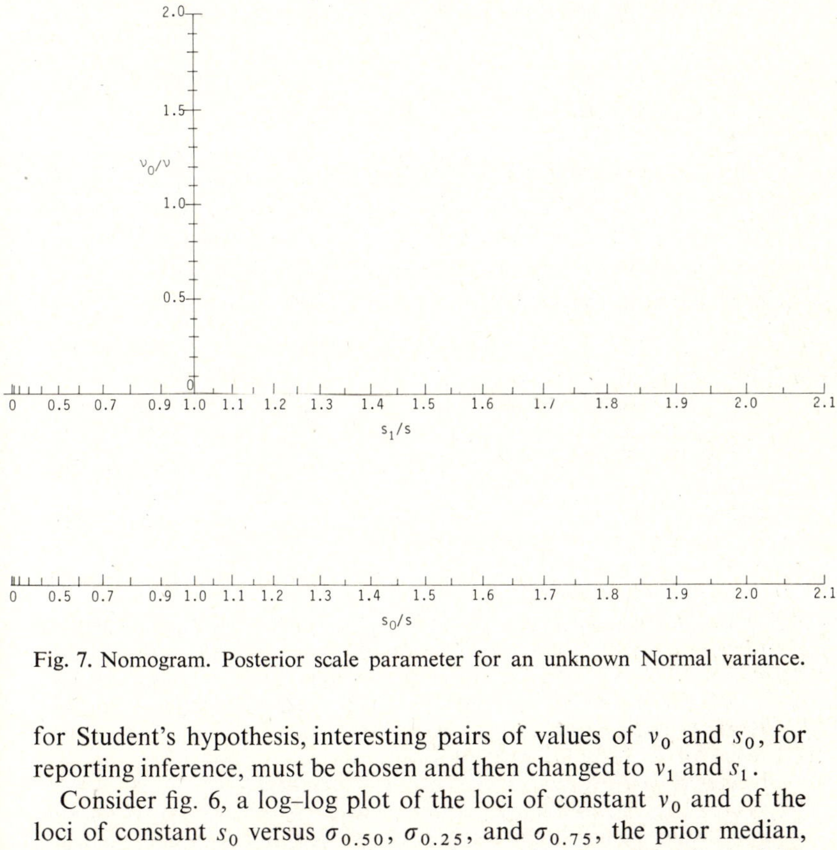

Fig. 7. Nomogram. Posterior scale parameter for an unknown Normal variance.

for Student's hypothesis, interesting pairs of values of ν_0 and s_0, for reporting inference, must be chosen and then changed to ν_1 and s_1.

Consider fig. 6, a log–log plot of the loci of constant ν_0 and of the loci of constant s_0 versus $\sigma_{0.50}$, $\sigma_{0.25}$, and $\sigma_{0.75}$, the prior median, 25%, and 75% points for σ. Since s_0 is a scale parameter, the loci for constant ν_0 are parallel 45-degree lines, labeled to read out ν_0. (The values $\nu_0 = 0$, 1, ∞ might be interpreted as representing "prior ignorance", "one prior observation", and known σ, respectively.) The loci for constant s_0/s cross $\nu_0 = \infty$ at $(s_0/s, s_0/s)$, by which s_0 can be read out. The parts of each such locus above and below $\nu_0 = \infty$ have points with coordinates $(\sigma_{0.50}/s, \sigma_{0.75}/s)$ and $(\sigma_{0.50}/s, \sigma_{0.25}/s)$, respectively. Fig. 6 was plotted from Pearson's [21] table of the incomplete gamma function and from Raiffa and Schlaifer's [17, fig. 7.5, p. 230].

Fig. 7 is a handy nomogram for the equation (from (47))

$$\frac{(s_1/s)^2-1}{(\nu_0/\nu)} = \frac{(s_0/s)^2-1}{(\nu_0/\nu)+1}. \tag{62}$$

The method of construction is immediate by similar triangles.

Acknowledgement

I am grateful for Leonard J. Savage's criticism of an early manuscript; and I thank Ron Goldman and Judy Walrath for their computer programming and other calculations, and various referees for various suggestions. This research was supported in part by the United States National Institutes of Health, Grant GM16557.

References

[18] F. J. Anscombe, Bayesian inference concerning many parameters with reference to supersaturated designs. Bull. Int. Statist. Inst. 40 (1963), 721–733.

[7] G. A. Barnard, unpublished series of lectures on statistical inference given at the National Institutes of Health, 1964.

[8] G. E. P. Box and George C. Tiao, A further look at robustness via Bayes' theorem. Biometrika 49 (1962), 419–432.

[22] Jerome Bracken and Arthur Schleifer, Jr., Tables for normal sampling with unknown variance. School of Business, Harvard University, 1964.

[2] Bruno DeFinetti, Foresight: its logical laws, its subjective sources. Reprinted in: Studies in subjective probability, edited by Henry E. Kyburg, Jr., and Howard E. Smokler. New York, Wiley, 1963.

[13] James Dickey, The weighted likelihood ratio, linear hypotheses on normal location parameters. Ann. Math. Statist. 42 (1971), 204–233.

[14] James Dickey, The Bayesian alternatives to the *F*-test and least-squares estimate in the Normal linear model. Research Report 50, Statistics Department, State University of New York at Buffalo. (Reprinted in the present volume.)

[20] James M. Dickey, Three multidimensional-integral identities with Bayesian applications. Annals of Mathematical Statistics 39 (1968), 1615–1627.

[12] James M. Dickey and B. P. Lientz, The weighted likelihood ratio, sharp hypotheses about chances, the order of a Markov chain. Annals of Mathematical Statistics 41 (1970), 214–226.

[6] Ward Edwards, Harold Lindman and Leonard J. Savage, Bayesian statistical inference for psychological research. Psychological Review 70 (1963), 193–242. Reprinted in: Readings in mathematical psychology, edited by Luce, Bush and Galanter. New York, Wiley.

[10] L. Ivan Epstein, Nomography. New York, Interscience, 1958.

[5] Ronald A. Fisher, Statistical methods and scientific inference, 2nd Ed. Edinburgh, Oliver and Boyd; New York, Hafner, 1959.

[9] Ronald A. Fisher, Design of experiments, 8th Ed. Edinburgh, Oliver and Boyd; New York, Hafner, 1966.

[15] Clifford Hildreth, Bayesian statisticians and remote clients. Econometrica 31 (1963), 422–438.

[11] D. V. Lindley, A statistical paradox. Biometrika 44 (1957), 187–192.

[16] D. V. Lindley, The use of prior probability in statistical inference and decision. In: Proc. of Fourth Berkeley Symp. on Math. Stat. and Prob. Univ. Calif. Press, 1961, 453–468.

[21] Karl Pearson, Tables of the incomplete gamma function. Cambridge University Press, 1965.

[1] Frank Plumpton Ramsey, Truth and probability. In: Studies in subjective probability, edited by Henry E. Kyburg, Jr., and Howard E. Smokler. New York, Wiley, 1926, 61–92. Reprinted from: The foundations of mathematics and other essays. London, Kegan, Paul, Trench, Trubner & Co., 1931, 156–198.

[4] John W. Pratt, Howard Raiffa and Robert Schlaifer, The foundations of decision under uncertainty: an elementary exposition. J. Am. Statist. Assoc. 59 (1964), 353–375.

[17] Howard Raiffa and Robert Schlaifer, Applied statistical decision theory. School of Business, Harvard University, Boston, 1961.

[3] Leonard J. Savage, The foundations of statistics. New York, Wiley, 1954.

[19] M. Stone, Comments on a posterior distribution of Geisser and Cornfield. J. Roy. Statist. Soc. (B) 26 (1964) 274–276.

PART 12

NEW DEVELOPMENTS IN ESTIMATION

S. E. Fienberg and A. Zellner (eds.)
Studies in Bayesian Econometrics and Statistics.

12.1 BAYESIAN ALTERNATIVES TO THE *F*-TEST AND LEAST-SQUARES ESTIMATE IN THE NORMAL LINEAR MODEL*

James DICKEY

State University of New York at Buffalo

1. *Introduction*

Coherent decision making is known to require the existence and use of personal probabilities and utilities ([1–4], and other references). Statistical inference, from data **D** from a parametrized statistical model $P(\mathbf{D}|\boldsymbol{\theta})$, can be defined as the process of updating one's prior probabilities $P_0(\boldsymbol{\theta}) = P(\boldsymbol{\theta})$ to posterior probabilities $P_1(\boldsymbol{\theta}) = P(\boldsymbol{\theta}|\mathbf{D})$, which then quantify beliefs usable for making any future decisions.

Scientific reports of statistical data should objectively exhibit as much as possible of the *inferential content* of the data and model, namely, the data-specific prior-to-posterior *transformation* of the class of all possible personal probability distributions $\{P_0(\boldsymbol{\theta})\}$. The complete likelihood function, $l_D(\boldsymbol{\theta}) = p(\mathbf{D}|\boldsymbol{\theta})$, the probability mass or density of **D** as a function of $\boldsymbol{\theta}$, would report this transformation in an indirect, disguised form.

Of course, it is usually not feasible to report directly the whole inferential-content transformation; but with some models $P(\mathbf{D}|\boldsymbol{\theta})$, a scientific reporter can summarize the application of Bayes' theorem to subclasses of prior distributions, carefully chosen to include approximations to the actual beliefs of his readers. His readers will be interested in the displayed sensitivity, to the prior distribution, of descriptive functions of the posterior distribution, for example aspects of the marginal posterior of a parameter of interest, or the predictive distribution of a future observation. It is here maintained that such summarizing can often be successfully accomplished for Normal-linear-model data.

* Research supported by NIGMS-NIH Grant GM-16557.

An important first task with the Normal linear model is to develop the prior-to-posterior transformation and descriptive posterior analysis for a wide and realistic class of tractible prior distributions. This paper proposes the use of prior distributions $P_0(\boldsymbol{\theta})$ on the parameter space Θ which are in the form of a mixture of two probability density functions, one, $p_0(\boldsymbol{\theta}|H)$, supported on a lower-dimensional linear manifold $H \subset \theta$, a "sharp null hypothesis", and the other, $p_0(\boldsymbol{\theta}|\bar{H})$, supported on the complement $\bar{H} = \theta - H$. The first density will be *induced* by a continuous version of the second (Savage's Density Condition); and hence we need only specify one density function $p_0(\boldsymbol{\theta}|\bar{H})$ and a mixing probability $P_0(H)$. The posterior distribution $P_1(\boldsymbol{\theta})$ then takes the same form, in terms of $p_1(\boldsymbol{\theta}|\bar{H})$, the posterior version of $p_0(\boldsymbol{\theta}|\bar{H})$, and the posterior probability $P_1(H) = P(H|\mathbf{D})$.

Such non-continuous-type prior distributions can be useful approximations to *density functions* which have relatively high density of probability near H, but would themselves be less tractible. Some readers may recognize this theory as an extension of work begun by Jeffreys [5]. Further extensions for nested or intersecting sharp hypotheses are easy.

The density functions on Θ, $p_0(\boldsymbol{\theta}|\bar{H})$, which we shall consider can be described as multivariate-t on the location parameters $\boldsymbol{\beta} = \boldsymbol{\beta}(\boldsymbol{\theta})$ (regression coefficients) and independently reciprocal-scaled-χ^2 on the scale parameter $\sigma^2 = \sigma^2(\boldsymbol{\theta})$ (common variance of the dependent variable). The usual, invariant, nonintegrable, "ignorance-prior densities" are special cases, though, by now, discredited as unrealistic in high dimensions [6; 7; 8, p. 223; 9]. Raiffa and Schlaifer's [10] integrable "natural conjugate prior" densities are here excluded: (i) their σ^2-known-case multivariate-Normal prior, although it is a limiting form of the prior densities here, is rejected for its strong prior knowledge about the distance of $\boldsymbol{\beta}$ from its mean [11]; and (ii) their σ^2-unknown-case version is excluded for its unrealistic prior stochastic dependence between σ^2 and $\boldsymbol{\beta}$ (in this paper, σ^2 will refer to a measuring instrument, and $\boldsymbol{\beta}$ to something measured).

Forms of the prior densities developed here, with $\boldsymbol{\beta}$ and σ^2 independent, seem first to have been used by Stein [7], Anscombe [11], Lindley [12], Tiao and Zellner [13], and Stone [14]; and since then, by Hill [16–17], Dickey [18–20], Cornfield [21–23], Lindley [24, 25], Novick et al. [26], and Zellner and Vandaele [27]. See also Efron and Morris [28, 29] and Strawderman [30]. In a significant new paper, Lindley and Smith [31] develop and extend such prior densities with an emphasis on exchangeability and on posterior modes as estimates.

It is customary in statistical practice to report estimates preceded by tests to establish the degree of generality needed in the statistical model. Tools are developed here appropriate for two Bayesian versions of this practice.

The first version is an informal one: calculate the posterior probability $P_1(H)$; if it is "practically unity", use a restricted estimate $\hat{\boldsymbol{\beta}}_H \in H$ based on the posterior density $p_1(\boldsymbol{\theta}|H)$; otherwise, use an estimate $\hat{\boldsymbol{\beta}}_*$ based on the posterior density $p_1(\boldsymbol{\theta}|\bar{H})$. In the second version, the two-decision problem, of whether to constrain the estimate by H, is formally imbedded in the estimation problem, by a modification of the estimation loss function.

A testing or estimation procedure from either version can be isolated from the full test-then-estimate context and advocated for use in its own limited context.

In the first version approach to hypothesis-testing, of Jeffreys [5], Good [32, 33], Savage [34], Lindley [35, 36], Edwards et al. [8], Cornfield [22, 23, 37, 38], Dickey and Lientz [39], and Dickey [19, 20], the hypothesis $H \subset \Theta$ is assigned positive probability, $0 < P_0(H) < 1$, under a prior distribution $P_0(\boldsymbol{\theta})$ on the parameter space Θ. The parameters $\boldsymbol{\theta}$ determine a statistical model for the observations $\mathbf{D}$, $P(\mathbf{D}|\boldsymbol{\theta})$, with (generalized) density $p(\mathbf{D}|\boldsymbol{\theta})$. A utility or income function $U(d, \boldsymbol{\theta})$ of $\boldsymbol{\theta}$ and of the two-valued decision variable, $d = d_H$ "for H", or $d_{\bar{H}}$ "for the alternative $\bar{H}$", may be assigned. Naturally, $U(d_{\bar{H}}, \theta) < U(d_H, \theta)$ for $\theta \in H$, and $U(d_H, \theta) < U(d_{\bar{H}}, \theta)$ for $\theta \in \bar{H}$. Then under the optimal decision procedure, one compares the posterior expected utilities, $E_1[U(d_H, \boldsymbol{\theta})]$ and $E_1[U(d_{\bar{H}}, \boldsymbol{\theta})]$, for the two possible decisions.

A prominent role in such a decision procedure is played by the weighted likelihood ratio or Bayes factor for H, $L_D(H)$, which is the ratio of the posterior odds for H, $O_1(H) = P_1(H)/P_1(\bar{H})$, to the prior odds for H, $O_0(H) = P_0(H)/P_0(\bar{H})$, and is a function of the data $\mathbf{D}$ and of the conditional prior distributions of the "nuisance" parameters given H and given $\bar{H}$, $P_0(\boldsymbol{\theta}|H)$ and $P_0(\boldsymbol{\theta}|\bar{H})$,

$$\begin{aligned} L_D(H) &= O_1(H)/O_0(H) = p(\mathbf{D}|H)/p(\mathbf{D}|\bar{H}) \\ &= \int p(\mathbf{D}|\boldsymbol{\theta})\, \mathrm{d}P_0(\boldsymbol{\theta}|H)/\int p(\mathbf{D}|\boldsymbol{\theta})\, \mathrm{d}P_0(\boldsymbol{\theta}|\bar{H}). \end{aligned} \tag{1}$$

For, if the utility function U is transformed to the canonical form of paying nothing for wrong decisions, $U(d_J, \boldsymbol{\theta}) = 0$ for $\boldsymbol{\theta} \notin J$,

$J = H, \bar{H}$, and $U(d_J, \theta) > 0$ for $\theta \in J$, then the decision procedure takes the form of choosing $d_{\bar{H}}$ or d_H, respectively, according to the criterion,

$$L_D(H) \cdot [P_0(H)/P_0(\bar{H})] \lessgtr E_1[U(d_{\bar{H}}, \boldsymbol{\theta})|\bar{H}]/E_1[U(d_H, \boldsymbol{\theta})|H]. \tag{2}$$

(For a derivation with discussion, see Dickey [20].) The threshold on the right-hand side of (2) depends on the data **D** through the posterior densities $p_1(\boldsymbol{\theta}|\bar{H})$ and $p_1(\boldsymbol{\theta}|H)$.

In scientific reports, the ratio of posterior odds to prior odds, $L_D(H)$, can be plotted as a function of the conditional prior densities.

Note that the posterior probabilities can also be expressed directly in terms of the prior probabilities and the weighted likelihood ratio: since $P_1(H) = O_1(H)/[O_1(H)+1]$,

$$\begin{aligned} P_1(H) &= P_0(H)/[P_0(H)+P_0(\bar{H})/L_D(H)] \\ &= 1/\{1+1/[L_D(H) \cdot P_0(H)/P_0(\bar{H})]\} \end{aligned} \tag{3}$$

and

$$\begin{aligned} P_1(\bar{H}) &= P_0(\bar{H})/[P_0(\bar{H})+P_0(H) \cdot L_D(H)] \\ &= 1/\{1+L_D(H) \cdot P_0(H)/P_0(\bar{H})\}. \end{aligned} \tag{4}$$

Geisel [40] and Lempers [41] calculate $P_1(H)$ directly, without recourse to odds and weighted likelihood ratios.

In the second-version approach, of Anscombe [11], Dickey [42], and Lindley [43, 44], the primary decision problem is the one of estimating an unknown parameter $\boldsymbol{\beta} = \boldsymbol{\beta}(\boldsymbol{\theta})$ by $\hat{\boldsymbol{\beta}} = \hat{\boldsymbol{\beta}}(\mathbf{D})$, subject to an estimation loss $\Delta(\hat{\boldsymbol{\beta}}, \boldsymbol{\beta})$, but with an additive "simplicity" reward $U_H > 0$ gained if $\hat{\boldsymbol{\beta}}$ is chosen in the subset H of $\boldsymbol{\beta}$ values. Hence we define another utility function $u(\hat{\boldsymbol{\beta}}, \boldsymbol{\beta})$,

$$u(\hat{\boldsymbol{\beta}}, \boldsymbol{\beta}) = U_H \cdot \delta_H(\hat{\boldsymbol{\beta}}) - \Delta(\hat{\boldsymbol{\beta}}, \boldsymbol{\beta}), \tag{5}$$

$\delta_H(\hat{\boldsymbol{\beta}}) = 1$ if $\hat{\boldsymbol{\beta}}$ is in H, and $\delta_H = 0$ for other arguments. As before, a prior distribution P_0 is assigned to the unknown parameters $\boldsymbol{\theta}$; and then, given the data **D**, the posterior utility,

$$\bar{u}(\hat{\boldsymbol{\beta}}) = E_1 u(\hat{\boldsymbol{\beta}}, \boldsymbol{\beta}), \tag{6}$$

is maximized in $\hat{\boldsymbol{\beta}}$. Clearly, this problem does not require the assignment of positive prior probability to H; but the general case will be treated here. Then the procedures will again involve $L_D(H)$.

The posterior expected utility $\bar{u}$ can be written,

$$\bar{u}(\hat{\boldsymbol{\beta}}) = U_H \cdot \delta_H(\hat{\boldsymbol{\beta}}) - \bar{\Delta}(\hat{\boldsymbol{\beta}}) , \tag{7}$$

where $\bar{\Delta}(\hat{\boldsymbol{\beta}}) = E_1 \Delta(\hat{\boldsymbol{\beta}}, \boldsymbol{\beta})$. Assume that the unconstrained minimum of $\bar{\Delta}(\hat{\boldsymbol{\beta}})$ is attained at $\hat{\boldsymbol{\beta}}_*$, namely,

$$\bar{\Delta}(\hat{\boldsymbol{\beta}}_*) = \min_{\hat{\beta}} \bar{\Delta}(\hat{\boldsymbol{\beta}}) . \tag{8}$$

And assume, as will be true in the following examples, that the constrained minimum of $\bar{\Delta}(\hat{\boldsymbol{\beta}})$ over $\hat{\boldsymbol{\beta}} \in H$ is actually attained in H, at $\hat{\boldsymbol{\beta}}_H$,

$$\bar{\Delta}(\hat{\boldsymbol{\beta}}_H) = \min_{\hat{\beta} \in H} \bar{\Delta}(\hat{\boldsymbol{\beta}}) , \qquad \hat{\beta}_H \in H . \tag{9}$$

Proposition 1.1. The above estimation problem is equivalent to the following two-decision problem. Choose $\hat{\boldsymbol{\beta}} = \hat{\boldsymbol{\beta}}_*$ or $\hat{\boldsymbol{\beta}}_H$, yielding, for $\hat{\boldsymbol{\beta}}_* \in \bar{H}$ (otherwise, $\hat{\boldsymbol{\beta}}_* = \hat{\boldsymbol{\beta}}_H$),

$$\bar{u}(\hat{\boldsymbol{\beta}}_*) = -\bar{\Delta}(\hat{\boldsymbol{\beta}}_*) \tag{10}$$

or

$$\bar{u}(\hat{\boldsymbol{\beta}}_H) = -\bar{\Delta}(\hat{\boldsymbol{\beta}}_H) + U_H , \tag{11}$$

respectively, according to the criterion $\bar{u}(\hat{\boldsymbol{\beta}}_*) \gtrless \bar{u}(\hat{\boldsymbol{\beta}}_H)$, namely

$$\bar{\Delta}(\hat{\boldsymbol{\beta}}_H) - \bar{\Delta}(\hat{\boldsymbol{\beta}}_*) \gtrless U_H . \tag{12}$$

These optimal estimates $\hat{\boldsymbol{\beta}}_H$ (9) and $\hat{\boldsymbol{\beta}}_*$ (8) (and hence the utilities $u(\hat{\boldsymbol{\beta}}_H, \boldsymbol{\beta})$ and $u(\hat{\boldsymbol{\beta}}_*, \boldsymbol{\beta})$) depend on the posterior probability structure. Hence, we cannot transform to a two-decision utility function of only $\boldsymbol{\theta}$ and d_H or $d_{\bar{H}}$, but not of the data $\mathbf{D}$, as we possessed in U below equation (1). Consequently, the two approaches to testing are distinct.

Note that an estimation problem can arise from a prediction problem, as in Lindley [43, 44] and Raiffa and Schlaifer [10, ch. 6], by taking $\Delta(\hat{\boldsymbol{\beta}}, \boldsymbol{\beta})$ to be the conditional expectation, over a future predictand y and any future predictors $\mathbf{x}$, of a prediction loss function $\tilde{\Delta}(\hat{y}, y)$, given the parameter $\boldsymbol{\beta}$ and past data $\mathbf{D}$,

$$\Delta(\hat{\boldsymbol{\beta}}, \boldsymbol{\beta}) = E_{y,\mathbf{x},\boldsymbol{\theta}|\boldsymbol{\beta},\mathbf{D}} \tilde{\Delta}(\hat{y}, y) , \tag{13}$$

where $\hat{\boldsymbol{\beta}}(\mathbf{D})$ and $\mathbf{x}$ yield a prediction $\hat{y} = \hat{y}(\hat{\boldsymbol{\beta}}, \mathbf{x})$.

Section 2 develops the prior-posterior density theory for the Normal linear model; section 3, the consequent weighted likelihood ratio for a linear hypothesis; section 4, the test, prediction, and estimation, with squared-error prediction and estimation loss; and section 5, the test and estimation, with error-threshold estimation loss. (Prediction and estimation with absolute-error loss, in general tractible only in one dimension, are developed by Raiffa and Schlaifer [10, ch. 6].)

2. *Distribution theory for the normal linear model*

Consider the familiar statistical model of a single response variable y conditionally Normally distributed with mean $\boldsymbol{\beta}^T\mathbf{x}$, a linear function of r regressor-variable values $\mathbf{x} = (x^{(1)}, \ldots, x^{(r)})^T$, and with variance σ^2.

Define the matrix of available data,

$$\mathbf{D} = (\mathbf{y}, \mathbf{X}), \qquad n \times (r+1)\,, \tag{14a}$$

where

$$\mathbf{y} = (y_1, \ldots, y_n)^T, \qquad \mathbf{X} = (\mathbf{x}_1, \ldots, \mathbf{x}_n)^T\,, \tag{14b}$$

the rows $(y_i, \mathbf{x}_i^T)$ of which, together with the future $(y_{n+1}, \mathbf{x}_{n+1}^T)$, have the statistical-model conditional Normal distribution independently in $i = 1, \ldots, n+1$,

$$y_i|\mathbf{X}, \mathbf{x}_i, \boldsymbol{\theta} \sim y_i|\mathbf{x}_i, \boldsymbol{\beta}, \sigma \sim \text{Normal}\,(\boldsymbol{\beta}^T\mathbf{x}_i, \sigma^2)\,. \tag{15}$$

We assume that $\boldsymbol{\beta}(\boldsymbol{\theta})$ and $\sigma(\boldsymbol{\theta})$ satisfy no known constraints and are functionally independent of $\mathbf{X}$ and $\mathbf{x}_{n+1}$.

The more-general-looking context, in which $\mathbf{y} \sim n$-variate-Normal $(\mathbf{X}\boldsymbol{\beta}, \sigma^2\mathbf{M})$, $\mathbf{M}$ known, is reduced to the above by transforming $\mathbf{y}$ to $\mathbf{M}^{-\frac{1}{2}}\mathbf{y}$, and $\mathbf{X}$ to $\mathbf{M}^{-\frac{1}{2}}\mathbf{X}$.

In case $\mathbf{X}$ and $\mathbf{x}_{n+1}$ obey an underlying statistical model with $\mathbf{X}$ and $\mathbf{x}_{n+1}$ conditionally independently distributed given parameters ξ (ξ functionally independent of $\boldsymbol{\beta}, \sigma$), then define

$$\boldsymbol{\theta} = (\boldsymbol{\beta}^T, \sigma, \xi^T)^T\,; \tag{16a}$$

but if $\mathbf{x}_{n+1}$ is assumed known, define

$$\boldsymbol{\theta} = (\boldsymbol{\beta}^T, \sigma)^T\,. \tag{16b}$$

If $(\boldsymbol{\beta}, \sigma)$ and ξ are prior independent (as then are $(\boldsymbol{\beta}, \sigma)$ independent of $\mathbf{X}$ and $\mathbf{x}_{n+1}$), then since the likelihood function factors into the product of the conditional probability density of $\mathbf{y}$ given $(\mathbf{X}, \boldsymbol{\beta}, \sigma)$ times the density of $\mathbf{X}$ given ξ, so also are $(\boldsymbol{\beta}, \sigma)$ and ξ posterior independent (as then are $(\boldsymbol{\beta}, \sigma)$ and $\mathbf{x}_{n+1}$) [43]. More generally, the conditional posterior distribution of $(\boldsymbol{\beta}, \sigma)$, given ξ, depends on the first factor, only, of the likelihood function and on the conditional prior distribution of $(\boldsymbol{\beta}, \sigma)$ given ξ. The unconditional posterior distribution of $(\boldsymbol{\beta}, \sigma)$ depends on the first factor, only, of the likelihood function and on the conditional distribution of $(\boldsymbol{\beta}, \sigma)$ given $\mathbf{X}$, prior to $\mathbf{y}$.

If the statistical model for the $\mathbf{x}_i$'s and $\mathbf{x}_{n+1}$ is, as is often assumed, independently identically r-variate-Normal, the maximum likelihood estimates $\hat{\xi}$ of the unknown parameters ξ are

$$\widehat{E(\mathbf{x}|\xi)} = \bar{\mathbf{x}}, \qquad \widehat{E(\mathbf{x}\mathbf{x}^T|\xi)} = \frac{1}{n}\mathbf{X}^T\mathbf{X}\,. \tag{17}$$

For large enough n, any nonpathological posterior distribution of ξ will be concentrated near these values. This author has little feeling for what sorts of uncertainty are modeled by the prior and posterior distributions for high-dimensional ξ in the literature, all of which involve Wishart and limiting Wishart distributions for the unknown inverse covariance matrix.

We turn now to inference on $(\boldsymbol{\beta}, \sigma)$ (conditional on $\mathbf{X}$ or on ξ). The sufficient statistics from $\mathbf{D}$, conditional on $\mathbf{X}$, are

$$\mathbf{N} = \mathbf{X}^T\mathbf{X}, \qquad \mathbf{b} \text{ satisfying } \mathbf{X}^T(\mathbf{y}-\mathbf{X}\mathbf{b}) = \mathbf{0}, \tag{18a}$$

$$\nu = n\text{-rank}(\mathbf{N})\,, \tag{18b}$$

and

$$\nu s^2 = SSW = \sum_1^n (y_i - \mathbf{b}^T\mathbf{x}_i)^2 = \sum_1^n (y_i - \boldsymbol{\beta}^T\mathbf{x}_i)^2 - \mathbf{N}((\mathbf{b}-\boldsymbol{\beta})), \quad \text{any } \boldsymbol{\beta}\,, \tag{18c}$$

where $\mathbf{M}((\mathbf{z}))$ denotes a quadratic form $\mathbf{z}^T\mathbf{M}\mathbf{z} = \Sigma_j\Sigma_k z^{(j)} m_{jk} z^{(k)}$. The usual least-squares projection $\mathbf{X}\mathbf{b}$ is unique, though the least-squares estimate $\mathbf{b}$ will be unique $(\mathbf{N}^{-1}\mathbf{X}^T\mathbf{y})$ only if $\mathbf{X}$ (and hence $\mathbf{N}$) is of full rank.

The likelihood function of $\boldsymbol{\beta}$ and σ is

$$p(\mathbf{y}|\mathbf{X}, \boldsymbol{\beta}, \sigma) \propto \exp\left[-\tfrac{1}{2}\sigma^{-2}\mathbf{N}((\boldsymbol{\beta}-\mathbf{b}))\right]\sigma^{-n} \exp\left[-\tfrac{1}{2}\sigma^{-2}\nu s^2\right]. \tag{19}$$

For the prior distribution of $(\boldsymbol{\beta}, \sigma)$, consider the following realistically rich class in which $\boldsymbol{\beta}$ and σ are independently distributed. If $\mathbf{X}$ (or ξ) is not prior independent of $(\boldsymbol{\beta}, \sigma)$, one should consider the following prior and posterior distributions (P_0 and P_1) taken conditional on

X (or ξ). Consider the two-parameter inverted-gamma family for σ^2,

$$\sigma^2 \sim \nu_0 s_0^2/\chi^2_{\nu_0} \sim s_0^2 \cdot F_{\infty,\nu_0}, \tag{20}$$

with density in σ,

$$p_0(\sigma) = k_{\nu_0}(\sigma; s_0) = [\Gamma(\tfrac{1}{2}\nu_0)(\tfrac{1}{2}\nu_0 s_0^2)^{\frac{1}{2}}]^{-1} \times 2(\tfrac{1}{2}\nu_0 s_0^2/\sigma^2)^{\frac{1}{2}(\nu_0+1)} \exp(-\tfrac{1}{2}\nu_0 s_0^2/\sigma^2), \tag{21}$$

with mode $s_0[\nu_0/(\nu_0+1)]^{\frac{1}{2}}$ for σ and $E_0\sigma^2 = (\nu_0-2)^{-1}\nu_0 s_0^2$. For further discussion of this family for σ, see Raiffa and Schlaifer [10, p. 227], or Dickey [19].

Consider the general multivariate-t family for $\boldsymbol{\beta}$ on the r-space-complement $\bar{H}$ of H, where H is a lower-dimensional manifold in r-space (H has zero r-volume),

$$\boldsymbol{\beta}|\bar{H} \sim \mathbf{t}_{\nu'}^{(r)}(\mathbf{b}', s'^2\mathbf{N}'^{-1}), \qquad \text{rank } N' = r, \tag{22}$$

with density,

$$\begin{aligned} p_0(\boldsymbol{\beta}|\bar{H}) &= h_{\nu'}^{(r)}(\boldsymbol{\beta}-\mathbf{b}'; s'^2\mathbf{N}'^{-1}) \\ &= [B(\tfrac{1}{2}\nu'; \tfrac{1}{2}\mathbf{1}_r)]^{-1}(\nu' s'^2)^{-\frac{1}{2}r}|\mathbf{N}'|^{\frac{1}{2}} \\ &\quad \times [1+\nu'^{-1}s'^{-2}\mathbf{N}'((\boldsymbol{\beta}-\mathbf{b}'))]^{-\frac{1}{2}(\nu'+r)}, \end{aligned} \tag{23}$$

where in a more general notation

$$\begin{aligned} B(c_1, \ldots, c_k; \mathbf{d}) &= B(c_1, \ldots, c_k, d^{(1)}, \ldots, d^{(r)}) \\ &= [\prod_1^k \Gamma(c_i)][\prod_1^r \Gamma(d^{(j)})]/[\Gamma(\sum_1^k c_i + \sum_1^r d^{(j)})], \end{aligned}$$

and $\mathbf{1}_r = (1, 1, \ldots, 1)^T$, r-dimensional. We have $E_0(\boldsymbol{\beta}|\bar{H}) = \mathbf{b}'$ and $\mathbf{V}_0(\boldsymbol{\beta}|\bar{H}) = (\nu'-2)^{-1}\nu' s'^2\mathbf{N}'^{-1}$. The marginal and conditional distributions of subvectors are again multivariate-t: in an obvious notation for partitioned vectors and matrices,

$$\boldsymbol{\beta}_{(1)}|\bar{H} \sim t_{\nu'}^{(r_1)}[\mathbf{b}'_{(1)}, s'^2(\mathbf{N}'^{-1})_{(11)}]$$

and

$$\boldsymbol{\beta}_{(2)}|\boldsymbol{\beta}_{(1)}, \bar{H} \sim \mathbf{t}_{\nu'+r_1}^{(r_2)}[\mathbf{b}'_{(2)} - \mathbf{N}'^{-1}_{(22)}\mathbf{N}'_{(21)}(\boldsymbol{\beta}_{(1)} - \mathbf{b}'_{(1)}), \; S'^2(\boldsymbol{\beta}_{(1)})\mathbf{N}'^{-1}_{(22)}],$$

where

$$\begin{aligned} S'^2(\boldsymbol{\beta}_{(1)}) &= (\nu'+r_1)^{-1}[\nu' s'^2 + \mathbf{N}'_{(11.2)}((\boldsymbol{\beta}_{(1)} - \mathbf{b}'_{(1)}))], \\ \mathbf{N}'_{(11.2)} &= \mathbf{N}'_{(11)} - \mathbf{N}'_{(12)}\mathbf{N}'^{-1}_{(22)}\mathbf{N}_{(21)} = [(\mathbf{N}'^{-1})_{(11)}]^{-1}. \end{aligned}$$

The quadratic form $\mathbf{N}'((\boldsymbol{\beta}-\mathbf{b}'))/r$ is distributed like $s'^2 \cdot F_{r,\nu'}$.

The distribution (22)–(23) is a generalization of the multivariate-Normal distribution; indeed, it will be convenient to introduce the prior mixing variable τ ($\boldsymbol{\beta}$ and τ independent of σ),

$$\tau^2|\bar{H} \sim s'^2 \cdot F_{\infty, v'} \,. \tag{24}$$

Given τ, we have the conditional r-variate-Normal distribution,

$$\boldsymbol{\beta}|\tau, \bar{H} \sim \mathbf{N}^{(r)}(\mathbf{b}', \tau^2\mathbf{N}'^{-1}) \sim \mathbf{t}_{\infty}^{(r)}(\mathbf{b}', \tau^2\mathbf{N}'^{-1}) \,. \tag{25}$$

Conditional on τ, $\mathbf{N}'((\boldsymbol{\beta}-\mathbf{b}'))/r$ is distributed like $\tau^2 \cdot F_{r,\infty}$.

So $\boldsymbol{\beta}$ can be loosely interpreted as coming from a prior "super population", multivariate-Normal with "unknown" variance scale τ^2 with the second-order prior distribution (24). This superpopulation could be either public (Model II) or private (Model I), depending on the practical context. As will be illustrated later, $\mathbf{b}'$ itself can be subjectively random to express "ignorance" about the superpopulation mean.

The joint prior density of $\boldsymbol{\beta}$, τ, and σ, then has parameters, $\mathbf{b}'$, $\mathbf{N}'$; v', s'; v_0, s_0, and satisfies the proportionality,

$$\begin{aligned} p_0(\boldsymbol{\beta}, \tau, \sigma|\bar{H}) \propto & \exp\left[-\tfrac{1}{2}\tau^{-2}\mathbf{N}'((\boldsymbol{\beta}-\mathbf{b}'))\right] \\ & \times \tau^{-(v'+r+1)} \exp\left[-\tfrac{1}{2}\tau^{-2}v's'^2\right] \\ & \times \sigma^{-(v_0+1)} \exp\left[-\tfrac{1}{2}\sigma^{-2}v_0 s_0^2\right] . \end{aligned} \tag{26}$$

The prior parameter values for some interesting special cases are listed in table 1. The usual joint "complete-ignorance" density for $\boldsymbol{\beta}$ and σ, $p_0(\boldsymbol{\beta}, \sigma|\bar{H}) \propto \sigma^{-1}$, arises as the limiting case of (26), $\mathbf{N}' \to \mathbf{O}_{r\times r}$, $v_0 \to 0$. Because of the factor τ^{-r} in (26), the usual joint "complete-ignorance" density for $\boldsymbol{\beta}$ and τ, $p_0(\boldsymbol{\beta}, \tau|\bar{H}) \propto \tau^{-1}$, is not a limiting case. Tiao and Tan [45] and Box and Tiao [46] chose to express joint "complete ignorance" of τ and σ by $p_0(\tau, \sigma|\bar{H}) = \tau\sigma^{-1}[\sigma^2 + (n/r)\tau^2]^{-1}$. For criticism, see Stone and Springer [47]. Novick [48] discusses a posterior nonintegrability resulting from the choice $v' = s' = 0$. This nonintegrability was first noted by Hill [15], who also proposed independent inverted gamma distributions on σ^2 and τ^2 and derived conditional posterior means and modes given τ/σ. See also Hill [49]. Dickey [20] used the family (26) for "known" $\tau(v' \to \infty)$, with no more tractability than what follows, for "unknown" τ.

TABLE 1
Special prior parameter values

Quantity	$\boldsymbol{\beta}$	τ	σ
Marginal "complete ignorance"	$\mathbf{N}' \to \mathbf{O}_{r\times r}$	$\nu' \to 0$	$\nu_0 \to 0$
Marginal certain knowledge	Not relevant	$\nu' \to \infty$	$\nu_0 \to \infty$

In the remainder of this section, we study the posterior density, $p_1(\boldsymbol{\beta}, \sigma, \tau|\bar{H}) = p(\boldsymbol{\beta}, \sigma, \tau|\bar{H}, \mathbf{D})$ (or $= p(\boldsymbol{\beta}, \sigma, \tau|\bar{H}, \mathbf{D}, \xi)$), which is proportional to the product of (26) and the likelihood function (19),

$$p_1(\boldsymbol{\beta}, \sigma, \tau|\bar{H}) \propto \exp\{-\tfrac{1}{2}[\sigma^{-2}\mathbf{N}((\boldsymbol{\beta}-\mathbf{b})) + \tau^{-2}\mathbf{N}'((\boldsymbol{\beta}-\mathbf{b}'))]\}$$
$$\times\sigma^{-(\nu_1+r+1)}\exp[-\tfrac{1}{2}\sigma^{-2}\nu_1 s_1^2]$$
$$\times\tau^{-(\nu'+r+1)}\exp[-\tfrac{1}{2}\tau^{-2}\nu' s'^2], \tag{27a}$$

$$\nu_1 = n-r+\nu_0 \quad \text{and} \quad \nu_1 s_1^2 = \nu s^2 + \nu_0 s_0^2 . \tag{27b}$$

Note that the factor in τ is unchanged by the likelihood, which is constant in τ.

Proposition 2.1. In terms of the transformed variable

$$\omega = \frac{\tau^2}{\tau^2+\sigma^2} = \frac{\sigma^{-2}}{\sigma^{-2}+\tau^{-2}}, \qquad 0 < \omega < 1 \tag{28}$$

(in the Model-II one-way analysis-of-variance, ω is the proportion of the variance of y accounted for by the Normal distribution of $\boldsymbol{\beta}$), define

$$\mathbf{N}_\omega = \omega\mathbf{N} + (1-\omega)\mathbf{N}', \tag{29a}$$

$$\mathbf{b}_\omega = \mathbf{N}_\omega^{-1}[\omega\mathbf{N}\mathbf{b} + (1-\omega)\mathbf{N}'\mathbf{b}']$$
$$= \mathbf{N}_\omega^{-1}\omega\mathbf{N}(\mathbf{b}-\mathbf{b}') + \mathbf{b}', \tag{29b}$$

$$\nu_* = \nu_1 + \nu' + r = n + \nu_0 + \nu', \tag{29c}$$

$$\nu_* s_\omega^2 = \omega\nu_1 s_1^2 + (1-\omega)\nu' s'^2 + \omega(1-\omega)(\mathbf{N}\mathbf{N}_\omega^{-1}\mathbf{N}')((\mathbf{b}-\mathbf{b}')) . \tag{29d}$$

Then, conditional on ω, we obtain the posterior densities,

$$p_1(\boldsymbol{\beta}|\omega, \bar{H}) = h_{\nu_*}^{(r)}(\boldsymbol{\beta}-\mathbf{b}_\omega\,;\, s_\omega^2\mathbf{N}_\omega^{-1})\,, \tag{31a}$$

$$p_1(\sigma|\omega, \bar{H}) = k_{\nu_*}(\sigma\,;\,\omega^{-\frac{1}{2}}s_\omega)\,. \tag{31b}$$

These conditional posterior densities of $\boldsymbol{\beta}$ and σ are not independent; indeed, $\boldsymbol{\beta}, \sigma^{-2}|\omega, \bar{H}, \mathbf{D} \sim$ (r-variate-Normal)-gamma $(\mathbf{b}_\omega, \omega^{-1}s_\omega^2, \omega^{-1}\mathbf{N}_\omega, \nu_*)$, the "conjugate" family of Raiffa and Schlaifer [10, p. 318]. The conditioning variable ω has the posterior density

$$p_1(\omega|\bar{H}) = |\mathbf{N}_\omega|^{-\frac{1}{2}}\omega^{\frac{1}{2}(\nu_1+r)-1}(1-\omega)^{\frac{1}{2}(\nu'+r)-1}(\nu_* s_\omega^2)^{-\frac{1}{2}\nu_*}/C_* \tag{32}$$

where C_* is the integral of the numerator over (0, 1).

Proof. In the joint density (27), complete the quadratic form in $\boldsymbol{\beta}$,

$$\sigma^{-2}\mathbf{N}((\boldsymbol{\beta}-\mathbf{b}))+\tau^{-2}\mathbf{N}'((\boldsymbol{\beta}-\mathbf{b}')) = \sigma^{-2}\omega^{-1}N_\omega((\boldsymbol{\beta}-\mathbf{b}_\omega)) + \sigma^{-2}(1-\omega)(\mathbf{N}\mathbf{N}_\omega^{-1}\mathbf{N}') \times ((\mathbf{b}-\mathbf{b}')).$$

Although $\mathbf{N}$, and even $\mathbf{N}'$ need not be positive definite, their sum (or $\mathbf{N}_\omega$) is assumed to be; hence this posterior density is integrable in $\boldsymbol{\beta}$. After a change of variable to $(\boldsymbol{\beta}, \sigma, \omega)$, $J(\partial\tau/\partial\omega) = \frac{1}{2}\sigma\omega^{-\frac{1}{2}}(1-\omega)^{-\frac{3}{2}}$, the Normal-gamma form of the conditional joint density of $(\boldsymbol{\beta}, \sigma)$ given ω can be recognized. Then integration of $(\boldsymbol{\beta}, \sigma)$ out of the joint density yields (32).

Forms of proposition 2.1 appear in Box and Tiao [50, sections 7.2.5 and 9.3.1], with nonintegrable prior densities termed "reference priors". Those authors' intention seems to be that the posterior density resulting from a "reference prior" can be used as a factor to obtain the posterior density from any more-realistic prior distribution. But then this cannot be done directly by using their marginal posterior density of $\boldsymbol{\beta}$ as a weighted likelihood function, unless their nonintegrable prior density of σ and τ happens to be an effective approximation to prior belief. One of their two choices [13], with $p_0(\sigma|\tau, \boldsymbol{\beta}, \bar{H}) = p_0(\sigma|\bar{H}) = \sigma^{-1}$, is a limiting member of our family of prior densities, $\nu_0 \to 0$; and the corresponding posterior density $p_1(\boldsymbol{\beta}, \tau \mid \bar{H})$ yields our weighted likelihood function under the substitution $\nu \leftarrow \nu_1$, $s^2 \leftarrow s_1^2$. Their other choice [46] and published objections to it were referenced above.

Corollary 1. By the symmetry between σ and τ in the form of the posterior density (27), obtain $\boldsymbol{\beta}, \tau^{-2}|\omega, \bar{H}, \mathbf{D} \sim$ (r-Normal)-gamma $(\mathbf{b}_\omega, (1-\omega)^{-1}s_\omega^2, (1-\omega)^{-1}\mathbf{N}_\omega, \nu_*)$,

$$p_1(\tau|\omega, \bar{H}) = k_{\nu_*}(\tau; (1-\omega)^{-\frac{1}{2}}s_\omega)\,. \tag{33}$$

Corollary 2. The posterior moments of $\boldsymbol{\beta}$, σ and τ can be obtained as one-dimensional integrals. For example,

$$E_1(\boldsymbol{\beta}|\bar{H}) = E_1(\mathbf{b}_\omega|\bar{H}) = E_1(\omega\mathbf{N}_\omega^{-1}|\bar{H})\mathbf{N}(\mathbf{b}-\mathbf{b}')+\mathbf{b}', \tag{34a}$$

$$\mathbf{V}_1(\boldsymbol{\beta}|\bar{H}) = \mathbf{V}_1(\mathbf{b}_\omega|\bar{H})+(v_*-2)^{-1}E_1(v_* s_\omega^2\mathbf{N}_\omega^{-1}|\bar{H}), \tag{34b}$$

$$E_1(\sigma^2|\bar{H}) = (v_*-2)^{-1}E_1(v_* s_\omega^2/\omega|\bar{H}),$$

$$E_1(\tau^2|\bar{H}) = (v_*-2)^{-1}E_1(v_* s_\omega^2/(1-\omega)|\bar{H}). \tag{34c}$$

The integrands can be expressed as rapidly electronically computable functions of ω by simultaneous diagonalization of $\mathbf{N}$ and $\mathbf{N}'$: as in Rao [51, p. 37], or Dickey [18, p. 1618], assuming $\mathbf{N}+\mathbf{N}' > 0$, write $\mathbf{N}+\mathbf{N}' = \mathbf{A}^T\mathbf{A}$, $\mathbf{N} = \mathbf{A}^T\boldsymbol{\Lambda}\mathbf{A}$, and $\mathbf{N}' = \mathbf{A}^T(\mathbf{I}_r-\boldsymbol{\Lambda})\mathbf{A}$; then

$$\begin{aligned}
\mathbf{N}_\omega &= \mathbf{A}^T[\omega\boldsymbol{\Lambda}+(1-\omega)(\mathbf{I}_r-\boldsymbol{\Lambda})]\mathbf{A},\\
|\mathbf{N}_\omega| &= |\mathbf{A}|^2\cdot|\omega\boldsymbol{\Lambda}+(1-\omega)(\mathbf{I}_r-\boldsymbol{\Lambda})|,\\
\mathbf{b}_\omega &= \mathbf{A}^{-1}[\omega\boldsymbol{\Lambda}+(1-\omega)(\mathbf{I}_r-\boldsymbol{\Lambda})]^{-1}\boldsymbol{\Lambda}\mathbf{A}(\mathbf{b}-\mathbf{b}')+\mathbf{b}',
\end{aligned} \tag{35}$$

and

$$\begin{aligned}
(\mathbf{N}\mathbf{N}_\omega^{-1}\mathbf{N}')((\mathbf{b}-\mathbf{b}')) = \{&\boldsymbol{\Lambda}[\omega\boldsymbol{\Lambda}+(1-\omega)(\mathbf{I}_r-\boldsymbol{\Lambda})]^{-1}\\
&\times(\mathbf{I}_r-\boldsymbol{\Lambda})\}((\mathbf{A}\mathbf{b}-\mathbf{A}\mathbf{b}')).
\end{aligned}$$

The "conjugate-posterior-mean" points $\mathbf{b}_\omega$, parametrized by ω, will be called *the curve décolletage*; and corollary 2 can be paraphrazed in part by saying, the posterior mean of $\boldsymbol{\beta}$ is an average of points on the curve décolletage.

Although the posterior density of $\boldsymbol{\beta}$ can be expressed as a one-dimensional mixture of (31a) by (32), its ordinates are more easily computed in practice by the following

Proposition 2.2.

$$\begin{aligned}
p_1(\boldsymbol{\beta}|\bar{H}) &= h_{v_1}^{(r)}(\boldsymbol{\beta}-\mathbf{b};\, s_1^2\mathbf{N}^{-1})h_{v'}^{(r)}(\boldsymbol{\beta}-\mathbf{b}';\, s'^2\mathbf{N}'^{-1})\\
&\qquad /h_{v_1,v'}^{(r)}(\mathbf{b}-\mathbf{b}';\, s_1^2\mathbf{N}^{-1},\, s'^2\mathbf{N}'^{-1})\\
&= [v_1 s_1^2+\mathbf{N}((\boldsymbol{\beta}-\mathbf{b}))]^{-\frac{1}{2}(v_1+r)}\\
&\qquad\times[v's'^2+\mathbf{N}'((\boldsymbol{\beta}-\mathbf{b}'))]^{-\frac{1}{2}(v'+r)}\Gamma[\tfrac{1}{2}(v_1+r)]\\
&\qquad\times\Gamma[\tfrac{1}{2}(v'+r)]\{\Gamma[\tfrac{1}{2}(v_1+v'+r)]\pi^{\frac{1}{2}r}\}^{-1}/C_*,
\end{aligned} \tag{36}$$

where C_* is the one-dimensional integral of the numerator of $p_1(\omega|\bar{H})$ (32), and $h_{v_1,v'}^{(r)}(\gamma;\, s_1^2\mathbf{N}^{-1},\, s'^2\mathbf{N}')$ is the density of the Behrens-Fisher random vector $\gamma = s_1 \mathbf{N}^{-\frac{1}{2}} \mathbf{t}_{v1}^{(r)} - s' \mathbf{N}'^{-\frac{1}{2}} \mathbf{t}_{v'}^{(r)}$, discussed in Dickey [18, 20]. Equation (36) implies

$$h_{v1,v'}^{(r)}(\mathbf{b}-\mathbf{b}'; s_1^2 \mathbf{N}^{-1}, s'^2 \mathbf{N}'^{-1})$$
$$= [B(\tfrac{1}{2}v_1, \tfrac{1}{2}v'; \tfrac{1}{2}\mathbf{1}_r)]^{-1} (v_1 s_1^2)^{\frac{1}{2}v_1} |\mathbf{N}|^{\frac{1}{2}} (v' s'^2)^{\frac{1}{2}v'} |\mathbf{N}'|^{\frac{1}{2}} \cdot C_*. \quad (37)$$

Proof. To obtain the proportionality of $p_1(\boldsymbol{\beta}|\bar{H})$ to the product of two multivariate-t densities $h_{v1}^{(r)} \cdot h_{v'}^{(r)}$, integrate τ and σ out of the joint density (27), using the fact that a density of the form (21) integrates to unity. Note that the factors $h_{v1}^{(r)}$ and $h_{v'}^{(r)}$ are proportional to the weighted likelihood function of $\boldsymbol{\beta}$ and its prior density, respectively. The denominator $h_{v1,v'}^{(r)}$ must equal the integral of the numerator-product, hence it is a convolution, and hence a density ordinate of the B–F vector γ. The third member of equation (36) follows from the second member by equation (37), which is essentially a rephrasing of Dickey's [18] second identity. But, alternatively, one can calculate

$$p_1(\boldsymbol{\beta}|\bar{H}) = \int_0^\infty \int_0^\infty p_1(\boldsymbol{\beta}, \tau, \sigma|\bar{H})\, \mathrm{d}\sigma\, \mathrm{d}\tau \Big/ \int_0^1 \int_0^\infty \int_{R^r} p_1(\boldsymbol{\beta}, \sigma, \omega|\bar{H})\, \mathrm{d}\boldsymbol{\beta}\, \mathrm{d}\sigma\, \mathrm{d}\omega$$

by using for integrands the right-hand side of proportionality (27) and its change-of-variable transformation as in the proof of proposition 2.1.

The conditional and marginal densities from the joint density $p_1(\boldsymbol{\beta}_{(1)}, \boldsymbol{\beta}_{(2)}|\bar{H})$ (36) are of interest. $p_1(\boldsymbol{\beta}_{(2)}|\boldsymbol{\beta}_{(1)}, \bar{H})$ is given by equation (36) with $r = r_1 + r_2$ replaced by r_2; with $\mathbf{b}$, $\mathbf{N}$, v_1, s_1^2 replaced by $\mathbf{B}(\boldsymbol{\beta}_{(1)})$, $\mathbf{N}_{(22)}$, $v_1 + r_1$, $S^2(\boldsymbol{\beta}_{(1)})$, where

$$\mathbf{B}(\boldsymbol{\beta}_{(1)}) = \mathbf{b}_{(2)} - \mathbf{N}_{(22)}^{-1} \mathbf{N}_{(21)} (\boldsymbol{\beta}_{(1)} - \mathbf{b}_{(1)}), \quad (38a)$$

$$S^2(\boldsymbol{\beta}_{(1)}) = (v_1 + r_1)^{-1} [v_1 s_1^2 + \mathbf{N}_{(11.2)}((\boldsymbol{\beta}_{(1)} - \mathbf{b}_{(1)}))]; \quad (38b)$$

and with $\mathbf{b}'$, $\mathbf{N}'$, v', s'^2 replaced by $\mathbf{B}'(\boldsymbol{\beta}_{(1)})$, $\mathbf{N}'_{(22)}$, $v' + r_1$, $S'^2(\boldsymbol{\beta}_{(1)})$, where

$$\mathbf{B}'(\boldsymbol{\beta}_{(1)}) = \mathbf{b}'_{(2)} - \mathbf{N}'^{-1}_{(22)} \mathbf{N}'_{(21)} (\boldsymbol{\beta}_{(1)} - \mathbf{b}'_{(1)}), \quad (38c)$$

$$S'^2(\boldsymbol{\beta}_{(1)}) = (v' + r_1)^{-1} [v' s'^2 + \mathbf{N}'_{(11.2)}((\boldsymbol{\beta}_{(1)} - \mathbf{b}'_{(1)}))]. \quad (38d)$$

The marginal density of $\boldsymbol{\beta}_{(1)}$ can be found by integrating this conditional density of $\boldsymbol{\beta}_{(2)}|\boldsymbol{\beta}_{(1)}$ out from the joint density of $\boldsymbol{\beta}_{(1)}, \boldsymbol{\beta}_{(2)}$ (36), and by utilizing the relationship between (36) and (32). We obtain the ratio of one-dimensional integrals,

$$p_1(\boldsymbol{\beta}_{(1)}|\overline{H}) = K_*(\boldsymbol{\beta}_{(1)})/[B(\tfrac{1}{2}v_*, \tfrac{1}{2}\mathbf{l}_{r_1})\cdot C_*], \tag{39a}$$

where

$$\begin{aligned} K_*(\boldsymbol{\beta}_{(1)}) = \int_0^1 &|\mathbf{N}_{\omega(22)}|^{-\frac{1}{2}}\omega^{\frac{1}{2}(v_1+r)-1}(1-\omega)^{\frac{1}{2}(v'+r)-1} \\ &\times[\omega(v_1+r)S^2(\boldsymbol{\beta}_{(1)})+(1-\omega)(v'+r)S'^2(\boldsymbol{\beta}_{(1)}) \\ &+\omega(1-\omega)(\mathbf{N}_{(22)}\mathbf{N}_{\omega(22)}^{-1}\mathbf{N}'_{(22)})((\mathbf{B}(\boldsymbol{\beta}_{(1)}) \\ &-\mathbf{B}'(\boldsymbol{\beta}_{(1)})))]^{-\frac{1}{2}(v_*+r_1)}\mathrm{d}\omega. \end{aligned} \tag{39b}$$

Alternatively, one can write, from (31a) and (32),

$$p_1(\boldsymbol{\beta}_{(1)}|\overline{H}) = \int_0^1 h_{v_*}^{(r_1)}(\boldsymbol{\beta}_{(1)}-\mathbf{b}_{\omega(1)};\, s_\omega^2\mathbf{N}_{\omega(11)}^{-1})\, p_1(\omega|\overline{H})\,\mathrm{d}\omega. \tag{40}$$

Corollary. In the special case when τ is "known" ($v' \to \infty$, $\tau \equiv s'$, $\omega = s'^2(s'^2+\sigma^2)^{-1}$), for which the prior distribution of $\boldsymbol{\beta}$ is multivariate-Normal (25), we have the analogues of equations (31), (32), (36), (37),

$$p_1(\boldsymbol{\beta}|\sigma, \overline{H}) = p_1(\boldsymbol{\beta}|\sigma, \tau, \overline{H}) = h_\infty^{(r)}(\boldsymbol{\beta}-\mathbf{b}_\omega;\, \sigma^2\omega\mathbf{N}_\omega^{-1}), \tag{41a}$$

$$\begin{aligned} p_1(\sigma|\overline{H}) = p_1(\sigma|\tau, \overline{H}) = &|\mathbf{N}_\omega|^{-\frac{1}{2}}(\sigma^2)^{-\frac{1}{2}(v_1+1)}(\sigma^2+\tau^2)^{-\frac{1}{2}r} \\ &\times \exp\{-\tfrac{1}{2}(\sigma^{-2}+\tau^{-2}) \\ &\times [v_*s_\omega^2-(1-\omega)v's'^2]\}/C_{**}, \end{aligned} \tag{41b}$$

where $C_{**} = \lim_{v'\to\infty} [\tau^{v'}v'^{\frac{1}{2}(n+v_0+v')}\tfrac{1}{2}C_*]$, the integral of the numerator of (41b), and

$$\begin{aligned} p_1(\boldsymbol{\beta}|\overline{H}) &= p_1(\boldsymbol{\beta}|\tau, \overline{H}) \\ &= [v_1s_1^2+\mathbf{N}((\boldsymbol{\beta}-\mathbf{b}))]^{-\frac{1}{2}(v_1+r)} \\ &\quad\times\exp[-\tfrac{1}{2}\tau^{-2}\mathbf{N}'((\boldsymbol{\beta}-\mathbf{b}'))]\tau^{-r}\Gamma[\tfrac{1}{2}(v_1+r)] \\ &\quad\times[(\tfrac{1}{2})^{\frac{1}{2}v_1}\pi^{\frac{1}{2}r}]^{-1}/(2C_{**}), \end{aligned} \tag{41c}$$

$$\begin{aligned} &h_{v_1,\infty}^{(r)}(\mathbf{b}-\mathbf{b}';\, s_1^2\mathbf{N}^{-1}, \tau^2\mathbf{N}'^{-1}) \\ &= (\tfrac{1}{2})^{\frac{1}{2}(v_1+r)}[\Gamma(\tfrac{1}{2}v_1)\pi^{\frac{1}{2}r}]^{-1}\cdot(v_1s_1^2)^{\frac{1}{2}v_1}|\mathbf{N}|^{\frac{1}{2}}|\mathbf{N}'|^{\frac{1}{2}}\cdot 2C_{**}. \end{aligned} \tag{41d}$$

The second equalities in (41a, b, c) hold whether or not $v' \to \infty$.

The posterior density of $\boldsymbol{\beta}$, (36) or (41c), can be either unimodal or bimodal. The modes may be obtained in practice by a one-dimensional search, according to the following.

Proposition 2.3. The local maxima of the density function $p_1(\boldsymbol{\beta}|\bar{H})$, (36) or (41c), are attained on the curve décolletage $\mathbf{b}_\omega$ (29b), (35).

Proof. Suppose $\hat{\boldsymbol{\beta}}$ is such an unfound point of local maximization. Then for fixed $\mathbf{N}'((\boldsymbol{\beta}-\mathbf{b}')) = \mathbf{N}'((\hat{\boldsymbol{\beta}}-\mathbf{b}'))$, $\mathbf{N}((\boldsymbol{\beta}-\mathbf{b}))$ is locally minimized at $\boldsymbol{\beta} = \hat{\boldsymbol{\beta}}$. But the solution to this new problem is well known and immediate by use of a Lagrange multiplier λ (λ strictly monotonic in ω).

Since

$$\frac{\mathrm{d}}{\mathrm{d}\omega}\mathbf{N}'((\mathbf{b}_\omega-\mathbf{b}')) = -\omega(1-\omega)^{-1}\frac{\mathrm{d}}{\mathrm{d}\omega}\mathbf{N}((\mathbf{b}_\omega-\mathbf{b}))$$
$$= 2\omega\mathbf{N}_\omega^{-1}((\mathbf{N}'\mathbf{N}_\omega^{-1}\mathbf{N}(\mathbf{b}-\mathbf{b}'))),$$

then the derivative, with respect to ω, of $p_1(\boldsymbol{\beta}|\bar{H})_{b_\omega}$ (36) is

$$p_1(\boldsymbol{\beta}|\bar{H})_{b_\omega}\cdot[(1-\omega)\hat{\sigma}_\omega^{-2}-\omega\hat{\tau}_\omega^{-2}]\cdot\mathbf{N}_\omega^{-1}((\mathbf{N}'\mathbf{N}_\omega^{-1}\mathbf{N}(\mathbf{b}-\mathbf{b}'))), \quad (42)$$

where

$$\hat{\sigma}_\omega^2 = (\nu_1+r)^{-1}[\nu_1 s_1^2+\mathbf{N}((\mathbf{b}_\omega-\mathbf{b}))], \quad (43a)$$
$$\hat{\tau}_\omega^2 = (\nu'+r)^{-1}[\nu' s'^2+\mathbf{N}'((\mathbf{b}_\omega-\mathbf{b}'))]. \quad (43b)$$

This derivative (42) takes the value zero for any $\omega = \hat{\omega}$ satisfying

$$\omega = \hat{\tau}_\omega^2(\hat{\tau}_\omega^2+\hat{\sigma}_\omega^2)^{-1}. \quad (44)$$

Hence, the (marginal) posterior modes of $\boldsymbol{\beta}$ take the form $\mathbf{b}_{\hat{\omega}}$, where $\hat{\omega}$ satisfies equation (44).

Equations such as (44) can be solved recursively in the form $\omega_{n+1} = f(\omega_n)$ for those fixed points $\hat{\omega}$ for which $|f'(\hat{\omega})| < 1$. For $f(\omega) = \hat{\tau}_\omega^2(\hat{\tau}_\omega^2+\hat{\sigma}_\omega^2)^{-1}$, we have

$$f'(\omega) = 2[(\nu_1+r)^{-1}(1-\omega)f+(\nu'+r)^{-1}\omega(1-f)]$$
$$\times\mathbf{N}_\omega^{-1}((\mathbf{N}'\mathbf{N}_\omega^{-1}\mathbf{N}(\mathbf{b}-\mathbf{b}'))).$$

If $\hat{\omega}$ is any fixed point for which $f'(\hat{\omega}) \neq 1$, redefine $g(\omega) = (1-u)\omega+uf(\omega)$, with u satisfying $|g'(\hat{\omega})| = |1+u[f'(\hat{\omega})-1]| < 1$, for example $u = -1/[f'(\hat{\omega})-1]$. Then $\omega_{n+1} = g(\omega_n)$ converges locally to $\hat{\omega}$. These modifications to the use of (44), requiring knowledge of bounds on $f'(\hat{\omega})$, were suggested to me by Howard Thaler.

Proposition 2.4. As in the prior distribution, the parameters σ, τ are conditionally posterior independent given $\boldsymbol{\beta}$; and

$$\sigma^2|\boldsymbol{\beta}, \bar{H}, \mathbf{D} \sim [\nu_1 s_1^2+\mathbf{N}((\boldsymbol{\beta}-\mathbf{b}))]/\chi^2_{\nu_1+r}, \quad (45a)$$

$$\tau^2|\boldsymbol{\beta}, \bar{H}, \mathbf{D} \sim \tau^2|\boldsymbol{\beta}, \bar{H} \sim [v's'^2+\mathbf{N}'((\boldsymbol{\beta}-\mathbf{b}'))]/\chi^2_{v'+r}\,. \tag{45b}$$

Lindley [53] notes that for any two variables such as σ, $\boldsymbol{\beta}$, a joint mode is a conditional mode for $\sigma|\boldsymbol{\beta}$ (for $\boldsymbol{\beta}|\sigma$) at a joint mode of $\boldsymbol{\beta}$ (of σ). The mode of $\sigma|\boldsymbol{\beta}, \bar{H}, \mathbf{D}$ is $\hat{\sigma} = \{(v_1+r+1)^{-1}[v_1 s_1^2+\mathbf{N}((\boldsymbol{\beta}-\mathbf{b}))]\}^{\frac{1}{2}}$, at which the joint density of σ, $\boldsymbol{\beta}|\bar{H}$, $\mathbf{D}$ resembles the density of $\boldsymbol{\beta}|\bar{H}$, $\mathbf{D}$ (36), but with the exponent $-\frac{1}{2}(v_1+r)$ replaced by $-\frac{1}{2}(v_1+r+1)$. Hence, a joint mode of σ, $\boldsymbol{\beta}|\bar{H}$, $\mathbf{D}$ has $\hat{\sigma}$ as immediately above with $\boldsymbol{\beta} = \hat{\boldsymbol{\beta}}$, and $\hat{\boldsymbol{\beta}} = \mathbf{b}_{\hat{\omega}}$ with $\hat{\omega}$ satisfying (43), (44) with v_1+r replaced by v_1+r+1.

By similar reasoning, a joint mode of τ, $\boldsymbol{\beta}|\bar{H}$, $\mathbf{D}$ has $\hat{\tau} = \{(v'+r+1)^{-1}[v's'^2+\mathbf{N}'((\hat{\boldsymbol{\beta}}-\mathbf{b}'))]\}^{\frac{1}{2}}$, and $\hat{\boldsymbol{\beta}} = \mathbf{b}_{\hat{\omega}}$ with $\hat{\omega}$ from (43), (44) with $v'+r$ replaced by $v'+r+1$.

Whereas, a fully joint mode of σ, τ, $\boldsymbol{\beta}|\bar{H}$, $\mathbf{D}$ has $\hat{\sigma}$ and τ the preceding functions of $\hat{\boldsymbol{\beta}}$, and $\hat{\boldsymbol{\beta}} = \mathbf{b}_{\hat{\omega}}$ with $\hat{\omega}$ from (43), (44) with both the preceding replacements, v_1+r by v_1+r+1, and $v'+r$ by $v'+r+1$. Lindley [53, eq. (8.15)], and Lindley and Smith [31] propose estimation by such joint posterior modes.

Context 1. *Simple Normal sample.* The reader may wish to compare the testing and estimation methodologies that follow in regard to the case of a simple Normal (μ, σ^2) sample $y_1, \ldots, y_n$, for which $r = 1$, $\beta = \mu$, $x_i \equiv 1$ (not random). It may suffice to model the low-dimensional prior knowledges by the 5-parameter $(\bar{y}', v', s'; v_0, s_0)$ family of independent $s' \cdot t_{v'}+\bar{y}'$ and $s_0^2/(\chi^2_{v_0}/v_0)$ distributions for μ and σ^2 (26) with $b' = \bar{y}'$, $N' = 1$). If a knowledge-modeler can tolerate the sharp normal tails, he might settle for the 4-parameter $(\bar{y}', \infty, \tau; v_0, s_0)$ subfamily with $N(\bar{y}', \tau^2)$ for μ.

We have $b = \bar{y}$, $N = n$, $v = n-1$, $vs^2 = \Sigma_{i=1}^n (y_i-\bar{y})^2$; and

$$N_\omega = n\omega+1-\omega,$$

$$b_\omega = \bar{y}_\omega = \bar{y}+(1-\omega)(n\omega+1-\omega)^{-1}(\bar{y}'-\bar{y})\,.$$

Context 2. *Uncontrolled-Variables Linear Model.* To realistically model knowledge prior to an uncontrolled experiment, redefine the linear mean of the model (15),

$$y_i|\mathbf{x}_i, \mu, \boldsymbol{\beta}, \sigma \sim \text{Normal}\,(\mu+\boldsymbol{\beta}^T\mathbf{x}_i, \sigma^2)\,. \tag{46}$$

If μ is a priori known with certainty (or if posterior distributions conditional on μ are sought), merely replace the y_i's by the residuals

$y_i - \mu$ in all the preceding expressions. Namely, in place of vs^2, write $vs^2(\mu) = \min_\beta \Sigma(y_i - \mu - \boldsymbol{\beta}^T \mathbf{x}_i)^2 = \Sigma(y_i - \mu - \mathbf{b}(\mu)^T \mathbf{x}_i)^2$ throughout. (Other known coefficients can be treated similarly.)

If μ is not known a priori, replace the sum of the arguments of the exponentials in the likelihood function (19), by

$$\sum_1^n [y_i - (\mu + \boldsymbol{\beta}^T \mathbf{x}_i)]^2 = n(\mu^* - u^*)^2 + \mathbf{N}((\boldsymbol{\beta} - \mathbf{b})) + vs^2 \,, \tag{47}$$

where

$$\mu^* = \mu + \boldsymbol{\beta}^T \bar{\mathbf{x}}, \qquad u^* = u + \mathbf{b}^T \bar{\mathbf{x}} = \bar{y} \,, \tag{48a}$$

$$\bar{\mathbf{x}} = n^{-1} \sum_1^n \mathbf{x}_i, \qquad \bar{y} = n^{-1} \sum_1^n y_i \,, \tag{48b}$$

and where the statistics $\mathbf{N}$, $\mathbf{b}$, v, s^2 are redefined, as in equations (18), but with each $\mathbf{x}_i$ and y_i replaced by $\mathbf{x}_i - \bar{\mathbf{x}}$ and $y_i - \bar{y}$, respectively: for $\mathbf{1}_n = (1, 1, \ldots, 1)^T$ (n copies),

$$\mathbf{N} = \sum_1^n (\mathbf{x}_i - \bar{\mathbf{x}})(\mathbf{x}_i - \bar{\mathbf{x}})^T$$

$$= (\mathbf{X} - \mathbf{1}_n \bar{\mathbf{x}}^T)^T (\mathbf{X} - \mathbf{1}_n \bar{\mathbf{x}}^T) = \mathbf{X}^T \mathbf{X} - n \bar{\mathbf{x}} \bar{\mathbf{x}}^T \,, \tag{49a}$$

$$(\mathbf{X} - \mathbf{1}_n \bar{\mathbf{x}}^T)^T [\mathbf{y} - \bar{y} \mathbf{1}_n - (\mathbf{X} - \mathbf{1}_n \bar{\mathbf{x}}^T) \mathbf{b}] = \mathbf{0} \,, \tag{49b}$$

or

$$\mathbf{N}\mathbf{b} = \sum_1^n (\mathbf{x}_i - \bar{\mathbf{x}})(y_i - \bar{y}) = (\mathbf{X} - \mathbf{1}_n \bar{\mathbf{x}}^T)^T (\mathbf{y} - \mathbf{1}_n \bar{y}) = \mathbf{X}^T \mathbf{y} - n \bar{\mathbf{x}} \bar{y} \,, \tag{49c}$$

$$vs^2 = \sum_1^n (y_i - u - \mathbf{b}^T \mathbf{x}_i)^2 \tag{49d}$$

(or vs^2 satisfying (47), any μ, $\boldsymbol{\beta}$).

Note that if $\Sigma_{j=1}^r x_i^{(j)}$ is the same for all $i = 1, \ldots, n$, then $\mathbf{N}\mathbf{1}_r = \mathbf{0}$, an undesirable, rare type of singularity for $\mathbf{N}$, which we hereby rule out by assumption.

If μ is a priori locally "completely unknown" ($p_0(\mu | \bar{H}) \doteq \text{Const}_1$) and prior independent of the other parameters, integrate μ out of the likelihood function to obtain the weighted likelihood function, (19) again, but with $\mathbf{N}$, $\mathbf{b}$, vs^2 as redefined above and with σ^{-n} replaced by $\sigma^{-(n-1)}$. With these modifications, obtain, as before, the posterior distributions of $\boldsymbol{\beta}$, σ, and ω marginal as regards μ.

Formally, make the foregoing modifications of $\mathbf{N}$, $\mathbf{b}$, and νs^2 in the definitions (29) of $\mathbf{N}_\omega$, $\mathbf{b}_\omega$, $\nu_* s_\omega^2$, with $\nu_* = n-1+\nu_0+\nu'$, and define the $(r+1)$-dimensional vector of coefficients $\boldsymbol{\alpha} = (\mu^*, \boldsymbol{\beta}^T)^T$, and $\mathbf{a}_\omega = (\bar{y}, \mathbf{b}_\omega^T)^T$. Then

$$\boldsymbol{\alpha}, \sigma^{-2}|\omega, \bar{H}, \mathbf{D} \sim ((r+1)\text{-Normal})\text{-gamma}$$
$$[\mathbf{a}_\omega, \omega^{-1}s_\omega^2, \omega^{-1}\,\mathbf{Diag}(n, \mathbf{N}_\omega), \nu_*],$$

$$p_1(\boldsymbol{\alpha}|\omega, \bar{H}) = h_{\nu_*}^{(r+1)}[\boldsymbol{\alpha}-\mathbf{a}_\omega; s_\omega^2\mathbf{Diag}(n^{-1}, \mathbf{N}_\omega^{-1})], \tag{50}$$

$$p_1(\mu^*|\boldsymbol{\beta}, \omega, \bar{H}) = h_{\nu_*+r}^{(1)}[\mu^*-\bar{y}; S_\omega^2(\boldsymbol{\beta})n^{-1}] \tag{51a}$$

where

$$S_\omega^2(\boldsymbol{\beta}) = (\nu_*+r)^{-1}[\nu_* s_\omega^2+\mathbf{N}_\omega((\boldsymbol{\beta}-\mathbf{b}_\omega))],$$

and

$$p_1(\mu^*|\omega, \bar{H}) = h_{\nu_*}^{(1)}(\mu^*-\bar{y}; s_\omega^2 n^{-1}). \tag{51b}$$

Note that $\bar{y}$ is the mean and mode for μ^*, regardless of $\boldsymbol{\beta}$ and ω. Since $\mu = \mu^*-\boldsymbol{\beta}^T\bar{\mathbf{x}} = (1, \bar{\mathbf{x}}^T)\boldsymbol{\alpha}$,

$$p_1(\mu|\boldsymbol{\beta}, \omega, \bar{H}) = h_{\nu_*+r}^{(1)}[\mu-(\bar{y}-\boldsymbol{\beta}^T\bar{\mathbf{x}}); S_\omega^2(\boldsymbol{\beta})n^{-1}], \tag{52a}$$

$$p_1(\mu|\omega, \bar{H}) = h_{\nu_*}^{(1)}[\mu-(\bar{y}-\bar{\mathbf{x}}^T\mathbf{b}_\omega); s_\omega^2(n^{-1}+\bar{\mathbf{x}}^T\mathbf{N}_\omega^{-1}\bar{\mathbf{x}})]. \tag{52b}$$

So μ should be estimated by $\bar{y}-\bar{\mathbf{x}}^T\hat{\boldsymbol{\beta}}$. Equations (31)–(45) hold again as read with the foregoing redefinitions and with $\nu_1 = n-(r+1)+\nu_0$.

Now, for an important special prior family, suppose that under some choice of scales for the $x^{(j)}$'s (perhaps even a full linear transformation), the prior distribution of the $\beta^{(j)}$'s is exchangeable, namely invariant under permutations of the variables $\beta^{(j)}$. In particular, consider the family of prior densities: (26) conditional on a scalar b', with equiangular vector $\mathbf{b}' = b'\mathbf{1}_r$ and identity matrix $\mathbf{N}' = \mathbf{I}_r$,

$$\boldsymbol{\beta}|\tau, b', \bar{H} \sim \mathbf{N}^{(r)}(b'\mathbf{1}_r, \tau^2\mathbf{I}_r), \qquad \boldsymbol{\beta}|b', \bar{H} \sim \mathbf{t}_{\nu'}^{(r)}(b'\mathbf{1}_r, s'^2\mathbf{I}_r), \tag{53}$$

hence

$$r^{-1}\sum_1^r(\beta^{(j)}-b')^2|\tau, b', \bar{H} \sim \tau^2\cdot F_{r,\infty},$$

$$\text{Same}\,|b', \bar{H} \sim s'^2\cdot F_{r,\nu'}. \tag{54}$$

but with b' a priori locally "completely unknown", $b' \sim$ "Uniform",

$$p_0(b'|\bar{H}) \doteq \text{Const}_2, \quad \text{just as} \quad p_0(\mu|\bar{H}) \doteq \text{Const}_1 \,. \tag{55}$$

Then, independently,

$$\sigma^2, \tau^2, \mu, b'|\bar{H} \sim s_0^2 \cdot F_{\infty\ \nu_0}, s_\tau^2 \cdot F_{\infty, \nu_\tau}, \text{"Unif."}, \text{"Unif."} \tag{56}$$

Since the likelihood function is constant in b', we can integrate b' out of the prior density, to obtain one-dimensionally nonintegrable prior densities, symbolically denoted

$$\boldsymbol{\beta}|\tau, \bar{H} \sim \text{"}\mathbf{N}^{(r)}(\mathbf{0}, \tau^2\mathbf{N}'^{-})\text{"}, \ \boldsymbol{\beta}|\bar{H} \sim \text{"}\mathbf{t}_{\nu'}^{(r)}(\mathbf{0}, s'^2\mathbf{N}'^{-})\text{"}, \tag{57a}$$

which with $\mathbf{N}' = \mathbf{I}_r - r^{-1}\mathbf{l}_r\mathbf{l}_r^T$, means

$$\beta^{(\cdot)}|\tau, \bar{H} \sim \beta^{(\cdot)}|\bar{H} \sim \text{"Unif."}, \quad \text{where} \quad \beta^{(\cdot)} = r^{-1}\sum_1^r \beta^{(j)}, \tag{57b}$$

$$\boldsymbol{\eta}|\tau, \bar{H} \sim \mathbf{N}^{(r-1)}(0, \tau^2\mathbf{J}_{r-1}^{-1}), \qquad \boldsymbol{\eta}|\bar{H} \sim \mathbf{t}_{\nu'}^{(r-1)}(\mathbf{0}, s'^2\mathbf{J}_{r-1}^{-1}), \tag{57c}$$

where

$$\boldsymbol{\eta} = (\beta^{(1)} - \beta^{(\cdot)}, \ldots, \beta^{(r-1)} - \beta^{(\cdot)})^T,$$
$$\mathbf{J}_{r-1}^{-1} = \mathbf{I}_{r-1} - r^{-1}\mathbf{l}_{r-1}\mathbf{l}_{r-1}^T,$$
$$\mathbf{J}_{r-1} = \mathbf{I}_{r-1} + \mathbf{l}_{r-1}\mathbf{l}_{r-1}^T, \quad \text{and}$$
$$\mathbf{J}_{r-1}((\boldsymbol{\eta})) = \sum_1^r (\beta^{(j)} - \beta^{(\cdot)})^2.$$

Hence,

$$(r-1)^{-1}\sum_1^r (\beta^{(j)} - \beta^{(\cdot)})^2|\tau, \bar{H} \sim \tau^2 \cdot F_{r-1,\infty},$$
$$\text{Same}\,|\bar{H} \sim s'^2 \cdot F_{r-1,\nu'}\,. \tag{57d}$$

The constant local approximation to the density of $\beta^{(\cdot)}|\tau, \bar{H} \sim \beta^{(\cdot)}|\bar{H}$ is again $\text{Const}_2 \doteq p_0(b'|\bar{H})$, which yields $p_0(\boldsymbol{\beta}|\tau, \bar{H})$ when multiplied by $r^{-1} \cdot h_\infty^{(r-1)}(\boldsymbol{\eta}; \tau^2\mathbf{J}_{r-1}^{-1})$, and yields $p_0(\boldsymbol{\beta}|\bar{H})$ when multiplied by $r^{-1} \cdot h_{\nu'}^{(r-1)}(\boldsymbol{\eta}; s'^2\mathbf{J}_{r-1}^{-1})$.

The prior density of $\boldsymbol{\beta}$, τ, $\sigma|\bar{H}$ is again of the form (26) with $\mathbf{b}' = \mathbf{0}$, $\mathbf{N}' = \mathbf{I}_r - r^{-1}\mathbf{l}_r\mathbf{l}_r^T$, and $\tau^{-(\nu'+r+1)}$ replaced by $\tau^{-[(\nu'-1)+r+1]}$. The posterior distributions for this 5-parameter $[p_0(\beta^{(\cdot)}|\bar{H}); \nu', s'; \nu_0, s_0]$ prior family can be read by applying the above modifications to the posterior distributions already modified in this context 2. Hence $\nu_* = n - 1 + \nu_0 + \nu' - 1$ and $\nu_1 = n - 1 - r + \nu_0$.

Context 3. *Analysis of variance.* Consider the one-way analysis of variance with an equal number m of observations $z_k^{(j)}$ $(k = 1, \ldots, m)$ in each of r groups $j\,(j = 1, \ldots, r)$, so $n = rm$. The coefficient $\beta^{(j)}$ is the unknown population-mean for the jth group of observations, $z_k^{(j)} = y_{(j-1)m+k}$, $k = 1, \ldots, m$. Hence the design matrix $\mathbf{X}$ is fixed (not random), and each row $\mathbf{x}_{(j-1)m+k}^T$ has zero extries except for 1 in its jth position. Then $\mathbf{N} = m\mathbf{I}_r$, $\mathbf{b} = \bar{\mathbf{z}} = (\bar{z}^{(1)}, \ldots, \bar{z}^{(r)})^T$, and $vs^2 = \text{SSW} = \Sigma_{j=1}^r \Sigma_{k=1}^m (z_k^{(j)} - \bar{z}^{(j)})^2$.

Take the same 5-parameter prior family for $\boldsymbol{\beta}$, τ, $\sigma|\bar{H}$ as in context 2. Of course, there is no μ; so n should not be replaced by $n-1$.

We have

$$\begin{aligned} v_* &= rm + v_0 + v' - 1; \\ \mathbf{N}_\omega &= (m\omega + 1 - \omega)\mathbf{I}_r - (1-\omega)r^{-1}\mathbf{1}_r\mathbf{1}_r^T; \\ \mathbf{N}_\omega^{-1} &= (m\omega + 1 - \omega)^{-1}[\mathbf{I}_r + (m\omega)^{-1}(1-\omega)r^{-1}\mathbf{1}_r\mathbf{1}_r^T]; \\ |\mathbf{N}_\omega| &= m\omega(m\omega + 1 - \omega)^{r-1}; \\ v_* s_\omega^2 &= \omega(SSW + v_0 s_0^2) + (1-\omega)v's'^2 + \omega(1-\omega) \\ &\quad \times (m\omega + 1 - \omega)^{-1} SSB, \end{aligned}$$

where

$$SSB = m\sum_1^r (\bar{z}^{(j)} - \bar{z}^{(\cdot)})^2 \quad \text{with} \quad \bar{z}^{(\cdot)} = \sum \bar{z}^{(j)}/r;$$

$\mathbf{b}_\omega$ has jth coordinate $\bar{z}^{(\cdot)} + m\omega(m\omega + 1 - \omega)^{-1}(\ ^{(j)} - \bar{z}^{(\cdot)})$;

$$\begin{aligned} \mathbf{N}_\omega((\boldsymbol{\beta} - \mathbf{b}_\omega)) &= m\omega \cdot r(\beta^{(\cdot)} - \bar{z}^{(\cdot)})^2 + (m\omega + 1 - \omega) \\ &\quad \times \sum_1^r [(\beta^{(j)} - \beta^{(\cdot)}) - m\omega(m\omega + 1 - \omega)^{-1} \\ &\quad \times (\bar{z}^{(j)} - \bar{z}^{(\cdot)})]^2; \end{aligned}$$

and

$$p_1(\beta^{(\cdot)}|\omega, \bar{H}) = h_{v_*}(\beta^{(\cdot)} - \bar{z}^{(\cdot)};\, s_\omega^2/(rm)). \tag{58}$$

Lindley [25] obtained the joint posterior modes in this context without the assumption of equal sample sizes, which we have made to simplify the illustration.

3. *The weighted likelihood ratio for a linear hypothesis*

In the r-space of $\boldsymbol{\beta}$ and of $\hat{\boldsymbol{\beta}}$, define the hypothesis H to be an $(r-q)$-dimensional linear manifold,

$$H: \boldsymbol{\eta} = \boldsymbol{\eta}_H, \qquad \bar{H}: \boldsymbol{\eta} \neq \boldsymbol{\eta}_H, \tag{59}$$

where $\boldsymbol{\eta}$ is a q-valued linear function of $\boldsymbol{\beta}$,

$$\boldsymbol{\eta} = \mathbf{C}_H\boldsymbol{\beta}, \qquad (q\times r)\mathbf{C}_H \quad \text{full rank} \quad q. \tag{60}$$

There is no loss of generality in assuming $\mathbf{C}_H$ is of full rank $q \leqq r$. Note the apparent generalization from the usual linear sub*space* $\boldsymbol{\eta} = \mathbf{0}$. This permits, for one thing, checking of a test-outcome against small variations in the definition of a possibly-unrealistically sharp null hypothesis. Of course, $\boldsymbol{\eta}_H$ could be replaced by $\mathbf{0}$, by having $\mathbf{y}$ replaced by $\mathbf{y}-\mathbf{X}\mathbf{C}^{-1}(\boldsymbol{\eta}_H^T, \mathbf{0}^T)^T$, where $\mathbf{C}$ is any $r\times r$ full-rank matrix of the form $\mathbf{C} = (\mathbf{C}_H^T, \mathbf{C}_{H*}^T)^T$.

We turn now to calculation of the weighted likelihood ratio for H based on $\mathbf{y}$ conditional on $\mathbf{X}$,

$$L_{y|X}(H) = [P(H|\mathbf{y}, \mathbf{X})/P(\bar{H}|\mathbf{y}, \mathbf{X})]/[P(H|\mathbf{X})/P(\bar{H}|\mathbf{X})]$$
$$= p(\mathbf{y}|\mathbf{X}, H)/p(\mathbf{y}|\mathbf{X}, \bar{H})$$

(or consider, conditional on ξ, $L_{y|X,\xi}$). If $(\boldsymbol{\beta}, \sigma)$ and $\mathbf{X}$ (or ξ) are prior independent,

$$L_{y|X}(H) = L_D(H) = p(\mathbf{D}|H)/p(\mathbf{D}|\bar{H}) \quad (\text{or} \quad L_{y|X,\xi} = L_D).$$

According to Savage's Density Ratio Theorem [20, 39], the weighted likelihood ratio can sometimes be calculated as the ratio of the posterior density of $\boldsymbol{\eta}$ under $\bar{H}$ at the boundary $\boldsymbol{\eta}_H$ to the corresponding prior density,

$$L_{y|X}(H) = p_1(\boldsymbol{\eta}|\bar{H})_{\eta_H}/p_0(\boldsymbol{\eta}|\bar{H})_{\eta_H}. \tag{61}$$

A sufficient condition is Savage's Density Condition, that the prior distribution of σ, $\boldsymbol{\beta}$ under H coincide with the corresponding boundary conditional prior distribution under $\bar{H}$,

$$p_0(\sigma|\zeta, H) = p_0(\sigma|\zeta, \boldsymbol{\eta}, \bar{H})_{\eta=\eta_H}, \tag{62a}$$

$$p_0(\zeta|H) = p_0(\zeta|\boldsymbol{\eta}, \bar{H})_{\eta=\eta_H}, \tag{62b}$$

where the $r-q$ nuisance parameters ζ (additional to the nuisance parameter σ) are defined by,

$$\zeta = \mathbf{C}_{H^*}\boldsymbol{\beta}, \qquad (r\times r)\mathbf{C} = (\mathbf{C}_H^T, \mathbf{C}_{H^*}^T)^T \quad \text{full rank.} \tag{63}$$

For our prior family of section 2, condition (62a) follows directly from the prior independence of σ and $\boldsymbol{\beta}$. We hereby assume condition (62b), and in so doing, we complete the specification of the prior family. $P_0(H)$ need not be specified, to report $L_{y|X}(H)$.

The denominator of $L_{y|X}(H)$ is just the marginal density of a linear function of a multivariate-t vector, immediate from the representation (24)–(25),

$$p_0(\boldsymbol{\eta}|\bar{H}) = h_{v'}^{(q)}(\boldsymbol{\eta}-\mathbf{C}_H\mathbf{b}'; s'^2\mathbf{C}_H\mathbf{N}'^{-1}\mathbf{C}_H^T). \tag{64}$$

The numerator is a ratio of one-dimensional integrals, according to the following generalization of (40).

Proposition 3.1.

$$p_1(\boldsymbol{\eta}|\bar{H}) = \int_0^1 p_1(\boldsymbol{\eta}|\omega, \bar{H})p_1(\omega|\bar{H})\,d\omega, \tag{65}$$

where $p_1(\omega|\bar{H})$ was given in (32) and

$$p_1(\boldsymbol{\eta}|\omega, \bar{H}) = h_{v*}^{(q)}(\boldsymbol{\eta}-C_H\mathbf{b}_\omega; s_\omega^2\mathbf{C}_H\mathbf{N}_\omega^{-1}\mathbf{C}_H^T). \tag{66}$$

Proof. Change variables in $p_1(\boldsymbol{\beta}|\omega, \bar{H})$ (31a) from $\boldsymbol{\beta}$ to $\boldsymbol{\eta}, \zeta$; partition the quadratic form by lemma 3.1 of Dickey [54]; then integrate out ζ.

Special cases of proposition 3.1 appear in Dickey [20, eq. (5.51)], and Box and Tiao [50, sect. 9.3.1].

To discuss practical computation of the densities of $\boldsymbol{\eta}$ for $L_{y|X}(H)$, and for other later use, consider the geometry of r-space under an arbitrary inner product $(\boldsymbol{\alpha}, \boldsymbol{\beta})^Q = \boldsymbol{\alpha}'\mathbf{Q}\boldsymbol{\beta}$, $(r\times r)\mathbf{Q} > 0$. The **Q**-inner-product projection operator B_H^Q onto H is given by

$$\begin{aligned} B_H^Q\boldsymbol{\beta} &= \boldsymbol{\beta}+\mathbf{Q}^{-1}\mathbf{C}_H^T(\mathbf{C}_H\mathbf{Q}^{-1}\mathbf{C}_H^T)^{-1}(\boldsymbol{\eta}_H-\mathbf{C}_H\boldsymbol{\beta}) \\ &= C^{-1}B_H^{\tilde{Q}}(\boldsymbol{\eta}^T, \zeta^T)^T, \end{aligned} \tag{67}$$

where $\tilde{\mathbf{Q}} = \mathbf{C}^{-T}\mathbf{Q}\mathbf{C}^{-1}$ [51, p. 49; 55, p. 254]. Then define the weighted Sum of Squares Between with weights **Q**,

$$\begin{aligned} \mathrm{SSB}_H^Q(\boldsymbol{\beta}) &= \mathbf{Q}((\boldsymbol{\beta}-B_H^Q\boldsymbol{\beta})) = (\mathbf{C}_H\mathbf{Q}^{-1}\mathbf{C}_H^T)^{-1}((\mathbf{C}_H\boldsymbol{\beta}-\boldsymbol{\eta}_H)) \\ &= \tilde{\mathbf{Q}}_{(11.2)}((\boldsymbol{\eta}-\boldsymbol{\eta}_H)). \end{aligned} \tag{68}$$

In this notation, the quadratic form to be divided by v' within

$p_0(\boldsymbol{\eta}|\bar{H})$ at $\boldsymbol{\eta} = \boldsymbol{\eta}_H$ (64) is $\mathrm{SSB}_H^{N'}(\mathbf{b}')/s'^2$, and the one to be divided by v_* within $p_1(\boldsymbol{\eta}|\omega, \bar{H})$ at $\boldsymbol{\eta} = \boldsymbol{\eta}_H$ (66) is $\mathrm{SSB}_H^{N_\omega}(\mathbf{b}_\omega)/s_\omega^2$.

Lindley [43] remarks that although the projection $B_H^N\mathbf{b}$ is also the *constrained* least squares solution for $\Sigma[y_i - \boldsymbol{\beta}^T\mathbf{x}_i]^2$, the projections of Bayesian conjugate posterior means, such as our $B_H^{N_\omega}\mathbf{b}_\omega$, have no such interpretation. Hence the equation,

$$\mathrm{SSB} \equiv \mathrm{SSB}_H^N(\mathbf{b}) = \sum_1^n [y_i - (B_H^N\mathbf{b})^T\mathbf{x}_i]^2 - \mathrm{SSW}, \tag{69}$$

has no analogue for general $\mathrm{SSB}_H^{N_\omega}(\mathbf{b}_\omega)$.

For Model-II contexts, the hypothesis of zero $\boldsymbol{\beta}$-population-variance is of interest. Define

$$\tilde{H}: \tau = 0.$$

Proposition 3.2. If $\mathbf{b}'$ is known,

$$\begin{aligned} L_{y|X}(\tilde{H}) &= L_{y|X}(H: \boldsymbol{\beta} = \mathbf{b}') \\ &= B[\tfrac{1}{2}(n+v_0), \tfrac{1}{2}v'] |\mathbf{N}'|^{-\frac{1}{2}}(v's'^2)^{-\frac{1}{2}v'} \\ &\quad \times [v_0 s_0^2 + vs^2 + \mathbf{N}((\mathbf{b}-\mathbf{b}'))]^{-\frac{1}{2}(n+v_0)}/C_*. \end{aligned} \tag{70}$$

Proof. To prove the first equality of (70), note that it suffices to prove that the symmetric-difference event, $\tilde{H}\Delta H = (\tilde{H} - H) \cup (H - \tilde{H})$, has zero prior probability, and hence zero posterior probability. For if it does, $\tilde{H}$ and H must have equal prior probabilities and equal posterior probabilities, thus implying equal Bayes factors. We have, $P_0(H - \underline{H}) = P_0(\tilde{H}) \cdot P_0(\bar{H}|H) = P_0(\tilde{H}) \cdot 0$; and $P_0(H - \tilde{H}) = P_0(H|\tilde{H}) \cdot P_0(\bar{\tilde{H}}) = 0 \cdot P_0(\bar{\tilde{H}})$. To prove the second equality of (70), use (61), and (23), (36), with $\boldsymbol{\eta} = \boldsymbol{\beta} - \mathbf{b}'$, $\boldsymbol{\eta}_H = \mathbf{0}$.

Note that if $v' > 0$, $p_0(\tau|\bar{H})_0 = 0 = p_1(\tau|\bar{H})_0$, with the second equality obtainable from the finiteness of $L_{y|X}$.

By straightforward calculations, we obtain, for general H (59), the following predictive densities for $\mathbf{y}|\mathbf{X}$ under $\bar{H}$ and under H.

Proposition 3.3. The denominator of $L_{y|X}(H) = p(\mathbf{y}|\mathbf{X}, H)/p(\mathbf{y}|\mathbf{X}, \bar{H})$ is

$$\begin{aligned} p(\mathbf{y}|\mathbf{X}, \bar{H}) &= [B(\tfrac{1}{2}v_0, \tfrac{1}{2}v'; \tfrac{1}{2}\mathbf{1}_n)]^{-1}(v_0 s_0^2)^{\frac{1}{2}v_0}(v's'^2)^{\frac{1}{2}v'} \\ &\quad \times |\mathbf{N}'|^{\frac{1}{2}} \cdot C_*. \end{aligned} \tag{71}$$

Then the numerator, $p(\mathbf{y}|\mathbf{X}, H) = p(\mathbf{y}|\mathbf{X}, \bar{H}) \cdot L_{y|X}(H)$.

Such a predictive density of $\mathbf{y}$, and the corresponding predictive density of $(\mathbf{b}, s)$, have the same proportionality relationship as the usual statistical-model densities conditional on $\boldsymbol{\theta}$: for $K = H$ or $\bar{H}$, $p(\mathbf{b}, \mathbf{s}|\mathbf{X}, K) = p(\mathbf{y}|\mathbf{X}, K)\cdot(|\mathbf{N}|^{\frac{1}{2}}\nu^{\frac{1}{2}}\cdot\text{Surf})$, where $\text{Surf} = B(1; \frac{1}{2}\mathbf{1}_\nu) \times \nu(\nu s^2)^{\frac{1}{2}(\nu-1)}$, the surface area of a ν-dimensional hyperball of radius $(\nu s^2)^{\frac{1}{2}}$.

The preceding proposition 3.3 and the restricted-dimensional version to follow (corollary to proposition 4.2) are useful for testing among several exhaustive and mutually exclusive hypotheses, $H_1, \ldots, H_M$ (for example $M = 2^r$), since

$$\begin{aligned} P_1(H_i) &= p(\mathbf{y}|\mathbf{X}, H_i)\cdot P_0(H_i) \Big/ \sum_1^M \text{Numerator } (i') \\ &= L_{y|X}(H_i \text{ versus } H_M)\cdot P_0(H_i) \\ &\quad \Big/ \sum_1^M \text{New Numerator } (i'). \end{aligned} \tag{72}$$

Context 1. *Student's hypothesis.* The weighted likelihood ratio for $H: \mu = \mu_H$ versus $\bar{H}: \mu \neq \mu_H$ is $L_{y|X}(H) = L_D(H)$,

$$L_D(H) = h_{\nu_1}(\bar{y}-\mu_H; s_1^2/n)/h_{\nu_1,\nu'}(\bar{y}-\bar{y}'; s_1^2/n, s'^2)\,, \tag{73}$$

in which the denominator is the density at $\bar{y}-\bar{y}'$ of the usual Behrens–Fisher random variable $(s_1/n)^{\frac{1}{2}}t_{\nu_1} - s't_{\nu'}$. In the case of a Normal $(\bar{y}', \tau^2)$ prior distribution for μ under $\bar{H}$, the denominator in (73) becomes $h_{\nu_1,\infty}(\bar{y}-\bar{y}'; s_1^2/n, \tau^2)$. Of special interest is the 4-parameter $(v', s'; \nu_0, s_0)$ prior subfamily with $\bar{y}' = \mu_H$; or such a 3-parameter $(\infty, \tau, \nu_0, s_0)$ Normal one, for which L_D appeared in Dickey [20, eq. (4.31)], and Dickey [19, eq. (4.15)].

Context 2. *The hypothesis of given effects for a subset of regressor variables.* Partition the vector of coefficients $\boldsymbol{\beta} = (\boldsymbol{\beta}_{(1)}^T, \boldsymbol{\beta}_{(2)}^T)^T$ into subvectors of dimensions $r_1 = q$ and $r_2 = r-q$, and consider the hypothesis $H: \boldsymbol{\beta}_{(1)} = \boldsymbol{\beta}_{(1)H}$ versus $\bar{H}: \boldsymbol{\beta}_{(1)} \neq \boldsymbol{\beta}_{(1)H}$. Then $\mathbf{C} = \mathbf{I}_r$, and

$$B_H^Q\boldsymbol{\beta} = \{\boldsymbol{\beta}_{(1)H}^T, [\boldsymbol{\beta}_{(2)} - (\mathbf{Q}_{(22)})^{-1}\mathbf{Q}_{(21)}(\boldsymbol{\beta}_{(1)H} - \boldsymbol{\beta}_{(1)})]^T\}^T, \tag{74a}$$

$$\text{SSB}_H^Q(\boldsymbol{\beta}) = \mathbf{Q}_{(11.2)}((\boldsymbol{\beta}_{(1)} - \boldsymbol{\beta}_{(1)H})), \tag{74b}$$

where

$$-(\mathbf{Q}_{(22)})^{-1}\mathbf{Q}_{(21)} = (\mathbf{Q}^{-1})_{(21)}[(\mathbf{Q}^{-1})_{(11)}]^{-1}.$$

Hence

$$L_{y|X}(H) = \int_0^1 h_{\nu_*}^{(q)}[\mathbf{b}_{\omega(1)} - \boldsymbol{\beta}_{(1)H};\, s_\omega^2(\mathbf{N}_\omega^{-1})_{(11)}]$$
$$\times |\mathbf{N}_\omega|^{-\frac{1}{2}}\omega^{\frac{1}{2}(\nu_1+r)-1}(1-\omega)^{\frac{1}{2}(\nu'-1+r)-1}$$
$$\times(\nu_* s_\omega^2)^{-\frac{1}{2}\nu_*}C_*^{-1}\mathrm{d}\omega/[p_0(\boldsymbol{\beta}_{(1)}|\bar{H}) \quad \text{at} \quad \boldsymbol{\beta}_{(1)H}],$$

in which the numerator has quadratic form

$$\mathrm{SSB}_H^{N_\omega}(\mathbf{b}_\omega)/s_\omega^2 = \mathbf{N}_{\omega(11.2)}((\mathbf{b}_{\omega(1)} - \boldsymbol{\beta}_{(1)H}))/s_\omega^2.$$

(Modify as before the definitions (29) of ν_*, $\mathbf{N}_\omega$, $\mathbf{b}_\omega$, s_ω^2, ν_1, and ν_*.)

The form (64) of the denominator of $L_{y/X}(H)$ is zero for $\mathbf{N}' = \mathbf{I}_r - r^{-1}\mathbf{1}_r\mathbf{1}_r^T$, since $\mathbf{N}'_{(11.2)} = \mathbf{I}_q - q^{-1}\mathbf{1}_q\mathbf{1}_q^T$ a singular matrix. But for our special prior family, the local approximate density is

$$p_0(\boldsymbol{\beta}_{(1)}|\bar{H}) \doteq h_{\nu'}^{(q-1)}(\boldsymbol{\iota};\, s'^2\mathbf{J}_{q-1}^{-1})\cdot p_0(b'|\bar{H})q^{-1},$$

with the latter two factors equal to $p_0(\Sigma_1^q\beta^{(j)}|\bar{H}) \doteq p_0(\beta^{(\cdot)}|\bar{H})q^{-1}$ and with the first factor equal to $p_0(\boldsymbol{\iota}|\bar{H})$, where $\boldsymbol{\iota}$ is defined in terms of q and $\boldsymbol{\beta}_{(1)}$ just as $\boldsymbol{\eta}$ was defined for (57) in terms of r and $\boldsymbol{\beta}$. Then

$$\mathbf{J}_{q-1}((\boldsymbol{\iota}))/s'^2 = \mathrm{SSB}_H^{N'}(0)/s'^2 = \sum_1^q (\beta^{(j)} - \sum_1^q \beta^{(i)}/q)^2/s'^2.$$

We have the modified form of (71)

$$p(\mathbf{y}|\mathbf{X}, \bar{H}) \doteq p_0(\mu|\bar{H})n^{-\frac{1}{2}}p_0(\beta^{(\cdot)}|\bar{H})r^{-\frac{1}{2}}[B(\tfrac{1}{2}\nu_0, \tfrac{1}{2}\nu', \tfrac{1}{2}\mathbf{l}_{n-2})]^{-1}$$
$$\times(\nu_0 s_0^2)^{\frac{1}{2}\nu_0}(\nu' s'^2)^{\frac{1}{2}\nu'}\cdot C_*, \tag{75}$$

with the modified definitions of ν_* and C_*.

Context 3. *The hypothesis of equal group means.* For the hypothesis of zero contrasts, $H\colon \boldsymbol{\eta} = 0$, versus $\bar{H}\colon \boldsymbol{\eta} \neq 0$, we have

$$B_H^{N_\omega}\boldsymbol{\beta} = \beta^{(\cdot)}\mathbf{1}_r,$$

$$\mathrm{SSB}_H^{N_\omega}(\boldsymbol{\beta}) = (m\omega + 1 - \omega)\sum_1 (\beta^{(j)} - \beta^{(\cdot)})^2; \tag{76}$$

in particular, $\mathrm{SSB}_H^{N_\omega}(\mathbf{b}_\omega) = m\omega^2(m\omega+1-\omega)^{-1}\mathrm{SSB}$.

The denominator of $L_{y|X}(H) = L_D(H)$ is

$$p_0(\boldsymbol{\eta}|\bar{H})_0 = [B(\tfrac{1}{2}\nu', \tfrac{1}{2}l_{r-1})]^{-1}(\nu' s'^2)^{-\frac{1}{2}(r-l)}r^{\frac{1}{2}}, \tag{77}$$

since $p_0(\boldsymbol{\eta}|\mu, \bar{H}) = h_{\nu'}^{(r-1)}(\boldsymbol{\eta};\, s'^2\mathbf{J}_{r-1}^{-1})$ independent of μ, and $|\mathbf{J}_{r-1}| = r$.

By integrating $\beta^{(\cdot)}$ out of the weighted-likelihood-function factor of $p_1(\boldsymbol{\beta}|\bar{H})$ (2.22), the numerator of $L_D(H)$ is seen to be

$$p_1(\boldsymbol{\eta}|\bar{H})_0 = (\text{SSB}+\text{SSW}+v_0 s_0^2)^{-\frac{1}{2}(rm+v_0-1)} \times (v' s'^2)^{-\frac{1}{2}(v'+r-1)}\Gamma[\tfrac{1}{2}(rm+v_0-1)] \times \Gamma[\tfrac{1}{2}(v'+r-1)][\Gamma(\tfrac{1}{2}v_*)\pi^{\frac{1}{2}(r-1)}]^{-1}(r/m)^{\frac{1}{2}}/C_* \tag{78}$$

where

$$p_1(\omega|\bar{H}) = m^{-\frac{1}{2}}\omega^{\frac{1}{2}(rm+v_0-1)-1}(1-\omega)^{\frac{1}{2}(v'-1+r)-1} \times (m\omega+1-\omega)^{-\frac{1}{2}(r-1)}(v_* s_\omega^2)^{-\frac{1}{2}v_*}/C_* .$$

The ratio of (78) to (77) is

$$L_D(H) = B[\tfrac{1}{2}(rm+v_0-1), \tfrac{1}{2}v'] \times (\text{SSB}+\text{SSW}+v_0 s_0^2)^{-\frac{1}{2}(rm+v_0-1)}(v' s'^2)^{-\frac{1}{2}v'} m^{-\frac{1}{2}}/C_*. \tag{79}$$

Also,

$$p(\mathbf{y}|\mathbf{X}, \bar{H}) \doteq p_0(\beta^{(\cdot)}|\bar{H})r^{-1}\cdot[B(\tfrac{1}{2}v_0, \tfrac{1}{2}v'; \tfrac{1}{2}\mathbf{1}_{rm-1})]^{-1} \times (v_0 s_0^2)^{\frac{1}{2}v_0}(v' s'^2)^{\frac{1}{2}v'}\cdot C_* .$$

The hypothesis of zero β-population variance component, $H: \tau = 0$, is probabilistically equivalent to the hypothesis of equal group means $H: \boldsymbol{\eta} = \mathbf{0}$. Hence, analogously to proposition 3.2, $L_D(H) = L_D(H)$, a' given by equation (79). I offer this ratio of densities, $p_1(\tau|\bar{H}')_0/p_0(\tau|\bar{H}')_0$ S as a contribution to the discussion by Hill [15].

4. *Squared-error prediction and estimation loss*

The reader will note that, to start, the following theory does not depend on the Normal model and prior family introduced in sections 2 and 3, but only on the consequent relations which hold for all possible $\mathbf{D}$, for $y = y_{n+1}$, $\mathbf{x} = \mathbf{x}_{n+1}$,

$$E(y|\mathbf{x}, \boldsymbol{\theta}) = E(y|\mathbf{x}, \boldsymbol{\theta}, \mathbf{D}) = E(y|\mathbf{x}, \boldsymbol{\beta}, \mathbf{D}) = \boldsymbol{\beta}^T\mathbf{x} , \tag{80a}$$

$$V(y|\mathbf{x}, \boldsymbol{\theta}) = V(y|\mathbf{x}, \boldsymbol{\theta}, \mathbf{D}) = \sigma^2 , \tag{80b}$$

σ^2 functionally independent of $\mathbf{x}$. Hence,

$$V(y|\mathbf{x}, \boldsymbol{\beta}, \mathbf{D}) = E_1(\sigma^2|\mathbf{x}, \boldsymbol{\beta}) . \tag{81}$$

Consider the squared-*prediction*-error loss, for $\hat{y} = \hat{y}_{n+1}$,

$$\tilde{\Delta}(\hat{y}, y) = (\hat{y} - y)^2 , \tag{82}$$

and the prediction functions $\hat{y} = \hat{y}(\mathbf{x}, \mathbf{D})$ linear in $\mathbf{x}$,

$$\hat{y} = \hat{\boldsymbol{\beta}}'\mathbf{x}, \qquad \hat{\boldsymbol{\beta}} = \hat{\boldsymbol{\beta}}(\mathbf{D}) . \tag{83}$$

Lindley [43] assumed some sufficient conditions for a squared-error loss under which the personally optimal prediction function $E(y|\mathbf{x}, \mathbf{D})$ is linear in $\mathbf{x}$, namely personal-stochastic independence of $\boldsymbol{\beta}$ and $\mathbf{x}$ given D. See Halpern [56] for implications of the theory in this section for polynomial fitting.

The notation $\hat{\boldsymbol{\beta}}$ anticipates the resulting *estimation* problem, as in (13),

$$\begin{aligned} \Delta(\hat{\boldsymbol{\beta}}, \boldsymbol{\beta}) &= E_{\mathbf{x},\boldsymbol{\theta}|\boldsymbol{\beta},\mathbf{D}} E_{y|\mathbf{x},\boldsymbol{\theta},\mathbf{D}} [(\hat{\boldsymbol{\beta}} - \boldsymbol{\beta})^T \mathbf{x} + (\boldsymbol{\beta}^T\mathbf{x} - y)]^2 \\ &= \mathbf{Q}_{\beta}((\hat{\boldsymbol{\beta}} - \boldsymbol{\beta})) + E_1(\sigma^2|\boldsymbol{\beta}) , \end{aligned} \tag{84}$$

$$\mathbf{Q}_{\beta} = E(\mathbf{x}\mathbf{x}^T|\boldsymbol{\beta}, \mathbf{D}), \qquad (\mathbf{Q}_{\beta} \geqq 0) . \tag{85}$$

It may be of interest, if prediction is not a goal, to omit the term $E_1(\sigma^2|\boldsymbol{\beta})$ and the requirement (85) from the estimation loss (84). If it is desired to estimate only *some* components of $\boldsymbol{\beta}$, say $\boldsymbol{\eta} = \mathbf{C}_H\boldsymbol{\beta}$, then one could choose $\mathbf{Q}_{\beta}$ singular, say $\mathbf{Q}_{\beta} = \mathbf{C}_H^T\mathbf{Q}_{\eta}\mathbf{C}_H$; but the following theory is more easily utilized with a positive- definite quadratic form, say $\mathbf{Q}_{\eta}$ $((\hat{\boldsymbol{\eta}} - \boldsymbol{\eta}))$ itself.

Note that if $\boldsymbol{\beta}$ and $\mathbf{x}$ are (second-order wide-sense) independent given $\mathbf{D}$, then $\mathbf{Q}_{\beta} \equiv \mathbf{Q}$, and the first term $\mathbf{Q}(\hat{\boldsymbol{\beta}} - \boldsymbol{\beta}))$ of Δ is just a constant-coefficient quadratic-error loss for estimation of $\boldsymbol{\beta}$. We shall find that for its effect on the expected penalty $\bar{\Delta}$, this is the most general quadratic estimation-error term.

Define,

$$\mathbf{Q} = E_1\mathbf{Q}_{\beta}\ [= E(\mathbf{x}\mathbf{x}^T|\mathbf{D})] , \tag{86a}$$

$$\bar{\boldsymbol{\beta}} = \mathbf{Q}^{-1}E_1\mathbf{Q}_{\beta}\boldsymbol{\beta}\ [= \mathbf{Q}^{-1}E(\mathbf{x}\mathbf{x}^T\boldsymbol{\beta}|\mathbf{D}) = \mathbf{Q}^{-1}E(\mathbf{x}\mathbf{y}|\mathbf{D})] , \tag{86b}$$

assuming $\mathbf{Q} > 0$, or, if $\mathbf{Q}_{\beta} \equiv \mathbf{Q}$ independent of $\boldsymbol{\beta}$, define $\bar{\boldsymbol{\beta}} = E_1\boldsymbol{\beta}$, whether or not $\mathbf{Q} > 0$.

Proposition 4.1. We have

$$\bar{\Delta}(\hat{\boldsymbol{\beta}}) = \mathbf{Q}((\hat{\boldsymbol{\beta}} - \bar{\boldsymbol{\beta}})) + E_1\mathbf{Q}_{\beta}((\boldsymbol{\beta} - \bar{\boldsymbol{\beta}})) + E_1\sigma^2 . \tag{87}$$

If

$$\hat{\boldsymbol{\beta}} \in H, \qquad \mathbf{Q}((\hat{\boldsymbol{\beta}} - \bar{\boldsymbol{\beta}})) = \mathbf{Q}((\hat{\boldsymbol{\beta}} - B_H^Q\bar{\boldsymbol{\beta}})) + \mathrm{SSB}_H^Q(\bar{\boldsymbol{\beta}}) . \tag{88}$$

Proof. Average $\mathbf{Q}_\beta((\hat{\boldsymbol\beta}-\boldsymbol\beta)) = \mathbf{Q}_\beta((\hat{\boldsymbol\beta}))-2\hat{\boldsymbol\beta}^T\mathbf{Q}_\beta\boldsymbol\beta+\mathbf{Q}_\beta((\boldsymbol\beta))$ term by term, and then complete the square in $\hat{\boldsymbol\beta}$. Note: $\mathbf{Q}((\bar{\boldsymbol\beta})) = \mathbf{Q}^{-1}((E_1\mathbf{Q}_\beta\boldsymbol\beta))$; and in the Hilbert space of random square-integrable r-vector functions of $\boldsymbol\beta$ with inner product $(\mathbf{f}_\beta, \mathbf{g}_\beta) = E_1\mathbf{f}_\beta^T\mathbf{Q}_\beta\,\mathbf{g}_\beta$, every constant vector is orthogonal to $\boldsymbol\beta-\bar{\boldsymbol\beta}$.

The first term of $\bar\Delta$ (87) can be interpreted as the penalty for not choosing $\hat{\boldsymbol\beta} = \hat{\boldsymbol\beta}_* = \bar{\boldsymbol\beta}$, the second term as the penalty for uncertainty concerning $\boldsymbol\beta$, and the third term as a necessity of the statistical prediction model, useful in practice to gage the sizes of the first two terms. Note that the second term has the interpretation, for prediction,

$$\begin{aligned} E_1\mathbf{Q}_\beta((\boldsymbol\beta-\bar{\boldsymbol\beta})) &= E[(\boldsymbol\beta^T\mathbf{x}-\bar{\boldsymbol\beta}^T\mathbf{x})^2|\mathbf{D}] \\ &= \operatorname{tr} E[(\mathbf{x}\mathbf{x}^T)\mathbf{V}(\boldsymbol\beta|\mathbf{x}, \mathbf{D})|\mathbf{D}]\,. \end{aligned} \tag{89}$$

Note, in passing, that if $\boldsymbol\beta$, $\mathbf{x}$ are independent given $\mathbf{D}$, then $\bar{\boldsymbol\beta} = E_1\boldsymbol\beta$, and the second and third terms combine to form the posterior expectation of the personal variance of y given $\mathbf{x}$, $\operatorname{tr}[\mathbf{Q}\mathbf{V}_1(\boldsymbol\beta)]+E_1\sigma^2$. Also, assuming (17), $\mathbf{Q} \doteq n^{-1}\mathbf{N}$.

Corollary 1. The optimal choice for $\hat{\boldsymbol\beta}$ is either

$$\hat{\boldsymbol\beta}_* = \bar{\boldsymbol\beta}, \quad \text{or} \quad \hat{\boldsymbol\beta}_H = B_H^Q\bar{\boldsymbol\beta}\,; \tag{90}$$

for which

$$\bar\Delta(\hat{\boldsymbol\beta}_*) = E_1\mathbf{Q}_\beta((\boldsymbol\beta-\bar{\boldsymbol\beta}))+E_1\sigma^2\,, \tag{91a}$$

and

$$\bar\Delta(\hat{\boldsymbol\beta}_H) = \bar\Delta(\hat{\boldsymbol\beta}_*)+\mathrm{SSB}_H^Q(\bar{\boldsymbol\beta})\,; \tag{91b}$$

the difference, $\mathrm{SSB}_H^Q(\bar{\boldsymbol\beta})$, of which is to be compared with U_H.

Note that, as in Lindley [43, 44], the constraint H will "almost always" be "rejected" if there is no "reward for simplicity", $U_H \to 0$. Compare the much different frequentist results in Anderson et al. [57] and Allen [58, 59].

Corollary 2. If $\boldsymbol\beta$ and $\mathbf{x}$ are independent given $\mathbf{D}$,

$$\hat{\boldsymbol\beta}_* = E_1\boldsymbol\beta = P_1(\bar H)\cdot E_1(\boldsymbol\beta|\bar H)+P_1(H)\cdot E_1(\boldsymbol\beta|H)\,, \tag{92a}$$

$$\hat{\boldsymbol\beta}_H = B_H^Q E_1\boldsymbol\beta = P_1(\bar H)\cdot B_H^Q E_1(\boldsymbol\beta|\bar H)+P_1(H)\cdot E_1(\boldsymbol\beta|H)\,, \tag{92b}$$

$$\begin{aligned} \bar\Delta(\hat{\boldsymbol\beta}_*) &= \operatorname{tr}[\mathbf{Q}\mathbf{V}_1(\boldsymbol\beta)]+E_1\sigma^2 \\ &= P_1(\bar H)\{\operatorname{tr}[\mathbf{Q}\mathbf{V}_1(\boldsymbol\beta|\bar H)]+E_1(\sigma^2|\bar H)\} \\ &\quad +P_1(H)\{\operatorname{tr}[\mathbf{Q}\mathbf{V}_1(\boldsymbol\beta|H)]+E_1(\sigma^2|H)\} \\ &\quad +P_1(H)P_1(\bar H)\cdot\mathbf{Q}((E_1(\boldsymbol\beta|\bar H)-E_1(\boldsymbol\beta|H))), \end{aligned} \tag{93a}$$

$$\bar{\Delta}(\hat{\boldsymbol{\beta}}_H) = \bar{\Delta}(\hat{\boldsymbol{\beta}}_*) + \mathrm{SSB}_H^Q(E_1\boldsymbol{\beta})$$
$$= \bar{\Delta}(\hat{\boldsymbol{\beta}}_*) + P_1(\bar{H})^2 \cdot \mathrm{SSB}_H^Q[E_1(\boldsymbol{\beta}|\bar{H})], \tag{93b}$$

and the optimal choice is $\hat{\boldsymbol{\beta}}_*$ or $\hat{\boldsymbol{\beta}}_H$ according as $P_1(\bar{H})^2 \gtrless U_H/\mathrm{SSB}_H^Q \times [E_1(\boldsymbol{\beta}|\bar{H})]$, namely as

$$L_{y|x}(H)\cdot O_0(H) \gtrless \{\mathrm{SSB}_H^Q[E_1(\boldsymbol{\beta}|\bar{H})]/U_H\}^{\frac{1}{2}} - 1 . \tag{94}$$

If $P_1(H) = 0$, the expressions simplify.

Now, for the Normal model and prior family introduced in sections 2 and 3, we have equations (3), (4), (61) for the posterior probabilities of H and $\bar{H}$, equations (34) for the posterior moments of $\boldsymbol{\beta}$ and σ conditional on $\bar{H}$, and the following for the moments conditional on H.

Proposition 4.2. The analogues of equations (31)–(34) and (41b)' conditioned on H, follow.

$$p_1(\boldsymbol{\zeta}|\omega, H) = h_{\nu_*+q}^{(r-q)}[\boldsymbol{\zeta} - \mathbf{C}_{H_*}B_H^{N_\omega}\mathbf{b}_\omega;\; s_{\omega H}^2(\tilde{\mathbf{N}}_\omega^{-1})_{(22.1)}], \tag{95a}$$

$$p_1(\sigma|\omega, H) = k_{\nu_*+q}(\sigma;\, \omega^{-\frac{1}{2}}s_{\omega H}), \tag{95b}$$

$$p_1(\omega|H) = |\tilde{\mathbf{N}}_{\omega(22)}|^{-\frac{1}{2}}\omega^{\frac{1}{2}(\nu_1+r)-1}(1-\omega)^{\frac{1}{2}(\nu'+r)-1}$$
$$\times[(\nu_*+q)s_{\omega H}^2]^{-\frac{1}{2}(\nu_*+q)}/C_{*H}, \tag{96}$$

$$p_1(\sigma|\tau, H) = |\tilde{\mathbf{N}}_{\omega(22)}|^{-\frac{1}{2}}(\sigma^2)^{-\frac{1}{2}(\nu_1+q+1)}(\sigma^2+\tau^2)^{-\frac{1}{2}(r-q)}$$
$$\times \exp\{-\tfrac{1}{2}(\sigma^{-2}+\tau^{-2})[(\nu_*+q)s_{\omega H}^2$$
$$-(1-\omega)\nu's'^2]\}/C_{**H}, \tag{97}$$

where

$$\mathbf{C}_{H*}B_H^{N_\omega}\mathbf{b}_\omega = \mathbf{C}_{H*}\mathbf{b}_\omega + (\mathbf{C}_{H*}\mathbf{N}_\omega^{-1}\mathbf{C}_H^T)(\mathbf{C}_H\mathbf{N}_\omega^{-1}\mathbf{C}_H^T)^{-1}(\boldsymbol{\eta}_H - \mathbf{C}_H\mathbf{b}_\omega)$$
$$= \omega(\tilde{\mathbf{N}}_\omega^{-1})_{(22.1)}\tilde{\mathbf{N}}_{(22)}\mathbf{C}_{H*}(B_H^N\mathbf{b} - B_H^{N'}\mathbf{b}') + \mathbf{C}_{H*}B_H^{N'}\mathbf{b}' \tag{98a}$$

where $\tilde{\mathbf{N}} = \mathbf{C}^{-T}\mathbf{N}\mathbf{C}^{-1}$ and similarly for $\tilde{\mathbf{N}}'$ and $\tilde{\mathbf{N}}_\omega$, and where

$$(\nu_*+q)s_{\omega H}^2 = \nu_* s_\omega^2 + \mathrm{SSB}_H^{N_\omega}(\mathbf{b}_\omega)$$
$$= \omega\{\mathrm{SSW} + \nu_0 s_0^2 + \mathrm{SSB}\}$$
$$+(1-\omega)\{\nu's'^2 + \mathrm{SSB}_H^{N'}(\mathbf{b}')\}$$
$$+\omega(1-\omega)[\tilde{\mathbf{N}}_{(22)}(\tilde{\mathbf{N}}_\omega^{-1})_{(22.1)}\tilde{\mathbf{N}}'_{(22)}]$$
$$\times((\mathbf{C}_{H*}B_H^N\mathbf{b} - \mathbf{C}_{H*}B_H^{N'}\mathbf{b}')); \tag{98b}$$

hence

$$E_1(\zeta|H) = E_1(\mathbf{C}_{H*}B_H^{N_\omega}\,\mathbf{b}_\omega|H), \tag{99a}$$

$$\mathbf{V}_1(\zeta|H) = \mathbf{V}_1(\mathbf{C}_{H*}B_H^{N_\omega}\mathbf{b}_\omega|H) + (\nu_*+q-2)^{-1}E_1[(\nu_*+q)s_{\omega H}^2\tilde{\mathbf{N}}_{\omega(22.1)}^{-1}|H], \tag{99b}$$

$$E_1(\sigma^2|H) = (\nu_*+q-2)^{-1}E_1[(\nu_*+q)s_{\omega H}^2/\omega|H], \tag{100a}$$

and

$$E_1(\boldsymbol{\beta}|H) = \mathbf{C}^{-1}[\boldsymbol{\eta}_H^T, E_1(\zeta|H)^T]^T \tag{100b}$$

and

$$\mathbf{V}_1(\boldsymbol{\beta}|H) = \mathbf{C}^{-1}\{\mathbf{Diag}[\mathbf{0}, \mathbf{V}_1(\zeta|H)]\}\mathbf{C}^{-T}. \tag{100c}$$

Proof. By Savage's Density Condition for the posterior distribution P_1, obtain the conditional density $p_1(\zeta, \sigma, \tau|H)$ from (27) by again completing the quadratic form in $\boldsymbol{\beta}$, as in the proof of proposition 2.1, and by then writing for $\boldsymbol{\beta} \in H$, $\mathbf{N}_\omega((\boldsymbol{\beta}-\mathbf{b}_\omega)) = \tilde{\mathbf{N}}_{\omega(22)}((\zeta-\mathbf{C}_{H*}B_H^{N_\omega}\mathbf{b}_\omega))+\mathrm{SSB}_H^{N_\omega}(\mathbf{b}_\omega)$, as in (88). Alternatively, condition on H both the prior density (26) and the likelihood function (19); then apply Bayes' theorem, thus obtaining the second equalities in (98a), (98b).

Also, we have, in generalization of (38), the conditional density $p_1(\zeta|H)$ given by (36) with r replaced by $r-q$; with $\mathbf{b}$, $\mathbf{N}$, ν_1, s_1^2 replaced by $\mathbf{C}_{H*}B_H^N\mathbf{b}$, $\tilde{\mathbf{N}}_{(22)}$, ν_1+q, $(\nu_1+q)^{-1}[\mathrm{SSW}+\nu_0 s_0^2+\mathrm{SSB}]$; and with $\mathbf{b}'$, $\mathbf{N}'$, ν', s'^2 replaced by

$$\mathbf{C}_{H*}B_H^{N'}\mathbf{b}', \quad \tilde{\mathbf{N}}'_{(22)}, \quad \nu'+q, \quad (\nu'+q)^{-1}[\nu's'^2+\mathrm{SSB}_H^{N'}(\mathbf{b}')].$$

Corollary. The numerator in proposition 3.3 for $L_{y|X}(H)$ takes the form,

$$p(\mathbf{y}|\mathbf{X}, H) = [B(\tfrac{1}{2}\nu_0, \tfrac{1}{2}\nu'+\tfrac{1}{2}q; \tfrac{1}{2}\mathbf{1}_n)]^{-1}(\nu_0 s_0^2)^{\frac{1}{2}\nu_0} \times [\nu's'^2+\mathrm{SSB}_H^{N'}(\mathbf{b}')]^{\frac{1}{2}(\nu'+q)}|\tilde{\mathbf{N}}'_{(22)}|^{\frac{1}{2}}\cdot C_{*H}. \tag{101}$$

Proof. This is essentially a lower-dimensional version of proposition 3.3. It can be checked that $L_{y|X}(H)$ (61) is equal to the ratio of (101) to (71).

Context 1. *Student's hypothesis*. For $\Delta(\hat{\mu}, \mu) = Q(\hat{\mu}-\mu)^2$, we have

$$\hat{\mu}_H = \mu_H,$$

$$\hat{\mu}_* = P_1(\overline{H})E_1(\mu|\overline{H})+P_1(H)\cdot\mu_H,$$

$$\bar{\Delta}(\hat{\mu}_*) = P_1(\bar{H})[QV_1(\mu|\bar{H})+E_1(\sigma^2|\bar{H})]$$
$$+P_1(H)\{E_1(\sigma^2|H)+P_1(\bar{H})Q[E_1(\mu|\bar{H})-\mu_H]^2\}\,,$$
$$\bar{\Delta}(\hat{\mu}_H) = \bar{\Delta}(\hat{\mu}_*)+P_1(\bar{H})^2 Q[E_1(\mu|\bar{H})-\mu_H]^2\,.$$

Hence the optimal choice is $\hat{\mu}_*$ or $\hat{\mu}_H$, chosen according to

$$L_D(H)\cdot O_0(H) \lesseqgtr |E_1(\mu|\bar{H})-\mu_H|/(U_H/Q)^{\frac{1}{2}}-1\,, \qquad (102)$$

which becomes for $P_1(H) = 0$, $|E_1(\mu)-\mu_H| \gtrless (U_H/Q)^{\frac{1}{2}}$. We have

$$E_1(\mu|\bar{H}) = \bar{y}+E_1[(1-\omega)/(n\omega+1-\omega)|\bar{H}](\bar{y}'-\bar{y})\,,$$
$$V_1(\mu|\bar{H}) = V_1[(1-\omega)/(n\omega+1-\omega)|\bar{H}](\bar{y}-\bar{y}')^2$$
$$+(\nu_*-2)^{-1}E_1[\nu_* s_\omega^2/(n\omega+1-\omega)|\bar{H}]\,,$$
$$E_1(\sigma^2|\bar{H}) = (\nu_*-2)^{-1}E_1[\nu_* s_\omega^2/\omega|\bar{H}]\,,$$

where

$$\nu_* = n+\nu_0+\nu'\,,$$
$$\nu_* s_\omega^2 = \omega(\nu s^2+\nu_0 s_0^2)+(1-\omega)\nu' s'^2$$
$$+\omega(1-\omega)(n\omega+1-\omega)^{-1}n(\bar{y}-\bar{y}')^2\,.$$
$$p_1(\omega|\bar{H}) = (n\omega+1-\omega)^{-\frac{1}{2}}\omega^{\frac{1}{2}(n+\nu_0)-1}(1-\omega)^{\frac{1}{2}(\nu'+1)-1}$$
$$\times(\nu_* s_\omega^2)^{-\frac{1}{2}\nu_*}/C_*\,,$$

or for $\nu' \to \infty$,

$$p_1(\sigma|\tau,\bar{H}) = (\sigma^2+n\tau^2)^{-\frac{1}{2}}(\sigma^2)^{-\frac{1}{2}(n+\nu_0)}$$
$$\times\exp\{-\tfrac{1}{2}[\sigma^{-2}(\nu s^2+\nu_0 s_0^2)$$
$$+(\sigma^2+n\tau^2)^{-1}(\bar{y}-\bar{y}')^2]\}/C_{**}\,.$$

If $P_1(H) > 0$, we require:

$$E_1(\sigma^2|H) = (\nu_*-1)^{-1}E_1[(\nu_*+1)s_{\omega H}^2/\omega|H]\,,$$
$$p_1(\omega|H) = \omega^{\frac{1}{2}(n+\nu_0)-1}(1-\omega)^{\frac{1}{2}(\nu'+1)-1}$$
$$\times[(\nu_*+1)s_{\omega H}^2]^{-\frac{1}{2}(\nu_*+1)}/C_{*H}\,,$$

where

$$(\nu_*+1)s_{\omega H}^2 = \omega[\nu s^2+\nu_0 s_0^2+n(\bar{y}-\mu_H)^2]$$
$$+(1-\omega)[\nu' s'^2+(\bar{y}'-\mu_H)^2]\,;$$

or for $\nu' \to \infty$,

$$p_1(\sigma|\tau,H) = k_{n+\nu_0}\{\sigma;\,(n+\nu_0)^{-\frac{1}{2}}[\nu s^2+\nu_0 s_0^2+n(\bar{y}-\mu_H)^2]^{\frac{1}{2}}\}\,,$$

as in (45a).

Context 2. *The hypothesis of given effects.* If $\mathbf{X}$ and $\mathbf{x}$ have an underlying Normal statistical model (16a), then by (17), approximately,

$$\mathbf{Q} \doteq n^{-1}\mathbf{X}^T\mathbf{X} = n^{-1}\mathbf{N}+\bar{\mathbf{x}}\bar{\mathbf{x}}^T . \tag{103}$$

There seems to be no additional simplification of the preceding general theory for this context; but for known σ^2, Normal prior $\boldsymbol{\beta}$ stochastically independent of $\mathbf{X}$, $\mathbf{x}$, and $P_1(H) = 0$, see Lindley [43, 44].

Obtain, with the Context-modified definition of C_{*H},

$$\begin{aligned} p(\mathbf{y}|\mathbf{X}, H) \doteq\; & p_0(\mu|\bar{H})n^{-\frac{1}{2}}[B(\tfrac{1}{2}\nu_0, \tfrac{1}{2}\nu'+\tfrac{1}{2}q-\tfrac{1}{2}; \tfrac{1}{2}\mathbf{I}_{n-2})]^{-1} \\ & \times(\nu_0 s_0^2)^{\frac{1}{2}\nu_0}[\nu' s'^2+\mathrm{SSB}_H^{N'}(\mathbf{0})]^{\frac{1}{2}(\nu'+q-1)} \\ & \times(q/r)^{\frac{1}{2}}\cdot C_{*H} . \end{aligned} \tag{104}$$

Context 3. *Equality of group means.* For $\mathbf{Q} = Q\mathbf{I}_r$, we have

$$\Delta(\hat{\boldsymbol{\beta}}, \boldsymbol{\beta}) = Q\sum_1^r(\hat{\beta}^{(j)}-\beta^{(j)})^2 , \qquad \hat{\beta}_H^{(j)} = \bar{z}^{(\cdot)},$$

$$\hat{\beta}_*^{(j)} = \bar{z}^{(\cdot)}+P_1(\bar{H})E_1[m\omega/(m\omega+1-\omega)|\bar{H}](\bar{z}^{(j)}-z^{(\cdot)}),$$

$$\begin{aligned} &\mathrm{tr}\,[\mathbf{QV}_1(\boldsymbol{\beta}|\bar{H})] \\ &\quad= Qm^{-1}\{V_1[m\omega/(m\omega+1-\omega)|\bar{H}]\mathrm{SSB} \\ &\qquad+(\nu_*-2)^{-1}E_1[(\nu_* s_\omega^2/\omega)(rm\omega+1-\omega)/(m\omega+1-\omega)|\bar{H}]\}, \end{aligned}$$

$$\begin{aligned} &\mathrm{tr}\,[\mathbf{QV}_1(\boldsymbol{\beta}|H)] \\ &\quad= Qm^{-1}(\nu_*+r-3)^{-1}E_1[(\nu_*+r-1)s_{\omega H}^2/\omega|H] , \end{aligned}$$

where

$$(\nu_*+r-1)s_{\omega H}^2 = \omega(\mathrm{SSW}+\nu_0 s_0^2+\mathrm{SSB})+(1-\omega)\nu' s'^2 ,$$

and in $p_1(\omega|H)$, $\tilde{\mathbf{N}}_{\omega(22)} = rm\omega$. Most importantly,

$$\begin{aligned} \mathrm{SSB}_H^Q[E_1(\boldsymbol{\beta}|\bar{H})] &= \mathbf{Q}((E_1(\boldsymbol{\beta}|\bar{H})-E_1(\boldsymbol{\beta}|H))) \\ &= Qm^{-1}\{E_1[m\omega/(m\omega+1-\omega)|\bar{H}]\}^2\mathrm{SSB} . \end{aligned} \tag{105}$$

Of greater practical interest than equation (58) of $p_1(\boldsymbol{\beta}^{(\cdot)}|\omega, \bar{H})$, is

$$p_1(\beta^{(\cdot)}|\omega, H) = h_{\nu_*+r-1}(\beta^{(\cdot)}-\bar{z}^{(\cdot)}; s_{\omega H}^2/(rm)) . \tag{106}$$

5. *Error-threshold estimation loss*

The *following* theory does not literally require the Normal model, but merely smoothness of the posterior densities. The *prediction* problem seems intractible for our prior family and will not be addressed.

Consider the *estimation*-error-threshold loss,

$$\Delta(\hat{\boldsymbol{\beta}}, \boldsymbol{\beta}) = \begin{cases} 0 & \text{if } \boldsymbol{\beta} \in K(\hat{\boldsymbol{\beta}}) \\ U_K & \text{otherwise} \quad (U_k > 0) \end{cases}$$
$$= -U_K \delta_{K(\beta)}(\boldsymbol{\beta}) + U_K, \tag{107}$$

where $K(\hat{\boldsymbol{\beta}})$ is a neighborhood of $\boldsymbol{\beta}$ values "centered" at $\hat{\boldsymbol{\beta}}$ having r-dimensional volume $\mathrm{Vol}_\beta(K)$. Take for $K(\hat{\boldsymbol{\beta}})$ an ellipsoidal solid,

$$K(\hat{\boldsymbol{\beta}}) = \{\boldsymbol{\beta}: \mathbf{Q}((\boldsymbol{\beta}-\hat{\boldsymbol{\beta}})) \leqq 1\}, \qquad (r \times r)\mathbf{Q} > 0\,. \tag{108}$$

Then

$$\mathrm{Vol}_\beta(K) = B(1; \tfrac{1}{2}\mathbf{l}_r) \cdot |\mathbf{Q}|^{-\frac{1}{2}} \tag{109}$$

[60, p. 44].

We consider estimation problems requiring high accuracy relative to the degree of posterior certainty; in other words, long shots.

Proposition 5.1.

$$\begin{aligned} \bar{\Delta}(\hat{\boldsymbol{\beta}})/U_K &= -P_1\{\boldsymbol{\beta} \in K\} + 1 \\ &= -[P_1(\bar{H})P_1(K|\bar{H}) + P_1(H)P_1(K|H)] + 1 \\ &\doteq -P_1(\bar{H})p_1(\boldsymbol{\beta}|\bar{H})_\beta \cdot \mathrm{Vol}_\beta(K) \\ &\quad -\delta_H(\hat{\boldsymbol{\beta}})P_1(H)p_1(\zeta|H)_{\zeta(\beta)} \cdot \mathrm{Vol}_\zeta(K \cap H) + 1\,. \end{aligned} \tag{110}$$

The approximation applies for small K (large latent roots of $\mathbf{Q}$) and for $\hat{\boldsymbol{\beta}}$ either in H or beyond K from H (all other $\hat{\boldsymbol{\beta}}$, obviously, will be nonoptimal, with increasing prior probability for decreasing K). $\mathrm{Vol}_\zeta(K \cap H)$ denotes the $(r-q)$-dimensional volume of $K(\hat{\boldsymbol{\beta}}) \cap H$ in the coordinates of $\boldsymbol{\zeta} = \mathbf{C}_{H*}\boldsymbol{\beta}$; then

$$\mathrm{Vol}_\zeta(K \cap H) = B(1; \tfrac{1}{2}\mathbf{1}_{r-q}) \cdot |\tilde{\mathbf{Q}}_{(22)}|^{-\frac{1}{2}}, \quad \text{for} \quad \hat{\boldsymbol{\beta}} \in H, \tag{111}$$

where

$$\tilde{\mathbf{Q}}^{-1} = \mathbf{C}\mathbf{Q}^{-1}\mathbf{C}^T = (\mathbf{C}_H^T, \mathbf{C}_{H*}^T)^T \mathbf{Q}^{-1} (\mathbf{C}_H^T, \mathbf{C}_{H*}^T).$$

Proof of (111). In the ζ-coordinates, for $\hat{\boldsymbol{\beta}} \in H$,

$$K(\hat{\boldsymbol{\beta}}) \cap H = \{\zeta(\boldsymbol{\beta}): \tilde{\mathbf{Q}}((\mathbf{C}\beta - \mathbf{C}\hat{\boldsymbol{\beta}})) \leqq 1, \quad \zeta(\boldsymbol{\beta}) = \mathbf{C}_{H^*}\boldsymbol{\beta},$$

$$\boldsymbol{\eta}(\boldsymbol{\beta}) \equiv \mathbf{C}_H\boldsymbol{\beta} = \mathbf{C}_H\hat{\boldsymbol{\beta}} = \boldsymbol{\eta}_H\}$$

$$= \{\zeta: \tilde{\mathbf{Q}}_{22}((\zeta - \hat{\zeta})) \leqq 1\}, \qquad \hat{\zeta} = \zeta(\hat{\boldsymbol{\beta}}).$$

Note:

$$|\tilde{\mathbf{Q}}_{(22)}| = |\mathbf{C}|^{-2} \cdot |\mathbf{Q}| \cdot |\mathbf{C}_H \mathbf{Q}^{-1} \mathbf{C}_H^T|,$$

since

$$|\tilde{\mathbf{Q}}| = |\tilde{\mathbf{Q}}_{(22)}| \cdot |\tilde{\mathbf{Q}}_{(11.2)}| \quad \text{and} \quad \tilde{\mathbf{Q}}_{(11.2)} = [(\tilde{\mathbf{Q}}^{-1})_{(11)}]^{-1}.$$

From $p_1(\boldsymbol{\beta}|\bar{H}) \propto p_1(\boldsymbol{\eta}|\bar{H})p_1(\zeta|\boldsymbol{\eta}, \bar{H})$ and $p_1(\zeta|\boldsymbol{\eta}_H, \bar{H}) = p_1(\zeta|H)$ (the posterior version of (62b), note that $p_1(\boldsymbol{\beta}|\bar{H})$ attains its constrained maximum over H simultaneously with $p_1(\zeta|H)$ $(\zeta = \mathbf{C}_{H^*}\beta)$.

Corollary. An approximately optimal choice for $\hat{\boldsymbol{\beta}}$, is either

$$\hat{\boldsymbol{\beta}}^*, \quad \text{where} \quad \max p_1(\boldsymbol{\beta}|\bar{H}) = p_1(\boldsymbol{\beta}|\bar{H})_{\beta^*} \quad \text{with} \quad \hat{\boldsymbol{\beta}}^* \in \bar{H} \tag{112a}$$

or

$$\hat{\boldsymbol{\beta}}_H, \quad \text{where} \quad \max{}_H p_1(\boldsymbol{\beta}|\bar{H}) = p_1(\boldsymbol{\beta}|\bar{H})_{\beta_H} \quad \text{with} \quad \hat{\boldsymbol{\beta}}_H \in H, \tag{112b}$$

for which

$$\bar{\Delta}(\hat{\boldsymbol{\beta}}^*)/U_K \doteq -P_1(\bar{H})p_1(\boldsymbol{\beta}|\bar{H})_{\beta^*} \cdot \mathrm{Vol}_{\beta}(K) + 1 \tag{113a}$$

$$\bar{\Delta}(\hat{\boldsymbol{\beta}}_H)/U_K \doteq \bar{\Delta}(\hat{\boldsymbol{\beta}}^*)/U_K$$
$$+ P_1(\bar{H})[p_1(\boldsymbol{\beta}|\bar{H})_{\beta^*} - p_1(\boldsymbol{\beta}|\bar{H})_{\beta_H}]\mathrm{Vol}_{\beta}(K)$$
$$- P_1(H)p_1(\zeta|\boldsymbol{\eta}_H, \bar{H})_{\zeta(\beta_H)}\mathrm{Vol}_{\zeta}(K \cap H). \tag{113b}$$

The choice is $\hat{\boldsymbol{\beta}}^*$ or $\hat{\boldsymbol{\beta}}_H$ according as $\bar{\Delta}(\hat{\boldsymbol{\beta}}_H)/U_K - \bar{\Delta}(\hat{\boldsymbol{\beta}}^*)/U_K \gtrless U_H/U_K$ (depending on U_H and U_K only through their ratio), namely approximately as

$$L_D(H) \cdot O_0(H) \lesseqgtr \{[p_1(\boldsymbol{\beta}|\bar{H})_{\beta^*} - p_1(\boldsymbol{\beta}|\bar{H})_{\beta_H}]\mathrm{Vol}_{\beta}(K) - U_H/U_K\}$$
$$/\{p_1(\zeta|\boldsymbol{\eta}_H, \bar{H})_{\zeta(\beta_H)}\mathrm{Vol}_{\zeta}(K \cap H) + U_H/U_K\}. \tag{114}$$

Note that the mode $\hat{\boldsymbol{\beta}}^*$ (with superior asterisk) is defined (112a) to yield an approximate constrained minimum of $\bar{\Delta}(\hat{\boldsymbol{\beta}})$ for $K(\hat{\boldsymbol{\beta}}) \cap H = \emptyset$. Either $\hat{\boldsymbol{\beta}}_* = \hat{\boldsymbol{\beta}}^*$ $(K(\hat{\boldsymbol{\beta}}_*) \cap H = \emptyset)$, or $\hat{\boldsymbol{\beta}}_* = \hat{\boldsymbol{\beta}}_H$ $(\hat{\boldsymbol{\beta}}_* \in H)$, regardless of U_H. Hence the constraint H can sometimes be "accepted" even if there is no "reward for simplicity", $U_H \to 0$. This partially answers the challenge raised by Lindley [44, pp. 46 and 115]. This holds only for $P_1(H) > 0$, as does the need for $K(\hat{\boldsymbol{\beta}}) \cap H = \emptyset$. The criterion (114) is an approximation to $U_K P_1[K(\hat{\boldsymbol{\beta}}^*)] - U_K P_1[K(\hat{\boldsymbol{\beta}}_H)] \gtrless U_H$. In case $P_1(H) = 0$, it takes the simple form, $[p_1(\boldsymbol{\beta})_{\hat{\boldsymbol{\beta}}^*} - p_1(\boldsymbol{\beta})_{\hat{\boldsymbol{\beta}}_H}] \cdot \mathrm{Vol}_{\beta}(K) \gtrless U_H/U_K$.

The mode $\hat{\boldsymbol{\beta}}^*$ was developed in proposition 2.3 for our family of posterior densities. The solution to the constrained maximization problem, $\max_H p_1(\boldsymbol{\beta}|\bar{H})$ for $\hat{\boldsymbol{\beta}}_H$, follows.

Proposition 5.2. The H-constrained local maxima of the density function $p_1(\boldsymbol{\beta}|\bar{H})$, (36) and (41c), are attained on the curve of projections $B_H^{N\omega}\mathbf{b}_\omega$ of the curve décolletage $\mathbf{b}_\omega$ (29b), (35): for some ω, $\hat{\boldsymbol{\beta}}_H = B_H^{N\omega}\mathbf{b}_\omega$ and $\boldsymbol{\zeta}(\hat{\boldsymbol{\beta}}_H) = \mathbf{C}_{H*}\, B_H^{N\omega}\mathbf{b}_\omega$ as given in (98a).

Proof. Apply proposition 2.3 to the conditional density $p_1(\boldsymbol{\zeta}|\boldsymbol{\eta}_H, \bar{H}) = p_1(\boldsymbol{\zeta}|H)$, given below the proof of proposition 4.2.

Hence, the operation $\max_H p_1(\boldsymbol{\beta}|\bar{H})$ for $\hat{\boldsymbol{\beta}}_H$ likewise can be carried out by a one-dimensional search using $p_1(\boldsymbol{\beta}|\bar{H})$, (36) and (41c). Then one can evaluate

$$p_1(\boldsymbol{\zeta}|H)_{\boldsymbol{\zeta}(\hat{\boldsymbol{\beta}}_H)} = p_1(\boldsymbol{\beta}|\bar{H})_{\hat{\boldsymbol{\beta}}_H}|\mathbf{C}|^{-1}/p_1(\boldsymbol{\eta}|\bar{H})_{\boldsymbol{\eta}_H}, \tag{115}$$

the denominator of which was developed (65) for use in the weighted likelihood ratio.

The approximate local probabilities in this section can be improved by using also second derivatives of the posterior density functions to obtain Normal local approximations.

Context 1. *Student's hypothesis.* For $\Delta(\hat{\mu}, \mu) = 0$ if $|\mu - \hat{\mu}| < 1/Q^{\frac{1}{2}}$ and $\Delta = U_K$ otherwise, we have $\hat{\mu}_H = \mu_H$, and $\hat{\mu}^*$ maximizes

$$p_1(\mu|\bar{H}) = h_{\nu_1}(\mu - \bar{y}; s_1^2/n) \cdot h_{\nu'}(\mu - \bar{y}'; s'^2) \\ /h_{\nu_1,\nu'}(\bar{y} - \bar{y}'; s_1^2/n, s'^2).$$

We have

$$\begin{aligned} \bar{\Delta}(\hat{\mu}^*)/U_K &\doteq -P_1(\bar{H})p_1(\mu|\bar{H})_{\hat{\mu}_*} 2/Q^{\frac{1}{2}} + 1, \\ \bar{\Delta}(\hat{\mu}_H)/U_K &\doteq \bar{\Delta}(\hat{\mu}^*)/U_K \\ &\quad + P_1(\bar{H})[p_1(\mu|\bar{H})_{\hat{\mu}_*} - p_1(\mu|\bar{H})_{\mu_H}]2/Q^{\frac{1}{2}} \\ &\quad - P_1(H)\,. \end{aligned}$$

Hence, the approximately optimal choice is $\hat{\mu}^*$ or $\hat{\mu}_H$ according to

$$L_D(H)\cdot O_0(H) \lesseqgtr \{[p_1(\mu|\bar{H})_{\hat{\mu}_*} - p_1(\mu|\bar{H})_{\mu H}]2/Q^{\frac{1}{2}} - U_H/U_K\} \\ /\{1+U_H/U_K\}\,, \tag{116}$$

which becomes for

$$P_1(H) = 0, \qquad [p_1(\mu|\bar{H})_{\hat{\mu}_*} - p_1(\mu|\bar{H})_{\mu H}] \gtrless U_H/(U_K Q^{\frac{1}{2}}/2).$$

Context 2. *Given effects.* Obtain $\hat{\boldsymbol{\beta}}^*$ by maximizing in ω, $p_1(\boldsymbol{\beta}|\bar{H})$ evaluated by $\mathbf{b}_\omega$, as given by (36) with ν' replaced by $\nu'-1$, n replaced by $n-1$, $\mathbf{N}((\mathbf{b}_\omega-\mathbf{b})) = (1-\omega)^2\mathbf{N}((\mathbf{N}_\omega^{-1}\,(\mathbf{b}-b^{(\cdot)}\mathbf{1}_r)))$, $\mathbf{N}'((\mathbf{b}_\omega-\mathbf{b}')) = \omega^2\mathbf{N}'((\mathbf{N}_\omega^{-1}\mathbf{N}\,\mathbf{b}))$, and with C_* given by (32) with $\nu_* = n-1+\nu_0+\nu'-1$. Obtain $\hat{\boldsymbol{\beta}}_H$ by maximizing in ω, $p_1(\boldsymbol{\beta}|\bar{H})$ evaluated at

$$B_H^{N_\omega}\mathbf{b}_\omega = \{\boldsymbol{\beta}_{(1)H}^T,\,[\mathbf{b}_\omega^{(2)} - (\mathbf{N}_{\omega(22)})^{-1}\mathbf{N}_{\omega(21)}(\boldsymbol{\beta}_{(1)H} - \mathbf{b}_{\omega(1)})]^T\}^T.$$

Then

$$p_1(\boldsymbol{\beta}_{(2)}|H)_{\beta(2)H} = p_1(\boldsymbol{\beta}|\bar{H})_{\beta H}/p_1(\boldsymbol{\beta}_{(1)}|\bar{H})_{\beta(1)H}$$

$$(\text{since } |\mathbf{C}| = |\mathbf{I}_r| = 1),$$

in which the denominator is the numerator in $L_{y|X}(H)$ below (74). In (111) for $\mathrm{Vol}_\zeta(K\cap H)$, $\tilde{\mathbf{Q}}_{(22)} = \mathbf{Q}_{(22)}$.

Context 3. *Equality of group means.* Obtain $\hat{\boldsymbol{\beta}}^*$ by maximizing in ω, $p_1(\boldsymbol{\beta}|\bar{H})$ evaluated at $\mathbf{b}_\omega$, as given by (36) with

$$\mathbf{N}((\mathbf{b}_\omega-\mathbf{b})) = [(1-\omega)/(m\omega+1-\omega)]^2\mathrm{SSB},$$

$$\mathbf{N}'((\mathbf{b}_\omega-\mathbf{b}')) = [m\omega/(m\omega+1-\omega)]^2 m^{-1}\,\mathrm{SSB},$$

and C_* as given by (32), all with ν' replaced by $\nu'-1$ and with the substitutions immediately preceding equation (58). The constrained estimate $\hat{\boldsymbol{\beta}}_H = \bar{z}^{(\cdot)}\mathbf{1}_r$, for which $p_1(\boldsymbol{\beta}|\bar{H})_{\beta H}$ has $\mathbf{N}((\hat{\boldsymbol{\beta}}_H-\mathbf{b})) = \mathrm{SSB}$ and $\mathbf{N}'((\hat{\boldsymbol{\beta}}_H-\mathbf{b}')) = 0$. Then $p_1(\boldsymbol{\beta}^{(\cdot)}|H)_{\bar{z}(.)} = p_1(\boldsymbol{\beta}|\bar{H})_{\beta H}\cdot r/p_1(\boldsymbol{\eta}|\bar{H})_0$, in which the denominator is (78). The choice $\mathbf{Q} = \mathbf{Q}\mathbf{I}_r$ implies in (109) for $\mathrm{Vol}_\beta(K)$, $|\mathbf{Q}| = Q^r$; and in (111) for $\mathrm{Vol}_\zeta(K\cap H)$, $q = r-1$, $|\tilde{\mathbf{Q}}_{(22)}| = Q\cdot r$.

Acknowledgement

This paper was much improved by the suggestions of various persons.

References

[58] David M. Allen, Mean square error of prediction as a criterion for selecting variables. Technometrics 13 (1971), 469–475.

[59] David M. Allen, The prediction sum of squares as a criterion for selecting predictor variables. Technical Report Number 23, Department of Statistics, University of Kentucky, Lexington, Kentucky, 1971.

[57] R. L. Anderson, D. M. Allen and F. B. Cady, Selection of predictor variables in linear multiple regression. Technical Report Number 5, Dept. of Statistics, Univ. of Kentucky, Lexington, Kentucky, 1970.

[11] F. J. Anscombe, Bayesian inference concerning many parameters with reference to supersaturated designs. Bull. Int. Statist. Inst. 40 (1963), 721–733.

[61] F. J. Anscombe, Topics in the investigation of linear relations fitted by the method of least squares. J. Roy. Statist. Soc. B 29 (1967), 1–52.

[62] M. S. Bartlett, A comment on D. V. Lindley's statistical paradox. Biometrika 44 (1957), 533–534.

[46] G. E. P. Box and George C. Tiao, Bayesian estimation of means for the random effect model. J. Amer. Statist. Assoc. 63 (1968), 174–181.

[50] G. E. P. Box and George C. Tiao, Bayesian inference in statistical analysis. Reading, Addison-Wesley, 1973.

[21] Jerome Cornfield, Bayesian estimation for higher order cross classifications. Millbank Memorial Quarterly, Oct. 1970, Part II, 57–70.

[22] Jerome Cornfield, A Bayesian approach to the one-way classification problem. Unpublished notes, 18 May 1971.

[23] Jerome Cornfield, RBO's for breakdowns. Handout for Advanced Science Seminar, Fort Collins, Colorado, 12 July 1971.

[37] Jerome Cornfield, A Bayesian test of some classical hypotheses—with applications to sequential clinical trials. J. Amer. Statist. Assoc. 61 (1966), 577–594.

[38] Jerome Cornfield, The Bayesian outlook and its application. Biometrics 25, No. 4 (1969), 617–657.

[2] Bruno DeFinetti, Foresight: its logical laws, its subjective sources. Reprinted in: Studies in subjective probability, edited by Henry E. Kyburg, Jr., and Howard E. Smokler, New York, Wiley, 1937.

[55] Morris H. DeGroot, Optimal statistical decisions. New York, McGraw-Hill, 1970.

[60] A. P. Dempster, Elements of continuous multivariate analysis. Reading, Addison-Wesley, 1969.

[18] James M. Dickey, Three multidimensional-integral identities with Bayesian applications. Ann. Math. Stat. 39 (1968), 1615–1627.

[19] James Dickey, Scientific reporting and personal probabilities: Student's hypothesis. Research Report 47. Statistics Department, State University of New York at Buffalo, 1970. J. Roy. Statist. Soc. B. (1973), part 2. (Reprinted in the present volume).

[20] James Dickey, The weighted likelihood ratio, linear hypotheses on normal location parameters. Ann. Math. Statist. 42 (1971), 204–223.

[39] James M. Dickey and B. P. Lientz, The weighted likelihood ratio, sharp hypotheses about chances, the order of a Markov chain. Ann. Math. Statist. 41 (1970), 214–226.

[42] James M. Dickey, A Bayesian hypothesis-decision procedure. Ann. Inst. of Stat. Math. 19 (1967), 367–369.

[54] James M. Dickey, Matricvariate generalizations of the multivariate t distribution and the inverted multivariate t distribution. Ann. Math. Statist. 38 (1967), 511–518.

[8] Ward Edwards, Harold Lindman and Leonard J. Savage, Bayesian statistical inference for psychological research. Psychological Review 70 (1963), 193–242. Reprinted in: Readings in mathematical psychology, edited by Duncan R. Luce, Robert R. Bush and Eugene Galanter. New York, Wiley, 1965, 2, 519–568.

[9] Bradley Efron, Comments on Blyth's paper. Ann. Math. Statist. 41 (1970), 1049–1054.

[28] Bradley Efron and Carl Morris, Limiting the risk of Bayes and empirical Bayes estimators—Part I: the Bayes case. J. Amer. Statist. Assoc. 66 (1971), 807–815.

[29] Bradley Efron and Carl Morris, Limiting the risk of Bayes and empirical Bayes estimators—Part II: the empirical Bayes case. J. Amer. Statist. Assoc. 67 (1972), 130–139.

[63] W. A. Ericson, Subjective Bayesian models in sampling finite populations (with Discussion). J. Roy. Statist. Soc. Ser. B 31 (1969), 195–233.

[40] M. S. Geisel, Comparing and choosing among parametric statistical models: a Bayesian analysis with macro-economic applications. Unpublished doctoral dissertation, University of Chicago, 1970.

[32] I. J. Good, Probability and the weighing of evidence. New York, Hafner, 1950.

[33] I. J. Good, The estimation of probabilities, Research Monograph No. 30, Cambridge, Mass., M.I.T. Press, 1965.

[56] Elkan F. Halpern, Polynomial regression from a Bayesian approach. Combination of two technical reports, Statistics Center, Rutgers—The State University, 1972. J. Amer. Statist. Assoc. 68 (1972), 137–143.

[15] Bruce M. Hill, Inference about variance components in the one-way model. J. Amer. Statist. Assoc. 60 (1965), 806–825.

[16] Bruce M. Hill, Foundations for the theory of least squares. J. Roy. Statist. Soc. B 31 (1969), 89–97.

[17] Bruce M. Hill, Inference about many parameters in the theory of least squares. Unpublished manuscript, Dept. of Statistics, Univ. of Michigan, Ann Arbor, Mich., 1969.

[49] Bruce M. Hill, Correlated errors in the random model, J. Amer. Statist. Assoc. 62 (1967), 1387–1400.

[64] A. E. Hoerl and R. W. Kennard, Ridge regression: biased estimation and applications for nonorthogonal problems. Technometrics 12 (1970), 55–82.

[5] Harold Jeffreys, Theory of probability, 3rd Ed. Oxford, Clarendon Press, 1961.

[52] Edward E. Leamer, personal communication, 1972.

[41] F. B. Lempers, Posterior probabilities of alternative linear models. Rotterdam University Press, 1971. (Contains relevant references to econometric literature.)

[12] D. V. Lindley, Discussion to C. M. Stein's, Confidence sets for the mean of a multivariate normal distribution. J. Roy. Statist. Soc. B 24 (1962), 285–287.

[24] D. V. Lindley, Bayesian least squares, Bull. Inst. Internat. Statist. 43, 2 (1969), 152–153.

[25] D. V. Lindley, The estimation of many parameters. In: Foundations of statistical inference, edited by V. P. Godambe and D. A. Sprott. Toronto, Holt, Rinehart and Winston, 1971, 435–455.

[31] D. V. Lindley and A. F. M. Smith, Bayes estimates for the linear model (with Discussion). J. Roy. Statist. Soc. B 34, No. 1 (1972).

[35] D. V. Lindley, A statistical paradox. Biometrika 44 (1957), 187–92. (See also Bartlett [62].)

[36] D. V. Lindley, The use of prior probability distributions in statistical inference and decision. In: Proc. of Fourth Berkeley Symp. on Math. Stat. and Prob., Univ. of Calif. Press, 1961, 453–468.

[43] D. V. Lindley, The choice of variables in multiple regression. J. Roy. Statist. Soc. B 30 (1968), 31–66.

[44] D. V. Lindley, Bayesian analysis in regression problems. In: Bayesian statistics, edited by Donald L. Meyer and Raymond O. Collier, Jr. Itasca, Ill., F. E. Peacock, 1970.

[53] D. V. Lindley, Bayesian statistics, a review. Philadelphia, Soc. Indust. Appl. Math., 1971.

[65] D. V. Lindley, Introduction to probability and statistics from a Bayesian viewpoint: part-two inference. Cambridge, Cambridge Univ. Press, 1965.

[26] Melvin R. Novick, Paul H. Jackson, Dorothy T. Thayer and Nancy S. Cole, Estimating multiple regressions in *m*-groups, a cross-validation study. Brit. J. Math. Statist. Psychol. 25 (1972) in press.

[48] M. R. Novick, Multiparameter Bayesian indifference procedures. J. Roy. Statist. Soc. B 31 (1969), 29–64.

[1] Frank Plumpton Ramsey, Truth and probability. In: Studies in subjective probability, edited by Henry E. Kyburg, Jr., and Howard E. Smokler. New York, Wiley, 1926, 61–92. Reprinted from: The foundations of mathematics and other essays. London, Kegan, Paul, Trench, Trubner, 1931, 156–198.

[4] John W. Pratt, Howard Raiffa and Robert Schlaifer, The foundations of decision under uncertainty: an elementary exposition. J. Amer. Statist. Assoc. 59 (1964), 353–375.

[51] C. Radhakrishna Rao, Linear statistical inference and its applications. New York, Wiley, 1965.

[10] Howard Raiffa and Robert Schlaifer, Applied statistical decision theory. Boston, Harvard Univ., 1961. Reprinted in paperback by M.I.T. Press.

[3] Leonard J. Savage, The foundations of statistics. New York, Wiley, 1954.

[34] Leonard J. Savage, The foundations of statistical inference. Joint Statistics Seminar at the University of London. London, Methuen, 1962.

[6] C. M. Stein, Inadmissibility of the usual estimator for the mean of a multivariate normal distribution. In: Proc. Third Berkeley Sympos., Vol. 1. Berkeley, University of California Press, 1956, 197–206.

[7] C. M. Stein, Confidence sets for the mean of a multivariate normal distribution, J. Roy. Statist. Soc., Series B 24 (1962), 265–296.

[14] M. Stone, Comments on a posterior distribution of Geisser and Cornfield. J. Roy. Statist. Soc. B 26 (1964), 274–276.

[47] M. Stone and B. G. F. Springer, A paradox involving quasi prior distributions. Biometrika 52 (1965), 623–627.

[30] William E. Strawderman, Proper Bayes minimax estimators of the multivariate normal mean. Ann. Math. Statist. 42 (1971), 385–388.

[13] George C. Tiao and Arnold Zellner, Bayes's theorem and the use of prior knowledge in regression analysis. Biometrika 51 (1964), 219–230.

[45] George C. Tiao and W. Y. Tan, Bayesian analysis of random-effect models in the analysis of variance. I. Posterior distributions of variance-components. Biometrika 52 (1965), 37–53.

[27] Arnold Zellner and Walter Vandaele, Bayes-Stein estimators for *k*-means, regression, and simultaneous equation models. Paper presented to the Third Seminar on Bayesian Inference in Econometrics, sponsored by the National Bureau of Economic Research, Harvard University, Oct. 15–16, 1971. (Reprinted in the present volume).

S. E. Fienberg and A. Zellner (eds.)
Studies in Bayesian Econometrics and Statistics.

12.2 ON COHERENCE, INADMISSIBILITY AND INFERENCE ABOUT MANY PARAMETERS IN THE THEORY OF LEAST SQUARES*

Bruce M. HILL

University of Michigan

1. Introduction

This article is concerned with inference about possibly many parameters, each of which may be of interest in its own right. The basic model supposes that a vector of observations Y in n-dimensional Euclidean space (R_n) is of the form $Y = \gamma + \xi$, with γ independent of ξ a priori, and with ξ thought of as a vector of "errors", while γ is the parameter vector about which it is desired to draw inference. If $f(\gamma)$ and $\rho(\xi)$ are prior densities for γ and ξ, respectively, and if $Y = y$ is observed, then the posterior density for γ is

$$f''(\gamma) \propto f(\gamma)\, \rho(y-\gamma).$$

When ξ is conditionally uniformly distributed on spheres centered at the origin, and with perhaps certain constraints satisfied, this model is identical with that of my article on the theory of least squares [1].

Jeffreys [2, ch. III] has proposed, at least when the errors are normally distributed, for purposes of estimation that the prior distribution of γ should be taken as the improper uniform distribution, or in other words, Lebesgue measure, on R_n. For purposes of hypothesis testing he suggests other prior distributions, and he views estimation as fundamentally different from hypothesis testing [2, ch. V]. Savage et al., [3] and Dickey [4] eliminate much of the rigidity of Jeffreys, but their approach is not basically different. In this article I present a unified approach to estimation and hypothesis testing, or in other words, to

* This work was presented at the Fifth NBER-NSF Seminar on Bayesian Inference in Econometrics, Univ. of Minnesota, October 1972.

inference about many parameters. My approach also falls in the general mode of Jeffreys and Savage, but is based more directly on the concept of subjective probability, and suggests and examines specific families of prior distributions and likelihood functions having those properties I consider appropriate for inference about many parameters.

In section 2 the subjectivistic notion of coherence is discussed, and in the light of this discussion the frequentistic notion of inadmissibility as a criterion is then considered. It is shown that the latter has some relevance even from the subjectivistic Bayesian point of view. In section 3 the inadmissibility of the usual multivariate estimator of a multivariate location vector, i.e., that based upon the uniform prior distribution of Jeffreys, is discussed, and it is shown that although here the inadmissibility is not nearly so serious a problem as is commonly thought, nevertheless certain improvements are possible and desirable. It is proved that the usual estimator Y is admissible (for rather general loss functions) within the class of estimators in which the estimate of γ_i is a function only of Y_i. When a subjectivist regards the γ_i as marginally independent he is automatically restricted to such estimators. On the other hand, when he regards the γ_i as dependent it is shown how this may be exploited to yield improved forms of inference about γ and $||\gamma||^2$. Finally, in the remainder of this article specific families of prior distributions which build in such improvements are suggested and examined in the framework of my theory of least squares [1].

2. *Coherence and inadmissibility*

Before considering explicitly the problem of inference about many parameters I would like to discuss in general terms the relevance of the criterion of admissibility from the subjectivistic Bayesian point of view. There is of course no single subjectivistic Bayesian theory, but the most prominent seems to be that associated with de Finetti and Savage. de Finetti ([5], p. 151) takes as fundamental the condition of coherency, which he says "... obliges us to take care in evaluating probabilities not to allow an adversary who bets against us the possibility of winning with certainty, whatever be the event that occurs, by a judicious combination of his stakes on the various events. The fundamental theorems (total probability, compound probability) are only the immediate corollaries of this fundamental condition." It will be useful in considering admissibility to first discuss this concept of coherency. Although as expressed above the concept relates to hypothetical bets, the force of the argument does not really depend upon whether or not one plans to make such bets. It is rather that if one wishes to use probabilities as a measure of belief, then it would be unfortunate if those probabilities could not be acted upon, with confidence, in a gambling situation, and most especially if one would be a sure loser whenever he did so act upon them. Expressed in this way the concept would be applicable whether or not one gambled, using his subjective probabilities, although it would of course be more vital to a gambler that he be coherent than to an armchair subjectivist who would not necessarily pay the consequences for incoherencies in his thought. Nonetheless the desire for coherence, even for one who acts upon his probabilities, is not primarily because he fears being made a sure loser by an intelligent opponent who chooses a judicious sequence of gambles (since this situation seems uncommon) but rather because he feels that incoherence is symptomatic of something basically unsound in his attitudes. To illustrate, suppose that I assess directly the probability p_i of each event A_i in a partition of the universe, and then discover that my $\Sigma p_i = P > 1$. If I were forced to bet simultaneously on each A_i occurring, paying p_i for the chance to win a unit if A_i occurs, and winning nothing if A_i does not occur, then no matter which A_i actually does occur, I would pay P units and receive one unit, and thus be a sure loser. But it is not necessary that I expect to make such a set of bets in order to desire coherence. Even if I were to bet upon A_1 alone, for example, in which case I would not be a sure loser (provided $0 \leqq p_1 \leqq 1$), I would feel that I should resolve the discrepancy, altering some or all of the p_i, until $P = 1$. This is because a method of evaluating probabilities which would make me a sure loser if I had to act upon them (whether or not I will have to so act) would seem suspicious and untrustworthy for any purposes whatsoever, even for casual thought, and it would seem desirable to remove such incoherencies wherever possible.

Now the reason I have dwelled upon the meaning of coherency is because I think a very similar situation occurs in regard to the admissibility of estimators and decision rules. Consider a decision problem in which the parameter is ϕ, the data is $X = x$, the loss function is $L(\phi, d)$, where d denotes an act in the action space, and $d_1(X)$, $d_2(X)$ are two decision rules under consideration. If the prior density for ϕ is $g(\phi)$, then the Bayesian with this prior distribution prefers d_1 to d_2 if $E[L(\phi, d_1(X))] < E[L(\phi, d_2(X))]$, where the expectation is taken with respect to the prior distribution of ϕ and the conditional distribution of X, given ϕ. Defining the risk functions $R_i(\phi) = E_{X|\phi}[L(\phi, d_i(X))]$, then provided Fubini's theorem holds, we have $E[L(\phi, d_i(X))] = \int R_i(\phi)g(\phi)\,\mathrm{d}\phi$, for $i = 1, 2$. The Bayes decision rule for the prior distribution with density $g(\phi)$ is to choose $d(X)$ such that when $X = x$ is observed, then $d(x)$ minimizes $E_{\phi|X=x}[L(\phi, d(x))]$, where this expectation is taken with respect to the posterior distribution of ϕ, given $X = x$. Although such a Bayes rule is ordinarily admissible, there are examples, such as that to be discussed below, where the prior density $g(\phi)$ is improper, where $EL(\phi, d(X))$ is infinite, and $d(X)$ is inadmissible, and in this situation it is not immediately clear whether $d(X)$ should be employed by a subjectivistic Bayesian, and also to what extent such inadmissibility may cast suspicion upon the prior density which gave rise to $d(X)$.

To consider these questions let us suppose that decision rule $d_2(X)$ is inadmissible but is the nominal Bayes rule using the improper prior density $g_2(\phi)$, i.e., $d = d_2(x)$ minimizes $E_{\phi|X=x}[L(\phi, d)]$, for each x, where ϕ has prior density $g_2(\phi)$, and the expectation is taken with respect to the posterior distribution of ϕ, given $X = x$. We suppose also that $d_1(X)$ (which may or may not be admissible) is such that $R_1(\phi) \leqq R_2(\phi)$, for all ϕ, with the inequality being strict on a set of positive Lebesgue measure, so that in the frequentist sense d_1 is better than d_2. However, it may well be that $\int R_2(\phi)g_2(\phi)\,\mathrm{d}\phi$ is infinite, so that if the loss function is non-negative then by Fubini's theorem it is easily shown that $E[L(\phi, d_1(X))] = E[L(\phi, d_2(X))] = \infty$ where ϕ has prior density $g_2(\phi)$ in both expectations, and thus on the basis of expected loss no choice is indicated between $d_1(X)$ and $d_2(X)$. Nonetheless I would like now to argue that $d_1(X)$ is better than $d_2(X)$ for very much the same reasons given above in defence of coherency.

Ideally a posterior distribution is supposed to represent one's total knowledge of the parameter, given both the current data $X = x$, and all prior knowledge. If $P(x|\phi)$ is the likelihood function, and if $g_2(\phi)$ were truly my prior density for ϕ, then $g_2''(\phi) \propto P(x|\phi)g_2(\phi)$ would be my posterior density for ϕ, and as such, would completely describe my present knowledge of ϕ, so that I could use it with complete confidence to choose an action in connection with *any* specified decision problem. In particular in the above decision problem I would necessarily choose $d = d_2(x)$ since this minimizes $\int g_2''(\phi)L(\phi, d)\,\mathrm{d}\phi$ with respect to d. But as will now be shown there is a disconcerting feature of this choice, namely, that there do exist sequences of decision problems such that I will be a sure loser if I use $d_2(X)$, and that this stems from the inadmissibility of $d_2(X)$.

Suppose that a sequence of such observations, $X_1, X_2, \ldots, X_J$, are to be given to me, and that I regard the pairs (ϕ_j, X_j) as independent, with each ϕ_j having improper prior density $g_2(\phi)$, and each X_j having conditional distribution $P(x_j|\phi_j)$ as specified by the likelihood function. Suppose further that I am called upon to make a decision $d(X_j)$ for each value of j, so that if $X_j = x_j, j = 1, \ldots, J$, is observed, then based upon the nominal Bayes rule my decisions are $d_2(x_1), \ldots, d_2(x_J)$, my actual losses are $L(\phi_1, d_2(x_1)), \ldots, L(\phi_J, d_2(x_J))$, and my total loss is $\Sigma_{j=1}^J (L(\phi_j, d_2(x_j))$. Because it is assumed that I regard the (ϕ_j, X_j) as independent, it is clear that with prior density $g_2(\phi)$, given the data $X_j = x_j$, $j = 1, \ldots, J$, the above decision rule minimizes not only my posterior expected loss in each component problem, but also my total posterior expected loss. Now suppose that in fact the (ϕ_j, X_j) have been selected independently as above, but with each ϕ_j having proper density $g(\phi)$ instead of my $g_2(\phi)$. For example they are selected by a computer using the density $g(\phi)$, all of which is unknown to me. It is easy to see that in this situation my total loss using $d_2(X)$ as above is virtually certain to be larger than if I had instead used $d_1(X)$ in each component problem. Thus if $Z_j = L(\phi_j, d_2(X_j)) - L(\phi_j, d_1(X_j)), j = 1, \ldots, J$, then the Z_j are independent, and identically distributed, with $E(Z_j) = \int[R_2(\phi) - R_1(\phi)]g(\phi)\mathrm{d}\phi > 0$, provided that $g(\phi)$ is positive where the inequality $R_1(\phi) < R_2(\phi)$ is strict, and that $\int R_2(\phi)g(\phi)\,\mathrm{d}\phi$ is finite. Note that the expectation here is taken with respect to the "true" prior density $g(\phi)$ rather than my $g_2(\phi)$.[1] If $Z^{(J)} = \Sigma_{j=1}^J Z_j$ then $Z^{(J)}$ is the actual increase in loss using $d_2(X)$ over that using $d_1(X)$, and so if for example the central limit theorem applies, then $Z^{(J)}$ is virtually certain to be positive for large J, with of course $EZ^{(J)}$ tending to infinity as J does. Thus in this situation use of the prior density $g_2(\phi)$ and the nominal Bayes rule makes me virtually a sure loser as compared to someone using $d_1(X)$. However, there are certain

[1] Although the notion of a "true" density $g(\cdot)$ is distasteful to a subjectivist such as myself, it can be given subjectivistic meaning by merely supposing that there is, for example, a computer which generates the data, and if I could only see the program for that computer then I too would use density $g(\cdot)$. Without seeing the program I am forced to use my $g_2(\cdot)$. Note also that $E(Z_j)$ might be positive even if my $R_2(\cdot)$ were the risk function of an admissible estimator, but this would be true only for certain choices of $g(\cdot)$, while in the inadmissible case $E(Z_j)$ is positive much more generally.

obvious objections to the relevance of this example, which must now be discussed.

The first objection is that such a sequence of repetitions is rather unusual, and that if it did occur, then a Bayesian would more realistically regard the ϕ_j as dependent and/or would not use the same prior density $g_2(\phi)$ for each ϕ_j. The second objection is that the particular loss function for which the inadmissibility occurs might be unreasonable in some sense. The final objection is that the properties of decision rules are completely irrelevant as far as inference is concerned.

In regard to the first two objections, I would argue that whether or not a particular loss function or decision framework is reasonable or typical in practice, it should be possible to use the posterior distribution to choose an appropriate act without fear of being a sure loser if such a decision framework and/or loss function should arise. This is essentially the same as the previous argument for coherence on the grounds that whether or not one has to act upon his subjective probabilities, it should be possible to do so without fear of being a sure loser if the occasion did arise. In the case of the repetitions it is quite true that in many applications one would regard the ϕ_j as dependent, perhaps via a prior distribution for some other underlying parameter, as in the random model discussed below, but there is nothing to stop one from taking them to be independent, and we could easily construct realistic examples of ϕ_j that one would view as approximately, if not exactly, independent, for surely all of us regard some apparently unrelated natural phenomena in this way. Again, if the ϕ_j were taken as independent, it would be rare that a subjectivistic Bayesian would employ the same prior density $g_2(\phi)$ for each of them. But there is nothing to stop this either, and of course it is precisely what does occur if one routinely uses the Jeffreys prior distribution for the parameter.[2] Thus despite the artificiality of the above situation, it can (and does) arise, and my argument is that a method of deriving a posterior distribution, which would make one a sure loser when it does arise, is itself suspect. Note that according to this line of reasoning certain sample space characteristics are relevant to the Bayesian. This does not depend upon the frequentistic notion of long run behavior, but rather stems from the fundamental subjectivistic concept of coherence. A Bayesian who uses a formal prior distribution,

[2] Furthermore the argument showing that one is virtually a sure loser can easily be extended to cover the case where different densities are employed in each component problem.

such as that of Jeffreys, which yields inadmissible decision rules, will be virtually a sure loser over a long sequence of such usages. On the other hand a subjectivistic Bayesian who employs whatever (proper) prior distribution seems appropriate to him, will not be subject to this phenomenon, since each component problem he faces is unique, and will be treated as such. This is the reason why Bayesians tend to ignore long run arguments, but it is a mistake to imagine that the performance in the long run of Bayesian methods is irrelevant, or that subjectivistic Bayesian methods are not optimal in the long run if repetition should arise.

As to the final objection it is difficult to discuss unless one knows precisely what is meant by inference. For a Bayesian all possible inference stems from the posterior distribution. But no matter what is meant by inference, if it is to be of any value, then somehow it must be used, or acted upon, and this does indeed lead back to the decision-theoretic framework. I suspect that for some "inference" is used as a shield to avoid discovery that their actions are incoherent. For others it is used to indicate quite realistically and honestly that the actions one might be called upon to make are manifold and complex, and that none of the usual decision framework and loss functions adequately measure up to the task. This is a valid point but is certainly not an argument against decision-theory, but rather only against its careless use. The complexity of the realistic decision problems that one might face also does not affect the argument that a posterior distribution should be capable of being acted upon, if necessary.

So far my discussion has been rather puritanical, referring to an ideal world in which the prior distribution and likelihood function are known precisely. Of course nothing could be further from the truth. Approximations are necessary, and as will be seen in the next section, may lead to inadmissible estimators, which must then be considered carefully as to the possible loss stemming from their use.

3. Many parameters, the Jeffreys prior distribution, and the random model

Return now to the model $Y = \gamma + \xi$, as in the introduction. If we use the formal Jeffreys prior density $f(\gamma) \equiv 1$, and if the components of ξ are independently normally distributed with mean 0 and variance 1, then for squared error loss the nominal Bayes estimator of γ is the vector of observations Y. Stein [6, 7], has shown that this estimator is inadmissible, when $n > 2$, and since the risk function is $R(\gamma) = E_{Y|\gamma}||Y-\gamma||^2 = n$, so $\int R(\gamma)f(\gamma)\, d\gamma = +\infty$, and we are in precisely the situation discussed in the previous section. Although the arguments given there suggest that the Jeffreys prior distribution is untrustworthy, the situation is extremely complex and will now be considered from several different points of view.

In the first place the stable estimation argument of Savage et al. [3] leads to use of the Jeffreys prior distribution as yielding an approximation to the "true" posterior distribution whenever the likelihood function is sufficiently sharp relative to the prior density in the vicinity of the observation y. Now Savage et al. [3] argue that such sharpness is rare when n is large, and suggest that the Jeffreys prior is appropriate as an approximation only for small n. But even if it were only used when n was 1, by simply combining a number of such usages on problems that the Bayesian regards as independent, as in the example of last section, the upshot would be that he would be using it for large n. Thus the issue is not so easily avoided. Indeed, an example will now be presented in which a Bayesian with a perfectly proper prior distribution finds himself virtually certain to behave as though he had the Jeffreys prior distribution.

Let $S_r(y)$ be the solid sphere of radius r centered at y in R_n, and let $f(\gamma)$ be the uniform density on $S_R(\mathbf{0})$. If $Y = y$ is observed with $0 < ||y|| < R-r < R$, then

$$f''(\gamma) \propto \rho(y-\gamma) \quad \text{for} \quad ||\gamma|| < R,$$

and in particular, for

$$\gamma \in S_r(y).$$

But a priori

$$\begin{aligned}\Pr\{||Y|| < R-r\} &\geqq \Pr\{||\gamma|| < R-2r,\, ||Y-\gamma|| < r\}\\ &= \Pr\{||\gamma|| < R-2r\}\Pr\{||\xi|| < r\}\\ &= (1-2r/R)^n \Pr\{||\xi|| < r\},\end{aligned}$$

and the right-hand side goes to 1 as both r and R go to ∞, with r/R going to 0. (Note that as n goes to ∞ for fixed r/R, $\Pr\{||\gamma|| < R-r\}$ goes to 0, which is sometimes referred to as the surface phenomenon; however, for fixed n a quite different phenomenon occurs.) Thus if the prior distribution of γ is uniform over a sufficiently large sphere, then one will regard it as nearly certain that he will observe y such that the posterior density of γ is proportional to $\rho(y-\gamma)$ for $\gamma \in S_r(y)$, with the integral of $\rho(y-\gamma)$ taken over $S_r(y)$ itself nearly 1. If ξ is normally distributed this would yield the inadmissible estimator y as the posterior expectation of γ, to an arbitrarily close approximation. Although such a broad uniform prior distribution may seem unreasonable in a particular application, nonetheless the mathematics of inadmissibility has nothing whatever to do with whether a prior distribution is reasonable, and the upshot is that a Bayesian who wishes his estimators to be admissible must be extremely cautious in invoking approximations such as $f''(\gamma) \propto \rho(y-\gamma)$, even when they seem most called for. This would seem to be an unfortunate state of affairs, since it demands that he be extremely precise both in the specification of his prior distribution and the evaluation of his posterior distribution. The situation, however, is not so bad as it at first seems. For there do exist specifications of $f(\gamma)$ which are both a priori compelling, lead to admissible estimators, and incorporate the desirable features of the Jeffreys prior distribution.

In discussing this phenomenon it is important to distinguish between a situation in which one regards the γ_i as dependent a priori, and a situation in which one regards them as (marginally) independent a priori. In the first situation the dependency can be exploited and the Bayes estimate of each γ_i will typically be a function of all the observations, while in the second situation the γ_i will remain independent a posteriori, and the Bayes estimate of γ_i will, for rather general loss functions, be a function only of the corresponding observation Y_i.

An insightful example of the first situation is provided by the usual random model for the analysis of variance, as is elegantly presented by Lindley and Smith [8]. (See also Box and Tiao [9], and Hill [10].) Suppose that the prior distribution of the γ_i is such that, given μ and σ_γ^2, the γ_i are independently normally distributed with mean μ and variance σ_γ^2, and that μ, σ_γ^2, σ^2 have a joint prior density that does not depend upon μ, so that μ has the univariate Jeffreys prior distribution. Note that the γ_i are only conditionally independent, given μ and σ_γ^2, and will be marginally dependent unless the prior distribution of μ and σ_γ^2 is degenerate. From standard Bayesian normal theory it follows that, given $\mu, \sigma_\gamma^2, \sigma^2$, and the data vector y, $\gamma_i \sim N(By_i+(1-B)\mu, (1/\sigma_\gamma^2+1/\sigma^2)^{-1})$, independently of one another, where $B \equiv B(\tau^2) = \tau^2(1+\tau^2)^{-1}$, and $\tau^2 = \sigma_\gamma^2/\sigma^2$. But it is immediate that the posterior distribution of μ, given σ^2 and σ_γ^2, is $\mu \sim N(y_{.}, (\sigma^2/n)(1+\tau^2))$, where

$y_. = (\Sigma y_i)/n$. It follows that, given σ^2 and σ_γ^2, the posterior distribution of γ_i is normal with mean $By_i + (1-B)y_.$, and with variance

$$(1/\sigma_\gamma^2 + 1/\sigma^2)^{-1} + (1-B)^2(\sigma^2/n)(1+\tau^2)$$
$$= (\sigma^2/n)(1+n\tau^2)(1+\tau^2)^{-1}.$$

These γ_i will be a posteriori marginally dependent by virtue of each involving the common term μ. Next, the posterior distribution of τ^2, as derived by Hill [10], can be used to obtain the marginal posterior expectation of γ_i, which is simply $y_. + (y_i - y_.)E[B(\tau^2)|y]$, and still a weighted average of y_i and $y_.$. For squared error loss this is the Bayes estimator of γ_i, and Lindley and Smith [8] state that such estimators are often admissible. They are also very reasonable from a common sense point of view. Thus if it were known that τ^2 was very small, and so $B(\tau^2)$ near 0, then clearly $y_.$ would be a very sensible estimate of each γ_i, since in this case the "bias" due to using $y_.$ would be very slight, and $y_.$ has greater precision than y_i. On the other hand, if it were known that τ^2 was very large, and so $B(\tau^2)$ near 1, then the "bias" would be large, and it would seem sensible to use y_i to estimate γ_i. This is precisely what the above model yields. One will obtain the usual inadmissible estimator y, of γ, as an approximation, only when there is reason to believe that B is nearly 1. Although in the present situation the data will provide little information about τ^2, in the random model with say J observations in each row, as shown in Hill [10], there can be substantial information about τ^2, and so the data itself will suggest an appropriate compromise between y_i and $y_.$. It might be objected that when $J = 1$, as above, then the choice of B is basically arbitrary, but it would seem even more arbitrary to automatically choose one of the extreme cases $B = 0$ or $B = 1$, the latter being equivalent to the Jeffreys prior distribution in R_n. It is perhaps worth noting that this approach also yields a new and interesting form of inference about $||\gamma||^2$. Thus, given μ, σ_γ^2, σ^2, and the data vector y, we have $\gamma_i \sim N(By_i + (1-B)\mu, \sigma_1^2)$, $i = 1, \ldots, n$, independently of one another, where $\sigma_1^2 = (1/\sigma_\gamma^2 + 1/\sigma^2)^{-1}$. Hence the corresponding conditional distribution of $||\gamma||^2$ will be of the noncentral chi-square form. The conditional expectation of $||\gamma||^2$ is

$$E\{||\gamma||^2|\mu, \sigma^2, \sigma_\gamma^2, y\}$$
$$= n\sigma_1^2 + B^2||y||^2 + (1-B)^2 n\mu^2 + 2B(1-B)\mu n y_.,$$

while

$$E\{||\gamma||^2|\sigma^2, \sigma_\gamma^2, y\}$$
$$= n\sigma_1^2 + B^2||y||^2 + n(y_.)^2(1-B^2) + (1-B)^2(\sigma^2+\sigma_\gamma^2).$$

Note that as $\sigma_\gamma^2 \to \infty$, which yields the Jeffreys prior distribution; the limit of the last expectation is $n\sigma^2 + ||y||^2$, while the usual unbiased estimate of $||\gamma||^2$ is $||Y||^2 - n\sigma^2$. Although both $||y||^2 + n\sigma^2$ and $||y||^2 - n\sigma^2$ seem unsatisfactory, the former tending to be too large, the latter too small, the general form above for the posterior expectation seems reasonable either for constant B or when a further expectation is taken with respect to the posterior distribution of σ^2 and σ_γ^2.

Our discussion of the random model clearly reveals how prior knowledge of the dependencies amongst the γ_i can be exploited to yield improved forms of inference and decision-making. In that example the type of prior knowledge that was crucial was knowledge of σ_γ^2/σ^2, and thereby also of $\Sigma_1^n(\gamma_i - \gamma_.)^2/\sigma^2$, but of course other forms of prior knowledge about the γ_i could similarly have been exploited. In practice one would obtain a probabilistic model for the γ_i that represents his actual prior knowledge as well as possible, and then obtain the appropriate posterior distribution and/or Bayes estimates. When that prior knowledge includes knowledge of dependencies amongst the γ_i it is of course neither novel nor surprising that improvements over the usual estimator Y are possible. However, there is nothing to prevent one from regarding the γ_i as a priori independent, and in this situation it is somewhat more disturbing that Y is inadmissible. Stein and James [11] showed that with squared error loss the estimator

$$\tilde{Y} = Y(1-(n-2)/||Y||^2)$$

has everywhere smaller risk than Y, provided $n > 2$. It is instructive to note that $\tilde{Y}$ is itself of the general form $BY + (1-B)\boldsymbol{\mu}$ that we derived from our analysis of the random model, with $\boldsymbol{\mu} = \mathbf{0}$ and $B = (1-(n-2)/||Y||^2)$. However, both $\tilde{Y}$, and our more general estimators, seem inappropriate in this situation, since both use data from problems regarded as completely unrelated to estimate a particular γ_i. Surely it is criminal to use data concerning the temperature on Venus to adjust census data, and yet in principle this is what $\tilde{Y}$ does.

To discuss this situation return to the general model $Y = \gamma + \xi$, and suppose that the ξ_i are independent of one another as well as of γ. If the γ_i are a priori independent, then they will be a posteriori independent also, so the posterior distribution of a particular γ_i will depend upon the data $Y = y$ only through the corresponding y_i. For any loss function of the form $L(\gamma, \hat{\gamma}) = \Sigma_{i=1}^{n} L_i(\gamma_i, \hat{\gamma}_i)$, where the $L_i(\gamma_i, \hat{\gamma}_i)$ are arbitrary, it follows that the Bayes estimator of γ will be of the form $d(Y) = (d_1(Y_1), \ldots, d_n(Y_n))'$, in which the estimate of γ_i is a function only of the corresponding Y_i. Thus a Bayesian who regards the γ_i as a priori independent, and who has a loss function of the above general form (and in fact this holds for even more general loss functions, although the above seems sufficiently general for most applications) can never be led to use estimators such as $\tilde{Y}$ or our random model estimators. The Bayesian analysis of this situation agrees with the common-sense loathing of statisticians to combine data from problems regarded as unrelated.

Now consider this situation from a non-Bayesian viewpoint. From such a viewpoint it does not really make sense to talk about the γ_i being independent, and the usual frequentistic criteria suggest that $\tilde{Y}$ is preferable to Y, despite the fact that $\tilde{Y}$ may seem unreasonable by virtue of combining data from perhaps apparently unrelated problems. Carried to its extreme this line of argument would suggest the absurd conclusion that all problems should be considered jointly in order to improve the risk function. A non-Bayesian can avoid this conclusion only by deciding, on some arbitrary basis, to restrict attention to certain subclasses of decision functions. For example, on the grounds that the problems are apparently unrelated, he might decide to restrict attention to the class of estimators in which the estimate of γ_i is a function only of the corresponding y_i, i.e., to the same class which a Bayesian is led to when he regards the γ_i as a priori independent. But within this class, as we shall now prove, the usual estimator Y is admissible with respect to squared error loss.

Theorem 1. *Let $Y = \gamma + \xi$, where $\xi \sim N(\mathbf{0}, I)$. For squared error loss, the estimator Y is admissible within the class of estimators of the form $d(Y) = (d_1(Y_1) \ldots, d_n(Y_n))'$.*

Proof: Suppose that Y is inadmissible, and that $d(Y)$ is better than Y, i.e.,

$$E\{\|\gamma - d(Y)\|^2\} \leqq n \quad \text{for all} \quad \gamma,$$

with strict inequality for some γ. Hence

$$E(\gamma_1 - d_1(Y_1))^2 \leqq 1 + \{n - 1 - \sum_{i=2}^{n} E(\gamma_i - d_i(Y))^2\}.$$

Hold $\gamma_2, \ldots, \gamma_n$ fixed. If the quantity in braces were negative, then letting γ_1 vary, it would follow that Y_1 is inadmissible, which contradicts the admissibility of the usual estimator in dimension 1 and 2. Hence

$$\sum_{i=2}^{n} E(\gamma_i - d_i(Y_i))^2 \leqq n-1 \quad \text{for all} \quad \gamma_2, \ldots, \gamma_n.$$

It is easily seen that the inequality must be strict for some $\gamma_2 \ldots, \gamma_n$, since otherwise we would again contradict the admissibility of Y_1. We have now showed that inadmissibility of Y in n dimensions implies inadmissibility in $n-1$ dimensions, and by induction we conclude that Y is inadmissible in dimension 2, which is once again a contradiction. Q.E.D.

Theorem 1 and our previous discussion suggest that a non-Bayesian who does not wish to combine data from apparently unrelated sources might just as well use the estimator Y, which is admissible within the appropriate subclass of estimators, despite its inadmissibility within the full class of possible estimators. In particular, the criterion of admissibility by itself does not rule out Y, since an arbitrary choice concerning the appropriate class of estimators to consider is also necessary (unless one is foolish enough to consider all conceivable problems jointly). In the Bayesian framework such a choice is implied by the prior distribution given to γ and the loss function. When we view the γ_i as a priori independent and our loss function is of a certain general form then we are led to estimators of γ_i which are functions only of the corresponding Y_i. When we view the γ_i as dependent and/or our loss function is not the sum of the separate loss functions for the individual problems, then we are led to other forms for the posterior distribution and for our estimators, forms which are appropriate under such circumstances. All this is quite reasonable. The only question that remains is whether Y itself is ever appropriate from the subjectivistic Bayesian viewpoint. My own view is that, taken literally, Y is not appropriate as an estimator, just as, taken literally, Lebesgue measure is not appropriate as a prior distribution. This has nothing to do with admissibility and is true even in dimension 1, where Y is in fact admissible. Thus even in dimension 1, where $Y_1 = \gamma_1 + \xi_1$, and $\xi_1 \sim N(0, 1)$, the use of Lebesgue measure as a prior distribution for γ_1 would imply that, given $Y_1 = y_1$, the posterior distribution of γ_1 is $\gamma_1 \sim N(y_1, 1)$, no matter what value y_1 has. But surely in any real problem, if y_1 were sufficiently extreme, then one would not have this posterior distribution, and thus by contemplating one's inference about γ_1 for various possible values of y_1 (before actually observing the data) one can easily see that Lebesgue measure is not a reasonable prior distribution. Similarly, the estimator Y, taken literally, seems unreasonable to me, and even in dimension 1 I would prefer an estimator of the form $By_1+(1-B)\mu_1$, say, where μ_1 is the prior expectation of γ_1. Of course often such an adjusted y_1 will be nearly equal to y_1, just as, for many values of y_1, the posterior distribution of γ_1 will be approximately $N(y_1, 1)$. Thus stable estimation may justify use of Lebesgue measure and the estimator Y, as approximations, for some values of y, but it is important to be aware that the approximation must be justified separately in each usage and that it cannot hold for all y.

This concludes our general discussion of admissibility. In the remainder of this article we formulate classes of prior distributions which generalize the usual random model and build in prior knowledge of a type that we regard as appropriate for inference about many parameters. In the next section certain technical questions concerning limiting posterior distributions of $\gamma - y$, as $||y|| \to \infty$, are considered.

4. *Limiting posterior distributions as* $\|y\| \to \infty$

In the last section it was argued that the Jeffreys prior distribution, taken literally, is unrealistic as a measure of prior knowledge in real problems of inference. Nonetheless, there exist many situations where it seems appropriate to represent one's prior distribution as a mixture of two or more components, at least one of which has some of the characteristics of the Jeffreys prior distribution. To take an example due to L. J. Savage, suppose that you are trying to create a chemical compound known as "statistic acid", and that γ is the melting point of the compound. You are given a measurement $Y = \gamma + \xi$ of the melting point and must now decide whether you have succeeded in creating statistic acid, and also what the melting point of your compound is. Let H_0 denote the hypothesis that the compound you have created is statistic acid, H_1 the hypothesis that it is not, and p be your prior probability for H_0. Given H_0, your prior density for γ is, say, $f_0(\gamma)$, while given H_1, it is $f_1(\gamma)$. Presumably $f_0(\gamma)$ is a much sharper function than $f_1(\gamma)$, e.g., more informative, since under H_1 the compound you have created might be virtually anything, while the melting point of statistic acid may be reasonably well known. As a first approximation it seems reasonable to consider using the Jeffreys prior density $f_1(\gamma) \equiv 1$ for the relatively flat function, but, unfortunately, as discovered by Jeffreys [2, ch. V], this is unsatisfactory because the posterior probability of H_1 will then be 0 no matter what data $Y = y$ is observed. Nonetheless we can try to choose a form of density for $f_1(\gamma)$ which has some of the appealing features of the uniform distribution but without the drawbacks. In particular, it seems appropriate that under H_1 one's prior knowledge of γ should be sufficiently weak so that for sufficiently large $|y|$ one should act approximately as though one had the Jeffreys prior density. It is therefore natural to examine the limiting posterior distribution of $\theta \equiv \gamma - y$, given H_1, as $||y|| \to \infty$. It will be proved in this section that essentially the only possible limiting distribution for θ is simply the initial prior distribution for the error ξ in the model $Y = \gamma + \xi$, and this is precisely the distribution one would obtain using the Jeffreys prior density for γ. We will return to the mixture model in the later sections. Throughout this section it is assumed that $Y = \gamma + \xi$, where γ and ξ are independent a priori with densities $f(\gamma)$ and $\rho(\xi)$, respectively, and all densities are proper relative to Lebesgue measure in R_n. We also assume $\rho(\xi) = \rho(-\xi)$, as is true in the most important case where $\rho(\xi)$ is spherically symmetric. We now prove

Lemma. *Let the probability density function $f(\gamma)$ and the non-negative function $g(y)$ be such that*

$$\lim_{|c| \to \infty} f(c\theta) \textit{ exists and is } 0 \textit{ for almost all } \theta,$$

$$\lim_{||y|| \to \infty} g(y) f(y+\theta) \textit{ exists and is equal to } h(\theta)$$

for almost all θ in the set H, where H contains the origin $\mathbf{0}$. *Then $h(\theta)$ is constant almost everywhere on H.*

Proof: Restrict attention to those θ for which $\lim f(c\theta) = 0$. For such a θ suppose that there exists $r = r(\theta) > 0$ such that $r\theta$ is in H. Then there exists a sequence $c_i = c_i(\theta) \to +\infty$ for which

$$f(c_i\theta) \geqq f(r\theta + c_i\theta), \quad \text{and so if} \quad ||\theta|| > 0,$$

$$h(\mathbf{0}) = \lim_i g(c_i\theta) f(c_i\theta) \geqq \lim_i g(c_i\theta) f(r\theta + c_i\theta) = h(r\theta).$$

Similarly, there exists a sequence $d_i = d_i(\theta) \to -\infty$ for which

$$f(d_i\theta) \leqq f(r\theta + d_i\theta),$$

and so in fact $h(r\theta) = h(\mathbf{0})$. This result holds for all $r\theta$ in H, with $r > 0$, on the set where $\lim_{|c| \to \infty} f(c\theta) = 0$, and so in fact $h(\theta) = h(\mathbf{0})$, almost everywhere on H.

Theorem. *Let $f(\gamma)$ be as in the lemma, $\rho(\mathbf{0}) > 0$, and $\rho''(\theta)$ be the posterior density of $\theta \equiv \gamma - y$, given $Y = y$. If there exists a limiting posterior distribution of θ as $||y|| \to \infty$, if this limiting distribution has density $\rho^*(\theta)$, and if $\rho''(\theta)$ converges to $\rho^*(\theta)$ for almost all θ for which $\rho(\theta) > 0$, then $\rho^*(\theta) = \rho(\theta)$ almost everywhere.*

Proof: Define $g(y) = (\int f(y+\theta)\, \rho(\theta)\, d\theta)^{-1}$ so $\rho''(\theta) = g(y) f(y+\theta)\, \rho(\theta)$. Since $\lim_{||y|| \to \infty} \rho''(\theta) = \rho^*(\theta)$ whenever $\rho(\theta) > 0$, the lemma implies that $\rho^*(\theta)/\rho(\theta)$ is constant almost everywhere on the set where $\rho(\theta) > 0$. But $\rho^*(\theta) = 0$ if $\rho(\theta) = 0$, and $\int \rho(\theta)\, d\theta = \int \rho^*(\theta)\, d\theta = 1$, so the constant must be unity. Note here that given $Y = y$, in fact $\theta = \gamma - y = -\xi$.

Remark 1. In general convergence in distribution does not imply convergence of densities, even when the latter exist, and so the theorem does not rule out the possibility of other limiting distributions. I doubt that this is of any relevance to the practice of statistics. The assumptions about $\rho(\xi)$ and $f(\gamma)$ will surely be appropriate in many practical situations.

Remark 2. Let $f(\gamma)$, $g(y)$ and $\rho(\xi)$ be as in the theorem, with $f(\gamma)$ strictly positive. Since $\{g(y)\}^{-1}$ is the marginal density of Y, I shall restrict attention to y such that $g(y)$ is both strictly positive and finite. For such y, and for θ such that $\rho(\theta) > 0$, $g(y) f(y+\theta)$ converges to 1 as $||y|| \to \infty$ iff $f(y+\theta)/f(y)$ and $g(y) f(y)$ each converge to 1. In this case the posterior distribution of $\theta = \gamma - y$ converges in distribution to the distribution with density $\rho(\cdot)$.

Remark 3. Let $f(\gamma)$, $g(y)$ and $\rho(\xi)$ be as in the theorem, with $f(\gamma)$ strictly positive, bounded and asymptotically as $||\gamma|| \to \infty$ a rational function of $||\gamma||$. Then $f(y+\theta)/f(y) \to 1$, and furthermore this convergence is uniform for $||\theta|| < c(||y||)$, where $c(||y||) \to +\infty$ and $c(||y||) = o(||y||)$. If also $\Pr\{\theta\colon ||\theta|| \geqq c(||y||)\}/f(y) \to 0$, then $g(y) f(y) \to 1$, and as in remark 2 there is convergence in distribution to the distribution with density $\rho(\cdot)$. This situation seems often appropriate in practice.

Remark 4. The roles of $f(\gamma)$ and $\rho(\xi)$ could, at least as far as the mathematics goes, be interchanged everywhere above. Thus as $\|y\| \to \infty$ one could have the posterior distribution of $\xi - y$ converging to the distribution with density $f(\cdot)$ instead of the posterior distribution of $\gamma - y$ converging to the distribution with density $\rho(\cdot)$. However, in practice an asymmetry of attitude towards ξ and γ is usually implicit in viewing the one as "error" and the other as more real in some usually ill-defined sense. Hopefully, errors are at least small relative to true effects, and the condition

$$\Pr\{\theta : \|\theta\| \geqq c(\|y\|)\}/f(y) \to 0 \quad \text{of remark 3}$$

specifies a sense in which this might hold. The situation is, unfortunately, further complicated by the fact that we are sometimes testing a hypothesis under which $\gamma = 0$, as will be discussed later. Despite philosophical ambiguity, I believe that in practice it is usually clear what kind of inferential procedure is appropriate.

Remark 5. If $f(\gamma) \propto \exp -\{\frac{1}{2}\|\gamma\|^2\}$ then $f(y+\theta)/f(y)$ does not converge for $\|\theta\| > 0$. This would not of itself rule out the use of such a prior distribution, but personally I think it is rarely appropriate in practice (at least as the overall prior distribution).

Remark 6. Let $f(\gamma)$ be uniform on $S_r(\mathbf{0})$ so the posterior density of γ is

$$f''(\gamma) = \rho(y-\gamma)/\Pr\{\xi : \|\xi - y\| \leqq r\} \quad \text{if} \quad \|\gamma\| \leqq r,$$

and is otherwise 0. From remarks 3 and 4 it follows that if $\rho(\xi)$ is strictly positive, bounded and asymptotically a rational function of $\|\xi\|$, then the posterior distribution of $\xi - y$ converges to the uniform distribution on $S_r(\mathbf{0})$ as $\|y\| \to \infty$. Thus the data supply virtually no information about γ for sufficiently large $\|y\|$.

Remark 7. In practice I believe it will often be appropriate to take $f(\gamma)$ of the form

$$f(\gamma) \propto \int_{\|\gamma\|^2}^{\infty} \phi(t)\,\mathrm{d}t, \quad \text{where} \quad \phi(\cdot) \geqq 0.$$

In effect this yields all densities which are non-increasing functions of $\|\gamma\|$. One way to arrive at such a form is to introduce a parameter R such that given $R^2 = r^2$, the prior distribution for γ is the uniform distribution on $S_r(\mathbf{0})$, as in remark 6. If R^2 has the probability density

$\phi_0(\cdot)$, then the marginal density for γ is

$$f(\gamma) \propto \int_{\|\gamma\|^2}^{\infty} t^{-n/2}\, \phi_0(t)\, \mathrm{d}t.$$

Note that when $\phi_0(\cdot)$ is tightly concentrated near 0, it is virtually as though $\gamma = \mathbf{0}$; when $\phi_0(\cdot)$ is non-zero only for sufficiently large values of R^2, then it is much the same situation as discussed in the Introduction in justifying the improper uniform prior distribution as an approximation. Apart from these extreme cases R has no natural interpretation I know of. Nonetheless it is interesting to study the posterior distribution of R^2, if only because of its relationship to the power function of the classical test of the hypothesis $\gamma = \mathbf{0}$. Writing for convenience $\Pr\{\cdot\}$ for a density function,

$$\begin{aligned}\Pr\{Y = y | R^2 = r^2\} &= \int \Pr\{\xi + \gamma = y | R^2 = r^2, \xi\} \\ &\qquad \Pr\{\xi | R^2 = r^2\}\, \mathrm{d}\xi \\ &\propto r^{-n} \Pr\{\xi \in S_r(y)\},\end{aligned}$$

where I have here assumed that ξ is independent of R, and that given R, γ is independent of ξ. The posterior density of R^2 is then

$$\phi_0''(r^2) \propto \phi_0(r^2)\, r^{-n} \Pr\{\xi \in S_r(y)\},$$

and $\Pr\{\xi \in S_r(y)\}$ is the probability of a non-central event of the kind that arises in the power function of the classical test of the hypothesis $\gamma = \mathbf{0}$.

Remark 8. Let $\phi(t) \propto t^{-\frac{1}{2}\lambda} \exp -(\beta/2t)$ in the general form of remark 7, where $t \geqq 0$, $\beta > 0$, $\lambda > 2$. Then

$$\begin{aligned}f(\gamma) &\propto \int_{\|\gamma\|^2}^{\infty} \phi(t)\, \mathrm{d}t \propto \int_0^{\|\gamma\|^{-2}} t^{\frac{1}{2}\lambda - 2} \exp -(\beta/2t)\, \mathrm{d}t \\ &\propto \Pr\{\chi^2_{(\lambda-2)} < \beta/\|\gamma\|^2\} \sim C\|\gamma\|^{-(\lambda-2)} \quad \text{as} \quad \|\gamma\| \to \infty,\end{aligned}$$

where C is constant and $\chi^2_{(z)}$ denotes a generic random variable having the gamma distribution with shape parameter $\frac{1}{2}z$ and scale parameter $\frac{1}{2}$. In order that $f(\gamma)$ be a proper density it is necessary that $\lambda > n+2$. This family of densities will play a central role in the analysis of section 7.

Remark 9. Convergence of the posterior distribution of $\theta = \gamma - y$ to the distribution with density $\rho(\cdot)$ typically occurs when $f(\gamma)$ is a mixture of densities, one of which is suitably diffuse or gentle relative to $\rho(\cdot)$ and the others. Thus suppose $f(\gamma) = p_1 f_1(\gamma) + \ldots + p_I f_I(\gamma)$ with $p_1 > 0, \Sigma p_i = 1$, $\rho(\cdot)$ and $f_1(\cdot)$ satisfying all the conditions of remark 3, and the other $f_i(\cdot)$ bounded densities such that $f_i(\gamma)/f_1(\gamma) \to 0$ as $||\gamma|| \to \infty$. Then

$$f(y+\theta)/f(y) \to 1 \quad \text{as} \quad ||y|| \to \infty.$$

Next, if $c(||y||)$ is the function of remark 3, then for $i > 1$,

$$[f_1(y)]^{-1} \int_{||\theta|| \geqq c(||y||)} f_i(y+\theta)\rho(\theta)\,\mathrm{d}\theta \to 0,$$

and

$$f_i(y+\theta)/f_1(y) \to 0 \quad \text{uniformly for} \quad ||\theta|| < c(||y||).$$

It follows that $g(y) f(y) \to 1$, so there is convergence in distribution as in remark 2. Note that the $f_i(\cdot)$ can be prior densities for γ under hypotheses H_i, in which case as $||y|| \to \infty$, the posterior probability of H_1 goes to 1. A version of this result which holds when the mixture includes atomic or otherwise singular distributions relative to Lebesgue measure in R_n plays a major role in the unified model of sections 6 and 7.

5. *Conditional uniformity on spheres for errors*

The most important application of the ideas in this article is to the theory of least squares based upon conditional uniformity on spheres centered at the origin for the error vector ξ, as set forth in my article [1]. In this case suppose $\rho(\xi) \propto ||\xi||^{-(n-2)}\rho_0(||\xi||^2)$ with $\int^\infty \rho_0(t)\,\mathrm{d}t = 1$. Although I shall not everywhere assume $\rho(\xi)$ to be of this form, or errors to be conditionally uniform on spheres, it seems well, because of its importance, to set forth the special properties of such a distribution for errors.

From remark 3 above it follows that if $f(\gamma)$ is strictly positive, bounded, and asymptotically (as $||\gamma|| \to \infty$) a rational function of $||\gamma||$, then a sufficient condition for convergence of the posterior distribution of $\xi = y - \gamma$ to the distribution with density $\rho(\cdot)$ is the existence of a function

$$c \equiv c(||y||) \quad \text{with} \quad c(||y||) \to \infty, \qquad c(||y||) = o(||y||),$$

and with

$$\Pr\{\theta: ||\theta|| > c(||y||)\}/f(y) = \frac{1}{f(y)} \int_{c^2}^{\infty} \rho_0(t)\,\mathrm{d}t \to 0 \quad \text{as} \quad ||y|| \to \infty.$$

Thus if $f(\gamma)$ is asymptotically proportional to $||\gamma||^{-(\lambda-2)}$ as $||\gamma|| \to \infty$, with $\lambda > n+2$, then the above condition is

$$||y||^{\lambda-2} \int_{c^2}^{\infty} \rho_0(t)\,\mathrm{d}t \to 0 \quad \text{as} \quad ||y|| \to \infty.$$

Three particularly important cases in which there will exist such a function $c(||y||)$ are

(a) $\rho_0(\cdot)$ identically 0 outside some bounded interval;

(b) $\rho_0(\cdot)$ of gamma form;

(c) $\rho_0(\cdot) \sim t^{-L/2}$ as $t \to \infty$, with $L > \lambda > n+2$.

Case (a) arises when it is known with certainty that $||\xi||$ cannot exceed a known constant. Case (b) includes the historically important case in which the errors are independently normally distributed with mean 0 and known variance. Case (c) includes the both historically and practically important case in which errors are conditionally independently normally distributed with mean 0 and variance σ^2, given σ^2, and with σ^2 itself having a prior density of inverted gamma form. In this case, if the prior density for σ^2 is proportional to

$$(\sigma^2)^{-\frac{1}{2}g-1} \exp -(\beta/2\sigma^2),$$

then

$$\rho(\xi) \propto \int_0^{\infty} (\sigma^2)^{-\frac{1}{2}(g+n)-1} \exp -\{(||\xi||^2+\beta)/2\sigma^2\}\,\mathrm{d}\sigma^2$$

$$\propto (||\xi||^2+\beta)^{-\frac{1}{2}(g+n)},$$

while

$$\rho_0(t) \propto t^{\frac{1}{2}(n-2)}(t+\beta)^{-\frac{1}{2}(g+n)}$$

For $\rho(\xi)$ to be a proper density function it is required that $g > 0$.

I believe that in practice it will ordinarily be appropriate to choose $\rho(\xi)$ of one of these three forms. Often the analysis will not be sensitive in critical respects to which of the three is used. When it is, then of course a choice must be made, as the price one pays for attempting inference.

The remainder of this article is devoted to a unified model for inference about many parameters. Although I shall not explicitly assume that $\rho(\xi)$ has one of the above three forms, or even that errors are conditionally uniform on spheres, it should be borne in mind that these are the most important cases in practice.

6. *A unified model for inference*

The usual model for Bayesian hypothesis testing assumes a mutually exclusive and exhaustive class of hypotheses H_i, having prior probabilities $\Pr\{H_i\}$. Data is then taken and the posterior probability of a hypothesis is evaluated as

$$\Pr\{H_i|\text{data}\} = \Pr\{\text{data}|H_i\} \Pr\{H_i\}/\Pr\{\text{data}\}.$$

The usual objection to this formulation is that the theory of personal probability does not tell how the prior probability of a hypothesis, or for that matter the probability of the data given a hypothesis when the hypothesis involves "unknown" parameters, is to be determined. This is true but no more implies that one should not utilize the Bayesian model than does the lack of a theory that tells us the "right" price to pay for our clothes imply that we should go naked. I do not mean to pretend, however, that there are not substantial difficulties in the use of the model. My purpose here is to implement the use of the Bayesian model, and to this end I shall propose a class of hypotheses which I believe must be an essential ingredient in any satisfactory theory of inference about many parameters. The philosophy underlying this approach will be discussed in section 8. The present section will be devoted to the mathematical model, which has singular aspects, and section 7 to its analysis. Since the generalization to R_n is relatively easy, for ease of exposition I shall begin with $\gamma = (\gamma_1, \gamma_2) \in R_2$.
Let

H_0 be any hypothesis under which $\gamma_1 = \gamma_2 = 0$,

H_1 be any hypothesis under which $\gamma_1 = 0$ and γ_2 has the univariate density $f_1(\gamma_2)$,

H_2 be any hypothesis under which $\gamma_2 = 0$ and γ_1 has the univariate density $f_2(\gamma_1)$,

H_3 be any hypothesis under which γ has the bivariate density $f_3(\gamma)$,

where all densities are proper, and let $p_i = \Pr\{H_i\}, \Sigma p_i = 1$.

This model defines a probability distribution for γ as follows. For any Borel set A in R_2 define

$$A_1 = \{\gamma_2 : \gamma \in A \text{ and } \gamma_1 = 0\},$$

$$A_2 = \{\gamma_1 : \gamma \in A \text{ and } \gamma_2 = 0\},$$

$$A_0 = \{\gamma : \gamma \in A \text{ and } \gamma_1 = \gamma_2 = 0\}.$$

Now consider the set function which assigns to A the number

$$m(A) = p_0 \delta(A_0) + p_1 \int_{A_1} f_1(\gamma_2)\, d\gamma_2 + p_2 \int_{A_2} f_2(\gamma_1)\, d\gamma_1 + p_3 \int_A f_3(\gamma)\, d\gamma,$$

where $\delta(A_0)$ is 0 or 1 according to whether A_0 is or is not empty. It is easy to verify that $m(R_2) = 1$, $m(A) \geqq 0$ wherever it is defined, and $m(\cdot)$ is a countably additive set function on the Borel sets of R_2, so that $m(\cdot)$ defines a probability distribution on these sets. This distribution is in effect a mixture of singular and non-singular distributions with respect to Lebesgue measure, and can be represented as

$$P(d\gamma) = p_0 \delta(||\gamma||) + p_1 f_1(\gamma_2) + p_2 f_2(\gamma_1) + p_3 f_3(\gamma),$$

where $\delta(||\gamma||)$ is 0 if $||\gamma|| = 0$ and is otherwise 1. Of course $P(d\gamma)$ cannot be interpreted as an ordinary density in the plane. (Alternatively, the same effect could be achieved working entirely with generalized functions in the sense of Schwartz, or approximated by means of ordinary densities in the plane suitably concentrated about points and axes. I find the above model more natural and simple.)

Now consider the distribution of $Y = \xi + \gamma$, where γ is as above, and ξ is independent of γ with the bivariate density $\rho(\xi)$. It is easy to verify that Y has an ordinary bivariate density in the plane given by the convolution

$$\begin{aligned} P(dy) &= \int \rho(y-\gamma) P(d\gamma) \\ &= p_0 \rho(y) + p_1 \int \rho_1(y-\gamma) f_1(\gamma_2)\, d\gamma_2 \\ &\quad + p_2 \int \rho_2(y-\gamma) f_2(\gamma_1)\, d\gamma_1 + p_3 \int \rho(y-\gamma) f_3(\gamma)\, d\gamma, \end{aligned}$$

where $\rho_i(y-\gamma)$ is by definition equal to $\rho(y-\gamma)$ with γ_i set equal to 0, $i = 1, 2$. Writing for convenience $\Pr\{Y = y\}$ in place of $P(dy)$, the posterior probabilities of the hypotheses are

$$\Pr\{H_0|Y=y\} = p_0\rho(y)/\Pr\{Y=y\},$$

$$\Pr\{H_1|Y=y\} = p_1(\int\rho_1(y-\gamma)f_1(\gamma_2)\,\mathrm{d}\gamma_2)/\Pr\{Y=y\},$$

$$\Pr\{H_2|Y=y\} = p_2(\int\rho_2(y-\gamma)f_2(\gamma_1)\,\mathrm{d}\gamma_1)/\Pr\{Y=y\},$$

$$\Pr\{H_3|Y=y\} = p_3(\int\rho(y-\gamma)f_3(\gamma)\,\mathrm{d}\gamma)/\Pr\{Y=y\},$$

while

$$\Pr\{\gamma_1=\gamma_2=0|Y=y\} = \Pr\{H_0|Y=y\},$$

$$\Pr\{\gamma_1=0|Y=y\} = \Pr\{H_0|Y=y\}+\Pr\{H_1|Y=y\},$$

$$\Pr\{\gamma_2=0|Y=y\} = \Pr\{H_0|Y=y\}+\Pr\{H_2|Y=y\}.$$

Finally, the posterior distribution of γ can be represented as

$$\begin{aligned} P(\mathrm{d}\gamma|Y=y) &= \Sigma P(\mathrm{d}\gamma|H_i, Y=y)\Pr\{H_i|Y=y\} \\ &= \Pr{}^{-1}\{Y=y\}[p_0\delta(||\gamma||)\rho(y) \\ &\quad +p_1f_1(\gamma_2)\rho_1(y-\gamma)+p_2f_2(\gamma_1)\rho_2(y-\gamma) \\ &\quad +p_3f_3(\gamma)\rho(y-\gamma)]. \end{aligned}$$

Like the prior distribution of γ, the posterior distribution is a mixture of singular and non-singular distributions, and cannot be interpreted as an ordinary bivariate density. In practice, of course, the data may make the posterior probability of one of the hypotheses nearly unity, and so the mixture may be adequately approximated by a single density, as will be discussed further below.

The extension of this model to R_n is immediate. There are now 2^n hypotheses corresponding to the 2^n subsets of the components of γ. Let $H_{i_1,\ldots,i_j}$ be a hypothesis under which $(\gamma_{i_1}\ldots,\gamma_{i_j})$ has density function $f_{i_1\ldots i_j}(\gamma_{i_1},\ldots,\gamma_{i_j})$, and the other $n-j$ components of γ are 0. The prior and posterior distributions of γ are then mixtures of singular and non-singular distributions with respect to Lebesgue measure in R_n, and the posterior probability of the hypothesis is

$$\Pr\{H_{i_1,\ldots,i_j} | Y = y\} = \frac{\Pr\{H_{i_1,\ldots,i_j}\}}{\Pr\{Y = y\}} \times \int f_{i_1,\ldots,i_j}(\gamma_{i_1}, \ldots, \gamma_{i_j}) \rho_{i_1,\ldots,i_j}(y-\gamma) \times d\gamma_{i_1} \times \ldots \times d\gamma_{i_j},$$

where $\rho_{i_1,\ldots,i_j}(y-\gamma) \equiv \rho(y-\gamma)$ with all components of γ other than $\gamma_{i_1}, \ldots, \gamma_{i_j}$ set equal to 0.

Since the model for R_n is such an obvious extension of that for R_2 I shall not elaborate further.

The most important question that remains is how to specify the densities given the various hypotheses. A family of densities that I believe often appropriate for practice will be suggested and examined in the next section.

7. *Analysis*

Excluding the hypothesis under which all components of γ are 0, there remain $2^n - 1$ hypotheses corresponding to the $2^n - 1$ possible subsets of the components of γ which are not set equal to 0. There is no necessary relationship between the densities for these non-zero components, and in particular, the density for a particular subset need not be either a conditional or marginal density derived from the n-dimensional density of γ. The most useful approach, here as everywhere in statistics, is to study families of distributions. A family which I believe leads to suitable forms of inference and decision making in practice, which provides substantial insight, and which is relatively tractable, is given by

$$f(\gamma) \propto \int_{||\gamma||^2}^{\infty} \phi(t)\, dt, \quad \text{with} \quad \phi(t) \propto t^{-\frac{1}{2}\lambda} \exp(-\beta/2t)$$

for $t \geqq 0$. As in remark 8 above,

$$f(\gamma) \propto \Pr\{\chi^2_{(\lambda-2)} < \beta/||\gamma||^2\} \sim C||\gamma||^{-(\lambda-2)} \text{ as } ||\gamma|| \to \infty.$$

Here I shall consider the case in which the densities under the various hypotheses are all of the above form, with common values $\beta_j > 0$ and

$\lambda_j > j+2$ for any hypothesis under which there are exactly j non-zero components of γ. Although not always appropriate I believe this case to be of most interest for practice.

Once the prior probabilities of the hypotheses and $\rho(\cdot)$ have been specified the posterior probabilities of the hypotheses and the posterior distribution of γ are completely determined by the choice of the β_j and λ_j. In practice it will often be convenient to choose the latter on the basis of desired moments for $||\gamma||$. If $H^{(j)}$ is any hypothesis under which there are exactly j non-zero components of γ, then

$$E(||\gamma||^r|H^{(j)}) = \frac{\int_0^\infty t^{r+j-1}\Pr\{\chi^2_{(\lambda_j-2)} < \beta_j/t^2\}\,dt}{\int_0^\infty t^{j-1}\Pr\{\chi^2_{(\lambda_j-2)} < \beta_j/t^2\}\,dt}$$

$$= \beta_j^{r/2}J(r+j-1, \lambda_j-2)/J(j-1, \lambda_j-2),$$

where

$$J(i, d) \equiv \int_0^\infty t^i\Pr\{\chi^2_{(d)} < 1/t^2\}\,dt.$$

Integration by parts yields

$$J(i, d) = (i+1)^{-1}\,\Gamma\{(d-i-1)/2\}/\Gamma\{d/2\}$$

for $d > i+1$, where $\Gamma\{t\}$ is the gamma function. With $\lambda_j > j+4$ these expressions allow β_j and λ_j to be chosen to yield desired first and second moments of $||\gamma||$ under $H^{(j)}$.

Now suppose the components of y are ordered so that $|y_1| \geqq |y_2| \geqq \ldots \geqq |y_n|$, and let $H^{(j)}$ denote the hypothesis under which all but the first j components of γ are 0, with these j components having density $f_j(\gamma_1, \ldots, \gamma_j) \propto \Pr\{\chi^2_{(\lambda_j-2)} < \beta_j/||\gamma||^2\}$. In practice such an ordering will be convenient because such $H^{(j)}$ will ordinarily have the highest posterior probability, and in any case there is no real loss in generality. The posterior probability of $H^{(j)}$ is then

$$\Pr\{H^{(j)}|Y = y\}$$

$$= \frac{\Pr\{H^{(j)}\}\int\rho_j(y-\gamma)\Pr\{\chi^2_{(\lambda_j-2)} < \beta_j/||\gamma||^2\}\,d\gamma_1\times\ldots\times d\gamma_j}{\Pr\{Y = y\}\int\Pr\{\chi^2_{(\lambda_j-2)} < \beta_j/||\gamma||^2\}\,d\gamma_1\times\ldots\times d\gamma_j}$$

$$= \frac{\Pr\{H^{(j)}\}}{\Pr\{Y = y\}\omega_j\beta_j^{j/2}J(j-1, \lambda_j-2)}$$

$$\times \int\rho_j(y-\gamma)\Pr\{\chi^2_{(\lambda_j-2)} < \beta_j/||\gamma||^2\}\,d\gamma_1\times\ldots\times d\gamma_j,$$

where $\rho_j(y-\gamma)$ is identically equal to $\rho(y-\gamma)$ with $\gamma_{j+1} = \ldots = \gamma_n = 0$, and $\omega_j = 2\pi^{j/2}\Gamma(j/2)$.

The asymptotic behavior of the posterior probabilities of the hypotheses as $||y|| \to \infty$ now follows easily from the results of sections 4 and 5. Thus if ξ has a conditionally uniform distribution of one of the three types discussed in section 5 (and in fact under substantially weaker conditions) then as $||y|| \to \infty$, with $y_{j+1}, \ldots, y_n$ fixed,

$$\begin{aligned} &\int \rho_j(y-\gamma) \Pr\{\chi^2_{(\lambda_j-2)} < \beta_j/||y||^2\} \, d\gamma_1 \times \ldots \times d\gamma_j \\ &\sim \Pr\{\chi^2_{(\lambda_j-2)} < \beta_j/||y||^2\} \int \rho_j(y-\gamma) \, d\gamma_1 \times \ldots \times d\gamma_j \\ &\sim [\Gamma(\tfrac{1}{2}\lambda_j)]^{-1} [2||y||^2/\beta_j]^{-(\lambda_j-2)/2} \int \rho_j(y-\gamma) \, d\gamma_1 \times \ldots \times d\gamma_j. \end{aligned}$$

In particular, if $|y_1| \geqq |y_2| \geqq \ldots \geqq |y_j| \to \infty$, $y_{j+1}, \ldots, y_n$ are held fixed, and $k > j$, then

$$\frac{\Pr\{H^{(j)} | Y = y\}}{\Pr\{H^{(k)} | Y = y\}} \sim ||y||^{\lambda_k - \lambda_j} \times 2^{\frac{1}{2}(\lambda_k - \lambda_j)} \times \frac{\Gamma\{\frac{1}{2}\lambda_k\}}{\Gamma\{\frac{1}{2}\lambda_j\}}$$

$$\times \frac{\Pr\{H^{(j)}\} \omega_k J(k-1, \lambda_k - 2) \times \beta_j^{(\lambda_j - j - 2)/2} \int \rho_j(y-\gamma) \, d\gamma_1 \times \ldots \times d\gamma_j}{\Pr\{H^{(k)}\} \omega_j J(j-1, \lambda_j - 2) \times \beta_k^{(\lambda_k - k - 2)/2} \int \rho_k(y-\gamma) \, d\gamma_1 \times \ldots \times d\gamma_k},$$

which holds for all $k \geqq 1$ when the errors are independently normally distributed. The factor $(\int \rho_j(y-\gamma) \, d\gamma_1 \times \ldots \times d\gamma_j)/\int \rho_k(y-\gamma) \, d\gamma_1 \times \ldots \times d\gamma_k$ is of critical importance in the above equation. For example, if $\rho(\xi) = (2\pi)^{-n/2} \exp -\frac{1}{2}||\xi||^2$, then this factor is equal to

$$(2\pi)^{(j-k)/2} \exp \tfrac{1}{2}\{y_{k+1}^2 + \ldots + y_n^2 - (y_{j+1}^2 + \ldots + y_n^2)\}.$$

For $k < j$, it follows that the posterior odds in favor of $H^{(j)}$ over $H^{(k)}$ go to ∞ as $|y_j|$ does. In other words sufficiently large values of $|y_{k+1}|, \ldots, |y_j|$, call for the introduction of the corresponding parameters $\gamma_{k+1}, \ldots, \gamma_j$. For $j < k$, on the other hand, the factor becomes $(2\pi)^{-(k-j)/2} \exp -\frac{1}{2}(y_{j+1}^2 + \ldots + y_k^2)$ and so the asymptotic behavior is determined by $||y||^{\lambda_k - \lambda_j}$. Clearly it would ordinarily be unreasonable to take $\lambda_k < \lambda_j$, since in this case, even with $y_{j+1} = y_{j+2} = \ldots = y_k = 0$, the data could make the posterior odds arbitrarily large in favour of $H^{(k)}$ over $H^{(j)}$. The behavior of the factor for the general conditionally uniform density $\rho(\xi) \propto ||\xi||^{-(n-2)} \rho_0(||\xi||^2)$ is qualitatively much the same as in the normal case.

It still remains to suggest choices for the λ_j, β_j and $\Pr\{H_{i_1}, \ldots, H_{i_j}\}$. Two ways of specifying the λ_j which I believe will often be appropriate in practice are either to take λ_j constant $= n+2+c$, or alternatively to take $\lambda_j = j+2+c$, where $c > 0$ determines the largest finite moment of $||\gamma||$. Ordinarily I think it will be appropriate to take a common value for the β_j, although in particular applications this will sometimes be undesirable. I shall not discuss further the selection of the β_j here because I do not think it will be a serious obstacle in practice. But the selection of the prior probabilities of the hypotheses is a more subtle matter. First recall that these hypotheses single out particular values, for some or all of the components of the parameter vector γ, which are to be given special consideration. In some applications such a hypothesis will be a genuine hypothesis about reality corresponding to a particular theory which specifies the distribution of the parameters allowed for by that theory. But it is no secret that such genuine hypotheses are relatively rare in practice. In spite of this I believe it will often be wise to behave as though such hypotheses were present. Furthermore, although ultimately the values singled out by a hypothesis are arbitrary, in many important applications (comparing treatments against a control, regression, etc.) zero does play a very special role. I believe that in such situations it is wise to consider inference in the light of the mixture model. If it is objected that this makes inference depend upon an arbitrary choice of the prior probabilities of the hypotheses, then I can only say that this degree of arbitrariness seems to me to be inherent in the problem, and that it is best not to cover it up by the cloak of a spurious objectivity. Note, for example, that some popular classical procedures are virtually equivalent to treating all parameters as "unknown", and this to me implies a concealed specification of the prior probabilities of all but one of the hypotheses as being zero. In comparing treatments against a control and in regression it may in particular applications be wise to take these probabilities as very tiny, but by considering suitably extreme data it becomes clear that zero is not ordinarily a wise choice. My point of view here is dualistic in that although I would prefer to specify the prior probabilities and all other quantities on the basis of introspection, I feel that when such choices are extremely difficult it is often useful to view them as simply parameters of a decision procedure which are chosen to achieve desired properties.

When all the choices have been made and the data is taken, the analysis will require integrals of the form

$$\int \rho_j(y-\gamma)\,\Pr\{\chi^2_{(\lambda_j-2)} < \beta_j/||\gamma||^2\}\,\mathrm{d}\gamma_1 \times \ldots \times \mathrm{d}\gamma_j,$$

which will probably have to be evaluated by techniques of numerical integration. Of course this need not ordinarily be done for all the hypotheses, since even casual inspection should allow a number to be discarded as having negligible posterior probability.

8. *Discussion*

I have suggested a general model for inference about many parameters, with particular reference to the theory of least squares. I should like now to discuss the philosophy underlying this approach. Although this philosophy is Bayesian, it is a personal rendering, and probably differs even from that of most Bayesians. I shall not go into the design of the experiment here, but simply suppose that the data will consist in the observation of the vector Y. In principle I would like to be able to specify my probability for any event A, given any conceivable observation $Y = y$, before actually observing y. The event A is any event of interest. It might concern the values of either real or mythical parameters (the proportion of rotten eggs in a dozen is a real parameter, the unknown probability that a coin flip yields heads is a mythical one), of future observations, or it might constitute a hypothesis about reality (e.g. "ESP does not exist"). Whatever the nature of A, Bayesian theory demands only that the specification be coherent in the sense of de Finetti [5]. Yet in practice much more is required, namely, that the specification reflect as well as possible the opinions and knowledge of the person or persons involved. Wisely chosen statistical models at present constitute the best way to achieve both ends. For the careful use of hypotheses, parameters (real or mythical), and prior distributions for the parameters under the various hypotheses, will lead to coherent inference, while if these are chosen judiciously, with a view to the global behaviour of the inferential procedure, then one can at least achieve approximately the kind of inference that best reflects his opinions and knowledge. But it should be understood that this is always only an approximation, and that one can unwittingly be led through the mechanical use of such models to inference, which although coherent, is not what one himself regards as wise. This danger is inherent in the use of so fine a tool as a Bayesian model. The way to gain the benefits and avoid the perils of Bayesian inference is to study carefully the behaviour of the inferential procedure for various data, before actually taking the data, until one is confident that the particular model he has chosen or is tempted to choose, yields to within an adequate approximation the kind of behavior he regards as wise. It is particularly useful in this connection to consider the form inference takes when the data is extreme in some fashion or other. If a model does not respond in a reasonable way to such extreme data, then it should be modified accordingly. This philosophy motivates my articles [1, 12], as well as the present one. Of course there are practical limitations to such a study, and one cannot literally consider the inference for all possible data. But it is surely advantageous to have as much foresight as possible.

The model for inference about many parameters proposed above represents an application of this philosophy. My underlying motivation was to provide a model which for sufficiently large magnitude of some of the components of y would lead to a posterior distribution for the corresponding components of γ, very nearly as though these components had the improper uniform prior distribution. At the same time I wanted the model to allow me to single out specific values for some or all components of γ, which might be given positive prior probability, and which the data might make either more or less probable, as the case may be. The model of section 6, with suitably chosen densities for the non-zero components of γ, does indeed yield this kind of inference. Of course I cannot prove that such inference is wise or even reasonable, but it does seem so to me, and others can judge for themselves. By studying the behavior of the posterior distribution for various choices of $\rho(\xi)$ and $P(\mathrm{d}\gamma)$ one can better learn to choose these distributions to yield the form of inference regarded as appropriate for the problem at hand. The procedure I have proposed is at least qualitatively in some harmony with certain classical traditions of hypothesis testing and estimation, for example, the tradition of first testing a null hypothesis, and then if this hypothesis is rejected, estimating parameters as though "nothing were known a priori". Despite philosophical differences, in practice such estimation is in some respects equivalent to the use of the improper uniform prior distribution for the parameter vector. However, a basic difference between classical pretesting and my procedure is reflected in the fact that for the latter the posterior distribution of γ is a mixture of the various posterior distributions corresponding to the different hypotheses. Often many of the hypotheses will have negligible posterior probability, and as an approximation one might delete such components of the posterior distribution, but typically there will be several hypotheses with non-negligible posterior probability. Classical pretesting, by contrast, acts as though a single hypothesis were true. Finally, in practice there will often be special information which neither classical procedures nor as yet the one I have proposed take into account. But mine can be relatively easily adjusted to do so. For example, in polynomial regression one might feel certain that the "true" curve was monotonic. Then $P(d\gamma)$ can be chosen to reflect such knowledge.

References

[1] Bruce M. Hill, Foundations for the theory of least squares. J. Roy. Statist. Soc. B 31 (1969), 89–97.

[2] H. Jeffreys, Theory of probability, 3rd ed. Oxford, Clarendon Press: Oxford, 1961.

[3] L. J. Savage, H. Lindman and W. Edwards, Bayesian statistical inference for psychological research. Psychological Review 70 (1963), 193–242.

[4] J. M. Dickey, The weighted likelihood ratio, linear hypotheses on Normal location parameters. Ann. Math. Statist. 42 (1971), 204–223.

[5] B. de Finetti, Foresight: its logical laws, its subjective sources. In: Studies in subjective probability, edited by H. E. Kyburg and H. E. Smokler. New York, John Wiley, 1964, 93–158.

[6] C. Stein, Inadmissibility of the usual estimator for the mean of a multivariate normal distribution. In: Third Berkeley Symposium, Vol. 1, 1956, 197–206.

[7] C. Stein, Confidence sets for the mean of a multivariate normal distribution. J. Roy. Statist. Soc. B 24 (1962), 265–296.

[8] D. Lindley and A. Smith, Bayes estimates for the linear model. J. Roy. Statist. Soc. B 34 (1972), 1–41.

[9] G. E. P. Box and G. C. Tiao, Bayesian estimation of means for the random effects model. J. Amer. Statist. Assoc. 63 (1968), 174–181.

[10] Bruce M. Hill, Inference about variance components in the one-way model. J. Amer. Statist. Assoc. 60 (1965), 806–825.

[11] C. Stein and W. James, Estimation with quadratic loss. In: Fourth Berkeley Symposium, 1960, 361–378.

[12] Bruce M. Hill, Posterior distribution of percentiles: Bayes' theorem for sampling from a population. J. Amer. Statist. Assoc. 63 (1968), 677–691.

[13] W. E. Strawderman, Proper Bayes minimax estimators of the multivariate normal mean. Ann. Math. Statist. 42 (1971), 385–388.

S. E. Fienberg and A. Zellner (eds.)
Studies in Bayesian Econometrics and Statistics.

12.3 COMBINING BAYES AND FREQUENCY APPROACHES TO ESTIMATE A MULTINOMIAL PARAMETER*

Michael SUTHERLAND

Hampshire College

Paul W. HOLLAND

NBER Computer Research Center

and

Stephen E. FIENBERG
University of Minnesota

1. *Introduction*

In this paper, we consider the problem of estimating the vector of cell probabilities associated with the multinomial distribution. We view this problem as one of simultaneously estimating a large number of parameters. One way to provide such estimators is to assume that these parameters themselves have a probability distribution which we can characterize by a smaller set of hyperparameters. By estimating the hyperparameters, we generate estimators for the original set, which often have superior frequency properties (i.e., smaller risk) when compared to estimators not based on the hyperparametric structure.

This approach to estimation is by no means new. It underlies much of Bayesian inference (see Good [1] and Lindley [2] for discussions) and has played a prominent role in the more classical statistical literature (e.g., see Mosteller [3]). We can also use this approach to generate estimators for the mean of the multivariate Normal distribution similar to those suggested by Stein and others from frequency considerations. We illustrate the application of this approach to estimating

* Presented to the Fifth NBER-NSF Seminar on Bayesian Inference in Econometrics, Univ. of Minnesota, October 1972. Research supported by NSF Grants GS-32327x and GJ-1154x.

the multivariate Normal mean in the next subsection before turning to the multinomial problem which is the main topic of this paper.

1.1. *Estimating a multivariate Normal mean*

This section summarizes some facts about estimating the mean of a multivariate Normal distribution that are analogous to the results on estimating the cell probabilities of a multinomial distribution we discuss in sections 3, 4 and 5. These Normal distribution results are instructive because while the two problems are similar the calculations for the Normal distribution are usually easier.

We consider only the simplest case when the covariance matrix is known. Let $\mathbf{X}' = (X_1, \ldots, X_t)$ have independent, Normally distributed coordinates with $E(X_i) = \theta_i$ and $\text{Var}(X_i) = N^{-1}$. Letting $\boldsymbol{\theta}' = (\theta_1, \ldots, \theta_t)$ be the vector of means, we denote this model by

$$\mathbf{X} \sim \mathcal{N}_t(\boldsymbol{\theta}, N^{-1}\mathbf{I}). \tag{1}$$

Taking the known value of the variance as N^{-1} allows us to interpret each X_i as an average, and results in expressions that are very similar to those appearing in sections 3, 4 and 5 for the multinomial distribution.

Throughout this paper, the loss function for an estimator, $\mathbf{T}(\mathbf{X})$, of $\boldsymbol{\theta}$ is squared-error given by

$$L(\boldsymbol{\theta}, \mathbf{T}) = N\|\mathbf{T}-\boldsymbol{\theta}\|^2 = N \sum_{i=1}^{t} (T_i-\theta_i)^2. \tag{2}$$

The maximum likelihood estimator of $\boldsymbol{\theta}$ is $\mathbf{T}(\mathbf{X}) \equiv \mathbf{X}$. This estimator is also the unique, minimum-risk (minimum expected loss), unbiased estimator relative to squared-error loss and is minimax with constant risk equal to t.

A simple Bayesian approach to this problem gives $\boldsymbol{\theta}$ a multivariate Normal prior distribution of the form

$$\boldsymbol{\theta} \sim \mathcal{N}_t(\boldsymbol{\lambda}, \tau^{-1}\mathbf{I}). \tag{3}$$

Thus the $\{\theta_i\}$ are regarded as independently drawn from Normal distributions each with mean λ_i and common inverse variance, τ. A judgmental Bayesian would specify $\boldsymbol{\lambda}$ and τ on the basis of his prior information. Relative to the loss function (2) the Bayes estimator of $\boldsymbol{\theta}$ is the posterior mean which is given by

$$E(\boldsymbol{\theta}|\mathbf{X}, \boldsymbol{\lambda}, \tau) = \frac{N}{N+\tau}\,\mathbf{X} + \frac{\tau}{N+\tau}\,\boldsymbol{\lambda}$$
$$= (1-w)(\mathbf{X}-\boldsymbol{\lambda})+\boldsymbol{\lambda}, \tag{4}$$

where

$$w = \frac{\tau}{N+\tau}. \tag{5}$$

If $\tau \to 0$, we see that $\mathbf{T}(\mathbf{X}) = \mathbf{X}$ is the limit of Bayes estimators. We regard $\boldsymbol{\lambda}$ as a specified, known vector (e.g., $\boldsymbol{\lambda} = 0$), so that the quantity $\mathbf{X}-\boldsymbol{\lambda}$ is observable. On the other hand, we regard τ as a parameter to be specified or estimated. The quantity w given in (5) is a one-to-one function of τ, and thus is an equivalent form of this parameter.

The marginal distribution of the observable quantity, $\sqrt{N}(\mathbf{X}-\boldsymbol{\lambda})$, is $\mathcal{N}_t(\mathbf{0}, w^{-1}\,\mathbf{I})$. The maximum likelihood estimate of τ relative to this marginal distribution is

$$\hat{\tau}_{\max} = \frac{t}{\|\mathbf{X}-\boldsymbol{\lambda}\|^2-(t/N)}, \tag{6}$$

or equivalently,

$$\hat{w}_{\max} = \frac{t}{N\|\mathbf{X}-\boldsymbol{\lambda}\|^2}. \tag{7}$$

Relative to the marginal distribution of $\sqrt{N}(\mathbf{X}-\boldsymbol{\lambda})$, $\hat{w}_{\max}$ in (7) is a biased estimator of w, but the estimator

$$\hat{w}_u = \frac{t-2}{N\|\mathbf{X}-\boldsymbol{\lambda}\|^2}, \tag{8}$$

is unbiased. If we replace w in (4) by $\hat{w}_u$ from (8) we obtain a new estimator for the vector parameter $\boldsymbol{\theta}$ given by

$$\mathbf{T}(\mathbf{X}) = \left(1-\frac{t-2}{N\|\mathbf{X}-\boldsymbol{\lambda}\|^2}\right)(\mathbf{X}-\boldsymbol{\lambda})+\boldsymbol{\lambda}. \tag{9}$$

While we have obtained this estimator by near-Bayesian methods (also called pseudo-, semi-, or bootstrap-Bayes) which are similar to the empirical Bayes approach of Robbins [4], this same estimator has been shown by James and Stein [5] to have good frequency properties in the sense of having uniformly smaller risk than the usual estimator, $\mathbf{X}$, if $t \geqq 3$. Our derivation of the James–Stein estimator is essentially that given by Lindley [6], and is discussed in Efron and Morris [7]. It is akin to the "method of marginal moments" estimator discussed in section 3.1 of this paper.

So far we have used a Bayesian approach and some heuristics to derive an estimator of $\boldsymbol{\theta}$ which has good frequency properties. Now we address the question of small risk directly. Regarding the right-hand side of (4) as defining a class of estimators of $\boldsymbol{\theta}$, we minimize the risk function over this class by differentiating it with respect to τ and solving for the optimal value, $\tau_0(\boldsymbol{\theta}, \boldsymbol{\lambda})$. This yields

$$\tau_0(\boldsymbol{\theta}, \boldsymbol{\lambda}) = \frac{t}{\|\boldsymbol{\theta}-\boldsymbol{\lambda}\|^2}, \tag{10}$$

which is a function of the unknown parameter, $\boldsymbol{\theta}$, so that we can only estimate this optimal value. Two alternative methods for estimating τ_0 suggest themselves. The first is the maximum likelihood estimate of τ_0 relative to the conditional distribution of $\mathbf{X}$ given $\boldsymbol{\theta}$. This method yields

$$\hat{\tau}_{0ml} = \frac{t}{\|\mathbf{X}-\boldsymbol{\lambda}\|^2} \tag{11}$$

or

$$\hat{w}_{0ml} = \frac{t}{N\|\mathbf{X}-\boldsymbol{\lambda}\|^2+t}. \tag{12}$$

While this choice of $\hat{\tau}$ has the nice property that it is never negative (a property not shared by $\hat{\tau}_{\max}$ in (6)), its denominator is a badly biased estimator of the denominator of τ_0 when t is large. If we estimate τ_0 by the reciprocal of an unbiased estimator of τ_0^{-1}, we obtain the "reciprocal unbiased" estimator, $\hat{\tau}_{0ru}$, given by

$$\hat{\tau}_{0ru} = \frac{t}{\|\mathbf{X}-\boldsymbol{\lambda}\|^2-(t/N)}. \tag{13}$$

We note that $\hat{\tau}_{0ru}$ and $\hat{\tau}_{\max}$ are identical so that using $\hat{\tau}_{0ru}$ to estimate τ_0 in (4) we arrive at the following estimator of $\boldsymbol{\theta}$:

$$\mathbf{T}(\mathbf{X}) = \left(1-\frac{t}{N\|\mathbf{X}-\boldsymbol{\lambda}\|^2}\right)(\mathbf{X}-\boldsymbol{\lambda})+\boldsymbol{\lambda}. \tag{14}$$

When t is large, this estimator has essentially the same frequency properties as the James–Stein estimator, whereas this is not true of the estimator that substitutes $\hat{\tau}_{0ml}$ for τ in (4).

A Bayesian who is not satisfied with the limited range of priors expressed by (3) might seek to enlarge it by compounding on λ, τ, or both (i.e., by giving $\boldsymbol{\lambda}$ or τ prior distributions). Not only does such compounding enlarge our range of prior distributions, but it also allows us to replace the assumption of independence of the θ_i a priori (see (3)), by the assumption that their distribution is exchangeable (see Lindley and Smith [8]). Suppose we do the compounding by letting $\boldsymbol{\lambda}$ and τ have joint density $\phi(\boldsymbol{\lambda}, \tau)$. This "two-stage" prior for $\boldsymbol{\theta}$ is a mixture of Normal distributions. We may now express the posterior mean of $\boldsymbol{\theta}$ as

$$E(\boldsymbol{\theta}|\mathbf{X}) = \int E(\boldsymbol{\theta}|\mathbf{X}, \boldsymbol{\lambda}, \tau)\phi(\boldsymbol{\lambda}, \tau|\mathbf{X})\, \mathrm{d}\boldsymbol{\lambda}\, \mathrm{d}\tau, \tag{15}$$

where $\phi(\boldsymbol{\lambda}, \tau|\mathbf{X})$ denotes the posterior density of $\boldsymbol{\lambda}$ and τ. Substituting the right-hand side of (4) into (15) and simplifying yields

$$E(\boldsymbol{\theta}|\mathbf{X}) = \frac{N}{N+\tau(\mathbf{X})}\mathbf{X} + \frac{\tau(\mathbf{X})}{N+\tau(\mathbf{X})}\boldsymbol{\lambda}(\mathbf{X}), \tag{16}$$

where

$$\tau(\mathbf{X}) = \frac{E\left(\dfrac{\tau}{N+\tau}\middle|\mathbf{X}\right)}{E\left(\dfrac{1}{N+\tau}\middle|\mathbf{X}\right)}, \tag{17}$$

and

$$\lambda_i(\mathbf{X}) = \frac{E\left(\lambda_i\dfrac{\tau}{N+\tau}\middle|\mathbf{X}\right)}{E\left(\dfrac{\tau}{N+\tau}\middle|\mathbf{X}\right)}. \tag{18}$$

The form of the right-hand side of (16) as an estimator of $\boldsymbol{\theta}$ is quite intriguing because it is of the same general form as the estimators given in (9) and (14). If $\phi(\boldsymbol{\lambda}, \tau)$ puts all of its mass on a specific value of $\boldsymbol{\lambda}$, then $\boldsymbol{\lambda}(\mathbf{X})$ will be identically equal to that value, and expression (16) reduces to

$$E(\boldsymbol{\theta}|\mathbf{X}) = (1-w(\mathbf{X}))(\mathbf{X}-\boldsymbol{\lambda})+\boldsymbol{\lambda} \tag{19}$$

where

$$w(\mathbf{X}) = E\left(\frac{\tau}{N+\tau}\middle|\mathbf{X}\right). \tag{20}$$

If ϕ puts zero mass on $\tau \leqq 0$, then $0 < w(\mathbf{X}) < 1$. This suggests modifying $\hat{w}_u$ in (8) and $\hat{w}_{0ru} = \hat{w}_{\max}$ in (7) to keep them in [0, 1], and therefore make the resulting estimators of $\boldsymbol{\theta}$ behave more like Bayes estimators of the same general form. Strawderman [9] has exhibited Bayes estimators of the form (19) that have good frequency properties (i.e., are admissible, minimax and therefore uniform improvements over $\mathbf{T}(\mathbf{X}) = \mathbf{X}$ when $t \geqq 5$). Unfortunately his expression for $w(\mathbf{X})$ is much more difficult to compute than $\hat{w}_u$ or $\hat{w}_{0ru}$ and the improvement in estimation over these choices of $w(\mathbf{X})$ is not substantial.

Lindley and Smith [8] consider a general multivariate Normal problem using compound priors, and they apply the general theory to examples in experimental design and regression analysis. Since explicit expressions for the posterior mean, (16), are quite difficult to derive for standard situations such as when τ^{-1} has a scaled inverted χ^2 distribution and $\boldsymbol{\lambda}$ has a multivariate Normal distribution, Lindley and Smith focus on modal estimates of $\boldsymbol{\theta}$, which are more or less of the same form as (16). Cleveland (this volume), and Zellner and Vandaele (this volume) have considered related problems.

1.2. *Outline of the remainder of the paper*

In section 2, we discuss one- and two-stage Bayesian estimators of the vector of cell probabilities of a multinomial variable. These estimators correspond to the smoothing of multinomial data by adding pseudo-counts. As in the case of the multivariate Normal mean discussed above, the form of the two-stage Bayes estimators suggests that the parameters in the one-stage Bayesian approach should be estimated from the data. Drawing an analogy with the James–Stein result, we would then expect the resulting estimators to have improved frequency properties when t is large.

In section 3, we derive several near-Bayesian estimators of the multinomial parameter using techniques similar to those discussed above for the multivariate Normal mean. The exact evaluation of the risk functions of these new estimators is not possible, and, in order to arrive at a basis for comparing them, in section 4 we derive asymptotic approximations to these risk functions that hold for large N, large t and moderate N/t—large, sparse multinomials. In section 5, we prove that one of the estimators in section 3 has the uniformly smallest dominant term in the expansion of its risk function, among all estimators of a particular class.

Section 6 deals briefly with possible extensions of this work, especially to the problem of incorporating the cross-classification structure of a two-way contingency table into the estimation of its cell probabilities. The final section applies 10 different estimators described in this paper to a 14 by 14 occupational mobility table.

2. *Bayesian estimation of a multinomial parameter*

We assume throughout the rest of this paper that $\mathbf{X}' = (X_1, \ldots, X_t)$ has a multinomial distribution with parameter $\mathbf{p}' = (p_1, \ldots, p_t)$, and we let $N = \Sigma X_i$. We denote this sampling model by

$$\mathbf{X} \sim \mathcal{M}_t(N, \mathbf{p}). \tag{21}$$

The parameter $\mathbf{p}$ takes values in the probability simplex

$$S_t = \{(p_1, \ldots, p_t) : p_i > 0, \sum p_i = 1\}.$$

We denote the center of S_t by $\mathbf{c} = (t^{-1}, \ldots, t^{-1})$. As usual, we use $\mathbf{X}$ to denote both the random variable and the value it assumes. The likelihood function for this multinomial problem is

$$l(\mathbf{p}|\mathbf{X}) = \binom{N}{\mathbf{X}} \prod_i p_i^{X_i}, \tag{22}$$

where $\binom{N}{\mathbf{X}}$ denotes the multinomial coefficient.

2.1. *A one-stage Bayesian approach*

The conjugate family of priors for the likelihood (22) is the Dirichlet family, whose densities have the form

$$f(\mathbf{p}|K, \boldsymbol{\lambda}) = G(K\boldsymbol{\lambda}) \prod_i p_i^{K\lambda_i - 1} \tag{23}$$

where $K > 0$, $\boldsymbol{\lambda} \in S_t$ and

$$G(K\boldsymbol{\lambda}) = \frac{\Gamma(K)}{\prod_i \Gamma(K\lambda_i)}. \tag{24}$$

We denote this model by

$$\mathbf{p} \sim \mathcal{D}_t(K, \boldsymbol{\lambda}). \tag{25}$$

If the prior is $\mathcal{D}_t(K, \boldsymbol{\lambda})$ then the posterior distribution is easily shown to be $\mathcal{D}_t(K+N, (\mathbf{X}+K\boldsymbol{\lambda})/(K+N))$. The moments of a Dirichlet distribution may be expressed as

$$E(\prod_i p_i^{a_i}|K, \boldsymbol{\lambda}) = \frac{G(K\boldsymbol{\lambda})}{G(K\boldsymbol{\lambda}+\mathbf{a})} \tag{26}$$

(see Wilks [10] or Watson [11]). From (26) we see that the prior and posterior means of $\mathbf{p}$ are given by

$$E(\mathbf{p}|K, \boldsymbol{\lambda}) = \boldsymbol{\lambda} \quad \text{(prior mean)} \tag{27}$$

$$E(\mathbf{p}|\mathbf{X}, K, \boldsymbol{\lambda}) = \frac{N}{N+K}(\mathbf{X}/N) + \frac{K}{N+K}\boldsymbol{\lambda} \quad \text{(posterior mean).} \tag{28}$$

For a geometrical discussion of K when $\boldsymbol{\lambda} = \mathbf{c}$, see Fienberg and Holland [12]. We note the similarity between (28) and the corresponding expression for the Normal distribution (4).

A Bayesian would specify K and $\boldsymbol{\lambda}$ on the basis of his prior information. This specification can be aided by the following interpretation of (28). We may regard the posterior mean of $\mathbf{p}$ as the estimate of $\mathbf{p}$ with prior information and regard $\mathbf{X}/N$ as the estimate without prior information. Thus, K may be interpreted as the prior sample size and $\boldsymbol{\lambda}$ as the prior estimate of $\mathbf{p}$. When we have prior information in the form of a previous set of data, we might wish to take as our prior a Dirichlet with K equal to the sample size of these data and $\boldsymbol{\lambda}$ equal to an estimate of $\mathbf{p}$. Watson [11] discusses some related cell estimation problems when the number of cells in the table is increased (or decreased) from one sample to the next.

In addition to its Bayesian interpretation, the right-hand side of (28) also illustrates a well-known method for "smoothing" multinomial data. The data, $\mathbf{X}/N$, is shrunk towards a "smooth" probability vector, $\boldsymbol{\lambda}$, by a convex weight, $N/(N+K)$. This is the same as adding $K\lambda_i$ "pseudo-counts" to X_i and normalizing by the new total, $N+K$. The latter device is a standard way of removing zero counts in contingency tables. A popular choice of parameters in this situation is $\boldsymbol{\lambda} = \mathbf{c}$ and $K = t/2$, which corresponds to "adding $\frac{1}{2}$" to each count. We call K the *smoothing constant* and regard $\boldsymbol{\lambda}$ as a device for allocating a fraction of K to each cell of the multinomial. "Adding $\frac{1}{2}$" is an example of a *data-independent* smoothing constant. Below we see that *data-dependent* smoothing constants, that are functions of $\mathbf{X}$, can also arise from Bayesian models.

2.2. *A two-stage Bayesian approach*

Just as the class of Normal priors can be widened by compounding on the prior parameters, so can the Dirichlet family be usefully expanded by compounding on K, $\boldsymbol{\lambda}$, or both. In this two-stage Bayesian approach we assume that

$$\mathbf{p}|K, \boldsymbol{\lambda} \sim \mathscr{D}_t(K, \boldsymbol{\lambda}) \tag{29}$$

and that $(K, \boldsymbol{\lambda})$ is given a joint density function, $\phi(K, \boldsymbol{\lambda})$. The resulting prior for $\mathbf{p}$ is a mixture of Dirichlets (or a compound Dirichlet distribution). If $\phi(K, \boldsymbol{\lambda})$ is a degenerate distribution concentrated on a single value of $(K, \boldsymbol{\lambda})$, then the prior reduces to the ordinary $\mathscr{D}_t(K, \boldsymbol{\lambda})$ distribution discussed above.

We observe that, because of the similarity between (28) and (4), the expression for the posterior mean of $\mathbf{p}$ is formally identical to that derived for $\boldsymbol{\theta}$ in the Normal example ((16), (17), and (18)). Thus, we have

$$E(\mathbf{p}|\mathbf{X}) = \int E(\mathbf{p}|\mathbf{X}, K, \boldsymbol{\lambda})\, \phi(K, \boldsymbol{\lambda}|\mathbf{X})\, \mathrm{d}K\, \mathrm{d}\boldsymbol{\lambda}, \tag{30}$$

where $\phi(K, \boldsymbol{\lambda}|\mathbf{X})$ denotes the posterior density of $(K, \boldsymbol{\lambda})$. Hence, we get

$$E(\mathbf{p}|\mathbf{X}) = \frac{N}{N+K(\mathbf{X})}(\mathbf{X}/N) + \frac{K(\mathbf{X})}{N+K(\mathbf{X})}\boldsymbol{\lambda}(\mathbf{X}), \tag{31}$$

where

$$K(\mathbf{X}) = \frac{E\left(\dfrac{K}{N+K}\middle|\mathbf{X}\right)}{E\left(\dfrac{1}{N+K}\middle|\mathbf{X}\right)} \tag{32}$$

and

$$\lambda_i(\mathbf{X}) = \frac{E\left(\lambda_i\dfrac{K}{N+K}\middle|\mathbf{X}\right)}{E\left(\dfrac{K}{N+K}\middle|\mathbf{X}\right)}. \tag{33}$$

As in the Normal example, $\lambda_i(\mathbf{X})$ does not depend on $\mathbf{X}$ if $\phi(K, \boldsymbol{\lambda})$ concentrates all of its mass on a specific value of $\boldsymbol{\lambda}$. Good [13] derives a similar expression for $E(\mathbf{p}|\mathbf{X})$ for the case of $\boldsymbol{\lambda} = \mathbf{c}$.

Under this two-stage prior the Bayes estimate of $\mathbf{p}$ (assuming quadratic loss) is again a convex shrinking of the data towards a probability vector. Unlike the one-stage case, both the smoothing constant K and the "smooth" probability vector, $\boldsymbol{\lambda}$, may now depend on the data, $\mathbf{X}$. In general this dependence on $\mathbf{X}$ through expressions (32) and (33) is computationally intractable. In the next section we consider data-dependent smoothing constants that are easier to compute than (32), but which nevertheless result in estimators of $\mathbf{p}$ that have generally better frequency properties than the "unsmoothed estimator", $\mathbf{X}/N$.

We end this section with a few comments about the estimator $\hat{\mathbf{p}} = \mathbf{X}/N$. It is the unique, minimum-risk, unbiased estimator of $\mathbf{p}$ but it is not minimax (relative to squared-error loss). The unique, minimax estimator is a Bayes estimator of the form (28) with $K = \sqrt{N}$ and $\boldsymbol{\lambda} = \mathbf{c}$ (see Steinhaus [14] and Trybula [15]). On the other hand, $\hat{\mathbf{p}}$ is admissible (see Johnson [16]) so that there is no hope of a uniform improvement over $\hat{\mathbf{p}}$. The reason $\hat{\mathbf{p}}$ is admissible, however, stems from how well it does for very extreme values of $\mathbf{p}$. If the loss function is

$$L(\mathbf{T}, \mathbf{p}) = N\|\mathbf{T}-\mathbf{p}\|^2, \tag{34}$$

then the risk (expected loss) of $\hat{\mathbf{p}}$ is easily shown to be

$$R(\hat{\mathbf{p}}, \mathbf{p}) = 1-\|\mathbf{p}\|^2. \tag{35}$$

If $\mathbf{p}$ has only one positive coordinate (an extreme value of $\mathbf{p}$), then the risk of $\hat{\mathbf{p}}$ is zero. In section 4, we show that when t is large and N/t is moderate then certain choices of data-dependent smoothing constants lead to improved estimators of $\mathbf{p}$, provided that $\mathbf{p}$ is not near the extreme values of S_t.

3. *Examples of data-dependent smoothing constants*

In this section we motivate three specific data-dependent smoothing constants for the case when $\boldsymbol{\lambda}$ is a fixed, specified vector (e.g., $\boldsymbol{\lambda} = \mathbf{c}$). We also introduce a general class of data-dependent smoothing constants $\mathscr{C}(\boldsymbol{\lambda})$. In section 4.3 we investigate the frequency properties of estimators of $\mathbf{p}$ based on these smoothing constants.

3.1. *The method of marginal moments*

Suppose we adopt the one-stage Bayesian approach of section 2.1. Then the marginal distribution of $\mathbf{X}$ is

$$\begin{aligned} P\{\mathbf{X} = \mathbf{x}|K, \boldsymbol{\lambda}\} &= \int P\{\mathbf{X} = \mathbf{x}|\mathbf{p}\} f(\mathbf{p}|K, \boldsymbol{\lambda})\, \mathrm{d}\mathbf{p} \\ &= \binom{N}{\mathbf{x}} \int \prod_i p_i^{x_i} G(K\boldsymbol{\lambda}) \prod_i p_i^{K\lambda_j - 1}\, \mathrm{d}\mathbf{p} \\ &= \binom{N}{\mathbf{x}} \frac{G(K\boldsymbol{\lambda})}{G(K\boldsymbol{\lambda}+\mathbf{x})}. \end{aligned} \tag{36}$$

Expressions (36) and (24) show the explicit dependence of the marginal distribution of **X** on the parameters $(K, \boldsymbol{\lambda})$ of the prior. For the remainder of this section we shall be concerned with this marginal distribution. As mentioned earlier, we assume $\boldsymbol{\lambda}$ to be a specified vector while we regard K as unknown and in need of estimation.

Good [1] suggests that the maximum likelihood estimate of K relative to the density (36) be used as a smoothing constant. It is evident from (36) that, in general, this maximum likelihood estimate of K can only be found using a computer.

Good also suggests basing a method of moments estimator of K on the sample "repeat rate" statistic

$$S = \frac{1}{N(N-1)} \sum_{i=1}^{t} X_i(X_i - 1). \tag{37}$$

As this approach results in a more tractable estimate of K than the previously mentioned maximum likelihood approach, we present further details here. Good calls S the "sample repeat rate" because its conditional expectation given $\mathbf{p}$ is $\Sigma_{i=1}^{t} p_i^2$, which is the probability that two independent observations drawn from the distribution $p_1, \ldots, p_t$ are tied (i.e., the population repeat rate). We are interested in the unconditional expected value of S. It is well known that

$$E(\mathbf{X}|\mathbf{p}) = N\mathbf{p}, \tag{38}$$

and

$$\mathrm{Cov}(\mathbf{X}|\mathbf{p}) = N(D_p - \mathbf{pp}'), \tag{39}$$

where D_p is the diagonal matrix based on $\mathbf{p}$. Now using (26), we get

$$E(\mathbf{p}|K, \boldsymbol{\lambda}) = \boldsymbol{\lambda}, \tag{40}$$

and

$$\mathrm{Cov}(\mathbf{p}|K, \boldsymbol{\lambda}) = \frac{1}{K+1}(D_\lambda - \boldsymbol{\lambda\lambda}'). \tag{41}$$

Combining (38) and (40) we have

$$E(\mathbf{X}|K, \boldsymbol{\lambda}) = E(E(\mathbf{X}|\mathbf{p})|K, \boldsymbol{\lambda})$$
$$= E(N\mathbf{p}|K, \boldsymbol{\lambda}), \tag{42}$$

or

$$E(\mathbf{X}|K, \boldsymbol{\lambda}) = N\boldsymbol{\lambda}. \tag{43}$$

To compute $\mathrm{Cov}(\mathbf{X}|K, \boldsymbol{\lambda})$ we make use of the rule that "the marginal covariance equals the mean of the conditional covariance plus the covariance of the conditional mean". Thus,

$$\begin{aligned}\operatorname{Cov}(\mathbf{X}|K, \boldsymbol{\lambda}) &= E(\operatorname{Cov}(\mathbf{X}|\mathbf{p})|K, \boldsymbol{\lambda}) + \operatorname{Cov}(E(\mathbf{X}|\mathbf{p})|K, \boldsymbol{\lambda}) \\ &= E(N(D_p - \mathbf{pp}')|K, \boldsymbol{\lambda}) + \operatorname{Cov}(N\mathbf{p}|K, \boldsymbol{\lambda}) \\ &= N\{D_\lambda - E(\mathbf{pp}'|K, \boldsymbol{\lambda}) + \frac{N}{K+1}(D_\lambda - \boldsymbol{\lambda\lambda}')\}. \end{aligned} \tag{44}$$

Since

$$E(\mathbf{pp}'|K, \boldsymbol{\lambda}) = \operatorname{Cov}(\mathbf{p}|K, \boldsymbol{\lambda}) + \boldsymbol{\lambda\lambda}', \tag{45}$$

if we combine terms in (44), we have

$$\operatorname{Cov}(\mathbf{X}|K, \boldsymbol{\lambda}) = \frac{K+N}{K+1} N(D_\lambda - \boldsymbol{\lambda\lambda}'). \tag{46}$$

Resolving (46) into the corresponding marginal variances and correlations we find:

$$\operatorname{Var}(X_i) = \frac{K+N}{K+1} N\lambda_i(1-\lambda_i) \tag{47}$$

and

$$\rho(X_i, X_j) = -\sqrt{\frac{\lambda_i}{1-\lambda_i}\,\frac{\lambda_j}{1-\lambda_j}}. \tag{48}$$

Hence we see that up to the level of second moments, the distribution of $\mathbf{X}$ depends on K only through the variances of the X_i. This suggests that the use of the sample repeat rate is quite reasonable, but it also suggests we consider other quadratic functions of the X_i. We do not examine the effect of other quadratic functions of the X_i in this paper.

We now compute the unconditional expected value of S, i.e.,

$$E(S|K, \boldsymbol{\lambda}) = \frac{1}{N(N-1)} \sum_i E(X_i^2|K, \boldsymbol{\lambda}) - \frac{1}{N-1}. \tag{49}$$

Using (43) and (47) in (49), after some algebraic manipulation we obtain

$$E(S) = \frac{1+K\sum \lambda_i^2}{1+K}. \tag{50}$$

Solving for K yields

$$K = \frac{1-E(S)}{E(S)-\sum \lambda_i^2}. \tag{51}$$

We obtain the "method of marginal moments" estimate of K, $\hat{K}_{MMM}$, by substituting S for $E(S)$ in (51) and simplifying the result. This yields

$$\hat{K}_{MMM} = \frac{N^2 - \sum X_i^2}{\sum X_i^2 - N(N-1)\sum \lambda_i^2 - N}. \tag{52}$$

When $\boldsymbol{\lambda} = \mathbf{c}$, the center of S_t, $\hat{K}_{MMM}$ reduces to

$$\hat{K}_{MMM} = \frac{N^2 - \sum X_i^2}{\sum X_i^2 - N(N+t-1)/t}. \tag{53}$$

3.2. *Estimates of the optimal smoothing constant*

We now derive two more data-dependent choices of K for $\boldsymbol{\lambda}$ specified. To do this we concentrate on the conditional distribution of $\mathbf{X}$ given $\mathbf{p}$, which we take, as usual, to be multinomial. We continue to use squared-error loss for evaluating different estimators of $\mathbf{p}$, so that the risk function of a generic estimator, $\mathbf{T}$, is

$$R(\mathbf{T}, \mathbf{p}) = NE\|\mathbf{T} - \mathbf{p}\|^2 = N\sum_{i=1}^{t} E(T_i - p_i)^2. \tag{54}$$

Suppose we consider estimates of $\mathbf{p}$ of the form,

$$\hat{\mathbf{q}}(K, \boldsymbol{\lambda}) = \frac{N}{N+K}(\mathbf{X}/N) + \frac{K}{N+K}\boldsymbol{\lambda}, \tag{55}$$

where $K \geqq 0$ and $\boldsymbol{\lambda}$ is a probability vector with non-zero components. The risk function for $\hat{\mathbf{q}} = \hat{\mathbf{q}}(K, \boldsymbol{\lambda})$ is

$$R(\hat{\mathbf{q}}, \mathbf{p}) = \left(\frac{N}{N+K}\right)^2 (1 - \|\mathbf{p}\|^2) + \left(\frac{K}{N+K}\right)^2 N\|\mathbf{p} - \boldsymbol{\lambda}\|^2, \tag{56}$$

(Fienberg and Holland [17]). If we differentiate (56) in K in order to minimize the risk for a specified $\boldsymbol{\lambda}$ and particular value of $\mathbf{p}$, we obtain

$$K_0(\mathbf{p}, \boldsymbol{\lambda}) = (1 - \|\mathbf{p}\|^2)/\|\mathbf{p} - \boldsymbol{\lambda}\|^2. \tag{57}$$

We regard $K_0(\mathbf{p}, \boldsymbol{\lambda})$ as the optimal choice of K for a given $\boldsymbol{\lambda}$. It is a parameter that we proceed to estimate in two different ways.

3.2.1. *Maximum likelihood estimate of $K_0(\mathbf{p}, \boldsymbol{\lambda})$*

Since $K_0(\mathbf{p}, \boldsymbol{\lambda})$ is a function of $\mathbf{p}$, its maximum likelihood estimate is found by replacing $\mathbf{p}$ by $\hat{\mathbf{p}} = \mathbf{X}/N$, yielding

$$\hat{K}_{0ml} = (1 - \|\hat{\mathbf{p}}\|^2)/\|\hat{\mathbf{p}} - \boldsymbol{\lambda}\|^2 \tag{58}$$

or

$$\hat{K}_{0ml} = \frac{N^2 - \sum X_i^2}{\sum X_i^2 - 2N \sum X_i \lambda_i + N^2 \sum \lambda_i^2}. \tag{59}$$

When $\boldsymbol{\lambda} = \mathbf{c}$, the center of S_t, $\hat{K}_{0ml}$ reduces to

$$\hat{K}_{0ml} = \frac{N^2 - \sum X_i^2}{\sum X_i^2 - N^2/t}. \tag{60}$$

This maximum likelihood estimate of $K_0(\mathbf{p}, \boldsymbol{\lambda})$ should not be confused with the maximum likelihood estimate of K mentioned in the previous section, because they maximize different likelihood functions.

3.2.2. *Ratio unbiased estimate of* $K_0(\mathbf{p}, \boldsymbol{\lambda})$

It is impossible to obtain an unbiased estimator of $K_0(\mathbf{p}, \boldsymbol{\lambda})$ since (57) is a rational function of $\mathbf{p}$, but it is possible to obtain unbiased estimators of both the numerator and denominator separately. Define $\theta_1(\mathbf{p})$ and $\theta_2(\mathbf{p})$ by

$$\theta_1 = \theta_1(\mathbf{p}) = \sum_i p_i^2 \tag{61}$$

and

$$\theta_2 = \theta_2(\mathbf{p}) = \sum_i p_i \lambda_i. \tag{62}$$

Thus, $K_0(\mathbf{p}, \boldsymbol{\lambda})$ depends on $\mathbf{p}$ only through θ_1 and θ_2 i.e.,

$$K_0(\mathbf{p}, \boldsymbol{\lambda}) = (1 - \theta_1)/(\theta_1 - 2\theta_2 + \sum \lambda_i^2). \tag{63}$$

As stated in section 3.1, an unbiased estimate of θ_1 is given by the sample repeat rate,

$$S = \frac{1}{N(N-1)} \sum_i X_i(X_i - 1). \tag{64}$$

Therefore, unbiased estimates of the numerator and denominator of $K_0(\mathbf{p}, \boldsymbol{\lambda})$ are given by

$$1 - S \tag{65}$$

and

$$S - (2/N) \sum X_i \lambda_i + \sum \lambda_i^2, \tag{66}$$

respectively. Taking the ratio of (65) to (66) and simplifying the result produces the "ratio unbiased" estimate of $K_0(\mathbf{p}, \boldsymbol{\lambda})$,

$$\hat{K}_{0ru} = \frac{N^2 - \sum X_i^2}{\sum X_i^2 - 2(N-1)\sum X_i\lambda_i + N(N-1)\sum \lambda_i^2 - N}. \tag{67}$$

When $\boldsymbol{\lambda} = \mathbf{c}$, the center of S_t, $\hat{K}_{0ru}$ reduces to

$$\hat{K}_{0ru} = \frac{N^2 - \sum X_i^2}{\sum X_i^2 - N(N+t-1)/t}, \tag{68}$$

so that $\hat{K}_{0ru}$ and $\hat{K}_{MMM}$ are identical in this case.

3.3. *A general class of data-dependent smoothing constants, $\mathscr{C}(\boldsymbol{\lambda})$*

The forms of $\hat{K}_{MMM}$, $\hat{K}_{0ml}$ and $\hat{K}_{0ru}$ given in (52), (59) and (67) are similar, and they suggest we consider a class of data-dependent smoothing constants that contains each as a special case. We define the class $\mathscr{C}(\boldsymbol{\lambda})$ as follows:

$$\mathscr{C}(\boldsymbol{\lambda}) = \left\{ K(\mathbf{X}) : K(\mathbf{X}) = \frac{a^{(N)} + b\sum X_i^2}{c\sum X_i^2 + d^{(N)}\sum X_i\lambda_i + e^{(N)}\sum \lambda_i^2 + f^{(N)}} \right\} \tag{69}$$

where b and c are constants and $a^{(N)}$, $d^{(N)}$, $e^{(N)}$ and $f^{(N)}$ are polynomials in N of the form

$$\begin{aligned} a^{(N)} &= a_1 N^2 + a_2 N + a_3 \\ d^{(N)} &= d_1 N + d_2 \\ e^{(N)} &= e_1 N^2 + e_2 N + e_3 \\ f^{(N)} &= f_1 N + f_2 . \end{aligned} \tag{70}$$

The constants $\{a_i\}$, b, c, $\{d_i\}$, $\{e_i\}$, and $\{f_i\}$ are all assumed to be independent of N, t, $\boldsymbol{\lambda}$ and $\mathbf{p}$. Table 1 shows the values of the constants in $\mathscr{C}(\boldsymbol{\lambda})$ that yield $\hat{K}_{MMM}$, $\hat{K}_{0ml}$ and $\hat{K}_{0ru}$.

We obtain a wide class of estimators of $\mathbf{p}$ by considering those of the form

$$\hat{\mathbf{q}} = \frac{N}{N+\hat{K}}(\mathbf{X}/N) + \frac{\hat{K}}{N+\hat{K}}\boldsymbol{\lambda} \tag{71}$$

where $\hat{K}$ is an element of $\mathscr{C}(\boldsymbol{\lambda})$. Unfortunately, we cannot show that any

TABLE 1

	$\hat{K}_{MMM}$	$\hat{K}_{Oml}$	$\hat{K}_{Oru}$
$a^{(N)}$	N	N	N
b	-1	-1	-1
c	1	1	1
$d^{(N)}$	0	$-2N$	$-2N+2$
$e^{(N)}$	$-N^2+N$	N^2	N^2-N
$f^{(N)}$	$-N$	0	$-N$

Values of constants in $\mathscr{C}(\lambda)$ that yield $\hat{K}_{MMM}$, $\hat{K}_{Oml}$ and $\hat{K}_{Oru}$.

of the Bayesian smoothing constants given by the two-stage Bayesian approach are elements of $\mathscr{C}(\boldsymbol{\lambda})$. It is probable that some elements of $\mathscr{C}(\boldsymbol{\lambda})$ approximate these Bayesian smoothing constants. In section 4.3 we give an approximation to the risk function of $\hat{\mathbf{q}}$ in (71) that holds when N and t are both large.

4. *The asymptotics of large sparse multinomials*

4.1. *Notation and assumptions*

We wish to compare the risk functions of various estimators of $\mathbf{p}$, including the highly non-linear estimators given by (71) where $\hat{K} \in \mathscr{C}(\boldsymbol{\lambda})$. Since the risk functions of these estimators are, in general, too complicated for useful exact comparisons, we shall make approximate risk calculations that hold for large N and t. We refer to the framework we shall presently introduce as the asymptotics of large, sparse multinomials (LSM), because it applies when N and t are both large but $\delta = N/t$ is moderate. By letting $\delta \to \infty$ we may also obtain results for the usual asymptotic framework, where $N \to \infty$ and t remains fixed; however, we do not focus on the usual asymptotic framework here (see Fienberg and Holland [17, 18] for details).

If we let $t \to \infty$ as well as N, then instead of a single probability vector, $\mathbf{p}$, we must consider an infinite sequence of probability vectors whose dimensions increase without bound. We choose to simplify the structure of the situation by "binding together" the probability vectors in this infinite sequence via the following device. Let $p(\cdot)$ be a probability density function on [0, 1]. For each t, we let

$$p_i = t^{-1}p\left(\frac{i-\frac{1}{2}}{t}\right) \quad i = 1, 2, \ldots, t. \tag{72}$$

Strictly speaking, the $\{p_i\}$ defined in (72) should depend explicitly on t (i.e., $p_{i,t}$) but we do not use this notation here. Furthermore, the vector $\mathbf{p}' = (p_1, \ldots, p_t)$ defined by (72) is not necessarily an element of the probability simplex, S_t, because the p_i need not sum to one. If $p(\cdot)$ is sufficiently smooth, however, standard results in numerical integration (Davis and Rabinowitz [19]) show that

$$\sum_{i=1}^{t} p_i = \sum_{i=1}^{t} p\left(\frac{i-\frac{1}{2}}{t}\right)t^{-1} = \int_0^1 p(x)\,\mathrm{d}x + o(t^{-1}) \tag{73}$$
$$= 1 + o(t^{-1}) \quad \text{as} \quad t \to \infty.$$

Hence the p_i sum to unity as $t \to \infty$ with an error that is small compared to t^{-1}. As will become evident, this is sufficient for our purposes.

Let $\lambda(\cdot)$ be a second probability density on [0, 1], and set

$$\lambda_i = t^{-1}\lambda\left(\frac{i-\frac{1}{2}}{t}\right) \quad i = 1, \ldots, t. \tag{74}$$

By assuming that both $p(\cdot)$ and $\lambda(\cdot)$ have continuous second-derivatives we have, for $\alpha, \beta \geqq 0$, that

$$\sum_{i=1}^{t} p_i^{\alpha}\lambda_i^{\beta} = t^{-(\alpha+\beta-1)}\int_0^1 p^{\alpha}(x)\lambda^{\beta}(x)\,\mathrm{d}x + o(t^{-(\alpha+\beta)})$$
$$\text{as} \quad t \to \infty. \tag{75}$$

Thus, we may replace summations involving p_i and λ_i by integrals involving $p(\cdot)$ and $\lambda(\cdot)$.

4.2. *Some examples of LSM asymptotics*

We now go through some simple calculations using the LSM asymptotics developed in section 4.1. Our purpose is to give the reader a feeling for the type of result that emerges in some simple situations.

4.2.1. *The risk of* $\mathbf{X}/N$

While we may easily calculate the exact risk function of the estimator

$$\hat{\mathbf{p}} = \mathbf{X}/N \tag{76}$$

we need its LSM approximation for later comparison with other estimators. This approximation is found as follows:

$$R(\hat{\mathbf{p}}, \mathbf{p}) = \sum_i p_i - \sum_i p_i^2$$

$$= \int_0^1 p(x)\,\mathrm{d}x + o(t^{-1}) - t^{-1}\int_0^1 p^2(x)\,\mathrm{d}x + o(t^{-2})$$

$$= 1 - t^{-1}\int p^2 + o(t^{-1}) \quad \text{as} \quad t \to \infty. \tag{77}$$

Henceforth, as in (77), we omit references to x, $\mathrm{d}x$ and the integration limits in all expressions using integrals. Equation (77) is prototypic of all LSM asymptotic expansions of risk functions in this paper. There is a dominant term (1 in this case), a t^{-1} term, and a remainder that is small compared to t^{-1}.

4.2.2. *The risk of adding α to each cell*

An often suggested method for smoothing contingency tables is to add a constant (e.g., 1, 0.5, or 0.25) to every cell. (See Good [1] and Fienberg and Holland [17] for references.) If we add α to each cell, we get the Bayes estimator

$$\hat{\mathbf{q}}(\alpha) = \hat{\mathbf{q}}(\alpha t, \mathbf{c}) = \frac{N}{N+\alpha t}\hat{\mathbf{p}} + \frac{\alpha t}{N+\alpha t}\mathbf{c}\,. \tag{78}$$

From (56) we may derive the LSM approximation to the exact risk function of $\hat{\mathbf{q}}(\alpha)$ as

$$R(\hat{\mathbf{q}}(\alpha), \mathbf{p}) = \left(\frac{\delta}{\delta+\alpha}\right)^2 (1 - \|\mathbf{p}\|^2) + \left(\frac{\alpha}{\delta+\alpha}\right)^2 (N\|\mathbf{p}\|^2 - \delta)$$

$$= \left(\frac{\delta}{\delta+\alpha}\right)^2 (1 - t^{-1}\textstyle\int p^2 + o(t^{-2})) \tag{79}$$

$$+ \left(\frac{\alpha}{\delta+\alpha}\right)^2 (\delta\textstyle\int p^2 - \delta + N o(t^{-2}))$$

$$= \left\{\left(\frac{\delta}{\delta+\alpha}\right)^2 + \left(\frac{\alpha}{\delta+\alpha}\right)^2 \delta[\textstyle\int p^2 - 1]\right\}$$

$$- t^{-1}\left(\frac{\delta}{\delta+\alpha}\right)^2 \textstyle\int p^2 + o(t^{-1})$$

$$\text{as} \quad t \to \infty. \tag{80}$$

The dominant term of (80), contained in braces, may be compared with the dominant term in (77). It is a linear function of the quantity

$$D_0 = \delta[\int p^2 - 1]. \tag{81}$$

D_0 is always non-negative and can be made as large as we please by a suitable choice of $p(\cdot)$. Hence the dominant term of (80) will exceed that of (77) if D_0 is large enough. This implies that the common practice of "adding $\frac{1}{2}$" is not a good idea in general, as it can produce an estimator whose risk is larger than that of $\hat{\mathbf{p}}$, even for values of $\mathbf{p}$ that are not extreme.

4.2.3. *Risk of the minimax estimator*

Trybula [15] shows that the estimator of $\mathbf{p}$ that minimizes the maximum value of its risk function under quadratic loss is

$$\hat{\mathbf{p}}_M = \hat{\mathbf{q}}(\sqrt{N}, \mathbf{c}) = \frac{\sqrt{N}}{\sqrt{N}+1}\hat{\mathbf{p}} + \frac{1}{\sqrt{N}+1}\mathbf{c}. \tag{82}$$

The estimator $\hat{\mathbf{p}}_M$ is the unique minimax estimator of $\mathbf{p}$ and has constant risk (over the entire simplex), i.e.,

$$R(\hat{\mathbf{p}}_M, \mathbf{p}) = (1+N^{-\frac{1}{2}})^{-2}(1-t^{-1}). \tag{83}$$

If we expand (83) out to order t^{-1} we obtain a formula analogous to (77) and (80):

$$R(\hat{\mathbf{p}}_M, \mathbf{p}) = 1 - \frac{2}{\sqrt{N}} + t^{-1}((3/\delta)-1) + o(t^{-1}). \tag{84}$$

We see that the dominant term of (84) is uniformly smaller than that of (77) but for large values of $\sqrt{N}$ the improvement is negligible.

4.2.4. *Risk of using the optimal smoothing constant*

In section 3.2, we introduced the optimal smoothing constant, $K_0(\mathbf{p}, \lambda)$, given in (57). We now put K_0 into the risk function (56) to obtain a lower bound on the risk function for any estimator of the form (55). A little algebra reveals that

$$\min_K R(\hat{\mathbf{q}}(K, \lambda), \mathbf{p}) = \left(\frac{N}{N+K_0}\right)(1 - \|\mathbf{p}\|^2). \tag{85}$$

Now the LSM expansion of $K_0(\mathbf{p}, \boldsymbol{\lambda})$ is given by

$$K_0(\mathbf{p}, \boldsymbol{\lambda}) = \frac{1-\|\mathbf{p}\|^2}{\|\mathbf{p}-\boldsymbol{\lambda}\|^2} = \frac{1-t^{-1}\int p^2+o(t^{-2})}{t^{-1}\int(p-\lambda)^2+o(t^{-2})}$$

$$= N\left(\frac{1-t^{-1}\int p^2+o(t^{-2})}{\delta\int(p-\lambda)^2+o(t^{-1})}\right)$$

$$= ND^{-1}-(Dt)^{-1}\int p^2+o(t^{-1}) \qquad (86)$$

where

$$D = \delta\int(p-\lambda)^2. \qquad (87)$$

Note that D in (87) is identical to D_0 in (81) when $\lambda \equiv 1$. We obtain a LSM expansion of the lower bound given in (85) by substituting (86) into (85) and simplifying. This yields

$$\min_K R(\hat{\mathbf{q}}(K, \boldsymbol{\lambda}), \mathbf{p}) = \frac{D}{D+1} - t^{-1}\left(\frac{D}{D+1}\right)^2 \int p^2 + o(t^{-1}). \qquad (88)$$

From (88) we see that the dominant term of this lower bound is $D/(D+1)$. In the next subsection we show that estimators of the form (55) with a data-dependent $\hat{K}$ can actually achieve this lower bound on the dominant term of the risk function.

4.3. *The risk of using the smoothing constants in $\mathscr{C}(\lambda)$*

Applying LSM asymptotics to the risk of estimators of the form

$$\hat{\mathbf{q}}(\hat{K}, \boldsymbol{\lambda}) = \frac{N}{N+\hat{K}}\hat{\mathbf{p}} + \frac{\hat{K}}{N+\hat{K}}\boldsymbol{\lambda} \qquad (89)$$

where $\hat{K} \in \mathscr{C}(\boldsymbol{\lambda})$, Sutherland [20] has proved the following result:

Theorem 1. *If $\boldsymbol{\lambda}$ is a fixed probability vector, $\hat{\mathbf{q}}(\hat{K}, \boldsymbol{\lambda})$ is of the form* (89), *and $\hat{K} \in \mathscr{C}(\boldsymbol{\lambda})$, then the risk of the estimator $\hat{\mathbf{q}} = \hat{\mathbf{q}}(\hat{K}, \boldsymbol{\lambda})$ is given by*:

$$R(\hat{\mathbf{q}}, \mathbf{p}) = S_0 + \frac{1}{N}S_1 - \frac{1}{t}\int p^2\, S_2 + \frac{1}{t}\int \lambda p\, S_3 + \frac{1}{t}\int \lambda^2\, S_4$$

$$+\frac{\delta}{t}\sigma^2(p)\, S_5 - \frac{\delta}{t}\sigma^2(\lambda)\, S_6 - \frac{\delta}{t}\sigma(\lambda, p) S_7 + o(t^{-1}), \qquad (90)$$

where

$$S_0 = (a_1^2 D + A^2)/(a_1 + A)^2, \tag{91}$$

$$S_1 = 2A((f_2 - a_2 - b)(A - a_1 D) + a_1(1 - a_1)f_2)/(a_1 + A)^3, \tag{92}$$

$$S_2 = (A^4 + 2a_1 A^3 + a_1(a_1 - 4c)A^2 - 2a_1 c(2a_1 + c)A - 2a_1^2 c^2 + 2(A - a_1 D)(bA^2 + a_1(b + c)A + a_1 c(a_1 + 3c)))/(a_1 + A)^4, \tag{93}$$

$$S_3 = 2a_1 d_2(A - a_1 D)/(a_1 + A)^3, \tag{94}$$

$$S_4 = 2a_1 e_2(A - a_1 D)/(a_1 + A)^3, \tag{95}$$

$$S_5 = 4a_1 c(A^2 + cA - 3c(A - a_1 D) + a_1(c - a_1))/(a_1 + A)^4, \tag{96}$$

$$S_6 = a_1 d_1(2A^2 - d_1 A + 3d_1(A - a_1 D) - a_1(2a_1 + d_1))/(a_1 + A)^4, \tag{97}$$

$$S_7 = 2a_1((2c - d_1)A^2 - 2cd_1 A + 6cd_1(A - a_1 D) - a_1(2c(a_1 + d_1) - a_1 d_1))/(a_1 + A)^4, \tag{98}$$

and

$$A = c + f_1 + c\delta \int p^2 + d_1 \delta \int \lambda p + e_1 \delta \int \lambda^2, \tag{99}$$

$$D = \delta \int (\lambda - p)^2, \tag{100}$$

$$\sigma^2(p) = \int p^3 - (\int p^2)^2, \tag{101}$$

$$\sigma^2(\lambda) = \int \lambda^2 p - (\int \lambda p)^2, \tag{102}$$

$$\sigma(\lambda, p) = \int \lambda p^2 - \int p^2 \int \lambda p. \tag{103}$$

The proof of this result is straightforward but tedious and is omitted.

To $o(1)$, theorem 1 shows that we may express the risk of estimators based on smoothing constants in the class $\mathscr{C}(\boldsymbol{\lambda})$ as

$$R(\hat{\mathbf{q}}, \mathbf{p}) = \frac{a_1^2 D + A^2}{(a_1 + A)^2} + o(1). \tag{104}$$

The class $\mathscr{C}(\boldsymbol{\lambda})$ contains the three special cases developed in section 3: $\hat{K}_{0ml}$, $\hat{K}_{0ru}$ and $\hat{K}_{MMM}$. Applying (104) using the values in table 1 we obtain:

$$(\hat{\mathbf{q}}(\hat{K}_{0ml}, \boldsymbol{\lambda}), \mathbf{p}) = \frac{D^2 + 3D + 1}{(D + 2)^2} + o(1), \tag{105}$$

$$R(\hat{\mathbf{q}}(\hat{K}_{0ru}, \boldsymbol{\lambda}), \mathbf{p}) = \frac{D}{\mathrm{D}+1} + o(1), \tag{106}$$

and

$$R(\hat{\mathbf{q}}(\hat{K}_{MMM}, \boldsymbol{\lambda}), \mathbf{p}) = \frac{D+A^2}{(A+1)^2} + o(1), \tag{107}$$

where $A = \delta\int p^2 - \delta\int \lambda^2$. We observe that the dominant term of (106) is identical to the dominant term of the lower bound given in (88). Thus the ratio unbiased estimator of K_0, $\hat{K}_{0ru}$ given by (67), leads to an estimator of $\mathbf{p}$ whose risk function possesses a LSM dominant term equal to the LSM dominant term of the optimal smoothing constant which minimizes the risk function among all Bayes estimators of the form (55).

5. *Optimal smoothing constants in the class, $\mathscr{C}(\boldsymbol{\lambda})$*

In this section we show that

$$S_0 = \frac{D}{D+1} \tag{108}$$

is the uniform lower bound on the dominant term of (90) over the class, $\mathscr{C}(\lambda)$.

Once again, our estimators are of the form:

$$\hat{\mathbf{q}} = \hat{\mathbf{q}}(\hat{K}, \boldsymbol{\lambda}) = \frac{N}{N+\hat{K}}\hat{\mathbf{p}} + \frac{\hat{K}}{N+\hat{K}}\boldsymbol{\lambda}, \tag{109}$$

where $\hat{K} \in \mathscr{C}(\boldsymbol{\lambda})$. In section 4.3 we showed that to $o(1)$, the risk of $\hat{\mathbf{q}}$ is expressible as

$$R(\hat{\mathbf{q}}, \mathbf{p}) = S_0 + o(1), \tag{110}$$

where

$$S_0 = \frac{a_1^2 D + A^2}{(a_1+A)^2}, \tag{111}$$

and

$$D = \delta\int(p-\lambda)^2, \tag{112}$$

$$A = c + f_1 + c\delta\int p^2 + d_1\delta\int \lambda p + e_1\delta\int \lambda^2 \tag{113}$$

$$= A^* + D,$$

with

$$A^* = c+f_1+(c-1)\delta\int p^2+(d_1+2)\delta\int p\lambda+(e_1-1)\delta\int\lambda^2. \quad (114)$$

S_0 is therefore dependent on $p(\cdot)$, $\lambda(\cdot)$, δ and the constants a_1, c, d_1, e_1 and f_1.

First we show that $D/(D+1)$ is a lower bound for S_0, and then we show that there is a unique choice of constants a_1, c, d_1, e_1 and f_1 that makes S_0 equal to this lower bound.

Theorem 2. S_0, *defined by* (110) *to* (114), *is uniformly bounded from below by* $D/(D+1)$.

Proof. Assume not. Then there is a choice of $p(\cdot)$, $\lambda(\cdot)$, δ and the constants a_1, c, d_1, e_1 and f_1 such that the following strict inequality holds

$$S_0 < \frac{D}{D+1}. \quad (115)$$

Then

$$\frac{a_1^2 D+(A^*+D)^2}{(a_1+A^*+D)^2} < \frac{D}{D+1},$$

or

$$(A^*+D(1-a_1))^2 < 0\,. \quad (116)$$

Thus, the strict inequality of (115) is impossible. ∎

We know from (106) that $\hat{K}_{0ru}$ is based on constants a_1, c, d_1, e_1 and f_1 such that

$$S_0 = \frac{D}{D+1}. \quad (117)$$

We wish to show this choice of the constants is the only one whose resulting risk has an LSM dominant term of $D/(D+1)$.

We start by examining the implications of (117) for a_1. Expression (117) demands for all $p(\cdot)$, $\lambda(\cdot)$ and δ that

$$\frac{a_1^2 D+A^2}{(a_1+A)^2} = \frac{D}{D+1},$$

which implies that

$$(a_1 D-A)^2 = 0. \quad (118)$$

Expression (118) is equivalent to

$$a_1 = a_1^*$$

where

$$a_1^* = \frac{A}{D}. \tag{119}$$

For a_1^* to be a possible value for a_1 in $\mathscr{C}(\lambda)$ it must be independent of $p(\cdot)$, $\lambda(\cdot)$ and δ. We show that if this is true, then it determines unique values for c, d_1, e_1 and f_1 and that a_1^* is a constant only if c, d_1, e_1 and f_1 assume these values.

Without loss of generality when assuming a_1^* is a constant we may assume it equals one, since any other non-zero value could be divided out of both the numerator and denominator of $\hat{K}$.

Theorem 3. *$a_1^* = 1$ for all $p(\cdot)$, $\lambda(\cdot)$ and δ if and only if $c = 1$, $d_1 = -2$, $e_1 = 1$ and $f_1 = -1$.*

Proof. (i) If $a_1^* = 1$ for all $p(\cdot)$, $\lambda(\cdot)$ and δ, then (119) implies that

$$1 = \frac{c+f_1+c\delta\int p^2+d_1\delta\int\lambda p+e_1\delta\int\lambda^2}{\delta\int p^2-2\delta\int\lambda p+\delta\int\lambda^2},$$

and hence

$$c+f_1+(c-1)\delta\int p^2+(d_1+2)\delta\int\lambda p+(e_1-1)\delta\int\lambda^2 = 0. \tag{120}$$

Equation (120) may be regarded as a linear polynomial in the three variables

$$\begin{aligned} x &= \delta\int p^2, \\ y &= \delta\int\lambda p, \\ z &= \delta\int\lambda^2. \end{aligned} \tag{121}$$

As δ ranges over the positive real numbers, and $p(\cdot)$ and $\lambda(\cdot)$ range over the set of all density functions on [0, 1], x, y and z are linearly independent. Therefore (120) holds only if the coefficients of this linear polynomial are all equal to zero. Hence we conclude

$$c = 1, \quad d_1 = -2, \quad e_1 = 1 \quad \text{and} \quad f_1 = -c = -1. \tag{122}$$

(ii) If (122) holds then by substituting these values in (119) we have $a_1^* = 1$ for all $p(\cdot)$, $\lambda(\cdot)$ and δ. ∎

The uniqueness of the constants a_1, c, d_1, e_1 and f_1 derives from the following considerations. First, in order for the lower bound to be obtained a_1 must equal a_1^* in (119). Second, in order for a_1^* to be a bona fide value of a_1 it must be independent of $p(\cdot)$, $\lambda(\cdot)$ and δ. Third, from theorem 3 this occurs if and only if c, d_1, e_1 and f_1 take on the specified values (implying that $a_1^* = 1$).

If we use the values $a_1 = 1$, $c = 1$, $d_1 = -2$, $e_1 = 1$ and $f_1 = -1$ in the risk expression (90) and the associated terms (91) to (98), then the LSM approximation to the risk function becomes

$$R(\hat{\mathbf{q}}, \mathbf{p}) = \frac{D}{D+1} - \frac{1}{t}\left[\frac{D^3+2D^2-4D-2}{(D+1)^3}\right]\int p^2 \qquad (123)$$

$$+\frac{\delta}{t}\frac{4D}{(D+1)^3}(\sigma^2(p)+\sigma^2(\lambda)-2\sigma(p,\lambda))+o(t^{-1}).$$

The $\hat{\mathbf{q}}$ in (123) is any $\hat{\mathbf{q}}$ of the form (71) with $\hat{K} \in \mathscr{C}(\boldsymbol{\lambda})$ and the constants a_1, c, d_1, e_1 and f_1 taking on the above values. The other constants in $\hat{K}$ are arbitrary.

In light of this result we may review the LSM dominant terms for $\hat{K}_{0ml}$, $\hat{K}_{0ru}$ and $\hat{K}_{MMM}$. Of the three, only $\hat{K}_{0ru}$ actually attains the first-order lower bound of $D/(D+1)$. Thus the estimator based on this smoothing constant will have smaller risk (when t is large) than any of the other estimators we have discussed. In section 6, we again turn to ratio unbiased estimation to choose a smoothing constant, $\hat{K}$, for the more complex case in which $\boldsymbol{\lambda}$ is no longer a fixed probability vector, but depends on the data.

Lastly, we should mention the need for a careful small sample analysis of the behavior of the various smoothing constants developed here so that something "less asymptotic" can be said about their respective behaviors. See Fienberg and Holland [17, 18] for a discussion of small sample properties of $\hat{K}_{0ml}$.

6. *Smoothing constants when $\boldsymbol{\lambda}$ is data-dependent*

A desire to smooth a given set of data towards a smoother fitted structure motivates our interest in using data-dependent values of $\boldsymbol{\lambda}$ as well as K. Examples of such procedures are smoothing a given histogram towards a fitted Normal or Poisson distribution, and smoothing a given two-way contingency table towards a smoother structure such as the fitted values under the assumption of independence. Such procedures create a value, $\hat{\boldsymbol{\lambda}}(\mathbf{X})$, for $\boldsymbol{\lambda}$ from the data.

An obvious method for incorporating a data-dependent $\hat{\boldsymbol{\lambda}}$ would be to substitute $\hat{\boldsymbol{\lambda}}$ for $\boldsymbol{\lambda}$ in our previously developed smoothing constants

$\hat{K}_{0ml}$, $\hat{K}_{0ru}$ and $\hat{K}_{MMM}$. Provided that the variability of $\hat{\boldsymbol{\lambda}}$ is small compared to that for $\hat{\mathbf{p}} = \mathbf{X}/N$, such a procedure should give results analogous to those of the corresponding fixed $\boldsymbol{\lambda}$ cases. While we have developed no theory to support this analogy, it appears that particularly when smoothing towards a fitted histogram with large N and t this "substitution method" provides an adequate, quick and easy method to generate smoothed estimators of $\mathbf{p}$. As a first approach to the case of a data-dependent $\hat{\boldsymbol{\lambda}}$ we define the following smoothing constants:

$$\hat{K}_{0ml}(\hat{\boldsymbol{\lambda}}) = \frac{N^2-\sum X_i^2}{\sum (X_i - N\hat{\lambda}_i)^2}, \tag{124}$$

$$\hat{K}_{0ru}(\hat{\boldsymbol{\lambda}}) = \frac{N^2-\sum X_i^2}{\sum X_i^2 - 2(N-1)\sum X_i\hat{\lambda}_i + N(N-1)\sum \hat{\lambda}_i^2 - N}, \tag{125}$$

$$\hat{K}_{MMM}(\hat{\boldsymbol{\lambda}}) = \frac{N^2-\sum X_i^2}{\sum X_i^2 - N(N-1)\sum \hat{\lambda}_i^2 - N}. \tag{126}$$

A second approach to choosing a $\hat{K}$ when using a data-dependent $\hat{\boldsymbol{\lambda}}$ is to estimate the value of K which minimizes the risk of resulting estimator of $\mathbf{p}$. The estimators are now of the form

$$\hat{\mathbf{q}} = \frac{N}{N+K}\hat{\mathbf{p}} + \frac{K}{N+K}\hat{\boldsymbol{\lambda}}, \tag{127}$$

and their risk is given by

$$R(\hat{\mathbf{q}}, \mathbf{p}) = NE\|\hat{\mathbf{q}} - \mathbf{p}\|^2. \tag{128}$$

Setting the derivative of (128) with respect to K equal to zero and solving gives the optimal K as

$$K_0(\mathbf{p}) = N\,\frac{E\left(\sum_{i=1}^{t} (\hat{p}_i - \hat{\lambda}_i)(\hat{p}_i - p_i)\right)}{E\left(\sum_{i=1}^{t} (p_i - \hat{\lambda}_i)(\hat{p}_i - \hat{\lambda}_i)\right)}. \tag{129}$$

Letting $E(\hat{\boldsymbol{\lambda}}(\mathbf{X})) = \boldsymbol{\lambda}(\mathbf{p}) = \boldsymbol{\lambda}$, and writing $\Sigma_i a_i b_i$ as $(\mathbf{a}, \mathbf{b})$ we may express (129) as

$$K_0(\mathbf{p}) = \frac{1 - \|\mathbf{p}\|^2 - NE(\hat{\mathbf{p}}, \hat{\boldsymbol{\lambda}}) + N(\mathbf{p}, \boldsymbol{\lambda})}{\|\mathbf{p} - \boldsymbol{\lambda}\|^2 + E\|\hat{\boldsymbol{\lambda}}\|^2 - \|\boldsymbol{\lambda}\|^2 - E(\hat{\mathbf{p}}, \hat{\boldsymbol{\lambda}}) + (\mathbf{p}, \boldsymbol{\lambda})}. \tag{130}$$

To actually use (130) we must be able to evaluate the quantities

$$\boldsymbol{\lambda}(\mathbf{p}) = (\lambda_1(\mathbf{p}), \lambda_2(\mathbf{p}), \ldots, \lambda_t(\mathbf{p})), \tag{131}$$

$$\|\mathbf{p}-\boldsymbol{\lambda}\|^2, \tag{132}$$

$$(\mathbf{p}, \boldsymbol{\lambda}), \tag{133}$$

$$\|\boldsymbol{\lambda}\|^2, \tag{134}$$

$$E(\hat{\mathbf{p}}, \hat{\boldsymbol{\lambda}}), \tag{135}$$

$$E\|\hat{\boldsymbol{\lambda}}\|^2. \tag{136}$$

In a two-dimensional contingency table (X_{ij}) we can evaluate the quantities in (131) through (136) when we take

$$\hat{\lambda}_{ij}(\mathbf{X}) = (X_{i+}X_{+j})/N^2 = \hat{p}_{i+}\hat{p}_{+j}\,.$$

In this case (131)–(136) become

$$\lambda_{ij} = E(\hat{\lambda}_{ij}) = \frac{1}{N}((N-1)p_{i+}p_{+j}+p_{ij}) \tag{137}$$

$$\|\mathbf{p}-\boldsymbol{\lambda}\|^2 = \left(\frac{N-1}{N}\right)^2 D, \tag{138}$$

$$(\mathbf{p}, \boldsymbol{\lambda}) = \left(\frac{N-1}{N}\right) H+\frac{1}{N} S, \tag{139}$$

$$\|\boldsymbol{\lambda}\|^2 = \left(\frac{N-1}{N}\right)^2 S_1S_2+\frac{N-1}{N^2}\, 2H+\frac{1}{N^2}\, S, \tag{140}$$

$$E(\hat{\mathbf{p}}, \hat{\boldsymbol{\lambda}}) = \frac{N^{(3)}}{N^3}\, H+\frac{N^{(2)}}{N^3}\,(S+S_1+S_2)+\frac{1}{N^2}, \tag{141}$$

$$E\|\hat{\boldsymbol{\lambda}}\|^2 = \frac{N^{(4)}}{N^4}\, S_1S_2+\frac{N^{(3)}}{N^4}\,(4H+S_1+S_2) + \frac{N^{(2)}}{N^4}\,(2S+2S_1+2S_2+1)+\frac{1}{N^3}, \tag{142}$$

where,

$$D = \sum_{i,j} (p_{ij}-p_{i+}p_{+j})^2, \tag{143}$$

$$H = \sum_{i,j} p_{ij}p_{i+}p_{+j}, \tag{144}$$

$$S = \sum_{i,j} p_{ij}^2 , \tag{145}$$

$$S_1 = \sum_{i} p_{i+}^2 , \tag{146}$$

$$S_2 = \sum_{j} p_{+j}^2 , \tag{147}$$

$$S_1 S_2 = \sum_{i,j} p_{i+}^2 p_{+j}^2 . \tag{148}$$

While $S_1 S_2$ is simply the product of S_1 and S_2 we list it separately, since we take an unbiased estimate of $S_1 S_2$ below. We note that

$$D = S + S_1 S_2 - 2H . \tag{149}$$

Substituting these values in (130), the optimal $K_0(\mathbf{p})$ for the two-dimensional table using the observed independence structure as $\hat{\lambda}$ is

$$K_2(\mathbf{p}) = \frac{1 - D - S_1 - S_2 + S_1 S_2}{D - \frac{1}{N}(3D - 2S + 2S_1 S_2) + \frac{1}{N^2}(4D - 2S + 2S_1 S_2)} . \tag{150}$$

Now (150) can be estimated in two ways. We can use maximum likelihood estimates for the components of (143) through (148) as we did for $\hat{K}_{0ml}$ and then use these in (150). This gives

$$\hat{D} = \sum_{i,j} (\hat{p}_{ij} - \hat{p}_{i+}\hat{p}_{+j})^2 , \tag{151}$$

$$\hat{H} = \sum_{i,j} \hat{p}_{ij}\hat{p}_{i+}\hat{p}_{+j} , \tag{152}$$

$$\hat{S} = \sum_{i,j} \hat{p}_{ij}^2 , \tag{153}$$

$$\hat{S}_1 = \sum_{i} \hat{p}_{i+}^2 , \tag{154}$$

$$\hat{S}_2 = \sum \hat{p}_{+j}^2 , \tag{155}$$

and

$$\hat{K}_{2ml} = \frac{1 - \hat{D} - \hat{S}_1 - \hat{S}_2 - \hat{S}_1 \hat{S}_2}{\hat{D} - \frac{1}{N}(3D - 2\hat{S} + 2\hat{S}_1 \hat{S}_2) + \frac{1}{N^2}(4\hat{D} - 2\hat{S} + 2\hat{S}_1 \hat{S}_2)} . \tag{156}$$

On the other hand, we may use the ratio unbiased approach as we did for $\hat{K}_{0ru}$ and replace the numerator and denominator of (150) with their unbiased estimates. This approach estimates (143) through (148) by

$$\tilde{D} = \frac{N}{N-3}\hat{S} - \frac{2N^2}{(N-2)(N-3)}\hat{H} + \frac{N^4}{N^{(4)}}\hat{S}_1\hat{S}_2 + \frac{N}{(N-1)(N-3)}(\hat{S}_1+\hat{S}_2) - \frac{N}{(N-1)(N-3)}, \tag{157}$$

$$\tilde{H} = \frac{N^2}{(N-1)(N-2)}\left(\hat{H} - \frac{1}{N}(\hat{S}+\hat{S}_1+\hat{S}_2) + \frac{2}{N^2}\right), \tag{158}$$

$$\tilde{S} = \frac{1}{N-1}(N\hat{S}-1), \tag{159}$$

$$\tilde{S}_1 = \frac{1}{N-1}(N\hat{S}_1-1), \tag{160}$$

$$\tilde{S}_2 = \frac{1}{N-1}(N\hat{S}_2-1), \tag{161}$$

$$\widetilde{S_1S_2} = \frac{N^4}{N^{(4)}}\left(\hat{S}_1\hat{S}_2 - \frac{4\hat{H}}{N} + \frac{2\hat{S}}{N^2} - \left(\frac{N-4}{N^2}\right)(\hat{S}_1+\hat{S}_2) + \frac{N-6}{N^3}\right) \tag{162}$$

Substituting (157)–(162) in (150) gives the ratio unbiased estimator of $K_2(\mathbf{p})$ which we denote as $\hat{K}_{2ru}$.

For the case of a two-way table, we now have five different ways of obtaining a data-dependent smoothing constant $\hat{K}$ when $\boldsymbol{\lambda} = \hat{\boldsymbol{\lambda}}$ is also data-dependent. These are $\hat{K}_{0ml}(\hat{\boldsymbol{\lambda}})$, $\hat{K}_{0ru}(\hat{\boldsymbol{\lambda}})$, $\hat{K}_{MMM}(\hat{\boldsymbol{\lambda}})$, $\hat{K}_{2ml}$ and $\hat{K}_{2ru}$. At the present time, we do not have any results that allow us to distinguish between them as we did for the class $\mathscr{C}(\boldsymbol{\lambda})$ in the specified $\boldsymbol{\lambda}$ case. However, the success of the ratio unbiased approach in that case suggests that this approach is a contender here. Unfortunately, it is the most complicated to calculate unless a computer is used. In the next and final section we compare all five smoothing constants on a set of real data.

7 An example

The data in table 2 come from Pearson [21] via Good [22], and give a cross-classification of the occupations of fathers and of their sons. The total number of cells is $t = 196$ and $N = 775$. Thus $\delta = N/t = 3.95$ and we would expect the LSM approximations of sections 4 and 5 to be valid.

TABLE 2

Contingency between the occupations of fathers and sons[a]

	(i)	(ii)	(iii)	(iv)	(v)	(vi)	(vii)	(viii)	(ix)	(x)	(xi)	(xii)	(xiii)	(xiv)	
(i)	28	0	4	0	0	0	1	3	3	0	3	1	5	2	50
(ii)	2	51	1	1	2	0	0	1	2	0	0	0	1	1	62
(iii)	6	5	7	0	9	1	3	6	4	2	1	1	2	7	54
(iv)	0	12	0	6	5	0	0	1	7	1	2	0	0	10	44
(v)	5	5	2	1	54	0	0	6	9	4	12	3	1	13	115
(vi)	0	2	3	0	3	0	0	1	4	1	4	2	1	5	26
(vii)	17	1	4	0	14	0	6	11	4	1	3	3	17	7	88
(viii)	3	5	6	0	6	0	2	18	13	1	1	1	8	5	69
(ix)	0	1	1	0	4	0	0	1	4	0	2	1	1	4	19
(x)	12	16	4	1	15	0	0	5	13	11	6	1	7	15	106
(xi)	0	4	2	0	1	0	0	0	3	0	20	0	5	6	41
(xii)	1	3	1	0	0	0	1	0	1	1	1	6	2	1	18
(xiii)	5	0	2	0	3	0	1	8	1	2	2	3	23	1	51
(xiv)	5	3	0	2	6	0	1	3	1	0	0	1	1	9	32
	84	108	37	11	122	1	15	64	69	24	57	23	74	86	775

[a] Presented originally by Karl Pearson [21]. The labels (i) to (xiv) have the following significance: (i) army, (ii) art, (iii) teacher, clerk, civil servant, (iv) crafts, (v) divinity, (vi) agriculture, (vii) landownership, (viii) law, (ix) literature, (x) commerce, (xi) medicine, (xii) navy, (xiii) politics and court, (xiv) scholarship and science. Rows indicate fathers' occupations, columns indicate sons' occupations.

7.1. *Smoothing all cells equally*

If we set $\boldsymbol{\lambda} = \mathbf{c}$ in this case, then smoothing consists of adding the same amount to each cell and renormalizing the result. Table 3 summarizes the values of K that would be used to smooth all cells equally for the five smoothing constants discussed in section 3.

TABLE 3

Values of $\hat{K}$ for data in table 2 when $\boldsymbol{\lambda} = \mathbf{c}$

Description	$\hat{K}$
Minimax $= \sqrt{N}$	27.8
"add one-half" $= t/2$	98.0
$\hat{K}_{0ml}$	65.3
$\hat{K}_{MMM} = \hat{K}_{0ru}$	71.4

We see quite a range of values in table 3. Given the approximate optimality of $\hat{K}_{0ru}$ we regard the minimax estimator as adding too few pseudo-counts while the traditional practice of "adding $\frac{1}{2}$" appears to add a few too many. The difference between $\hat{K}_{0ml}$ and $\hat{K}_{0ru}$ does not appear to be very important in this example.

7.2. *Smoothing to preserve row and column totals*

In order to preserve the row and column totals in the smoothed table, it is necessary to require that $\boldsymbol{\lambda}$ be data-dependent in such a way that

$$N\hat{\lambda}_{i+} = X_{i+} \quad \text{and} \quad N\hat{\lambda}_{+j} = X_{+j}\,.$$

One choice of $\hat{\boldsymbol{\lambda}}$ that does this is

$$\hat{\lambda}_{ij}(\mathbf{X}) = X_{i+}X_{+j}/N^2\,. \tag{163}$$

Fienberg and Holland [17] discuss other possible choices of $\hat{\boldsymbol{\lambda}}$ in this case.

In table 4 we give values for $\hat{K}$ for $\hat{\lambda}$ given in (163). In this example we see that there is general agreement among the various choices of $\hat{K}$ suggesting that it may sometimes be possible to avoid calculating estimates of the optimal value $K_2(\mathbf{p})$.

TABLE 4

Values of $\hat{K}$ for data in table 2 when $\hat{\lambda}$ is given by (163)

Description	$\hat{K}$
$\hat{K}_{0ml}(\hat{\lambda})$	95.2
$\hat{K}_{0ru}(\hat{\lambda})$	108.6
$\hat{K}_{MMM}(\hat{\lambda})$	103.1
$\hat{K}_{2ml}$	78.3
$\hat{K}_{2ru}$	87.3

References

[19] P. J. Davis and P. Rabinowitz, Numerical integration. Waltham, Mass., Blaisdell, 1967, 15–17 and 108–112.

[7] B. Efron and C. Morris, Limiting the risk of Bayes and empirical Bayes estimators—Part II: the empirical Bayes case. Journal of the American Statistical Association 67 (1972), 130–139.

[12] S. E. Fienberg and P. W. Holland, On the choice of flattening constants for estimating multinomial probabilities. Journal of Multivariate Analysis 2 (1972), 127–134.

[17] S. E. Fienberg and P. W. Holland, Methods for eliminating zero counts in contingency tables. In: Random counts on models and structures, edited by G. P. Patil. University Park, Pa., Pennsylvania State University Press, 1970, 233–260.

[18] S. E. Fienberg and P. W. Holland, Simultaneous estimation of multinomial cell probabilities. Journal of the American Statistical Association 68 (1973), 683–691.

[1] I. J. Good, The estimation of probabilities. Cambridge, Mass., Massachusetts Institute of Technology Press, 1965.

[13] I. J. Good, A Bayesian significance test for multinomial distributions (with discussion). Journal of The Royal Statistical Society (B) 29 (1967), 399–431.

[22] I. J. Good, On the estimation of small frequencies in contingency tables. Journal of The Royal Statistical Society (B) 18 (1956), 113–124.

[5] W. James and C. Stein, Estimation with quadratic loss. In: Proceedings of the Fourth Berkeley Symposium on Mathematical Statistics and Probability, 1. University of California Press, 1961, 361–379.

[16] B. McK. Johnson, On the admissible estimation for certain fixed sample binomial problems. Annals of Mathematical Statistics 42 (1971), 1579–1587.

[2] D. V. Lindley, Bayesian statistics, a review. Philadelphia, Society for Industrial and Applied Mathematics, 1971.

[6] D. V. Lindley, Discussion of a paper by C. Stein. Journal of The Royal Statistical Society (B), 24 (1962), 285–287

[8] D. V. Lindley and A. F. M. Smith, Bayes estimates for the linear model (with discussion). Journal of The Royal Statistical Society (B) 34 (1972), 1–42.

[3] F. Mosteller, On pooling data. Journal of the American Statistical Association 43 (1948), 231–242.

[21] K. Pearson, On the theory of contingency and its relation to association and normal correlation. Drapers Co. Res. Mem. Biometrics Series, 1904. (Reprinted in Karl Pearson's Early Papers, Cambridge University Press, 1948.)

[4] H. Robbins, An empirical Bayes approach to statistics. In: Proceedings of the Third Berkeley Symposium on Statistics and Probability, 1. University of California Press, 1955, 157–164.

[23] C. Stein, Confidence sets for the mean of a multivariate normal distribution (with discussion). Journal of The Royal Statistical Society (B) 24 (1962), 265–296.

[14] H. Steinhaus, The problem of estimation. Annals of Mathematical Statistics 28 (1957), 633–648.

[9] W. E. Strawderman, Proper Bayes minimax estimators of the multivariate Normal mean. Annals of Mathematical Statistics 42 (1971), 385–388.

[20] M. Sutherland, Estimation in large sparse multinomials. Unpublished Ph.D. dissertation, Department of Statistics, Harvard University, 1974.

[15] S. Trybula, Some problems of simultaneous minimax estimation. Annals of Mathematical Statistics 29 (1958), 245–253.

[11] G. S. Watson, Some Bayesian methods related to χ^2. Bulletin of International Statistical Institute (Proc. of 35th Session) 61, No. 1 (1965), 64–76.

[10] S. S. Wilks, Mathematical statistics. New York, Wiley, 1962.

S. E. Fienberg and A. Zellner (eds.)
Studies in Bayesian Econometrics and Statistics.

12.4 SOME COMMENTS ON "BAYES" ESTIMATORS*

G. C. TIAO and G. E. P. BOX

1. Introduction

In the Bayesian approach to inference, a posterior distribution of unknown parameters is produced as the normalized product of the likelihood and a prior distribution. Inferences about the unknown parameters are then based on the entire posterior distribution resulting from the one specific data set **y** which has actually occurred.

In recent years, some workers have used a dual approach which, in a sense, goes part of the way along the Bayes route and then switches to sampling theory. Specifically, the Bayesian posterior distribution is obtained, but not used to make inferences directly. Instead, it is employed to suggest appropriate functions of the data for use as estimators. Inference is then attempted by considering the *sampling properties* of these "Bayes" estimators.

We believe that while such an estimator undoubtedly serves some useful purpose, consideration of its sampling properties has little relevance to inference.

2. Description of posterior distributions

While the *whole* of the appropriate posterior distribution provides the means of making all relevant inference about a parameter or a set of parameters,[1] nevertheless a description of the posterior distribution

* Reprinted from the American Statistician 27, No. 1 (1973), 12–14, with permission of the American Statistical Association. This work was reported to the Fourth NBER-NSF Seminar on Bayesian Inference in Econometrics, Univ. of Chicago, May 1972. Research supported by the U.S. Army Research Office No. DA-ARO-D-31-124-G917.

[1] When, as is frequently the case for scientific problems, it is desired to represent the inferential situation appropriate when there is very little prior information relative to that expected from the data, Jeffreys and others [1], [2] have suggested how a suitable non-informative prior may be chosen.

may be needed in terms of a few numerical values. These should be chosen, *after* the posterior distribution has been calculated, in any way which will give a realistic impression of it.

For example, the description of the posterior distribution of fig. 1(a) is simple. It is a Normal distribution with mean 22 and standard

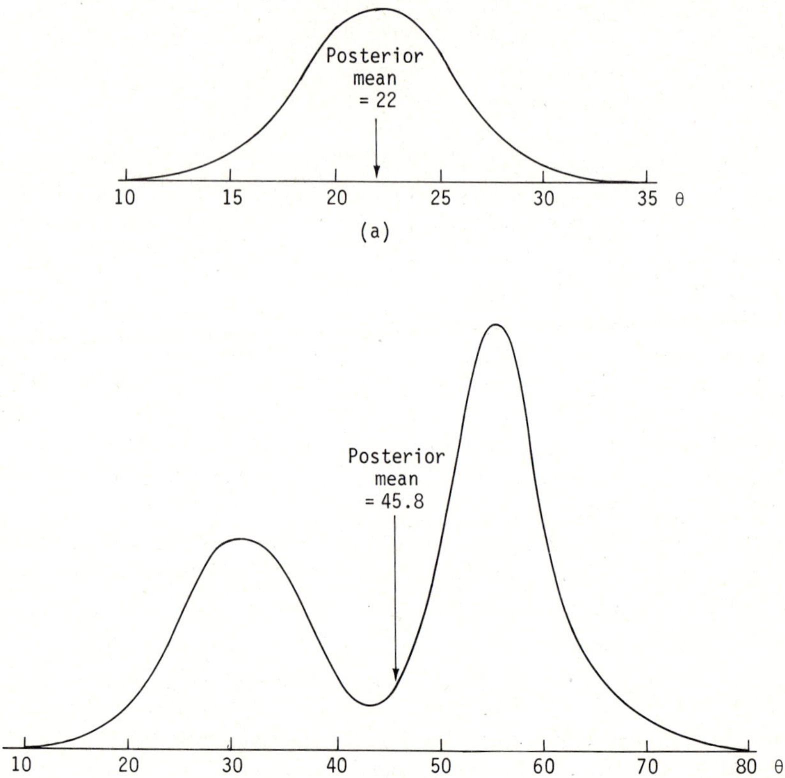

Fig. 1. Two posterior distributions.

deviation 3.7. Thus, with knowledge of the Normal form, two quantities suffice to provide an exact description.

Description of the posterior distribution in fig. 1(b) would be more difficult.[2] We might describe it thus: "It is a bimodal distribution with

[2] There is nothing unrealistic about this posterior distribution. Practical examples of this sort often occur in the analysis of scientific data [3].

most of its mass between 10 and 80. The left-hand mode at 30 is about half the height of the right-hand mode at 57. A local minimum occurs at 43 and is of about one-tenth the height of the right-hand mode." Alternatively, we could quote, say, 10 suitably spaced densities to give the main features of the distribution. However we proceeded, we would need to quote five or more different numbers to provide even an approximate idea of what we were talking about. On a Bayesian view, then, it would be the *descriptive power* and *not* the *sampling properties* of the quantities quoted which would be of interest.

Posterior distributions met in practice are usually neither so simple as the Normal nor so complex as the bimodal form just considered. But what *is* clear is (i) that the number and kinds of measures needed to approximately describe the posterior distribution must depend on the type and complexity of that distribution; (ii) questions about how to describe a posterior distribution cannot be (and do not need to be) decided until after that distribution is computed; (iii) they are decided simply on the basis of what would best convey the right picture to the investigator's mind; (iv) on this view the sampling properties of the quantities are irrelevant.

3. *Point estimators*

For many years, attempts of one sort or another have been made to obtain by some *prior* recipe a *single number* calculated from the data which in some way best represents a parameter under study. Historically there has been considerable ambivalence about objectives. Classical writers would for example sometimes state the problem as that of finding the most "probable" value and sometimes that of finding a best "average" or "mean" value.

3.1. *A measure of goodness of estimators*

Modern sampling theorists refer to the single representative quantity calculated from a sample $\mathbf{y}$ as a *point estimator* and they argue as follows. Suppose interest is centered on a particular parameter ϕ. Then any function $\hat{\phi}(\mathbf{y})$ of the data which provides some idea of the value of ϕ may be called an *estimator*. Usually a very large number of possible estimators can be devised. Therefore, a criterion of goodness is needed. It is argued that goodness of an estimator ought to be measured by its average closeness over all possible samples to the true value of ϕ. Many different criteria of average closeness could be used. For definitiveness we will illustrate our argument only in terms of one particular criterion—the mean squared error (m.s.e.). However, it is to be noted that difficulties which exactly parallel those we will discuss are encountered whichever measure is picked.

In terms of this criterion, among a number of possible candidate estimators $\hat{\phi}_1(\mathbf{y}), \ldots, \hat{\phi}_j(\mathbf{y}), \ldots, \hat{\phi}_k(\mathbf{y})$, the goodness of a particular one, say $\hat{\phi}_j(\mathbf{y})$, would be measured by the size of

$$\text{m.s.e.}(\hat{\phi}_j) = E_{\mathbf{y}}[\hat{\phi}_j(\mathbf{y}) - \phi]^2 = f_j(\phi) \tag{1}$$

where the expectation is taken over the sampling distribution of $\mathbf{y}$. The estimator $\hat{\phi}_i(\mathbf{y})$ would be "best" in the class considered if

$$f_i(\phi) \leqq f_j(\phi) \qquad (j = 1, \ldots, k; j \neq i)$$

for all values of ϕ, and would be called a minimum mean squared error (m.m.s.e.) estimator.

This m.s.e. criterion, like any other arbitrarily selected measure, is easily shown to be unreliable. For example, consider the estimation of the reciprocal of the mean of the Normal distribution $N(\theta, 1)$ from a random sample $\mathbf{y}$. There is surely something to be said for the maximum likelihood estimator $\bar{y}^{-1}$ which is, after all, sufficient for θ^{-1}, but the m.s.e. of $\bar{y}^{-1}$ is infinite.

The real weakness of the argument is that it requires a prior choice of a universal measure of goodness applicable to all problems.

4. *How to find estimators*

The m.s.e. criterion, if it could be accepted, would provide a means of ranking a set of estimators *once they were presented.* But how can one cope with the embarrassingly large variety of functions of the data which might have relevance?

One early attempt (arbitrarily) limited the estimators to be (a) *linear* functions of $\mathbf{y}$ which were (b) *unbiased.* Among these, the estimator which had (c) *minimum variance* was selected. For example, as is well known, in a random sample from a population with finite mean ϕ and finite variance the sample mean $\bar{y}$ is the minimum variance linear unbiased (m.v.l.u.) estimator of ϕ. These (m.v.l.u.) estimators had the advantage that they could often be derived by a simple minimization process.

However, after it was realized that biased estimators often had smaller mean squared errors than unbiased ones, the requirement of unbiasedness was usually dropped and the criterion of minimum m.s.e. substituted for that of minimum variance. As an example, for the exponential distribution $f(y) = \phi^{-1} e^{-y/\phi}$, $(y > 0, \phi > 0)$, the linear function $n\bar{y}/(n+1)$, although biased, has smaller m.s.e. than that of the unbiased sample mean $\bar{y}$ in estimating ϕ. When it was further shown that non-linear estimators could be readily produced (from Bayesian sources) having smaller m.s.e. than the "best" linear ones for all values of ϕ, linear estimators could no longer be wholeheartedly supported either [4]. Possible ways of obtaining non-linear estimators which *still might minimize m.s.e.* are now being examined. The logical final step would seem to be that the idea of a universal criterion of closeness (be it m.s.e. or something else) should be abandoned also.

One method employed for finding estimators which are not necessarily linear in the observations is Fisher's method of maximum likelihood. Here the estimator is the mode of the likelihood, and, under certain conditions, asymptotically at least, the estimators have Normal distribution and smallest m.s.e. However, some investigators have felt that the likelihood method itself was too limited for their purposes and in recent years various features of Bayesian posterior distributions have been considered as estimators. Whereas it was the *mode* of the likelihood function which had received almost exclusive attention, it has been the mean of the posterior distribution which became the main center of interest. Nevertheless, the posterior mode and the median have also come in for examination and each has been recommended or condemned in one context or another.

4.1. *A decision-theoretic approach*

Justification for this way of choosing estimators is sometimes attempted on formal decision-theoretic grounds. In this argument, let $\hat{\phi}$ be the decision-maker's "best" guess of the unknown ϕ relative to a loss function $L(\hat{\phi}, \phi)$. Then the so-called "Bayes estimator" is the guessed value which minimizes the posterior expected loss

$$\underset{\phi|y}{E}\; L(\hat{\phi}, \phi) = \int_{-\infty}^{\infty} L(\hat{\phi}, \phi) p(\phi|\mathbf{y})\, d\phi \tag{2}$$

where $p(\phi|\mathbf{y})$ is the posterior distribution of ϕ given a specific data set **y.** In particular, if the loss function is the squared error

$$L_1(\hat{\phi}, \phi) = (\hat{\phi}-\phi)^2 \tag{3}$$

then, provided it exists, the Bayes estimator is the posterior mean,[3] but by suitably changing the loss function the posterior mode or median can also be produced. Thus, see [5], for

$$L_2(\hat{\phi}, \phi) = \begin{cases} 1 & \text{for} \quad |\hat{\phi}-\phi| > \varepsilon \\ 0 & \phantom{\text{for}} \quad |\hat{\phi}-\phi| < \varepsilon \end{cases} \tag{4}$$

where $\varepsilon > 0$ is an arbitrarily small positive constant, the Bayes estimator is a mode of the posterior distribution. Also, for

$$L_3(\hat{\phi}, \phi) = |\hat{\phi}-\phi| \tag{5}$$

the Bayes estimator is the posterior median. Indeed, if the loss function were

$$L_4(\hat{\phi},\phi) = \begin{cases} a_1|\hat{\phi}-\phi| & \text{for} \quad \hat{\phi} < \phi \\ a_2|\hat{\phi}-\phi| & \phantom{\text{for}} \quad \hat{\phi} > \phi \end{cases}, \tag{6}$$

then by appropriately choosing a_1 and a_2, *any* fractile of the posterior distribution could be an estimator for ϕ.

Now it is undoubtedly true on the one hand that situations exist where the loss function is at least approximately known (for example certain problems in business and sampling inspection are of this sort). But in this case the resulting estimator would be best for *minimizing this loss* and this would be so *whether its m.s.e. in repeated sampling was large or small.*

On the other hand, a vast number of inferential problems occur, particularly in the analysis of scientific data, where there is no way of knowing in advance to what use the results of research will subsequently be put. In this case, if we are to follow the decision-theoretic approach, we should publish the function which would allow us to compute the estimator which minimizes the posterior expected loss *whatever the loss function*. This function is the complete posterior distribution.

[3] When authors write of the "Bayes estimator" without further qualification they usually refer to the posterior mean. The naming of the "Bayes" estimator seems particularly unfair to the Rev. Thomas Bayes who already has had much to answer for and should be cleared of all responsibility for fathering this particular foundling.

Thus, except for rather special cases, the decision-theoretic approach provides little motivation for the use of specific point estimators and, in particular, of the posterior mean.

The question remains as to why some sampling theorists cling tenaciously to the m.s.e. criterion and the quadratic loss function. The reasons seem to be that (a) for their theory to work, they must cling to something, arbitrarily though it be and (b) given this, it is best to cling to something that works well for the Normal distribution. The implied belief in universal near-Normality of distributions may be traced, perhaps, to early feeding on a diet of asymptotic Normality of maximum likelihood and other estimators. The resulting loyalty to the posterior mean has led to a curious reaction in cases where that estimator has been found to have large m.s.e. [6], [7]. The Bayes approach or the Bayesian prior have been blamed and not the arbitrary features, curious mechanism and dubious relevance of the point estimation method itself.

5. *Inferences based on distributions of "Bayes" estimators*

It seems to us particularly inappropriate that having obtained the estimator from the posterior distribution the sampling theorist should then use its *sampling distribution* to make inferences about the unknown parameter. The bimodal example in fig. 1(b) perhaps makes it easy to see why for inferential purposes the relevance of the sampling distribution of the posterior mean might be questioned. It also illustrates rather dramatically the appeal of using the posterior distribution itself to arrive at appropriate inferences.[4]

[4] While it is easy to demonstrate examples for which there can be no satisfactory point estimate, yet the idea is very strong among people in general, and some statisticians in particular, that there is a need for such a quantity. To the idea that people like to have a single number we answer that usually they shouldn't get it. Most people know they live in a statistical world and common parlance is full of phrases implying uncertainty represented by more than one number. The sentence "Probably 50,000 to 70,000 people visit the museum in a single week" implies that the exact number is unknown, but that there is a high probability that this number falls within the stated limits. The trend towards more realistic specification of uncertain quantities ought to be encouraged.

More elementary examples may be used to illustrate further this point. For the problem of the Normal mean θ with known variance σ^2, the posterior distribution of θ based on the locally uniform reference prior, $p(\theta) \doteq$ constant, is $N(\bar{y}, \sigma^2/n)$. This distribution allows us to make inferences not only about θ but also about any transformation of θ such as e^θ or θ^{-1}. However, while the posterior mean $\bar{y}$ of θ is a useful and important characteristic of the Normal posterior, the posterior mean $\exp[\bar{y}+(1/2n)\sigma^2]$ of e^θ tells us much less about the log Normal posterior, and the posterior mean of θ^{-1}, since it fails to exist, tells us *nothing* about the distribution of θ^{-1}!

Thus, so far as inference is concerned, it is irrelevant whether some *arbitrarily chosen* characteristic of the posterior has or does not have some allegedly desirable sampling property. In particular, it is irrelevant that $\bar{y}$ is the m.m.s.e. estimator for θ, nor is it relevant that for e^θ, as can be readily shown, the posterior mean $\exp[\bar{y}+(1/2n)\sigma^2]$ has larger m.s.e. than that of the mode $\exp[\bar{y}-(\sigma^2/n)]$ (which, for that matter, has larger m.s.e. than that of the estimator $\exp[\bar{y}-(3/2n)\sigma^2]$). It follows that a locally uniform prior for θ should not be blamed for causing the large m.s.e. of the posterior mean of e^θ nor should the *same* prior be praised for producing the minimum m.s.e. of the posterior mean of θ.

Inferences about a parameter are provided not by the sampling properties of some arbitrarily chosen characteristic of the posterior distribution, but by that distribution itself.

References

[1] G. E. P. Box and G. C. Tiao, Bayesian inference in statistical analysis. Reading, Addison-Wesley, 1973.

[3] G. E. P. Box and N. R. Draper, The Bayesian estimation of common parameters from several responses. Biometrika 52 (1965), 355.

[4] W. James and C. M. Stein, Estimation with quadratic loss function, Proc. 4th Berkeley Symposium Math. Statist. Prob. 1 (1961), 361.

[2] H. Jeffreys, Theory of probability, 3rd Ed. Oxford, Clarendon Press, 1961.

[6] J. H. Klotz, R. C. Milton and S. Zacks, Mean square efficiency of estimators of variance component. J. Amer. Statist. Assoc. 64 (1969), 1383.

[7] S. Portnoy, Formal Bayes estimation with application to a random effect model. Ann. Math. Statist. 42 (1971), 1379.

[5] J. W. Pratt, H. Raiffa and R. Schlaifer, Introduction to statistical decision theory. New York, McGraw-Hill, 1965.

S. E. Fienberg and A. Zellner (eds.)
Studies in Bayesian Econometrics and Statistics.

12.5 BAYES–STEIN ESTIMATORS FOR *k*-MEANS, REGRESSION AND SIMULTANEOUS EQUATION MODELS*

Arnold ZELLNER and Walter VANDAELE

University of Chicago

1. *Introduction*

In this paper we provide a brief review of Stein and Stein-like estimators for normal k-means and regression models. As is well-known, such estimators dominate usual estimators, maximum likelihood (ML) and least squares (LS) estimators, in the sense that the latter estimators have been shown to be inadmissible relative to a quadratic loss function under often-encountered conditions. Then we present and discuss Bayesian and non-Bayesian methods of generating Stein and Stein-like estimators that have appeared in the literature and a few of our own. The latter include a minimal mean square error (MMSE) approach and a Bayesian minimum expected loss (BMEL) approach. Also, we consider the problem of generating improved estimators for parameters of simultaneous equation models. Last, we present a summary of results and some concluding remarks.

2. *Analysis of k-means model*[1]

2.1. *Estimators for k-means model*

The model for k independent observations, $\mathbf{y}' = (y_1, y_2, \ldots, y_k)$, is

* Research financed in part by NSF Grant GS-2347 and by the H. G. B. Alexander Endowment Fund, Graduate School of Business, Univ. of Chicago. The second author being on leave from the Katholieke Universiteit, Leuven, Belgium, received fellowship support from the Ford Foundation, the Belgium-American Educational Foundation and the Belgian NSF (Nationaal Fonds voor Wetenschappelijk Onderzoek). This work was reported to the NBER-NSF Seminar on Bayesian Inference in Econometrics, October 15–16, 1971. Participants in the seminar, in particular J. M. Dickey, J. H. Drèze, S. Fienberg, J. Kadane, J. W. Pratt and S. J. Press, provided helpful comments for which we are grateful.

[1] See Stein [27], James and Stein [13], Lindley [16], and Box and Tiao [4].

assumed to be:

$$y_i = \xi_i + u_i, \qquad i = 1, 2, \ldots, k \tag{1}$$

with the u_i's normally and independently distributed, each with zero mean and common finite variance, σ^2, that is NID$(0, \sigma^2)$. Further, $E(y_i|\xi_i) = \xi_i$, $-\infty < \xi_i < \infty$, $i = 1, 2, \ldots, k$. Then the likelihood function is:

$$p(\mathbf{y}|\xi, \sigma) = (2\pi\sigma^2)^{-k/2} \exp\left\{-\frac{1}{2\sigma^2}(\mathbf{y}-\xi)'(\mathbf{y}-\xi)\right\}, \tag{2}$$

where $\xi' = (\xi_1, \xi_2, \ldots, \xi_k)$. From (2), it is apparent that the maximum likelihood (ML) (and least squares (LS)) estimator for ξ is given by

$$\hat{\xi}_0 = \mathbf{y}. \tag{3}$$

Given that we employ the following quadratic loss function,

$$L(\xi, \tilde{\xi}) = (\xi - \tilde{\xi})'(\xi - \tilde{\xi}) \tag{4}$$

the risk function for the ML (and LS estimator) is

$$R(\xi, \hat{\xi}_0) = EL(\xi, \hat{\xi}_0) = k\sigma^2. \tag{5}$$

In the case that σ^2 has a known value, say $\sigma^2 = 1$, Stein [27] put forward the following estimator for ξ:

$$\hat{\xi}_1 = \left(1 - \frac{b}{\mathbf{y}'\mathbf{y}}\right)\mathbf{y} \qquad b > 0 \tag{6a}$$

where b is a positive scalar. On evaluating the risk associated with $\hat{\xi}_1$, we have[2]

$$\begin{aligned} R(\xi, \hat{\xi}_1) = EL(\xi, \hat{\xi}_1) &= \underset{y}{E}\left[\left(1 - \frac{b}{\mathbf{y}'\mathbf{y}}\right)\mathbf{y} - \xi\right]'\left[\left(1 - \frac{b}{\mathbf{y}'\mathbf{y}}\right)\mathbf{y} - \xi\right] \\ &= k - [2(k-2)b - b^2]\underset{K}{E}[1/(k-2+2K)] \end{aligned} \tag{7a}$$

where K is a random variable having a Poisson distribution with mean $\xi'\xi/2$. For $k \geqq 3$, taking $b = k-2$ provides minimal risk for $\hat{\xi}_1$ for all ξ; thus Stein's estimator for ξ, when $\sigma^2 = 1$, is

[2] See e.g. James and Stein [13, p. 365].

$$\hat{\xi}_2 = \left(1 - \frac{k-2}{\mathbf{y}'\mathbf{y}}\right)\mathbf{y}, \tag{6b}$$

with risk, from (7a) with $b = k-2$,

$$R(\xi, \hat{\xi}_2) = k-(k-2)^2 \underset{K}{E}[1/(k-2+2K)]. \tag{7b}$$

Some comments regarding the usual estimator in (3) and the Stein estimator in (6b) follow:

(1) When $k \leqq 2$, Stein [27] has shown that $\hat{\xi}_0 = \mathbf{y}$ is admissible.

(2) When $k \geqq 3$, $\hat{\xi}_2$ has smaller risk than the ML for all ξ. Therefore the ML (and LS) estimator, $\hat{\xi}_0$, is inadmissible.

(3) James and Stein [13, p. 363] state that the risk associated with $\hat{\xi}_2$ is 2 at $\xi = 0$ and increases gradually with $\xi'\xi$ to the value k, the risk associated with $\hat{\xi}_0$ when $\sigma^2 = 1$, as $\xi'\xi \to \infty$.

(4) James and Stein [13, p. 363] note that although $\hat{\xi}_2$ is not admissible, "... it seems unlikely that there are spherically symmetrical estimators which are appreciably better ...".

(5) We note that while $\hat{\xi}_0$ is a linear, unbiased estimator, Stein's estimator $\hat{\xi}_2$ is obviously a non-linear, biased estimator. Stein has thus gone outside the class of linear, unbiased estimators to produce an estimator with lower risk. That the risk associated with Stein's estimator is lower over the entire parameter space, given $k \geqq 3$ and $\xi'\xi < \infty$, is indeed surprising.

When σ^2 has an unknown value, James and Stein [13] consider the following estimator

$$\hat{\xi}_3 = \left(1 - \frac{az}{\mathbf{y}'\mathbf{y}}\right)\mathbf{y} \qquad a > 0 \tag{8a}$$

where a is a positive scalar and z is a random variable, independent of $\mathbf{y}$, distributed as σ^2 times a χ^2 variable with n degrees of freedom. For example, an independent experiment could yield z.[3] The risk associated with $\hat{\xi}_3$ relative to the quadratic loss function in (4), given in James Stein [13], is:

$$R(\xi, \hat{\xi}_3) = \sigma^2\{k-[2an(k-2)-a^2n(n+2)]\underset{K}{E}[1/(k-2+2K)]\} \tag{9}$$

[3] Note that it is not possible to estimate both ξ and σ^2 from just data $\mathbf{y}$ in the k-means model.

where here K is a random variable with a Poisson distribution with mean $\xi'\xi/2\sigma^2$. From (9), the optimal value for the parameter a is $a = (k-2)/(n+2)$, for $k \geqq 3$, leading to

$$\hat{\hat{\xi}}_4 = \left[1-\left(\frac{k-2}{n+2}\right)\frac{z}{\mathbf{y}'\mathbf{y}}\right]\mathbf{y} \tag{8b}$$

and

$$R(\xi, \hat{\hat{\xi}}_4) = \sigma^2\left\{k-\frac{n}{n+2}(k-2)^2 \underset{K}{E}[1/(k-2+2K)]\right\}. \tag{10}$$

It is seen from (10) that the risk associated with $\hat{\hat{\xi}}_4$ will, for $k \geqq 3$, be smaller than that associated with $\hat{\hat{\xi}}_0$, the ML (and LS) estimator, namely $k\sigma^2$.[4]

2.2. *Interpretations of Stein's estimators for k-means*

Stein [29, p. 266] has provided the following Bayesian interpretation of his estimator for k-means given $\sigma^2 = 1$. He assumed that the ξ_i's in (1) are NID(0, τ^2), i.e., $g(\xi|\tau) \propto \tau^{-k}\exp\{-\xi'\xi/2\tau^2\}$. Then the posterior probability density function (pdf) for ξ given $\sigma^2 = 1$ and τ is:

$$\begin{aligned} p(\xi|\mathbf{y}, \sigma^2 = 1, \tau) &\propto \exp\{-\tfrac{1}{2}[(\mathbf{y}-\xi)'(\mathbf{y}-\xi)+\xi'\xi/\tau^2]\} \\ &\propto \exp\left\{-\left(\frac{1+\tau^2}{2\tau^2}\right)\left(\xi-\frac{\tau^2}{1+\tau^2}\mathbf{y}\right)' \right. \\ &\qquad \left. \times\left(\xi-\frac{\tau^2}{1+\tau^2}\mathbf{y}\right)\right\}. \end{aligned} \tag{11}$$

From the second line of (11), where we have completed the square on ξ, it is apparent that the conditional posterior mean of ξ given $\sigma^2 = 1$ and τ is[5]

$$E(\xi|\mathbf{y}, \sigma^2 = 1, \tau) = \frac{\tau^2}{1+\tau^2}\mathbf{y} = \left(1-\frac{1}{1+\tau^2}\right)\mathbf{y}. \tag{12}$$

[4] On comparing (10) and (7b), it is seen that a factor $n/(n+2)$ appears in (10) that is not present in (7b). Thus when σ^2 is estimated, there is a smaller reduction in risk.

[5] Fienberg and Holland [10] have recently employed a similar argument in connection with estimating parameters of multinomial distributions.

From (1), $\sigma_y^2 = \tau^2+1$, or $\tau^2 = \sigma_y^2-1$. To approximate (12), Stein inserts $\tau^2 = \hat{\tau}^2 = \mathbf{y}'\mathbf{y}/k-1$ to obtain

$$\hat{\xi} = \left(1-\frac{k}{\mathbf{y}'\mathbf{y}}\right)\mathbf{y}\,, \tag{13}$$

which is in the form of (6b). Note that (13) is an approximation to the mean of the posterior pdf for ξ that is optimal for a quadratic loss function.[6]

In Stein [29, p. 281] the following interpretation is attributed to I. J. Good and D. L. Wallace. Assume that for given $r = (\xi'\xi)^{\frac{1}{2}}$, ξ is distributed uniformly over a sphere of radius r and that the marginal prior pdf for r is uniform, $0 < r < \infty$. Then on transforming from spherical to rectangular coordinates, the implied prior pdf for the elements of ξ is:

$$p(\xi)\,\mathrm{d}\xi \propto 1/(\xi'\xi)^{(k-1)/2}\,\mathrm{d}\xi \qquad -\infty < \xi_i < \infty \qquad i = 1, 2, \ldots, k \tag{14a}$$

which is close to the prior pdf

$$p(\xi)\,\mathrm{d}\xi \propto 1/(\xi'\xi)^{(k-3)/2}\,\mathrm{d}\xi \qquad -\infty < \xi_i < \infty \qquad i = 1, 2, \ldots, k \tag{14b}$$

that Stein [29, p. 281] states will produce his estimator for ξ.

Note further that if our prior for ξ given τ is $g(\xi|\tau) \propto \tau^{-k}\exp\{-\xi'\xi/2\tau^2\}$ and if the marginal prior on τ is $h(\tau) \propto \tau^{-c}$, $0 < \tau < \infty$, then the joint prior for ξ and τ is

$$p(\xi, \tau) \propto \tau^{-(k+c)}\exp\{-\xi'\xi/2\tau^2\} \tag{15}$$

and the marginal prior for ξ, obtained by integrating (1 5) with respe c to τ, $0 < \tau < \infty$, is[7]

$$p(\xi) \propto 1/(\xi'\xi)^{(k+c-1)/2}\,. \tag{16}$$

Clearly, if $c = 0$, we have the Good–Wallace prior for ξ shown in (14a). For $c = -2$, (16) takes the form of the Stein prior in (14b).

[6] Stein presents (13) just to illustrate the nature of his estimators. Given a proper prior pdf for τ, it would, of course, be possible in principle to obtain the exact posterior mean for ξ that is optimal for a quadratic loss function.

[7] The implied pdf for $r = (\xi'\xi)^{\frac{1}{2}}$ is: $p(r)\mathrm{d}r \propto r^{-(c+1)}\mathrm{d}r$, $0 < r < \infty$.

Lindley's interpretation [16] of Stein's results proceeds in terms of the following system:

$$\begin{aligned} y_i &= \xi_i + u_i \\ \xi_i &= \mu + \varepsilon_i \end{aligned} \qquad i = 1, 2, \ldots, k. \tag{17}$$

It is assumed that the y_i's given the ξ_i's are NID(ξ_i, σ^2) while the ξ_i's given μ are NID(μ, τ^2). The parameter μ represents a "general effect" that links the y_i's and provides a rationale for an estimate of a particular ξ_i to depend not only on y_i but also on other y's as well. From (17), the joint pdf for $\mathbf{y}$ and ξ is given by

$$\begin{aligned} p(\mathbf{y}, \xi|\sigma, \mu, \tau) &= p(\mathbf{y}|\xi, \sigma)p(\xi|\mu, \tau) \\ &\propto \sigma^{-k} \exp\left\{-\frac{1}{2\sigma^2}(\mathbf{y}-\xi)'(\mathbf{y}-\xi)\right\} \\ &\quad \times \tau^{-k} \exp\left\{-\frac{1}{2\tau^2}(\xi-\mu\iota)'(\xi-\mu\iota)\right\}. \end{aligned} \tag{18}$$

To parallel Stein's analysis and for illustrative purposes, Lindley considers σ and τ given and uses a spread-out normal prior for μ (effectively $p(\mu) \propto$ const). On integrating out μ, $-\infty < \mu < \infty$, Lindley gives the mean for ξ, $\tilde{\xi}$, as

$$\tilde{\xi} = (\mathbf{y}/\sigma^2 + \bar{y}\iota/\tau^2)/(1/\sigma^2 + 1/\tau^2),$$

where $\iota' = (1, 1, \ldots, 1)$, a $1 \times k$ vector with all elements equal to 1. Then assuming $\sigma^2 = 1$ and $\tau^2 = \hat{\tau}^2 = (\mathbf{y}-\bar{y}\iota)'(\mathbf{y}-\bar{y}\iota)/a' - 1$, where a' is a positive scalar, Lindley obtains

$$\begin{aligned} \tilde{\xi} &= \mathbf{y} - \frac{a'(\mathbf{y}-\bar{y}\iota)}{(\mathbf{y}-\bar{y}\iota)'(\mathbf{y}-\bar{y}\iota)} \\ &= \bar{y}\iota + \left[1 - \frac{a'}{(\mathbf{y}-\bar{y}\iota)'(\mathbf{y}-\bar{y}\iota)}\right](\mathbf{y}-\bar{y}\iota) \end{aligned} \tag{19}$$

which is in the form of (6a) if $\bar{y} = 0$. With $\bar{y} \neq 0$, Lindley's suggestion [16, p. 286] is to use Stein's methods for the deviations, $\mathbf{y}-\bar{y}\iota$, rather than for $\mathbf{y}$ and he notes that this avoids the difficulty of the arbitrary origin that appears in Stein's argument (see Stein's interpretation above in which he assumes that the mean of the ξ_i's is 0). Lindley computed the risk associated with the estimator in (19) relative to a quadratic loss function in the manner of James and Stein [13] and found that taking $a' = k-3$ (rather than $k-2$) is optimal and states that the risk associated with his estimator is lower than that associated with the estimator in (6b).

A problem for which the Stein estimator in (6) is optimal is the following. Introduce a "relative squared error" loss function in place of (4). That is, let our loss function be

$$L(\xi, \tilde{\xi}) = (\xi-\tilde{\xi})'(\xi-\tilde{\xi})/(\xi'\xi)^m \qquad m > 0. \tag{20}$$

Then, if the prior pdf for ξ is $p(\xi) \propto$ const, $-\infty < \xi_i < \infty$, $i = 1, 2, \ldots, k$, with $\sigma^2 = 1$, the posterior pdf for ξ is $p(\xi|\mathbf{y}) \propto \exp\{-(\mathbf{y}-\xi)'(\mathbf{y}-\xi)/2\}$ and posterior expected loss is given by

$$EL = \int_{\xi} [(\xi-\tilde{\xi})'(\xi-\tilde{\xi})/(\xi'\xi)^m] p(\xi|\mathbf{y})\, \mathrm{d}\xi. \tag{21}$$

With $m = (k-3)/2$, the value of $\tilde{\xi}$ that minimizes (21) is equal to Stein's estimate in (6b). Note that the factor $(\xi'\xi)^{-(k-3)/2}$ enters the integrand of (21) via the loss function rather than via the prior pdf as in (14b).

Next, we turn to a "minimal mean square error" (MMSE) interpretation of Stein-like estimators.[8] Consider the class of estimators $\tilde{\xi} = q\mathbf{y}$, where q is a scalar. Then the risk associated with this estimator is

$$E(\xi-q\mathbf{y})'(\xi-q\mathbf{y}) = \xi'\xi-2q\xi'\xi+q^2(\xi'\xi+k\sigma^2). \tag{22}$$

The value of q that minimizes (22), say q_*, is:

$$q_* = \frac{\xi'\xi}{\xi'\xi+k\sigma^2} = 1-\frac{k\sigma^2}{\xi'\xi+k\sigma^2} \tag{23}$$

and the MMSE estimator in the class $q\mathbf{y}$ is:

$$\tilde{\xi}^* = q_*\mathbf{y} = \left(1-\frac{k\sigma^2}{\xi'\xi+k\sigma^2}\right)\mathbf{y} \tag{24}$$

which depends on the parameters σ^2 and $\xi'\xi$. For $\sigma^2 = 1$, we can approximate (24) by

$$\tilde{\xi}^*_a = \left(1-\frac{k}{\mathbf{y}'\mathbf{y}}\right)\mathbf{y} \tag{25}$$

which is precisely (6a) with risk given by (7a) with $b = k$.

[8] See Savage [21, p. 224] where this approach has been used in generating a minimal mean square error (MMSE) estimator for a variance and Zellner [34] where it has been employed to generate MMSE estimators for parameters of the two-parameter log-normal distribution.

When σ^2 has an unknown value, (24) can be approximated by

$$\tilde{\xi}_a^* = \left(1 - \frac{k}{n+2}\frac{z}{\mathbf{y}'\mathbf{y}}\right)\mathbf{y} \tag{26}$$

where z is distributed independently of $\mathbf{y}$ as σ^2 times a χ^2 variable with n degrees of freedom.[9] It is seen that (25) and (26) are very similar to (6b) and (8b). Further (25) is identical to (13).

In (25) and (26), there is no Lindley-like correction for the overall mean—see (19). To obtain an overall mean correction in the MMSE approach, consider the class of estimators given by $q_0\boldsymbol{\iota}+q_1\mathbf{y}$ where q_0 and q_1 are scalars and $\boldsymbol{\iota}$ is a $k \times 1$ vector with all elements equal to one. Values of q_0 and q_1 that minimize risk, that is minimize $E(q_0\boldsymbol{\iota}+q_1\mathbf{y}-\xi)'(q_0\boldsymbol{\iota}+q_1\mathbf{y}-\xi)$ are given by $q_0^* = \sigma^2\bar{\xi}/\Delta$ and $q_1^* = (m-\bar{\xi}^2)/\Delta$ where $\bar{\xi} = \Sigma_{i=1}^k \xi_i/k$, $m = \Sigma_{i=1}^k \xi_i^2/k$ and $\Delta = m+\sigma^2-\bar{\xi}^2$. Then the MMSE estimator, $\check{\xi}$, is:

$$\check{\xi} = q_0^*\boldsymbol{\iota}+q_1^*\mathbf{y}, \tag{27a}$$

which depends on the parameters σ^2, $\bar{\xi}$ and m. As above, we can approximate (27a) when $\sigma^2 = 1$ by

$$\check{\xi}_a = \frac{a_k'\bar{y}}{(\mathbf{y}-\bar{y}\boldsymbol{\iota})'(\mathbf{y}-\bar{y}\boldsymbol{\iota})}\boldsymbol{\iota}+\left(1-\frac{a_k'}{(\mathbf{y}-\bar{y}\boldsymbol{\iota})'(\mathbf{y}-\bar{y}\boldsymbol{\iota})}\right)\mathbf{y}$$

$$= \bar{y}\boldsymbol{\iota}+\left(1-\frac{a_k'}{(\mathbf{y}-\bar{y}\boldsymbol{\iota})'(\mathbf{y}-\bar{y}\boldsymbol{\iota})}\right)(\mathbf{y}-\bar{y}\boldsymbol{\iota}) \tag{27b}$$

where a_k is a positive scalar. Taking $a_k' = k-3$ in (27b) yields exactly Lindley's proposed estimator (19) that incorporates an over-all mean correction. Note however that in obtaining (27b), it was not necessary to resort to the random effects model in (17) that Lindley employed.

Last we note that Stein-like estimators can be generated as usual Bayesian minimum expected loss (BMEL) when we average over two hypotheses. To demonstrate this, let p_0 denote the posterior probability on the hypothesis H_0: $\xi = \xi_0$, where ξ_0 is a given vector. Further let $1-p_0$ be the posterior probability on the hypothesis $H_1 : \xi \neq \xi_0$. With a quadratic loss function, $L(\xi, \tilde{\xi}) = (\xi-\tilde{\xi})'(\xi-\tilde{\xi})$, posterior

[9] In (26), as in (8b), z is divided by $n+2$ to provide a MMSE estimate for σ^2. Also $z/(n+2)$ is an optimal estimate from the Bayesian point of view for the relative squared error loss function, $L(\sigma^2,\hat{\sigma}^2) = [(\sigma^2-\hat{\sigma}^2)/\sigma^2]^2$.

expected loss is given by

$$\begin{aligned} \underset{\xi}{E}L(\xi, \tilde{\xi}) &= p_0 E(L|\xi = \xi_0) + (1-p_0)E(L|\xi \neq \xi_0) \\ &= p_0(\xi_0 - \tilde{\xi})'(\xi_0 - \tilde{\xi}) + (1-p_0)E(\xi - \tilde{\xi})'(\xi - \tilde{\xi}) \\ &= p_0(\xi_0 - \tilde{\xi})'(\xi_0 - \tilde{\xi}) + (1-p_0)[E(\bar{\xi} - \xi)'(\xi - \bar{\xi}) \\ &\quad + (\tilde{\xi} - \bar{\xi})'(\tilde{\xi} - \bar{\xi})] \end{aligned} \tag{27}$$

where $\bar{\xi}$ = posterior mean vector for ξ given $\xi \neq \xi_0$. On minimizing (27) with respect to $\tilde{\xi}$, the minimizing value for $\tilde{\xi}$ is found to be

$$\begin{aligned} \tilde{\xi}^* &= p_0 \xi_0 + (1-p_0)\bar{\xi} \\ &= \xi_0 + \left(1 - \frac{K_{01}}{1+K_{01}}\right)(\bar{\xi} - \xi_0) \end{aligned} \tag{28}$$

where $K_{01} = p_0/(1-p_0)$ is the posterior odds ratio for the hypotheses $H_0\colon \xi = \xi_0$ and $H_1 : \xi \neq \xi_0$. It is seen that (28) is in a form similar to that of (19). Further, note that $1 - K_{01}/(1+K_{01}) > 0$ and thus the elements of $\tilde{\xi}^* - \xi_0$ have the same algebraic signs as the corresponding elements of $\bar{\xi} - \xi_0$. Also, it should be observed that (a) $\tilde{\xi}^*$ is a continuous function of the observations, (b) prior information about ξ under H_1 can be introduced through use of an appropriate prior pdf, and (c) there is no arbitrariness in choice of a significance level as is the case with sampling theory "pre-test" estimators (see below).

In summary, the k-means problem has three components: (i) a model for the observations, (ii) a prior pdf for the parameters and (iii) a loss function. When the model is (1), with likelihood function (2), and the prior is $p(\xi, \sigma) \propto 1/\sigma$, $-\infty < \xi_i < \infty$, $i = 1, 2, \ldots, k$, then the mean of the posterior pdf for ξ is $\hat{\xi}_0 = \mathbf{y}$, the usual estimate. This estimate provides a minimal value to the *posterior mean* of the quadratic loss function in (4). However, with the prior $p(\xi, \sigma) \propto 1/\sigma$, the average risk of $\hat{\xi}_0 = \mathbf{y}$ is not finite; thus $\hat{\xi}_0 = \mathbf{y}$ does not minimize average risk and there is no assurance that $\hat{\xi}_0 = \mathbf{y}$ will in general be admissible. A Bayesian who seeks an estimator with good sampling properties can employ the MMSE approach described above, an approach that does not involve changes in the prior, the model or the loss function.[10] In all the other approaches, one of the components of the problem has been changed. Stein, Good and Wallace introduced a different prior pdf. Lindley resorted to a change in the model. Use of a relative squared

[10] There is, however, a restriction on the class of estimators considered.

error loss function, as shown in (20), obviously involves a modification of the loss function of the original problem. Last in the BMEL approach, a proper prior pdf for ξ under the hypothesis $\xi \neq \xi_0$ must be introduced. In particular applications these modified assumptions may be eminently reasonable and the fact that they all lead to Stein-like estimates for ξ is of great interest. Finally, while the discussion has emphasized point estimation, it is important to recognize that point estimates alone are not adequate for most inference problems.

3. *Analysis of regression models*

3.1. *Univariate regression*

We now consider improved estimators[11] for the $k \times 1$ regression coefficient vector, $\boldsymbol{\beta}$, of a multiple regression equation,

$$\mathbf{y} = X\boldsymbol{\beta} + \mathbf{u} \tag{29}$$

where $\mathbf{y}' = (y_1, y_2, \ldots, y_T)$, X is a $T \times k$ nonstochastic matrix with rank k, and $\mathbf{u}' = (u_1, u_2, \ldots, u_T)$ is a random disturbance vector. We shall assume that the elements of $\mathbf{u}$ are normally and independently distributed, each with mean zero and common variance, σ^2; that is $\mathbf{u}$: $N(\mathbf{0}, \sigma^2 I_T)$.

Following Sclove [24], we first observe that if $X'X = TI_k$, then we can write

$$X'\mathbf{y}/T = \boldsymbol{\beta} + X'\mathbf{u}/T \tag{30a}$$

or

$$\mathbf{w} = \boldsymbol{\beta} + \boldsymbol{\varepsilon} \tag{30b}$$

where $\mathbf{w} = X'\mathbf{y}/T$ and $\boldsymbol{\varepsilon} = X'\mathbf{u}/T$ are each $k \times 1$ vectors with $\boldsymbol{\varepsilon}$: $N[\mathbf{0}, (\sigma^2/T)I_k]$. Note that (30b) is precisely in the form of a k-means model with $\mathbf{w}$ the ML (and LS) estimator for $\boldsymbol{\beta}$. Therefore all the results for the k-means problem apply directly[12] to the estimation of $\boldsymbol{\beta}$ with $\boldsymbol{\beta}$ replacing ξ and $\mathbf{w}$ replacing $\mathbf{y}$. For example, the form of (8b) for the

[11] These were first put forward in Stein [28] and exposited by Sclove [24]. See also Dickey [5] and Lindley and Smith [17].

[12] The prior pdf's, used above, may not be appropriate for $\boldsymbol{\beta}$. See below for a generalization of these prior assumptions leading to a posterior pdf with a mean exhibiting Stein-like properties.

present problem is:

$$\hat{\boldsymbol{\beta}}_4 = \left[1 - \left(\frac{k-2}{n+2}\right)\frac{z}{\mathbf{w}'\mathbf{w}}\right]\mathbf{w} \tag{31}$$

where $z = \hat{\mathbf{u}}'\hat{\mathbf{u}} = (y - X\mathbf{w})'(y - X\mathbf{w})$ and $n = T - k$, the d.f. associated with z. Then relative to a quadratic loss function, $L(\boldsymbol{\beta}, \tilde{\boldsymbol{\beta}}) = (\boldsymbol{\beta} - \tilde{\boldsymbol{\beta}})'(\boldsymbol{\beta} - \tilde{\boldsymbol{\beta}})$, $\hat{\boldsymbol{\beta}}_4$ has lower risk for $\boldsymbol{\beta}$, when $k \geqq 3$, $\boldsymbol{\beta}'\boldsymbol{\beta} < \infty$ and $T < \infty$, than $\mathbf{w}$, the ML (and LS) estimator.

If the assumption that $X'X = TI_k$ is dropped, we can reparameterize the model as follows: $\mathbf{y} = XPP^{-1}\boldsymbol{\beta} + \mathbf{u} = W\boldsymbol{\theta} + \mathbf{u}$ where $\boldsymbol{\theta} = P^{-1}\boldsymbol{\beta}$, a $k \times 1$ vector, P is a $k \times k$ non-singular matrix such that $P'X'XP = W'W = TI_k$. Then, as pointed out by Sclove [24], the results for the k-means problem apply to the estimation of $\boldsymbol{\theta}$ and from $\boldsymbol{\beta} = P\boldsymbol{\theta}$ to the estimation of $\boldsymbol{\beta}$. However, note that in applying the Bayesian interpretations of Stein's estimators, presented above, to the present problem, the prior pdf's relate directly to $\boldsymbol{\theta}$ and only indirectly, via $\boldsymbol{\beta} = P\boldsymbol{\theta}$, to $\boldsymbol{\beta}$. In view of this complication, we present a slight generalization of Lindley's interpretation—see equation (17) and what follows.

Let the prior pdf for $\boldsymbol{\beta}$ be given by

$$p(\boldsymbol{\beta}|\bar{\boldsymbol{\beta}}, \tau, A) \propto \tau^{-k} \exp\left\{-\frac{1}{2\tau^2}(\boldsymbol{\beta} - \bar{\boldsymbol{\beta}})'A(\boldsymbol{\beta} - \bar{\boldsymbol{\beta}})\right\}$$

$$A\text{: p.d.s.}$$

$$-\infty < \beta_i < \infty, \quad i = 1, 2, \ldots, k \tag{32}$$

where $\bar{\boldsymbol{\beta}}$ is the prior mean vector and $A^{-1}\tau^2$ is the prior covariance matrix. We can then combine (32) with the likelihood function,

$$p(\mathbf{y}|\boldsymbol{\beta}, \sigma) \propto \sigma^{-T} \exp\left\{-\frac{1}{2\sigma^2}[vs^2 + (\boldsymbol{\beta} - \hat{\boldsymbol{\beta}})'X'X(\boldsymbol{\beta} - \hat{\boldsymbol{\beta}})]\right\}, \tag{33}$$

where $\hat{\boldsymbol{\beta}} = (X'X)^{-1}X'\mathbf{y}$, $v = T - k$, and $vs^2 = (\mathbf{y} - X\hat{\boldsymbol{\beta}})'(\mathbf{y} - X\hat{\boldsymbol{\beta}})$, to obtain a posterior pdf for $\boldsymbol{\beta}$ given σ. The mean, $\bar{\bar{\boldsymbol{\beta}}}$, of this conditional posterior pdf is[13]:

$$\bar{\bar{\boldsymbol{\beta}}} = (A/\tau^2 + X'X/\sigma^2)^{-1}(A\bar{\boldsymbol{\beta}}/\tau^2 + X'X\hat{\boldsymbol{\beta}}/\sigma^2)$$

$$= \bar{\boldsymbol{\beta}} + \left[I - \left(A + \frac{\tau^2}{\sigma^2}X'X\right)^{-1} A\right](\hat{\boldsymbol{\beta}} - \bar{\boldsymbol{\beta}}). \tag{34}$$

[13] After this result was obtained, J. Dickey informed us that a similar result appears in Lindley and Smith [17]. The result in the first line of (34) appears in Tiao and Zellner [32] and Zellner [33, p. 100].

It is seen that the second line of (34) is in a form resembling that of (19).[14] And, of course, with a proper prior pdf, the mean of the posterior pdf is optimal relative to a quadratic loss function. That is, the posterior mean vector is the estimator that minimizes average risk. Also, it is admissible and consistent.

Just as Lindley's interpretation of Stein's k-means estimator generalizes to relate to the regression model, the "relative squared error" loss function, the MMSE and the BMEL interpretations do so also. Below we extend the last two interpretations to apply to the regression model.

In the MMSE approach, we utilize the following loss function,[15]

$$L(\boldsymbol{\beta}, \tilde{\boldsymbol{\beta}}) = (\boldsymbol{\beta}-\tilde{\boldsymbol{\beta}})'Q(\boldsymbol{\beta}-\tilde{\boldsymbol{\beta}}) \qquad Q: k\times k, \text{ p.d.s.} \tag{35}$$

The risk associated with the ML estimator, $\hat{\boldsymbol{\beta}} = (X'X)^{-1}X'\mathbf{y}$, is

$$R_1 = EL(\boldsymbol{\beta}, \hat{\boldsymbol{\beta}}) = \sigma^2 \mathrm{tr}(X'X)^{-1}Q. \tag{36}$$

Now consider the class of estimators, $\tilde{\boldsymbol{\beta}} = q\hat{\boldsymbol{\beta}}$, where q is a scalar. The risk associated with $\tilde{\boldsymbol{\beta}}$ is

$$\begin{aligned} EL(\boldsymbol{\beta}, \tilde{\boldsymbol{\beta}}) &= E(\tilde{\boldsymbol{\beta}}-\boldsymbol{\beta})'Q(\tilde{\boldsymbol{\beta}}-\boldsymbol{\beta}) \\ &= E[q(\hat{\boldsymbol{\beta}}-\boldsymbol{\beta})-(1-q)\boldsymbol{\beta}]'Q[q(\hat{\boldsymbol{\beta}}-\boldsymbol{\beta})-(1-q)\boldsymbol{\beta}] \\ &= q^2E(\hat{\boldsymbol{\beta}}-\boldsymbol{\beta})'Q(\hat{\boldsymbol{\beta}}-\boldsymbol{\beta})+(1-q)^2\boldsymbol{\beta}'Q\boldsymbol{\beta}. \end{aligned} \tag{37}$$

Minimization of this last expression with respect to q provides the following optimal value, q_*:

$$q_* = 1-\frac{R_1}{R_1+\boldsymbol{\beta}'Q\boldsymbol{\beta}} \tag{38}$$

with R_1 as given in (36). It is the case that q_* satisfies $0 \leqq q_* \leqq 1$. Then the MMSE estimator, $\tilde{\boldsymbol{\beta}}^*$, is

$$\tilde{\boldsymbol{\beta}}^* = \left[1-\frac{g\sigma^2}{g\sigma^2+\boldsymbol{\beta}'Q\boldsymbol{\beta}}\right]\hat{\boldsymbol{\beta}} \tag{39}$$

[14] We have pursued the analysis conditional on σ being given merely for illustrative purposes. See Tiao and Zellner [32] for an unconditional analysis of the regression model using the prior for $\boldsymbol{\beta}$ shown in (32) and a diffuse prior pdf for σ. This analysis can easily be extended to accommodate a proper inverted gamma prior on σ.

[15] As an aside, write $Q = C'C$, where C is a $k\times k$ matrix of rank k with ith row $\mathbf{c}_i'$. Then $L = (\boldsymbol{\beta}-\tilde{\boldsymbol{\beta}})'C'C\,(\boldsymbol{\beta}-\tilde{\boldsymbol{\beta}}) = \Sigma_{i=1}^{k}\mathbf{c}_i'(\boldsymbol{\beta}-\tilde{\boldsymbol{\beta}})(\boldsymbol{\beta}-\tilde{\boldsymbol{\beta}})'\mathbf{c}_i$. Thus use of the expected loss criterion is, as is well known, equivalent to the use of $E(\boldsymbol{\beta}-\tilde{\boldsymbol{\beta}})(\boldsymbol{\beta}-\tilde{\boldsymbol{\beta}})'$ as a criterion for evaluating alternative estimators.

where $g = \text{tr}\,(X'X)^{-1}Q > 0$ and the risk associated with $\tilde{\beta}^*$, R_{q_*}, is given by

$$R_{q_*} = E[L(\tilde{\beta}, \beta)|q = q_*] = R_1[\beta'Q\beta/(R_1+\beta'Q\beta)]. \tag{40}$$

For $0 < \beta'Q\beta < \infty$ and $R_1 \neq 0$, $R_{q_*} < R_1$.[16] Further, as $T \to \infty$, $R_{q_*} \to R_1$.

Since $\tilde{\beta}^*$ in (39) involves parameters with unknown values, it can not be computed. However, it can be approximated as follows[17]

$$\tilde{\beta}_a^* = \left[1 - \frac{g\hat{\sigma}^2}{\hat{\beta}'Q\hat{\beta}}\right]\hat{\beta} \tag{41}$$

where $\hat{\beta} = (X'X)^{-1}X'\mathbf{y}$, $\hat{\sigma}^2 = (\mathbf{y}-X\hat{\beta})'(\mathbf{y}-X\hat{\beta})/(v+2)$, and $v = T-k$. The estimator in (41) is in a form analogous to the Stein-like k-means estimator shown in (26) above. Further, if $Q = X'X$, then $g\hat{\sigma}^2/\hat{\beta}'Q\hat{\beta} = k\hat{\sigma}^2/\hat{\beta}'X'X\hat{\beta}$ is related to an F statistic for testing the hypothesis $\beta = \mathbf{0}$; see Baranchik [1, 2], Sclove [24, 25] and Bock et al. [3] for work on sampling theory pre-test estimators that resemble (41) in form. For example Baranchik [1, 2] and Sclove [24] present the following "positive part" estimator,

$$\tilde{\beta}^+ = \left(1 - \frac{a(F)}{F}\right)\hat{\beta}, \tag{42}$$

where F is an F statistic associated with the pre-test of the hypothesis $\beta = \mathbf{0}$, $a(F) \equiv c$ for $F > c$, $a(F) \equiv F$ for $F \leqq c$, and c is the critical value of F employed in the pre-test. With the function $a(F)$ defined as it is, the algebraic signs of the elements of $\tilde{\beta}^+$ are the same as those of the corresponding elements of $\hat{\beta}$. However, it should be noted that $\tilde{\beta}^+$'s properties depend on the value chosen for c.

[16] If, for example, all variables in the regression equation are expressed in terms of deviations from sample means, the population squared multiple correlation coefficient, ϕ^2, is given by: $\phi^2 = \beta'X'X\beta/(\beta'X'X\beta+T\sigma^2)$. For the special case $Q = X'X$, $R_1 = k\sigma^2$ and thus from (40),

$$R_{q_*} = R_1[\beta'X'X\beta/(\beta'X'X\beta+k\sigma^2)]$$
$$= R_1\left[1/\left(1+\frac{k}{T}\frac{1-\phi^2}{\phi^2}\right)\right].$$

From this we see that if $k = 5$, $T = 20$, and $\phi^2 = 0.5$, $R_{q_*} = 0.8R_1$.

[17] Note that $E\hat{\beta}'Q\hat{\beta} = g\sigma^2+\beta'Q\beta$. Below we investigate the risk associated with approximate MMSE estimators.

The MMSE and pre-test estimators above do not incorporate a Lindley-like mean correction—see (19). In the MMSE approach such a correction is obtained if we consider the class of estimators $\tilde{\boldsymbol{\beta}} = q_0\boldsymbol{\iota}+q_1\hat{\boldsymbol{\beta}}$, where q_0 and q_1 are scalar parameters and $\hat{\boldsymbol{\beta}} = (X'X)^{-1}X'\mathbf{y}$.[18] On inserting $\tilde{\boldsymbol{\beta}} = q_0\boldsymbol{\iota}+q_1\hat{\boldsymbol{\beta}}$ in the loss function (35) and taking its expectation, the values of q_0 and q_1 that minimize expected loss are:

$$q_0^* = \frac{R_1\bar{\beta}}{R_1+(\boldsymbol{\beta}-\bar{\beta}\boldsymbol{\iota})'Q(\boldsymbol{\beta}-\bar{\beta}\boldsymbol{\iota})} \quad \text{and}$$

$$q_1^* = \frac{(\boldsymbol{\beta}-\bar{\beta}\boldsymbol{\iota})'Q(\boldsymbol{\beta}-\bar{\beta}\boldsymbol{\iota})}{R_1+(\boldsymbol{\beta}-\bar{\beta}\boldsymbol{\iota})'Q(\boldsymbol{\beta}-\bar{\beta}\boldsymbol{\iota})}, \tag{43}$$

where $R_1 = E(\hat{\boldsymbol{\beta}}-\boldsymbol{\beta})'Q(\hat{\boldsymbol{\beta}}-\boldsymbol{\beta})$ and $\bar{\beta} = \boldsymbol{\iota}'Q\boldsymbol{\beta}/\boldsymbol{\iota}'Q\boldsymbol{\iota}$. Then the MMSE estimator in the above class is:

$$\begin{aligned}\tilde{\boldsymbol{\beta}}^* &= q_0^*\boldsymbol{\iota}+q_1^*\hat{\boldsymbol{\beta}} \\ &= \bar{\beta}\boldsymbol{\iota}+\left[1-\frac{R_1}{R_1+(\boldsymbol{\beta}-\bar{\beta}\boldsymbol{\iota})'Q(\boldsymbol{\beta}-\bar{\beta}\boldsymbol{\iota})}\right](\hat{\boldsymbol{\beta}}-\bar{\beta}\boldsymbol{\iota}).\end{aligned} \tag{44}$$

The risk associated with $\tilde{\boldsymbol{\beta}}^*$ in (44), denoted R^*, is:

$$R^* = R_1(\boldsymbol{\beta}-\bar{\beta}\boldsymbol{\iota})'Q(\boldsymbol{\beta}-\bar{\beta}\boldsymbol{\iota})/[R_1+(\boldsymbol{\beta}-\bar{\beta}\boldsymbol{\iota})'Q(\boldsymbol{\beta}-\bar{\beta}\boldsymbol{\iota})]. \tag{45}$$

Since $(\boldsymbol{\beta}-\bar{\beta}\boldsymbol{\iota})'Q(\boldsymbol{\beta}-\bar{\beta}\boldsymbol{\iota}) = \boldsymbol{\beta}'Q\boldsymbol{\beta}-\bar{\beta}^2\boldsymbol{\iota}'Q\boldsymbol{\iota} < \boldsymbol{\beta}'Q\boldsymbol{\beta}$ for $\bar{\beta} \neq 0$, R^* in (45) is smaller than R_{q_*} in (40). Of course, since $\tilde{\boldsymbol{\beta}}^*$ in (44) depends on $\boldsymbol{\beta}$ and σ, it can not be used in practice. However, just as with (39), it is possible to construct an operational estimator, $\tilde{\boldsymbol{\beta}}_a^* = \hat{q}_0^*\boldsymbol{\iota}+\hat{q}_1^*\hat{\boldsymbol{\beta}}$, where $\hat{q}_0^*$ and $\hat{q}_1^*$ are estimates of q_0^* and q_1^*, respectively.[19] See section 4 for consideration of the properties of such approximate MMSE estimators.

As in the case of the k-means problem, a BMEL estimator can be generated for the regression coefficient vector $\boldsymbol{\beta}$. Let p_0 be the posterior probability on the hypothesis H_0: $\boldsymbol{\beta} = \boldsymbol{\beta}_0$, where $\boldsymbol{\beta}_0$ is a given vector.

[18] It is possible to consider a broader class of estimators $\tilde{\boldsymbol{\beta}} = q_0\boldsymbol{\iota}+A\hat{\boldsymbol{\beta}}$, where A is a $k\times n$ matrix but we do not analyze this broader class at present.

[19] Explicitly,

$$\tilde{\boldsymbol{\beta}}_a^* = \hat{\bar{\beta}}\boldsymbol{\iota}+\left[1-\frac{g\hat{\sigma}^2}{(\hat{\boldsymbol{\beta}}-\hat{\bar{\beta}}\boldsymbol{\iota})'Q(\hat{\boldsymbol{\beta}}-\hat{\bar{\beta}}\boldsymbol{\iota})+h\hat{\sigma}^2}\right](\hat{\boldsymbol{\beta}}-\hat{\bar{\beta}}\boldsymbol{\iota}),$$

where $\hat{\bar{\beta}} = \boldsymbol{\iota}'Q\hat{\boldsymbol{\beta}}/\boldsymbol{\iota}'Q\boldsymbol{\iota}$, $g = \operatorname{tr}(X'X)^{-1}Q$, $h = \boldsymbol{\iota}'Q(X'X)^{-1}Q\boldsymbol{\iota}/\boldsymbol{\iota}'Q\boldsymbol{\iota}$, and $\hat{\sigma}^2 = (\mathbf{y}-X\hat{\boldsymbol{\beta}})'(\mathbf{y}-X\hat{\boldsymbol{\beta}})/(\nu+2)$.

Let $1-p_0$ be the posterior probability on the hypothesis $H_1: \boldsymbol{\beta} \neq \boldsymbol{\beta}_0$. With a quadratic loss function, $L(\boldsymbol{\beta}, \tilde{\boldsymbol{\beta}}) = (\boldsymbol{\beta}-\tilde{\boldsymbol{\beta}})'Q(\boldsymbol{\beta}-\tilde{\boldsymbol{\beta}})$, where Q is p.d.s., posterior expected loss is given by:

$$\begin{aligned} EL &= p_0 E(L|\boldsymbol{\beta} = \boldsymbol{\beta}_0)+(1-p_0)E(L|\boldsymbol{\beta} \neq \boldsymbol{\beta}_0) \\ &= p_0(\tilde{\boldsymbol{\beta}}-\boldsymbol{\beta}_0)'Q(\tilde{\boldsymbol{\beta}}-\boldsymbol{\beta}_0)+(1-p_0)[(\tilde{\boldsymbol{\beta}}-\bar{\boldsymbol{\beta}})'Q(\tilde{\boldsymbol{\beta}}-\bar{\boldsymbol{\beta}}) \\ &\quad +E(\boldsymbol{\beta}-\bar{\boldsymbol{\beta}})'Q(\boldsymbol{\beta}-\bar{\boldsymbol{\beta}})], \end{aligned} \tag{46}$$

where $\bar{\boldsymbol{\beta}}$ = posterior mean of $\boldsymbol{\beta}$ under $H_1: \boldsymbol{\beta} \neq \boldsymbol{\beta}_0$. The value of $\tilde{\boldsymbol{\beta}}$ that minimizes (46), say $\tilde{\boldsymbol{\beta}}_B^*$, is given by[20]:

$$\begin{aligned} \tilde{\boldsymbol{\beta}}_B^* &= p_0\boldsymbol{\beta}_0+(1-p_0)\bar{\boldsymbol{\beta}} \\ &= \boldsymbol{\beta}_0+\left[1-\frac{K_{01}}{1+K_{01}}\right](\bar{\boldsymbol{\beta}}-\boldsymbol{\beta}_0), \end{aligned} \tag{47}$$

where $K_{01} = p_0/(1-p_0)$, the posterior odds ratio for the hypotheses H_0 and H_1. With average risk assumed finite, the estimator in (47) is optimal in the usual Bayesian senses; that is, it is consistent, admissible and minimizes average risk. Further, $\tilde{\boldsymbol{\beta}}_B^*$ is a continuous function of the observations and since $0 < K_{01}/(1+K_{01}) < 1$, the elements of $\boldsymbol{\beta}_B^*-\boldsymbol{\beta}_0$ have the same algebraic signs as the corresponding elements of $\bar{\boldsymbol{\beta}}-\boldsymbol{\beta}_0$. Therefore there is no need to introduce a "positive part" estimator.

3.2. *Multivariate regression*

The estimation principles discussed above can be readily applied to estimate regression coefficients in the "seemingly unrelated" multivariate regression model,[21]

$$\mathbf{y}_\alpha = X_\alpha\boldsymbol{\beta}_\alpha+\mathbf{u}_\alpha \qquad \alpha = 1, 2, \ldots, m \tag{48}$$

[20] Independent work by Dickey [5] contains analysis similar in certain respect to ours. Also, in Geisel [11], optimal point predictions, derived from two alternative regression models, were found to be a weighted average of predictions from each with the weights being posterior probabilities on the models.

[21] Sclove [23] presents a Stein-like estimator for the model in (49a) for the case $X_1 = X_2 = \ldots = X_m$. He did not employ the approaches that we utilize below to generate improved estimators and our estimators differ somewhat from his.

which we express as

$$\begin{bmatrix} \mathbf{y}_1 \\ \mathbf{y}_2 \\ \vdots \\ \mathbf{y}_m \end{bmatrix} = \begin{bmatrix} X_1 & & & \\ & X_2 & & \\ & & \ddots & \\ & & & X_m \end{bmatrix} \begin{bmatrix} \boldsymbol{\beta}_1 \\ \boldsymbol{\beta}_2 \\ \vdots \\ \boldsymbol{\beta}_m \end{bmatrix} + \begin{bmatrix} \mathbf{u}_1 \\ \mathbf{u}_2 \\ \vdots \\ \mathbf{u}_m \end{bmatrix} \tag{49a}$$

or

$$y = Z\boldsymbol{\eta} + \mathbf{u} \tag{49b}$$

where, for $\alpha = 1, 2, \ldots, m$, $\mathbf{y}_\alpha$ is a $T \times 1$ vector, X_α is a $T \times k_\alpha$ non-stochastic matrix of rank k_α, $\mathbf{u}_\alpha$ is a $T \times 1$ vector of disturbance terms, and $\boldsymbol{\beta}_\alpha$ is a $k_\alpha \times 1$ vector of regression coefficients. Further, in (49b) $\mathbf{y}' = (\mathbf{y}_1', \mathbf{y}_2', \ldots, \mathbf{y}_m')$, $\boldsymbol{\eta}' = (\boldsymbol{\beta}_1', \boldsymbol{\beta}_2', \ldots, \boldsymbol{\beta}_m')$, $\mathbf{u}' = (\mathbf{u}_1', \mathbf{u}_2', \ldots, \mathbf{u}_m')$ and Z represents the block diagonal matrix on the right-hand side of (49a). Regarding $\mathbf{u}$ in (49b), we assume $E\mathbf{u} = \mathbf{0}$ and

$$E\mathbf{u}\mathbf{u}' = \Omega = \Sigma \otimes I_T \tag{50}$$

where the $m \times m$ matrix Σ is p.d.s.

Let H be a non-singular matrix such that $H'\Omega H = I_{mT}$. Then on premultiplying both sides of (49b) by H, we obtain

$$\dot{y} = \dot{Z}\boldsymbol{\eta} + \dot{u} \tag{51}$$

where $\dot{y} = H\mathbf{y}$, $\dot{Z} = HZ$ and $\dot{u} = H\mathbf{u}$. We have $E\dot{u} = \mathbf{0}$ and $E\dot{u}\,\dot{u}' = I_{mT}$. Thus (51) is in the form of a standard univariate regression model to which the results of the preceding section can be applied.

For example, let $\hat{\boldsymbol{\eta}} = (\dot{Z}'\dot{Z})^{-1}\dot{Z}'\dot{y} = (Z'\Omega^{-1}Z)^{-1}Z'\Omega^{-1}\mathbf{y}$. The risk associated with this estimator, relative to a quadratic loss function, is

$$\begin{aligned} R_1 &= E(\hat{\boldsymbol{\eta}} - \boldsymbol{\eta})'Q(\hat{\boldsymbol{\eta}} - \boldsymbol{\eta}) \\ &= \operatorname{tr}(\dot{Z}'\dot{Z})^{-1}Q = \operatorname{tr}(Z'\Omega^{-1}Z)^{-1}Q \end{aligned} \tag{52}$$

where Q is a p.d.s. matrix. If, as above, we consider estimators of the form $q\hat{\boldsymbol{\eta}}$, where q is a scalar,[22] the optimal value for q relative to a quadratic loss function with matrix Q, is

$$q_* = 1 - \frac{R_1}{R_1 + \boldsymbol{\eta}'Q\boldsymbol{\eta}}, \tag{53}$$

with R_1 as shown in (52). Then the MMSE estimator in the class

[22] Obviously, estimators of the forms $q_0\iota + q_1\hat{\boldsymbol{\eta}}$ or $q_0\iota + A\hat{\boldsymbol{\eta}}$ could also be considered.

$q\hat{\boldsymbol{\eta}}$ is $\tilde{\boldsymbol{\eta}}^* = q_*\hat{\boldsymbol{\eta}}$, or

$$\tilde{\boldsymbol{\eta}}^* = \left[1 - \frac{g}{g + \boldsymbol{\eta}'Q\boldsymbol{\eta}}\right]\hat{\boldsymbol{\eta}} \tag{54}$$

where here $g = \text{tr}\,(Z'\Omega^{-1}Z)^{-1}Q$. Note that (54) involves $\boldsymbol{\eta}$ and Ω and thus is not operational. We can however approximate (54) as follows:

$$\tilde{\boldsymbol{\eta}}_a^* = \left[1 - \frac{\hat{g}}{\hat{\boldsymbol{\eta}}_a'Q\hat{\boldsymbol{\eta}}_a}\right]\hat{\boldsymbol{\eta}}_a \tag{55}$$

where $\hat{g} = \text{tr}\,(Z'\hat{\Omega}^{-1}Z)^{-1}Q$, $\hat{\boldsymbol{\eta}}_a = (Z'\hat{\Omega}^{-1}Z)^{-1}Z'\hat{\Omega}^{-1}\mathbf{y}$, and $\hat{\Omega} = \hat{\Sigma} \otimes I_T$. The elements of $\hat{\Sigma}$, say $s_{\alpha l}$, are given by: $s_{\alpha l} = (\mathbf{y}_\alpha - X_\alpha\hat{\boldsymbol{\beta}}_\alpha)'(\mathbf{y}_l - X_l\hat{\boldsymbol{\beta}}_l)/T$, with $\hat{\boldsymbol{\beta}}_\alpha = (X_\alpha'X_\alpha)^{-1}X_\alpha'\mathbf{y}_\alpha'$ and $\hat{\boldsymbol{\beta}}_l = (X_l'X_l)^{-1}X_l'\mathbf{y}_l$, $\alpha, l = 1, 2, \ldots, m$. The risk associated with $\tilde{\boldsymbol{\eta}}^*$ in (54), R_{q_*}, is:

$$R_{q_*} = R_1[\boldsymbol{\eta}'Q\boldsymbol{\eta}/(R_1 + \boldsymbol{\eta}'Q\boldsymbol{\eta})] \tag{56}$$

with R_1 given in (52). Additional work is required to establish the risk associated with the approximately optimal estimator $\tilde{\boldsymbol{\eta}}_a^*$, shown in (55).[23]

Further, as noted in Zellner [33, p. 242], given Σ, the posterior pdf for $\boldsymbol{\eta}$ is given by

$$p(\boldsymbol{\eta}|\Sigma, \mathbf{y}, PI) \propto p(\boldsymbol{\eta}) \exp\{-\tfrac{1}{2}(\boldsymbol{\eta} - \hat{\boldsymbol{\eta}})'\dot{Z}'\dot{Z}(\boldsymbol{\eta} - \hat{\boldsymbol{\eta}})\} \tag{57}$$

where PI denotes prior information and $p(\boldsymbol{\eta})$ is the prior pdf for $\boldsymbol{\eta}$. If $p(\boldsymbol{\eta})$ is in the multivariate normal form with mean $\bar{\boldsymbol{\eta}}$ and covariance matrix Φ^{-1}, then the posterior mean for $\boldsymbol{\eta}$ given Σ, denoted $\bar{\bar{\boldsymbol{\eta}}}$, is given by

$$\begin{aligned}\bar{\bar{\boldsymbol{\eta}}} &= (\dot{Z}'\dot{Z} + \Phi)^{-1}(\dot{Z}'\dot{Z}\hat{\boldsymbol{\eta}} + \Phi\bar{\boldsymbol{\eta}}) \\ &= [Z'(\Sigma^{-1} \otimes I_T)Z + \Phi]^{-1}[Z'(\Sigma^{-1} \otimes I_T)Z\hat{\boldsymbol{\eta}} + \Phi\bar{\boldsymbol{\eta}}]\end{aligned} \tag{58}$$

which is the multivariate analogue of (34).[24] Last, given Σ, the BMEL estimator analysis can be readily applied to yield an optimal estimator for $\boldsymbol{\eta}$.

4. *General discussion of improved estimators and the simultaneous equation model*

4.1. *General results for improving biased estimators*

In the preceding sections, we have discussed the improvement of usual

[23] Some work on this problem appears in Srivastava [26].

[24] Given $\Sigma = \hat{\Sigma}$, (39) can be employed in practice. The conditional covariance matrix is $[Z'(\hat{\Sigma}^{-1} \otimes I_T)Z + \Phi]^{-1}$.

estimators all of which had the property of being unbiased. For many problems, when we work with diffuse prior pdf's, the mean of a posterior pdf, viewed as an estimator, will not be unbiased.[25] Also, many usual sampling theory estimators, including maximum likelihood estimators, two- and three-stage least squares estimators, etc., are usually biased in finite-sized samples. It is thus natural to ask whether it is possible to improve biased, as well as unbiased, estimators.[26] The results below bear on this issue.

Theorem. *Let $\hat{\boldsymbol{\theta}}$ be an estimator, not necessarily unbiased, for $\boldsymbol{\theta}$, a $p \times 1$ vector, and assume that the first and second moments of the elements of $\hat{\boldsymbol{\theta}}$ exist and have finite values. Then if $\tilde{\boldsymbol{\theta}} = q_0\boldsymbol{\iota} + q_1\hat{\boldsymbol{\theta}}$, where q_0 and q_1 are scalars, there are values of q_0 and q_1 such that the risk associated with $\tilde{\boldsymbol{\theta}}$ relative to a positive definite quadratic loss function is equal to or less than that associated with the estimator $\hat{\boldsymbol{\theta}}$.*

Proof. The risk associated with $\tilde{\boldsymbol{\theta}}$, denoted R, is:

$$R = E(q_0\boldsymbol{\iota}+q_1\hat{\boldsymbol{\theta}}-\boldsymbol{\theta})'Q(q_0\boldsymbol{\iota}+q_1\hat{\boldsymbol{\theta}}-\boldsymbol{\theta}), \tag{59}$$

where Q is a p.d.s. matrix. On differentiating R partially with respect to q_0, the value of q_0 that sets this partial derivative equal to zero must satisfy

$$\begin{aligned} q_0 &= (\boldsymbol{\iota}'Q\boldsymbol{\theta}-q_1\boldsymbol{\iota}'Q\boldsymbol{\alpha})/\boldsymbol{\iota}'Q\boldsymbol{\iota} \\ &= \bar{\theta}-q_1\bar{\alpha}, \end{aligned} \tag{60}$$

where $\boldsymbol{\alpha} \equiv E\hat{\boldsymbol{\theta}}$, $\bar{\theta} \equiv \boldsymbol{\iota}'Q\boldsymbol{\theta}/\boldsymbol{\iota}'Q\boldsymbol{\iota}$, and $\bar{\alpha} \equiv \boldsymbol{\iota}'Q\boldsymbol{\alpha}/\boldsymbol{\iota}'Q\boldsymbol{\iota}$. On substituting from (60) in (59), we have:

$$R = E[q_1(\hat{\boldsymbol{\theta}}-\bar{\alpha}\boldsymbol{\iota})-(\boldsymbol{\theta}-\bar{\theta}\boldsymbol{\iota})]'Q[q_1(\hat{\boldsymbol{\theta}}-\bar{\alpha}\boldsymbol{\iota})-(\boldsymbol{\theta}-\bar{\theta}\boldsymbol{\iota})]. \tag{61}$$

The value of q_1 that minimizes (61), q_1^*, is given by:

$$q_1^* = (\boldsymbol{\alpha}-\bar{\alpha}\boldsymbol{\iota})'Q(\boldsymbol{\theta}-\bar{\theta}\boldsymbol{\iota})/E(\hat{\boldsymbol{\theta}}-\bar{\alpha}\boldsymbol{\iota})'Q(\hat{\boldsymbol{\theta}}-\bar{\alpha}\boldsymbol{\iota}). \tag{62}$$

Let R_1 be the risk associated with $\hat{\boldsymbol{\theta}}$, namely, $R_1 = E(\hat{\boldsymbol{\theta}}-\boldsymbol{\theta})'Q(\hat{\boldsymbol{\theta}}-\boldsymbol{\theta}) = E[(\hat{\boldsymbol{\theta}}-\bar{\alpha}\boldsymbol{\iota})-(\boldsymbol{\theta}-\bar{\alpha}\boldsymbol{\iota})]'Q[(\hat{\boldsymbol{\theta}}-\bar{\alpha}\boldsymbol{\iota})-(\boldsymbol{\theta}-\bar{\alpha}\boldsymbol{\iota})]$. Then after substituting $q_1 = q_1^*$ in (61), we have

[25] For example, in univariate regression analysis with a diffuse prior pdf, the posterior mean for σ^2, the common variance of the error terms, is not unbiased.

[26] In a Bayesian context, we assume that sampling properties are of concern to an investigator.

$$R_1 - R = E(\hat{\theta}-\bar{\alpha}\iota)'Q(\hat{\theta}-\bar{\alpha}\iota) - 2(\alpha-\bar{\alpha}\iota)'Q(\theta-\bar{\theta}\iota) + \frac{[(\alpha-\bar{\alpha}\iota)'Q(\theta-\bar{\theta}\iota)]^2}{E(\hat{\theta}-\bar{\alpha}\iota)'Q(\hat{\theta}-\bar{\alpha}\iota)} + (\theta-\bar{\alpha}\iota)'Q(\theta-\bar{\alpha}\iota) - (\theta-\bar{\theta}\iota)'Q(\theta-\bar{\theta}\iota) + 2[(\alpha-\bar{\alpha}\iota)'Q(\alpha-\bar{\alpha}\iota) - (\alpha-\bar{\alpha}\iota)'Q(\theta-\bar{\alpha}\iota)],$$

or, on collecting terms,

$$R_1 - R = \frac{[E(\hat{\theta}-\bar{\alpha}\iota)'Q(\hat{\theta}-\bar{\alpha}\iota) - (\alpha-\bar{\alpha}\iota)'Q(\alpha-\bar{\theta}\iota)]^2}{E(\hat{\theta}-\bar{\alpha}\iota)'Q(\hat{\theta}-\bar{\alpha}\iota)} + \iota'Q\iota(\bar{\alpha}-\bar{\theta})^2. \tag{63}$$

Since the r.h.s. of (63) is non-negative, $R_1 \geqq R$.

The following lemmas provide specializations of the theorem.

Lemma 1. *If $\hat{\theta}$ is unbiased, i.e. $E\hat{\theta} = \alpha = \theta$, then $\bar{\alpha} = \bar{\theta}$ and* (63) *specializes to:*

$$R = R_1 - \frac{[E(\hat{\theta}-\bar{\theta}\iota)'Q(\hat{\theta}-\bar{\theta}\iota) - (\theta-\bar{\theta}\iota)'Q(\theta-\bar{\theta}\iota)]^2}{E(\hat{\theta}-\bar{\theta}\iota)'Q(\hat{\theta}-\bar{\theta}\iota)}$$

$$= R_1(\theta-\bar{\theta}\iota)'Q(\theta-\bar{\theta}\iota)/[R_1 + (\theta-\bar{\theta}\iota)'Q(\theta-\bar{\theta}\iota)]. \tag{63a}$$

The result in (45) *for the univariate regression model is a special case of this lemma with $\hat{\theta} = \hat{\beta}$.*

Lemma 2. *If we restrict the class of estimators considered to $\tilde{\theta} = q\hat{\theta}$ with $\hat{\theta}$ not necessarily unbiased, then the optimal value for q is $q_* = E\hat{\theta}'Q\theta/E\hat{\theta}'Q\hat{\theta}$ and the risk of $\tilde{\theta}^* = q_*\hat{\theta}$ is given by $R_* = E(q_*\hat{\theta}-\theta)'Q(q_*\hat{\theta}-\theta)$. With $R_1 = E(\hat{\theta}-\theta)'Q(\hat{\theta}-\theta)$, we have*

$$R_1 - R_* = [E\hat{\theta}'Q\hat{\theta} - E\hat{\theta}'Q\theta]^2/E\hat{\theta}'Q\hat{\theta} \tag{63b}$$

Thus, $R_1 - R_* \geqq 0$ or $R_1 \geqq R_*$. On comparing (63) *and* (63b), *$R_1 - R_* \leqq R_1 - R$ and thus $R \leqq R_*$.* If *$E\hat{\theta} = \theta$,* (63b) *specializes to $R_* = R_1[\theta'Q\theta/(R_1+\theta'Q\theta)]$, a result that includes* (40) *as a special case.*

Since the optimal q's in the theorem and in the lemmas all depend on population parameters, estimators depending on them are not operational. In the following lemma, we indicate some of the effects of using an estimate of an optimal q in forming an approximation to an optimal estimator.

Lemma 3. *If in lemma 2, we approximate the optimal estimator $\tilde{\theta}^* = q_*\hat{\theta}$ by $\tilde{\theta}^*_a = \hat{q}_*\theta$, where $\hat{q}_*$ is an estimator for q_* such that $1-\hat{q}_* =$*

$1-q_*+\Delta_1$, *where* $E\Delta_1$ *is* $0(T^{-m})$ with $m>1$, *then the risk associated with* $\tilde{\theta}_*$, R_*, *and that associated with* $\tilde{\theta}_a^*$, R_a *are related as follows:* $R_a-R_* = E\Delta_2$, *where* $E\Delta_2$ *is* $0(Q/T^{2+\delta})$, $\delta>0$ *given that* $R_1 = E(\hat{\theta}-\theta)'Q(\hat{\theta}-\theta)$ *and* $E(\hat{\theta}-\theta)'Q\theta$ *are each* $0(Q/T)$.

Proof. Note that $\tilde{\theta}_a^*-\theta = \hat{q}_*\hat{\theta}-\theta = \tilde{\theta}^*-\theta-\Delta_1\hat{\theta}$. Then

$$\begin{aligned} R_a &= E(\tilde{\theta}^*-\theta-\Delta_1\hat{\theta})'Q(\tilde{\theta}^*-\theta-\Delta_1\hat{\theta}) \\ &= R_*+E[\Delta_1^2\hat{\theta}'Q\hat{\theta}-2\Delta_1\hat{\theta}'Q(\tilde{\theta}^*-\theta)] \\ &= R_1-\frac{[R_1+E\theta'Q(\hat{\theta}-\theta)]^2}{E\hat{\theta}'Q\hat{\theta}}+E\Delta_2 \end{aligned} \tag{64}$$

where the first two terms on the r.h.s. of the last line of (64) come from (63b) and $\Delta_2 = \Delta_1^2\hat{\theta}'Q\hat{\theta}-2\Delta_1\hat{\theta}'Q(\tilde{\theta}^*-\theta)$. With respect to orders of the terms in the last line of (64), R_1 is $0(Q/T)$, the second term is $0(Q/T^2)$ and $E\Delta_2$ is $0(Q/T^{2+\delta})$, $\delta>0$.[27]

4.2. *Estimators for parameters of simultaneous equation model*

In this section, we consider estimation of the parameters of a single equation in a linear simultaneous equation model. Let the first equation of a system, with parameters assumed identified, be:

$$\mathbf{y}_1 = Y_1\boldsymbol{\gamma}_1+X_1\boldsymbol{\beta}_1+\mathbf{u}_1 \tag{65}$$

where $\mathbf{y}_1$ is a $T\times 1$ vector of observations on an endogenous variable, Y_1 is a $T\times m_1$ matrix of observations on m_1 other endogenous variables appearing in the first equation, X_1 is a $T\times k_1$ non-stochastic matrix with rank k_1, $\boldsymbol{\gamma}_1$ and $\boldsymbol{\beta}_1$ are, respectively, $m_1\times 1$ and $k_1\times 1$ coefficient vectors, and $\mathbf{u}_1$ is a $T\times 1$ disturbance vector. We assume that $E\mathbf{u}_1 = \mathbf{0}$ and $E\mathbf{u}_1\mathbf{u}_1' = \sigma_{11}I_T$.

The reduced form equations for $\mathbf{y}_1$ and Y_1 are

$$(\mathbf{y}_1 \vdots Y_1) = (X_1 \vdots X_0)\begin{bmatrix}\pi_{11} & \Pi_{10}\\ \pi_{01} & \Pi_{00}\end{bmatrix}+(\mathbf{v}_1 \vdots V_1) \tag{66}$$

[27] $E\Delta_2 = E\Delta_1{}^2\hat{\theta}'Q\hat{\theta}-2E\Delta_1\hat{\theta}'Q(q_*\hat{\theta}-\theta)$. Given that $E\Delta_1$ is $0(T^{-m})$, $m>1$, the first term is $0(Q/T^{2+\delta})$, $\delta>0$. For the second term, we have:

$$\begin{aligned} E\Delta_1\hat{\theta}'Q(q_*\hat{\theta}-\theta) = E\Delta_1\{&q_*[(\hat{\theta}-\theta)'Q\theta+(\hat{\theta}-\theta)'Q(\hat{\theta}-\theta)] \\ &-(1-q_*)[\theta'Q\theta+(\hat{\theta}-\theta)'Q\theta]\}. \end{aligned}$$

Since q_* is $0(1)$, $1-q_*$ is $0(T^{-1})$, $E\Delta_1$ is $0(T^{-m})$, $m>1$, and $E(\hat{\theta}-\theta)'Q\theta$ is $0(Q/T)$, $E\Delta_1\hat{\theta}'Q(q_*\hat{\theta}-\theta)$ is $0(Q/T^{2+\delta})$, $\delta>0$.

where $X = (X_1 \vdots X_0)$, a $T \times k$ non-stochastic matrix with rank k, with X_0 representing variables that are excluded from (59), and $(\mathbf{v}_1 \vdots V_1)$ is a $T \times (m_1 + 1)$ matrix of reduced form disturbance terms. The rows of $(\mathbf{v}_1 \vdots V_1)$, say $(v_{1t} \vdots \mathbf{v}'_{0t})$, $t = 1, 2, \ldots, T$, are assumed to have zero mean vectors, to be uncorrelated, and to have common covariance matrix,

$$E\begin{pmatrix} v_{1t} \\ \hdashline \mathbf{v}_{0t} \end{pmatrix}(v_{1t} \vdots \mathbf{v}'_{0t}) = \left(\begin{array}{c:c} \omega_{11} & \omega'_1 \\ \hdashline \omega_1 & \Omega_{11} \end{array}\right). \tag{67}$$

From (65) and (66), we have the following well-known restrictions

$$\pi_{11} - \Pi_{10}\gamma_1 = \boldsymbol{\beta}_1 \tag{68}$$

$$\pi_{01} - \Pi_{00}\gamma_1 = \mathbf{0}. \tag{69}$$

Using (68)–(69) to substitute for π_{11} and π_{01} in (66), we obtain

$$\begin{aligned} \mathbf{y}_1 &= X\Pi_0\gamma_1 + X_1\boldsymbol{\beta}_1 + \mathbf{v}_1 \\ &= \bar{Z}_1\boldsymbol{\delta}_1 + \mathbf{v}_1 \end{aligned} \tag{70}$$

and

$$Y_1 = X\Pi_0 + V_1 \tag{71}$$

where $\Pi'_0 = (\Pi'_{10} \vdots \Pi'_{00})$, $\bar{Z}_1 = (X\Pi_0 \vdots X_1)$, and $\boldsymbol{\delta}'_1 = (\gamma'_1 \vdots \boldsymbol{\beta}'_1)$. The system (70)–(71) is a nonlinear (in the parameters) multivariate regression model.

If the elements of Π_0 in (70) had known values, the usual estimator for $\boldsymbol{\delta}_1$ is

$$\boldsymbol{\delta}_1^* = (\bar{Z}'_1\bar{Z}_1)^{-1}\bar{Z}'_1\mathbf{y}_1 \tag{72}$$

and the methods of section 3 are directly applicable for obtaining improved estimators. Since the elements of Π_0 rarely have known values in practice, Π_0 is often replaced by $\hat{\Pi}_0 = (X'X)^{-1}X'Y_1$ to yield

$$\hat{\boldsymbol{\delta}}_1 = (\hat{Z}'_1\hat{Z}_1)^{-1}\hat{Z}'_1\mathbf{y}_1, \tag{73}$$

the 2SLS estimator, where $\hat{Z}_1 = (X\hat{\Pi}_0 \vdots X_1)$. While $\hat{\boldsymbol{\delta}}_1$ is known to be an asymptotically efficient estimator, it is not clear in what sense, if any, (73) is optimal in finite samples. For example, in certain instances the second moments of the elements of $\hat{\boldsymbol{\delta}}_1$ do not exist and thus $\hat{\boldsymbol{\delta}}_1$ can not be optimal relative to a quadratic loss function. Further, given that $E\hat{\boldsymbol{\delta}}_1$ is finite, it is known that $\hat{\boldsymbol{\delta}}_1$ can be biased—see for example Richardson [19, 20] and Sawa [22]. Since $\hat{\boldsymbol{\delta}}_1$ is usually biased, the theorem in the preceding section can be applied to obtain an improved estimator when the second moments of the elements of $\hat{\boldsymbol{\delta}}_1$ exist and are finite. To illustrate, we consider $\tilde{\boldsymbol{\delta}} = q\hat{\boldsymbol{\delta}}$, where q is a scalar and where, for convenience we have suppressed the subscript 1; that is $\hat{\boldsymbol{\delta}}_1 \equiv \hat{\boldsymbol{\delta}}$, $\boldsymbol{\delta}_1 \equiv \boldsymbol{\delta}$, etc. Then, for this problem, application of lemma 2 yields the following optimal value for q relative to a quadratic loss function, $(\tilde{\boldsymbol{\delta}}-\boldsymbol{\delta})'Q(\tilde{\boldsymbol{\delta}}-\boldsymbol{\delta})$:

$$\begin{aligned} q_* &= E\hat{\delta}'Q\delta/E\hat{\delta}'Q\hat{\delta} \\ &= [\delta'Q\delta+E(\hat{\delta}-\delta)'Q\delta]/[R_1+\delta'Q\delta+2E(\hat{\delta}-\delta)'Q\delta] \end{aligned} \tag{74}$$

where $R_1 = E(\hat{\boldsymbol{\delta}}-\boldsymbol{\delta})'Q(\hat{\boldsymbol{\delta}}-\boldsymbol{\delta})$, the risk function for the 2SLS estimator. Then, using the results of lemma 2,

$$R_* = R_1 - \frac{[R_1+E(\hat{\delta}-\delta)'Q\delta]^2}{E\hat{\delta}'Q\hat{\delta}} \tag{75}$$

where $R_* = E(\tilde{\boldsymbol{\delta}}^*-\boldsymbol{\delta})'Q(\tilde{\boldsymbol{\delta}}^*-\boldsymbol{\delta})$ with $\tilde{\boldsymbol{\delta}}^* = q_*\hat{\boldsymbol{\delta}}$.

Since q_* in (74) depends on population parameters, $\tilde{\boldsymbol{\delta}}^* = q_*\hat{\boldsymbol{\delta}}$ is not an operational estimator. However, pursuing the approach set forth in lemma 3 above, we can approximate $\tilde{\boldsymbol{\delta}}^*$ by $\tilde{\boldsymbol{\delta}}^*_a = \hat{q}_*\hat{\boldsymbol{\delta}}$ where

$$\hat{q}_* = 1 - \frac{\hat{\sigma}_{11}\,\mathrm{tr}\,(\hat{Z}_1'\hat{Z}_1)^{-1}Q + r_1\hat{\mathbf{d}}'(\hat{Z}_1'\hat{Z}_1)^{-1}Q\hat{\delta}_1}{\hat{\delta}_1'Q\hat{\delta}_1} \tag{76}$$

with $\hat{\boldsymbol{\delta}}_1 = (\hat{Z}_1'\hat{Z}_1)^{-1}\hat{Z}_1'\mathbf{y}_1$, $\hat{\sigma}_{11} = (\mathbf{y}_1 - Z_1\hat{\boldsymbol{\delta}}_1)'(\mathbf{y}_1 - Z_1\hat{\boldsymbol{\delta}}_1)/T$, $r_1 = k-k_1-m_1-1$,[28] $\hat{\mathbf{d}}' = (\hat{\boldsymbol{\omega}}_1' - \hat{\boldsymbol{\gamma}}_1'\hat{\Omega}_{11} \vdots \mathbf{0}_{k_1}')$, a $1\times(m_1+k_1)$ vector where $\hat{\boldsymbol{\gamma}}_1'$ is a subvector of $\hat{\boldsymbol{\delta}}_1' = (\hat{\boldsymbol{\gamma}}_1' \vdots \hat{\boldsymbol{\beta}}_1')$ and $\hat{\omega}_{11}$ and $\hat{\Omega}_{11}$ are consistent estimates of $\boldsymbol{\omega}_1$ and Ω_{11}, defined in (67). The form of $\hat{q}_*$ in (76) is similar to that in (41) for a regression coefficient vector except that (76) has an additional term in the numerator that arises because $\hat{\boldsymbol{\delta}}_1$ is usually biased. Then, since the conditions of lemma 3 are satisfied,[29]

$$R_a = R_1 - \frac{[R_1+E\delta'Q(\hat{\delta}-\delta)]^2}{E\hat{\delta}'Q\hat{\delta}} + E\Delta_2 \tag{77}$$

where R_a is the risk associated with $\tilde{\boldsymbol{\delta}}^*_a$ and R_1 with $\hat{\boldsymbol{\delta}}$. As stated in lemma 3, R_1 is $0(Q/T)$, the second term on the r.h.s. is $0(Q/T^2)$ and the third term is $0(Q/T^{2+\delta})$, $\delta > 0$.

[28] Note that r_1+1 is the "degree of overidentification" of the coefficients in (65).

[29] Previous work, see e.g. Nagar [18] and Kadane [14] has established that $E(\hat{\boldsymbol{\delta}}-\boldsymbol{\delta})$ and $E(\hat{\boldsymbol{\delta}}-\boldsymbol{\delta})(\hat{\boldsymbol{\delta}}-\boldsymbol{\delta})'$ have elements of $0(T^{-1})$. Thus $E(\hat{\boldsymbol{\delta}}-\boldsymbol{\delta})'Q\boldsymbol{\delta}$ is $0(Q/T)$ and $R_1 = E(\hat{\boldsymbol{\delta}}-\boldsymbol{\delta})'Q(\hat{\boldsymbol{\delta}}-\boldsymbol{\delta})$ is $0(Q/T)$. That $1-\hat{q}_* = 1-q_*+\Delta_1$ with $E\Delta_1$ of $0(T^{-m})$ with $m = 3/2$ can be established using the results of Nagar [18] and Kadane [14].

The Bayesian analysis of (70), conditional upon given Π_0, is the same as the analysis of a univariate regression model, as pointed out in Zellner [33, p. 266]. That is, assume that the elements of $\mathbf{v}_1$ are normally and independently distributed, each with zero mean and common variance ω_{11}. The likelihood function is:

$$p(\mathbf{y}_1|\Pi_0, \boldsymbol{\delta}_1, \omega_{11})$$
$$\propto \omega_{11}^{-T/2} \exp\left\{-\frac{1}{2\omega_{11}}(\mathbf{y}_1 - \bar{Z}_1\boldsymbol{\delta}_1)'(\mathbf{y}_1 - \bar{Z}_1\boldsymbol{\delta}_1)\right\}$$
$$\propto \omega_{11}^{-T/2} \exp\left\{-\frac{1}{2\omega_{11}}(\nu_1 s_{11} + (\boldsymbol{\delta}_1 - \hat{\boldsymbol{\delta}}_1^*)'\bar{Z}_1'\bar{Z}_1(\boldsymbol{\delta}_1 - \hat{\boldsymbol{\delta}}_1^*)\right\} \tag{78}$$

where $\bar{Z}_1 = (X\Pi_0 \vdots X_1)$, $\hat{\boldsymbol{\delta}}_1^* = (\bar{Z}_1'\bar{Z}_1)^{-1}\bar{Z}_1'\mathbf{y}_1$, $\nu_1 s_{11} = (\mathbf{y}_1 - \bar{Z}_1'\hat{\boldsymbol{\delta}}_1^*)'(\mathbf{y}_1 - \bar{Z}_1\hat{\boldsymbol{\delta}}_1^*)$, and $\nu_1 = T - m_1 - k_1$. If we employ a diffuse prior for Π_0, $\boldsymbol{\delta}_1$ and ω_{11}, $p(\Pi_0, \boldsymbol{\delta}_1, \omega_{11}) \propto 1/\omega_{11}$, the conditional posterior pdf for $\boldsymbol{\delta}_1$ given Π_0 is multivariate normal with mean vector $\hat{\boldsymbol{\delta}}^*$. If in this conditional posterior pdf, we take $\Pi_0 = \hat{\Pi}_0 = (X'X)^{-1}X'Y_1$—see (71), the conditional posterior mean for $\boldsymbol{\delta}_1$ is $\hat{\boldsymbol{\delta}}_1 = (\hat{Z}_1'\hat{Z}_1)^{-1}\hat{Z}_1'\mathbf{y}_1$, the 2SLS estimate—see (73). On the other hand, if our prior is $p(\Pi_0, \boldsymbol{\delta}_1 \omega_{11}) \propto p_1(\boldsymbol{\delta}_1)\omega_{11}$ with

$$p_1(\boldsymbol{\delta}_1|\bar{\boldsymbol{\delta}}_1, \tau, A) \propto \tau^{-(m_1+k_1)} \exp\left\{-\frac{1}{2\tau^2}(\boldsymbol{\delta}_1 - \bar{\boldsymbol{\delta}}_1)'A(\boldsymbol{\delta}_1 - \bar{\boldsymbol{\delta}}_1)\right\}, \tag{79}$$

where $\bar{\boldsymbol{\delta}}_1$ is the prior mean vector and $\tau^2 A^{-1}$ the prior covariance matrix, then given ω_{11} and $\Pi_0 = \hat{\Pi}_0 = (X'X)^{-1}X'Y_1$, the posterior pdf for $\boldsymbol{\delta}_1$ is[30]:

$$p(\boldsymbol{\delta}_1|\omega_{11}, \Pi_0 = \hat{\Pi}_0, \mathbf{y}_1) \propto \exp\{-\tfrac{1}{2}(\boldsymbol{\delta}_1 - \bar{\bar{\boldsymbol{\delta}}}_1)'M(\boldsymbol{\delta}_1 - \bar{\bar{\boldsymbol{\delta}}}_1)\} \tag{80}$$

where $\bar{\bar{\boldsymbol{\delta}}}_1$ and M^{-1}, the posterior mean vector and covariance matrix, respectively, are given by:

$$\bar{\bar{\boldsymbol{\delta}}}_1 = (A/\tau^2 + \hat{Z}_1'\hat{Z}_1/\omega_{11})^{-1}(A\bar{\boldsymbol{\delta}}_1/\tau^2 + \hat{Z}_1'\hat{Z}_1\hat{\boldsymbol{\delta}}_1/\omega_{11}) \tag{81}$$

[30] Here, for illustrative purposes, we go ahead conditional on given ω_{11}. With a prior for ω_{11}, say in an inverted gamma form, it is possible to integrate out ω_{11} to obtain the marginal conditional posterior pdf for $\boldsymbol{\delta}_1$ given Π_0. See Drèze [6, 7, 8], Drèze and Morales [9] and Zellner [33, ch. 9] for results not conditional on Π_0 assumed given.

and

$$M^{-1} = (A/\tau^2 + \hat{Z}_1'\hat{Z}_1/\omega_{11})^{-1}. \tag{82}$$

The posterior mean in (81) is precisely in the form of (34) and thus all that has been said above applies to (81).

It should be noted that we can generate a BMEL for $\boldsymbol{\delta}_1$ as follows. Let p_0 = posterior probability on the hypothesis $H_0: \boldsymbol{\delta}_1 = \boldsymbol{\delta}_1^0$, a given vector, and $1-p_0$ = posterior probability on $H_1: \boldsymbol{\delta}_1 \neq \boldsymbol{\delta}_1^0$. Then, as in section 3, the optimal point estimate relative to a quadratic loss function is:

$$\begin{aligned}\hat{\boldsymbol{\delta}}_1^{\text{opt}} &= p_0\boldsymbol{\delta}_1^0 + (1-p_0)\boldsymbol{\delta}_1^m \\ &= \boldsymbol{\delta}_1^0 + \left(1 - \frac{K_{01}}{1+K_{01}}\right)(\boldsymbol{\delta}_1^m - \boldsymbol{\delta}_1^0),\end{aligned} \tag{83}$$

where $K_{01} = p_0/(1-p_0)$, the posterior odds ratio relating to H_0 and H_1, and $\boldsymbol{\delta}_1^m$ = posterior mean of $\boldsymbol{\delta}_1$ under H_1. If this analysis is performed conditional upon $\bar{Z}_1 = \hat{Z}_1$, it will be precisely the same as that for the univariate regression model.

Last, while we do not pursue the matter here, it should be appreciated that the above methods can be employed to improve "systems estimators" such as the full-information maximum likelihood and 3SLS estimators.

5. *Summary and concluding remarks*

In this paper we have provided a review of Stein and Stein-like estimators for widely-used models. Several methods of generating such estimators have been described and applied. For example, approximate minimal mean square error estimators for parameters of several models have been derived and found to be in a Stein-like form. While these MMSE estimators have lower risk than usual estimators under a wide range of conditions, it should be appreciated that they are not in general admissible. Further, the degree of improvement relative to properties of usual estimators deserves further study and more analysis is needed to provide standard errors, confidence intervals, etc.

In the Bayesian approach, it was seen that use of standard proper prior pdf's led to posterior pdf's having means in the form of Stein or Stein-like estimators. Since such means are optimal relative to quadratic loss functions, they have much to recommend them. Further, it is the case that in the Bayesian approach, complete posterior pdf's are available and can be used to construct interval estimates, etc. Of course, this analysis requires that prior information about parameters be available and represented suitably by a prior pdf.

Last certain Bayesian minimum expected loss estimators were derived and found to be in a Stein-like form. These estimators incorporate prior information, are optimal relative to a quadratic loss function, and reflect uncertainty regarding the validity of different hypotheses. Since this uncertainty is often present, it is thought that such Bayesian estimators may be found useful in practice. Also, in this approach the complete posterior pdf for the parameter vector is available and thus there is no need to rely on only a point estimate in making inferences about parameters' probable values.

References

[1] A. J. Baranchik, Multiple regression and estimation of the mean of a multivariate normal distribution. Tech. Report No. 51, Stanford, Stanford U., 1964.

[2] A. J. Baranchik, A family of minimax estimators of the mean of a multivariate normal distribution. Annals of Mathematical Statistics 41 (1970), 642–645.

[3] M. E. Bock, T. A. Yancey and G. G. Judge, Properties of some preliminary test estimators in regression using a quadratic loss criterion. October 1971, mimeographed.

[4] G. E. P. Box and G. C. Tiao, Bayesian estimation of means for the random effect model. Journal of the American Statistical Association 63 (1968), 174–181.

[5] J. Dickey, The Bayesian alternatives to the *F* test and the least squares estimates in the normal linear model. Research Report No. 50, Buffalo, State U. of New York, 1971.

[6] J. H. Drèze, Bayesian limited information analysis of the simultaneous equations model. CORE Discussion Paper No. 7111, U. of Louvain, July 1971.

[7] J. H. Drèze, Limited information estimation from a Bayesian viewpoint. CORE Discussion Paper No. 6816, U. of Louvain, August 1968.

[8] J. H. Drèze, The Bayesian approach to simultaneous equation estimation. ONR Research Memorandum No. 67, Northwestern U., 1962.

[9] J. H. Drèze and J. A. Morales, Bayesian full information analysis of the simultaneous equation model. CORE Discussion Paper No. 7031, U. of Louvain, August 1970.

[10] S. Fienberg and P. W. Holland, Simultaneous estimation of multinomial probabilities. Journal of American Statistical Association 68 (1973), 683–691.

[11] M. S. Geisel, Comparing and choosing among parametric statistical models: a Bayesian analysis with macroeconomic applications. Unpublished doctoral dissertation, Graduate School of Business, U. of Chicago, 1970.

[12] D. V. Huntsberger, A generalization of a preliminary testing procedure for pooling data. Annals of Mathematical Statistics 26 (1955), 734–743.

[13] W. James and C. Stein, Estimation with quadratic loss. In: Proceedings of the Fourth Berkeley Symposium on Mathematical Statistics and Probability, Vol. I, Berkeley, U. of California Press, 1961.

[14] J. B. Kadane, Comparison of k-class estimators when the disturbances are small. Tech. Report No. 64, Stanford, Stanford U., 1969.

[15] J. B. Kadane, The distribution, when the residuals are small, of statistics testing overidentifying restrictions. Journal of the American Statistical Association 65 (1970), 182–185.

[16] D. V. Lindley, Discussion on Professor Stein's paper. Journal of the Royal Statistical Society, Series B 24 (1962), 285–287.

[17] D. V. Lindley and A. F. M. Smith, Bayes estimates for the linear model. Journal of the Royal Statistical Society, Series B 34 (1972), 1–41.

[18] A. L. Nagar, The bias and moment matrix of the general k-class estimators of the parameters in simultaneous equations, Econometrica 27 (1959), 575–595.

[19] D. H. Richardson, The exact distribution of a structural coefficient estimator. Journal of the American Statistical Association 63 (1968), 1214–1226.

[20] D. H. Richardson and R. T. Rohr, Distribution of a structural t-statistic for the case of two included endogenous variables. Journal of the American Statistical Association 66 (1971), 375–382.

[21] L. J. Savage, The foundations of statistics. New York, John Wiley, 1954.

[22] T. Sawa, The exact sampling distribution of ordinary least squares and two-stage least squares. Journal of the American Statistical Association 64 (1969), 923–937.

[23] S. L. Sclove, Improved estimation of parameters in multivariate regression. Sankhyā, Series A 33 (1971), 61–66.

[24] S. L. Sclove, Improved estimators for coefficients in linear regression. Journal of the American Statistical Association 63 (1968), 596–606.

[25] S. L. Sclove, C. Morris and R. Radhakrishnan, Non-optimality of preliminary-test estimators for the mean of a multinomial normal distribution. Annals of Mathematical Statistics 43 (1972), 1481–1490.

[26] V. K. Srivastava, The efficiency of an improved method of estimating seemingly unrelated regression equations. Journal of Econometrics 1 (1973), 341–350.

[27] C. Stein, Inadmissibility of the usual estimator for the Mean of a multivariate normal distribution. In: Proceedings of the Third Berkeley Symposium on Mathematical Statistics and Probability, Vol. I, Berkeley, U. of California Press, 1956.

[28] C. Stein, Multiple regression. In: Contributions to probability and statistics, essays in honor of Harold Hotelling, edited by I. Olkin. Stanford, Stanford U. Press, 1960.

[29] C. Stein, Confidence sets for the mean of a multivariate normal distribution. Journal of the Royal Statistical Society, Series B 24 (1962), 265–296.

[30] W. E. Strawderman and A. Cohen, Admissibility of estimators of the mean vector of a multivariate normal distribution with quadratic loss. Annals of Mathematical Statistics 42 (1971), 270–296.

[31] J. R. Thompson, Some shrinkage techniques for estimating the mean. Journal of the American Statistical Association 63 (1968), 113–122.

[32] G. C. Tiao and A. Zellner, Bayes' theorem and the use of prior knowledge in regression analysis. Biometrika 51 (1964), 219–230.

[33] A. Zellner, An introduction to Bayesian inference in econometrics. New York, John Wiley, 1971.

[34] A. Zellner, Bayesian and non-Bayesian analysis of the log-normal distribution and log-normal regression. Journal of the American Statistical Association 66 (1971), 327–330.

APPENDIX

PROGRAM OF THE FIRST NBER-NSF SEMINAR ON BAYESIAN INFERENCE IN ECONOMETRICS

Meeting Place: University of Chicago
Date of Seminar: December 11–12, 1970

Program

(arranged by A. Zellner, U. of Chicago)

Friday, December 11

10.00–12.00:	*Organizational Meeting*
Chairman:	Arnold Zellner, U. of Chicago
2.00–5.00:	*Bayesian Methods for Comparing and Choosing Among Models*
Chairman:	Harry V. Roberts, U. of Chicago
Speakers:	"Comparison of Alternative Lag Distributions and Error Structures", Martin S. Geisel, Carnegie-Mellon U.
	"Distributed Lag Analysis with Informative Prior Distributions", Edward E. Leamer, Harvard U.
	"Comments on Distributed Lag Models", G. S. Maddala, Rochester U.

Saturday, December 12

9.00–12.00:	*Bayesian Analyses of Regression and Simultaneous Equation Models*
Chairman:	John W. Pratt, Harvard U.

Speakers:	"Foundations for the Theory of Least Squares", Bruce M. Hill, U. of Michigan
	"On Constraints Often Overlooked in Analyses of Simultaneous Equation Models" and "Bayesian and Non-Bayesian Analyses of Simultaneous Equation Models", Arnold Zellner, U. of Chicago
2.00–4.00:	*Bayesian Adaptive Control Problems*
Chairman:	Joseph Kadane, Center for Naval Analyses and Carnegie-Mellon U.
Speakers:	"The Multi-Period Control Problem Under Uncertainty", Edward C. Prescott, U. of Pennsylvania
	"Bayesian Control Theory and the St. Louis Model", H. Woods Bowman, Chicago Federal Reserve Bank

PROGRAM OF THE SECOND NBER-NSF SEMINAR ON BAYESIAN INFERENCE IN ECONOMETRICS

Meeting Place: Carnegie-Mellon University, Pittsburgh
Date of Seminar: May 14–15, 1971

Program

(arranged by M. S. Geisel, Carnegie-Mellon U.)

Friday, May 14

10.00–12.00:	*Organizational Meeting*
Chairmen:	A. Zellner, U. of Chicago and M. S. Geisel, Carnegie-Mellon U.
	Reporting the Results of a Bayesian Analysis:
	Discussion led by J. M. Dickey, State U. of New York at Buffalo
2.00–5.00:	*Bayesian Applications and Analysis of Specific Models*
Chairman:	Martin S. Geisel, Carnegie-Mellon U.

Speakers: "Use of Prior Information in the Analysis and Estimation of Cobb-Douglas Production Function Models",
Arnold Zellner, U. of Chicago
"Tests of the Adaptive Regression Model",
Edward C. Prescott, U. of Pennsylvania
"Multiperiod Predictions from Stochastic Difference Equations by Bayesian Methods",
Gregory C. Chow, Princeton U.

6.30– Pittsburgh Pirates *vs.* N.Y. Mets cum Dinner
Chairman and Host: Richard M. Cyert

Saturday, May 15

9.30–12.00: *Bayesian Methods in Economic Theory*
Chairman: Thomas Rothenberg, U. of California, Berkeley
Speakers: "Analysis of Cooperation and Learning in a Duopoly Context",
Richard M. Cyert and Morris H. DeGroot, Carnegie-Mellon U.

1.30–4.00: *Analysis of Specific Models*
Chairman: Arnold Zellner, U. of Chicago
Speakers: "Bayesian Analysis of Error Components Regression Models",
P. A. V. B. Swamy, Ohio State U.
"Estimation of Multinomial Process when only the Sum and the Number Governed by Each Process is Observed",
Joseph B. Kadane, Center for Naval Analyses and Carnegie-Mellon U.

PROGRAM OF THE THIRD NBER-NSF SEMINAR ON BAYESIAN INFERENCE IN ECONOMETRICS

Meeting Place: Harvard University
Date of Seminar: October 15–16, 1971

Program

(arranged by Edward Leamer, Harvard U.)

Friday, October 15

9.30–12.45: "Properties of Preliminary Test Estimators in Regression,"
George Judge, U. of Illinois
"Pseudo-Bayes Multinomial Estimators",
Stephen Fienberg, U. of Chicago

2.15–5.15: "Bayes–Stein Estimators for *k*-Means, Regression and Simultaneous Equation Models",
Arnold Zellner and Walter Vandaele, U. of Chicago
"Identification in Probability: A Bayesian Concept and Application to Simultaneous Equations",
Jacques Drèze, U. of Louvain and Cornell U.
"Bayesian Estimation with a Time Series of Cross Sections",
Gary Chamberlain, Harvard U.

Saturday, October 16

9.30–12.15: "Demonstration of an Interactive Computer Program for Simple-Minded Data Analysis",
Robert Schlaifer and Robert Glauber, Harvard U.

1.30–4.00: "Elicitation of Personal Probabilities and Expectations",
Leonard J. Savage, Yale U.

PROGRAM OF THE FOURTH NBER-NSF SEMINAR ON BAYESIAN INFERENCE IN ECONOMETRICS

Meeting Place: University of Chicago
Date of Seminar: May 19–20, 1972

Program

(arranged by A. Zellner, U. of Chicago)

Friday, May 19

9.00–12.00: *Bayesian Analyses of Missing Observation Problems*
"Multiple Regression Analysis with Incomplete Observations from a Bayesian Viewpoint",
M. G. Dagenais, U. of Montreal
"Missing Independent Variables in Bayesian Regression",
S. J. Press, U. of Chicago
"Bayesian Analysis of the Bivariate Normal Distribution when Some Observations are Missing",
P. A. V. B. Swamy, Ohio State U.

2.00–5.00: *Time Series Problems*
"Estimation of Parameters in Distributed Lag Econometric Models",
W. S. Cleveland, U. of North Carolina
"A Distributed Lag Estimator Derived from Smoothness Priors",
R. J. Shiller, M.I.T.
"False Models and Post-Data Model Construction",
E. E. Leamer, Harvard U.

Saturday, May 20

9.00–12.00: *Prior Distributions in Bayesian Analyses*
"Bayesian Inference and Non-informative Priors",
G. E. P. Box, U. of Wisconsin
"Behrens–Fisher Problems: Bayesian Solutions",
H. H. Dayal, State U. of New York at Buffalo

1.30–4.30: *Analyses of Selected Problems*
"Bayesian Factor Analysis",
G. M. Kaufman, M.I.T. and S. J. Press, U. of Chicago
"Some Comments on 'Bayes' Estimators",
G. C. Tiao, U. of Wisconsin
"Comparing Linear Regression Models with Serially Correlated Errors",
M. S. Geisel, Carnegie-Mellon U.

PROGRAM OF THE FIFTH NBER-NSF SEMINAR ON BAYESIAN INFERENCE IN ECONOMETRICS

Meeting Place: University of Minnesota
Date of Seminar: October 27–28, 1972

Program

(arranged by S. E. Fienberg)

Friday, October 27

9.00–12.00: "Ventures, Bets, and Current Prospects",
C. Hildreth, U. of Minnesota
"Optimal Gambling Systems",
D. Berry, U. of Minnesota
"On Coherence, Inadmissibility and Inference About Many Parameters in the Theory of Least Squares",
B. Hill, U. of Michigan

1.30–5.00: *Linear Model Problems*
"Multicollinearity: A Bayesian Interpretation",
E. E. Leamer, Harvard U.
"Weak Priors and Strong Posteriors in Simultaneous Equations Models",
G. S. Maddala, U. of Rochester
"Bayesian Analysis of Regression Errors",
A. Zellner, U. of Chicago
"Post-Data Model Construction as Estimation with Infinitely Many Parameters",
C. A. Sims, U. of Minnesota

Saturday, October 28

9.00–12.00: *Bayesian Analysis of Econometric Models*
"A Bayesian Approach to Static, Stochastic Equilibria",
S. Grossman, U. of Chicago
"Effect of Uncertainty on Optimal Control Policies",
G. C. Chow, Princeton U.

"Doing What Comes Naturally: Interpreting a Tail Area as a Posterior Probability" and "Reaching a Consensus: How a Group of Econometricians Can Pool Their Opinions to Reach Agreement About Subjective Probabilities",
M. H. DeGroot, Carnegie-Mellon U.

1.30–4.00: *Discussion of SBIE Volume*
Sampling Properties of Bayes Estimators
"More on the Risk of Multinomial Estimators",
P. W. Holland, NBER Computer Research Center
"A Comparison of Preliminary Test, and New and Old Stein-like Estimators",
G. G. Judge, U. of Illinois

REPORT OF THE SIXTH NBER-NSF SEMINAR ON BAYESIAN INFERENCE IN ECONOMETRICS

Meeting Place: University of Wisconsin, Madison
Date of Seminar: May 4–5, 1973

Program

(arranged by R. B. Miller and G. C. Tiao, U. of Wisconsin)

Friday, May 4
9.00–11.55: *Regression Analysis*
"Regression Prediction",
H. V. Roberts, U. of Chicago
"Comparing Linear Regression Models with Diffuse Prior Information",
M. G. Dagenais, U. of Montreal
"Bayesian Factor Analysis",
G. M. Kaufman, M.I.T., and S. J. Press, Yale U. and U. of Chicago

2.00–5.00: *Potpourri*

"Rational Expectations and Bayesian Analysis",
R. M. Cyert and M. H. DeGroot, Carnegie-Mellon U.

"Bayesian Tolerance Regions",
I. Guttman, U. of Toronto

"Classification-Information-Loss Cluster Analysis",
W. S. Cleveland, Bell Laboratories, and D. A. Relles, Rand Corporation

Saturday, May 5

9.00–12.30: (a) *Panel Discussion on Bayesian Inference in Econometrics*

Panel: G. E. P. Box, U. of Wisconsin
Jacques H. Drèze, U. of Louvain
Seymour Geisser, U. of Minnesota
Arnold Zellner, U. of Chicago, Moderator

(b) *Time Series*

"A Bayesian Approach to the Smoothing of Seasonal Time Series with Special Reference to the Census Bureau X–11 Method",
W. P. Cleveland, Duke U., and G. C. Tiao, U. of Wisconsin

"Some Remarks on Parameter Shifts in Time Series Models",
R. B. Miller, U. of Wisconsin

AUTHOR INDEX

SUBJECT INDEX